FINITE TEMPERATURE FIELD THEORY

Second Edition

FINITE TEMPERATURE FIELD THEORY

Second Edition

Ashok Das
University of Rochester, USA

World Scientific

NEW JERSEY · LONDON · SINGAPORE · BEIJING · SHANGHAI · HONG KONG · TAIPEI · CHENNAI · TOKYO

Published by

World Scientific Publishing Co. Pte. Ltd.

5 Toh Tuck Link, Singapore 596224

USA office: 27 Warren Street, Suite 401-402, Hackensack, NJ 07601

UK office: 57 Shelton Street, Covent Garden, London WC2H 9HE

Library of Congress Control Number: 2023007025

British Library Cataloguing-in-Publication Data
A catalogue record for this book is available from the British Library.

FINITE TEMPERATURE FIELD THEORY
Second Edition

ISBN 978-981-127-234-9 (hardcover)
ISBN 978-981-127-293-6 (paperback)
ISBN 978-981-127-235-6 (ebook for institutions)
ISBN 978-981-127-236-3 (ebook for individuals)

For any available supplementary material, please visit
https://www.worldscientific.com/worldscibooks/10.1142/13308#t=suppl

To

Charu, Gayatri, Namrata, Pushpa,

Shanusree, Smitu, Subhashree,

and

to

Jackie (BaaLungaa)

Preface

Preface to the First Edition

In the past several years, there have been a lot of interest in various phenomena, in quantum field theories, at finite temperature. The formalism, for such field theoretic investigations, of course, dates back to Matsubara. However, over the years, there have been alternate methods of doing calculations in field theories at finite temperature which shed light on the structure of theories at finite temperature. And, as more and more studies are carried out in these formalisms, we become more and more aware of the subtleties at finite temperature and we understand the structure of these theories much better.

While many people are quite familiar with the Matsubara formalism or the imaginary time formalism, the alternate, real time formalisms are not as widely appreciated. However, the analysis of many phenomena of current interest need such alternate formalisms. Keeping this in mind, I have tried to describe all the three formalisms – imaginary time formalism, closed time path formalism and thermofield dynamics – right in the beginning chapters of this book. There are numerous examples that are worked out to give a sense of which formalism may be better suited for a given kind of problem. It is, of course, left completely to the reader to choose from among these formalisms, to carry out any calculation.

I also develop, in detail, various other concepts that are not as widely stressed in the literature. For example, the study of gauge theories at finite temperature, the cutting rules at finite temperature, the gauge dependence of the effective potential are just a few topics that are discussed at great lengths. Most of these topics are basic field theory questions and, therefore, to keep the material self-complete, many of the zero temperature topics on field theory are also worked out in detail. The idea here is to bring out, from these finite temperature calculations, the similarities with and differences from what we know at zero temperature. Although, I had originally planned to include a discussion of the hard thermal loops in this book, neither space nor time allowed this. Moreover, I recognize that there exist, by now, several excellent sources of discussion of the hard thermal loops in the literature. The references, at the end of each chapter, only represent those which I have found useful in presenting the material. There are many more papers on the various topics being discussed and I apologize for not being able to include an exhaustive list of references.

The material in the first few chapters of this book were presented as a series of lectures at the Mehta Research Institute, Allahabad and I would like to thank Prof. H. S. Mani for his hospitality. I have benefited and learned a lot from all of my collaborators on the subject – K. S. Babu, P. F. Bedaque, M. Hott, A. Karev, S. Naik, S. Panda, P. Panigrahi and S. Vokos – and I would like to thank them all. I also acknowledge numerous suggestions (and help in simplifying many difficult derivations) from Drs. P. F. Bedaque and S. Pernice as well as from Profs. S. Okubo and H. A. Weldon. I would also like to thank M. Begel and T. Guptill for much needed assistance on computer related matters. All the figures in this book were drawn using PSTricks.

I appreciate all the help that many of my friends have provided during this project. But, most of all, I would like to thank my family members, especially Jhilli, for their support and understanding, particularly during the last part of the project.

Ashok Das
Rochester

Preface to the Second Edition

There has been a lot of progress in the understanding of thermal field theories since the first edition of this book was published in 1997. As a result, the new edition has several new chapters and has expanded chapters that were already there in the previous edition, but now with additional newer material. Several appendices have been included, particularly, in the beginning chapters mainly to point out the intricacies in the calculations at finite temperature. The new chapters include thermal field theories on an arbitrary path in the complex t plane, both in the operator as well as in the path integral formulations, thermal operator representation, thermal lightfront field theories as well as the fluctuation-dissipation theorem, both in equilibrium and out of equilibrium theories. The expanded chapters include material on gauge independence of physical parameters at finite temperature through Nielsen identities as well as effective actions for the fermions in the proper time formalism. In addition, several examples have been added in various chapters to clarify things in a better manner. The appendices added, at the end of initial chapters, include rotations to Euclidean space, periodic/antiperiodic delta functions, Green's functions for bosonic as well as $0+1$ dimensional fermionic theories, Gaussian wave packets, coherent and squeezed states, bosonic oscillator on an arbitrary path, propagator for a fermionic oscillator on an arbitrary path, KMS condition on an arbitrary path as well as $0+1$ dimensional Chern-Simons QED. Two important topics of interest, namely, hard thermal loops and infrared divergences at finite temperature, have not been intentionally addressed in this new edition since they have been covered in great detail in other books and review articles.

I would like to thank Prof. Josif Frenkel and Dr. Pushpa Kalauni for going through the expanded material of this new edition carefully and Prof. Fernando Brandt for help with some of the figures in the book.

Ashok Das

Rochester

Contents

CHAPTER 1

Imaginary time formalism

1.1 Introduction

Let us begin by reviewing some of the general concepts from equilibrium statistical mechanics. The statistical behavior of a quantum mechanical system, in thermal equilibrium, is normally studied through an appropriate ensemble (an ensemble is a collection of an infinitely large number of identical and independent systems). In general, one defines a density matrix for the ensemble as

$$\rho(\beta) = \frac{e^{-\beta\mathcal{H}}}{Z(\beta)}, \tag{1.1}$$

where β represents the inverse of the equilibrium temperature (we are assuming that the Boltzmann constant $k = 1$, otherwise, $\beta = \frac{1}{kT}$) and $\mathcal{H}$ can be thought of as the appropriate Hamiltonian for the particular choice of the ensemble. $Z(\beta)$ which is a normalization factor, to be determined shortly, is known as the partition function of the ensemble.

The exact form of the Hamiltonian $\mathcal{H}$ in the Boltzmann factor in (1.1) depends on the type of the ensemble under study. For example, if a system is allowed to exchange only energy with the heat bath (environment), then such an ensemble is known as a canonical ensemble and, in this case, we have

$$\mathcal{H} = H, \tag{1.2}$$

where H denotes the (conventional) dynamical Hamiltonian of the individual systems. In a canonical ensemble, temperature T denotes the only characteristic of equilibrium. On the other hand, if the system is allowed to exchange not only energy but also particles with the external reservoir, then such an ensemble is known as a grand canonical ensemble. In this case, we write

$$\mathcal{H} = H - \mu N, \tag{1.3}$$

where H and N represent, respectively, the dynamical Hamiltonian and the number operators for the particular system under study and μ, which is a constant, corresponds to the chemical potential. In such an ensemble, the chemical potential μ also defines a characteristic of equilibrium like the temperature T. (Chemical potential, in very simple terms, represents the energy required to change the number of particles in a system by unity.) If there is another conserved quantity (charge), say Q, in a system besides the number of particles, then, one can define a grand canonical ensemble as described by the Hamiltonian

$$\mathcal{H} = H - \mu N - \mu_Q Q, \tag{1.4}$$

where μ_Q would represent the "chemical potential" for the corresponding conserved quantity (charge) Q. A general ensemble may have several conserved quantities besides the number operator and, in this case, a grand canonical ensemble is represented by the Hamiltonian

$$\mathcal{H} = H - \mu N - \sum_i \mu_{Q_i} Q_i, \tag{1.5}$$

and, for such a grand canonical ensemble, μ as well as all the "chemical potentials" μ_{Q_i} represent the totality of all the characteristics of equilibrium besides the temperature T (known as the equilibrium state properties). Therefore, from (1.2)-(1.5), we see that we can think of the canonical ensemble as a grand canonical ensemble where all the "chemical potentials" vanish, $\mu = \mu_{Q_i} = 0$ for all i. In our general discussions, we will keep the choice of the ensemble arbitrary, since the qualitative properties of a system at finite temperature do not depend on the nature of the ensemble. In working out specific examples, however, we will choose the ensemble that is appropriate for the particular physical system under study.

Let us recall from quantum mechanics[1] that, in a quantum system in thermal equilibrium with a heat bath (environment), the density matrix, in the energy basis, is defined as

$$\rho = \sum_n p_n |n\rangle\langle n|, \tag{1.6}$$

where p_n denotes the probability of finding a quantum mechanical system, in the ensemble, in the energy state $|n\rangle$ (recall that the energy basis is orthonormal, $\langle n|m\rangle = \delta_{nm}$ and that $|n\rangle\langle n|$ corresponds to the projection operator onto the state vector $|n\rangle$). Being probabilities, p_ns satisfy

$$0 \le p_n \le 1, \quad \sum_n p_n = 1. \tag{1.7}$$

It follows from the orthonormality of the energy basis as well as from (1.6)-(1.7) that

$$\mathrm{Tr}\,\rho = \sum_m \langle m|\rho|m\rangle = \sum_{n,m} p_n \langle m|n\rangle\langle n|m\rangle$$

$$= \sum_{m,n} p_n \delta_{nm} = \sum_m p_m = 1. \tag{1.8}$$

Namely, the density matrix has a unit trace and, using this in (1.1), we can obtain the form of the partition function of the ensemble to be (recall that $Z(\beta)$ is a normalization factor, a scalar function of temperature and other equilibrium state properties, and not a matrix)

$$\mathrm{Tr}\,\rho(\beta) = \frac{1}{Z(\beta)}\,\mathrm{Tr}\,e^{-\beta\mathcal{H}} = 1,$$

$$\text{or,}\quad Z(\beta) = \mathrm{Tr}\,e^{-\beta\mathcal{H}} = e^{-\beta F(\beta)}, \tag{1.9}$$

where "Tr" stands for the trace over the matrix indices or the sum over the expectation values of the operator in any complete basis (and not necessarily in the energy basis). Here we have defined another scalar function of temperature (and other equilibrium state

---

[1]See, for example, section **5.6** in A. Das, *Lectures on Quantum Mechanics* (second edition), Hindustan Book Agency (New Delhi) and World Scientific Publishing (Singapore), 2011.

properties) which is known as the free energy of the ensemble. Both the partition function $Z(\beta)$ and the free energy $F(\beta)$ are very useful in calculating ensemble averages of various observables. We see from (1.9) that we can write the density matrix in (1.1) in a more complete form as

$$\rho(\beta) = \frac{e^{-\beta\mathcal{H}}}{\operatorname{Tr} e^{-\beta\mathcal{H}}} = Z^{-1}(\beta) \, e^{-\beta\mathcal{H}}. \tag{1.10}$$

The ensemble average of any observable (operator) A is defined as

$$\begin{aligned}
\langle A \rangle_\beta = \operatorname{Tr}(\rho(\beta)A) &= \frac{\operatorname{Tr}(e^{-\beta\mathcal{H}}A)}{\operatorname{Tr} e^{-\beta\mathcal{H}}} \\
&= Z^{-1}(\beta) \operatorname{Tr}(e^{-\beta\mathcal{H}}A),
\end{aligned} \tag{1.11}$$

where we have used (1.10). (The ensemble average in (1.11) is normalized such that when $A = \mathbb{1}$, the average reduces to unity as it should, see (1.9).) The thermal average of the correlation function of any two operators, A and B, with different coordinates can similarly be written as

$$\begin{aligned}
\langle AB \rangle_\beta = \operatorname{Tr}(\rho(\beta)AB) &= \frac{\operatorname{Tr}(e^{-\beta\mathcal{H}}AB)}{\operatorname{Tr} e^{-\beta\mathcal{H}}} \\
&= Z^{-1}(\beta) \operatorname{Tr}(e^{-\beta\mathcal{H}}AB),
\end{aligned} \tag{1.12}$$

where we have suppressed the coordinate dependence of the operators for simplicity. Similarly, the ensemble average of a product of any number of observables is defined as

$$\begin{aligned}
\langle A_1 A_2 \cdots \rangle_\beta = \operatorname{Tr}(\rho(\beta)A_1 A_2 \cdots) &= \frac{\operatorname{Tr}(e^{-\beta\mathcal{H}}A_1 A_2 \cdots)}{\operatorname{Tr} e^{-\beta\mathcal{H}}} \\
&= Z^{-1}(\beta) \operatorname{Tr}(e^{-\beta\mathcal{H}}A_1 A_2 \cdots),
\end{aligned} \tag{1.13}$$

where, once again, the coordinate dependences of the operators are suppressed for simplicity.

1.2 Different pictures

In quantum mechanics, we know that the system can be described in one of the three equivalent pictures. (In this section, a superscript on an operator or on a state denotes the picture in which it is being described.[2] In the Schrödinger picture states carry all the time dependence while the operators, in general, are time independent $(O^{(S)} \neq O^{(S)}(t))$. The time evolution of the states is governed by the time dependent Schrödinger equation (we set $\hbar = 1$ throughout)

$$i \frac{\mathrm{d}|\psi(t)\rangle^{(S)}}{\mathrm{d}t} = \mathcal{H}^{(S)}|\psi(t)\rangle^{(S)}, \tag{1.14}$$

where $\mathcal{H}^{(S)}$ denotes the time independent Hamiltonian governing the dynamics of the system and, for such Hamiltonians, (1.14) leads to the solution

$$|\psi(t)\rangle^{(S)} = e^{-i\mathcal{H}^{(S)}t}|\psi(0)\rangle^{(S)} = U^{(S)}(t)|\psi(0)\rangle^{(S)}. \tag{1.15}$$

Here $U^{(S)}(t) = e^{-i\mathcal{H}^{(S)}t}$ denotes the time evolution operator in the Schrödinger picture. (Note that, unlike other operators in this picture which do not carry time dependence, time evolution operator does carry time dependence and, in fact, it has to carry time dependence in any picture since it evolves the state from a given reference time.) If the Hamiltonian becomes time dependent (for example, if a system is in a time dependent external field), the time evolution operator in (1.15), which solves (1.14), can be written as the time ordered exponential

$$U^{(S)}(t) = T\left(e^{-i\int_0^t \mathrm{d}t'\, \mathcal{H}^{(S)}(t')}\right), \tag{1.16}$$

where T denotes time ordering of the operators inside the parenthesis such that an operator with a larger time stands to the left of those with earlier times. However, let us continue our discussion with the time independent Hamiltonian for simplicity.

In general, the time evolution operator takes a state at time t_1 to a state at time t_2

$$|\psi(t_2)\rangle^{(S)} = U^{(S)}(t_2, t_1)|\psi(t_1)\rangle^{(S)}$$
$$= e^{-i\mathcal{H}^{(S)}(t_2 - t_1)}|\psi(t_1)\rangle^{(S)}, \tag{1.17}$$

---

[2]See, for example, section **6.3** in A. Das, *Lectures on Quantum Field Theory* (second edition), World Scientific Publishing, Singapore (2020).

and satisfies the properties

$$U^{(S)}(t,t) = \mathbb{1},$$

$$U^{(S)}(t_1, t_2) U^{(S)}(t_2, t_3) = U^{(S)}(t_1, t_3),$$

$$\left(U^{(S)}\right)^\dagger (t_1, t_2) = \left(U^{(S)}\right)^{-1}(t_1, t_2) = U^{(S)}(t_2, t_1). \tag{1.18}$$

(The Hamiltonian is assumed to be Hermitian in the last relation.) Viewed in this way, we can identify (this is true in any picture)

$$U^{(S)}(t) = U^{(S)}(t, 0),$$

$$U^{(S)}(-t) = U^{(S)}(0, t) = \left(U^{(S)}\right)^{-1}(t),$$

$$U^{(S)}(t_1)(U^{(S)})^{-1}(t_2) = U^{(S)}(t_1, 0) U^{(S)}(0, t_2) = U^{(S)}(t_1, t_2), \tag{1.19}$$

where we have used the second and the third properties in (1.18).

In the Heisenberg picture, on the other hand, the states are assumed to be time independent with the operators carrying all the time dependence so that if we identify $|\psi\rangle^{(H)} = |\psi(0)\rangle^{(S)}$ and $O^{(S)} = O^{(H)}(0)$, the two pictures can be related as

$$|\psi\rangle^{(H)} = |\psi(0)\rangle^{(S)} = U^{(S)}(0, t)|\psi(t)\rangle^{(S)} = (U^{(S)})^{-1}(t)|\psi(t)\rangle^{(S)}$$

$$= e^{i\mathcal{H}^{(S)} t}|\psi(t)\rangle^{(S)},$$

$$O^{(H)}(t) = (U^{(S)})^{-1}(t) O^{(S)} U^{(S)}(t) = e^{i\mathcal{H}^{(S)} t} O^{(S)} e^{-i\mathcal{H}^{(S)} t}. \tag{1.20}$$

Here we have used (1.19) as well as the form of $U^{(S)}(t)$ following from (1.17). Equation (1.20) shows that $\mathcal{H}^{(H)} = \mathcal{H}^{(S)} = \mathcal{H}$ (if we choose $O^{(S)} = \mathcal{H}^{(S)}$ in the second relation in (1.20)). We also note, from (1.20) that the expectation value of any operator (observable) is independent of the picture, namely,

$$^{(S)}\langle\psi(t)|O^{(S)}|\psi(t)\rangle^{(S)} = {}^{(H)}\langle\psi|e^{i\mathcal{H}^{(S)} t} O^{(S)} e^{-i\mathcal{H}^{(S)} t}|\psi\rangle^{(H)}$$

$$= {}^{(H)}\langle\psi|O^{(H)}(t)|\psi\rangle^{(H)}. \tag{1.21}$$

The time evolution of operators in the Heisenberg picture is given by the Heisenberg equation of motion which follows from (1.20)

$$i\frac{dO^{(H)}(t)}{dt} = i\frac{d}{dt}(e^{i\mathcal{H}^{(S)} t} O^{(S)} e^{-i\mathcal{H}^{(S)} t}) = \left[O^{(H)}(t), \mathcal{H}^{(S)}\right]$$

$$= \left[O^{(H)}(t), \mathcal{H}\right]. \tag{1.22}$$

(Note that the time evolution operator for the operators in the Heisenberg picture is determined to be the inverse of the time evolution operator for states in the Schrödinger picture, $U^{(H)}(t) = (U^{(S)})^{-1}(t) = e^{i\mathcal{H}t}$, see, for example, the second relation in (1.20).)

There is yet another description of quantum mechanical systems which is quite useful in the presence of nontrivial interactions in quantum field theories. This is known as the interaction picture, and here both the states as well as operators carry partial time dependence. To introduce this picture, let us decompose the Hamiltonian in the Schrödinger picture as

$$\mathcal{H}^{(S)} = \mathcal{H}_0^{(S)} + \mathcal{H}_I^{(S)}, \tag{1.23}$$

where $\mathcal{H}_0^{(S)}$ denotes the free part of the Hamiltonian while $\mathcal{H}_I^{(S)}$ contains all the nontrivial interactions. Let us next define

$$|\psi(t)\rangle^{(I)} = (U_0^{(S)})^{-1}(t)|\psi(t)\rangle^{(S)} = (U_0^{(S)})^{-1}(t)U^{(S)}(t)|\psi\rangle^{(H)}$$
$$= U^{(I)}(t)|\psi\rangle^{(H)} = U^{(I)}(t)|\psi(0)\rangle^{(I)}, \tag{1.24}$$

if we identify

$$|\psi(0)\rangle^{(I)} = |\psi\rangle^{(H)} = |\psi(0)\rangle^{(S)}. \tag{1.25}$$

Furthermore, we have identified

$$U_0^{(S)}(t) = e^{-i\mathcal{H}_0^{(S)}t}, \tag{1.26}$$

so that the time evolution operator (for the states), in the interaction picture, is seen from (1.24) to be given by

$$U^{(I)}(t) = (U_0^{(S)})^{-1}(t)U^{(S)}(t) = e^{i\mathcal{H}_0^{(S)}t}e^{-i\mathcal{H}^{(S)}t}, \tag{1.27}$$

The operators, in this picture, are related to those in the other two pictures as (so that the expectation value is independent of the picture)

$$O^{(I)}(t) = (U_0^{(S)})^{-1}(t)O^{(S)}U_0^{(S)}(t)$$
$$= e^{i\mathcal{H}_0^{(S)}t}O^{(S)}e^{-i\mathcal{H}_0^{(S)}t}$$
$$= (U_0^{(S)})^{-1}(t)U^{(S)}(t)O^{(H)}(t)(U^{(S)})^{-1}(t)U_0^{(S)}(t)$$
$$= U^{(I)}(t)O^{(H)}(t)(U^{(I)})^{-1}(t), \tag{1.28}$$

which follows from (1.20) as well as (1.24) and implies that

$$\mathcal{H}_0^{(I)} = e^{i\mathcal{H}_0^{(S)}t} \mathcal{H}_0^{(S)} e^{-i\mathcal{H}_0^{(S)}t} = \mathcal{H}_0^{(S)},$$
$$\mathcal{H}_I^{(I)}(t) = e^{i\mathcal{H}_0^{(S)}t} \mathcal{H}_I^{(S)} e^{-i\mathcal{H}_0^{(S)}t},$$
$$\mathcal{H}^{(I)}(t) = e^{i\mathcal{H}_0^{(S)}t} \mathcal{H}^{(S)} e^{-i\mathcal{H}_0^{(S)}t}. \tag{1.29}$$

Taking the time derivative of (1.28), the time evolution of operators in the interaction picture follows to be

$$i\frac{dO^{(I)}(t)}{dt} = i\frac{d}{dt}(e^{i\mathcal{H}_0^{(S)}t}O^{(S)}e^{-i\mathcal{H}_0^{(S)}t}) = [O^{(I)}(t), \mathcal{H}_0^{(S)}]$$
$$= [O^{(I)}(t), \mathcal{H}_0^{(I)}], \tag{1.30}$$

where we have used the first relation in (1.29) in the last step. This is the simplicity of the interaction picture, namely, the operators evolve with the free Hamiltonian and, therefore, the operator quantization conditions do not change because of interactions (operators only acquire phases). The time evolution for the states in the interaction picture in (1.24), on the other hand, is governed by the interaction Hamiltonian in the interaction picture since we can write (1.27) also in the compact form

$$U^{(I)}(t) = e^{i\mathcal{H}_0^{(S)}t}e^{-i\mathcal{H}^{(S)}t} = T\left(e^{-i\int_0^t dt'\mathcal{H}_I^{(I)}(t')}\right). \tag{1.31}$$

This is best seen by taking the time derivative of the time evolution operator (1.27) (or (1.31))

$$i\frac{dU^{(I)}(t)}{dt} = e^{i\mathcal{H}_0^{(S)}t}(-\mathcal{H}_0^{(S)} + \mathcal{H}^{(S)})e^{-i\mathcal{H}^{(S)}t}$$
$$= e^{i\mathcal{H}_0^{(S)}t} \mathcal{H}_I^{(S)} e^{-i\mathcal{H}_0^{(S)}t} e^{i\mathcal{H}_0^{(S)}t}e^{-i\mathcal{H}^{(S)}t}$$
$$= \mathcal{H}_I^{(I)}(t)U^{(I)}(t), \tag{1.32}$$

which is solved by the right hand side of (1.31) (see, for example, (1.16)). Note that we have used the second relation of (1.29) in the final step in (1.32).

Since the three pictures are related by the unitary time evolution operator and since cyclicity holds under "Trace", it follows that the ensemble averages of products of operators (in any fixed picture) has the same value when evaluated in the basis of states in any picture. For example, suppose the operators (whose ensemble average is being calculated) are in the Schrödinger picture

$$
\mathrm{Tr}^{(S)} \left(e^{-\beta \mathcal{H}^{(S)}} A^{(S)}(t_1) B^{(S)}(t_2) \cdots \right)
$$

$$
= \sum_n {}^{(S)}\langle \psi_n(t) | e^{-\beta \mathcal{H}^{(S)}} A^{(S)}(t_1) B^{(S)}(t_2) \cdots |\psi_n(t)\rangle^{(S)}
$$

$$
= \sum_n {}^{(H)}\langle \psi_n | e^{i\mathcal{H}^{(S)}t} e^{-\beta \mathcal{H}^{(S)}} A^{(S)}(t_1) B^{(S)}(t_2) \cdots e^{-i\mathcal{H}^{(S)}t} |\psi_n\rangle^{(H)}
$$

$$
= \mathrm{Tr}^{(H)} \left(e^{i\mathcal{H}^{(S)}t} e^{-\beta \mathcal{H}^{(S)}} A^{(S)}(t_1) B^{(S)}(t_2) \cdots e^{-i\mathcal{H}^{(S)}t} \right)
$$

$$
= \mathrm{Tr}^{(H)} \left(e^{-\beta \mathcal{H}^{(S)}} A^{(S)}(t_1) B^{(S)}(t_2) \cdots \right), \tag{1.33}
$$

where we have used the cyclicity of the trace in the last step. The same analysis can be extended to interaction picture as well. Namely, the thermal ensemble average can be evaluated in the basis of any picture that is suitable for the question under study. The cyclicity of trace (as well as the fact that $\mathcal{H}^{(S)} = \mathcal{H}^{(H)}$) also implies that,

$$
\mathcal{Z}(\beta) = \mathrm{Tr}\, e^{-\beta \mathcal{H}^{(S)}}
$$

$$
= \mathrm{Tr}\, e^{-\beta \mathcal{H}^{(H)}}
$$

$$
= \mathrm{Tr} \left(e^{i\mathcal{H}_0^{(S)}t} e^{-\beta \mathcal{H}^{(S)}} e^{-i\mathcal{H}_0^{(S)}t} \right) = \mathrm{Tr}\, e^{-\beta \mathcal{H}^{(I)}(t)}, \tag{1.34}
$$

where the last relation follows from the second line in (1.28) (as well as (1.29)). This shows that the partition function has the same value when evaluated in any picture.

1.3 Kubo-Martin-Schwinger (KMS) condition

Let us next note that, given an ensemble and a set of Schrödinger operators (operators in the Schrödinger picture are time indepen-dent), we can define operators in a (modified) Heisenberg picture as follows. For any arbitrary Schrödinger operator A, we have the Heisenberg operator $A_H(t)$ defined as (see (1.20), here we have writ-ten the Heisenberg picture label as a subscript unlike in the last section and operators without a picture label are assumed to be in the Schrödinger picture for simplicity)

$$
A_H(t) = e^{i\mathcal{H}t} A e^{-i\mathcal{H}t}. \tag{1.35}
$$

It is clear now that, for a general thermal correlation function (ensemble average) of two Heisenberg operators $A_H(t)$ and $B_H(t')$, we can write (see (1.13))

$$
\begin{aligned}
\langle A_H(t) B_H(t') \rangle_\beta &= \text{Tr}\left(\rho(\beta) A_H(t) B_H(t') \right) \\
&= Z^{-1}(\beta)\,\text{Tr}\left(e^{-\beta\mathcal{H}} A_H(t) B_H(t') \right) \\
&= Z^{-1}(\beta)\,\text{Tr}\left(e^{-\beta\mathcal{H}} A_H(t) e^{\beta\mathcal{H}} e^{-\beta\mathcal{H}} B_H(t') \right) \\
&= Z^{-1}(\beta)\,\text{Tr}\left(A_H(t+i\beta) e^{-\beta\mathcal{H}} B_H(t') \right) \\
&= Z^{-1}(\beta)\,\text{Tr}\left(e^{-\beta\mathcal{H}} B_H(t') A_H(t+i\beta) \right) \\
&= \langle B_H(t') A_H(t+i\beta) \rangle_\beta,
\end{aligned}
\tag{1.36}
$$

where we have used the fact that the Hamiltonians in the two pictures (Schrödinger and Hamiltonian) are the same, namely, $\mathcal{H}_H = \mathcal{H}$ (see discussion after (1.20)), the cyclicity property of the trace operation as well as (1.35), namely,

$$
e^{-\beta\mathcal{H}} A_H(t)\, e^{\beta\mathcal{H}} = e^{i\mathcal{H}(t+i\beta)} A\, e^{-i\mathcal{H}(t+i\beta)} = A_H(t+i\beta). \tag{1.37}
$$

It is important to note that (1.36) holds independent of the Grassmann parities of the operators A and B, namely, it holds for both bosonic as well as fermionic operators. Relations of the type (1.36) are known as the Kubo-Martin-Schwinger (KMS) relations. Let us note, in particular, from (1.36), that

$$
\langle A_H(t) A_H(t') \rangle_\beta = \langle A_H(t') A_H(t+i\beta) \rangle_\beta. \tag{1.38}
$$

As we will see later, this relation leads to periodicity and antiperiodicity properties in bosonic and fermionic two point Green's functions at finite temperature. The KMS conditions are also extremely useful in studying the asymptotic behavior of various matrix elements as well as dispersion relations at finite temperature. (It is worth emphasizing here that cyclicity of the trace may fail if the trace diverges. Such a situation, as we will see later in chapter **3**, may arise when there is a spontaneous breakdown of a symmetry.)

1.4 Matsubara formalism

It is clear that the main idea in the study of equilibrium statistical mechanics is to evaluate the partition function for the system.

In general, however, the partition function of an interacting system cannot be evaluated exactly. From the form of (1.9), we see that even a perturbative expansion in powers of a coupling constant seems formidable, primarily because we have a sum over the expectation values in all possible states in the Hilbert space and there is an infinite number of such states in any quantum field theory. The Matsubara formalism provides a way of evaluating the partition function perturbatively using a diagrammatic method which is analogous to what is used in conventional quantum field theories at zero temperature. There are several ways to introduce the Matsubara formalism, the most popular being the path integral method. Let us, instead, describe the operatorial method first and then go over to the path integral method. In either of the approaches, the crucial observation lies in the fact that the Boltzmann factor in (1.1) has the form of a time evolution operator for a negative imaginary time.

Since our goal is to calculate the partition function $Z(\beta)$, let us recall from (1.9) that we have

$$Z(\beta) = \mathrm{Tr}\ e^{-\beta \mathcal{H}} = \mathrm{Tr}\ \tilde{\rho}(\beta), \tag{1.39}$$

where we have identified

$$\tilde{\rho}(\beta) = e^{-\beta \mathcal{H}} = Z(\beta)\rho(\beta). \tag{1.40}$$

Namely, we can think of $\tilde{\rho}(\beta)$ as the Boltzmann factor which is the unnormalized density matrix or the scaled density matrix. Furthermore, we note that if the total (dynamical) Hamiltonian, H, of the system can be separated into a free part and an interaction part (earlier we had used H_I for the interaction part, but here we are writing the picture label as a subscript and, therefore, denoting the interaction Hamiltonian with a prime)

$$H = H_0 + H', \tag{1.41}$$

then, we can write (all the heat bath characteristics in (1.5) go into $\mathcal{H}_0$ while H' contains only the interaction terms beyond the quadratic terms in the Hamiltonian)

$$\mathcal{H} = \mathcal{H}_0 + H'. \tag{1.42}$$

In this case, the scaled density matrix in (1.40) can also be written as

$$\tilde{\rho}(\beta) = e^{-\beta \mathcal{H}} = \tilde{\rho}_0(\beta)U(\beta), \tag{1.43}$$

where we have defined

$$\tilde{\rho}_0(\beta) \equiv e^{-\beta \mathcal{H}_0}, \tag{1.44}$$

which allows us to identify

$$U(\beta) = (\tilde{\rho}_0)^{-1}(\beta)\tilde{\rho}(\beta) = e^{\beta \mathcal{H}_0} e^{-\beta \mathcal{H}}. \tag{1.45}$$

(Remember that $\mathcal{H}_0$ and H' and, therefore, $\mathcal{H}$, need not commute so that the two exponents in (1.45) cannot be added trivially.)

The scaled density matrix, as we see from (1.43)-(1.44), can be thought of as an initial value problem for the Bloch equation in the finite time interval $0 \leq \tau \leq \beta$ (we use partial derivative here to allow for the fact that the density matrices may depend on other quantities such as the equilibrium characteristics, see, for example, (1.5))

$$\frac{\partial \tilde{\rho}_0(\tau)}{\partial \tau} = -\mathcal{H}_0 \, \tilde{\rho}_0(\tau),$$

$$\frac{\partial \tilde{\rho}(\tau)}{\partial \tau} = -\mathcal{H}\tilde{\rho}(\tau) = -(\mathcal{H}_0 + H')\tilde{\rho}(\tau). \tag{1.46}$$

From equations (1.45)-(1.46), it follows that the evolution equation satisfied by $U(\tau)$, in this interval, is given by (recall that $\frac{\partial (\tilde{\rho}_0)^{-1}(\tau)}{\partial \tau} = -(\tilde{\rho}_0)^{-1}(\tau)\frac{\partial \tilde{\rho}_0(\tau)}{\partial \tau}(\tilde{\rho}_0)^{-1}(\tau)$)

$$\begin{aligned}
\frac{\partial U(\tau)}{\partial \tau} &= \frac{\partial (\tilde{\rho}_0)^{-1}(\tau)}{\partial \tau}\tilde{\rho}(\tau) + (\tilde{\rho}_0)^{-1}(\tau)\frac{\partial \tilde{\rho}(\tau)}{\partial \tau} \\
&= \left((\tilde{\rho}_0)^{-1}(\tau)\mathcal{H}_0\tilde{\rho}_0(\tau)(\tilde{\rho}_0)^{-1}(\tau)\right)\tilde{\rho}(\tau) - (\tilde{\rho}_0)^{-1}(\tau)\mathcal{H}\tilde{\rho}(\tau) \\
&= (\tilde{\rho}_0)^{-1}(\tau)(\mathcal{H}_0 - \mathcal{H})\tilde{\rho}(\tau) \\
&= -(\tilde{\rho}_0)^{-1}(\tau)H'\tilde{\rho}_0(\tau)\left((\tilde{\rho}_0)^{-1}(\tau)\tilde{\rho}(\tau)\right) \\
&= -H_I'(\tau)\left((\tilde{\rho}_0)^{-1}\tilde{\rho}(\tau)\right) = -H_I'(\tau)U(\tau),
\end{aligned} \tag{1.47}$$

where we have defined

$$H_I'(\tau) = \tilde{\rho}_0^{-1}(\tau)H'\tilde{\rho}_0(\tau) = e^{\tau \mathcal{H}_0} H' e^{-\tau \mathcal{H}_0}. \tag{1.48}$$

The relations (1.45), (1.47) and (1.48) are quite familiar from our earlier discussions of interaction picture at zero temperature (see, for example, (1.29), (1.31) and (1.32)) for negative imaginary times (within a finite interval). We can think of $U(\tau)$ as the time evolution operator in the interaction picture (see (1.31)) for $t \to -i\tau$ (or for Euclidean time) with its imaginary time (or Euclidean time) evolution given by (1.47) (which can be compared with (1.32)). The Matsubara formalism is also sometimes known as the imaginary time formalism for this reason. Relation (1.48) simply defines a modified interaction picture for operators through the relation (see the second line in (1.28) with $t = -i\tau$ and remember that operators without any subindex are in the Schrödinger picture)

$$A_I(\tau) = e^{\tau \mathcal{H}_0} A e^{-\tau \mathcal{H}_0}. \tag{1.49}$$

Such a transformation is not unitary for real τ with a Hermitian (self-adjoint) Hamiltonian $\mathcal{H}_0$. On the other hand, we note that it is unitary in the Euclidean space where the Hamiltonian (energy) also rotates (together with time) to the imaginary axis, but in a counter clockwise direction, namely, $P_0 = \mathcal{H}_0 \to i P_0^{\mathrm{E}} = i \mathcal{H}_0^{\mathrm{E}}$.

As we have already pointed out, from our studies of zero temperature field theory, the evolution equation (1.47) is nothing other than the evolution equation for the time evolution operator in the interaction picture, see (1.32) with $t = -i\tau$. Equation (1.47) can be formally integrated to give

$$U(\beta) = P_\tau \left(e^{-\int_0^\beta d\tau H_I'(\tau)} \right), \tag{1.50}$$

where P_τ stands for (time) ordering in the τ variable (such that the larger τ values appear to the left of the smaller ones and, we use the notation P_τ as opposed to the more conventional T_τ to avoid any possible confusion with "Tr" used in the definition of ensemble averages). Equation (1.50) is reminiscent of the S-matrix (scattering matrix) in zero temperature quantum field theory except for the fact that the time integration is over a finite interval along the negative imaginary axis. As in the zero temperature case, we can expand the exponential and each term in the expansion would give rise to a (modified) Feynman diagram and, therefore, thermal quantities (averages) can also have a diagrammatic representation. This is basically the essence of the Matsubara formalism or the imaginary time formalism.

In fact, the discussion of diagrams at finite temperature is quite parallel to that at zero temperature. In particular, we note that Wick's theorem can be naturally generalized to finite temperature in this formalism. Furthermore, if we define (see (1.50) and we assume $\tau_1 > \tau_2$)

$$U(\tau_1, \tau_2) = P_\tau \left(e^{-\int_{\tau_2}^{\tau_1} d\tau\, H_I'(\tau)} \right) = U(\tau_1)U^{-1}(\tau_2), \tag{1.51}$$

then, this operator satisfies the semi-group properties of the zero temperature time evolution operator (see, for example, (1.19) in the Schrödinger picture), namely,

$$U(\tau) = U(\tau, 0),$$

$$U^{-1}(\tau) = U(0, \tau),$$

$$U(\tau_1, \tau_2)U(\tau_2, \tau_3) = U(\tau_1, \tau_3). \tag{1.52}$$

Let us also note that the two point Green's functions can be defined in the Heisenberg picture as (here we are assuming that the field ϕ_H is complex and $0 \leq \tau, \tau' \leq \beta$)

$$\mathcal{G}_\beta(\tau, \tau') = \langle P_\tau(\phi_H(\tau)\phi_H^\dagger(\tau')) \rangle_\beta$$

$$= Z^{-1}(\beta)\, \mathrm{Tr}\left(e^{-\beta\mathcal{H}} P_\tau(\phi_H(\tau)\phi_H^\dagger(\tau')) \right). \tag{1.53}$$

Here ϕ_H can represent either a bosonic or a fermionic field. We have suppressed all dependence of the fields on spatial coordinates as well as any spinorial indices which are not relevant for our discussion. The τ ordering in (1.53), like the time ordering at zero temperature, is sensitive to the Grassmann parity of the field variables and is defined as

$$P_\tau(\phi_H(\tau)\phi_H^\dagger(\tau'))$$

$$= \theta(\tau - \tau')\phi_H(\tau)\phi_H^\dagger(\tau') \pm \theta(\tau' - \tau)\phi_H^\dagger(\tau')\phi_H(\tau), \tag{1.54}$$

where the minus sign in the second term is for fermionic fields which anti-commute.

Incidentally, we note that, by definition (see (1.35) with $t = -i\tau$ and we recall that ϕ and $\phi^\dagger$ are operators in the Schrödinger picture),

$$\phi_H(\tau) = e^{\tau\mathcal{H}}\phi e^{-\tau\mathcal{H}},$$

$$\phi_H^\dagger(\tau') = e^{\tau'\mathcal{H}}\phi^\dagger e^{-\tau'\mathcal{H}}. \tag{1.55}$$

The relation between the interaction picture and the Heisenberg picture, in the present context, takes the form (see (1.35), (1.45) and (1.49))

$$A_H(\tau) = e^{\tau\mathcal{H}} A e^{-\tau\mathcal{H}}$$

$$= e^{\tau\mathcal{H}} e^{-\tau\mathcal{H}_0} A_I(\tau) e^{\tau\mathcal{H}_0} e^{-\tau\mathcal{H}}$$

$$= U^{-1}(\tau) A_I(\tau) U(\tau). \tag{1.56}$$

Using (1.54) and (1.56), it is easy to show that the two point Green's function in (1.53) can also be written as ($0 \leq \tau, \tau' \leq \beta$)

$$\mathcal{G}_\beta(\tau, \tau') = \frac{\text{Tr}\left(e^{-\beta\mathcal{H}} P_\tau(U^{-1}(\tau)\phi_I(\tau)U(\tau)U^{-1}(\tau')\phi_I^\dagger(\tau')U(\tau'))\right)}{\text{Tr}\, e^{-\beta\mathcal{H}}}$$

$$= \frac{\text{Tr}\left(e^{-\beta\mathcal{H}_0}U(\beta) P_\tau(U^{-1}(\tau)\phi_I(\tau)U(\tau)U^{-1}(\tau')\phi_I^\dagger(\tau')U(\tau'))\right)}{\text{Tr}\left(e^{-\beta\mathcal{H}_0}U(\beta)\right)}$$

$$= \frac{\text{Tr}\left(e^{-\beta\mathcal{H}_0} P_\tau(U(\beta)U^{-1}(\tau)\phi_I(\tau)U(\tau)U^{-1}(\tau')\phi_I^\dagger(\tau')U(\tau'))\right)}{\text{Tr}\left(e^{-\beta\mathcal{H}_0}U(\beta)\right)}$$

$$= \frac{\text{Tr}\left(e^{-\beta\mathcal{H}_0} P_\tau(\phi_I(\tau)\phi_I^\dagger(\tau')U(\beta))\right)}{\text{Tr}\left(e^{-\beta\mathcal{H}_0}U(\beta)\right)}$$

$$= \frac{\langle P_\tau(\phi_I(\tau)\phi_I^\dagger(\tau')U(\beta))\rangle_{\beta,0}}{\langle U(\beta)\rangle_{\beta,0}}, \tag{1.57}$$

where we have used the fact that, since β has the largest (τ) value, $U(\beta)$ can be taken inside the τ-ordering P_τ and, furthermore, the order of the bosonic factors (such as U, U^{-1}) does not matter inside P_τ. Here the subscript "0" represents the fact that the averages are calculated in a non-interacting (free) ensemble. This is exactly like the zero temperature results.

In addition to the similarities, we should also note the differences from the zero temperature case. The most significant, from our point of view, is the fact that unlike at zero temperature, in the present case, the "time" variable is integrated over a finite interval. As a result, as we go to higher and higher orders of expansion, the number of such integration regions grows rapidly and it is not clear whether the diagrammatic evaluations can even be carried out in a manner which would be useful to perform calculations.

1.5 Matsubara frequencies

To tackle the problem of evaluating finite temperature diagrams, let us note that even at zero temperature, the calculation of Feynman diagrams is much easier in the momentum space. Correspondingly, we can ask whether going over to the momentum space will help in the evaluation of diagrams at finite temperature. Let us first note some of the essential properties of the two point Green's functions at finite temperature. From the definition in (1.53) as well as from the relation between the Heisenberg and the Schrödinger pictures in (1.55), it is straightforward to show that the two point Green's functions defined in (1.53) depend only on the difference $\tau - \tau'$, namely, $\mathcal{G}_\beta(\tau, \tau') = \mathcal{G}_\beta(\tau - \tau')$. Second, each of the time variables can lie between $0 \leq \tau, \tau' \leq \beta$. Consequently, the argument of the two point function has the range $-\beta \leq (\tau - \tau') \leq \beta$. Furthermore, from the cyclicity properties of the trace as well as from the definitions of time ordering in (1.54), it is easy to show that (here τ stands for the argument of the Green's function, namely, for $(\tau - \tau')$ in the earlier discussion)

$$\mathcal{G}_\beta(-\beta \leq \tau \leq 0) = \pm \mathcal{G}_\beta(\tau + \beta). \tag{1.58}$$

More explicitly, we note that for $-\beta \leq \tau \leq 0$, we can write from (1.53)-(1.55),

$$
\begin{aligned}
\mathcal{G}_\beta(\tau) = \mathcal{G}_\beta(\tau, 0) &= \langle P_\tau(\phi_H(\tau)\phi_H^\dagger(0)))\rangle_\beta = \pm\langle \phi_H^\dagger(0)\phi_H(\tau)\rangle_\beta \\
&= \pm Z^{-1}(\beta)\,\mathrm{Tr}\left(e^{-\beta\mathcal{H}}\,\phi_H^\dagger(0)\phi_H(\tau)\right) \\
&= \pm Z^{-1}(\beta)\,\mathrm{Tr}\left(\phi_H(\tau)\,e^{-\beta\mathcal{H}}\,\phi_H^\dagger(0)\right) \\
&= \pm Z^{-1}(\beta)\,\mathrm{Tr}\left(e^{-\beta\mathcal{H}}e^{\beta\mathcal{H}}\phi_H(\tau)e^{-\beta\mathcal{H}}\phi_H^\dagger(0)\right) \\
&= \pm Z^{-1}(\beta)\,\mathrm{Tr}\left(e^{-\beta\mathcal{H}}\phi_H(\tau + \beta)\phi_H^\dagger(0)\right)
\end{aligned}
$$

$$= \pm \langle \phi_H(\tau + \beta) \phi_H^\dagger(0) \rangle_\beta$$

$$= \pm \langle P_\tau (\phi_H(\tau + \beta) \phi_H^\dagger(0)) \rangle_\beta$$

$$= \pm \mathcal{G}_\beta(\tau + \beta, 0) = \pm \mathcal{G}_\beta(\tau + \beta), \tag{1.59}$$

which is the result given in (1.58). This result can also be seen to follow from the general KMS relation in (1.36) or (1.38) with appropriate identification of times (namely, rotation to negative imaginary times). It simply relates the value of the two point function with a negative argument to that with a positive argument.

We note next that since the Green's functions are defined within a finite time interval (remember that $-\beta \le \tau \le \beta$), it can be written as a Fourier series involving only discrete frequencies, namely, we can write

$$\mathcal{G}_\beta(\tau) = \frac{1}{\beta} \sum_n e^{i\omega_n \tau} \mathcal{G}_\beta(\omega_n), \tag{1.60}$$

where we can write the general form of the frequencies, ω_n, from the (anti) periodicity properties of the Green's function given in (1.58) and (1.59). For example, for periodic boundary condition (for bosons), the Green's function has to satisfy (with $\tau = 0$)

$$\mathcal{G}_\beta(0) = \mathcal{G}_\beta(\beta)$$

$$\text{or,} \quad \frac{1}{\beta} \sum_n \mathcal{G}_\beta(\omega_n) = \frac{1}{\beta} \sum_n e^{i\omega_n \beta} \mathcal{G}_\beta(\omega_n), \tag{1.61}$$

which determines (assuming $\mathcal{G}_\beta(\omega_n)$ is nontrivial)

$$\omega_n = \frac{2n\pi}{\beta}, \quad n = 0, \pm 1, \pm 2, \cdots. \tag{1.62}$$

On the other hand, for Green's functions (for fermions) satisfying anti-periodic boundary condition (see (1.58) and (1.59)), we have

$$\mathcal{G}_\beta(0) = -\mathcal{G}_\beta(\beta)$$

$$\text{or,} \quad \frac{1}{\beta} \sum_n \mathcal{G}_\beta(\omega_n) = -\frac{1}{\beta} \sum_n e^{i\omega_n \beta} \mathcal{G}_\beta(\omega_n), \tag{1.63}$$

leading to

$$\omega_n = \frac{(2n+1)\pi}{\beta}, \quad n = 0, \pm 1, \pm 2, \cdots. \tag{1.64}$$

In general, we note from (1.62) and (1.64) that we can write $\omega_n = \frac{n\pi}{\beta}$, $n = 0, \pm 1, \pm 2, \cdots$ with even integer values (which include 0) corresponding to bosons and odd integer values corresponding to fermions. These frequencies are conventionally called the Matsubara frequencies. In general, therefore, we can write

$$\mathcal{G}_\beta(\tau) = \frac{1}{\beta} \sum_{-\infty}^{\infty} e^{i\omega_n \tau} \mathcal{G}_\beta(\omega_n), \quad \omega_n = \frac{n\pi}{\beta}, \, n = 0, \pm 1, \pm 2, \cdots .$$

$$(1.65)$$

The inverse Fourier transform, $\mathcal{G}_\beta(\omega_n)$, can be obtained to be

$$\mathcal{G}_\beta(\omega_n) = \int_{-\beta}^{\beta} d\tau \, e^{-i\omega_n \tau} \mathcal{G}_\beta(\tau).$$

$$(1.66)$$

This is easily seen in the following way. For example, for bosons with $\omega_n = \frac{2n\pi}{\beta}$, let us note the following identity (the derivation of which is given in the second appendix at the end of this chapter)

$$\frac{1}{\beta} \sum_{-\infty}^{\infty} e^{\frac{2in\pi(\tau-\tau')}{\beta}} = \lim_{N \to \infty} S_N(\tau - \tau') = \lim_{N \to \infty} \frac{1}{\beta} \sum_{-N}^{N} e^{\frac{2in\pi(\tau-\tau')}{\beta}}$$

$$= \sum_{m=-\infty}^{\infty} \delta(\tau - \tau' + m\beta)$$

$$= \delta_\beta^{(\text{per.})}(\tau - \tau'),$$

$$(1.67)$$

where $\delta_\beta^{(\text{per.})}(\tau - \tau')$ is known as the periodic delta function with period β (namely, it remains the same under $(\tau - \tau') \to (\tau - \tau') + \beta$) and is quite useful in the study of periodic potentials in quantum mechanics. On the other hand, for fermions with $\omega_n = \frac{(2n+1)\pi}{\beta}$, one can show that (see the second appendix at the end of this chapter)

$$\frac{1}{\beta} \sum_{-\infty}^{\infty} e^{\frac{(2n+1)i\pi(\tau-\tau')}{\beta}} = \sum_{m=-\infty}^{\infty} (-1)^m \delta(\tau - \tau' + m\beta)$$

$$= \delta_\beta^{(\text{anti-per.})}(\tau - \tau'),$$

$$(1.68)$$

which is known as the anti-periodic delta function with a period β (it changes sign under $(\tau - \tau') \to (\tau - \tau') + \beta$). However, in either case, if both τ, τ' are restricted to lie in the same fixed interval (in the case of Green's functions, $-\beta \le \tau, \tau' \le \beta$), the sum on the right hand side in (1.67) or (1.68) picks out only the $m = 0$ term so that, in this case, we can write the general result

$$\frac{1}{\beta} \sum_{-\infty}^{\infty} e^{\frac{in\pi(\tau - \tau')}{\beta}} = \delta(\tau - \tau'). \tag{1.69}$$

Using this identity, we obtain from (1.60) and (1.66)

$$\mathcal{G}_\beta(\tau) = \frac{1}{\beta} \sum_n e^{i\omega_n \tau} \mathcal{G}_\beta(\omega_n) = \frac{1}{\beta} \int_{-\beta}^{\beta} d\tau' \sum_n e^{i\omega_n(\tau - \tau')} \mathcal{G}_\beta(\tau')$$

$$= \int_{-\beta}^{\beta} d\tau' \frac{1}{\beta} \sum_{-\infty}^{\infty} e^{\frac{in\pi(\tau - \tau')}{\beta}} \mathcal{G}_\beta(\tau')$$

$$= \int_{-\beta}^{\beta} d\tau' \, \delta(\tau - \tau') \mathcal{G}_\beta(\tau') \equiv \mathcal{G}_\beta(\tau). \tag{1.70}$$

This shows that (1.66) indeed denotes the inverse Fourier transform of the Green's function in (1.60).

The quantization conditions for the boson and fermion energies, given respectively in (1.62) and (1.64), can also be alternatively derived from the inverse Fourier transform of the Green's function as follows. Let us denote, as before, $\omega_n = \frac{n\pi}{\beta}, n = 0, \pm 1, \pm 2, \cdots$. Decomposing the integral in (1.66) into positive and negative arguments of τ and using the periodicity for bosons and fermions in (1.58), we can write the inverse Fourier transform as

$$\mathcal{G}_\beta(\omega_n) = \int_{-\beta}^{\beta} d\tau \, e^{-i\omega_n \tau} \mathcal{G}_\beta(\tau)$$

$$= \int_{-\beta}^{0} d\tau \, e^{-i\omega_n \tau} \mathcal{G}_\beta(\tau) + \int_{0}^{\beta} d\tau \, e^{-i\omega_n \tau} \mathcal{G}_\beta(\tau)$$

$$= \pm \int_{-\beta}^{0} d\tau\, e^{-i\omega_n \tau} \mathcal{G}_\beta(\tau + \beta) + \int_{0}^{\beta} d\tau\, e^{-i\omega_n \tau} \mathcal{G}_\beta(\tau)$$

$$= \pm \int_{0}^{\beta} d\tau\, e^{-i\omega_n(\tau - \beta)} \mathcal{G}_\beta(\tau) + \int_{0}^{\beta} d\tau\, e^{-i\omega_n \tau} \mathcal{G}_\beta(\tau)$$

$$= (1 \pm e^{i\omega_n \beta}) \int_{0}^{\beta} d\tau\, e^{-i\omega_n \tau} \mathcal{G}_\beta(\tau)$$

$$= (1 \pm (-1)^n) \int_{0}^{\beta} d\tau\, e^{-i\omega_n \tau} \mathcal{G}_\beta(\tau), \tag{1.71}$$

where we have used $\omega_n = \frac{n\pi}{\beta}$ in the last step. This shows that for bosons (the upper sign), $\mathcal{G}_\beta(\omega_n)$ vanishes for odd n, while for fermions (the lower sign) it vanishes when n is even. Thus we conclude, from (1.60) and (1.71), that the Fourier transform of the two point function $\mathcal{G}_\beta(\tau)$ and the inverse Fourier transform $\mathcal{G}_\beta(\omega_n)$ can be written, in general, as

$$\mathcal{G}_\beta(\tau) = \frac{1}{\beta} \sum_n e^{i\omega_n \tau} \mathcal{G}_\beta(\omega_n),$$

$$\mathcal{G}_\beta(\omega_n) = \int_{-\beta}^{\beta} d\tau\, e^{-i\omega_n \tau} \mathcal{G}_\beta(\tau), \tag{1.72}$$

where

$$\omega_n = \begin{cases} \frac{2n\pi}{\beta} & \text{for bosons,} \\ \frac{(2n+1)\pi}{\beta} & \text{for fermions.} \end{cases} \tag{1.73}$$

These are commonly referred to as the Matsubara frequencies.

In contrast to the imaginary time variable τ at finite temperature (in the Green's function $\mathcal{G}_\beta(\tau)$) taking values in a finite interval, $-\beta \leq \tau \leq \beta$, the spatial coordinates have an infinite range just as in the case of zero temperature field theory and, therefore, Fourier transformation of the spatial coordinates does not give rise to any new feature. Thus, putting in all the coordinates, we can write (see (1.72))

$$\mathcal{G}_\beta(\tau, \mathbf{x}) = \frac{1}{\beta} \sum_n \int \frac{d^3k}{(2\pi)^3} \, e^{i(\omega_n \tau + \mathbf{k} \cdot \mathbf{x})} \, \mathcal{G}_\beta(\omega_n, \mathbf{k}),$$

$$\mathcal{G}_\beta(\omega_n, \mathbf{k}) = \int_{-\beta}^{\beta} d\tau \int d^3x \, e^{-i(\omega_n \tau + \mathbf{k} \cdot \mathbf{x})} \, \mathcal{G}_\beta(\tau, \mathbf{x}), \tag{1.74}$$

where we have assumed that we are in four space-time dimensions (the spatial integrals can, otherwise, be appropriately defined in other dimensions) and the allowed frequencies are as defined in (1.73). Equation (1.74) sets the rules for taking the Fourier transformation of the propagator in going from Minkowski space to finite temperature Euclidean space, namely,

$$\int_{-\infty}^{\infty} \frac{dk^0}{2\pi} \rightarrow \frac{1}{\beta} \sum_{n=-\infty}^{\infty},$$

$$\int_{-\infty}^{\infty} \frac{d^3k}{(2\pi)^3} \rightarrow \int_{-\infty}^{\infty} \frac{d^3k}{(2\pi)^3}. \tag{1.75}$$

We are now ready to derive the form of the propagator for a bosonic (or fermionic) theory at finite temperature.

For example, let us consider the bosonic (free) Klein-Gordon theory of a massive, real scalar field ϕ described by the Lagrangian density ($\mu = 0, 1, 2, 3$ in four space-time dimensions)

$$\mathcal{L} = \frac{1}{2} \partial^\mu \phi \partial_\mu \phi - \frac{m^2}{2} \phi^2, \quad \partial_\mu = \frac{\partial}{\partial x^\mu}. \tag{1.76}$$

In this case, we know that the zero temperature Green's function satisfies (our metric in the Minkowski space is $(+,-,-,-)$)

$$(\partial_\mu \partial^\mu + m^2)G(x) = (\partial_t^2 - \boldsymbol{\nabla}^2 + m^2)G(x) = -\delta^4(x). \tag{1.77}$$

The momentum space Green's function, in this case, has the form

$$G(k) = G(k_0, \mathbf{k}) = \frac{1}{k_0^2 - \mathbf{k}^2 - m^2}. \tag{1.78}$$

Going over to Euclidean space (when rotating to Euclidean space or imaginary time, we let $x^0 = t \rightarrow -i\tau$, $k^0 = k_0 \rightarrow ik_4 = i\omega_n$ and

$G \to -\mathcal{G}_\beta$, we will discuss briefly the rotation to Euclidean space at the end of this chapter in a short appendix), we note from (1.77) that the finite temperature Green's function (in the absence of a chemical potential) would satisfy the equation

$$\left(\frac{\partial^2}{\partial \tau^2} + \boldsymbol{\nabla}^2 - m^2\right)\mathcal{G}_\beta(\tau, \mathbf{x}) = -\delta(\tau)\,\delta^3(x). \tag{1.79}$$

There are two equivalent ways we can determine the momentum space Green's function for the scalar boson at finite temperature. First, we can take the zero temperature Euclidean Green's function given in (1.78) and rotate it to Euclidean space which gives

$$G(k_0, \mathbf{k}) = \frac{1}{k_0^2 - \mathbf{k}^2 - m^2}$$

$$\to \frac{1}{-\omega_n^2 - \mathbf{k}^2 - m^2} = -\mathcal{G}_\beta(\omega_n, \mathbf{k}),$$

$$\text{or,} \quad \mathcal{G}_\beta(\omega_n, \mathbf{k}) = \frac{1}{\omega_n^2 + \mathbf{k}^2 + m^2}. \tag{1.80}$$

Alternatively, we can take the Fourier transformation of the equation (Fourier transformations in Minkowski space and in Euclidean space will be briefly discussed in the first appendix at the end of this chapter) satisfied by the Green's function in (1.79) using (1.74) and solve for it in momentum space, namely,

$$\left(\frac{\partial^2}{\partial \tau^2} + \boldsymbol{\nabla}^2 - m^2\right)\frac{1}{\beta}\sum_n \int \frac{\mathrm{d}^3 k}{(2\pi)^3}\, e^{i(\omega_n \tau + \mathbf{k}\cdot\mathbf{x})}\mathcal{G}_\beta(\omega_n, \mathbf{k})$$

$$= -\frac{1}{\beta}\sum_n \int \frac{\mathrm{d}^3 k}{(2\pi)^3}\, e^{i(\omega_n \tau + \mathbf{k}\cdot\mathbf{x})}, \tag{1.81}$$

which leads to

$$\frac{1}{\beta}\sum_n \int \frac{\mathrm{d}^3 k}{(2\pi)^3}e^{i(\omega_n \tau + \mathbf{k}\cdot\mathbf{x})}\left((-\omega_n^2 - \mathbf{k}^2 - m^2)\mathcal{G}_\beta(\omega_n, \mathbf{k}) + 1\right) = 0,$$

$$\text{or,} \quad \mathcal{G}_\beta(\omega_n, \mathbf{k}) = \frac{1}{\omega_n^2 + \mathbf{k}^2 + m^2}, \tag{1.82}$$

and this coincides with (1.80). In this derivation (namely, on the right hand side in (1.81)), we have used the representation of $\delta(\tau)$

as a sum given in (1.69) with the appropriate ω_n given in (1.73). The Green's function for the fermions can similarly be obtained in a straightforward manner. The important thing to note from (1.82) (or (1.80)) is that, unlike the zero temperature case, here the Green's functions do not have singularities for real values of energy and momentum variables.

The analysis of the propagators (Green's functions), in the momentum space, simplifies the diagrammatic calculations at finite temperature considerably. In fact, it makes the finite temperature calculations completely parallel (at least qualitatively) to the zero temperature case. The interaction vertices of the theory are defined as in the zero temperature (Minkowski space) case (but with a multiplicative factor of (-1)). Only the forms of the propagators are different from the zero temperature (Minkowski) case and carry the temperature dependence. As we will see, this is sufficient to introduce new features into the quantum mechanical systems at finite temperature.

1.6 Path integral formulation

The Matsubara formalism or the imaginary time formalism can also be very easily understood within the path integral context. Let us recall that the transition amplitude in a quantum field theory at zero temperature (in Minkowski space) has the functional representation given by (we use $\hbar = c = 1$)

$$\langle \phi(\mathbf{x}_f, t_f) | \phi(\mathbf{x}_i, t_i) \rangle = \langle \phi_f | e^{-iH(t_f - t_i)} | \phi_i \rangle$$

$$= N' \int \mathcal{D}\phi \, e^{iS}, \tag{1.83}$$

where ϕ is the basic quantum field variable, N' is an irrelevant normalization constant and the subindices i, f stand for the initial and final coordinates of the path. The action S is defined to be (normally we identify $t_i \to -\infty$ (infinite past) and $t_f \to \infty$ (infinite future))

$$S[\phi] = \int\limits_{t_i}^{t_f} dt \int d^3x \, \mathcal{L}, \tag{1.84}$$

with $\mathcal{L}$ representing the appropriate Lagrangian density for the system under study. The functional integral (path integral), in this case, is defined over paths which satisfy

$$\phi(\mathbf{x}_f, t_f) = \phi_f,$$
$$\phi(\mathbf{x}_i, t_i) = \phi_i, \tag{1.85}$$

namely, the end points are held fixed (and not integrated over in (1.83)).

Given (1.83)-(1.85), it is now clear that if we define

$$t_f - t_i = -i\beta, \tag{1.86}$$

and identify the field variables at the end points (upto a phase) $\phi_f = \pm\phi_i$ as well and integrate over the fields at the end points (which is equivalent to taking the trace) then, we can write the partition function for any quantum system as

$$Z(\beta) = \text{Tr}\, e^{-\beta\mathcal{H}} = \int d\phi_f \, \langle\phi_f|e^{-\beta\mathcal{H}}|\phi_f\rangle$$

$$= N' \int \mathcal{D}\phi\, e^{-S_E}. \tag{1.87}$$

The path integral measure $\mathcal{D}\phi$ in (1.87) includes integration over the end point field configurations. Here S_E is related to the Euclidean (imaginary time) action as (if the chemical potential is nonzero)

$$\mathcal{S}_E = S_E + \beta\mu N$$

$$= \int_0^\beta d\tau \left[\left(\int d^3x\, \mathcal{L}_E\right) + \mu N\right], \tag{1.88}$$

and N denotes the number operator. The (anti) periodicity conditions (end points are identified up to a phase) which the field variables are assumed to satisfy are given by

$$\phi(t_f) = \pm\phi(t_i),$$

$$\text{or,} \quad \phi(t_i - i\beta) = \pm\phi(t_i), \tag{1.89}$$

and, with the identification $t_i = 0$ (remember $t \to -i\tau$ in the imaginary time formalism), leads to

$$\phi(\mathbf{x}, \beta) = \pm\phi(\mathbf{x}, 0), \tag{1.90}$$

depending on whether the field variables are bosonic or fermionic. We emphasize again that the end points, for a path integral representation of the partition function, are integrated over in (1.87) as well unlike in the case of the genrating functional in zero temperature field theories (see (1.83)). (Periodicity condition for the bosons is obvious from (1.85) and (1.87). Namely, the partition function involves a trace so that the initial and the final states in (1.87) must be the same and integrated over for the trace operation. Anti-periodicity for fermions, on the other hand, is not so obvious and we will discuss this in chapter **5**, more precisely in section **5.5**.) We note here that the contour of integration in the complex time plane, in the present formalism, is along the negative imaginary axis as shown in Fig. 1.1 (recall that $t \to -i\tau$ and $0 \leq \tau \leq \beta$).

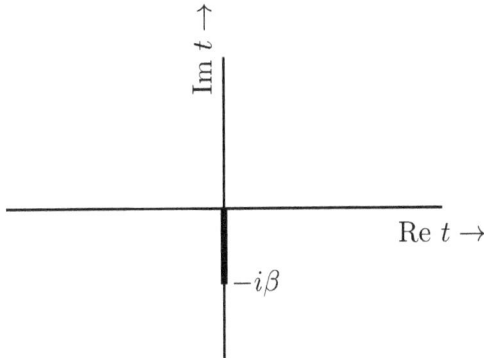

Figure 1.1: Time interval, in the complex t-plane, in the imaginary time formalism.

We make a short digression here to explain the derivation of (1.88). The chemical potential can be thought of as arising from a constant electrostatic potential for charged particles. Therefore, with minimal coupling, we can write the Lagrangian density for a complex scalar field in Minkowski space with a chemical potential, for example, as (the same can be done for a fermion field as well)

$$\mathcal{L}^{(\mu)} = ((\partial_t + i\mu)\phi)^*(\partial_t + i\mu)\phi - \boldsymbol{\nabla}\phi^* \cdot \boldsymbol{\nabla}\phi - m^2\phi^*\phi$$

$$= \partial_\mu \phi^* \partial^\mu \phi - (m^2 - \mu^2)\phi^*\phi + \mu N(\mathbf{x}, t)$$

$$= \mathcal{L} + \mu N(\mathbf{x}, t), \tag{1.91}$$

where

$$N(\mathbf{x}, t) = i\,\phi^* \overset{\leftrightarrow}{\partial_t}\,\phi, \tag{1.92}$$

denotes the number density operator. As a result, the Lagrangian has the form

$$L^{(\mu)} = \int d^3x\,(\mathcal{L} + \mu N(\mathbf{x}, t)) = L + \mu N, \tag{1.93}$$

with $N = \int d^3x\,N(\mathbf{x}, t)$ is the number operator. Under a Euclidean rotation $t \to -i\tau$, $\partial_t \to i\partial_\tau$. Furthermore, thinking of the chemical potential as a constant electrostatic potential, we see that it must rotate as the time derivative so that $\mu \to i\mu$ and N, being a conserved charge, must also rotate likewise, namely, $N \to iN$ (see also (1.92)). Therefore, the Lagrangian with a chemical potential (see, for example, (1.91) and (1.93)) rotates to Euclidean space as

$$L^{(\mu)} \to -(L_E + \mu N), \tag{1.94}$$

so that the action behaves as

$$S^{(\mu)} \to \int_0^\beta (-i)d\tau\,(-)(L_E + \mu N)$$

$$= i(S_E + \beta\mu N) = i\mathcal{S}_E, \tag{1.95}$$

which can be compared with (1.88). Here we have used the fact that, at finite temperature, the Euclidean time interval is finite as well as the fact that the number operator, being a conserved charge, is independent of time (constant) so that the second term (in the first line of (1.95)) can be trivially integrated over (Euclidean) time. (The overall phase i in (1.95) combines with the phase in the exponent of (1.83) to give a negative sign in the exponent in (1.87).)

The path integral formulation of the partition function is quite interesting in the sense that it shows how parallel the zero temperature and the finite temperature descriptions of quantum field theories are, at least qualitatively. The definitions of the diagrams – 1PI, connected etc. – are exactly the same and the respective graphs can be

obtained from an expansion of the path integral. The only difference is that the field variables now have to satisfy (anti) periodicity
conditions (see (1.90)) and correspondingly should be expanded in
an appropriate basis. This leads to the energy values being discrete
as in (1.73). (This is completely analogous to quantizing a quantum
mechanical system in a box – it is, however, a one dimensional box
in the τ direction. The topology of space-time, in this description,
correspondingly becomes that of $\mathbf{R}^3 \times S^1$.)

It is also worth noting at this point that the Matsubara (imaginary time) formalism has been developed completely within the context of equilibrium systems. We have given up the time variable in
favor of the equilibrium temperature. This method is ideal for studying static, equilibrium properties of a quantum system. Time dependence can be introduced into Green's functions through a nontrivial
analytic continuation which we will discuss later. However, that
would only describe slow time developments of quantum mechanical
systems in thermal equilibrium. But this formalism is completely
unsuitable for studying non-equilibrium phenomena such as phase
transitions. Furthermore, dynamical relations, such as the Ward
identities of gauge theories, are hard to study in this formalism. But
this is the formalism that is most widely used in the calculation of
equilibrium thermodynamic quantities.

To summarize, then, in the Matsubara formalism, the partition
function is given a path integral representation where the action for
the path integral corresponds to the Euclidean action of the original
system with the time integration over a finite interval. Furthermore,
the path integral is carried over paths where the fundamental field
variables of this Euclidean action are supposed to satisfy (anti) periodic boundary conditions with a period of β. The Feynman rules for
this theory can be read out from the path integral. The vertices are
the same as that of the zero temperature Euclidean theory. However,
the propagators now correspond to the inverses of the operators of
the quadratic part of the Lagrangian defined in a space of (anti) periodic functions. That gives the propagators a nontrivial dependence
on the temperature (see, for example, (1.72)-(1.74)). In this formalism, the propagator does not have a simple decomposition into a zero
temperature part and a genuinely temperature dependent part (although graphs calculated with them do) which, as we will see later, is
a useful property of the real time formulations of finite temperature
field theories.

1.7 Applications

As we have seen earlier, in the imaginary time formalism, the only difference between the zero temperature and the finite temperature field theories lies in the form of the propagator which carries all the temperature dependence. The vertices at finite temperature are exactly the same as those at zero temperature (in a Euclidean theory). Thus, given any quantum field theory, we can carry out calculations of thermodynamic interest perturbatively by calculating Feynman diagrams. Let us note here that the propagators for the spin 0 and spin $\frac{1}{2}$ fields are given respectively by (see, for example, (1.73) and (1.82) for the spin zero case)

$$
\mathcal{G}_\beta(\mathbf{k}, \omega_n) = \frac{1}{\omega_n^2 + \mathbf{k}^2 + m^2}
$$

$$
= \frac{1}{(4n^2\pi^2/\beta^2) + \mathbf{k}^2 + m^2},
$$

$$
\mathcal{S}_\beta(\mathbf{k}, \omega_n) = \frac{\gamma^0 \omega_n + \gamma \cdot \mathbf{k} - m}{\omega_n^2 + \mathbf{k}^2 + m^2}
$$

$$
= \frac{\gamma^0((2n+1)\pi/\beta) + \gamma \cdot \mathbf{k} - m}{((2n+1)^2\pi^2/\beta^2) + \mathbf{k}^2 + m^2}. \tag{1.96}
$$

With these, let us next calculate the one loop mass correction in a self-interacting bosonic theory.

1.7.1 One loop mass correction. Let us consider the self-interacting ϕ^4 theory described by the Lagrangian density

$$
\mathcal{L}(\phi) = \frac{1}{2}\partial_\mu\phi\partial^\mu\phi - \frac{m^2}{2}\phi^2 - \frac{\lambda}{4!}\phi^4. \tag{1.97}
$$

According to our discussion so far, if we want to calculate quantities at finite temperature, we should treat time as an imaginary parameter in which case the theory leads to a Euclidean field theory (there are no upper and lower indices in Euclidean field theory unlike in Minkowski space)

$$
\mathcal{L}_E(\phi) = \frac{1}{2}\partial_\mu^E\phi\partial_\mu^E\phi + \frac{m^2}{2}\phi^2 + \frac{\lambda}{4!}\phi^4. \tag{1.98}
$$

The propagator for the theory can be read out from (1.96) and the vertices are those of the Euclidean theory (the vertices are obtained from $(-S_E)$). The diagrammatic calculations can, then, be carried out analogous to the zero temperature case. The only difference is that since the energy values are now quantized, the internal energy integrals have to be replaced by sums over discrete values, namely,

$$\int \frac{\mathrm{d}^4 k_E}{(2\pi)^4} \longrightarrow \frac{1}{\beta} \sum_n \int \frac{\mathrm{d}^3 k}{(2\pi)^3}. \tag{1.99}$$

In this theory, the only nontrivial contribution, that arises at one loop, is to the self-energy for which the Feynman diagram is shown in Fig. 1.2. The self-energy leads only to a mass correction which has the form (the symmetry factor for this diagram is $\frac{1}{2}$)

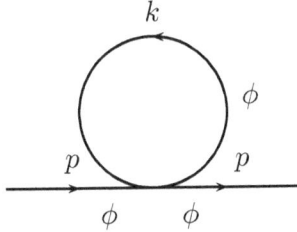

Figure 1.2: One loop correction to the self-energy in the ϕ^4 theory.

$$-\Delta m^2 = -\frac{\lambda}{2\beta} \sum_n \int \frac{\mathrm{d}^3 k}{(2\pi)^3} \frac{1}{(4n^2\pi^2/\beta^2) + \omega_k^2}$$

$$= -\frac{\lambda}{2\beta} \left(\frac{\beta}{2\pi}\right)^2 \sum_n \int \frac{\mathrm{d}^3 k}{(2\pi)^3} \frac{1}{n^2 + (\beta\omega_k/2\pi)^2}, \tag{1.100}$$

where we have used the fact that the quartic interaction vertex (in the Euclidean theory), in momentum space, is $(-\lambda)$ (with a momentum conserving delta function) and have introduced the notation

$$\omega_k = (\mathbf{k}^2 + m^2)^{\frac{1}{2}}. \tag{1.101}$$

The sum in (1.100) can be easily evaluated using the method of residues which gives[3]

$$\sum_{n=-\infty}^{\infty} \frac{1}{n^2 + y^2} = \frac{\pi}{y} \coth \pi y, \quad \text{for} \quad y > 0, \tag{1.102}$$

which can also be written equivalently as

$$\sum_{n=1}^{\infty} \frac{1}{n^2 + y^2} = -\frac{1}{2y^2} + \frac{\pi}{2y} \coth \pi y, \quad \text{for} \quad y > 0. \tag{1.103}$$

For later use, let us note that

$$\sum_{n=-\infty}^{\infty} \frac{1}{n + iy} = \sum_{n=-\infty}^{\infty} \frac{1}{-n + iy}$$

$$= \frac{1}{2} \sum_{n=-\infty}^{\infty} \left(\frac{1}{n + iy} + \frac{1}{-n + iy} \right)$$

$$= \frac{1}{2} \times (-2iy) \sum_{n=-\infty}^{\infty} \frac{1}{n^2 + y^2}$$

$$= (-iy) \times \frac{\pi}{y} \coth \pi y = -i\pi \coth \pi y, \tag{1.104}$$

where we have used (1.102). Using (1.102) in (1.100), we obtain

$$\Delta m^2 = \frac{\lambda \beta}{8\pi^2} \int \frac{d^3k}{(2\pi)^3} \frac{2\pi^2}{\beta \omega_k} \coth \left(\frac{\beta \omega_k}{2} \right)$$

$$= \frac{\lambda}{4} \int \frac{d^3k}{(2\pi)^3} \frac{1}{\omega_k} \coth \left(\frac{\beta \omega_k}{2} \right). \tag{1.105}$$

There are several things to note from (1.105). First, using the identity

$$\coth \beta x = 1 + 2n_B(2x), \tag{1.106}$$

where $n_B(x)$ represents the bosonic distribution function

$$n_B(x) = \frac{1}{e^{\beta x} - 1}, \tag{1.107}$$

---

[3]See, for example, problem 26 in chapter 7 of M. R. Spiegel, *Theory and Problems of Complex Variables*, McGraw-Hill (1964).

it is clear that the mass correction in (1.105) can be written as a sum of two terms – one corresponding to a zero temperature contribution and the other explicitly temperature dependent,

$$\Delta m^2 = \Delta m_0^2 + \Delta m_T^2$$

$$= \frac{\lambda}{4} \int \frac{d^3k}{(2\pi)^3} \frac{1}{\omega_k} + \frac{\lambda}{2} \int \frac{d^3k}{(2\pi)^3} \frac{1}{\omega_k} \frac{1}{e^{\beta\omega_k} - 1}. \tag{1.108}$$

Second, the divergence in the expression for the mass correction in (1.108) is entirely contained in the zero temperature part. The temperature dependent part is free from ultraviolet divergences (the exponential in the distribution function suppresses the ultraviolet contributions coming from large values of $|\mathbf{k}|$). Therefore, the zero temperature counter terms are sufficient to renormalize the theory. We will come to a further discussion of this point in the next chapter. Furthermore, unlike the zero temperature case, the integral in (1.105), even for this simple process, cannot be evaluated in a closed form.

Since finite temperature calculations are notoriously difficult, we discuss in some detail how the temperature dependent integral is evaluated at high temperature ($\beta m \ll 1$). Let us note that, since the integrand in the finite temperature part of (1.108) (as well as the zero temperature part) is radially symmetric, we can do the angular integral trivially which leads to

$$\Delta m_T^2 = \frac{\lambda}{2} \int_0^\infty \frac{(4\pi)}{(2\pi)^3} \frac{dk\, k^2}{\omega_k} \frac{1}{e^{\beta\omega_k} - 1}$$

$$= \frac{\lambda}{4\pi^2} \int_m^\infty d\omega_k\, (\omega_k^2 - m^2)^{\frac{1}{2}} \frac{1}{e^{\beta\omega_k} - 1}, \tag{1.109}$$

where we have used the definition in (1.101) and changed the variable of integration $k \to \omega_k$. Redefining the variable of integration to $x = \beta\omega_k$ and recalling that, at high temperature, $\beta m \ll 1$, we obtain

$$\Delta m_T^2 = \frac{\lambda}{4\pi^2\beta^2} \int_{\beta m}^\infty dx\, (x^2 - (\beta m)^2)^{\frac{1}{2}} \frac{1}{e^x - 1}$$

$$= \frac{\lambda}{4\pi^2\beta^2} \int_{\beta m}^\infty dx \left(x - \frac{(\beta m)^2}{2x} + O((\beta m)^4) \right) \frac{1}{e^x - 1}$$

$$= \frac{\lambda}{4\pi^2\beta^2} \left(\int_{\beta m}^\infty dx\, \frac{x}{e^x - 1} + O((\beta m)^2) \right). \tag{1.110}$$

To the leading order, at high temperature (β small), therefore, the higher order terms in (1.110) can be neglected. Furhermore, since $\beta m \ll 1$, to leading order, the lower limit can be extended to 0, namely,

$$\int_{\beta m}^{\infty} dx\, f(x) = \int_{0}^{\infty} dx\, f(x + \beta m)$$

$$= \int_{0}^{\infty} dx\, \left(f(x) + (\beta m) f'(x) + \cdots \right)$$

$$= \int_{0}^{\infty} dx\, f(x) + O(\beta m). \tag{1.111}$$

As a result, to leading order at high temperature, (1.110) gives

$$\Delta m_T^2 = \frac{\lambda}{4\pi^2 \beta^2} \int_0^\infty dx\, \frac{x}{e^x - 1} = \frac{\lambda}{4\pi^2 \beta^2} \zeta(2), \tag{1.112}$$

where $\zeta(n)$ denotes Riemann's zeta function[4] and its value for $n = 2$ was already calculated (earlier) by Euler to be $\zeta(2) = \frac{\pi^2}{6}$. As a result, to leading order at high temperature ($\beta m \ll 1$), the temperature dependent part of the mass correction in (1.112) has the form (recall that $\beta = \frac{1}{T}$ in our units of $k = 1$)

$$\Delta m_T^2 = \frac{\lambda}{4\pi^2 \beta^2} \times \frac{\pi^2}{6} + O(\beta m) = \frac{\lambda T^2}{24} + O\left(\frac{m}{T}\right). \tag{1.113}$$

This shows that finite temperature induces a (finite) mass correction for the mass of the boson analogous to that of a particle moving in a (thermal) medium. Furthermore, this mass correction is positive which is at the heart of studies in the restoration of spontaneously broken symmetries, a topic which we will discuss later.

1.7.2 Self-energy at finite temperature. Let us next analyze the general structure of the self-energy which depends on external momentum (unlike the mass correction discussed in the last section). For this purpose we study a (particularly) simple bosonic theory to bring

---

[4]See, for example, page 436 in A. Das, *Field Theory: A Path Integral Approach* (third edition), World Scientific Publishing (2019), for a definition of the Riemann zeta function and the value of $\zeta(2)$ is given in page 434.

out some specific features that arise at finite temperature. Let us consider, as a toy model, an interacting theory of two spin zero bosons described by the Lagrangian density

$$\mathcal{L} = \frac{1}{2}\partial_\mu\phi\partial^\mu\phi - \frac{m^2}{2}\phi^2 + \frac{1}{2}\partial_\mu B\partial^\mu B - \frac{M^2}{2}B^2 - \frac{g}{2}\phi B^2. \quad (1.114)$$

We can think of B as describing a heavy mass particle ($M \gg m$). We can integrate this (heavy) field out since the Lagrangian density depends on it only quadratically and ask for its contribution to the self-energy of the lighter mass field. We emphasize that the Lagrangian density in (1.114) is not a realistic model in any sense, and should only be treated as a toy model to bring out some particular features at finite temperature.

The propagator for the B-field can be read out from (1.96). The vertices are the same as in the zero temperature Euclidean theory (obtained from $(-S_E)$). However, since the energy values are now quantized, as we have discussed earlier, we have to replace the energy integrals in the intermediate loops by sums (see (1.99)). With these rules, the self-energy diagram for the ϕ field at one loop, shown in Fig. 1.3, leads to (the internal loop consists of the B-propagator and the symmetry factor for the diagram is $\frac{1}{2}$)

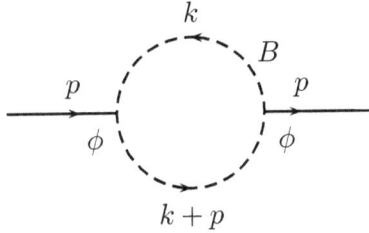

Figure 1.3: Self-energy contribution for the ϕ field coming from an internal loop of the heavy B field.

$$\Pi(\mathbf{p}, p^0) = \frac{(-g)^2}{2\beta} \sum_n \int \frac{\mathrm{d}^3 k}{(2\pi)^3} \frac{1}{\frac{4n^2\pi^2}{\beta^2} + \mathbf{k}^2 + M^2}$$

$$\times \frac{1}{(\frac{2n\pi}{\beta} + p^0)^2 + (\mathbf{k} + \mathbf{p})^2 + M^2}, \quad (1.115)$$

where we have denoted the energy of the external ϕ line as p^0 which is assumed to take discrete values at finite temperature in the imaginary time formalism.

Let us next introduce the notations

$$\omega_k = (\mathbf{k}^2 + M^2)^{\frac{1}{2}},$$

$$\omega_{k+p} = ((\mathbf{k} + \mathbf{p})^2 + M^2)^{\frac{1}{2}}, \tag{1.116}$$

and note the identity

$$\frac{1}{\frac{4n^2\pi^2}{\beta^2} + \omega_k^2} = \left(\frac{\beta}{2\pi}\right)^2 \frac{1}{n^2 + \left(\frac{\beta\omega_k}{2\pi}\right)^2}$$

$$= \frac{i\beta}{2\pi} \frac{1}{2\omega_k} \left(\frac{1}{n + \frac{i\beta\omega_k}{2\pi}} - \frac{1}{n - \frac{i\beta\omega_k}{2\pi}}\right). \tag{1.117}$$

Using this identity for each of the two propagators in (1.115), the self-energy in (1.115) can be rewritten as

$$\Pi(\mathbf{p}, p^0) = \left(\frac{i\beta}{2\pi}\right)^2 \left(\frac{g^2}{2\beta}\right)$$

$$\times \sum_n \int \frac{d^3k}{(2\pi)^3} \frac{1}{4\omega_k\omega_{k+p}} \left(\frac{1}{n + i\frac{\beta\omega_k}{2\pi}} - \frac{1}{n - i\frac{\beta\omega_k}{2\pi}}\right)$$

$$\times \left(\frac{1}{n + \frac{\beta p^0}{2\pi} + i\frac{\beta\omega_{k+p}}{2\pi}} - \frac{1}{n + \frac{\beta p^0}{2\pi} - i\frac{\beta\omega_{k+p}}{2\pi}}\right). \tag{1.118}$$

There are four terms in the integrand in (1.118) and it can be checked that the integrand is symmetric under $p^0 \leftrightarrow -p^0$ (with $n \leftrightarrow -n$ since we are summing over all integers). The sum over n for each term in (1.118) can be simply evaluated using the relation

$$\sum_n \frac{1}{n + ix} \frac{1}{n + iy} = \frac{i}{x - y} \sum_n \left(\frac{1}{n + ix} - \frac{1}{n + iy}\right)$$

$$= \frac{i}{x - y} (-i\pi) (\coth \pi x - \coth \pi y)$$

$$= \frac{\pi}{x - y} (\coth(\pi x) - \coth(\pi y)), \tag{1.119}$$

where we have used (1.104) in the intermediate (second) step. The sum of the four terms in (1.118) can, then, be rearranged using the symmetry $p^0 \leftrightarrow -p^0$ and leads to

$$
-\frac{2\pi^2}{\beta}\left[\left(\frac{1}{\omega_{k+p}-(\omega_k-ip^0)}+\frac{1}{\omega_{k+p}+(\omega_k-ip^0)}\right)\coth\left(\frac{\beta\omega_k}{2}\right)\right.
$$
$$
+\left(\frac{1}{\omega_k-(\omega_{k+p}-ip^0)}+\frac{1}{\omega_k+(\omega_{k+p}-ip^0)}\right)
$$
$$
\left.\times\coth\left(\frac{\beta(\omega_{k+p}-ip^0)}{2}\right)+(p^0\leftrightarrow-p^0)\right]
$$
$$
=-\frac{2\pi^2}{\beta}\left[\frac{2\omega_{k+p}}{\omega_{k+p}^2-(\omega_k-ip^0)^2}\coth\left(\frac{\beta\omega_k}{2}\right)\right.
$$
$$
\left.+\frac{2\omega_k}{\omega_k^2-(\omega_{k+p}-ip^0)^2}\coth\left(\frac{\beta(\omega_{k+p}-ip^0)}{2}\right)+(p^0\leftrightarrow-p^0)\right].
$$
$$
(1.120)
$$

Substituting this into the integrand in (1.118), we obtain

$$
\Pi(\mathbf{p},p^0)=\frac{g^2}{8}\int\frac{d^3k}{(2\pi)^3}\left[\frac{1}{\omega_k}\coth\left(\frac{\beta\omega_k}{2}\right)\frac{1}{\omega_{k+p}^2-(\omega_k-ip^0)^2}\right.
$$
$$
+\frac{1}{\omega_{k+p}}\coth\left(\frac{\beta(\omega_{k+p}-ip^0)}{2}\right)\frac{1}{\omega_k^2-(\omega_{k+p}-ip^0)^2}
$$
$$
\left.+(p^0\leftrightarrow-p^0)\right].
$$
$$
(1.121)
$$

We can now use the fact that the energy for the external lines takes the discrete value $p^0=\frac{2l\pi}{\beta}$ (where l is an integer since the external lines represent bosons). Furthermore, using the periodicity of "coth" for such values of ip^0 (namely, $\coth(x+il\pi)=\coth x$, for l integer), we obtain

$$
\Pi(\mathbf{p},p^0)=\frac{g^2}{8}\int\frac{d^3k}{(2\pi)^3}\left[\frac{1}{\omega_k}\coth\left(\frac{\beta\omega_k}{2}\right)\frac{1}{\omega_{k+p}^2-(\omega_k-ip^0)^2}\right.
$$
$$
+\frac{1}{\omega_{k+p}}\coth\left(\frac{\beta\omega_{k+p}}{2}\right)\frac{1}{\omega_k^2-(\omega_{k+p}-ip^0)^2}
$$
$$
\left.+(p^0\leftrightarrow-p^0)\right].
$$
$$
(1.122)
$$

This simple calculation already brings out the subtleties associated with finite temperature quantum field theories. First, we note that the one loop integral in (1.122) cannot be evaluated in a closed form. We can only evaluate this quantity in the limits of high and low temperatures. But even without evaluating the integral, it is not hard to see that the self-energy at finite temperatures is a non-analytic function of the external momentum and energy at the origin. To see this, let us pretend that p^0 is a continuous variable and analytically continue $\Pi(\mathbf{p}, p^0)$ to continuous values of p^0 which would allow us to take the limit $p^0 \to 0$. The analytic continuation of a function to continuous values of its argument is, in general, not well defined if we know the values of the function only at a set of discrete points. However, if the function is well behaved at infinity (convergent), then this can be done. With this then, we now see from (1.122) that, if we set the external momentum to zero, namely, $\mathbf{p} = 0$ we obtain (note that, in this limit, $\omega_{k+p} = \omega_k$)

$$
\Pi(0, p^0) = \frac{g^2}{8} \int \frac{d^3 k}{(2\pi)^3} \frac{2}{\omega_k} \coth\left(\frac{\beta \omega_k}{2}\right)
$$

$$
\times \left(\frac{1}{2ip^0 \omega_k + (p^0)^2} + \frac{1}{-2ip^0 \omega_k + (p^0)^2} \right)
$$

$$
= \frac{g^2}{8} \int \frac{d^3 k}{(2\pi)^3} \frac{4}{\omega_k} \frac{1}{4\omega_k^2 + (p^0)^2} \coth\left(\frac{\beta \omega_k}{2}\right)
$$

$$
\xrightarrow{p^0 \to 0} \frac{g^2}{8} \int \frac{d^3 k}{(2\pi)^3} \frac{1}{\omega_k^3} \coth\left(\frac{\beta \omega_k}{2}\right). \tag{1.123}
$$

On the other hand, taking the reverse limit (namely, setting $p^0 = 0$ and then taking the limit $\mathbf{p} \to 0$) in (1.122), we obtain

$$
\Pi(\mathbf{p}, 0) = -\frac{g^2}{8} \int \frac{d^3 k}{(2\pi)^3} \frac{2}{\omega_k^2 - \omega_{k+p}^2}
$$

$$
\times \left(\frac{1}{\omega_k} \coth\left(\frac{\beta \omega_k}{2}\right) - \frac{1}{\omega_{k+p}} \coth\left(\frac{\beta \omega_{k+p}}{2}\right) \right)
$$

$$
= -\frac{g^2}{8} \int \frac{d^3 k}{(2\pi)^3} \frac{1}{\omega_k \omega_{k+p}}
$$

$$
\times \left[\frac{1}{\omega_k - \omega_{k+p}} \left(\coth\left(\frac{\beta \omega_k}{2}\right) - \coth\left(\frac{\beta \omega_{k+p}}{2}\right) \right) \right.
$$

$$-\frac{1}{\omega_k + \omega_{k+p}}\left(\coth\left(\frac{\beta\omega_k}{2}\right) + \coth\left(\frac{\beta\omega_{k+p}}{2}\right)\right)\Bigg]$$

$$\xrightarrow{\mathbf{p}\to 0} \frac{g^2}{8}\int\frac{d^3k}{(2\pi)^3}\frac{1}{\omega_k^3}\left(\coth\left(\frac{\beta\omega_k}{2}\right) + \frac{\beta\omega_k}{2}\operatorname{csch}^2\left(\frac{\beta\omega_k}{2}\right)\right).$$

$$(1.124)$$

Here we have used the identity

$$\frac{2}{\omega_k^2 - \omega_{k+p}^2} = \frac{1}{\omega_{k+p}}\left(\frac{1}{\omega_k - \omega_{k+p}} - \frac{1}{\omega_k + \omega_{k+p}}\right)$$

$$= \frac{1}{\omega_k}\left(\frac{1}{\omega_k - \omega_{k+p}} + \frac{1}{\omega_k + \omega_{k+p}}\right),$$

$$(1.125)$$

in going from the first to the second step and

$$\coth\left(\frac{\beta\omega_k}{2}\right) - \coth\left(\frac{\beta\omega_{k+p}}{2}\right) \xrightarrow{\mathbf{p}\to 0} -\frac{\beta(\omega_k - \omega_{k+p})}{2}\operatorname{csch}^2\left(\frac{\beta\omega_k}{2}\right),$$

$$(1.126)$$

in going from the second to the third (last) step. This shows that the limits $p^0 = 0$, $\mathbf{p}\to 0$ and $\mathbf{p} = 0$, $p^0 \to 0$ do not commute and, in fact, the self-energy is non-analytic at the origin in the (energy) momentum space.

Such a non-analyticity is unexpected from our experience with zero temperature field theory. That is because Lorentz invariance forces the zero temperature amplitudes to be covariant functions of the external momentum, p^μ, and that makes them analytic at the origin. At finite temperature, however, Lorentz invariance is broken by the choice of a specific frame (say, for example, the rest frame of the heat bath) and, therefore, the amplitudes can depend independently on p^0 and $\mathbf{p}$. The limits $p^0 = 0$, $\mathbf{p}\to 0$ and $\mathbf{p} = 0, p^0 \to 0$, consequently, need not be the same. The non-commuting nature of limits is not merely of academic interest – rather, they represent physical effects. In fact, the limit $\mathbf{p}\to 0$, $p^0 = 0$ corresponds to taking the static limit and leads to the dynamical screening mass for the electric fields. The other limit, namely, $p^0 \to 0$, $\mathbf{p} = 0$, on the other hand, gives the plasmon mass associated with the damping of the oscillations in a plasma and the two masses do not have to coincide.

That is physically the origin of the non-analyticity of the self-energy at the origin in the momentum space at finite temperature.

An alternate way to look at the nonanalyticity in the self-energy at finite temperature is as follows. We describe the calculation with some details since it will be of use later as well. If we introduce the number densities (here we are dealing with bosons only and, therefore, we will ignore the subscript representing a bosonic distribution for simplicity)

$$n(\omega_k) = \frac{1}{e^{\beta\omega_k} - 1},$$

$$n(\omega_{k+p}) = \frac{1}{e^{\beta\omega_{k+p}} - 1}, \tag{1.127}$$

and note, as pointed out earlier in (1.106), that

$$\coth\left(\frac{\beta\omega_k}{2}\right) = 1 + 2n(\omega_k)$$

$$= 1 + n(\omega_k) + n(\omega_{k+p}) + n(\omega_k)n(\omega_{k+p}) - n(\omega_k)n(\omega_{k+p})$$
$$+ (n(\omega_k) - n(\omega_{k+p})),$$

$$= \left((1 + n(\omega_k))(1 + n(\omega_{k+p})) - n(\omega_k)n(\omega_{k+p})\right)$$
$$+ \left(n(\omega_k)(1 + n(\omega_{k+p})) - n(\omega_{k+p})(1 + n(\omega_k))\right). \tag{1.128}$$

Similarly, we can also show that

$$\coth\left(\frac{\beta\omega_{k+p}}{2}\right) = 1 + 2n(\omega_{k+p})$$

$$= \left((1 + n(\omega_k))(1 + n(\omega_{k+p})) - n(\omega_k)n(\omega_{k+p})\right)$$
$$- \left(n(\omega_k)(1 + n(\omega_{k+p})) - n(\omega_{k+p})(1 + n(\omega_k))\right). \tag{1.129}$$

The next thing that we note from (1.122) is that, if we identify $\omega = ip^0$, then using (the first form of) (1.125) we can write

$$\frac{1}{\omega_{k+p}^2 - (\omega_k - \omega)^2} + \omega \leftrightarrow -\omega = -\frac{1}{(\omega - \omega_k)^2 - \omega_{k+p}^2} + \omega \leftrightarrow -\omega$$

$$= \frac{1}{2\omega_{k+p}}\left(\frac{1}{\omega - \omega_k + \omega_{k+p}} - \frac{1}{\omega - \omega_k - \omega_{k+p}}\right) + \omega \leftrightarrow -\omega$$

$$= \frac{1}{2\omega_{k+p}} \left[\left(\frac{1}{\omega - \omega_k + \omega_{k+p}} - \frac{1}{\omega + \omega_k - \omega_{k+p}} \right) \right.$$

$$\left. + \left(\frac{1}{\omega + \omega_k + \omega_{k+p}} - \frac{1}{\omega - \omega_k - \omega_{k+p}} \right) \right]. \qquad (1.130)$$

Similarly, using (the second form of) (1.125) we can show that

$$\frac{1}{\omega_k^2 - (\omega_{k+p} - \omega)^2} + \omega \leftrightarrow -\omega$$

$$= \frac{1}{2\omega_k} \left(\frac{1}{\omega + \omega_k - \omega_{k+p}} - \frac{1}{\omega - \omega_k - \omega_{k+p}} \right) + \omega \leftrightarrow -\omega$$

$$= \frac{1}{2\omega_k} \left[- \left(\frac{1}{\omega - \omega_k + \omega_{k+p}} - \frac{1}{\omega + \omega_k - \omega_{k+p}} \right) \right.$$

$$\left. + \left(\frac{1}{\omega + \omega_k + \omega_{k+p}} - \frac{1}{\omega - \omega_k - \omega_{k+p}} \right) \right]. \qquad (1.131)$$

Substituting (1.128)-(1.131) into the expression for the self-energy at finite temperature given in (1.122) (we suppress the argument $\mathbf{p}$ in the self-energy which allows us to denote $\Pi(\mathbf{p}, p^0) = \Pi(\omega)$ for simplicity)

$$\Pi(\omega) = \frac{g^2}{8} \int \frac{d^3k}{(2\pi)^3} \frac{1}{\omega_k \omega_{k+p}}$$

$$\times \left[((1 + n(\omega_k))(1 + n(\omega_{k+p})) - n(\omega_k)n(\omega_{k+p})) \right.$$

$$\times \left(\frac{1}{\omega + \omega_k + \omega_{k+p}} - \frac{1}{\omega - \omega_k - \omega_{k+p}} \right)$$

$$+ (n(\omega_k)(1 + n(\omega_{k+p})) - n(\omega_{k+p})(1 + n(\omega_k)))$$

$$\left. \times \left(\frac{1}{\omega - \omega_k + \omega_{k+p}} - \frac{1}{\omega + \omega_k - \omega_{k+p}} \right) \right]. \qquad (1.132)$$

Although the self-energy in (1.132) is defined only for imaginary discrete values of ω, we can extend it to the full complex plane satisfying

$$\Pi^*(\omega) = \Pi(\omega^*). \qquad (1.133)$$

With this, it is easy to see that the analytic extension has cuts along the real axis and that the discontinuity across these cuts is purely imaginary for real ω and is given by

$$\text{Disc } \Pi(\omega) = \lim_{\eta \to 0} \left(\Pi(\omega + i\eta) - \Pi(\omega - i\eta) \right) = 2i \, \text{Im} \, \Pi(\omega).$$

$$(1.134)$$

The discontinuity can be read out directly from (1.132) using (1.134) and is determined to be (recall that $\lim_{\eta \to 0} \frac{1}{x \pm i\eta} = P\frac{1}{x} \mp i\pi\delta(x)$)

$$
\text{Im} \, \Pi(\omega) = -\frac{g^2}{16} \int \frac{\mathrm{d}^3 k}{(2\pi)^2} \frac{1}{\omega_k \omega_{k+p}}
$$

$$
\times \Big[\big((1 + n(\omega_k))(1 + n(\omega_{k+p})) - n(\omega_k) n(\omega_{k+p}) \big)
$$

$$
\times \big(\delta(\omega + \omega_k + \omega_{k+p}) - \delta(\omega - \omega_k - \omega_{k+p}) \big)
$$

$$
+ \big(n(\omega_k)(1 + n(\omega_{k+p})) - n(\omega_{k+p})(1 + n(\omega_k)) \big)
$$

$$
\times \big(\delta(\omega - \omega_k + \omega_{k+p}) - \delta(\omega + \omega_k - \omega_{k+p}) \big) \Big].
$$

$$(1.135)$$

This brings out an interesting feature of field theories at finite temperature. We note that, at zero temperature, $\text{Im} \, \Pi$ represents a decay rate which is the (absolute) square of the decay amplitude integrated over the appropriate phase space factors. At finite temperature, however, the available density of states has to be appropriately weighted by statistical factors (as seen in (1.135)). Furthermore, there are real particles present in a medium and consequently, there are more channels of reaction available at finite temperature that are not possible at zero temperature. For example, in the present example of the scalar self-energy, we note that the direct process

$$\phi \longrightarrow B + B,$$

$$(1.136)$$

would, of course, occur with a statistical weight factor $(1+n(\omega_k))(1+n(\omega_{k+p}))$ (think of the stimulated emission in a medium for a system such as the laser). But more importantly, there are real B's in the thermal medium and, therefore, the inverse process

$$B + B \longrightarrow \phi,$$

$$(1.137)$$

is also possible. However, this will take place with a statistical weight factor $n(\omega_k)n(\omega_{k+p})$. The total rate for the decay of ϕ, therefore, should correspond to the difference of these two rates and we note that both these processes are indeed included in the first term in (1.135). The other terms, similarly, represent the processes

$$\phi + B \longrightarrow B,$$

$$B \longrightarrow \phi + B. \tag{1.138}$$

It is worth noting here that if the decay products were fermions, as opposed to bosons as in the present example, then the statistical factors would have the form $(1-n)$ corresponding to a Pauli blocking representing the fact that there is a real distribution of fermions in the medium of density n and, therefore, the available density of states for the decay products is suppressed.

In general, the location of the branch cuts (that are responsible for the non-analyticity) can also be determined from (1.135). Let us assume that the masses of the two particles in the propagators of the loop are given by m_1 and m_2 respectively (for generality) so that we can define (see (1.116))

$$\omega_1 = \omega_k = (\mathbf{k}^2 + m_1^2)^{\frac{1}{2}},$$

$$\omega_2 = \omega_{k+p} = ((\mathbf{k} + \mathbf{p})^2 + m_2^2)^{\frac{1}{2}}. \tag{1.139}$$

It follows now that

$$\begin{aligned}
\omega_1\omega_2 &= (|\mathbf{k}|^2 + m_1^2)^{\frac{1}{2}}(|\mathbf{k} + \mathbf{p}|^2 + m_2^2)^{\frac{1}{2}} \\
&= (|\mathbf{k}|^2|\mathbf{k} + \mathbf{p}|^2 + m_1^2|\mathbf{k} + \mathbf{p}|^2 + m_2^2|\mathbf{k}|^2 + m_1^2 m_2^2)^{\frac{1}{2}} \\
&= ((|\mathbf{k}||\mathbf{k} + \mathbf{p}| + m_1 m_2)^2 + (m_2|\mathbf{k}| - m_1|\mathbf{k} + \mathbf{p}|)^2)^{\frac{1}{2}} \\
&\geq |\mathbf{k}||\mathbf{k} + \mathbf{p}| + m_1 m_2,
\end{aligned}$$

or, $\omega_1\omega_2 - m_1 m_2 \geq |\mathbf{k}||\mathbf{k} + \mathbf{p}| \geq 0. \tag{1.140}$

Let us next define a variable

$$s = \omega^2 - \mathbf{p}^2. \tag{1.141}$$

It is now clear that, if in (1.135) the delta function constraint that is satisfied is $\omega = \omega_1 + \omega_2$ (or $\omega = -(\omega_1 + \omega_2)$), then, using (1.139), we obtain

$$s = \omega^2 - \mathbf{p}^2 = (\omega_1 + \omega_2)^2 - \mathbf{p}^2 = \omega_1^2 + \omega_2^2 + 2\omega_1\omega_2 - \mathbf{p}^2$$
$$= m_1^2 + m_2^2 + 2\mathbf{k} \cdot (\mathbf{k} + \mathbf{p}) + 2\omega_1\omega_2$$
$$= (m_1 + m_2)^2 + 2\mathbf{k} \cdot (\mathbf{k} + \mathbf{p}) + 2(\omega_1\omega_2 - m_1 m_2)$$
$$\geq (m_1 + m_2)^2 + 2(|\mathbf{k}||\mathbf{k} + \mathbf{p}| + \mathbf{k} \cdot (\mathbf{k} + \mathbf{p}))$$
$$\geq (m_1 + m_2)^2, \tag{1.142}$$

where we have used (1.140) in the last but one step as well as $-1 \leq \cos\theta \leq 1$ in the last step. This gives the first branch cut of the two point function lying between $(m_1 + m_2)^2 \leq s < \infty$ which coincides with the zero temperature cut that describes the standard threshold for particle decays (through pair production, see Fig. 1.4).

However, a second branch cut develops at finite temperature because of the presence of additional processes (channels of reaction). This is easy to see in the following way. Let us assume that the delta function constraint in (1.135) that is satisfied is $\omega = \omega_1 - \omega_2$ (or the one with $\omega = \omega_2 - \omega_1$). Then, using (1.139) and (1.140), we now see that

$$s = \omega^2 - \mathbf{p}^2 = (\omega_1 - \omega_2)^2 - \mathbf{p}^2 = \omega_1^2 + \omega_2^2 - 2\omega_1\omega_2 - \mathbf{p}^2$$
$$= m_1^2 + m_2^2 + 2\mathbf{k} \cdot (\mathbf{k} + \mathbf{p}) - 2\omega_1\omega_2$$
$$= (m_1 - m_2)^2 + 2\mathbf{k} \cdot (\mathbf{k} + \mathbf{p}) - 2(\omega_1\omega_2 - m_1 m_2)$$
$$\leq (m_1 - m_2)^2 - 2(|\mathbf{k}||\mathbf{k} + \mathbf{p}| - \mathbf{k} \cdot (\mathbf{k} + \mathbf{p}))$$
$$\leq (m_1 - m_2)^2. \tag{1.143}$$

This shows that, at finite temperature, there is an additional branch cut along $-\infty < s \leq (m_1 - m_2)^2$ where the lower limit is obtained for space like s. This new cut corresponds to particle absorption from the medium (or due to thermal scattering in the medium) and pictorially, the cut structure, at finite temperature, is shown in Fig. 1.4. We note that the first branch cut, of course, does not lead to any non-analyticity at the origin. However, the second branch cut is the one responsible for the non-commuting limits of $\mathbf{p} \to 0, p^0 = 0$ and $p^0 \to 0, \mathbf{p} = 0$ when $m_1 = m_2$ which we have discussed earlier in (1.123)

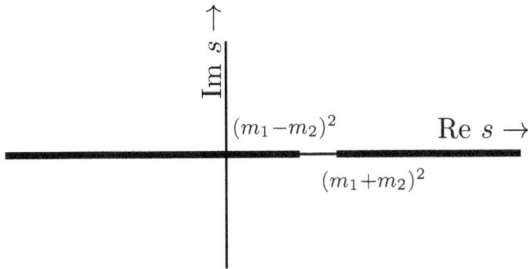

Figure 1.4: Branch cuts for the self-energy at finite temperature.

and (1.124). This analysis also makes it clear that for processes where the loop would involve particles with distinct masses, the non-analyticity would be absent. This can, in fact, be explicitly checked by calculating the self-energy of the gauge boson in a spontaneously broken gauge theory in the unitary gauge. It is also worth noting here that the non-analyticity is not a special feature of the self-energy of a boson. It is present even for a fermion. In fact, this non-analyticity manifests itself in the calculation of effective actions as well. In particular, one has to define what one means by an effective potential when the effective action is non-analytic at the origin in the momentum space.

1.8 Covariant description

Normally, one believes that the very definition of a system at finite temperature breaks Lorentz invariance because we conventionally define a thermal system in the rest frame of the heat bath and, thereby select out a specific Lorentz frame. On the other hand, it is also possible to formulate statistical mechanics in a manifestly Lorentz covariant manner in the following way. This is particularly useful if we want to take advantage of concepts such as tensor decomposition of amplitudes in a manifestly covariant theory. (We cannot help but point out here that this is very much like what we happens in gauge theories. For example, the temporal gauge condition, $A_0 = 0$, as well as the axial gauge condition, $A_3 = 0$, naively break Lorentz invariance. However, choosing a space-time independent four vector

n^μ, we can also formulate such noncovariant gauge conditions in a covariant (Lorentz invariant) manner by writing $n^\mu A_\mu = 0$.)

Let us consider a heat bath (or a fluid) moving with a proper four velocity u^μ satisfying

$$u^2 = u_\mu u^\mu = 1. \tag{1.144}$$

In the rest frame of the heat bath, the four velocity has the form $u^\mu = (1, 0, 0, 0)$ and the inverse temperature, β, is defined in this rest frame.

Given the vector u^μ, we can decompose any other four vector into parallel and perpendicular components with respect to the proper velocity. Thus, for example, an arbitrary four vector A^μ will have components parallel and perpendicular to u^μ given by

$$A^\mu = A^\mu_\parallel + A^\mu_\perp,$$

$$A^\mu_\parallel = (A \cdot u) u^\mu,$$

$$A^\mu_\perp \equiv \widetilde{A}^\mu = A^\mu - A^\mu_\parallel = A^\mu - (A \cdot u) u^\mu, \tag{1.145}$$

so that we have

$$u \cdot A_\parallel = u \cdot A, \qquad u \cdot A_\perp = u \cdot \widetilde{A} = u \cdot A - u \cdot A = 0. \tag{1.146}$$

This shows explicitly that the longitudinal and the perpendicular components of any vector are indeed orthogonal to each other, namely,

$$A_\parallel \cdot A_\perp = (A \cdot u) \, u \cdot A_\perp = 0, \tag{1.147}$$

which follows from (1.146).

In particular, we note that if p^μ represents the four momentum of a particle, then its components parallel and perpendicular to the four velocity are given by

$$p^\mu_\parallel = (p \cdot u) u^\mu \equiv \omega u^\mu,$$

$$p^\mu_\perp = \widetilde{p}^\mu = p^\mu - (p \cdot u) u^\mu = p^\mu - \omega u^\mu, \qquad u_\mu \widetilde{p}^\mu = 0, \tag{1.148}$$

where we have defined

$$\omega = (p \cdot u). \tag{1.149}$$

It is easy to see from the two definitions in (1.148) that in the rest frame of the heat bath ($u^\mu = (1, 0, 0, 0)$)

$$\omega = p^0 = p_0,$$
$$\widetilde{p}^\mu = (0, \mathbf{p}), \tag{1.150}$$

so that we can think of ω as the Lorentz invariant energy. It can also be easily checked now using (1.144)-(1.150) that (k is some positive number which has the value $k = |\mathbf{p}|$ in the rest frame)

$$\widetilde{p}^2 = \widetilde{p}^\mu \widetilde{p}_\mu = -\mathbf{p}^2 \equiv -k^2 < 0, \tag{1.151}$$

where we have used the second relation in (1.150). Thus, we can think of $\widetilde{p}$ as the Lorentz invariant spatial momentum. Clearly, then, we can write, from (1.144)-(1.151),

$$p^2 = \eta_{\mu\nu} p^\mu p^\nu = \eta_{\mu\nu} (p_\parallel^\mu + p_\perp^\mu)(p_\parallel^\nu + p_\perp^\nu)$$
$$= (\omega u^\mu + \widetilde{p}^\mu)(\omega u_\mu + \widetilde{p}_\mu) = \omega^2 + \widetilde{p}^2 = \omega^2 - k^2, \tag{1.152}$$

where we have used $u \cdot \widetilde{p} = 0$ (see (1.146)) and this shows that we can think of k as the length of the momentum (three) vector (remember that $k = |\mathbf{p}|$ is non-negative).

The density matrix for the system can be covariantly written as (compare with (1.1), the form in the rest frame)

$$\rho(\beta) = \frac{1}{Z(\beta)} e^{-\beta u \cdot p}, \quad Z(\beta) = \mathrm{Tr}\, e^{-\beta u \cdot p}, \quad \mathrm{Tr}\, \rho(\beta) = 1, \tag{1.153}$$

which shows that the bosonic and the fermionic distribution functions can now be generalized to an arbitrary boosted frame of the heat bath simply as

$$n = \frac{1}{e^{\beta|\omega|} \mp 1} = \frac{1}{e^{\beta|u \cdot p|} \mp 1}, \tag{1.154}$$

where $\mp$ specify respectively the signs in the denominator for the bosonic and fermionic distributions. We can also carry out the Lorentz tensor decomposition of various amplitudes at finite temperature, in an arbitrary frame of the heat bath, with respect to the (heat bath) velocity four vector u^μ. Thus, for example, let us note that the projection operator onto the space of vectors, which are transverse to u^μ, is given by (the longitudinal projection operator is simply $u^\mu u^\nu$)

$$\widetilde{\eta}^{\mu\nu} = \eta^{\mu\nu} - u^\mu u^\nu, \quad u_\mu \widetilde{\eta}^{\mu\nu} = 0 = \widetilde{\eta}^{\mu\nu} u_\nu, \qquad (1.155)$$

so that, for an arbitrary vector A^μ, we can write the transverse part with respect to u^μ as

$$A^\mu_\perp = \widetilde{A}^\mu = \widetilde{\eta}^{\mu\nu} A_\nu = A^\mu - (A \cdot u)u^\mu, \quad u_\mu \widetilde{A}^\mu = 0, \qquad (1.156)$$

which can be compared with the last relation in (1.145).

Just as we can decompose any vector into components parallel and perpendicular to u^μ, we can similarly decompose any vector into components which are parallel and perpendicular to a given momentum vector, say p^μ. Thus, let A^μ represent an arbitrary four vector. Then, its component parallel to p^μ is, of course, given by $\frac{1}{p^2}(p \cdot A)p^\mu$. And the component perpendicular to p^μ has the form

$$\overline{A}^\mu = A^\mu - \frac{1}{p^2}(p \cdot A)p^\mu, \quad p \cdot \overline{A} = (p \cdot A) - (p \cdot A) = 0. \quad (1.157)$$

Similarly, the orthogonal components of u^μ and $\eta^{\mu\nu}$ with respect to p^μ are given respectively by (remember that $\omega = p \cdot u$, see (1.149))

$$\overline{u}^\mu = u^\mu - \frac{\omega}{p^2}p^\mu, \qquad\qquad p \cdot \overline{u} = 0,$$

$$\overline{\eta}^{\mu\nu} = \eta^{\mu\nu} - \frac{p^\mu p^\nu}{p^2}, \qquad\qquad p_\mu \overline{\eta}^{\mu\nu} = \overline{\eta}^{\mu\nu} p_\nu = 0. \qquad (1.158)$$

Given these relations, it is easy to check that there are only three possible second rank, symmetric tensors that one can construct at finite temperature from p^μ, u^μ and $\eta^{\mu\nu}$ which are orthogonal to p^μ, namely,

$$A^{\mu\nu} = \widetilde{\eta}^{\mu\nu} - \frac{\widetilde{p}^\mu \widetilde{p}^\nu}{\widetilde{p}^2} = A^{\nu\mu}, \qquad p_\mu A^{\mu\nu} = 0 = A^{\mu\nu} p_\nu,$$

$$B^{\mu\nu} = \frac{p^2}{\widetilde{p}^2} \overline{u}^\mu \overline{u}^\nu = B^{\nu\mu}, \qquad p_\mu B^{\mu\nu} = 0 = B^{\mu\nu} p_\nu,$$

$$C^{\mu\nu} = \eta^{\mu\nu} - \frac{p^\mu p^\nu}{p^2} = C^{\nu\mu}, \qquad p_\mu C^{\mu\nu} = 0 = C^{\mu\nu} p_\nu. \qquad (1.159)$$

We note that of the three structures, only the first structure ($A^{\mu\nu}$) is also transverse to the four velocity of the heat bath, in addition to being transverse to the momentum (p^{μ}), namely,

$$u_{\mu}A^{\mu\nu} = 0 = A^{\mu\nu}u_{\nu}, \tag{1.160}$$

which follows from the fact that $u_{\mu}\widetilde{\eta}^{\mu\nu} = 0 = \widetilde{\eta}^{\mu\nu}u_{\nu}$ as well as $u_{\mu}\widetilde{p}^{\mu} = 0$ (see (1.155) and (1.156)). On the other hand, for the other two structures, we have

$$u_{\mu}B^{\mu\nu} = \overline{u}^{\nu} = u_{\mu}C^{\mu\nu}, \tag{1.161}$$

where we have used

$$u \cdot \overline{u} = \frac{\widetilde{p}^2}{p^2} = \overline{u} \cdot \overline{u}, \tag{1.162}$$

as well as the definition of $\overline{u}^{\mu}$ in (1.158). (Note also that $u_{\mu}B^{\mu\nu} = \overline{u}_{\mu}B^{\mu\nu}$ and $u_{\mu}C^{\mu\nu} = \overline{u}_{\mu}C^{\mu\nu}$ since the structures are transverse to p_{μ}.) Each of the structures in (1.159) is normalized so that they act like projection operators. However, they are not linearly independent. In fact, let us show the linear dependence in some detail

$$
\begin{aligned}
C^{\mu\nu} &= \eta^{\mu\nu} - \frac{p^{\mu}p^{\nu}}{p^2} \\
&= \eta^{\mu\nu} - u^{\mu}u^{\nu} - \frac{\widetilde{p}^{\mu}\widetilde{p}^{\nu}}{\widetilde{p}^2} + u^{\mu}u^{\nu} + \frac{\widetilde{p}^{\mu}\widetilde{p}^{\nu}}{\widetilde{p}^2} - \frac{p^{\mu}p^{\nu}}{p^2} \\
&= \widetilde{\eta}^{\mu\nu} - \frac{\widetilde{p}^{\mu}\widetilde{p}^{\nu}}{\widetilde{p}^2} \\
&\quad + \frac{1}{p^2\widetilde{p}^2}\left(p^2\widetilde{p}^2 u^{\mu}u^{\nu} + p^2(p^{\mu} - \omega u^{\mu})(p^{\nu} - \omega u^{\nu}) - \widetilde{p}^2 p^{\mu}p^{\nu}\right) \\
&= A^{\mu\nu} + \frac{1}{p^2\widetilde{p}^2}\left((p^2)^2 u^{\mu}u^{\nu} - p^2\omega(u^{\mu}p^{\nu} + u^{\nu}p^{\mu}) + \omega^2 p^{\mu}p^{\nu}\right) \\
&= A^{\mu\nu} + \frac{p^2}{\widetilde{p}^2}\left(u^{\mu} - \frac{\omega}{p^2}p^{\mu}\right)\left(u^{\nu} - \frac{\omega}{p^2}p^{\nu}\right) \\
&= A^{\mu\nu} + \frac{p^2}{\widetilde{p}^2}\overline{u}^{\mu}\overline{u}^{\nu} = A^{\mu\nu} + B^{\mu\nu}, \tag{1.163}
\end{aligned}
$$

where we have used the definitions in (1.148), (1.151)-(1.152), (1.155) as well as (1.158)-(1.159). Normally, one takes $A^{\mu\nu}$ and $B^{\mu\nu}$ as the two independent structures since they are also orthogonal to each other. (This follows from the fact that $\widetilde{\eta}^{\mu\nu}\bar{u}_\nu = -\frac{\omega}{p^2}\,\widetilde{p}^\mu = \frac{\widetilde{p}^\mu\widetilde{p}^\nu}{\widetilde{p}^2}\bar{u}_\nu$.)

The reason behind discussing second rank symmetric tensors which are orthogonal to the momentum four vector (p^μ) is that the self-energy of a gauge field (also known as the polarization tensor) is a second rank symmetric tensor (in the external vector indices). Furthermore, it is known from the Ward-Takahasi (or Slavnov-Taylor) identities to be transverse to the (external) momentum four vector because of gauge invariance (see, for example, (5.49)). At zero temperature the self-energy can depend only on the external momentum four vector and, in such a case, we can construct only one second rank symmetric tensor which is transverse to the momentum. Therefore, the self-energy of a gauge field at zero temperature has the form

$$\Pi_{\mu\nu}(p) = \left(\eta_{\mu\nu} - \frac{p_\mu p_\nu}{p^2}\right)\Pi(p), \tag{1.164}$$

where we recognize the tensor structure to coincide with $C^{\mu\nu}$ defined in (1.159). We can take the trace of the polarization tensor and it leads to

$$\eta^{\mu\nu}\Pi_{\mu\nu}(p) = (D-1)\Pi(p), \tag{1.165}$$

namely, the form factor $\Pi(p)$ can be determined from the trace of the polarization tensor when $D \neq 1$. Here D denotes number of spacetime dimensions and we see that the right hand side vanishes for $D = 1$ simply reflecting the fact that there is no transverse direction in one dimension and, therefore, a transverse polarization tensor cannot exist, $\Pi_{\mu\nu}(p) = 0$.

On the other hand, since at finite temperature, there is a second four vector, namely, the velocity of the heat bath (u^μ), we see that the number of independent second rank symmetric tensors which are orthogonal to the momentum increases to two. As a result, we can write down the general decomposition of the self-energy of a vector (gauge) field, at finite temperature, which is transverse to the external momentum as

$$\Pi_{\mu\nu}^{(\beta)}(p) = A_{\mu\nu}\Pi_T^{(\beta)}(p) + B_{\mu\nu}\Pi_L^{(\beta)}(p)$$

$$= \left(\widetilde{\eta}_{\mu\nu} - \frac{\widetilde{p}_\mu\widetilde{p}_\nu}{\widetilde{p}^2}\right)\Pi_T^{(\beta)}(p) + \frac{p^2}{\widetilde{p}^2}\bar{u}_\mu\bar{u}_\nu\Pi_L^{(\beta)}(p). \tag{1.166}$$

We know that the structures, $A^{\mu\nu}$ and $B^{\mu\nu}$ (as well as $C^{\mu\nu}$), are transverse to the external momentum. However, we also know that, while $A^{\mu\nu}$ is transverse to the four velocity of the heat bath, $B^{\mu\nu}$ is not. This reflects in the subindices T and L in the two form factors (coefficients) $\Pi_T^{(\beta)}(p), \Pi_L^{(\beta)}(p)$ associated with the tensor structures which are respectively transverse and longitudinal to the velocity of the heat bath (we emphasize that both structures are transverse to the momentum as the self-energy should be). Comparing with the structure of the polarization tensor at zero temperature given in (1.164), we conclude that the two form factors in (1.166) must become equal at zero temperature, namely,

$$\Pi_T(p) = \Pi_L(p) = \Pi(p), \tag{1.167}$$

where we have identified $\Pi_T^{(\beta)}(p) \to \Pi_T(p), \Pi_L^{(\beta)}(p) \to \Pi_L(p)$ as $\beta \to \infty$, so that (1.166) can reduce to (1.164) (see (1.163)).

Furthermore, since $A^{\mu\nu}$ is transverse to the four velocity (in addition to being transverse to momentum), contracting (1.166) with $u^\mu u^\nu$ we obtain that, at finite temperature,

$$u^\mu u^\nu \Pi_{\mu\nu}^{(\beta)}(p) = \frac{p^2}{\widetilde{p}^2} (u \cdot \overline{u})^2 \, \Pi_L^{(\beta)}(p),$$

$$\text{or,} \quad \Pi_L^{(\beta)}(p) = \frac{p^2}{\widetilde{p}^2} u^\mu u^\nu \Pi_{\mu\nu}^{(\beta)}(p), \tag{1.168}$$

where we have used (1.162). This shows that the form factor $\Pi_L^{(\beta)}(p)$, at finite temperature, can be determined from the transverse polarization tensor $\Pi_{\mu\nu}^{(\beta)}(p)$ by contracting both the indices with the four velocity in all space-time dimensions (except $D = 1$ where, as mentioned earlier following (1.165), a transverse $\Pi_{\mu\nu}^{(\beta)}(p)$ does not exist). Similarly, taking the trace of the polarization tensor in (1.166), we obtain

$$\eta^{\mu\nu} \Pi_{\mu\nu}^{(\beta)}(p) = \eta^{\mu\nu} A_{\mu\nu} \Pi_T^{(\beta)}(p) + \frac{p^2}{\widetilde{p}^2} (\overline{u} \cdot \overline{u}) \Pi_L^{(\beta)}(p),$$

$$\text{or,} \quad (D - 2)\Pi_T^{(\beta)}(p) = \eta^{\mu\nu} \Pi_{\mu\nu}^{(\beta)} - \Pi_L^{(\beta)}. \tag{1.169}$$

Here we have used the fact that $\eta_{\mu\nu} \widetilde{\eta}^{\mu\nu} = D - 1$ (see (1.155) and recall from (1.144) that $u_\mu u^\mu = 1$) with D, as pointed out earlier, denoting the number of space-time dimensions and we have used (1.162). At zero temperature, where the longitudinal and the transverse form factors are equal (as already pointed out in (1.167)), this relation reduces to (1.165). At finite temperature, (1.169) determines the transverse form factor in all dimensions (if we ignore $D = 1$ where a transverse self-energy $\Pi_{\mu\nu}$ can not be defined) except when $D = 2$. Of course, the coefficient $(D - 2)$ on the left hand side in (1.169) vanishes in two space-time dimensions (independent of the value of $\Pi_T^{(\beta)}(p)$). Let us now show that the right hand side of (1.169) vanishes identically in two dimensions so that $\Pi_T^{(\beta)}(p)$ remains undetermined.

Using (1.168) in the right hand side of (1.169), we note that we can write

$$(D - 2)\Pi_T^{(\beta)}(p) = \left(\eta^{\mu\nu} - \frac{p^2}{\overline{p}^2}u^\mu u^\nu\right)\Pi_{\mu\nu}^{(\beta)}(p). \tag{1.170}$$

In two dimensions, $D = 2$, we have two natural second rank tensors, the usual symmetric metric tensor $\eta^{\mu\nu} = \eta^{\nu\mu}$ as well as the anti-symmetric Levi-Civita tensor $\epsilon^{\mu\nu} = -\epsilon^{\nu\mu}$. This leads to a simpler definition of the transverse direction to the momentum as $\overline{p}^\mu = \epsilon^{\mu\nu}p_\nu$. As a result, we can have a natural decomposition of a vector into its components parallel and transverse to the momentum, namely,

$$A^\mu = \frac{(p \cdot A)}{p^2}p^\mu + \frac{(\overline{p} \cdot A)}{\overline{p}^2}\overline{p}^\mu = \frac{(p \cdot A)}{p^2}p^\mu - \frac{(\overline{p} \cdot A)}{p^2}\epsilon^{\mu\nu}p_\nu, \tag{1.171}$$

where we have used $\overline{p}^2 = -p^2$. In particular, for the four velocity u^μ, this implies that it can be decomposed as

$$u^\mu = \frac{p \cdot u}{p^2}p^\mu - \frac{\overline{p} \cdot u}{p^2}\epsilon^{\mu\nu}p_\nu = \frac{\omega}{p^2}p^\mu - \frac{(-\widetilde{p}^2)^{\frac{1}{2}}}{p^2}\epsilon^{\mu\nu}p_\nu, \tag{1.172}$$

which follows from the identity

$$(\overline{p} \cdot u)^2 = (\epsilon^{\mu\nu}p_\nu u_\mu)^2 = -p^2 + (p \cdot u)^2 = -p^2 + \omega^2 = -\widetilde{p}^2. \tag{1.173}$$

Note that the decomposition in (1.172) naturally satisfies

$$p \cdot u = \frac{\omega}{p^2}p^2 = \omega, \quad u \cdot u = \frac{\omega^2}{p^2} - \frac{(-\widetilde{p}^2)}{p^2} = 1, \tag{1.174}$$

as it should.

The decomposition in (1.172) leads to

$$u^\mu u^\nu = \frac{1}{(p^2)^2} \left(\omega^2 p^\mu p^\nu - \omega(-\tilde{p}^2)^{\frac{1}{2}}(\epsilon^{\mu\lambda}p^\nu + \epsilon^{\nu\lambda}p^\mu) \right.$$

$$\left. -\tilde{p}^2 \epsilon^{\mu\lambda}\epsilon^{\nu\rho}p_\lambda p_\rho \right)$$

$$= \frac{\tilde{p}^2}{p^2}\eta^{\mu\nu} - \frac{\omega(-\tilde{p}^2)^{\frac{1}{2}}}{(p^2)^2}(\epsilon^{\mu\lambda}p^\nu + \epsilon^{\nu\lambda}p^\mu)p_\lambda$$

$$+ \frac{1}{(p^2)^2}(\omega^2 - \tilde{p}^2)p^\mu p^\nu. \tag{1.175}$$

Here we have used the two dimensional identity

$$\epsilon^{\mu\lambda}\epsilon^{\nu\rho} = -\eta^{\mu\nu}\eta^{\lambda\rho} + \eta^{\mu\rho}\eta^{\nu\lambda}. \tag{1.176}$$

Using (1.175) in the right hand side of (1.170), we see that, in two dimensions,

$$\left(\eta^{\mu\nu} - \frac{p^2}{\tilde{p}^2}u^\mu u^\nu \right) \Pi^{(\beta)}_{\mu\nu}(p)$$

$$= \frac{1}{p^2\tilde{p}^2} \left(\omega(-\tilde{p}^2)^{\frac{1}{2}}(\epsilon^{\mu\lambda}p^\nu + \epsilon^{\nu\lambda}p^\mu)p_\lambda - (\omega^2 - \tilde{p}^2)p^\mu p^\nu \right) \Pi^{(\beta)}_{\mu\nu}(p)$$

$$= 0, \tag{1.177}$$

where we have used the transversality of the self-energy to the momentum in the last step. This shows that, in two dimensions, the right hand side of (1.170) identically vanishes. Since the coefficient $(D-2)$ on the left hand side also vanishes for $D = 2$, the two sides are consistent independent of the value of $\Pi^{(\beta)}_T(p)$ which can not be determined and has to be calculated.

Such a decomposition of tensors, at finite temperature where an additional vector besides the momentum is available, can be extended to other amplitudes as well and is often quite useful.

1.9 Appendices

In these two short appendices, we discuss briefly the question of rotation to Euclidean space, in general, as well as the question of periodic and anti-periodic delta functions.

1.9.1 Rotation to Euclidean space. The Minkowski space is characterized by its metric which, in our convention, is given by $\eta^{\mu\nu} = (+,-,-,-) = \eta_{\mu\nu}$, where $\mu,\nu = 0,1,2,3$. Namely, the space and time coordinates as well as the (space-time) indices of a vector behave differently under raising and lowering. For example, if the contravariant coordinates are chosen to be $x^\mu = (x^0, x^i) = (x^0, \mathbf{x}) = (t, \mathbf{x})$, then the covariant coordinates have the form $x_\mu = (x_0, x_i) = (x^0, -x^i) = (t, -\mathbf{x})$. Similarly, for a general contravariant vector $A^\mu = (A^0, A^i) = (A^0, \mathbf{A})$, the corresponding covariant vector is given by $A_\mu = (A_0, A_i) = (A^0, -\mathbf{A})$. On the other hand, Euclidean space is characterized by the simple metric $\delta^{\mu\nu} = (+,+,+,+) = \delta_{\mu\nu}$, where $\mu,\nu = 1,2,3,4$, so that the space and time coordinates are treated on a equal footing and raising and lowering of indices is not very relevant.

We can go to Euclidean space by rotating the time coordinate clockwise to the (negative) imaginary axis, namely,

$$x^0 = x_0 = t \to -i\tau = -ix_4^{\mathrm{E}}, \quad \mathbf{x} \to \mathbf{x}^{\mathrm{E}}, \tag{1.178}$$

so that the length of a vector transforms as (repeated indices are summed)

$$x^2 = \eta_{\mu\nu}x^\mu x^\nu = t^2 - \mathbf{x}^2$$
$$\to -\tau^2 - \mathbf{x}^2 = -\delta_{\mu\nu}x_\mu^{\mathrm{E}}x_\nu^{\mathrm{E}} = -x_{\mathrm{E}}^2. \tag{1.179}$$

This shows that the Minkowski invariant, length squared, transforms (up to an overall sign) to the Euclidean invariant, length squared. As a result, we can think of the Minkowski metric rotating (transforming) to the Euclidean metric as $\eta_{\mu\nu} \to -\delta_{\mu\nu}$. We note that, for any any vector $A^\mu = (A^0, \mathbf{A})$ in Minkowski space, $A^2 \to -A_{\mathrm{E}}^2$ under $A^0 = A_0 \to \pm iA_4^{\mathrm{E}}, \mathbf{A} \to \mathbf{A}_{\mathrm{E}}$. This does not pick out whether the time component of A^μ should rotate clockwise or anti-clockwise under a transformation to Euclidean space. Similarly, for any two vectors A^μ, B^μ in Minkowski space, if the time components rotate in the same way to Euclidean space, namely, $A^0 = A_0 \to \pm iA_4^{\mathrm{E}}, B^0 = B_0 \to \pm iB_4^{\mathrm{E}}$, not only would their lengths rotate to Euclidean lengths (up to a sign), their Minkowski invariant scalar product will also rotate to the Euclidean invariant scalar product (up to a sign), namely, $A \cdot B \to -A_{\mathrm{E}} \cdot B_{\mathrm{E}}$. On the other hand, for two vectors whose time components rotate to Euclidean space in an opposite manner, namely, $A^0 = A_0 \to \pm i A_4^{\mathrm{E}}, B^0 = B_0 \to \mp i B_4^{\mathrm{E}}$, the Minkowski invariant scalar product of the two will not go into the Euclidean invariant product, namely, $A \cdot B \not\to A_{\mathrm{E}} \cdot B_{\mathrm{E}}$, even up to signs. This is very important to note. For example, from (1.178) we see that

$$\partial_0 = \frac{\partial}{\partial x^0} = \frac{\partial}{\partial t}$$
$$\to \frac{\partial}{\partial(-i\tau)} = i\frac{\partial}{\partial x_4^{\mathrm{E}}} = i\partial_4^{\mathrm{E}}. \tag{1.180}$$

Therefore, while $x^0 \to -ix_4^{\mathrm{E}}$, the energy (recall that $\hbar = 1$ in our convention) $k_0 = i\partial_0 \to i(i\partial_4^{\mathrm{E}}) = ik_4^{\mathrm{E}}$. Note that this consistently preserves the commutation relation (quantization condition) in rotating to Euclidean space,

$$1 = [\partial_0, x^0] \to [i\partial_4^{\mathrm{E}}, -ix_4^{\mathrm{E}}] = [\partial_4^{\mathrm{E}}, x_4^{\mathrm{E}}] = 1. \tag{1.181}$$

This clarifies why when the time coordinate is rotated clockwise in going to Euclidean space, the energy rotates anti-clockwise to the imaginary axis (so that the commutation relations are preserved under the rotation). Furthermore, this allows the Feynman Green's functions to be analytically continued to the Euclidean space (the anti-clockwise rotation of energy does not cross any of the singularities of the Feynman Green's function which is not true for the retarded and advanced Green's functions). We would like to emphasize here that, under such a rotation,

$$k \cdot x = k_0 x^0 - \mathbf{k} \cdot \mathbf{x} \to k_4^{\mathrm{E}} x_4^{\mathrm{E}} - \mathbf{k}^{\mathrm{E}} \cdot \mathbf{x}^{\mathrm{E}} \not\to k^{\mathrm{E}} \cdot x^{\mathrm{E}}. \tag{1.182}$$

This has the consequence that the Minkowski (invariant) plane wave solution does not go into the corresponding Euclidean invariant,

$$e^{ik \cdot x} \not\to e^{\pm i k^{\mathrm{E}} \cdot x^{\mathrm{E}}}, \tag{1.183}$$

so that the Fourier transforms need to be defined carefully.

Fourier transformation is defined in the n-dimensional Euclidean space as a map $\mathbb{R}^n \to \mathbb{R}^n$ and it is defined together with its inverse transformation as

$$F(x) = \int \frac{\mathrm{d}^n k}{(2\pi)^n} e^{ik \cdot x} G(k), \quad G(k) = \int \mathrm{d}^n x \, e^{-ik \cdot x} F(x), \tag{1.184}$$

where k, x are conjugate vectors in the n-dimensional Euclidean space and $k \cdot x = k_1 x_1 + k_2 x_2 + \cdots + k_n x_n$ which is the Euclidean invariant scalar product. In contrast, in the n-dimensional Minkowski space is non-Euclidean (or pseudo-Euclidean) generally denoted as $\mathbb{R}^{(1,n-1)}$ where the $(n-1)$ spatial coordinates are considered to be Euclidean. In such a case, the Fourier transformation denotes a map $\mathbb{R}^{(1,n-1)} \rightarrow \mathbb{R}^{(1,n-1)}$ and is defined together with its inverse as

$$F(x) = \int \frac{d^n k}{(2\pi)^n} e^{-ik \cdot x} G(k), \quad G(k) = \int d^n x \, e^{ik \cdot x} F(x), \quad (1.185)$$

where k, x are n-dimensional Minkowski vectors and $k \cdot x = k_0 x_0 - k_1 x_1 - k_2 x_2 - \cdots - k_{n-1} x_{n-1}$ denotes the Minkowski invariant scalar product. (Note that the Fourier transforms in (1.184) and (1.185) do not rotate into each other because of (1.183) and should be thought of as independent definitions.) The relative negative signs in the exponents in (1.185) compared to those in (1.184) arise in order to keep the (Euclidean) space parts in the exponents to have the same Euclidean sign as in (1.184). We note here that the definitions of Fourier transformations used in (1.74) as well as in (1.81) are consistent with (1.184) (together with (1.75)).

1.9.2 Periodic/Anti-periodic delta function. Let us first consider the bosonic case and write (see, for example, (1.67))

$$\frac{1}{\beta} \sum_{-\infty}^{\infty} e^{\frac{2i\pi n\tau}{\beta}} = \lim_{N \to \infty} \frac{1}{\beta} \sum_{n=-N}^{N} e^{\frac{2i\pi n\tau}{\beta}} = \lim_{N \to \infty} S_N(\tau), \quad (1.186)$$

where we have identified

$$S_N(\tau) = \frac{1}{\beta} \sum_{n=-N}^{N} e^{\frac{2i\pi n\tau}{\beta}}. \quad (1.187)$$

We note that this is a periodic function of τ with a period π, namely,

$$S_N(\tau + m\pi) = S_N(\tau), \quad m = 0, \pm 1, \pm 2, \cdots. \quad (1.188)$$

Furthermore, this is a geometric series with a common ratio $e^{\frac{2i\pi\tau}{\beta}}$ so that it can be summed easily and can be written in the compact form

$$S_N(\tau) = \frac{1}{\beta} \frac{\sin \frac{(2N+1)\pi\tau}{\beta}}{\sin \frac{\pi\tau}{\beta}}. \quad (1.189)$$

The periodicity discussed in (1.188) is manifest here since the numerator and the denominator of the function change by the same phase under $\tau \to \tau + m\beta$ so that the function remains invariant, namely,

$$S_N(\tau) = \frac{1}{\beta} \frac{\sin \frac{(2N+1)\pi\tau}{\beta}}{\sin \frac{\pi\tau}{\beta}} \to \frac{1}{\beta} \frac{\sin \frac{(2N+1)\pi(\tau+m\beta)}{\beta}}{\sin \frac{\pi(\tau+m\beta)}{\beta}}$$

$$= \frac{1}{\beta} \frac{(-1)^{(2N+1)m} \sin \frac{(2N+1)\pi\tau}{\beta}}{(-1)^m \sin \frac{\pi\tau}{\beta}} = S_N(\tau). \tag{1.190}$$

For finite, but large values of N (remember that we are supposed to take the limit $N \to \infty$), the numerator of $S_N(\tau)$ in (1.187) oscillates rapidly and will average out to zero if $\tau \neq \tau_m = m\beta, m = 0, \pm 1, \pm 2, \cdots$. On the other hand, near τ_m, both the numerator and the denominator vanish and can lead to a non-negligible contribution. In fact, let us note from (1.190), that for finitely large values of N,

$$S_N(\tau_m) = \lim_{\tau \to 0} S_N(\tau + m\pi) = \lim_{\tau \to 0} \frac{1}{\beta} \frac{\sin \frac{(2N+1)\pi(\tau+m\pi)}{\beta}}{\sin \frac{\pi(\tau+m\pi)}{\beta}}$$

$$= \lim_{\tau \to 0} \frac{1}{\beta} \frac{\sin \frac{(2N+1)\pi\tau}{\beta}}{\sin \frac{\pi\tau}{\beta}} = \frac{1}{\beta} (2N+1), \tag{1.191}$$

which is large for large N and the important thing to note is that it is independent of the value of m. This shows that, for large values of N, the function $S_N(\tau)$ has negligible value everywhere except at points $\tau = m\beta, m = 0, \pm 1, \pm 2, \cdots$ where it peaks with the same height (independent of the value of m). This is precisely the behavior of a sum of Dirac delta functions located at $\tau = m\beta$. To see this manifestly, let us look at the function $S_N(\tau)$ near a small interval around zero, namely, $-\epsilon \leq \tau \leq \epsilon$ where the function can be approximated by (see (1.189))

$$S_N(\tau) = \frac{1}{\beta} \frac{\sin \frac{(2N+1)\pi\tau}{\beta}}{\frac{\pi\tau}{\beta}} = \frac{1}{\pi} \frac{\sin \frac{(2N+1)\pi\tau}{\beta}}{\tau}, \tag{1.192}$$

and the sequence of functions $\lim_{N\to\infty} S_N(\tau)$ is known to provide a representation for the Dirac delta function[5] $\delta(\tau)$. Moreover, such a behavior would arise around each of $\tau_m = m\beta$. Therefore, for arbitrary τ, we can write

$$\lim_{N\to\infty} \frac{1}{\beta} \sum_{n=-N}^{N} e^{\frac{2i\pi n\tau}{\beta}} = \delta_\beta^{(\text{per.})}(\tau) = \sum_{m=-\infty}^{\infty} \delta(\tau + m\beta), \quad (1.193)$$

where $m = 0, 1, 2, \cdots$. This is the periodic delta function described in (1.67) which is manifestly periodic with a period of β.

The fermion case can also be discussed much along the same lines with some essential differences which we would like to describe briefly. For the fermions, we can write

$$\frac{1}{\beta} \sum_{-\infty}^{\infty} e^{\frac{i\pi(2n+1)\tau}{\beta}} = \lim_{N\to\infty} S_N(\tau), \quad (1.194)$$

where we have identified

$$S_N = \frac{1}{\beta} \sum_{n=-(N+1)}^{n=N} e^{-\frac{i\pi(2n+1)\tau}{\beta}}. \quad (1.195)$$

The seemingly asymmetric limits in the sum in (1.195) arises from the fact that while, in the bosonic case, the infinite series in (1.186) is symmetric under $n \leftrightarrow -n$, for the fermionic case, the infinite series in (1.194) is symmetric under $(2n+1) \leftrightarrow -(2n+1)$ which translates to $n \leftrightarrow -(n+1)$. We note from (1.194) and (1.195) that, unlike the bosonic case, the infinite series as well as $S_N(\tau)$ are anti-periodic under $\tau \to \tau + \beta$. In general, however, we have

$$S_N(\tau + m\beta) = (-1)^m S_N(\tau), \quad m = 0, \pm 1, \pm 2, \cdots. \quad (1.196)$$

Namely, the geometric series (in (1.195) or (1.196)) alternate in sign as τ changes by multiples of β successively.

The geometric series in (1.195) has a common ratio $e^{\frac{2i\pi\tau}{\beta}}$ much like the bosonic case and can be summerd easily leading to

$$S_N(\tau) = \frac{1}{\beta} \frac{\sin\frac{2(N+1)\pi\tau}{\beta}}{\sin\frac{\pi\tau}{\beta}}, \quad (1.197)$$

---

[5]See, for example, Eq. (2.138) in A. Das, *Lectures on Quantum Mechanics* (second edition), Hindustan Book Agency and World Scientific Publishing (2011).

where the symmetry (1.196) is manifest (the numerator is invariant under the shift while the denominator picks up alternating signs). The analysis from the bosonic case, from (1.191)-(1.193) can now be taken over entirely with the difference that the series would now consist of a sum of alternating Dirac delta functions (or with alternating peaks) and we can write

$$\frac{1}{\beta} \sum_{-\infty}^{\infty} e^{\frac{i\pi(2n+1)\tau}{\beta}} = \lim_{N\to\infty} S_N(\tau) = \delta_\beta^{(\text{anti}-\text{per.})}(\tau)$$

$$= \sum_{m=-\infty}^{\infty} (-1)^m \delta(\tau + m\beta). \tag{1.198}$$

This is manifestly anti-periodic with a period β and is known as the anti-periodic delta function. Note, however, that within a fixed finite interval, say $-\beta \leq \tau \leq \beta$, only the term $m = 0$ can contribute to the sum in both (1.193) as well as in (1.198) so that, in this case, we can write the general result

$$\frac{1}{\beta} \sum_{-\infty}^{\infty} e^{\frac{i\pi n\tau}{\beta}} = \delta(\tau), \quad n = 0, \pm 1, \pm 2, \cdots . \tag{1.199}$$

1.10 References

A. A. Abrikosov, L. P. Gorkov and I. E. Dzyaloshinski, *Methods of Quantum Field Theory in Statistical Physics*, Dover Publication (1975).

A. A. Abrikosov, L. P. Gorkov and I. E. Dzyaloshinski, Soviet Physics JETP **9**, 636 (1959).

P. Arnold, P. F. Bedaque, A. Das and S. Vokos, Physical Review **D47**, 498 (1993).

A. Das, *Lectures on Quantum Mechanics* (second edition), Hindustan Book Agency and World Scientific Publishing (2011).

A. Das, *Field Theory: A Path Integral Approach* (third edition), World Scientific Publishing (2019).

A. Das, *Lectures on Quantum Field Theory* (second edition), World Scientific Publishing (2020).

A. L. Fetter and J. D. Walecka, *Quantum Theory of Many Particle Systems*, McGraw-Hill (1971).

D. Gross, R. Pisarski and L. Yaffe, Reviews of Modern Physics **53**, (1981) 43.

R. Kubo, Journal of Physics Society of Japan **12**, 570 (1957).

P. Martin and J. Schwinger, Physical Reviews **115**, 1342 (1959).

T. Matsubara, Progress of Theoretical Physics **14**, 351 (1955).

M. R. Spiegel, *Theory and Problems of Complex Variables*, McGraw-Hill (1964).

H. Umezawa, Y. Tomozawa and H. Ezawa, Nuovo Cimento **5**, 810 (1957).

H. A. Weldon, Physical Review **D26**, 1394 (1982); *ibid* **D28**, 2007 (1983).

CHAPTER 2

Real time formalism

2.1 Closed time path formalism

Our discussion, in the last chapter, has concentrated on the imaginary time formalism applied to statistical systems. As we have seen, this formalism is very convenient and calculationally simple if we are interested in static, equilibrium properties. In fact, by definition, it starts out with an equilibrium distribution of particles where the time variable is traded in for temperature. Time dependence can, of course, be brought into the imaginary time formalism through analytic continuation, although this is nontrivial. We will discuss the analytic continuation later, but let us note here that this can only describe a slow time dependence. A system far away from equilibrium or out of equilibrium cannot be described in the imaginary time formalism.

On the other hand, there are systems where we may be interested in dynamical questions. This may involve systems evolving through a phase transition. Study of the evolution of the universe or the study of chiral symmetry breaking in QCD are just a few examples of questions of this kind. Or, we may simply be interested in the Ward identities in a gauge theory in equilibrium. In either case, the time variable plays a fundamental role and cannot be traded in for an equilibrium temperature. The real time formalism handles such questions in a natural manner. Here one has a time variable which describes the dynamics of the system as well as a temperature variable which represents the equilibrium temperature if we are studying a system in thermal equilibrium. While the real time formalism naturally answers dynamical questions (and, of course, also gives static equilibrium properties), it is conceptually (as well as calculationally) more complicated than the imaginary time formalism. There are various reasons for this. First, there are various subtleties that show up in this formalism which one should be careful about. Second, the real time formalism has not been as widely used as the imaginary time formalism and, therefore, not many things are standardly known about it. But as we will see, the real time formalism provides a much more physical and powerful description of what happens at finite temperature.

There are several formulations of the real time formalism and they all lead to equivalent physical results for systems in thermal equilibrium. In this chapter, we will describe only one formulation, commonly known as the closed time path formalism because we believe that this is potentially richer and unifies both equilibrium and non-equilibrium processes in a natural way. However, other approaches have their own advantages and we will return to another approach known as thermofield dynamics in the next chapter which allows for a nice operatorial, real time description of equilibrium systems at finite temperature. Following that, in chapter **4**, we will describe a general real time description, still for equilibrium processes for completeness, which reduces to both these two earlier cases in special limits. We will discuss non-equilibrium phenomena only in later chapters of the book.

To describe the closed time path formalism in a simple manner, let us consider a general, quantum mechanical system in interaction with an environment (for example, a heat bath). In the Schrödinger picture description where the state vector carries all the time dependence (see section **1.2**), the system is described by a mixed state (consisting of the wave function of the quantum system as well as that of the environment) which is not easy to calculate. A simpler approach, instead, is to describe the system by a density matrix ρ where the degrees of freedom of the environment have already been summed out.[1] In general, the system, embedded in the external surrounding, may or may not be in thermal equilibrium. In such a case, we know that the density matrix can be written as

$$\rho(t) = \sum_n p_n |\psi_n(t)\rangle \langle \psi_n(t)|. \tag{2.1}$$

---

[1]See, for example, section **5.6** in A. Das, *Lectures on Quantum Mechanics* (second edition), Hindustan Book Agency (New Delhi) and World Scientific Publishing (Singapore), 2011.

Here, p_n describes the probability of finding the quantum mechanical system in the state $|\psi_n(t)\rangle$ (which is not necessarily an energy eigenstate as was assumed in (1.6)) in the ensemble and, for simplicity, we are assuming that the states of the system are discrete. We note that we have not explicitly put the superscript (S) in the states in (2.1) denoting the Schrödinger picture description for simplicity as well as to avoid any possible confusion with the entropy of the system (which is usually denoted by S). Being a probability, p_n satisfies

$$p_n \geq 0, \quad \sum_n p_n = 1. \tag{2.2}$$

The average of any operator associated with this system can be calculated from a knowledge of the density matrix and, in the Schrödinger picture, takes the form (see also (1.11)-(1.13))

$$\langle A \rangle(t) = \mathrm{Tr}\, \rho(t) A = \sum_n p_n \langle \psi_n(t)|A|\psi_n(t)\rangle = \sum_n p_n A_n, \tag{2.3}$$

where we have defined

$$A_n = \langle \psi_n(t)|A|\psi_n(t)\rangle. \tag{2.4}$$

It is clear, therefore, that the average of any operator, in such a case, develops a time dependence which is determined entirely from the time dependence of the density matrix (recall that the operators, in the Schrödinger picture are time independent). We also note that we can define an entropy associated with the system in a standard way (see (2.3)-(2.4)), namely,

$$S = -\langle \ln p \rangle \equiv -\sum_n p_n \ln p_n, \tag{2.5}$$

which is a measure of the disorder in the system and is easily seen to be positive semi-definite in view of (2.2).

The time evolution of the density matrix can be obtained from the evolution of the states of the system. Since the quantum states of the system satisfy the Schrödinger equation ($\hbar = 1$)

$$i\frac{\partial|\psi(t)\rangle}{\partial t} = H|\psi(t)\rangle, \quad i\frac{\partial\langle\psi(t)|}{\partial t} = -\langle\psi(t)|H, \tag{2.6}$$

where H is the (Hermitian) Hamiltonian of the system, it is clear from (2.1) that the evolution of the density matrix is given by the quantum Liouville equation (which differs from the Heisenberg equation only in an overall sign, see (1.22))

$$i \frac{\partial \rho(t)}{\partial t} = [H, \rho(t)] = -[\rho(t), H]. \tag{2.7}$$

Here we have assumed that the probabilities, p_n, do not change appreciably with time (the system is near equilibrium) which also amounts to saying that the entropy is almost a constant in time. The primary reason for such an assumption is the lack of knowledge, on our part, as to the dynamical evolution of the external system (such as the heat bath). However, adiabatic evolutions do arise frequently in our study of physical systems and we will continue with this assumption.

It now follows that if the Hamiltonian, H, governing the dynamics of the quantum system is time independent, then we can integrate (2.7) to write

$$\rho(t) = e^{-iHt} \rho(0) e^{iHt}. \tag{2.8}$$

Furthermore, if the Hamiltonian commutes with $\rho(0)$, then the density matrix will be constant in time and will describe a system in equilibrium. This is true, for example, if the states of the system are stationary states. This is also true if the probability, $\rho(0)$, corresponds to a Boltzmann distribution. The second case is what we describe as thermal equilibrium. In general, however, the system may not be in equilibrium and the Hamiltonian may have a time dependence in which case, we can express the time evolution of the density matrix (solution of (2.7)) through the time evolution operator as (see, for example, (1.15)-(1.19))

$$\rho(t) = U(t, 0) \rho(0) U^\dagger(t, 0) = U(t, 0) \rho(0) U(0, t), \tag{2.9}$$

where the time evolution operator is defined in the standard manner

$$i \frac{\partial U(t, t')}{\partial t} = H(t) U(t, t'),$$

$$\text{or,} \quad U(t, t') = T \left(e^{-i \int_{t'}^{t} dt'' \, H(t'')} \right), \tag{2.10}$$

for a time dependent Hamiltonian (see (1.16)). If the Hamiltonian is time-independent, it has the simpler form

$$U(t, t') = e^{-iH(t-t')},\tag{2.11}$$

which follows from (2.10) (see also (1.17)). In either case, the time evolution operator satisfies the condition that

$$U(t, t) = \mathbb{1}.\tag{2.12}$$

Furthermore, the time evolution operator also satisfies (see also (1.18)-(1.19))

$$U(t_1, t_2)\, U(t_2, t_1) = 1,$$
$$U(t_1, t_2)U(t_2, t_3) = U(t_1, t_3).\tag{2.13}$$

Since the density matrix, $\rho(0)$, is a positive Hermitian matrix with unit trace (see (2.2) or (2.3) with $A = \mathbb{1}$), without loss of generality we can assume that it has the form

$$\rho(0) = \frac{e^{-\beta H_i}}{\operatorname{Tr} e^{-\beta H_i}},\tag{2.14}$$

for some Hermitian operator H_i. Let us next suppose that the quantum system, which we are studying, is governed by a Hamiltonian of the form

$$H(t) = \begin{cases} H_i & \text{for} \quad \operatorname{Re} t \leq 0, \\ \mathcal{H}(t) & \text{for} \quad \operatorname{Re} t \geq 0. \end{cases}\tag{2.15}$$

The physical meaning of this is quite clear. For negative times, we prepare the system in a equilibrium state with temperature β and let the system evolve for positive times with the true Hamiltonian of the system, $\mathcal{H}(t)$, which may be time dependent. It is clear that if $\mathcal{H}(t) = H_i$, then the system will evolve in equilibrium, but not otherwise. From (2.9)-(2.15), we note that we can write

$$e^{-\beta H_i} = e^{-i(T-i\beta-T)H_i} = U(T - i\beta, T),\tag{2.16}$$

where T represents a large negative time (for which the "Hamiltonian" of the system, H_i, is time independent) which allows us to write

$$\rho(0) = \frac{U(T - i\beta, T)}{\operatorname{Tr} U(T - i\beta, T)}.\tag{2.17}$$

Furthermore from (2.9) we, then, obtain

$$\rho(t) = U(t,0)\,\rho(0)\,U(0,t)$$
$$= \frac{U(t,0)U(T - i\beta, T)U(0,t)}{\operatorname{Tr} U(T - i\beta, T)}. \tag{2.18}$$

The time evolution of the average of any operator can now be calculated and since this is at the heart of the closed time path formalism, we will go through the derivation in some detail. We note from (2.3) and (2.18) that we can write (recall that the operators in the Schrödinger picture are time independent)

$$\langle A \rangle(t) = \operatorname{Tr} \rho(t) A$$
$$= \frac{\operatorname{Tr} U(t,0)U(T - i\beta, T)U(0,t)A}{\operatorname{Tr} U(T - i\beta, T)}$$
$$= \frac{\operatorname{Tr} U(T - i\beta, T)U(0,t)AU(t,0)U(0,T)U(T,0)}{\operatorname{Tr} U(T - i\beta, T)}$$
$$= \frac{\operatorname{Tr} U(T - i\beta, T)U(T,0)U(0,t)AU(t,0)U(0,T)}{\operatorname{Tr} U(T - i\beta, T)}$$
$$= \frac{\operatorname{Tr} U(T - i\beta, T)U(T,t)AU(t,T)}{\operatorname{Tr} U(T - i\beta, T)}. \tag{2.19}$$

Here we have used the cyclicity of the trace as well as (2.13) and the fact that for $T < 0$, (see, for example, (2.11), (2.15) and (2.16)) $U(T - i\beta, T) = e^{-\beta H_i}$ and $U(T,0) = e^{-iTH_i}$ commute.

There is one final step in bringing the expression in (2.19) into the standard form. Let T' represent a large positive time. Then, using the property in (2.13), we can write

$$\langle A \rangle(t) = \frac{\operatorname{Tr} U(T - i\beta, T)U(T,T')U(T',T)U(T,t)AU(t,T)}{\operatorname{Tr} U(T - i\beta, T)}$$
$$= \frac{\operatorname{Tr} U(T - i\beta, T)U(T,T')U(T',t)AU(t,T)}{\operatorname{Tr} U(T - i\beta, T)U(T,T')U(T',T)}. \tag{2.20}$$

The physical meaning of this expression is quite clear. The system evolves (read from right to left) from a large negative time T to some time t where an operator insertion of A is introduced. Then, the system evolves from t to a large positive time T' after which it evolves

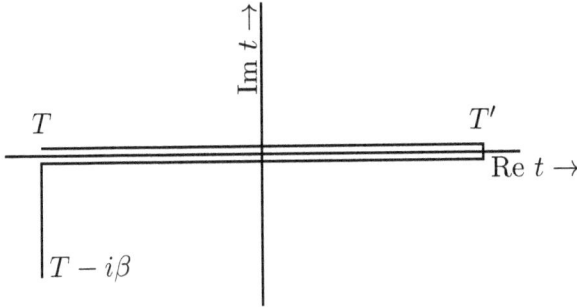

Figure 2.1: Contour for the closed time path in the complex t-plane.

backward in time to T and finally along the imaginary time axis from T to $(T - i\beta)$. Thus, the contour C, representing the evolution of the system in the complex t-plane, has the form shown in Fig. 2.1 (the paths along the real axis have been infinitesimally displaced along the imaginary axis only in order to distinguish them). Eventually, of course, we would like to take the limits $T \to -\infty$ and $T' \to \infty$ for real time descriptions, (for the imaginary time description $T = 0 = T'$, see Fig. 1.1) in which case the form of the average in (2.20) suggests that we should define a generating functional of the form

$$Z[J_c] = \text{Tr}\, U_{J_c}(T - i\beta, T) U_{J_c}(T, T') U_{J_c}(T', T), \tag{2.21}$$

where J_c denote sources which are defined along the three branches of the time contour (in the complex t-plane in Fig. 2.1) that we have described. We should also note that U_{J_c} represents the evolution operator in the presence of the external sources defined along the contour. For a time-independent Hamiltonian and for $J_c = J$, namely, if the source is the same along the entire contour, it is clear that the generating functional in (2.21) reduces to the one that we have already discussed for the imaginary time formalism since, in this case,

$$U_{J_c}(T, T') U_{J_c}(T', T) = \mathbb{1}, \tag{2.22}$$

which follows from the properties in (2.13) (for the case of imaginary time formalism, see (1.86)-(1.87) where $T = 0 = T'$ and $U_{J_c}(0,0) = \mathbb{1}$ trivially). In such a case, we will have a real time description of a system in equilibrium. On the other hand, the sources need not coincide all along the time path (in particular, along the two branches going forward and backward in time) just as the Hamiltonian can be different for different times. If that is the case, then the generating functional can describe a system not in equilibrium.

We conclude this section by noting that the analysis in (2.19)-(2.21) can be generalized leading to a more general real time path contour. We will see an example of this in the next chapter when we discuss the thermofield dynamics formalism. We will further generalize the contour 2.1 in the complex t-plane in chapter **4**.

2.2 Propagators

It is clear now that if we take (2.21) as defining the generating functional of our theory, then we can give it a path integral representation much like at zero temperature. Thus, for example, for a free, real Klein-Gordon field interacting only with an external source, we can write the generating functional as (see also the discussions in (1.83)-(1.87))

$$Z[J_c] = N \int \mathcal{D}\phi \, e^{i \int_c \mathrm{d}t \int \mathrm{d}^3x \, (\mathcal{L} + J_c\phi)}, \tag{2.23}$$

where N is a normalization constant and we can choose the free Lagrangian density to have the form (see (1.76))

$$\mathcal{L} = \frac{1}{2}\partial_\mu \phi \partial^\mu \phi - \frac{m^2}{2}\phi^2. \tag{2.24}$$

Here the (bosonic) field, ϕ, has to satisfy periodic boundary condition in the path integration (because of the trace, see (1.90)) and the time integral is defined along the contour we have described earlier in Fig. 2.1 (with $T \to -\infty, T' \to \infty$) which has mainly three branches. We will denote the branch going from $-\infty$ to ∞ (slightly) above the real axis as C_+, the return path from ∞ to $-\infty$ (slightly) below the real axis as C_- and the final segment along the imaginary axis as C_3. We will show shortly that the last segment contributes in a trivial way to the generating functional (namely, it does not contribute in the calculation of amplitudes) and hence can be ignored. For all practical purposes, therefore, we can think of the contour as effectively consisting of C_+ and C_- which defines a closed contour in the complex time plane. This is why this formalism is also commonly referred to as the closed time path formalism. We note here that the time integration along this closed contour can also be written, with our convention, as

$$
\int_c dt = \int_{-\infty}^{\infty} dt_+ - \int_{-\infty}^{\infty} dt_-. \tag{2.25}
$$

Given a path integral (representation) for the partition function as in (2.23), we can obtain various Green's functions by taking functional derivatives with respect to the sources and these would correspond to appropriate ensemble averages. However, these Green's functions, unlike the zero temperature ones, will be time ordered along the contour, C, in the complex time plane. Let us denote such contour ordered Green's functions by $\langle T_c(A(t_1)B(t_2)\cdots)\rangle$. Thus, for example, for the real Klein-Gordon theory in (2.23)-(2.24), we can define the propagator as

$$
\begin{aligned}
iG_c(t - t') &= \langle T_c(\phi(t)\phi(t')) \rangle \\
&= \theta_c(t - t')\langle \phi(t)\phi(t') \rangle + \theta_c(t' - t)\langle \phi(t')\phi(t) \rangle \\
&= (-i)^2 \frac{1}{Z[J_c]} \frac{\delta^2 Z[J_c]}{\delta J_c(t)\delta J_c(t')}\bigg|_{J_c=0},
\end{aligned} \tag{2.26}
$$

where t and t' are any two points on the contour and we have suppressed the spatial coordinates of the field variables for simplicity. The theta function on the contour is defined in a natural way such that

$$
\theta_c(t - t') = \begin{cases}
\theta(t - t'), & t, t' \text{ on } C_+, \\
\theta(t' - t), & t, t' \text{ on } C_- \text{ or } C_3, \\
0, & t \text{ on } C_+, \ t' \text{ on } C_- \text{ or } C_3, \\
& \text{or } t \text{ on } C_-, \ t' \text{ on } C_3, \\
1, & t' \text{ on } C_+, \ t \text{ on } C_- \text{ or } C_3, \\
& \text{or } t' \text{ on } C_-, \ t \text{ on } C_3,
\end{cases} \tag{2.27}
$$

which automatically satisfies

$$
\theta_c(t - t') + \theta_c(t' - t) = 1. \tag{2.28}
$$

This also leads naturally to a definition of the delta function on the contour as

$$\delta_c(t - t') = \frac{d\theta_c(t - t')}{dt}$$

$$= \begin{cases} \delta(t - t') & t, t' \text{ on } C_+, \\ -\delta(t - t') & t, t' \text{ on } C_- \text{ or } C_3, \\ 0 & \text{otherwise}, \end{cases} \qquad (2.29)$$

which clearly satisfies (see (2.25))

$$\int_c dt' \, \delta_c(t - t') \, f(t') = f(t). \qquad (2.30)$$

The generating functional in (2.23) can be easily evaluated (it is a Gaussian integral) and has the form (here we have restored the spatial coordinate dependences)

$$Z[J_c] = N \, e^{-\frac{i}{2} \int_c dt dt' \int d^3x d^3y \, J_c(\mathbf{x},t) G_c(\mathbf{x}-\mathbf{y},t-t') J_c(\mathbf{y},t')}, \qquad (2.31)$$

where G_c satisfies the equation on the contour given by (compare, for example, with (1.77))

$$\left(\partial_\mu \partial^\mu + m^2\right) G_c(\mathbf{x} - \mathbf{y}, t - t') = -\delta^3(x - y)\delta_c(t - t'). \qquad (2.32)$$

Taking the Fourier transform of the spatial coordinates, we can rewrite this equation also as

$$\left(\frac{\partial^2}{\partial t^2} + \omega_p^2\right) G_c(t - t', \omega_p) = -\delta_c(t - t'), \qquad (2.33)$$

where, as in (1.101),

$$\omega_p = (\mathbf{p}^2 + m^2)^{\frac{1}{2}}, \qquad (2.34)$$

which we are going to denote as ω for simplicity.

The solution of (2.33), subject to the periodicity condition (see (1.38)), can be easily determined to be (note that the solution of this second order equation with a delta potential has to be continuous at

$t = t'$ with a discontinuity in the first (time) derivative (across $t = t'$) given by (-1))

$$G_c(t - t', \omega) = \frac{n_B(\omega)}{2i\omega} \left[\theta_c(t - t') \left(e^{\beta\omega - i\omega(t-t')} + e^{i\omega(t-t')} \right) \right.$$
$$\left. + \theta_c(t' - t) \left(e^{-i\omega(t-t')} + e^{\beta\omega + i\omega(t-t')} \right) \right], \quad (2.35)$$

where the bosonic distribution function, $n_B(\omega)$, is defined as in (1.127). The derivation of this result is given in the appendix at the end of this chapter. Note that this propagator can also be written, alternatively, as

$$G_c(t - t', \omega)$$
$$= \frac{1}{2i\omega} \left[\theta_c(t - t') \left((1 + n_B(\omega))e^{-i\omega(t-t')} + n_B(\omega)e^{i\omega(t-t')} \right) \right.$$
$$\left. + \theta_c(t' - t) \left(n_B(\omega)e^{-i\omega(t-t')} + (1 + n_B(\omega))e^{i\omega(t-t')} \right) \right], \quad (2.36)$$

where we have used the identity

$$n_B(\omega)e^{\beta\omega} = \frac{e^{\beta\omega}}{e^{\beta\omega} - 1} = 1 + n_B(\omega). \quad (2.37)$$

The form of the Green's function in (2.36) will also be useful later in our discussions as well as in some calculations.

There are several things to note from the expression for the propagator in (2.35) or (2.36) (recall that the propagator is i times the Green's function). First, in the zero temperature limit, $\beta \to \infty$, the Green's function, for example, reduces to the conventional one on C_+, namely,

$$G(t - t', \omega) = \frac{1}{2i\omega} \left(\theta(t - t')e^{-i\omega(t-t')} + \theta(t' - t)e^{i\omega(t-t')} \right), \quad (2.38)$$

where we have used the identity ((2.38) follows more directly from (2.36) using the following identity)

$$n_B(\omega) = \frac{1}{e^{\beta\omega} - 1} \xrightarrow{\beta \to \infty} 0. \quad (2.39)$$

Second, the Green's function (for a scalar field) is an even function (on the contour) very much like at zero temperature, namely,

$$G_c(t - t') = G_c(t' - t). \quad (2.40)$$

It satisfies the KMS condition $G_c(-\infty - t) = G_c(-\infty - i\beta - t)$ (see (1.89) with $t_i = -\infty$). Finally, let us note that if t is on the real branch C_+ (or C_-) of the time contour and t' is on C_3 (or vice versa), then the Green's function (and, therefore, the propagator) (see, for example, (2.35) together with (2.27))

$$
\begin{aligned}
G_c(t - t', \omega) &= \lim_{T \to -\infty} G_c(t - (T - is), \omega) \\
&= \frac{n_B(\omega)}{2i\omega} \lim_{T \to -\infty} \left(e^{-i\omega(t-T+is)} + e^{\beta\omega + i\omega(t-T+is)} \right) \to 0, \quad (2.41)
\end{aligned}
$$

in this limit. This shows that the propagator cannot connect two distinct points when one of them lies on C_+ (or C_-) and the other on C_3. In other words, the generating functional in (2.31) cannot have mixed terms involving the sources J_3 and J_+ (or J_-). Therefore, the contribution coming from the branch C_3 to the path integral factorizes so that we can safely set the sources on the branch C_3 to zero. This contribution can be absorbed into a temperature dependent normalization factor which is unimportant as far as dynamical questions are concerned and we will ignore this branch of the contour from now on. Consequently, the relevant contour, in this formalism, really corresponds to a closed path in the complex t-plane. We also note here without going into details that the Green's function for the fermionic fields can be similarly derived in a straightforward manner (see the appendix at the end of this chapter for a derivation).

2.3 Matrix structure for propagators

It is clear from our discussion of the earlier sections that the propagator for any theory can be derived in a straightforward manner very much like the zero temperature case, once we impose the proper periodicity conditions. It is also clear from (2.29) and (2.35) that at nonzero temperatures, the propagator has more structure than is present at zero temperature. Thus, for example, on the two branches C_+ and C_- of the closed time path (we are going to ignore the branch C_3 from now on for reasons alluded to above), there are four non-trivial propagator structures possible and are given by (propagators are simply "i" times the Green's functions, $G_{\pm\pm}$)

$$G_{++}^{\beta}(t - t', \omega) = -\frac{i}{2\omega}\Big((\theta(t - t') + n_B(\omega))\, e^{-i\omega(t-t')}$$
$$+ (\theta(t' - t) + n_B(\omega))\, e^{i\omega(t-t')}\Big),$$

$$G_{+-}^{\beta}(t - t', \omega) = -\frac{i}{2\omega}\Big(n_B(\omega)e^{-i\omega(t-t')} + (1 + n_B(\omega))e^{i\omega(t-t')}\Big),$$

$$G_{-+}^{\beta}(t - t', \omega) = -\frac{i}{2\omega}\Big((1 + n_B(\omega))e^{-i\omega(t-t')} + n_B(\omega)e^{i\omega(t-t')}\Big),$$

$$G_{--}^{\beta}(t - t', \omega) = -\frac{i}{2\omega}\Big((\theta(t' - t) + n_B(\omega))\, e^{-i\omega(t-t')}$$
$$+ (\theta(t - t') + n_B(\omega))\, e^{i\omega(t-t')}\Big), \quad (2.42)$$

where we have used the second form of the Green's function given in (2.36). Here the subscripts for the Green's functions denote the branches of the contour in the complex time plane on which the appropriate time coordinates reside. In terms of ensemble averages, these have the definitions, from (2.26) and (2.27),

$$\begin{aligned}
iG_{++}^{\beta}(t - t') &= \langle T(\phi(t)\phi(t'))\rangle, & t, t' &\in C_+, \\
iG_{+-}^{\beta}(t - t') &= \langle \phi(t')\phi(t)\rangle, & t &\in C_+, t' \in C_-, \\
iG_{-+}^{\beta}(t - t') &= \langle \phi(t)\phi(t')\rangle, & t &\in C_-, t' \in C_+, \\
iG_{--}^{\beta}(t - t') &= \langle \widetilde{T}(\phi(t)\phi(t'))\rangle, & t, t' &\in C_-,
\end{aligned} \quad (2.43)$$

where $\widetilde{T}$ denotes anti-time ordering which is defined on the C_+ branch, for example, as

$$\widetilde{T}(\phi(t)\phi(t')) = \theta(t - t')\phi(t')\phi(t) + \theta(t' - t)\phi(t)\phi(t'), \quad (2.44)$$

which can be compared with (2.26) (on the C_+ branch).

This form of the propagator suggests that we can define a doublet of field variables and sources as

$$\phi_a = \begin{pmatrix} \phi_+ \\ \phi_- \end{pmatrix}, \quad J_a = \begin{pmatrix} J_+ \\ J_- \end{pmatrix}, \quad (2.45)$$

where $a = +, -$ and $\phi_+, J_+ \in C_+$ while $\phi_-, J_- \in C_-$. Furthermore, if we introduce a diagonal metric $\eta^{ab} = (1, -1), a, b = +, -$ in this two dimensional space with the diagonal inverse $\eta_{ab} = (1, -1)$, then we can write the generating functional for the free theory (interacting only with sources) given in (2.31) can be written as (indices a, b are raised and lowered with the metrics η^{ab} and η_{ab} respectively)

$$Z[J_+, J_-] = Z[J^a] = N \, e^{-\frac{i}{2} \int d^4x \, d^4y \, J^a(x) G^{\beta}_{ab}(x,y) J^b(y)}. \qquad (2.46)$$

This would represent the generating functional corresponding to an action (see (2.23)-(2.24))

$$\begin{aligned}
S[\phi_a, J_a] &= S(\phi_+, J_+) - S(\phi_-, J_-) \\
&= \int d^4x \left(\mathcal{L}(\phi_+, J_+) - \mathcal{L}(\phi_-, J_-) \right) \\
&= \int d^4x \, \mathcal{L}(\phi_a, J_a) \\
&= \int d^4x \, \eta^{ab} \left(\frac{1}{2} \left(\partial_\mu \phi_a \partial^\mu \phi_b - m^2 \phi_a \phi_b \right) + J_a \phi_b \right), \quad (2.47)
\end{aligned}$$

so that the components of the propagator in (2.43) would follow from the generating functional in (2.46) simply as

$$iG^{\beta}_{ab}(x - y) = (-i)^2 \, \frac{1}{Z[J]} \frac{\delta^2 Z[J]}{\delta J^a(x) \delta J^b(y)} \bigg|_{J^a=0}. \qquad (2.48)$$

We note here that the time integration (in (2.46) and (2.47)), in this case, is over the usual interval, $-\infty \leq t \leq \infty$, and that we have gotten rid of the branch C_- of the closed time path contour by doubling the field degrees of freedom. It is clear from the definition in (2.43) that G^{β}_{++} corresponds to the conventional time ordered Green's function whereas G^{β}_{--} is the anti-time ordered Green's function. (Contour ordering on the branch C_- corresponds to anti-time ordering in the conventional sense since time runs backwards along this branch.) Thus, intuitively, it is quite clear that the finite temperature Green's functions would display very different analyticity properties compared to the zero temperature ones. We will come to this later. But for the present, let us note that we can write the four possible Green's functions as the elements of a 2×2 matrix as (either on the two branches of the closed time path contour or in the space of the doubled degrees of freedom)

$$G^{(CT)}(p) = \begin{pmatrix} G^\beta_{++}(p) & G^\beta_{+-}(p) \\ G^\beta_{-+}(p) & G^\beta_{--}(p) \end{pmatrix}. \tag{2.49}$$

So far, we have talked about the Green's functions in the co-ordinate space. However, as we know from our experience at zero temperature, calculations simplify considerably in momentum space. In the real time formalism, (unlike in the imaginary time formalism) we can take the Fourier transform of the Green's functions in a simple manner and the components take the forms

$$G^\beta_{++}(p) = \frac{1}{p^2 - m^2 + i\epsilon} - 2i\pi n_B(|p^0|)\delta(p^2 - m^2),$$

$$G^\beta_{+-}(p) = -2i\pi \left(\theta(-p^0) + n_B(|p^0|)\right)\delta(p^2 - m^2)$$

$$= -2i\pi n_B(|p^0|)e^{\frac{\beta(|p^0|-p^0)}{2}}\delta(p^2 - m^2),$$

$$G^\beta_{-+}(p) = -2i\pi \left(\theta(p^0) + n_B(|p^0|)\right)\delta(p^2 - m^2)$$

$$= -2i\pi n_B(|p^0|)e^{\frac{\beta(|p^0|+p^0)}{2}}\delta(p^2 - m^2),$$

$$G^\beta_{--}(p) = -\frac{1}{p^2 - m^2 - i\epsilon} - 2i\pi n_B(|p^0|)\delta(p^2 - m^2), \tag{2.50}$$

where we have used the identities (see also the identity in (2.37))

$$\theta(\mp p^0) + n_B(|p^0|) = \theta(\mp p^0)(1 + n_B(|p^0|)) + \theta(\pm p^0)n_B(|p^0|)$$

$$= \left(\theta(\mp p^0)e^{\beta|p^0|} + \theta(\pm p^0)\right)n_B(|p^0|)$$

$$= e^{\frac{\beta|p^0|}{2}}\left(\theta(\mp p^0)e^{\frac{\beta|p^0|}{2}} + \theta(\pm p^0)e^{-\frac{\beta|p^0|}{2}}\right)n_B(|p^0|)$$

$$= e^{\frac{\beta|p^0|}{2}}\left(\theta(\mp p^0)e^{\mp\frac{\beta p^0}{2}} + \theta(\pm p^0)e^{\mp\frac{\beta p^0}{2}}\right)n_B(|p^0|)$$

$$= e^{\frac{\beta(|p^0|\mp p^0)}{2}}n_B(|p^0|). \tag{2.51}$$

Note that ϵ in (2.50) is an infinitesimal constant parameter which we take to zero from the positive side, namely, $\epsilon \to 0^+$, at the end (although, for simplicity, we will write $\epsilon \to 0$). As a result, the matrix $G^{(CT)}(p)$ can also be written as

$$G^{(CT)}(p) = \begin{pmatrix} \frac{1}{p^2-m^2+i\epsilon} & 0 \\ 0 & -\frac{1}{p^2-m^2-i\epsilon} \end{pmatrix}$$

$$- 2i\pi n_B(|p^0|)\delta(p^2-m^2) \begin{pmatrix} 1 & e^{\frac{\beta(|p^0|-p^0)}{2}} \\ e^{\frac{\beta(|p^0|+p^0)}{2}} & 1 \end{pmatrix},$$

$$(2.52)$$

There are several things to note from the structures in (2.50). We note here that even though the (bosonic) distribution functions in the above expressions can also be written as $n_B(\omega_p)$ (see (2.34)), we write them in a form which is the most convenient for carrying out calculations. It is quite clear now that the Green's functions, even for the free theory, do not have well defined analyticity properties. This is because, by definition,

$$\delta(x) = \lim_{\epsilon \to 0} \frac{1}{\pi} \frac{\epsilon}{x^2+\epsilon^2} = \lim_{\epsilon \to 0} \frac{1}{2i\pi} \left(\frac{1}{x-i\epsilon} - \frac{1}{x+i\epsilon} \right), \qquad (2.53)$$

and, therefore, the delta function is not analytic in either half of the complex plane. Furthermore, from the definition of the Green's functions in (2.42) and (2.50) we note that not all the components of the matrix in (2.49) are independent. In fact, it is easy to check from (2.50) that the matrix components of the Green's function satisfy the constraint relation

$$G_{++}^\beta(p) + G_{--}^\beta(p) = G_{+-}^\beta(p) + G_{-+}^\beta(p). \qquad (2.54)$$

Thus, there are only three independent components of the matrix Green's function.

It is interesting to note here also that the components of the real time propagators (iG_{ab}^β) in (2.42) (or (2.50)) can be given an intuitive physical picture, unlike the imaginary time ones. First, we note that the real time propagators consist of a sum of two parts – one corresponding to the propagator at zero temperature and the other corresponding to a temperature dependent part. It is to be noted that the temperature dependent part is the same for all the components of the Green's function in this formalism. Second, if we look at G^{β}_{++}, we note that while the zero temperature part of the propagator can be thought of in the conventional manner as representing the exchange of a virtual particle, the temperature dependent part, as is clear from (2.50), represents an on-shell contribution because of the delta function enforcing the mass-shell condition. In a plasma of hot particles, there is a distribution of real particles and these real particles can also participate in the absorption and emission process in addition to the exchange of virtual particles. And the temperature dependent part of the Feynman propagator describes precisely this.

The other important thing to note here, from the structure of the propagator in (2.50), is the fact that since the temperature dependent part corresponds to an on-shell contribution, it is unlikely to change the ultraviolet properties of the theory. Namely, a quantum system cannot develop any additional ultraviolet divergences at finite temperature. Thus, from the point of view of renormalization of a theory, the zero temperature counter terms will be sufficient to renormalize the theory. (This, however, does not say anything about the infrared divergences in a theory which can become quite severe at finite temperature.) This fact can also be understood in a physical manner in the following way. The thermal contribution comes with a Boltzmann distribution factor associated with the real particles in a hot medium. This distribution, however, is highly suppressed for high energies and thus temperature contributions are negligible in the ultraviolet region.

It is clear from this description that the fields as well as the propagators acquire a natural two dimensional matrix structure at finite temperature. This is a characteristic property of all real time formalisms and is commonly referred to as the (thermal) doubling of degrees of freedom. We will discuss the meaning of this doubling in a bit more detail in the next chapter, but let us simply note here, heuristically, that one can think of this doubling as follows. At finite temperature, the entire system really consists of two parts – one the quantum system which we would like to study and the other the heat bath. Doubling of degrees of freedom merely represents this aspect in some sense. It is, in fact, required for the self-consistency of the formalism as we will see shortly.

Let us also note that the Green's functions that we have discussed so far, can be calculated perturbatively in the presence of interactions in a diagrammatic form. However, often, we may not be interested in these causal Green's functions. For example, in statistical systems, we are more interested in the physical Green's functions which are defined as (for bosonic fields)

$$iG_R(x, x') = \theta(t - t')\langle [\phi(x), \phi(x')] \rangle,$$
$$iG_A(x, x') = \theta(t' - t)\langle [\phi(x), \phi(x')] \rangle,$$
$$iG_C(x, x') = \langle [\phi(x), \phi(x')]_+ \rangle. \tag{2.55}$$

These are respectively known as the retarded, the advanced and the correlated Green's functions. It would be interesting, therefore, to see how the diagrammatic technique of finite temperature can be used to study dynamical questions which depend on the physical Green's functions. A typical example of such a study, for instance, would involve the study of the formation of domains and the growth of domain sizes during a phase transition.

The interesting point to note from the definitions of the causal Green's functions in (2.43) and the physical Green's functions in (2.55) is that the physical Green's functions, in the closed time path formalism, can be written as linear combinations of the components of the causal Green's functions as (see also the identity (2.54))

$$G_R^{(\mathrm{CT})} = G_{++}^\beta - G_{+-}^\beta = G_{-+}^\beta - G_{--}^\beta,$$
$$G_A^{(\mathrm{CT})} = G_{++}^\beta - G_{-+}^\beta = G_{+-}^\beta - G_{--}^\beta,$$
$$G_C^{(\mathrm{CT})} = G_{++}^\beta + G_{--}^\beta = G_{+-}^\beta + G_{-+}^\beta. \tag{2.56}$$

Thus, one way to calculate the physical Green's functions using the diagrammatic techniques is as follows. We know that $G_{++}^\beta(p)$, $G_{+-}^\beta(p)$, $G_{-+}^\beta(p)$ and $G_{--}^\beta(p)$ can be calculated using Feynman diagrams. One can, then, obtain the physical Green's functions from these calculations using (2.56). (It is worth noting here that there have also been attempts at calculating the physical Green's functions using Feynman diagrams directly.) In fact, let us note that if we define a 2×2 matrix of physical Green's functions as

$$\widehat{G}^{(\mathrm{CT})}(p) = \begin{pmatrix} 0 & G_A^{(\mathrm{CT})}(p) \\ G_R^{(\mathrm{CT})}(p) & G_C^{(\mathrm{CT})}(p) \end{pmatrix}, \tag{2.57}$$

then, from (2.56) we see that we can write

$$G^{(\mathrm{CT})}(p) = Q^{-1}\, \widehat{G}^{(\mathrm{CT})}(p)\, Q,$$

$$\widehat{G}^{(\mathrm{CT})}(p) = Q\, G^{(\mathrm{CT})}(p)\, Q^{-1}, \qquad (2.58)$$

where the matrix $G^{(\mathrm{CT})}(p)$ is defined in (2.49) and the unitary matrix relating the two is given by

$$Q = \frac{1}{\sqrt{2}} \begin{pmatrix} 1 & -1 \\ 1 & 1 \end{pmatrix},$$

$$Q^{-1} = \frac{1}{\sqrt{2}} \begin{pmatrix} 1 & 1 \\ -1 & 1 \end{pmatrix} = Q^{\dagger}. \qquad (2.59)$$

Thus, we note that once the matrix $G^{(\mathrm{CT})}(p)$ has been calculated diagrammatically, the physical Green's functions, $\widehat{G}^{(\mathrm{CT})}(p)$, can be obtained in a simple manner through a change of basis defined by (2.58). (Namely, the retarded, advanced and the correlated Green's function then follow automatically and need not be calculated again.)

For the sake of completeness, we record here the forms of the retarded, advanced and correlated Green's functions at finite temperature for the free Klein-Gordon theory which can be obtained from (2.50) and (2.56),

$$G_R^{(\mathrm{CT})}(p) = \frac{1}{p^2 - m^2 + i\epsilon p^0} \simeq \frac{1}{(p^0 + \frac{i\epsilon}{2})^2 - \omega^2},$$

$$G_A^{(\mathrm{CT})}(p) = \frac{1}{p^2 - m^2 - i\epsilon p^0} \simeq \frac{1}{(p^0 - \frac{i\epsilon}{2})^2 - \omega^2},$$

$$G_C^{(\mathrm{CT})}(p) = -2i\pi \left(1 + 2n_B(|p^0|)\right) \delta(p^2 - m^2), \qquad (2.60)$$

where ω (denoted as ω_p) is defined in (2.34). It should be noted that the retarded and the advanced Green's functions are independent of the temperature at the lowest order as they should be. (The retarded and the advanced Green's functions are described in terms of ensemble averages of commutators (see (2.55)) which are c–number functions.) The effect of the boundary conditions or the initial state conditions is carried completely by the correlated Green's function. Furthermore, we see from (2.60) that while the retarded Green's function is analytic in the upper half of the complex p^0–plane, the advanced Green's function is analytic in the lower half plane much like at zero temperature. This is clear from the last forms of these two Green's functions in (2.60). For example, the two poles of the retarded Green's function are at

$$p^0 + \frac{i\epsilon}{2} = \pm\omega, \quad \text{or,} \quad p^0 = \pm\omega - \frac{i\epsilon}{2}, \qquad (2.61)$$

both of which lie in the lower half of the complex p^0-plane. Similarly, it follows from (2.60) that the two poles of the advanced Green's function lie in the upper half of the complex p^0-plane. On the other hand, the correlated Green's function does not have any well defined analytic behavior in either halves of the complex plane (because of the delta function, see (2.53)).

Let us also note here, without going into details, that the forms of the finite temperature Green's functions for a free massive Dirac field can also be worked out in a straightforward manner leading to the forms (see section **6.2.5** for more details)

$$S^{\beta}_{++}(p) = (\not{p} + m)\left(\frac{1}{p^2 - m^2 + i\epsilon} + 2i\pi n_F(|p^0|)\delta(p^2 - m^2)\right),$$

$$S^{\beta}_{+-}(p) = 2i\pi(\not{p} + m)\left(-\theta(-p^0) + n_F(|p^0|)\right)\delta(p^2 - m^2),$$

$$S^{\beta}_{-+}(p) = 2i\pi(\not{p} + m)\left(-\theta(p^0) + n_F(|p^0|)\right)\delta(p^2 - m^2), \qquad (2.62)$$

$$S^{\beta}_{--}(p) = (\not{p} + m)\left(-\frac{1}{p^2 - m^2 - i\epsilon} + 2i\pi n_F(|p^0|)\delta(p^2 - m^2)\right),$$

where $\not{p} = \gamma^\mu p_\mu$ and the fermion distribution function is defined as

$$n_F(|p^0|) = \frac{1}{e^{\beta|p^0|} + 1}. \qquad (2.63)$$

The physical Green's functions can, then, be obtained from (2.62) in a straightforward manner using the (constant and dynamics independent) unitary transformation matrix, Q, defined in (2.59).

2.4 Analyticity properties

As we have discussed earlier, at finite temperature, the Green's functions do not have very well defined analyticity properties (for example, recall the delta function argument in (2.53) and the discussions around there). Let us investigate this in some more detail with the simple example of the propagators of the free scalar theory although similar conclusions can be easily derived for the ensemble averages of any two arbitrary operators which we will do in the next chapter. We note from (2.50) that (remember that $\epsilon \to 0$ is understood and we will drop the subindex β, for simplicity, although all the Green's functions are at finite temperature)

$$G^{\beta}_{++}(p) = \frac{1}{p^2 - m^2 + i\epsilon} - 2i\pi n_B(|p^0|)\delta(p^2 - m^2)$$

$$= \frac{p^2 - m^2 - i\epsilon}{(p^2 - m^2)^2 + \epsilon^2} - 2i\pi n_B(|p^0|)\delta(p^2 - m^2), \qquad (2.64)$$

so that we can identify (in the limit $\epsilon \to 0$)

$$\operatorname{Re} G^{\beta}_{++}(p) = \frac{(p^2 - m^2)}{(p^2 - m^2)^2 + \epsilon^2} = P\frac{1}{p^2 - m^2},$$

$$\operatorname{Im} G^{\beta}_{++}(p) = -\frac{\epsilon}{(p^2 - m^2)^2 + \epsilon^2} - 2\pi n_B(|p^0|)\delta(p^2 - m^2)$$

$$= -\pi(1 + 2n_B(|p^0|))\delta(p^2 - m^2),$$

$$= -\pi \coth\left(\frac{\beta|p^0|}{2}\right)\delta(p^2 - m^2), \qquad (2.65)$$

where P stands for the principal value of the expression defined as

$$P\left(\frac{1}{x}\right) = \lim_{\epsilon \to 0} \frac{x}{x^2 + \epsilon^2} = \lim_{\epsilon \to 0} \frac{1}{2}\left(\frac{1}{x - i\epsilon} + \frac{1}{x + i\epsilon}\right), \qquad (2.66)$$

and we have used (2.53) as well in the calculation of the imaginary part. Furthermore, we note that

$$1 + 2n_B(x) = 1 + \frac{2}{e^{\beta x} - 1} = \frac{e^{\beta x} + 1}{e^{\beta x} - 1} = \coth\frac{\beta x}{2}, \qquad (2.67)$$

which we have used in the last step in (2.65). Equation (2.65) leads to the dispersion relation

$$\operatorname{Re} G^{\beta}_{++}(p^0, \omega_p)$$

$$= -\frac{1}{\pi}P\int\limits_{-\infty}^{\infty} dp'^0\, \frac{\operatorname{Im} G^{\beta}_{++}(p'^0, \omega_p)}{p^0 - p'^0}\tanh\left(\frac{\beta p'^0}{2}\right). \qquad (2.68)$$

This can be simply checked as follows. The integrand on the right hand side can be simplified to give

$$-\frac{1}{\pi}\left(\frac{-\pi\coth\left(\frac{\beta|p'^0|}{2}\right)\delta((p'^0)^2-\omega_p^2)}{p^0-p'^0}\right)\tanh\left(\frac{\beta p'^0}{2}\right)$$

$$=\frac{\mathrm{sgn}(p'^0)}{p^0-p'^0}\delta((p'^0)^2-\omega_p^2)$$

$$=\frac{1}{p^0-p'^0}\frac{\mathrm{sgn}(p'^0)}{2|p'^0|}\left(\delta(p'^0-\omega_p)+\delta(p'^0+\omega_p)\right)$$

$$=\frac{1}{p^0-p'^0}\frac{1}{2p'^0}\left(\delta(p'^0-\omega_p)+\delta(p'^0+\omega_p)\right).\qquad(2.69)$$

Substituting this back into the integral on the right hand side in (2.68) and doing the integral over p'^0 using the delta functions leads to the desired result, namely, the right hand side of (2.68) becomes

$$=P\left[\frac{1}{p^0-\omega_p}\frac{1}{2\omega_p}+\frac{1}{p^0+\omega_p}\left(-\frac{1}{2\omega_p}\right)\right]$$

$$=P\left[\frac{1}{2\omega_p}\left(\frac{1}{p^0-\omega_p}-\frac{1}{p^0+\omega_p}\right)\right]$$

$$=P\frac{1}{(p^0)^2-\omega_p^2}=P\frac{1}{p^2-m^2},\qquad(2.70)$$

where we have used the definition in (2.34) in the last step.

Similarly, we note from (2.60) that

$$G_R^{(\mathrm{CT})}(p)=\frac{1}{p^2-m^2+i\epsilon p^0}=\frac{p^2-m^2-i\epsilon p^0}{(p^2-m^2)^2+(\epsilon p^0)^2},\qquad(2.71)$$

so that we can determine

$$\mathrm{Re}\,G_R^{(\mathrm{CT})}(p)=\frac{(p^2-m^2)}{(p^2-m^2)^2+(\epsilon p^0)^2}=\frac{1}{p^0}\frac{\frac{p^2-m^2}{p^0}}{\left(\frac{p^2-m^2}{p^0}\right)^2+\epsilon^2}$$

$$=\frac{1}{p^0}P\frac{1}{\frac{p^2-m^2}{p^0}}=P\frac{1}{p^2-m^2},\qquad(2.72)$$

$$\operatorname{Im} G_R^{(\mathrm{CT})}(p) = -\frac{\epsilon p^0}{(p^2 - m^2)^2 + (\epsilon p^0)^2} = -\frac{1}{p^0} \frac{\epsilon}{\left(\frac{p^2 - m^2}{p^0}\right)^2 + \epsilon^2}$$

$$= -\frac{\pi}{p^0} \delta\left(\frac{p^2 - m^2}{p^0}\right) = -\pi \operatorname{sgn}(p^0)\, \delta(p^2 - m^2). \tag{2.73}$$

Here we have used (2.66) in the last line of (2.72) while both (2.53) as well as the delta function identity $\delta(ax) = |a|\delta(x)$ have been used in the last line of (2.73). In the last line of (2.72), the factor $\frac{1}{p^0}$ can be moved inside the principal value because there is really no singularity at $p^0 = 0$. Of course, relation (2.72) can also be obtained simply by noting that, since $\epsilon \to 0$, with a finite p^0, we can define a new infinitesimal parameter $\epsilon' = \epsilon |p^0| \to 0$ leading directly to (2.72). (We emphasize that, since the infinitesimal parameter is assumed to approach 0 from the positive side, the meaningful definition of the new parameter should be with $|p^0|$ and, of course, $(p^0)^2 = |p^0|^2$.) Equations (2.72) and (2.73) lead to the simpler dispersion relation (compare with (2.68))

$$\operatorname{Re} G_R^{(\mathrm{CT})}(p) = -\frac{1}{\pi} P \int dp'^0 \, \frac{\operatorname{Im} G_R^{(\mathrm{CT})}(p'^0, \omega_p)}{p^0 - p'^0}, \tag{2.74}$$

which can be easily checked, using (2.72) and (2.73), as before.

Let us note here that, although $G_R^{(\mathrm{CT})}(p)$ is analytic in the upper half of the complex p^0-plane, neither $\operatorname{Re} G_R^{(\mathrm{CT})}(p)$ nor $\operatorname{Im} G_R^{(\mathrm{CT})}(p)$ is. This follows from the definitions of the principal part as well as that of the delta function (given respectively in (2.66) and (2.53)) and leads to

$$\operatorname{Re} G_R^{(\mathrm{CT})}(p) = P \frac{1}{p^2 - m^2}$$

$$= \lim_{\epsilon \to 0} \frac{1}{2}\left(\frac{1}{p^2 - m^2 - i\epsilon} + \frac{1}{p^2 - m^2 + i\epsilon}\right),$$

$$\operatorname{Im} G_R^{(\mathrm{CT})}(p) = -\pi \operatorname{sgn}(p^0)\delta(p^2 - m^2)$$

$$= \lim_{\epsilon \to 0} -\frac{\operatorname{sgn}(p^0)}{2i}\left(\frac{1}{p^2 - m^2 - i\epsilon} - \frac{1}{p^2 - m^2 + i\epsilon}\right). \tag{2.75}$$

This shows clearly that neither $\operatorname{Re} G_R^{(\mathrm{CT})}(p)$ nor $\operatorname{Im} G_R^{(\mathrm{CT})}(p)$ is analytic in either half of the complex p^0-plane. On the other hand, we see from (2.75) that

$$G_R^{(\mathrm{CT})}(p) = \mathrm{Re}\, G_R^{(\mathrm{CT})}(p) + i\mathrm{Im}\, G_R^{(\mathrm{CT})}(p)$$

$$= \frac{1}{2}\left(\frac{1 - \mathrm{sgn}(p^0)}{p^2 - m^2 - i\epsilon} + \frac{1 + \mathrm{sgn}(p^0)}{p^2 - m^2 + i\epsilon}\right)$$

$$= \frac{\theta(-p^0)}{p^2 - m^2 - i\epsilon} + \frac{\theta(p^0)}{p^2 - m^2 + i\epsilon}$$

$$= \frac{\theta(-p^0)}{p^2 - m^2 + i\epsilon' p^0} + \frac{\theta(p^0)}{p^2 - m^2 + i\epsilon' p^0}, \qquad (\epsilon = \epsilon'|p^0|),$$

$$= \frac{1}{p^2 - m^2 + i\epsilon' p^0}, \tag{2.76}$$

as it should be (see (2.60)) and it was already shown in (2.60)-(2.61) that this is analytic in the upper half of the complex p^0-plane. In the derivation, given in (2.76), we have used the standard relations

$$\mathrm{sgn}(x) = \theta(x) - \theta(-x), \qquad \theta(x) + \theta(-x) = 1. \tag{2.77}$$

It follows now from (2.65), (2.72)-(2.73) that (we note that $\coth x = (\coth|x|)\mathrm{sgn}(x)$)

$$\mathrm{Re}\, G_{++}^{\beta}(p) = \mathrm{Re}\, G_R^{(\mathrm{CT})}(p),$$

$$\mathrm{Im}\, G_{++}^{\beta}(p) = \coth\left(\frac{\beta p^0}{2}\right)\mathrm{Im}\, G_R^{(\mathrm{CT})}(p). \tag{2.78}$$

Since $\mathrm{Re}\, G_R^{(\mathrm{CT})}(p)$ and $\mathrm{Im}\, G_R^{(\mathrm{CT})}(p)$ are not analytic in either half of the complex p^0-plane, this shows that $\mathrm{Re}\, G_{++}^{\beta}(p)$ as well as $\mathrm{Im}\, G_{++}^{\beta}(p)$ (in addition to $G_{++}^{\beta}(p)$)) are also not analytic in either half of the planes. Furthermore, $\mathrm{Im}\, G_{++}^{\beta}(p)$ has additional singularities coming from an infinite number of poles of $\coth x$ along the imaginary axis in both the upper and the lower half planes (these poles are there in addition to those coming from $\delta(x)$ present in $\mathrm{Im}\, G_R^{(\mathrm{CT})}(p)$, see (2.75)). We note here that although, in this section, we have derived dispersion relations for the free Green's functions (propagators), they continue to hold even in an interacting theory as the derivation in the next chapter will show.

We can derive relations similar to (2.78) involving the advanced Green's function as well. However, the more interesting question to ask is the relation between the Green's functions in the imaginary and the real time formalisms. This can be obtained in the following way. From the form of the Green's function in (1.96), we note that

$$\mathcal{G}_\beta(p^0, \mathbf{p}) = \frac{1}{\omega^2 + \mathbf{p}^2 + m^2}, \tag{2.79}$$

where ω is assumed to take discrete values as discussed earlier. Let us recall that the Euclidean rotation, which takes us to Euclidean space (imaginary time) corresponds to $p^0 = p_0 \to i\omega$ (see discussion immediately following (1.180)). Therefore, expressing ω in terms of the continuous real variable p^0 (namely, $\omega = -ip^0$) and analytically continuing it to the complex p^0-plane, we obtain

$$\mathcal{G}_\beta(-i(p^0 + i\epsilon), \mathbf{p}) = -\frac{1}{(p^0 + i\epsilon)^2 - \mathbf{p}^2 - m^2}$$

$$= -\frac{1}{p^2 - m^2 + 2i\epsilon p^0} = -G_R(p), \tag{2.80}$$

which can be compared with (2.60) (note that, since $\epsilon \to 0$, scaling it by a finite positive number 2 does not modify anything). This shows that if we know the values of the imaginary time Green's functions at a set of discrete points, then by analytic continuation, we can naturally obtain the real time, retarded Green's function which would then lead to the real time Feynman Green's function through the relation in (2.78). The analytic continuation of a function, whose values are known only at a set of discrete values on the imaginary axis in the complex energy plane, to the real axis is not unambiguous in general. For example, we can always multiply any given function by a factor $e^{i\beta\omega}$ which has a unit value at $\omega = \frac{2n\pi}{\beta}$ for integer n and this will lead to a different function upon analytic continuation. However, if the functions are well behaved and convergent at infinity, then according to a theorem in complex analysis, the analytic continuation is unique. This, therefore, establishes the connection between the Green's functions in the two formalisms.

2.5 One loop propagator

Let us now work out the propagator for a self-interacting ϕ^4-theory up to one loop order at finite temperature in the closed time path formalism. As we have discussed earlier, the complete action that enters into the generating functional in (2.23) and (2.25) can be written as (see (2.46))

$$S = \int d^4x \, (\mathcal{L}(\phi_+) - \mathcal{L}(\phi_-)),$$ (2.81)

where

$$\mathcal{L}(\phi) = \frac{1}{2} \partial_\mu \phi \partial^\mu \phi - \frac{m^2}{2} \phi^2 - \frac{\lambda}{4!} \phi^4.$$ (2.82)

As before we can think of ϕ_+ and ϕ_- to correspond to the values of the fields on the two branches of the contour or simply as two components of a thermal doublet of scalar fields.

We have already derived the propagator for the free theory in (2.50). The interaction vertices can also be read out from the action in (2.81)

$$= -i\lambda,$$

$$= i\lambda.$$ (2.83)

It is clear that there are two kinds of vertices in the theory now – one involving the ϕ_+ fields while the other involves ϕ_- fields. The strength of interaction is the same for the two, only the signs are opposite. Given this, we can immediately calculate the one loop corrections to the two point functions of the theory which would involve only a mass correction

$$(2.84)$$

In fact, we note that

$$-i\Delta m_+^2 = \frac{(-i\lambda)}{2} \int \frac{\mathrm{d}^4k}{(2\pi)^4} \, iG_{++}(k)$$

$$= \frac{\lambda}{2} \int \frac{\mathrm{d}^4k}{(2\pi)^4} \left(\frac{1}{k^2 - m^2 + i\epsilon} - 2i\pi n_B(|k^0|)\delta(k^2 - m^2) \right)$$

$$= -i \left(\Delta m_0^2 + \Delta m_T^2 \right). \tag{2.85}$$

The zero temperature and the finite temperature contributions to the one loop mass correction decompose explicitly. It is also clear from (2.85) that the only divergent integral is the temperature independent one and can be taken care of with a zero temperature counter term.

Let us note next that if we use the method of residues to do the k^0 integral in the first term, then the zero temperature part takes the form ($\omega_k = (\mathbf{k}^2 + m^2)^{\frac{1}{2}}$)

$$-i\Delta m_0^2 = \frac{\lambda}{2} \int \frac{\mathrm{d}^3k}{(2\pi)^3} \left(\frac{(-i)}{2\omega_k} \right) = -\frac{i\lambda}{4} \int \frac{\mathrm{d}^3k}{(2\pi)^3} \frac{1}{\omega_k}, \tag{2.86}$$

whereas the k^0 integral in the temperature dependent part (second term) can be done trivially using the delta function and gives

$$-i\Delta m_T^2 = -\frac{i\lambda}{2} \int \frac{\mathrm{d}^3k}{(2\pi)^3} \frac{n_B(\omega_k)}{\omega_k}. \tag{2.87}$$

Using (2.67) (namely, $1 + 2n_B(x) = \coth \frac{\beta x}{2}$), the total one loop mass correction at finite temperature becomes

$$\Delta m_+^2 = \Delta m_0^2 + \Delta m_T^2 = \frac{\lambda}{4} \int \frac{\mathrm{d}^3k}{(2\pi)^3} \frac{1}{\omega_k} \coth \left(\frac{\beta \omega_k}{2} \right). \tag{2.88}$$

This is precisely the value obtained for the mass correction using the imaginary time formalism (see (1.118)). For the ϕ_- field, we can similarly calculate and show that

$$\tag{2.89}$$

$$\text{or,} \quad \Delta m_-^2 = \frac{i\lambda}{2} \int \frac{d^4k}{(2\pi)^4} G_{--}(k) = \Delta m_+^2. \tag{2.90}$$

The equality of the two mass corrections, on the two branches C_+ and C_-, follows since both the vertex as well as the propagator change signs on the C_- branch (see (2.83) and (2.50)).

Given this, we can now calculate the propagator including the one loop correction, which would involve both $+$ and $-$ type vertices,

$$\tag{2.91}$$

and will have the form

$$iG_{++}^{(1)}(p) = iG_{++}(p) + (-i\Delta m_+^2)(iG_{++}(p))^2$$
$$+ (i\Delta m_-^2)(iG_{+-}(p))(iG_{-+}(p)),$$
$$\text{or,} \quad G_{++}^{(1)}(p) = G_{++}(p) + \Delta m_+^2 \left((G_{++}(p))^2 - G_{+-}(p)G_{-+}(p) \right), \tag{2.92}$$

where we have used the relation in (2.90) and $G_{\pm\pm}(p)$ denote the tree level propagators given in (2.50). Let us note from the structure of the Green's functions in (2.50) that each propagator has a term with a delta function, the square of which is an ill defined quantity. Consequently, (2.92) would appear to be ill defined. However, such ill defined quantities cancel out in the difference in the second term in (2.92) giving rise to a meaningful perturbative expansion. (In fact, note from (2.50) that the delta function term in each of the components of the propagator is identical leading to the cancellation of the squared terms in (2.92).) This is another way to see that a doubling of degrees of freedom is really necessary for the consistency of a meaningful diagrammatic description. It is worth noting here that some earlier attempts at formulating real time formalisms, which did not involve a doubling of fields, failed for this reason (among others). Finally we note that, using a regularized expression for the delta function as given in (2.53) (we will discuss more on this later),

(2.92) can also be written as

$$G_{++}^{(1)}(p) = G_{++}(p) + \Delta m_+^2 \frac{\partial G_{++}(p)}{\partial m^2}, \qquad (2.93)$$

which shows that the mass correction, merely shifts the position of the pole in the propagator much like at zero temperature.

2.6 Appendices

In these two short appendices, we discuss the derivations of the Green's functions (propagators) for the scalar field theory as well as for the fermion theory (in $0 + 1$ dimension, for simplicity) on the closed time path contour in some detail.

2.6.1 Green's function for the scalar field theory. We note from (2.33) that the homogeneous equation (without the delta function on the right) has two linearly independent solutions $e^{\pm i\omega_p(t-t')}$. Therefore, the contour ordered Green's function, $G_c(t-t', \omega_p)$, can be expressed as a linear superposition of the two homogeneous solutions as

$$G_c(t - t', \omega_p) = \theta_c(t - t')\left(Ae^{-i\omega_p(t-t')} + Be^{i\omega_p(t-t')}\right)$$

$$+ \theta_c(t' - t)\left(Ce^{-i\omega_p(t-t')} + De^{i\omega_p(t-t')}\right), \quad (2.94)$$

where A, B, C, D are constant coefficients of expansion and the step functions $\theta_c(t - t')$ and $\theta_c(t' - t)$ denote the contour ordered step functions which arise in the definition of the contour ordered Green's function in (2.26) and are defined in (2.27). Action of the quadratic operator in (2.33) on $G_c(t - t', \omega_p)$ given in (2.94) leads to

$$\left(\frac{\partial^2}{\partial t^2} + \omega_p^2\right)G_c(t - t', \omega_p) = -\delta_c(t - t')\left(i\omega_p(A - C - B + D)\right)$$

$$+ \delta_c'(t - t')\left(A - C + B - D\right), \qquad (2.95)$$

where we have used the identity

$$\delta_c'(t)f(t) = (\delta_c(t)f(t))' - \delta_c(t)f'(t) = (\delta_c(t)f(0))' - \delta_c(t)f'(t)$$

$$= \delta_c'(t)f(0) - \delta_c(t)\left(f'(t)\big|_{t=0}\right), \qquad (2.96)$$

in the intermediate steps. Therefore, we see that, for $G_c(t - t', \omega_p)$ in (2.94) to satisfy the differential equation in (2.33), the constant coefficients of expansion must satisfy

$$A + B = C + D, \quad i\omega_p(A - B - C + D) = 1, \tag{2.97}$$

which determine

$$C = A - \frac{1}{2i\omega_p}, \quad D = B + \frac{1}{2i\omega_p}. \tag{2.98}$$

Let us next note that, for a second order differential equation with a delta function term (potential) as in (2.33), the solution must be continuous across the boundary $t = t'$ while the first derivative (of the solution) must be discontinuous with the discontinuity across the boundary given by (-1). (The discontinuity is easily determined by integrating the differential equation (2.33) across the boundary $t = t'$.) With the relations derived in (2.98), these conditions are easily seen to be automatically satisfied. The solution in (2.94) would still seem to depend on two arbitrary constants A, B and, therefore, not unique.

However, let us recall that a bosonic Green's function, at finite temperature, has to satisfy periodic boundary conditions (see (1.38) or (1.89))

$$G_c(t_f - t, \omega_p) = G_c(t_i - t, \omega_p), \tag{2.99}$$

where $t_i \leq t \leq t_f$ and which we are yet to impose. For the closed time path formalism, $t_i = T$ and $t_f = T - i\beta$ (see Fig. 2.1), so that the bosonic Green's function has to satisfy

$$G_c(T - i\beta - t, \omega_p) = G_c(T - t, \omega_p). \tag{2.100}$$

Using (2.98) in (2.94), we obtain (remember that, on the contour, the ordering is $T \leq t \leq T - i\beta$)

$$G_c(T - t, \omega_p)$$

$$= \theta_c(t - T) \left(C e^{-i\omega_p(T-t)} + D e^{i\omega_p(T-t)} \right),$$

$$= \left(A - \frac{1}{2i\omega_p} \right) e^{-i\omega_p(T-t)} + \left(B + \frac{1}{2i\omega_p} \right) e^{i\omega_p(T-t)},$$

$$G_c(T - i\beta - t, \omega_p)$$

$$= \theta_c(T - i\beta - t) \left(A e^{-i\omega_p(T - i\beta - t)} + B e^{i\omega_p(T - i\beta - t)} \right)$$

$$= \left(A e^{-\beta\omega_p} e^{-i\omega_p(T - t)} + B e^{\beta\omega_p} e^{i\omega_p(T - t)} \right), \tag{2.101}$$

and equating the two we determine (see, for example, the relations in (2.37) and (2.98))

$$A = \frac{1}{2i\omega_p} (1 + n_B(\omega_p)), \qquad B = \frac{1}{2i\omega_p} n_B(\omega_p),$$

$$C = \frac{1}{2i\omega_p} n_B(\omega_p), \qquad D = \frac{1}{2i\omega_p} (1 + n_B(\omega_p)). \tag{2.102}$$

Here $n_B(\omega_p)$ is the bosonic distribution function defined in (1.127). This completes the derivation of the Green's function given in (2.35) (or (2.36)).

2.6.2 Green's function for the $0+1$ dimensional fermionic theory. In this appendix, we describe, along the lines of the previous appendix, the derivation of Green's function for a free fermionic field theory. There are two main differences between a bosonic and a fermionic theory at zero temperature. First, the dynamical equations (and, therefore, the equation for the Green's function) in a fermionic theory are first order in nature unlike in the bosonic theory. Second, the fermionic theories involve Dirac matrices so that the equations become matrix equations which can make it technically a little bit involved. However, at zero temperature, the calculation of the Green's function for the free Dirac theory is almost trivial because of certain identities which the Dirac matrices satisfy. Let us indicate this briefly here before moving on to the finite temperature case.

The Lagrangian density, for the free Dirac theory of a massive fermion, is given by

$$\mathcal{L} = \bar{\psi} \left(i\gamma^\mu \partial_\mu - m \right) \psi = \bar{\psi} \left(i\partial\!\!\!/ - m \right) \psi, \tag{2.103}$$

where, for example, in $3+1$ dimensions, $\partial_\mu = \frac{\partial}{\partial x^\mu}$, $\mu = 0, 1, 2, 3$ (see, for example, (1.76)) define the four space-time derivatives (gradients) and γ^μ denote the four linearly independent Dirac matrices which satisfy the Clifford algebra (involving the anti-commutator)

$$[\gamma^\mu, \gamma^\nu]_+ = 2\eta^{\mu\nu} \mathbb{1}, \quad ([A, B]_+ = AB + BA). \tag{2.104}$$

The fermion Green's function, $S(x - x')$, satisfies the equation

$$(i\partial\!\!\!/ - m)S(x - x') = \delta^4(x - x'),$$ (2.105)

which can be compared with the equation for the Green's function for the scalar field theory given in (1.77) (where x' is assumed to be at the origin). Using the Clifford algebra given in (2.104), it is easy to show that $(i\partial\!\!\!/ + m)(i\partial\!\!\!/ - m) = -(\partial^\mu \partial_\mu + m^2)$ and, therefore, comparing (1.77) and (2.105), we conclude that at zero temperature, we can write

$$S(x - x') = (i\partial\!\!\!/ + m)G(x - x').$$ (2.106)

In other words, once we know the Green's function for the free, massive scalar field theory at zero temperature, we can easily determine the Green's function for the free, massive Dirac theory at zero temperature using (2.106) (namely, we do not need to calculate it afresh). At finite temperature, however, the two Green's functions satisfy different periodicity conditions – scalar Green's function is periodic while the fermion Green's function is anti-periodic – which makes any such relation between the two becomes untenable. Therefore, we need to calculate the Green's function for the free, massive Dirac theory carefully which we do next. To avoid the technicalities arising from the matrix structure of the Dirac equation in (2.105) and, in order to bring out and emphasize the effect of the anti-periodicity as clearly as is possible, we look at the $0 + 1$ dimensional (quantum mechanical) free, massive fermionic theory where there are no nontrivial matrices. (Furthermore, this will also come in handy in later chapters when we discuss $0 + 1$ dimensional fermionic theories in other contexts.)

In $0+1$ dimension, the free, massive Dirac (fermion) Lagrangian has the form

$$L = \bar{\psi}\left(i\frac{\mathrm{d}}{\mathrm{d}t} - m\right)\psi,$$ (2.107)

where $\psi = \psi(t)$ and $\bar{\psi} = \bar{\psi}(t)$ denote the two anti-commuting fermion field variables of the theory and m denotes the mass of the free fermion. On the closed time path contour (see Fig. 2.1), the Green's function, for this theory, satisfies the equation

$$(i\partial_t - m)\,S_c(t - t', m) = \delta_c(t - t'),$$ (2.108)

which can be compared with (2.105) (there is no spatial coordinate in $0+1$ dimension, only time). We note that the theta function as well as the delta function, on the closed time path contour are already defined in (2.27)-(2.30). This is a first order differential equation with a delta potential. So, unlike (2.33), which is a second order equation, here the solution itself has to be discontinuous across the boundary $t = t'$ with a discontinuity $-i$ (which can be seen by simply integrating the differential equation in (2.108) across the boundary). Furthermore, since (2.108) is a first order equation, the homogeneous equation (without the delta function term on the right) has only one independent solution, e^{-imt}, unlike the second order equation (see (2.94) and the preceding discussion). Therefore, the general solution of the inhomogeneous equation (2.108) can be written as

$$S_c(t - t', m) = -i \left(\theta_c(t - t')A - \theta_c(t' - t)B \right) e^{-im(t-t')}. \quad (2.109)$$

Here the relative negative sign between the two terms arises from the definition of time ordering for anti-commuting fields when we change the order of the two fields in the second term (see, for example, (2.26)). Substituting this into the equation (2.108), we obtain

$$(i\partial_t - m) S_c(t - t', m) = \delta_c(t - t') (A + B) = \delta_c(t - t'), \quad (2.110)$$

which determines

$$A + B = 1, \quad \text{or,} \quad A = 1 - B, \quad (2.111)$$

so that we can write (see (2.109))

$$S_c(t - t', m) = -i \left(\theta_c(t - t')(1 - B) - \theta_c(t' - t)B \right) e^{-im(t-t')}$$
$$= -i \left(\theta_c(t - t') - B \right) e^{-im(t-t')}, \quad (2.112)$$

where we have used (2.28). This Green's function is clearly discontinuous across the boundary $t = t'$ with a discontinuity $-i$ as we would expect. Finally, let us note that this is the Green's function for a fermion so that it has to satisfy anti-periodic boundary condition (1.89) (recall that, in the closed time path formalism, $t_i = T, t_f = T - i\beta$, see Fig. 2.1)

$$S_c(t_f - t, m) = -S_c(t_i - t, m),$$

$$\text{or,} \ S_c(T - i\beta - t, m) = -S_c(T - t), \quad T - i\beta \le t \le T, \quad (2.113)$$

which determines

$$(1 - B)e^{-\beta m} = B, \tag{2.114}$$

and leads to

$$B = \frac{e^{-\beta m}}{e^{-\beta m} + 1} = \frac{1}{e^{\beta m} + 1} = n_F(m),$$

$$A = (1 - B) = (1 - n_F(m)) = \frac{e^{\beta m}}{e^{\beta m} + 1}. \tag{2.115}$$

Here $n_F(m)$ denotes the fermion distribution function and the unique (parameter free) fermion Green's function in (2.112) has the form

$$S_c(t - t', m) = -i \left(\theta_c(t - t') - n_F(m) \right) e^{-im(t-t')}. \tag{2.116}$$

These two derivations show how the (anti) periodicity conditions play an important role in determining the Green's function (propagator) at finite temperature. We will use this observation and, in particular, the form of the $0 + 1$ dimensional fermion Green's function given in (2.116) in a later chapter when we discuss the thermal operator representation (in chapter **6**, see section **6.5.1**) as well as effective actions at finite temperature.

2.7 References

P. Aurenche and T. Becherray, Nuclear Physics **B379**, 259 (1992).

P. M. Bakshi and K. T. Mahanthappa, Journal of Mathematical Physics **4**, 1 (1963).

G. Baym and N. Mermin, Journal Mathematical Physics **2**, 232 (1961).

P. F. Bedaque and A. Das, Modern Physics Letters **A8**, 3151 (1993).

K. Chou, Z. Su, B. Hao and L. Yu, Physics Reports **118**, 1 (1985).

M. A. van Eijck and Ch. G. van Weert, Physics Letters **B278**, 305 (1992).

M. A. van Eijck, R. Kobes and Ch. G. van Weert, Physical Review **D50**, 4097 (1994).

L. V. Keldysh, Soviet Physics JETP **20**, 1018 (1965).

J. Schwinger, Journal of Mathematical Physics **2**, 407 (1961).

J. Schwinger, *Lecture Notes of Brandeis Summer Institute in Theoretical Physics* (1960).

CHAPTER 3

Thermofield dynamics

Another very interesting approach to thermal field theories within the framework of real time formalism is thermofield dynamics. Thermofield dynamics has already been widely used in many areas of physics. We can essentially think of thermofield dynamics as an operator formalism (although a path integral representation does exist for it and will be described later in this chapter) and complementary to the closed time path formalism. It can answer such questions as the structure of the thermal vacuum, the nature of the Goldstone states and so on that are not possible within the framework of the closed time path formalism. It can also describe quite naturally the time development of quantities near equilibrium.

3.1 General formalism

To introduce thermofield dynamics, let us recall (see (1.11)) that the ensemble average of any operator in thermal equilibrium is given by ($Z(\beta)$ denotes the partition function of the system)

$$\langle A \rangle_\beta = Z^{-1}(\beta) \, \mathrm{Tr} \, (e^{-\beta \mathcal{H}} A), \tag{3.1}$$

and can be written in the energy basis as

$$\langle A \rangle_\beta = Z^{-1}(\beta) \sum_n e^{-\beta E_n} \langle n|A|n \rangle. \tag{3.2}$$

Here we are assuming that the eigenvalues of $\mathcal{H}$ are discrete for simplicity and the energy eigenstates are assumed to be orthonormal and complete

$$\mathcal{H}|n\rangle = E_n|n\rangle,$$

$$\langle n|m\rangle = \delta_{nm}, \quad \sum_n |n\rangle\langle n| = \mathbb{1}, \tag{3.3}$$

where the inner product is assumed to be a Dirac inner product.

The partition function which has the form (see (1.9))

$$Z(\beta) = \text{Tr}\, e^{-\beta\mathcal{H}}, \tag{3.4}$$

resembles the vacuum to vacuum transition amplitude (or the generating functional) of zero temperature field theory, but does not quite coincide with it (because of the trace). We can take over the machinery of the zero temperature field theory completely if we can express the ensemble average of any operator as an expectation value in a vacuum (say, a thermal vacuum). Thus, if we can define a state $|0,\beta\rangle$ such that we can write

$$\langle A\rangle_\beta = \langle 0,\beta|A|0,\beta\rangle = Z^{-1}(\beta)\sum_n e^{-\beta E_n}\langle n|A|n\rangle, \tag{3.5}$$

for any arbitrary operator A, then, the finite temperature formalism will be completely parallel to the zero temperature one.

If the state $|0,\beta\rangle$ belongs to our Hilbert space, then, we can express it as a linear superposition of any (complete) basis states of our Hilbert space. Since energy eigenstates define a complete orthonormal basis of the Hilbert space (see (3.3)), we can write

$$|0,\beta\rangle = \sum_n |n\rangle\langle n|0,\beta\rangle = \sum_n f_n(\beta)\,|n\rangle, \quad f_n(\beta) = \langle n|0,\beta\rangle, \tag{3.6}$$

where we have used the completeness of the states of our Hilbert space (see (3.3)) and this will lead to

$$\langle 0,\beta|A|0,\beta\rangle = \sum_{n,m} f_n^*(\beta)f_m(\beta)\langle n|A|m\rangle. \tag{3.7}$$

This will agree with (3.5) provided

$$f_n^*(\beta)f_m(\beta) = Z^{-1}(\beta)\,e^{-\beta E_n}\delta_{nm}. \tag{3.8}$$

However, as we see from (3.6), $f_n(\beta)$'s are merely coefficients of expansion and correspond to ordinary scalar functions. Therefore, it is not possible for them to satisfy (3.8), namely, product of two ordinary scalar functions cannot produce a Kronecker delta. This shows that, as long as we restrict ourselves to the original Hilbert space, we cannot define a finite temperature vacuum which would satisfy (3.5).

On the other hand, we note that the condition, (3.8), is quite analogous to an orthonormality condition for state vectors (see (3.3)) except that $f_n(\beta)$'s are not state vectors as long as we remain in the same Hilbert space. Therefore, to carry out this program, let us introduce an additional fictitious system which is an identical copy of the original system we are interested in and denote it as the tilde system, namely,

$$\widetilde{\mathcal{H}}|\tilde{n}\rangle = E_n|\tilde{n}\rangle,$$

$$\langle\tilde{n}|\tilde{m}\rangle = \delta_{nm}, \quad \sum_n |\tilde{n}\rangle\langle\tilde{n}| = \mathbb{1}. \tag{3.9}$$

In other words, the state $|\tilde{n}\rangle$ is a tilde state corresponding to the energy state $|n\rangle$ of the original system and is an energy eigenstate of this fictitious tilde system with the same energy eigenvalue E_n, see, for example, (3.3). Let us consider the product space of the two systems as

$$|n, \tilde{m}\rangle = |n\rangle \otimes |\tilde{m}\rangle, \tag{3.10}$$

and assume that the thermal vacuum state is a vacuum state in this product (Hilbert) space. Let us write the thermal vacuum in this doubled Hilbert space as a superposition of states (we note here that the reason for expanding the thermal vacuum as a diagonal superposition in the product space will become clear shortly, basically it is to conform with (3.8))

$$|0, \beta\rangle = \sum_n f_n(\beta)|n, \tilde{n}\rangle = \sum_n f_n(\beta)|n\rangle \otimes |\tilde{n}\rangle, \tag{3.11}$$

where the coefficients of expansion are given by (see (3.6))

$$f_n(\beta) = \langle n, \tilde{n}|0, \beta\rangle = (\langle n| \otimes \langle\tilde{n}|)\,|0, \beta\rangle. \tag{3.12}$$

In this case, we note that the thermal vacuum average of an operator A in the original Hilbert space can be calculated to give

$$\langle 0, \beta | A | 0, \beta \rangle = \sum_{n,m} f_n^*(\beta) f_m(\beta) \langle n, \tilde{n} | A | m, \tilde{m} \rangle$$

$$= \sum_{n,m} f_n^*(\beta) f_m(\beta) \langle n | A | m \rangle \, \delta_{nm}$$

$$= \sum_n f_n^*(\beta) f_n(\beta) \langle n | A | n \rangle. \tag{3.13}$$

Here, in the intermediate step, we have used the fact that an operator of the original physical system does not act on the Hilbert space of the tilde system (doubled system) and *vice versa* so that using the orthonormality of the states, we can write (note from (3.3) that $\langle n | m \rangle = \delta_{nm} = \langle \tilde{n} | \tilde{m} \rangle$)

$$\langle n, \tilde{n} | A | m, \tilde{m} \rangle = \langle n | A | m \rangle \, \langle \tilde{n} | \tilde{m} \rangle = \delta_{nm} \langle n | A | m \rangle. \tag{3.14}$$

Similarly, for a tilde operator, we can write

$$\langle n, \tilde{n} | \tilde{A} | m, \tilde{m} \rangle = \langle n | m \rangle \, \langle \tilde{n} | \tilde{A} | \tilde{m} \rangle = \delta_{nm} \langle \tilde{n} | \tilde{A} | \tilde{m} \rangle. \tag{3.15}$$

In fact, in this doubled space, it is more appropriate to write the operators A and $\tilde{A}$ as

$$A = A \otimes \mathbb{1}, \quad \tilde{A} = \mathbb{1} \otimes \tilde{A}, \tag{3.16}$$

but we will continue with our simpler (nonrigorous) notations only for simplicity.

We note now from (3.13) that the thermal vacuum average can represent an ensemble average if (see (3.5))

$$f_n^*(\beta) f_n(\beta) = Z^{-1}(\beta) e^{-\beta E_n}, \tag{3.17}$$

which can have the simple real solution

$$f_n(\beta) = f_n^*(\beta) = Z^{-\frac{1}{2}}(\beta) \, e^{-\frac{\beta E_n}{2}}. \tag{3.18}$$

This, therefore, shows that we can, in fact, introduce a temperature dependent vacuum state (thermal vacuum) so that the statistical ensemble average of any operator can be identified with the expectation value of the operator in this vacuum state. This, however, entails a doubling of the Hilbert space. (This doubling is quite reminiscent of our discussion of doubling of paths and degrees of freedom in the last chapter within the context of the closed time path formalism.) The advantage here, on the other hand, lies in the fact that the operator techniques of zero temperature quantum field theory can also be naturally carried over to the finite temperature case. We would like to note here that choices other than the simple one given in (3.18) are, indeed, possible. For example, besides the trivial phase ambiguity in (3.18), it is possible to define the duals and, therefore, the inner products in such a doubled Hilbert space nontrivially. We will address these issues in a later chapter. For the moment, however, we continue with our simple choice given in (3.18) in this chapter.

3.2 Fermionic oscillator

Let us now go into the details of the formalism through simple examples. To start with, let us consider the simplest example of a quantum mechanical fermionic oscillator with frequency ω described by the Hamiltonian (we have neglected the zero point energy and we have set $\hbar = 1$)

$$H = \omega\, a^\dagger a = \omega N, \quad N = a^\dagger a, \tag{3.19}$$

where N is the number operator and we are assuming that there is no chemical potential which allows us to identify $\mathcal{H} = H$. The creation and the annihilation operators of the theory satisfy the anticommutation relations

$$[a, a^\dagger]_+ = (aa^\dagger + a^\dagger a) = \mathbb{1},$$
$$[a, a]_+ = [a^\dagger, a^\dagger]_+ = 0. \tag{3.20}$$

The second relation in (3.20) implies that the annihilation and the creation operators are nilpotent, namely, $a^2 = (a^\dagger)^2 = 0$. Consequently, the Hilbert space of the theory becomes two dimensional (in the energy eigenspace) with the two energy eigenstates given by $|0\rangle$ and $|1\rangle = a^\dagger|0\rangle$ corresponding to the energy eigenvalues 0 and ω (and occupation numbers 0 and 1) respectively. Namely, (recall that

$a|0\rangle = 0)$

$$H|0\rangle = E_0|0\rangle = 0, \qquad H|1\rangle = E_1|1\rangle = \omega|1\rangle,$$
$$N|0\rangle = 0, \qquad N|1\rangle = |1\rangle. \tag{3.21}$$

According to our general discussion of thermofield dynamics, in order to consider this system at a finite temperature, we are next supposed to introduce a fictitious tilde system which is identical to the original system. Thus, we define a system described by the Hamiltonian

$$\widetilde{H} = \omega\widetilde{a}^\dagger\widetilde{a} = \omega\widetilde{N}, \quad \widetilde{N} = \widetilde{a}^\dagger\widetilde{a}, \tag{3.22}$$

where $\widetilde{a}, \widetilde{a}^\dagger$ are respectively the annihilation and creation operators of the tilde system and they satisfy identical anti-commutation relations (as in (3.20)) given by

$$[\widetilde{a}, \widetilde{a}^\dagger]_+ = \mathbb{1},$$
$$[\widetilde{a}, \widetilde{a}]_+ = [\widetilde{a}^\dagger, \widetilde{a}^\dagger]_+ = 0. \tag{3.23}$$

Note that the non-tilde and the tilde oscillators have the same frequency which is consistent with the fact the tilde system is an identical fictitious system. It follows now that the Hilbert space of this fictitious system is also two dimensional (in the energy space) as is the case with the original (fermion) system and (note that $\widetilde{a}|\widetilde{0}\rangle = 0$)

$$\widetilde{H}|\widetilde{0}\rangle = \widetilde{E}_0|\widetilde{0}\rangle = 0, \qquad \widetilde{H}|\widetilde{1}\rangle = \widetilde{E}_1|\widetilde{1}\rangle = \omega|\widetilde{1}\rangle,$$
$$\widetilde{N}|\widetilde{0}\rangle = 0, \qquad \widetilde{N}|\widetilde{1}\rangle = |\widetilde{1}\rangle. \tag{3.24}$$

Furthermore, we assume that the creation and the annihilation operators for the tilde and the non-tilde systems anti-commute so that observables like N, H and $\widetilde{N}, \widetilde{H}$ can commute (the two systems are independent). For example, we note that

$$[N, \widetilde{N}] = [a^\dagger a, \widetilde{a}^\dagger\widetilde{a}] = a^\dagger[a, \widetilde{a}^\dagger\widetilde{a}] + [a^\dagger, \widetilde{a}^\dagger\widetilde{a}]a$$
$$= a^\dagger\left(-\widetilde{a}^\dagger[a, \widetilde{a}]_+ + [a, \widetilde{a}^\dagger]_+\widetilde{a}\right)$$
$$+ \left(-\widetilde{a}^\dagger[a^\dagger, \widetilde{a}]_+ + [a^\dagger, \widetilde{a}^\dagger]_+\widetilde{a}\right)a, \tag{3.25}$$

and this can vanish only if the tilde and non-tilde creation and annihilation operators anti-commute. Namely,

$$[a, \tilde{a}]_+ = 0 = [a, \tilde{a}^\dagger]_+, \quad [a^\dagger, \tilde{a}]_+ = 0 = [a^\dagger, \tilde{a}^\dagger]_+, \tag{3.26}$$

which we assume, in addition to the relations given in (3.20) and (3.23).

The product Hilbert space of states for the two systems now becomes four dimensional (in the energy space) with the four basis states given by $(n, \tilde{m} = 0, 1$, see (3.21) and (3.24))

$$|n, \tilde{m}\rangle = |n\rangle \otimes |\tilde{m}\rangle : \quad |0, \tilde{0}\rangle; |0, \tilde{1}\rangle; |1, \tilde{0}\rangle; |1, \tilde{1}\rangle. \tag{3.27}$$

We assume the energy eigenstates to be orthonormal (the Hamiltonians are Hermitian) and according to our general discussion, we can choose the thermal vacuum for this system as in (3.11) which takes the simple form

$$|0, \beta\rangle = \sum_{n=0}^{1} f_n(\beta)|n, \tilde{n}\rangle = \sum_{n=0}^{1} f_n(\beta)|n\rangle \otimes |\tilde{n}\rangle$$

$$= f_0(\beta)|0\rangle \otimes |\tilde{0}\rangle + f_1(\beta)|1\rangle \otimes |\tilde{1}\rangle. \tag{3.28}$$

The normalization of the state $|0, \beta\rangle$ imposes a condition on the coefficients of expansion (recall that the energy states are orthonormal)

$$\langle 0, \beta|0, \beta\rangle = |f_0(\beta)|^2 + |f_1(\beta)|^2 = 1. \tag{3.29}$$

We derive a second condition on the coefficients by requiring that the thermal vacuum state $|0, \beta\rangle$ gives the correct thermal distribution function for fermions, namely,

$$\langle 0, \beta|N|0, \beta\rangle = \langle 0, \beta|a^\dagger a|0, \beta\rangle = |f_1(\beta)|^2 = \frac{1}{e^{\beta\omega} + 1}. \tag{3.30}$$

As a result, the coefficients of expansion in (3.28) can be uniquely determined (of course, up to phases, we choose the coefficients to be real for simplicity) from (3.29) and (3.30),

$$f_0(\beta) = f_0^*(\beta) = \frac{1}{\sqrt{1 + e^{-\beta\omega}}} = Z^{-\frac{1}{2}}(\beta),$$

$$f_1(\beta) = f_1^*(\beta) = \frac{e^{-\frac{\beta\omega}{2}}}{\sqrt{1 + e^{-\beta\omega}}} = Z^{-\frac{1}{2}}(\beta) e^{-\frac{\beta\omega}{2}}, \tag{3.31}$$

where $Z(\beta) = (1 + e^{-\beta\omega})$ is the partition function (see (1.9)) for a two level fermionic oscillator with vanishing zero point energy and this can be compared with (3.17). Thus, we see from (3.28) that the thermal vacuum, in this case, can be written as the simple superposition

$$|0, \beta\rangle = Z^{-\frac{1}{2}}(\beta) \left(|0, \tilde{0}\rangle + e^{-\frac{\beta\omega}{2}} |1, \tilde{1}\rangle \right)$$

$$= \frac{1}{\sqrt{1 + e^{-\beta\omega}}} \left(|0, \tilde{0}\rangle + e^{-\frac{\beta\omega}{2}} |1, \tilde{1}\rangle \right). \tag{3.32}$$

Let us also note from (3.32) (or from (3.28) and (3.31)) that we can identify (remember $T = \frac{1}{\beta}$, so that $T \to 0$ corresponds to $\beta \to \infty$)

$$|0, \tilde{0}\rangle = |0\rangle \otimes |\tilde{0}\rangle = |0, \beta\rangle|_{T=0}. \tag{3.33}$$

We also observe that, in this case, it follows trivially from the anti-commutation relations of the creation and annihilation operators in (3.20), (3.23) and (3.26) that (recall that these fermion operators are nilpotent)

$$\left(\tilde{a}a - a^\dagger \tilde{a}^\dagger \right)^{2n} |0, \tilde{0}\rangle = (-1)^n |0, \tilde{0}\rangle$$

$$\left(\tilde{a}a - a^\dagger \tilde{a}^\dagger \right)^{2n+1} |0, \tilde{0}\rangle = (-1)^{n+1} |1, \tilde{1}\rangle. \tag{3.34}$$

Therefore, if we define a Hermitian operator

$$Q(\theta(\beta)) = -i\theta(\beta)\left(\tilde{a}a - a^\dagger \tilde{a}^\dagger \right), \tag{3.35}$$

where $\theta(\beta)$ is a real parameter, then the exponential operator

$$U(\theta(\beta)) = e^{-iQ(\theta(\beta))} = e^{-\theta(\beta)(\tilde{a}a - a^\dagger \tilde{a}^\dagger)},$$

$$U^\dagger(\theta(\beta))U(\theta(\beta)) = \mathbb{1} = U(\theta(\beta))U^\dagger(\theta(\beta)), \tag{3.36}$$

would be formally unitary. Furthermore, we note that, acting on the vacuum state in the doubled space, $|0, \tilde{0}\rangle$, this operator will lead to

$$U(\theta(\beta))|0, \tilde{0}\rangle = \cos\theta(\beta) |0, \tilde{0}\rangle + \sin\theta(\beta) |1, \tilde{1}\rangle, \tag{3.37}$$

which follows from (3.34). And, this state can be identified with the thermal vacuum state, if $\theta(\beta)$ is chosen as (compare also with (3.32))

$$\cos\theta(\beta) = Z^{-\frac{1}{2}}(\beta) = \frac{1}{\sqrt{1 + e^{-\beta\omega}}},$$

$$\sin\theta(\beta) = Z^{-\frac{1}{2}}(\beta) e^{-\frac{\beta\omega}{2}} = \frac{e^{-\frac{\beta\omega}{2}}}{\sqrt{1 + e^{-\beta\omega}}}, \tag{3.38}$$

which is easily seen to satisfy $\cos^2 \theta(\beta) + \sin^2 \theta(\beta) = 1$. This choice is also equivalent to saying that $\tan \theta(\beta) = e^{-\frac{\beta\omega}{2}}$. Thus, we conclude that the (formally) unitary operator $U(\theta(\beta))$ generates the thermal vacuum, $|0, \beta\rangle$ acting on the vacuum of the doubled space, $|0, \widetilde{0}\rangle$, namely,

$$|0, \beta\rangle = U(\theta(\beta))|0, \widetilde{0}\rangle = \cos \theta(\beta)|0, \widetilde{0}\rangle + \sin \theta(\beta)|1, \widetilde{1}\rangle$$

$$= \frac{1}{\sqrt{1 + e^{-\beta\omega}}} \left(|0, \widetilde{0}\rangle + e^{-\frac{\beta\omega}{2}}|1, \widetilde{1}\rangle\right), \tag{3.39}$$

which can be compared with (3.32). Another way of saying this is that the two states, $|0, \beta\rangle$ and $|0, \widetilde{0}\rangle$, are related by the (formally) unitary transformation of (3.36). (We say "formally" because to be truly unitary, the operator must exist (be finite) and acting on a state in a Hilbert space, must take it to another state in the same Hilbert space. In quantum mechanics, this is not generally an issue. However, in continuum field theories, sometimes an operator, which formally satisfies $U^\dagger U = \mathbb{1}$, may not, in fact, exist (be finite) and acting on a state in a given Hilbert space may take it to a state in a different (inequivalent) Hilbert space. The simplest example of such a phenomenon arises when there is spontaneous breaking of a continuous symmetry in a quantum field theory. In such a case, the "formally" unitary transformation (operator) is known as a Bogoliubov transformation (operator).)

In general, we can say that, under such a unitary transformation $U(\theta(\beta))$ (given in (3.36)), any state $|\psi\rangle$ in the doubled Hilbert space would transform to the state $|\psi(\beta)\rangle$ as

$$|\psi(\beta)\rangle = U(\theta(\beta))|\psi\rangle. \tag{3.40}$$

It now follows that, under this unitary transformation, any operator O, in this doubled Hilbert space, would transform as

$$O(\beta) = U(\theta(\beta)) \, O \, U^\dagger(\theta(\beta)) = e^{-iQ(\theta(\beta))} \, O \, e^{iQ(\theta(\beta))}$$

$$= O + [(-iQ(\theta(\beta))), O] + \frac{1}{2!}[(-iQ(\theta(\beta))), [(-iQ(\theta(\beta))), O]]$$

$$+ \frac{1}{3!}[(-iQ(\theta(\beta))), [(-iQ(\theta(\beta))), [(-iQ(\theta(\beta))), O]]] + \cdots, \tag{3.41}$$

so that the expectation values would remain invariant under the unitary transformation. Namely,

$$\langle \psi(\beta)|O(\beta)|\psi(\beta)\rangle$$
$$= \langle \psi|U^{\dagger}(\theta(\beta))\,(U(\theta(\beta))OU^{\dagger}(\theta(\beta)))\,U(\theta(\beta))|\psi\rangle$$
$$= \langle \psi|O|\psi\rangle, \tag{3.42}$$

where the last relation follows from the unitarity property in (3.36). We see, from (3.41), that the basic quantity needed for calculating $O(\beta)$ is the commutator $[(-iQ(\theta(\beta))), O]$ and, for the annihilation and creator operators $a, a^{\dagger}, \tilde{a}, \tilde{a}^{\dagger}$, these can be easily calculated using (3.20), (3.23) and (3.26) and take the simple forms (remember that these are fermionic (anti-commuting) operators)

$$[(-iQ(\theta(\beta))), a] = [-\theta(\beta)(\tilde{a}a - a^{\dagger}\tilde{a}^{\dagger}), a] = -\theta(\beta)\tilde{a}^{\dagger},$$
$$[(-iQ(\theta(\beta))), a^{\dagger}] = [-\theta(\beta)(\tilde{a}a - a^{\dagger}\tilde{a}^{\dagger}), a^{\dagger}] = -\theta(\beta)\tilde{a},$$
$$[(-iQ(\theta(\beta))), \tilde{a}] = [-\theta(\beta)(\tilde{a}a - a^{\dagger}\tilde{a}^{\dagger}), \tilde{a}] = \theta(\beta)a^{\dagger},$$
$$[(-iQ(\theta(\beta))), \tilde{a}^{\dagger}] = [-\theta(\beta)(\tilde{a}a - a^{\dagger}\tilde{a}^{\dagger}), \tilde{a}^{\dagger}] = \theta(\beta)a. \tag{3.43}$$

These relations show that $Q(\theta(\beta))$ rotates $(a, a^{\dagger})$ into $(\tilde{a}^{\dagger}, \tilde{a})$ respectively and back in a very simple manner.

Therefore, using (3.43), the nested commutators in (3.41) can be easily calculated leading to the transformation of the creation and the annihilation operators under the unitary (Bogoliubov) transformation to be

$$a(\beta) = a\cos\theta(\beta) - \tilde{a}^{\dagger}\sin\theta(\beta),$$
$$\tilde{a}(\beta) = \tilde{a}\cos\theta(\beta) + a^{\dagger}\sin\theta(\beta),$$
$$a^{\dagger}(\beta) = a^{\dagger}\cos\theta(\beta) - \tilde{a}\sin\theta(\beta),$$
$$\tilde{a}^{\dagger}(\beta) = \tilde{a}^{\dagger}\cos\theta(\beta) + a\sin\theta(\beta), \tag{3.44}$$

where $\theta(\beta)$ is given in (3.38). We also note here, for later use, that since both (3.43) as well as (3.44) show that, under the unitary (Bogoliubov) transformation a rotates to $\tilde{a}^{\dagger}$ and back (and correspondingly, the Hermitian conjugates $a^{\dagger}$ to $\tilde{a}$ and back), it is useful to define a doublet of operators as

$$A = \begin{pmatrix} a \\ \tilde{a}^\dagger \end{pmatrix}, \quad A^\dagger = \begin{pmatrix} a^\dagger & \tilde{a} \end{pmatrix}. \tag{3.45}$$

Then, we can write the thermal transformations of (3.44) in the matrix form as

$$A(\beta) = \begin{pmatrix} a(\beta) \\ \tilde{a}^\dagger(\beta) \end{pmatrix} = \begin{pmatrix} \cos\theta(\beta) & -\sin\theta(\beta) \\ \sin\theta(\beta) & \cos\theta(\beta) \end{pmatrix} \begin{pmatrix} a \\ \tilde{a}^\dagger \end{pmatrix}$$

$$= \overline{U}(\theta(\beta))\, A,$$

$$A^\dagger(\beta) = \begin{pmatrix} a^\dagger(\beta) & \tilde{a}(\beta) \end{pmatrix} = \begin{pmatrix} a^\dagger & \tilde{a} \end{pmatrix} \begin{pmatrix} \cos\theta(\beta) & \sin\theta(\beta) \\ -\sin\theta(\beta) & \cos\theta(\beta) \end{pmatrix}$$

$$= A^\dagger \overline{U}^\dagger(\theta), \tag{3.46}$$

where

$$\overline{U}(\theta(\beta)) = \begin{pmatrix} \cos\theta(\beta) & -\sin\theta(\beta) \\ \sin\theta(\beta) & \cos\theta(\beta) \end{pmatrix}, \tag{3.47}$$

denotes the (unitary) rotation matrix in this two dimensional ($a - \tilde{a}^\dagger$) plane (note that $\det\overline{U}(\theta(\beta)) = \cos^2\theta(\beta) + \sin^2\theta(\beta) = 1$ and $\overline{U}^\dagger(\theta(\beta))\overline{U}(\theta(\beta)) = \mathbb{1} = \overline{U}(\theta(\beta))\overline{U}^\dagger(\theta(\beta))$, as it should be for a unitary rotation $SO(2)$). Note that the unitarity condition can also be written as

$$\overline{U}^\dagger(\theta(\beta))\mathbb{1}U(\theta(\beta)) = \mathbb{1}, \tag{3.48}$$

where $\mathbb{1}$ denotes the metric of the $SO(2)$ space (in components $\delta^{ij}, i, j = 1.2$).

It follows now from (3.37)-(3.44) that

$$a(\beta)|0,\beta\rangle = (a\cos\theta(\beta) - \tilde{a}^\dagger\sin\theta(\beta))\, |0,\beta\rangle$$

$$= (a\cos\theta(\beta) - \tilde{a}^\dagger\sin\theta(\beta))$$

$$\times (\cos\theta(\beta)|0,\tilde{0}\rangle + \sin\theta(\beta)|1,\tilde{1}\rangle)$$

$$= \cos\theta(\beta)\sin\theta(\beta)\, (a|1,\tilde{1}\rangle - \tilde{a}^\dagger|0,\tilde{0}\rangle)$$

$$= \cos\theta(\beta)\sin\theta(\beta)(|0,\tilde{1}\rangle - |0,\tilde{1}\rangle) = 0,$$

$$\tilde{a}(\beta)|0,\beta\rangle = (\tilde{a}\cos\theta(\beta) + a^\dagger \sin\theta(\beta))|0,\beta\rangle$$

$$= (\tilde{a}\cos\theta(\beta) + a^\dagger \sin\theta(\beta))$$

$$\times (\cos\theta(\beta)|0,\tilde{0}\rangle + \sin\theta(\beta)|1,\tilde{1}\rangle)$$

$$= \cos\theta(\beta)\sin\theta(\beta)(\tilde{a}|1,\tilde{1}\rangle + a^\dagger|0,\tilde{0}\rangle)$$

$$= \cos\theta(\beta)\sin\theta(\beta)(-|1,\tilde{0}\rangle + |1,\tilde{0}\rangle) = 0, \qquad (3.49)$$

where we have used $a|0,\tilde{0}\rangle = 0 = \tilde{a}|0,\tilde{0}\rangle$, $a^\dagger|1,\tilde{1}\rangle = 0 = \tilde{a}^\dagger|1,\tilde{1}\rangle$ (the two second relations reflect the nilpotency of the fermion operators physically saying that there can, at the most, be a single fermion of a kind in a state) as well as the fact that the tilde and the nontilde operators anti-commute. This shows that the thermal vacuum is truly the vacuum state of the thermal creation and annihilation operators. Such a condition is also known as the thermal state condition. Incidentally, such relations also follow directly from the definitions in (3.40) and (3.41) (without explicit evaluations given in (3.49)) as

$$a(\beta)|0,\beta\rangle = (U(\theta(\beta))aU^\dagger(\theta(\beta)))U(\theta(\beta))|0,\tilde{0}\rangle$$

$$= U(\theta(\beta))(a|0,\tilde{0}\rangle) = 0,$$

$$\tilde{a}(\beta)|0,\beta\rangle = (U(\theta(\beta))\tilde{a}U^\dagger(\theta(\beta)))U(\theta(\beta))|0,\tilde{0}\rangle$$

$$= U(\theta(\beta))(\tilde{a}|0,\tilde{0}\rangle) = 0, \qquad (3.50)$$

where we have used (3.36).

The thermal creation and annihilation operators in (3.44) can be easily shown to satisfy the same anti-commutation relations as the original ones given in (3.20) and (3.23) as well as the anti-commutativity of the two sets since they are unitarily related. We can now construct the thermal Hilbert space and the four dimensional space would consist of the states (see also (3.27))

$$|0,\beta\rangle; \quad a^\dagger(\beta)|0,\beta\rangle; \quad \tilde{a}^\dagger(\beta)|0,\beta\rangle; \quad a^\dagger(\beta)\tilde{a}^\dagger(\beta)|0,\beta\rangle. \qquad (3.51)$$

It is worth noting from the definition of the thermal vacuum in (3.37) that it is like a squeezed state. (We will discuss coherent states and squeezed states in an appendix at the end of this chapter.) However, there is an essential difference. A little bit of analysis will show that the thermal Hilbert space and the original Hilbert space will not become (unitarily) equivalent when the theory contains an infinite number of degrees of freedom or modes (namely, in a quantum field theory). In other words, the transformation in (3.36), even though formally unitary, is more like a Bogoliubov transformation which can take us (out of the Hilbert space) to an inequivalent Hilbert space when infinite number of degrees of freedom are involved. It is also worth noting from the two equations in (3.49) that annihilating a particle quantum in the thermal vacuum is equivalent to creating a tilde quantum and *vice versa*, namely,

$$a|0, \beta\rangle = \tan \theta(\beta) \, \tilde{a}^\dagger |0, \beta\rangle,$$

$$\tilde{a}|0, \beta\rangle = -\tan \theta(\beta) \, a^\dagger |0, \beta\rangle. \tag{3.52}$$

This allows us to intuitively think of the tilde particles as kind of hole states of the particle or particle states of the heat bath which gives an intuitive meaning to the doubling of the degrees of freedom in thermofield dynamics.

The states in the thermal Hilbert space are not eigenstates of either H or $\tilde{H}$ as can be easily checked. However, they are eigenstates of $(a^\dagger(\beta)a(\beta) - \tilde{a}^\dagger(\beta)\tilde{a}(\beta))$. Furthermore, from (3.44) as well as the anti-commutation relations in (3.20) and (3.23), it follows that

$$a^\dagger(\beta)a(\beta) - \tilde{a}^\dagger(\beta)\tilde{a}(\beta) = (a^\dagger a \cos^2 \theta(\beta) - aa^\dagger \sin^2 \theta(\beta))$$

$$+ (\tilde{a}\tilde{a}^\dagger \sin^2 \theta(\beta) - \tilde{a}^\dagger \tilde{a} \cos^2 \theta(\beta))$$

$$= a^\dagger a - \tilde{a}^\dagger \tilde{a}. \tag{3.53}$$

Therefore, if we define a total Hamiltonian for the combined system as

$$\widehat{H} = H - \tilde{H} = \omega(a^\dagger a - \tilde{a}^\dagger \tilde{a}), \tag{3.54}$$

then, it is clear that the thermal states will be eigenstates of $\widehat{H}$. Note also that $\widehat{H}$ is invariant under the arbitrary unitary transformation $U(\theta)$ (where θ is an arbitrary constant and not identified with $\theta(\beta)$ as given in (3.38)), namely, $U(\theta)\widehat{H}U^\dagger(\theta) = \widehat{H}$, as (3.53) explicitly shows. This, therefore, defines a rotational symmetry of the doubled system (in the two dimensional space) and leads to

$$\widehat{H}|0, \beta\rangle = 0. \tag{3.55}$$

However, when we fix the arbitrary parameter θ as in (3.38), this rotational symmetry is spontaneously broken. Furthermore, we note that, because of the relative negative sign between the two terms in (3.54), although $|0, \beta\rangle$ has zero eigenvalue for $\widehat{H}$, it is not the lowest eigenvalue state. In fact, the eigenvalues of $\widehat{H}$ can be negative, for example, the state $|0, \widetilde{1}\rangle$ is an eigenstate of $\widehat{H}$ with eigenvalue $(-\omega)$.

3.3 Bosonic oscillator

Let us next summarize, without going into too much detail, the results of thermofield dynamics when applied to a bosonic oscillator. The Hamiltonian, in this case, has the same form as in (3.19) (if we neglect the zero point energy)

$$H = \omega \, a^\dagger a, \tag{3.56}$$

except that the creation and the annihilation operators satisfy commutation relations

$$[a, a^\dagger] = aa^\dagger - a^\dagger a = \mathbb{1},$$
$$[a, a] = [a^\dagger, a^\dagger] = 0. \tag{3.57}$$

The Hilbert space is infinite dimensional and the eigenstates of the Hamiltonian satisfy

$$H|n\rangle = n\omega \, |n\rangle, \qquad n = 0, 1, 2, \cdots . \tag{3.58}$$

Following the general discussion of thermofield dynamics (see section **3.1**), we now introduce the tilde system described by the Hamiltonian

$$\widetilde{H} = \omega \, \widetilde{a}^\dagger \widetilde{a}, \tag{3.59}$$

where the tilde creation and annihilation operators satisfy identical commutation relations as in (3.57) and are assumed to commute with the original operators, namely,

$$[\widetilde{a}, \widetilde{a}^\dagger] = \widetilde{a}\widetilde{a}^\dagger - \widetilde{a}^\dagger\widetilde{a} = \mathbb{1},$$
$$[\widetilde{a}, \widetilde{a}] = 0 = [\widetilde{a}^\dagger, \widetilde{a}^\dagger] = [a, \widetilde{a}] = [a, \widetilde{a}^\dagger] = [a^\dagger, \widetilde{a}]. \tag{3.60}$$

As in the original oscillator system, the doubled Hamiltonian, $\widetilde{H}$, also will have identical energy levels (see (3.9)),

$$\widetilde{H}|\widetilde{n}\rangle = n\omega\,|\widetilde{n}\rangle, \quad n = 0, 1, 2, \cdots . \tag{3.61}$$

We can now represent a generic state in the product space as (see (3.12))

$$|n, \widetilde{m}\rangle = |n\rangle \otimes |\widetilde{m}\rangle, \qquad n, m = 0, 1, 2, \cdots , \tag{3.62}$$

where the states are assumed to be normalized. The thermal vacuum can now be constructed as in the last section and has the form

$$|0, \beta\rangle = (1 - e^{-\beta\omega})^{\frac{1}{2}} \sum_{n=0}^{\infty} e^{-\frac{n\beta\omega}{2}}\,|n, \widetilde{n}\rangle$$

$$= Z^{-\frac{1}{2}} \sum_{n=0}^{\infty} e^{-\frac{n\beta\omega}{2}}\,|n, \widetilde{n}\rangle, \tag{3.63}$$

which can be compared with (3.11) and (3.18). (Recall that $Z^{-1}(\beta) = (1 - e^{-\beta\omega})$ for a bosonic oscillator with vanishing zero point energy, see (1.9).) We note here once again, as in (3.33), that we can identify

$$|0, \widetilde{0}\rangle = |0\rangle \otimes |\widetilde{0}\rangle = |0, \beta\rangle|_{T=0}. \tag{3.64}$$

This, therefore, seems to be the general feature, in thermofield dynamics, that

$$|0, \beta\rangle|_{T=0} = |0, \widetilde{0}\rangle = |0\rangle \otimes |\widetilde{0}\rangle, \tag{3.65}$$

namely, independent of whether it is a bosonic or a fermionic theory, in the zero temperature limit, the thermal vacuum in thermofield dynamics reduces to the first term of the series $|0, \widetilde{0}\rangle$.

Once again, as in the fermionic case (see (3.35)), we can define a Hermitian operator

$$Q(\theta) = -i\theta(\beta)\left(\widetilde{a}a - a^{\dagger}\widetilde{a}^{\dagger}\right), \tag{3.66}$$

and show that the formally unitary operator (see (3.36))

$$U(\theta) = e^{-iQ(\theta)} = e^{-\theta(\beta)\left(\widetilde{a}a - a^{\dagger}\widetilde{a}^{\dagger}\right)}, \tag{3.67}$$

takes us to the thermal vacuum through a Bogoliubov transformation of the form

$$|0, \beta\rangle = U(\theta)|0, \widetilde{0}\rangle = e^{-\theta(\beta)(\widetilde{a}a - a^\dagger \widetilde{a}^\dagger)}|0, \widetilde{0}\rangle, \tag{3.68}$$

provided we identify

$$\cosh \theta(\beta) = \frac{1}{\sqrt{1 - e^{-\beta\omega}}}, \qquad \sinh \theta(\beta) = \frac{e^{-\frac{\beta\omega}{2}}}{\sqrt{1 - e^{-\beta\omega}}}, \tag{3.69}$$

which is equivalent to $\tanh \theta(\beta) = e^{-\frac{\beta\omega}{2}}$.

The thermal operators can also be obtained in a standard manner. We simply note here (see (3.45)-(3.46)) that if we define a doublet as

$$A = \begin{pmatrix} a \\ \widetilde{a}^\dagger \end{pmatrix}, \qquad A^\dagger = \begin{pmatrix} a^\dagger & \widetilde{a} \end{pmatrix}, \tag{3.70}$$

then, we can write

$$A(\beta) = \begin{pmatrix} a(\beta) \\ \widetilde{a}^\dagger(\beta) \end{pmatrix} = \begin{pmatrix} U(\theta)aU^\dagger(\theta) \\ U(\theta)\widetilde{a}^\dagger U^\dagger(\theta) \end{pmatrix}$$

$$= \begin{pmatrix} \cosh \theta(\beta) & -\sinh \theta(\beta) \\ -\sinh \theta(\beta) & \cosh \theta(\beta) \end{pmatrix} \begin{pmatrix} a \\ \widetilde{a}^\dagger \end{pmatrix} = \overline{U}(\theta) A,$$

$$A^\dagger(\beta) = \begin{pmatrix} a^\dagger(\beta) & \widetilde{a}(\beta) \end{pmatrix} = A^\dagger \overline{U}^\dagger(\theta) = \begin{pmatrix} a^\dagger & \widetilde{a} \end{pmatrix} \overline{U}^\dagger(\theta), \tag{3.71}$$

with $\theta(\beta)$ given in (3.69) and

$$\overline{U}(\theta) = \begin{pmatrix} \cosh \theta(\beta) & -\sinh \theta(\beta) \\ -\sinh \theta(\beta) & \cosh \theta(\beta) \end{pmatrix} = \overline{U}^\dagger(\theta). \tag{3.72}$$

There are two things to note from the structure of the matrix $\overline{U}(\theta)$ in (3.72). First, as in (3.47), it rotates the operators a to $\widetilde{a}^\dagger$ and $\widetilde{a}$ to $a^\dagger$ and as a rotation matrix in the two dimensional plane satisfies $\det \overline{U}(\theta) = \cosh^2 \theta(\beta) - \sinh^2 \theta(\beta) = 1$. However, the relative negative sign between the two terms in the determinant suggests that it is not a standard compact rotation. Second, the matrix (3.72) is not naively unitary, $\overline{U}^\dagger(\theta)\overline{U}(\theta) \neq \mathbb{1}$. On the other hand, we note that if we introduce a two dimensional metric as

$$\sigma_3 = \begin{pmatrix} 1 & 0 \\ 0 & -1 \end{pmatrix}, \tag{3.73}$$

then, we can easily check that

$$\overline{U}^\dagger(\theta)\sigma_3\overline{U}(\theta) = \sigma_3 = \overline{U}^\dagger(-\theta)\sigma_3\overline{U}(-\theta). \tag{3.74}$$

In this case, we see that the system has a two dimensional rotational invariance given by the noncompact group $SO(1,1)$ where the metric in the group space is σ_3, which, in components is given by $\epsilon^{ij}, i = 1, 2$ (namely, $\overline{U}(\theta(\beta))$ leaves the metric invariant) and $\overline{U}(\theta(\beta))$ is unitary in that sense (even though it is not naively unitary). This should be contrasted with the fermionic case discussed in (3.48).

From the definition of the thermal vacuum as well as the thermal creation and annihilation operators, it can be readily verified that the thermal vacuum satisfies

$$a(\beta)|0, \beta\rangle = (U(\theta)aU^\dagger(\theta))U(\theta)|0, \widetilde{0}\rangle = U(\theta)(a|0, \widetilde{0}\rangle) = 0,$$

$$\text{or,} \quad a(\beta)|0, \beta\rangle = (a\cosh\theta(\beta) - \widetilde{a}^\dagger \sinh\theta(\beta))|0, \beta\rangle = 0,$$

$$\widetilde{a}(\beta)|0, \beta\rangle = (U(\theta)\widetilde{a}U^\dagger(\theta))(U(\theta)|0, \widetilde{0}\rangle) = U(\theta)(\widetilde{a}|0, \widetilde{0}\rangle) = 0,$$

$$\text{or,} \quad \widetilde{a}(\beta)|0, \beta\rangle = (\widetilde{a}\cosh\theta(\beta) - a^\dagger \sinh\theta(\beta))|0, \beta\rangle = 0. \tag{3.75}$$

We see again that the state $|0, \beta\rangle$ is annihilated by the thermal annihilation operators, $a(\beta), \widetilde{a}(\beta)$, and, therefore, can be thought of as the thermal vacuum. Furthermore, as in the fermionic case, we note that the annihilation of a particle quantum in the thermal vacuum is equivalent to the creation of a tilde quantum and $vice\ versa$. In fact, we can explicitly write the thermal state condition from (3.75) as

$$a|0, \beta\rangle = \tanh\theta(\beta)\widetilde{a}^\dagger|0, \beta\rangle,$$

$$\widetilde{a}|0, \beta\rangle = \tanh\theta(\beta)a^\dagger|0, \beta\rangle, \tag{3.76}$$

which should be compared with (3.52). Finally, let us note that we can build up the thermal Hilbert space by applying thermal creation operators on the state $|0, \beta\rangle$ and, unlike the fermion case, the thermal Hilbert space will be infinite dimensional. However, the states in this space will not be eigenstates of either H or $\widetilde{H}$. Rather, they can be identified with the eigenstates of the operator

$$\widehat{H} = H - \widetilde{H} = \omega(a^\dagger a - \widetilde{a}^\dagger \widetilde{a}), \tag{3.77}$$

which can be thought of as the total Hamiltonian governing the dynamics of the combined system (see (3.53)-(3.54)). Note that, although the state $|0, \beta\rangle$ is an eigenstate of $\widehat{H}$ with zero eigenvalue, the eigen spectrum of this total Hamiltonian can not only be negative as is the case in (3.53), but is unbounded from below (in the bosonic case) since the particle numbers for the tilde particles can become infinitely large.

Let us now observe the following. For either the bosonic or the fermionic case, we note that

$$\widehat{H}|0, \beta\rangle = 0. \tag{3.78}$$

Moreover, in either case, it can be checked that

$$[\widehat{H}, a] = -\omega a, \qquad\qquad [\widehat{H}, a^\dagger] = \omega a^\dagger,$$
$$[\widehat{H}, \widetilde{a}] = \omega \widetilde{a}, \qquad\qquad [\widehat{H}, \widetilde{a}^\dagger] = -\omega \widetilde{a}^\dagger. \tag{3.79}$$

We also note from (3.38) and (3.69) that for the fermionic and the bosonic cases, we have, respectively

$$\tan \theta(\beta) = e^{-\frac{\beta\omega}{2}}, \qquad \tanh \theta(\beta) = e^{-\frac{\beta\omega}{2}}. \tag{3.80}$$

Furthermore, we note that, using (3.79), we can evaluate the nested commutators in (3.41) (with $Q = \mp i\widehat{H}$ respectively) to obtain

$$e^{-\frac{\beta\widehat{H}}{2}} a^\dagger = \left(e^{-\frac{\beta\widehat{H}}{2}} a^\dagger e^{\frac{\beta\widehat{H}}{2}} \right) e^{-\frac{\beta\widehat{H}}{2}} = e^{-\frac{\beta\omega}{2}} a^\dagger e^{-\frac{\beta\widehat{H}}{2}},$$

$$e^{\frac{\beta\widehat{H}}{2}} \widetilde{a}^\dagger = \left(e^{\frac{\beta\widehat{H}}{2}} \widetilde{a}^\dagger e^{-\frac{\beta\widehat{H}}{2}} \right) e^{\frac{\beta\widehat{H}}{2}} = e^{-\frac{\beta\omega}{2}} \widetilde{a}^\dagger e^{\frac{\beta\widehat{H}}{2}}. \tag{3.81}$$

Using these relations, we note that we can write the conditions in (3.49) and (3.76) in a unified way as

$$a|0, \beta\rangle = e^{-\frac{\beta\omega}{2}} \widetilde{a}^\dagger |0, \beta\rangle = e^{\frac{\beta\widehat{H}}{2}} \widetilde{a}^\dagger |0, \beta\rangle,$$

$$\widetilde{a}|0, \beta\rangle = (-1)^{|a|} e^{-\frac{\beta\omega}{2}} a^\dagger |0, \beta\rangle = (-1)^{|a|} e^{-\frac{\beta\widehat{H}}{2}} a^\dagger |0, \beta\rangle, \tag{3.82}$$

where we have used (3.78) (namely, $\widehat{H}|0, \beta\rangle = 0$). Here $|a|$ represents the Grassmann parity of the operator a which is also the same as that of $\widetilde{a}$ and takes values 0 or 1 for bosonic and fermionic operators respectively. There are also the adjoint conditions which follow in a straightforward manner. Together, they lead to what are known as the thermal state conditions (also known as tilde conjugation relations) that are essential for thermofield dynamics as we will see later in this chapter. Before moving on to field theories at finite temperature, let us only point out here that there are other quantum mechanical systems besides bosons and fermions which go under the names of parabosons or parafermions. These are quantum mechanical systems which are not defined by operators satisfying commutation or anti-commutation relations, rather they involve triple brackets (namely, the basic operators satisfy combination of three brackets, commutators and anti-commutators, as they case may be). Particles described by such systems obey a generalized statistics and one can describe such a system through thermofield dynamics as well.[1] We will not go into details of such systems here.

3.4 Free Schrödinger field theory

In the last two sections, we studied simple quantum mechanical systems at finite temperature in the formalism of thermofield dynamics. We will now look at some of the field theories at finite temperature in this formalism. Let us consider the free Schrödinger field theory in $3 + 1$ dimensions as a first example. As we know, the Lagrangian density is given by (this is a non-relativistic quantum field theory)

$$\mathcal{L} = i\psi^\dagger \dot{\psi} - \frac{1}{2m} \boldsymbol{\nabla}\psi^\dagger \cdot \boldsymbol{\nabla}\psi, \tag{3.83}$$

where ψ denotes a complex scalar field. According to thermofield dynamics, we should introduce a tilde system and let us take the Lagrangian density for the tilde system to be

$$\widetilde{\mathcal{L}} = -i\widetilde{\psi}^\dagger \dot{\widetilde{\psi}} - \frac{1}{2m} \boldsymbol{\nabla}\widetilde{\psi}^\dagger \cdot \boldsymbol{\nabla}\widetilde{\psi}. \tag{3.84}$$

The reason for the negative sign in the first term in $\widetilde{\mathcal{L}}$ will become clear shortly (as well as when we discuss the tilde conjugation rules). From the discussion of the last chapter, we recognize that we can

---

[1] See, for example, S. N. Biswas and A. Das, Modern Physics Letters **A3**, 549 (1988).

take the total Lagrangian density for the system to be

$$\widehat{\mathcal{L}} = \mathcal{L} - \widetilde{\mathcal{L}}$$

$$= i\psi^\dagger\dot\psi + i\widetilde\psi^\dagger\dot{\widetilde\psi} - \frac{1}{2m}\boldsymbol{\nabla}\psi^\dagger\cdot\boldsymbol{\nabla}\psi + \frac{1}{2m}\boldsymbol{\nabla}\widetilde\psi^\dagger\cdot\boldsymbol{\nabla}\widetilde\psi. \qquad (3.85)$$

The momenta conjugate to the field variables ψ and $\widetilde\psi$ are given by

$$\Pi(\mathbf{x}, t) = \frac{\partial\widehat{\mathcal{L}}}{\partial\dot\psi(\mathbf{x}, t)} = i\psi^\dagger(\mathbf{x}, t),$$

$$\widetilde\Pi(\mathbf{x}, t) = \frac{\partial\widehat{\mathcal{L}}}{\partial\dot{\widetilde\psi}(\mathbf{x}, t)} = i\widetilde\psi^\dagger(\mathbf{x}, t). \qquad (3.86)$$

This explains why we chose the sign of the first term in $\widetilde{\mathcal{L}}$ to be different from that of $\mathcal{L}$. Namely, with this choice, the relations between the field variables and their conjugate momenta become identical for the tilde and the nontilde systems. Correspondingly, they satisfy similar equal-time canonical commutation relations (as is assumed in thermofield dynamics)

$$[\psi(\mathbf{x}, t), \psi^\dagger(\mathbf{y}, t)] = \delta^3(x - y),$$

$$[\psi(\mathbf{x}, t), \psi(\mathbf{y}, t)] = 0 \ = [\psi^\dagger(\mathbf{x}, t), \psi^\dagger(\mathbf{y}, t)],$$

$$[\widetilde\psi(\mathbf{x}, t), \widetilde\psi^\dagger(\mathbf{y}, t)] = \delta^3(x - y),$$

$$[\widetilde\psi(\mathbf{x}, t), \widetilde\psi(\mathbf{y}, t)] = 0 \ = [\widetilde\psi^\dagger(\mathbf{x}, t), \widetilde\psi^\dagger(\mathbf{y}, t)], \qquad (3.87)$$

and the two systems become identical. (Recall that the tilde system should be an identical copy of the original system and that the two nontrivial equal-time commutation relations should lead to $[\psi(\mathbf{x}, t), \Pi(\mathbf{y}, t)] = i\delta^3(x - y) = [\widetilde\psi(\mathbf{x}, t), \widetilde\Pi(\mathbf{y}, t)]$.) We assume further that the tilde fields commute with the nontilde ones. This brings out one of the rules for "tilde conjugation", namely, the Lagrangian density for the tilde system is obtained from that of the original system by replacing the field variables by the tilde ones and complex conjugating all the coefficients in the Lagrangian density. This is also commonly represented by saying that

$$\widetilde{c\psi} = c^*\widetilde\psi. \qquad (3.88)$$

This is quite crucial, even when the Lagrangian density does not involve any complex parameters (manifestly), in bringing out the correct analytic structure of the finite temperature Green's functions (for example, in the $i\epsilon$ prescription in the propagators). We would like to emphasize here that the tilde conjugation rules are to be taken as postulates within the framework of thermofield dynamics.

The Hamiltonian densities for the tilde and the nontilde systems, following from (3.83)-(3.85), are given by

$$\mathcal{H} = \frac{1}{2m} \boldsymbol{\nabla}\psi^\dagger(\mathbf{x}, t) \cdot \boldsymbol{\nabla}\psi(\mathbf{x}, t),$$

$$\widetilde{\mathcal{H}} = \frac{1}{2m} \boldsymbol{\nabla}\widetilde{\psi}^\dagger(\mathbf{x}, t) \cdot \boldsymbol{\nabla}\widetilde{\psi}(\mathbf{x}, t), \tag{3.89}$$

leading to the total Hamiltonian for the combined system to be

$$\widehat{H} = H - \widetilde{H}. \tag{3.90}$$

It is straightforward to check using (3.87) that the dynamical equations for the system can be written as Heisenberg equations of the form (recall that the original and the tilde operators commute)

$$i\dot{\psi}(\mathbf{x}, t) = [\psi(\mathbf{x}, t), \widehat{H}] = -\frac{1}{2m} \boldsymbol{\nabla}^2 \psi(\mathbf{x}, t),$$

$$i\dot{\widetilde{\psi}}(\mathbf{x}, t) = [\widetilde{\psi}(\mathbf{x}, t), \widehat{H}] = \frac{1}{2m} \boldsymbol{\nabla}^2 \widetilde{\psi}(\mathbf{x}, t). \tag{3.91}$$

This shows that $\widehat{H}$ can indeed be thought of as the generator of time evolution for the combined system.

Let us next expand the Schrödinger fields in a plane wave basis (of the free Schrödinger equation)

$$\psi(\mathbf{x}, t) = \frac{1}{\sqrt{V}} \sum_{\mathbf{p}} e^{i\mathbf{p}\cdot\mathbf{x}} \psi_{\mathbf{p}}(t) = \frac{1}{\sqrt{V}} \sum_{\mathbf{p}} e^{-i\omega_p t + i\mathbf{p}\cdot\mathbf{x}} a_{\mathbf{p}},$$

$$\widetilde{\psi}(\mathbf{x}, t) = \frac{1}{\sqrt{V}} \sum_{\mathbf{p}} e^{-i\mathbf{p}\cdot\mathbf{x}} \widetilde{\psi}_{\mathbf{p}}(t) = \frac{1}{\sqrt{V}} \sum_{\mathbf{p}} e^{i\omega_p t - i\mathbf{p}\cdot\mathbf{x}} \widetilde{a}_{\mathbf{p}}. \tag{3.92}$$

There are two things to note here. We have quantized the system in a finite, cubic box of volume V for simplicity which leads to discrete values for $\mathbf{p}$ and the energy of a quantum state (namely, the dispersion relation) is given by

$$\omega_p = \frac{\mathbf{p}^2}{2m}. \tag{3.93}$$

Second, the expansion for $\widetilde{\psi}$ in (3.92) is consistently obtained from that of ψ using the tilde conjugation rule of (3.88) (namely, the arbitrariness in the signs in the exponential is settled by the tilde conjugation rules). Moreover, we can think of $a_\mathbf{p}$ and $\widetilde{a}_\mathbf{p}$ as representing the annihilation operators for the original system and the tilde system respectively although, from the structure of the exponential factors multiplying the two expansions in (3.92), it may seem that $\widetilde{a}_\mathbf{p}$ is more like a creation operator. In terms of these operators, we can write down the Hamiltonians (for the two component systems as well as the total system) and they take the forms

$$H = \sum_\mathbf{p} \omega_p a_\mathbf{p}^\dagger a_\mathbf{p},$$

$$\widetilde{H} = \sum_\mathbf{p} \omega_p \widetilde{a}_\mathbf{p}^\dagger \widetilde{a}_\mathbf{p},$$

$$\widehat{H} = H - \widetilde{H} = \sum_\mathbf{p} \omega_p \left(a_\mathbf{p}^\dagger a_\mathbf{p} - \widetilde{a}_\mathbf{p}^\dagger \widetilde{a}_\mathbf{p} \right), \tag{3.94}$$

where $(a_\mathbf{p}^\dagger, \widetilde{a}_\mathbf{p}^\dagger)$ are the respective creation operators. (These can be compared, respectively, with the quantum mechanical Hamiltonians in (3.56), (3.59) and (3.77). As pointed out earlier, a field theory can be thought of as a quantum mechanical system with infinitely many modes.) The annihilation and the creation operators satisfy the standard commutation relations for a harmonic oscillator with an infinite number of degrees of freedom. Namely, we have (and, of course, the original and the tilde operators commute)

$$[a_\mathbf{p}, a_{\mathbf{p}'}^\dagger] = \delta_{\mathbf{p}\mathbf{p}'}, \quad [a_\mathbf{p}, a_{\mathbf{p}'}] = 0 = [a_\mathbf{p}^\dagger, a_{\mathbf{p}'}^\dagger],$$

$$[\widetilde{a}_\mathbf{p}, \widetilde{a}_{\mathbf{p}'}^\dagger] = \delta_{\mathbf{p}\mathbf{p}'}, \quad [\widetilde{a}_\mathbf{p}, \widetilde{a}_{\mathbf{p}'}] = 0 = [\widetilde{a}_\mathbf{p}^\dagger, \widetilde{a}_{\mathbf{p}'}^\dagger]. \tag{3.95}$$

The Hilbert space for the combined (doubled) system can now be constructed in a straightforward manner as described earlier. Let us simply note here, following the discussion of the last section, that we can now define the thermal vacuum as (see also (3.63))

$$|0, \beta\rangle = Z^{-\frac{1}{2}}(\beta) \sum_{\mathbf{p}} \sum_{n_{\mathbf{p}}} e^{-\frac{\beta \omega_p n_{\mathbf{p}}}{2}} |n_{\mathbf{p}}, \tilde{n}_{\mathbf{p}}\rangle, \tag{3.96}$$

where $n_{\mathbf{p}}$ ($\tilde{n}_{\mathbf{p}}$) denotes the occupation number in a state with momentum $\mathbf{p}$ and the partition function for the system is given by

$$Z(\beta) = \sum_{\mathbf{p}} \frac{1}{1 - e^{-\beta \omega_p}}. \tag{3.97}$$

We also note, following the discussion of (3.68), that we can define a Hermitian generator of a Bogoliubov transformation as (see (3.66) and (3.67))

$$Q(\theta) = -i \sum_{\mathbf{p}} \theta_p(\beta) \left(\tilde{a}_{\mathbf{p}} a_{\mathbf{p}} - a_{\mathbf{p}}^\dagger \tilde{a}_{\mathbf{p}}^\dagger \right), \tag{3.98}$$

which would lead to a formally unitary operator

$$U(\theta) = e^{-iQ(\theta)} = e^{-\sum_{\mathbf{p}} \theta_p(\beta) \left(\tilde{a}_{\mathbf{p}} a_{\mathbf{p}} - a_{\mathbf{p}}^\dagger \tilde{a}_{\mathbf{p}}^\dagger \right)}. \tag{3.99}$$

This (formally) unitary operator will connect the thermal vacuum to the usual one (in the doubled space) in the standard manner, namely,

$$|0, \beta\rangle = U(\theta)|0, \tilde{0}\rangle, \tag{3.100}$$

provided we identify (see (3.69))

$$\cosh \theta_p(\beta) = \frac{1}{\sqrt{1 - e^{-\beta \omega_p}}}, \quad \sinh \theta_p(\beta) = \frac{e^{-\frac{\beta \omega_p}{2}}}{\sqrt{1 - e^{-\beta \omega_p}}}. \tag{3.101}$$

This can also be written equivalently as $\tanh \theta_p(\beta) = e^{-\frac{\beta \omega_p}{2}}$.

Given these, the thermal operators can also be calculated in a straightforward manner (see (3.71)) and we have

$$\begin{pmatrix} a_{\mathbf{p}}(\beta) \\ \tilde{a}_{\mathbf{p}}^\dagger(\beta) \end{pmatrix} = \overline{U}(\theta_p) \begin{pmatrix} a_{\mathbf{p}} \\ \tilde{a}_{\mathbf{p}}^\dagger \end{pmatrix}$$

$$= \begin{pmatrix} \cosh \theta_p(\beta) & -\sinh \theta_p(\beta) \\ -\sinh \theta_p(\beta) & \cosh \theta_p(\beta) \end{pmatrix} \begin{pmatrix} a_{\mathbf{p}} \\ \tilde{a}_{\mathbf{p}}^\dagger \end{pmatrix},$$

$$\left(a_{\mathbf{p}}^\dagger(\beta) \quad \tilde{a}_{\mathbf{p}}(\beta) \right) = \left(a_{\mathbf{p}}^\dagger \quad \tilde{a}_{\mathbf{p}} \right) \overline{U}^\dagger(\theta_p)$$

$$= \left(a_{\mathbf{p}}^\dagger \quad \tilde{a}_{\mathbf{p}} \right) \begin{pmatrix} \cosh \theta_p(\beta) & -\sinh \theta_p(\beta) \\ -\sinh \theta_p(\beta) & \cosh \theta_p(\beta) \end{pmatrix}. \tag{3.102}$$

They satisfy the same commutation relations as the ones at zero temperature since unitary transformations preserve commutation relations. Alternatively, we can write these in terms of the field variables as

$$
\begin{pmatrix} \psi_{\mathbf{p}}(t, \beta) \\ \widetilde{\psi}_{\mathbf{p}}^{\dagger}(t, \beta) \end{pmatrix} = \overline{U}(\theta_p) \begin{pmatrix} \psi_{\mathbf{p}}(t) \\ \widetilde{\psi}_{\mathbf{p}}^{\dagger}(t) \end{pmatrix}
$$

$$
= \begin{pmatrix} \cosh \theta_p(\beta) & -\sinh \theta_p(\beta) \\ -\sinh \theta_p(\beta) & \cosh \theta_p(\beta) \end{pmatrix} \begin{pmatrix} \psi_{\mathbf{p}}(t) \\ \widetilde{\psi}_{\mathbf{p}}^{\dagger}(t) \end{pmatrix}, \quad (3.103)
$$

as well as its adjoint. With these relations, it is easy to see that the thermal vacuum satisfies, as in (3.75),

$$
a_{\mathbf{p}}(\beta)|0, \beta\rangle = \left(a_{\mathbf{p}} \cosh \theta_p(\beta) - \widetilde{a}_{\mathbf{p}}^{\dagger} \sinh \theta_p(\beta) \right)|0, \beta\rangle = 0,
$$

$$
\widetilde{a}_{\mathbf{p}}(\beta)|0, \beta\rangle = \left(\widetilde{a}_{\mathbf{p}} \cosh \theta_p(\beta) - a_{\mathbf{p}}^{\dagger} \sinh \theta_p(\beta) \right)|0, \beta\rangle = 0, \quad (3.104)
$$

which bring out the thermal state conditions

$$
a_{\mathbf{p}}|0, \beta\rangle = \tanh \theta_p(\beta)\, \widetilde{a}_{\mathbf{p}}^{\dagger}|0, \beta\rangle,
$$

$$
\widetilde{a}_{\mathbf{p}}|0, \beta\rangle = \tanh \theta_p(\beta)\, a_{\mathbf{p}}^{\dagger}|0, \beta\rangle \quad (3.105)
$$

which can be compared with (3.76). This also shows, as before, that annihilating a quantum of the original particle in the thermal vacuum is equivalent to creating a quantum of the tilde particle (which, therefore, can be thought of as a hole of the original system) and *vice versa* and that the thermal vacuum contains an equal number of particles and holes which can also be seen from

$$
\widehat{N}|0, \beta\rangle = 0, \qquad \widehat{H}|0, \beta\rangle = 0. \quad (3.106)
$$

Let us note here that a Bogoliubov transformation is normally associated with a spontaneously broken symmetry. It is interesting, therefore, to ask what is the symmetry that is spontaneously broken at finite temperature. To see that, let us note that the total Hamiltonian, $\widehat{H}$, in (3.90) or (3.94) is invariant under the transformations

$$
a_{\mathbf{p}} \rightarrow a_{\mathbf{p}} \cosh \alpha_{\mathbf{p}} - \widetilde{a}_{\mathbf{p}}^{\dagger} \sinh \alpha_{\mathbf{p}},
$$

$$
\widetilde{a}_{\mathbf{p}} \rightarrow \widetilde{a}_{\mathbf{p}} \cosh \alpha_{\mathbf{p}} - a_{\mathbf{p}}^{\dagger} \sinh \alpha_{\mathbf{p}}, \quad (3.107)
$$

where $\alpha_{\mathbf{p}}$ is an arbitrary (real) global parameter. As we have mentioned in earlier sections, this can be thought of as a rotation in the $a_{\mathbf{p}} - \tilde{a}_{\mathbf{p}}^{\dagger}$ plane (or, equivalently in the $\tilde{a}_{\mathbf{p}} - a_{\mathbf{p}}^{\dagger}$ plane) which is non-compact (the parameter is not restricted to take values in a finite range, for example, $(0, 2\pi)$ in conventional rotations). This is a symmetry that is not present in the original system (H does not have this symmetry) – rather it develops for the combined system once the tilde field is introduced. It can be easily checked, with the standard commutation relations given in (3.95), that the generator of the infinitesimal form of this symmetry transformations is given by (see also (3.98))

$$Q(\epsilon) = -i \sum_{\mathbf{p}} \epsilon_{\mathbf{p}} \left(\tilde{a}_{\mathbf{p}} a_{\mathbf{p}} - a_{\mathbf{p}}^{\dagger} \tilde{a}_{\mathbf{p}}^{\dagger} \right), \tag{3.108}$$

where $\epsilon_{\mathbf{p}}$ is the constant infinitesimal parameter of transformation (namely, $\alpha_{\mathbf{p}} = \epsilon_{\mathbf{p}} = $ infinitesimal in (3.107)). Furthermore, it is quite clear from the structure of the thermal vacuum in (3.96) that

$$Q(\alpha)|0, \tilde{0}\rangle \neq 0, \quad Q(\alpha)|0, \beta\rangle \neq 0. \tag{3.109}$$

The generator of the symmetry does not annihilate either the original doubled vacuum or the thermal vacuum. (In fact, if it did, the thermal vacuum will coincide with the vacuum of the original theory with doubled degrees of freedom.) In other words, the rotational symmetry between the original and the tilde system (see (3.107)) which is present in the Hamiltonian for the doubled system is spontaneously broken by the vacuum. However, we do not expect a massless, Goldstone boson in this case because at finite temperature Lorentz invariance is not manifest and the spectrum of $\hat{H}$ is indefinite (as a result of the indefinite metric in the original and the space, see, for example, (3.73)-(3.74)).

3.5 Free Klein-Gordon theory

As a second example of quantum field theories at finite temperature, let us look at the free Klein-Gordon theory of real (scalar) fields at finite temperature. We continue to work in $3 + 1$ dimensions, but this is a relativistic field theory unlike the Schrödinger field theory in the last section and the Lagrangian density for the system is given by

$$\mathcal{L} = \frac{1}{2} \partial_\mu \phi \partial^\mu \phi - \frac{m^2}{2} \phi^2. \tag{3.110}$$

The Lagrangian density for the tilde system can now be written down in a straightforward manner

$$\widetilde{\mathcal{L}} = \frac{1}{2} \partial_\mu \widetilde{\phi} \partial^\mu \widetilde{\phi} - \frac{m^2}{2} \widetilde{\phi}^2, \tag{3.111}$$

so that the complete Lagrangian density for the combined system can be written as

$$\widehat{\mathcal{L}} = \mathcal{L} - \widetilde{\mathcal{L}}. \tag{3.112}$$

We note here that because the fields are real in this case, the tilde conjugation rule does not seem to have any apparent effect. That is not true, however, if we remember that the Feynman boundary condition is equivalent to adding an infinitesimal, imaginary term to the quadratic part of the Lagrangian density. The sign of that term would change under complex conjugation implying that the $i\epsilon$-prescription will be different for the tilde fields compared to the original fields. (We have already alluded to this earlier.)

We can now go on and quantize the system, construct the combined Hilbert space as well as the thermal vacuum. Let us instead ask how one can construct the thermal propagator of the theory from all the information we have. To that end, let us introduce the doublet of fields (from our earlier discussion in (3.45) or (3.70) or (3.103), we recognize that we should group ϕ and $\phi^\dagger$ into a doublet; in this case, however, the fields are real)

$$\Phi = \begin{pmatrix} \phi \\ \widetilde{\phi} \end{pmatrix}. \tag{3.113}$$

Then, we can write the propagator for the combined system at zero temperature to be

$$iG_{T=0}^{(\mathrm{TFD})}(x - y) = \langle 0, \widetilde{0} | T(\Phi(x)\Phi(y)) | 0, \widetilde{0} \rangle. \tag{3.114}$$

We should recognize that the propagator, in the present case, is a 2×2 matrix whose Fourier transform has the form (note that the

Feynman prescription for the tilde field as well as the overall negative sign in its propagator simply reflect the tilde conjugation rules which we have already discussed in (3.88))

$$iG_{T=0}^{(\text{TFD})}(p) = \begin{pmatrix} \frac{i}{p^2-m^2+i\epsilon} & 0 \\ 0 & -\frac{i}{p^2-m^2-i\epsilon} \end{pmatrix}. \tag{3.115}$$

This is what we will expect for two decoupled propagators except for the $i\epsilon$ term in the 22 element (which actually results from anti-time ordering in that term). We will give a brief derivation of the momentum space propagator matrix in (3.115) starting from its definition as a vacuum expectation value of the time ordered product given in (3.114) in the coordinate space in an appendix at the end of this chapter.

We can now determine the thermal propagator quite easily from the following heuristic observation. (For a more detailed derivation, we refer the reader to section **4.3** in the next chapter.) By definition, given in (3.114), (We note here that the symbol T, in the following equation, has been unfortunately used for time ordering, temperature as well as transpose of a matrix. However, we hope that the meaning is clear within the context and will not cause any confusion.)

$$iG^{(\text{TFD})}(x-y) = \langle 0, \beta | T(\Phi(x)\Phi(y)) | 0, \beta \rangle$$

$$= \langle 0, \tilde{0} | T(U^\dagger(\theta)\Phi(x)\Phi(y)U(\theta)) | 0, \tilde{0} \rangle$$

$$= \langle 0, \tilde{0} | T(U(-\theta)\Phi(x)U^\dagger(-\theta)U(-\theta)\Phi(y)U^\dagger(-\theta)) | 0, \tilde{0} \rangle$$

$$= \langle 0, \tilde{0} | T(\overline{U}(-\theta)\Phi(x)\Phi(y)\overline{U}^T(-\theta)) | 0, \tilde{0} \rangle$$

$$= \overline{U}(-\theta)\langle 0, \tilde{0} | T(\Phi(x)\Phi(y)) | 0, \tilde{0} \rangle \overline{U}^T(-\theta)$$

$$= \overline{U}(-\theta)iG_{T=0}^{(\text{TFD})}(x-y)\overline{U}^T(-\theta). \tag{3.116}$$

Here $U(\theta)$ denotes the (formally) unitary Bogoliubov transformation (operator) connecting the thermal vacuum and the product of vacuum states in the doubled space, namely, $|0, \beta\rangle = U(\theta)|0, \tilde{0}\rangle$, and $\overline{U}(\theta)$ is the transformation matrix in the two dimensional component space (see, for example, (3.68) and (3.71)) and satisfies, as in (3.74),

$$\overline{U}^\dagger(-\theta)\sigma_3\overline{U}(-\theta) = \sigma_3 = \overline{U}^\dagger(\theta)\sigma_3\overline{U}(\theta). \tag{3.117}$$

Furthermore, we note here that, since the operator $U(\theta)$ and, therefore, the matrix $\overline{U}(\theta)$ are time independent, we can take them inside the time ordering. Taking the Fourier transformation of (3.115) and using (3.116), we can now obtain the thermal propagator in the momentum space in the formalism of thermofield dynamics to be (the superscript T denotes the transpose of the matrix)

$$
iG^{(\text{TFD})}(p) = \overline{U}(-\theta_p)\, iG^{(\text{TFD})}_{T=0}(p)\, \overline{U}^T(-\theta_p)
$$

$$
= \begin{pmatrix} \cosh\theta_p(\beta) & \sinh\theta_p(\beta) \\ \sinh\theta_p(\beta) & \cosh\theta_p(\beta) \end{pmatrix} \begin{pmatrix} \frac{i}{p^2-m^2+i\epsilon} & 0 \\ 0 & -\frac{i}{p^2-m^2-i\epsilon} \end{pmatrix}
$$

$$
\times \begin{pmatrix} \cosh\theta_p(\beta) & \sinh\theta_p(\beta) \\ \sinh\theta_p(\beta) & \cosh\theta_p(\beta) \end{pmatrix}. \tag{3.118}
$$

Using the standard form for $\theta_p(\beta)$ for bosonic theories (see (3.101)), we can now simplify and write the finite temperature propagator for the free Klein-Gordon theory in momentum space in thermofield dynamics to be

$$
iG^{(\text{TFD})}(p) = \begin{pmatrix} \frac{i}{p^2-m^2+i\epsilon} & 0 \\ 0 & -\frac{i}{p^2-m^2-i\epsilon} \end{pmatrix}
$$

$$
+ 2\pi n_B(|p^0|)\delta(p^2-m^2) \begin{pmatrix} 1 & e^{\frac{\beta|p^0|}{2}} \\ e^{\frac{\beta|p^0|}{2}} & 1 \end{pmatrix}. \tag{3.119}
$$

Here $n_B(|p^0|)$ denotes the bosonic distribution function (defined, for example, in (1.107)). Note from (3.101) that, as (temperature) $T \to 0$ or $\beta \to \infty$, $\cosh\theta_p \to 1$, $\sinh\theta_p \to 0$ so that the Bogoliubov matrix

$$
\overline{U}(-\theta_p) \to \mathbb{1}, \tag{3.120}
$$

and the propagator in (3.118) reduces to the zero temperature propagator given in (3.115). This can also be seen directly in (3.119).

There are several things to note here. First, the 2×2 matrix propagator in (3.119) (at finite temperature) separates into a sum of two 2×2 matrices, the first of which is diagonal and corresponds to the zero temperature part of the propagator and the second, which is not diagonal, explicitly depends on temperature and vanishes in the zero temperature limit. This is similar to the case of the propagator in the closed time path formalism which we had discussed earlier (see, for example, (2.51)), but not exactly. Namely, the finite temperature propagator matrix, in the closed time path formalism given in (2.51), also separates into two parts, the first of which is diagonal and coincides with the first term in (3.119). But, it is not exactly the zero temperature propagator in the closed time path formalism because, even though the first term has no temperature dependence, the second matrix in (2.51) leads to nontrivial off-diagonal terms in the zero temperature limit. Therefore, the two matrices in (2.51) do not lead to a clean separation of zero temperature and finite temperature parts of the propagator as is the case in (3.119) (namely, in thermofield dynamics). Second, the propagator has a 2×2 matrix structure very much like the closed time path formalism. However, the structure of the propagator in thermofield dynamics is not quite the same as that in (2.50). In fact, we note that whereas the diagonal terms in the propagator matrices in (2.50) and (3.119) exactly coincide, the off-diagonal terms are different. In a contour space in the real time formalism, say in the closed time path, we have already noted earlier that the off-diagonal terms in the propagator simply correspond to propagation between the two real branches of the contour. Therefore, this suggests that, may be one can define the second real branch slightly differently from the closed time path formalism to accommodate the results of thermofield dynamics. In that case, in addition to the operator description, one can also give a contour description (and, therefore, a path integral description) to thermofield dynamics.

Without going into details, we simply note here that, indeed, we can describe thermofield dynamics in a path integral formulation as well where the time contour in the complex time plane has the form shown in Fig. 3.1. (We will discuss about this in more detail in the next chapter.) Here T denotes a large negative time (and not the temperature as in Fig. 2.1) and T' is a large positive time. In this case, the average value of an operator insertion, given in (2.19)-(2.20), can be understood as

$$\langle A \rangle (t) = \frac{\mathrm{Tr}\, \mathcal{N}}{\mathrm{Tr}\, \mathcal{D}}, \tag{3.121}$$

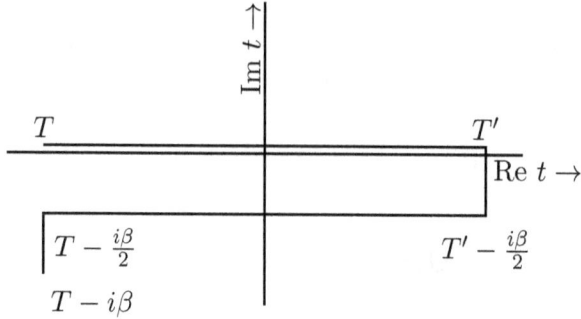

Figure 3.1: Contour in the complex t-plane for thermofield dynamics.

where

$$\mathcal{N} = U(T - i\beta, T - \frac{i\beta}{2})U(T - \frac{i\beta}{2}, T' - \frac{i\beta}{2})U(T' - \frac{i\beta}{2}, T')$$
$$\times U(T', t)AU(t, T),$$

$$\mathcal{D} = U(T - i\beta, T - \frac{i\beta}{2})U(T - \frac{i\beta}{2}, T' - \frac{i\beta}{2})U(T' - \frac{i\beta}{2}, T')$$
$$\times U(T', T). \tag{3.122}$$

We can identify the tilde fields as living on the second real branch of this contour in the complex plane and this is how the matrix structure for the propagator in (3.119) arises. To distinguish this contour from the closed time path contour in Fig. 2.1, we label the two real branches as 1 and 2 (as opposed to $+$ and $-$ which is generally reserved for closed time path formalism) so that the 2×2 matrix propagator can be written as

$$iG^{(\text{TFD})}(p) = \begin{pmatrix} iG^{(\text{TFD})}_{\beta,11}(p) & iG^{(\text{TFD})}_{\beta,12}(p) \\ iG^{(\text{TFD})}_{\beta,21}(p) & iG^{(\text{TFD})}_{\beta,22}(p) \end{pmatrix}. \tag{3.123}$$

Let us note here the identity (see also (2.51)),

$$e^{\frac{\beta|p^0|}{2}} n_B(|p^0|) = e^{\frac{\beta|p^0|}{2}} \left(\theta(-p^0) + \theta(p^0)\right) n_B(|p^0|)$$

$$= e^{-\frac{\beta|p^0|}{2}} \theta(-p^0) \left(1 + n_B(|p^0|)\right) + e^{\frac{\beta|p^0|}{2}} \theta(p^0) n_B(|p^0|)$$

$$= e^{\frac{\beta p^0}{2}} \theta(-p^0)\left(1 + n_B(|p^0|)\right) + e^{\frac{\beta p^0}{2}} \theta(p^0) n_B(|p^0|)$$

$$= e^{\frac{\beta p^0}{2}} \left(\theta(-p^0) + n_B(|p^0|)\right), \tag{3.124}$$

where we have used the relation

$$\left(1 + n_B(|p^0|)\right) = e^{\beta|p^0|} n_B(|p^0|),$$

$$\text{or,} \quad e^{-\frac{\beta|p^0|}{2}}\left(1 + n_B(|p^0|)\right) = e^{\frac{\beta|p^0|}{2}} n_B(|p^0|), \tag{3.125}$$

and, similarly, the identity

$$e^{\frac{\beta|p^0|}{2}} n_B(|p^0|) = e^{-\frac{\beta p^0}{2}}\left(\theta(p^0) + n_B(|p^0|)\right). \tag{3.126}$$

Using these identities, we can write from (2.50) and (3.119)

$$iG^{\beta}_{++}(p) = iG^{(\text{TFD})\,\beta}_{11}(p),$$

$$iG^{\beta}_{+-}(p) = 2\pi(\theta(-p^0) + n_B(|p^0|))\delta(p^2 - m^2)$$

$$= 2\pi n_B(|p^0|)\, e^{\frac{\beta(|p^0|-p^0)}{2}}\, \delta(p^2 - m^2)$$

$$= e^{-\frac{\beta p^0}{2}}\, iG^{(\text{TFD})\,\beta}_{12}(p),$$

$$iG^{\beta}_{-+}(p) = 2\pi(\theta(p^0) + n_B(|p^0|))\delta(p^2 - m^2)$$

$$= 2\pi n_B(|p^0|)\, e^{\frac{\beta(|p^0|+p^0)}{2}}\, \delta(p^2 - m^2)$$

$$= e^{\frac{\beta p^0}{2}}\, iG^{(\text{TFD})\,\beta}_{21}(p),$$

$$iG^{\beta}_{--}(p) = iG^{(\text{TFD})\,\beta}_{22}(p). \tag{3.127}$$

Here $iG^{\beta}_{\pm\pm}$ denote the components of the finite temperature propagator in the closed time path formalism (see (2.49)-(2.50)). Note also that in the context of thermofield dynamics, the elements of the propagator matrix do not satisfy any simple natural relations (without additional factors) as in (2.50). Thus, while there are similarities between the two formalisms, one should not take the analogies too far. Finally, let us note here that if there are interactions present, we can now introduce the corresponding interactions involving the tilde fields as well through the tilde conjugation relations. The complete interacting Lagrangian density, then, gives rise to the interaction vertices where the interaction vertices involving the tilde fields (in general) will have a relative negative sign compared to those for the original system (as is also true for the theory in the closed time path formalism). One can use the matrix form of the propagator in (3.119) (or its components, now labelled as $iG_{11}^{(\text{TFD})\,\beta}$, $iG_{12}^{(\text{TFD})\,\beta}$, $iG_{21}^{(\text{TFD})\,\beta}$ and $iG_{22}^{(\text{TFD})\,\beta}$) and carry out perturbative calculations in a standard manner. As we discussed in chapter **2**, the doubling of the fields, among other things, leads to cancellation of ill-defined terms in the loop calculations and the zero temperature counter terms are sufficient to renormalize the theory (see, for example, (2.92) and the discussion in the following paragraph).

3.6 KMS condition

As we have seen in the first chapter, the KMS condition (see section **1.3**, equations (1.89) or (1.90) and (2.100) or (2.113)) is at the heart of equilibrium statistical mechanics and arises from the fact that we are evaluating a trace when we are calculating an ensemble average. In thermofield dynamics, the KMS condition also follows quite easily from the thermal state condition (see (3.49) or (3.75) or (3.104)). We note, from (3.82), (we are going to suppress the spatial coordinates for simplicity) that the thermal state condition says that for any operator A we can write (the second relation below is obtained by taking the adjoint of the second relation in (3.82) and rearranging factors)

$$A(t)|0,\beta\rangle = e^{\frac{\beta \widehat{H}}{2}}\, \widetilde{A}^{\dagger}(t)|0,\beta\rangle,$$

$$\langle 0,\beta|A(t) = (-1)^{|A|}\langle 0,\beta|\widetilde{A}^{\dagger}(t)e^{\frac{\beta \widehat{H}}{2}}, \tag{3.128}$$

where $|A| = 0,1$ denotes the Grassmann parity of the operator A (discussed earlier below (3.82)). This can be understood by recognizing that the operator A can be written as a combination of annihilation and creation operators and each such operator acting on the thermal vacuum gives rise to the creation or annihilation operator (respectively) of the tilde system upto an energy dependent factor. This can then be reexpressed, as in (3.82), in terms of the operator $\widetilde{A}^{\dagger}$ and an exponent that involves $\widehat{H}$. Let us use (3.128) to calculate the ensemble average of the correlation of any two arbitrary operators

$$\langle 0, \beta | A(t) B(t') | 0, \beta \rangle = (-1)^{|A|} \langle 0, \beta | \widetilde{A}^\dagger(t) e^{\frac{\beta \widehat{H}}{2}} B(t') | 0, \beta \rangle$$

$$= (-1)^{|A|} \langle 0, \beta | e^{\frac{\beta \widehat{H}}{2}} \left(e^{-\frac{\beta \widehat{H}}{2}} \widetilde{A}^\dagger(t) e^{\frac{\beta \widehat{H}}{2}} \right) B(t') | 0, \beta \rangle$$

$$= (-1)^{|A|} \langle 0, \beta | e^{\frac{\beta \widehat{H}}{2}} \widetilde{A}^\dagger(t + \frac{i\beta}{2}) B(t') | 0, \beta \rangle$$

$$= (-1)^{|A|} \langle 0, \beta | \widetilde{A}^\dagger(t + \frac{i\beta}{2}) B(t') | 0, \beta \rangle$$

$$= (-1)^{|A|(1+|B|)} \langle 0, \beta | B(t') \widetilde{A}^\dagger(t + \frac{i\beta}{2}) | 0, \beta \rangle$$

$$= (-1)^{|A|(1+|B|)} \langle 0, \beta | B(t') e^{-\frac{\beta \widehat{H}}{2}} \left(e^{\frac{\beta \widehat{H}}{2}} \widetilde{A}^\dagger(t + \frac{i\beta}{2}) \right) | 0, \beta \rangle$$

$$= (-1)^{|A|(1+|B|)} \langle 0, \beta | B(t') e^{-\frac{\beta \widehat{H}}{2}} A(t + \frac{i\beta}{2}) | 0, \beta \rangle$$

$$= (-1)^{|A|(1+|B|)} \langle 0, \beta | B(t') \left(e^{-\frac{\beta \widehat{H}}{2}} A(t + \frac{i\beta}{2}) e^{\frac{\beta \widehat{H}}{2}} \right) e^{-\frac{\beta \widehat{H}}{2}} | 0, \beta \rangle$$

$$= (-1)^{|A|(1+|B|)} \langle 0, \beta | B(t') A(t + i\beta) e^{-\frac{\beta \widehat{H}}{2}} | 0, \beta \rangle$$

$$= \langle 0, \beta | B(t') A(t + i\beta) | 0, \beta \rangle. \tag{3.129}$$

This is the KMS condition which holds for both bosonic and fermionic operators (see (1.36) for bosonic operators) and to obtain this, we have used the thermal state conditions in (3.106) (namely, $\widehat{H}|0, \beta\rangle = 0$) as well as in (3.128). (Note that $|A|, |B| = 0, 1$ define the Grassmann parities of the operators A, B as emphasized earlier.) We have also used the fact that tilde and non-tilde operators either commute or anti-commute depending on their Grassmann parities which gives rise to the relevant phase factors in the intermediate steps. In addition, in the last step, we have used the fact that the thermal vacuum expectation value vanishes if both the operators do not have the same Grassmann parity and when they are the same, 0 or 1, the overall phase factor becomes unity. Furthermore, we recognize that $\widehat{H}$ is the generator of time translation for the system. It is clear that (3.129) holds for both bosonic as well as fermionic operators.

3.7 Dispersion relations

In chapter **2** (see section **2.4**), we derived explicitly the dispersion relations for the free, bosonic Green's functions (propagators) in the closed time path formalism. Here we will derive the same relations for two arbitrary operators within the thermofield dynamics formalism. Let us consider two general operators, $A(t)$ and $B(t)$, both of which are assumed to be either bosonic or fermionic. For simplicity, we are suppressing the spatial coordinates of the operators. Let $I_{AB}(p^0)$ represent the Fourier transform of the thermal correlation function (namely, the thermal ensemble average) of $A(t)$ and $B(t')$, namely,

$$\langle 0, \beta | A(t) B(t') | 0, \beta \rangle = \int_{-\infty}^{\infty} \frac{dp^0}{2\pi} e^{-ip^0(t-t')} I_{AB}(p^0), \qquad (3.130)$$

where time translation invariance has been assumed. Once again, we emphasize that we are suppressing the dependence of the operators on other coordinates. We note that, for real thermal averages, we can choose $I_{AB}(p^0)$ to be a real function, which we will assume in the following.

If we now use the KMS condition in (3.129), namely,

$$\langle 0, \beta | A(t) B(t') | 0, \beta \rangle = \langle 0, \beta | B(t') A(t + i\beta) | 0, \beta \rangle, \qquad (3.131)$$

we can write, using (3.131) followed by (3.130),

$$\langle 0, \beta | B(t') A(t) | 0, \beta \rangle = \langle 0, \beta | A(t - i\beta) B(t') | 0, \beta \rangle$$

$$= \int_{-\infty}^{\infty} \frac{dp^0}{2\pi} e^{-ip^0((t-i\beta)-t')} I_{AB}(p^0)$$

$$= \int_{-\infty}^{\infty} \frac{dp^0}{2\pi} e^{-ip^0(t-t')} e^{-\beta p^0} I_{AB}(p^0). \quad (3.132)$$

Given these two relations, (3.130) and (3.131), we can now calculate various other correlation functions. In particular, we note from the standard definition (see (2.55)) that the retarded correlation function of the two operators is given by

$$iR_{AB}(t - t') = \theta(t - t') \langle 0, \beta | [A(t), B(t')]_{\mp} | 0, \beta \rangle$$

$$= \theta(t - t') \int \frac{dp^0}{2\pi} e^{-ip^0(t-t')} (1 \mp e^{-\beta p^0}) I_{AB}(p^0), \quad (3.133)$$

where the subscript $\mp$ refers to the cases where both A and B are bosonic or fermionic. Defining the Fourier transform as

$$R_{AB}(t - t') = \int\limits_{-\infty}^{\infty} \frac{dp^0}{2\pi} e^{-ip^0(t-t')} R_{AB}(p^0), \quad (3.134)$$

and using a standard integral representation for the θ–function, namely,

$$\theta(t - t') = \int\limits_{-\infty}^{\infty} \frac{dp^0}{2i\pi} \frac{e^{ip^0(t-t')}}{p^0 - i\epsilon}, \quad (3.135)$$

it is now easy to show, using (3.130)-(3.135), that

$$R_{AB}(t - t') = \int\limits_{-\infty}^{\infty} \frac{dp^0}{2\pi} e^{-ip^0(t-t')} R_{AB}(p^0)$$

$$= \int\limits_{-\infty}^{\infty} \frac{dp^0}{2i\pi} \frac{e^{ip^0(t-t')}}{p^0 - i\epsilon} (-i) \int\limits_{-\infty}^{\infty} \frac{dp'^{\,0}}{2\pi} e^{-ip'^0(t-t')} (1 \mp e^{-\beta p'^0}) I_{AB}(p'^0)$$

$$= - \int\limits_{-\infty}^{\infty} \frac{dp^0}{2\pi} \int\limits_{-\infty}^{\infty} \frac{dp'^{\,0}}{2\pi} \frac{e^{i(p^0 - p'^0)(t-t')}}{p^0 - i\epsilon} (1 \mp e^{-\beta p'^{\,0}}) I_{AB}(p'^{\,0})$$

$$= - \int\limits_{-\infty}^{\infty} \frac{dp^0}{2\pi} \int\limits_{-\infty}^{\infty} \frac{dp'^{\,0}}{2\pi} \frac{e^{-ip^0(t-t')}}{p'^{\,0} - p^0 - i\epsilon} (1 \mp e^{-\beta p'^{\,0}}) I_{AB}(p'^{\,0})$$

$$= \int\limits_{-\infty}^{\infty} \frac{dp^0}{2\pi} e^{-ip^0(t-t')} \int\limits_{-\infty}^{\infty} \frac{dp'^{\,0}}{2\pi} \frac{I_{AB}(p'^{\,0})}{p^0 - p'^{\,0} + i\epsilon} (1 \mp e^{-\beta p'^0}), \quad (3.136)$$

where we have redefined $p^0 \to p'^{\,0} - p^0$ in the last but one step. Therefore, comparing with (3.134) (or the first line of (3.136)), we obtain

$$R_{AB}(p^0) = \int\limits_{-\infty}^{\infty} \frac{dp'^0}{2\pi} \frac{I_{AB}(p'^0)}{p^0 - p'^0 + i\epsilon} (1 \mp e^{-\beta p'^0}).$$

(3.137)

Similarly, the causal correlation function can also be calculated in a straightforward manner from the definitions

$$iG_{AB}(t - t') = \langle 0, \beta | T(A(t)B(t')) | 0, \beta \rangle,$$

$$G_{AB}(t - t') = \int\limits_{-\infty}^{\infty} \frac{dp^0}{2\pi} e^{-ip^0(t-t')} G_{AB}(p^0),$$

(3.138)

where the time ordering is defined to be

$$T(A(t)B(t')) = \theta(t - t') A(t)B(t') \pm \theta(t' - t) B(t')A(t), \quad (3.139)$$

with $+$ for bosonic operators and $-$ for the fermionic ones. It follows now from (3.130), (3.132) and (3.135) that

$$G_{AB}(t - t') = \int\limits_{-\infty}^{\infty} \frac{dp^0}{2\pi} e^{-ip^0(t-t')} \int\limits_{-\infty}^{\infty} \frac{dp'^0}{2\pi} I_{AB}(p'^0)$$

$$\times \left(\frac{1}{p^0 - p'^0 + i\epsilon} \mp \frac{e^{-\beta p'^0}}{p^0 - p'^0 - i\epsilon} \right). \quad (3.140)$$

Therefore, comparing with (3.138), we conclude that

$$G_{AB}(p^0) = \int\limits_{-\infty}^{\infty} \frac{dp'^0}{2\pi} I_{AB}(p'^0) \left(\frac{1}{p^0 - p'^0 + i\epsilon} \mp \frac{e^{-\beta p'^0}}{p^0 - p'^0 - i\epsilon} \right).$$

(3.141)

It now follows from (3.137) and (3.141) that (PV stands for the principal value and remember that we are assuming $I_{AB}(p^0)$ is a real function)

$$\mathrm{Re}\,R_{AB}(p^0) = \int_{-\infty}^{\infty} \frac{dp'^0}{2\pi}\, PV\left(\frac{1}{p^0 - p'^0}\right) I_{AB}(p'^0)(1 \mp e^{-\beta p'^0}),$$

$$= \mathrm{Re}\,G_{AB}(p^0)$$

$$\mathrm{Im}\,R_{AB}(p^0) = -\frac{1}{2} I_{AB}(p^0)(1 \mp e^{-\beta p^0})$$

$$= \frac{1 \mp e^{-\beta p^0}}{1 \pm e^{-\beta p^0}}\left(-\frac{1}{2} I_{AB}(p^0)(1 \pm e^{-\beta p^0})\right)$$

$$= \frac{e^{\frac{\beta p^0}{2}} \mp e^{-\frac{\beta p^0}{2}}}{e^{\frac{\beta p^0}{2}} \pm e^{-\frac{\beta p^0}{2}}}\left(-\frac{1}{2} I_{AB}(p^0)(1 \pm e^{-\beta p^0})\right)$$

$$= \frac{e^{\frac{\beta p^0}{2}} \mp e^{-\frac{\beta p^0}{2}}}{e^{\frac{\beta p^0}{2}} \pm e^{-\frac{\beta p^0}{2}}}\,\mathrm{Im}\,G_{AB}(p^0)$$

$$= \begin{cases} \tanh(\frac{\beta p^0}{2})\,\mathrm{Im}\,G_{AB}(p^0) & \text{for bosons,} \\ \coth(\frac{\beta p^0}{2})\,\mathrm{Im}\,G_{AB}(p^0) & \text{for fermions.} \end{cases} \tag{3.142}$$

Equation (3.142) coincides with the dispersion relation for the basic bosonic field variables obtained in (2.78) and generalizes it to arbitrary operators. We note here (as also pointed out in the last chapter) that the derivation of these relations here is general enough that they are true even in an interacting field theory. By choosing the operators A and B to be fundamental field variables, we can obtain conditions on the spectral function from these relations. Furthermore, the LSZ reduction of zero temperature can now be extended in a standard manner to nonzero temperatures.

3.8 Goldstone theorem

The thermal field theories are particularly interesting from the point of view of the study of spontaneous symmetry breaking and restoration of symmetries. However, a naive application of the ideas of symmetry breaking into finite temperature runs into difficulties. This can be seen as follows. Let H denote the Hamiltonian governing the dynamics of the system and let U denote a unitary symmetry of the dynamical system. Then, by definition,

$$UHU^\dagger = H, \tag{3.143}$$

If we write the unitary operator in terms of the generator of the infinitesimal symmetry, Q, then it takes the form (α is the constant parameter of symmetry transformation)

$$U = e^{-i\alpha Q}. \tag{3.144}$$

In such a case, the invariance of the Hamiltonian can be written as

$$[Q, H] = 0. \tag{3.145}$$

We say that a symmetry is spontaneously broken if the vacuum is not invariant under the symmetry transformations even when the Hamiltonian for the system is. Explicitly, if

$$Q|0\rangle \neq 0, \tag{3.146}$$

then, we say that the symmetry is spontaneously broken. An alternative way of characterizing spontaneous symmetry breaking is to note that in such a case, there always exists an operator A such that (we will discuss this in more detail in chapter **6** and we assume here that both Q and A are bosonic)

$$\langle 0|[Q, A]|0\rangle \neq 0. \tag{3.147}$$

The field A is conventionally known as the Goldstone field which, for manifestly Lorentz invariant theories with a positive semi-definite Hilbert space, is known to give rise to massless quanta.

At finite temperature, on the other hand, we have to work with ensemble averages. We note that for any arbitrary operator A, the ensemble average of $[Q, A]$ is given by

$$\begin{aligned}
\langle 0, \beta|[Q, A]|0, \beta\rangle &= \operatorname{Tr} e^{-\beta H} [Q, A] = \operatorname{Tr} e^{-\beta H} (QA - AQ) \\
&= \operatorname{Tr}\left(e^{-\beta H} QA - Q e^{-\beta H} A\right) \\
&= \operatorname{Tr}\left(e^{-\beta H} QA - e^{-\beta H} QA\right) = 0, \tag{3.148}
\end{aligned}$$

where we have used the cyclicity of the trace operation followed by the fact that Q and H commute (see (3.145)). This would, therefore, seem to suggest that, at finite temperature, there cannot be any spontaneous symmetry breaking since (3.148) must hold for any arbitrary operator A. The flaw in this argument, of course, lies in the fact that for infinite dimensional operators, cyclicity of trace may not necessarily hold as we had also noted in chapter **1**. The simplest example of this phenomenon arises in quantum mechanics where

$$[x, p] = i\hbar \mathbb{1}, \tag{3.149}$$

and, if we were to take the trace of both sides in (3.149) and use the cyclicity of trace, it leads to the inconsistent result that

$$\text{Tr}\,[x, p] = 0, \qquad \text{whereas} \quad \text{Tr}\,i\hbar \mathbb{1} \neq 0. \tag{3.150}$$

In addition, the cyclicity property may also be violated for singular operators. Therefore, one has to be careful in the study of spontaneous symmetry breaking at finite temperature.

As in zero temperature, there are only two possibilities that can arise even at finite temperature.

1. For all possible operators A, we have

$$\langle 0, \beta | [Q, A] | 0, \beta \rangle = 0. \tag{3.151}$$

In this case, we say that the symmetry is a true symmetry of the system (if (3.145) holds) and that

$$Q | 0, \beta \rangle = 0. \tag{3.152}$$

2. There exists an operator A for which

$$\langle 0, \beta | [Q, A] | 0, \beta \rangle \neq 0, \tag{3.153}$$

even when (3.145) is satisfied. In such a case, we say that the symmetry is spontaneously broken and that the thermal vacuum is not annihilated by the symmetry generator Q. To see the consequences of symmetry breaking at finite temperature, let us note the following. If the symmetry is a continuous global symmetry, then when matter is sufficiently localized and if there are no long range forces, we have a continuity equation following from Nöther's theorem

$$\partial_\mu J^\mu(\mathbf{x}, t) = 0. \tag{3.154}$$

The generator of the infinitesimal symmetry is related to the conserved current through the relation

$$Q = \int d^3x \, J^0(\mathbf{x}, t), \tag{3.155}$$

and we note that the charge Q is, in fact, independent of time which follows from the continuity equation.

Given these we note now, following (3.130)-(3.132), that

$$\langle 0, \beta | [J^0(y), A(x)] | 0, \beta \rangle = \int \frac{d^4p}{(2\pi)^4} \, e^{-ip \cdot (y-x)} \, I(p^0, \mathbf{p})(1 - e^{-\beta p^0}). \tag{3.156}$$

Upon integrating (3.156) over d^3y (see (3.155)), we obtain

$$\langle 0, \beta | [Q, A(x)] | 0, \beta \rangle = \int\int \frac{d^4p \, d^3y}{(2\pi)^4} e^{-ip \cdot (x-y)} I_A(p^0, \mathbf{p})(1 - e^{-\beta p^0})$$

$$= \int \frac{d^4p}{(2\pi)^4} \, (2\pi)^3 \, \delta^3(p) \, e^{-ip^0(y^0 - x^0) - i\mathbf{p} \cdot \mathbf{x}} \, I_A(p^0, \mathbf{p})(1 - e^{-\beta p^0})$$

$$= \int \frac{dp^0}{2\pi} \, e^{-ip^0(y^0 - x^0)} \, I_A(p^0, \mathbf{0})(1 - e^{-\beta p^0}). \tag{3.157}$$

The conserved charge Q is a constant independent of time y^0. Therefore, differentiating (3.157) with respect to y^0 and setting it to zero, we obtain

$$\int \frac{dp^0}{2\pi} \, e^{-ip^0(y^0 - x^0)} \, p^0 I_A(p^0, \mathbf{0})(1 - e^{-\beta p^0}) = 0. \tag{3.158}$$

This must hold true for any operator A in general. On the other hand, if A is an operator for which (3.153) holds (signalling spontaneous symmetry breaking), then the right hand side of (3.157) cannot be zero and this can be compatible with (3.158) only if

$$I_A(p^0, \mathbf{0})(1 - e^{-\beta p^0}) = a \, \delta(p^0), \tag{3.159}$$

where a is a constant. This shows that, in this case, certain correlation functions will develop a delta function singularity (and as we pointed out earlier, cyclicity may not hold when there are singularities).

This leads to a reformulation of the Goldstone theorem at finite temperature as follows. If in the absence of long range forces, a theory with sufficiently localized fields has a spontaneous breakdown of a continuous global symmetry at finite temperature, then, there must appear quanta with zero energy. These quanta need not be massless (as is the case at zero temperature) since we are no longer in a covariant framework at finite temperature. (In fact, as we have seen before, mass is not a uniquely defined concept at finite temperature.) We can think of these quanta with zero energy as quasi-particle states. In fact, a little analysis, following from our earlier discussions, shows that the Goldstone states can be thought of as composites of tilde and non-tilde states.

3.9 Appendices

In the first appendix in this section, we discuss the behavior of a general Gaussian wave packet first so that understanding the coherent and the squeezed states (both single and double mode) becomes easy. After the discussion of the general Gaussian wave packet, in the second and third appendices in this section, we take up the discussion of coherent states and squeezed states respectively. Such states, of course, play an important role in quantum optics, particularly in the study of lasers, but they are also very essential in the study of spontaneous symmetry breaking in quantum field theories as well as in thermal field theories, as we have mentioned earlier in the context of thermofield dynamics in this chapter. The other two appendices will discuss briefly thermofield dynamical descriptions of parafermions and parabosons respectively within the context of a quantum harmonic oscillator just for completeness.

3.9.1 A general Gaussian wave packet. Let us consider a general one dimensional wave packet of the form

$$\psi_{\alpha,\beta}(x) = \left(\frac{\beta}{\pi}\right)^{\frac{1}{4}} e^{\frac{\beta}{8}(\alpha-\alpha^*)^2} e^{-\frac{\beta}{2}(x-\alpha)^2}, \tag{3.160}$$

where, for simplicity, we assume the constant parameter β to be real and positive while we allow α to be complex, in general. As we will see later, reality of β would correspond to the physical cases that we will study and positivity of β is necessary for normalization of the wave packet. (If β were complex, then normalization would require $\mathrm{Re}\,\beta > 0$.) The prefactors are there for the normalization of the wave function, namely,

$$\int_{-\infty}^{\infty} \mathrm{d}x\, \psi_{\alpha,\beta}^*(x)\psi_{\alpha,\beta}(x)$$

$$= \left(\frac{\beta}{\pi}\right)^{\frac{1}{2}} e^{\frac{\beta}{4}(\alpha-\alpha^*)^2} \int_{-\infty}^{\infty} \mathrm{d}x\, e^{-\frac{\beta}{2}(x-\alpha^*)^2}\, e^{-\frac{\beta}{2}(x-\alpha)^2}$$

$$= \left(\frac{\beta}{\pi}\right)^{\frac{1}{2}} e^{\frac{\beta}{4}(\alpha-\alpha^*)^2}\, e^{-\frac{\beta}{4}(\alpha-\alpha^*)^2} \int_{-\infty}^{\infty} \mathrm{d}x\, e^{-\beta(x-\frac{1}{2}(\alpha+\alpha^*))^2}$$

$$= \left(\frac{\beta}{\pi}\right)^{\frac{1}{2}} \int_{-\infty}^{\infty} \mathrm{d}x\, e^{-\beta x^2} = \left(\frac{\beta}{\pi}\right)^{\frac{1}{2}} \left(\frac{\pi}{\beta}\right)^{\frac{1}{2}} = 1, \qquad (3.161)$$

where we have shifted the variable of integration in the last line. Such a Gaussian probability density $\psi_{\alpha,\beta}^*(x)\psi_{\alpha,\beta}(x) \sim e^{-\beta(x-\frac{1}{2}(\alpha+\alpha^*)^2)}$, therefore, peaks away from the origin at $x = \frac{1}{2}(\alpha + \alpha^*)$ and has a full width at half the maximum (FWHM) given by

$$\mathrm{FWHM} = 2\sqrt{\frac{\ln 2}{\beta}}, \qquad (3.162)$$

which will be useful in the following. (Note that it is independent of the value of α.)

In such a Gaussian state, we can calculate the average value of the coordinate as

$$\langle X \rangle = \int_{-\infty}^{\infty} \mathrm{d}x\, x\, \psi_{\alpha,\beta}^*(x)\psi_{\alpha,\beta}(x)$$

$$= \left(\frac{\beta}{\pi}\right)^{\frac{1}{2}} \int_{-\infty}^{\infty} \mathrm{d}x\, x e^{-\beta(x-\frac{1}{2}(\alpha+\alpha^*))^2}$$

$$= \left(\frac{\beta}{\pi}\right)^{\frac{1}{2}} \times \frac{1}{2}(\alpha + \alpha^*) \left(\frac{\pi}{\beta}\right)^{\frac{1}{2}} = \frac{1}{2}(\alpha + \alpha^*). \qquad (3.163)$$

Similarly, we can calculate

$$\langle X^2 \rangle = \int_{-\infty}^{\infty} dx \, x^2 \, \psi_{\alpha,\beta}^*(x) \psi_{\alpha,\beta}(x)$$

$$= \left(\frac{\beta}{\pi}\right)^{\frac{1}{2}} \int_{-\infty}^{\infty} dx \, x^2 \, e^{-\beta(x-\frac{1}{2}(\alpha+\alpha^*))^2}$$

$$= \left(\frac{\beta}{\pi}\right)^{\frac{1}{2}} \left(\frac{\pi}{\beta}\right)^{\frac{1}{2}} \left(\frac{1}{2\beta} + \frac{1}{4}(\alpha+\alpha^*)^2\right). \tag{3.164}$$

As a result, we obtain the root mean square deviation (or the uncertainty) of the coordinate, in this state, to be

$$(\Delta x) = \sqrt{\langle X^2 \rangle - \langle X \rangle^2} = \frac{1}{\sqrt{2\beta}}, \tag{3.165}$$

namely, it is proportional to the FWHM of the Gaussian (see (3.162)) and is independent of α although both $\langle X \rangle$ and $\langle X^2 \rangle$ depend on α.

The momentum space wave function, for such a state, is obtained by taking the Fourier transformation of the state in (3.160) and leads to

$$\psi_{\alpha,\beta}(p) = \frac{1}{\sqrt{2\pi}} \int dx \, e^{-ipx} \, \psi_{\alpha,\beta}(x)$$

$$= \left(\frac{1}{\pi\beta}\right)^{\frac{1}{4}} e^{\frac{\beta}{8}(\alpha-\alpha^*)^2} \, e^{-\frac{\beta}{2}\alpha^2} \, e^{-\frac{1}{2\beta}(p+i\alpha\beta)^2}. \tag{3.166}$$

Using this, we can calculate $\langle P \rangle, \langle P^2 \rangle$ in the state in (3.160) (alternatively, we can use the coordinate representation of the momentum operator $P \to -i\frac{d}{dx}$ on the state in the coordinate space in (3.160)) to obtain

$$\langle P \rangle = \int dp \, p \, \psi_{\alpha,\beta}^*(p) \psi_{\alpha,\beta}(p)$$

$$= \frac{1}{\sqrt{\pi\beta}} e^{\frac{\beta}{4}(\alpha-\alpha^*)^2} \, e^{-\frac{\beta}{2}(\alpha^2+(\alpha^*)^2)}$$

$$\times \int dp \, p \, e^{-\frac{1}{2\beta}((p-i\alpha^*\beta)^2+(p+i\alpha\beta)^2)}$$

$$= \frac{1}{\sqrt{\pi\beta}} \int dp \, p \, e^{-\frac{1}{\beta}(p+\frac{i\beta}{2}(\alpha-\alpha^*))^2}$$

$$= \frac{1}{\sqrt{\pi\beta}} \left(-\frac{i\beta}{2}(\alpha - \alpha^*) \right) \sqrt{\pi\beta} = -\frac{i\beta}{2}(\alpha - \alpha^*),$$

$$\langle P^2 \rangle = \int dp\, p^2\, \psi^*_{\alpha,\beta}(p)\psi_{\alpha,\beta}(p)$$

$$= \frac{1}{\sqrt{\pi\beta}} \int dp\, p^2\, e^{-\frac{1}{\beta}\left(p + \frac{i\beta}{2}(\alpha-\alpha^*)\right)^2}$$

$$= \frac{1}{\sqrt{\pi\beta}} \left(\frac{\beta}{2} + \left(-\frac{i\beta}{2}(\alpha - \alpha^*) \right)^2 \right) \sqrt{\pi\beta}$$

$$= \frac{\beta}{2} + \left(-\frac{i\beta}{2}(\alpha - \alpha^*) \right)^2. \tag{3.167}$$

This determines the root mean square deviation (or the uncertainty) of the momentum, in this state to be

$$(\Delta p) = \sqrt{\langle P^2 \rangle - \langle P \rangle^2} = \sqrt{\frac{\beta}{2}}, \tag{3.168}$$

which is, again, independent of the shift in the center of the Gaussian.

The uncertainty in the momentum (or the root mean square deviation of the momentum) is again proportional to the width of the Gaussian in the momentum space given in (3.166). However, it is related to the inverse of the uncertainty in the coordinate given in (3.165). Therefore, from (3.165) and (3.168), we obtain

$$(\Delta x)(\Delta p) = \frac{1}{\sqrt{2\beta}} \times \sqrt{\frac{\beta}{2}} = \frac{1}{2}. \tag{3.169}$$

Namely, a general Gaussian of the form in (3.160) (or its Fourier transform (3.166)) is a minimum uncertainty state, much like the ground state of the harmonic oscillator. However, we note that, for the general Gaussian of the form (3.160),

$$(\Delta x) = \frac{1}{\sqrt{2\beta}} \neq (\Delta p) = \sqrt{\frac{\beta}{2}}, \tag{3.170}$$

They are equal if $\beta = 1$, independent of the value of α. This includes the ground state wave function of the harmonic oscillator which has the form $\psi_0(x) \sim e^{-\frac{1}{2}x^2}$ or even a shifted Gaussian of the form $\psi_\alpha(x) \sim e^{-\frac{1}{2}(x-\alpha)^2}$. Namely, while the uncertainties in the coordinate as well as the momentum are the same for the ground state of the harmonic oscillator (shifted or unshifted), for a general Gaussian wave function, the uncertainty may be smaller (than that in the ground state of the harmonic oscillator) in one variable while it is larger in the conjugate variable in a compensating manner so that the product remains a minimum. We can say it even differently as the Gaussian wave function may be squeezed in one variable while it is more spread out in the conjugate variable. This feature will be helpful in understanding the following two subsections.

3.9.2 Coherent states. Let us consider a bosonic quantum harmonic oscillator described in section **3.3** with the Hamiltonian and the commutation relations given in (3.56)-(3.58). Normally, we work with the energy eigenbasis states. However, coherent states define an alternative set of basis states which are also complete and very useful to study various phenomena. To introduce the coherent states, let us consider the Hermitian operator

$$Q(\alpha) = i(\alpha a^\dagger - \alpha^* a), \tag{3.171}$$

where α is a complex parameter in general. This allows us to define a Unitary operator as

$$U(\alpha) = e^{-iQ(\alpha)} = e^{(\alpha a^\dagger - \alpha^* a)}, \quad (U(\alpha))^\dagger = e^{iQ(\alpha)},$$

$$U(\alpha)(U(\alpha))^\dagger = \mathbb{1} = (U(\alpha))^\dagger U(\alpha). \tag{3.172}$$

It follows trivially from the basic commutation relations in (3.130) that

$$[Q(\alpha), a] = -i\alpha\mathbb{1}, \qquad [Q(\alpha), [Q(\alpha), a]] = 0, \quad \cdots,$$

$$[Q(\alpha), a^\dagger] = -i\alpha^*\mathbb{1}, \qquad [Q(\alpha), [Q(\alpha), a^\dagger]] = 0, \quad \cdots, \tag{3.173}$$

namely, all the higher nested commutators (other than the first) vanish and lead to

$$(U(\alpha))^\dagger a U(\alpha) = e^{iQ(\alpha)} a \, e^{-iQ(\alpha)} = a + i[Q(\alpha), a]$$

$$= a + i(-i\alpha)\mathbb{1} = a + \alpha\mathbb{1},$$

$$(U(\alpha))^\dagger a^\dagger U(\alpha) = e^{iQ(\alpha)} a^\dagger e^{-iQ(\alpha)} = a^\dagger + i[Q(\alpha), a^\dagger]$$

$$= a^\dagger + i(-i\alpha^*)\mathbb{1} = a^\dagger + \alpha^*\mathbb{1}. \tag{3.174}$$

Namely, the unitary operators simply translate the creation and the annihilation operators by constants and correspondingly $U(\alpha)$ is also known as the displacement operator. As a result, if we define a new state related to the vacuum state of the harmonic oscillator by this unitary operator, namely,

$$|\alpha\rangle = U(\alpha)|0\rangle, \tag{3.175}$$

then, it follows from (3.174) that

$$a|\alpha\rangle = aU(\alpha)|0\rangle = U(\alpha)((U(\alpha))^\dagger aU(\alpha))|0\rangle$$
$$= U(\alpha)(a + \alpha\mathbb{1})|0\rangle = \alpha U(\alpha)|0\rangle = \alpha|\alpha\rangle. \tag{3.176}$$

Here we have used the fact that the annihilation operator, a, annihilates the vacuum state of the oscillator. Equation (3.176) shows that the new state $|\alpha\rangle$, obtained from the vacuum state of the oscillator through this unitary operator (see (3.175)), is an eigenstate of the annihilation operator which is also known as a coherent state. Coherent states are very useful in the study of spontaneous symmetry breaking in quantum field theories. It is worth noting here that $Q(\alpha)$ is not a symmetry generator since it does not commute with the Hamiltonian, namely, using (3.173) we obtain (we assume $\hbar = m = \omega = 1$)

$$[Q(\alpha), H] = [Q(\alpha), a^\dagger a] = [Q(\alpha), a^\dagger]a + a^\dagger[Q(\alpha), a]$$
$$= -i\alpha^* a - i\alpha a^\dagger = -i(\alpha a^\dagger + \alpha^* a) \neq 0. \tag{3.177}$$

It is merely a generator of transformation from the vacuum state $|0\rangle$ to the coherent state $|\alpha\rangle$. This same construction can also be carried out in the case of a fermionic oscillator (see section **3.2**) which we leave to the readers.

To bring out another interesting feature of the coherent states, we note that, since $|\alpha\rangle$ and $|0\rangle$ are related by a unitary transformation, the coherent state is normalized, namely,

$$\langle\alpha|\alpha\rangle = \langle 0|(U(\alpha))^\dagger U(\alpha)|0\rangle = \langle 0|\mathbb{1}|0\rangle = 1. \tag{3.178}$$

Let us further recall that the coordinate and momentum operators, associated with this one dimensional oscillator (remember that we are assuming $\hbar = m = \omega = 1$) are given by[2]

$$X = \frac{1}{\sqrt{2}}(a + a^\dagger), \quad P = -\frac{i}{\sqrt{2}}(a - a^\dagger). \tag{3.179}$$

As a result, using (3.174), we obtain

$$\langle \alpha|X|\alpha \rangle = \frac{1}{\sqrt{2}} \langle \alpha|(a + a^\dagger)|\alpha \rangle = \frac{1}{\sqrt{2}} \langle 0|(U(\alpha))^\dagger (a + a^\dagger) U(\alpha)|0 \rangle$$

$$= \frac{1}{\sqrt{2}} \langle 0|(a + a^\dagger + \alpha + \alpha^*)|0 \rangle = \frac{1}{\sqrt{2}}(\alpha + \alpha^*),$$

$$\langle \alpha|P|\alpha \rangle = -\frac{i}{\sqrt{2}} \langle \alpha|(a - a^\dagger)|\alpha \rangle = -\frac{i}{\sqrt{2}} \langle 0|(U(\alpha))^\dagger (a - a^\dagger) U(\alpha)|0 \rangle$$

$$= -\frac{i}{\sqrt{2}} \langle 0|(a - a^\dagger + \alpha - \alpha^*)|0 \rangle = -\frac{i}{\sqrt{2}}(\alpha - \alpha^*), \tag{3.180}$$

where we have used (3.174) and (3.175). Similarly, we can easily calculate (using (3.174) repeatedly as well as the properties of the ground state of the oscillator) that

$$\langle \alpha|X^2|\alpha \rangle = \frac{1}{2} \langle \alpha|(a + a^\dagger)^2|\alpha \rangle = \frac{1}{2} \langle 0|(U(\alpha))^\dagger (a + a^\dagger)^2 U(\alpha)|0 \rangle$$

$$= \frac{1}{2} \langle 0|(a + a^\dagger + \alpha + \alpha^*)^2|0 \rangle = \frac{1}{2}(1 + (\alpha + \alpha^*)^2),$$

$$\langle \alpha|P^2|\alpha \rangle = -\frac{1}{2} \langle \alpha|(a - a^\dagger)^2|\alpha \rangle = -\frac{1}{2} \langle 0|(U(\alpha))^\dagger (a - a^\dagger)^2|0 \rangle$$

$$= -\frac{1}{2} \langle 0|(a - a^\dagger + \alpha - \alpha^*)^2|0 \rangle = \frac{1}{2}(1 - (\alpha - \alpha^*)^2). \tag{3.181}$$

As a result, the root mean square deviations, in the coherent state, follow from (3.180) and (3.181) to correspond to

$$(\Delta x)^2 = \langle \alpha|X^2|\alpha \rangle - (\langle \alpha|X|\alpha \rangle)^2 = \frac{1}{2},$$

$$(\Delta p)^2 = \langle \alpha|P^2|\alpha \rangle - (\langle \alpha|P|\alpha \rangle)^2 = \frac{1}{2}, \tag{3.182}$$

leading to

$$(\Delta x)(\Delta p) = \frac{1}{2}. \tag{3.183}$$

---

[2]See, for example, Eq. (5.37) in A. Das, *Lectures on Quantum Mechanics* (second edition), Hindustan Book Agency and World Scientific Publishing (2011).

In other words, the coherent state is not only the eigenstate of the annihilation operator, it is also a state with the minimum uncertainty (remember $\hbar = 1$). From (3.182) we also note that, in a coherent state, the uncertainty in the coordinate is the same as that in the momentum, namely, $(\Delta x) = (\Delta p)$.

To understand this in view of our discussions in the earlier subsection, let us compare (3.180) and (3.181) with (3.163)-(3.164) and (3.167) and we see that they coincide if we let $\beta = 1$ and $\alpha \to \sqrt{2}\alpha$ in (3.163)-(3.164) and (3.167). Therefore, we suspect, from (3.160) that the coordinate space wave function for the coherent state is likely to be of the form

$$\psi_{\text{coherent}}(x) = \left(\frac{1}{\pi}\right)^{\frac{1}{4}} e^{\frac{1}{4}(\alpha-\alpha^*)^2} e^{-\frac{1}{2}(x-\sqrt{2}\alpha)^2}. \tag{3.184}$$

Let us, in fact, derive this wave function from the Unitary operator $U(\alpha)$ in (3.172) which generates the coherent state from the ground state of the harmonic oscillator as given in (3.175). Let us note from (3.179) that we can write

$$\alpha a^\dagger - \alpha^* a = \frac{1}{2}(\alpha - \alpha^*)(a + a^\dagger) - \frac{1}{2}(\alpha + \alpha^*)(a - a^\dagger)$$

$$= \frac{1}{2}(\alpha - \alpha^*)\sqrt{2}X - \frac{1}{2}(\alpha + \alpha^*)\sqrt{2}iP, \tag{3.185}$$

which, in the coordinate basis takes the form

$$(\alpha a^\dagger - \alpha^* a) \to \frac{1}{\sqrt{2}}\left((\alpha - \alpha^*)x - (\alpha + \alpha^*)\frac{\mathrm{d}}{\mathrm{d}x}\right). \tag{3.186}$$

As a result, in the coordinate basis, the unitary operator $U(\alpha)$ in (3.172) takes the form

$$U(\alpha) = e^{(\alpha a^\dagger - \alpha^* a)}$$

$$\to e^{\frac{1}{\sqrt{2}}\left((\alpha-\alpha^*)x - (\alpha+\alpha^*)\frac{\mathrm{d}}{\mathrm{d}x}\right)} = U(\alpha, x). \tag{3.187}$$

Let us recall that when we have exponentials of two non-commuting operators, A, B, then the Baker-Campbell-Hausdorff (BCH) formula gives a way of combining them, namely,

$$e^A e^B = e^{A+B+\frac{1}{2}[A,B]+\frac{1}{12}[A,[A,B]]-\frac{1}{12}[B,[A,B]]+\cdots}, \tag{3.188}$$

which takes a particularly simple form when $[A, B] = c\mathbb{1}$, namely,

$$e^A e^B = e^{A+B+\frac{c}{2}\mathbb{1}} = e^{\frac{c}{2}\mathbb{1}} e^{A+B}. \tag{3.189}$$

In this case, the inverse Baker-Campbell-Hausdorff formula is simply obtained to be

$$e^{A+B} = e^{-\frac{c}{2}\mathbb{1}} e^A e^B = e^{\frac{c}{2}\mathbb{1}} e^B e^A. \tag{3.190}$$

Using these, in particular, the inverse relation given in (3.189), we can write the $U(\alpha)$ in the coordinate representation given in (3.187) also as

$$U(\alpha, x) \rightarrow e^{\frac{1}{2} \times \frac{1}{2}(\alpha^2 - (\alpha^*)^2)} \, e^{-\frac{(\alpha+\alpha^*)}{\sqrt{2}} \frac{d}{dx}} \, e^{\frac{(\alpha-\alpha^*)}{\sqrt{2}} x}. \tag{3.191}$$

Therefore, we can now obtain the coherent state from the ground state of the harmonic oscillator as

$$\psi_{\text{coherent}}(x) = \langle x|\alpha \rangle = \langle x|U(\alpha)|0 \rangle = U(\alpha, x)\psi_0(x)$$

$$= \left(\frac{1}{\pi}\right)^{\frac{1}{4}} e^{\frac{1}{4}(\alpha^2 - (\alpha^*)^2)} e^{-\frac{(\alpha+\alpha^*)}{\sqrt{2}} \frac{d}{dx}} e^{\frac{(\alpha-\alpha^*)}{\sqrt{2}} x} e^{-\frac{1}{2}x^2}$$

$$= \left(\frac{1}{\pi}\right)^{\frac{1}{4}} e^{i\theta} e^{-\frac{(\alpha+\alpha^*)}{\sqrt{2}} \frac{d}{dx}} e^{-\frac{1}{2}\left((x - \frac{1}{\sqrt{2}}(\alpha-\alpha^*))^2 - \frac{1}{2}(\alpha-\alpha^*)^2\right)}$$

$$= \left(\frac{1}{\pi}\right)^{\frac{1}{4}} e^{i\theta} e^{\frac{1}{4}(\alpha-\alpha^*)^2} e^{-\frac{1}{2}\left(x - \frac{1}{\sqrt{2}}(\alpha+\alpha^*) - \frac{1}{\sqrt{2}}(\alpha-\alpha^*)\right)^2}$$

$$= \left(\frac{1}{\pi}\right)^{\frac{1}{4}} e^{i\theta} e^{\frac{1}{4}(\alpha-\alpha^*)^2} e^{-\frac{1}{2}(x - \sqrt{2}\alpha)^2}. \tag{3.192}$$

There are a couple of comments that need to be made here. First, we have identified $i\theta = \frac{1}{4}(\alpha^2 - (\alpha^*)^2) = i(\operatorname{Re}\alpha)(\operatorname{Im}\alpha)$ and second, we have used the fact that $\frac{d}{dx}$ is the generator of translations of the x coordinate. As a result, we see that the coherent state in (3.192) coincides with the state in (3.184) except for a constant phase factor which is not physically relevant.

Since the energy eigenstates define a complete basis, the coherent state can be expanded in terms of them as

$$|\alpha\rangle = \sum_{n=0}^{\infty} f_n |n\rangle, \tag{3.193}$$

and the coefficients of expansion, f_n, can be determined in various ways. We indicate here only two methods of determining these constant coefficients. The direct method determines the coefficients from (3.193) to be

$$f_n = \langle n|\alpha\rangle = \langle 0|\frac{a^n}{\sqrt{n!}}|\alpha\rangle = \frac{a^n}{\sqrt{n!}}\langle 0|\alpha\rangle, \tag{3.194}$$

leading to

$$|\alpha\rangle = \langle 0|\alpha\rangle \sum_n \frac{a^n}{\sqrt{n!}}|n\rangle, \tag{3.195}$$

where, in addition to the standard representation of the energy eigenstates in terms of creation/annihilation operators acting on the vacuum, we have used (3.176). Substituting this into (3.193) and using the normalization of the coherent state in (3.178), we obtain

$$\langle\alpha|\alpha\rangle = 1 = \sum_{m,n} f_m^* f_n \langle m|n\rangle = \sum_n |f_n|^2 = |\langle 0|\alpha\rangle|^2 \sum_n \frac{|\alpha|^{2n}}{n!}$$

$$= |\langle 0|\alpha\rangle|^2 e^{|\alpha|^2}$$

$$\text{or,} \quad \langle 0|\alpha\rangle = \langle\alpha|0\rangle = e^{-\frac{1}{2}|\alpha|^2}, \tag{3.196}$$

which gives (see (3.195))

$$|\alpha\rangle = e^{-\frac{1}{2}|\alpha|^2} \sum_n \frac{a^n}{\sqrt{n!}}|n\rangle \tag{3.197}$$

where we have set the phase in the solution in (3.196) to zero for simplicity.

The alternative derivation is quite short and elegant. We see, from the definitions in (3.172) that

$$U(\alpha) = e^{\alpha a^\dagger - \alpha^* a} = e^{\alpha a^\dagger} e^{-\alpha^* a} e^{\frac{1}{2}|\alpha|^2[a^\dagger,a]}$$

$$= e^{-\frac{1}{2}|\alpha|^2} e^{\alpha a^\dagger} e^{-\alpha^* a}, \tag{3.198}$$

where we have used the inverse Baker-Campbell-Hausdorff formula and which leads, upon using (3.175), to

$$|\alpha\rangle = U(\alpha)|0\rangle = e^{-\frac{1}{2}|\alpha|^2} e^{\alpha a^\dagger} e^{-\alpha^* a}|0\rangle = e^{-\frac{1}{2}|\alpha|^2} e^{\alpha a^\dagger}|0\rangle$$

$$= e^{-\frac{1}{2}|\alpha|^2} \sum_n \frac{(\alpha a^\dagger)^n}{n!}|0\rangle = e^{-\frac{1}{2}|\alpha|^2} \sum_n \frac{a^n}{\sqrt{n!}}|n\rangle, \tag{3.199}$$

which is the result in (3.197).

As we have seen in (3.178), the coherent state is normalized. However, they are not orthogonal for different values of $\alpha, \beta \in \mathbb{C}$. This can be seen as follows. Using (3.197) (or (3.199)), we have

$$\langle\alpha|\beta\rangle = e^{-\frac{1}{2}(|\alpha|^2+|\beta|^2)} \sum_{n,m} \frac{(\alpha^*)^n}{\sqrt{n!}} \frac{\beta^m}{\sqrt{m!}} \langle n|m\rangle$$

$$= e^{-\frac{1}{2}(|\alpha|^2+|\beta|^2)} \sum_n \frac{(\alpha^*\beta)^n}{n!}$$

$$= e^{-\frac{1}{2}(|\alpha|^2+|\beta|^2-2\alpha^*\beta)}, \tag{3.200}$$

so that we obtain

$$|\langle\alpha|\beta\rangle|^2 = e^{-(|\alpha|^2+|\beta|^2)+(\alpha^*\beta+\alpha\beta^*)}$$

$$= e^{-(\alpha-\beta)(\alpha^*-\beta^*)} = e^{-|\alpha-\beta|^2}. \tag{3.201}$$

It is clear, therefore, either from (3.200) or (3.201) that if $\alpha = \beta$, the state is normalized, but if $\alpha \neq \beta$, the inner product does not vanish signifying that such states are not orthogonal. (Note from (3.201), however, that the coherent states become asymptotically orthogonal as $|\alpha - \beta| \to \infty$.) The coherent states, nonetheless, provide a complete basis which, however, is not orthonormal like the energy basis. This can be seen as follows. Note that α is a complex parameter so that integration over α is a two dimensional integration – we have to integrate over the real as well as the imaginary parts. Therefore, writing (recall the two dimensional polar coordinates)

$$\alpha = |\alpha|e^{i\theta}, \quad \mathrm{d}^2\alpha = |\alpha|\mathrm{d}|\alpha|\mathrm{d}\theta, \tag{3.202}$$

we obtain,

$$\int \frac{\mathrm{d}^2\alpha}{\pi} |\alpha\rangle\langle\alpha| = \int_0^\infty \mathrm{d}|\alpha| \int_0^{2\pi} \frac{\mathrm{d}\theta}{\pi} |\alpha|e^{-|\alpha|^2} \sum_{m,n} \frac{\alpha^m}{\sqrt{m!}} \frac{(\alpha^*)^n}{\sqrt{n!}} |m\rangle\langle n|$$

$$= \int_0^\infty \mathrm{d}|\alpha| e^{-|\alpha|^2} \sum_{m,n} \int_0^{2\pi} \frac{\mathrm{d}\theta}{\pi} e^{i(m-n)\theta} \frac{|\alpha|^{m+n+1}}{\sqrt{(m!)(n!)}} |m\rangle\langle n|$$

$$= \int_0^\infty \mathrm{d}|\alpha| e^{-|\alpha|^2} \sum_{m,n} 2\delta_{mn} \frac{|\alpha|^{m+n+1}}{\sqrt{(m!)(n!)}} |m\rangle\langle n|$$

$$= \sum_n \left(2 \int_0^\infty \mathrm{d}|\alpha|\, e^{-|\alpha|^2} |\alpha|^{2n+1} \right) \frac{1}{n!} |n\rangle\langle n|$$

$$= \sum_n \Gamma(n+1) \frac{1}{n!} |n\rangle\langle n| = \sum_n |n\rangle\langle n| = \mathbb{1}. \qquad (3.203)$$

As a result, it follows that the coherent states are not linearly independent, namely,

$$|\alpha\rangle = \mathbb{1}|\alpha\rangle = \int \frac{\mathrm{d}^2\beta}{\pi} |\beta\rangle\langle\beta|\alpha\rangle$$

$$= \int \frac{\mathrm{d}^2\beta}{\pi} e^{-\frac{1}{2}(|\alpha|^2+|\beta|^2-2\beta^*\alpha)} |\beta\rangle, \qquad (3.204)$$

where we have used (3.200) as well as (3.203). Nonetheless, any vector can be uniquely expanded in the basis of coherent states.

3.9.3 Squeezed states. Let us continue with the discussion of a single bosonic harmonic oscillator described in the last subsection and consider the following Hermitian operator quadratic in the creation and annihilation operators

$$S(\xi) = i(\xi(a^\dagger)^2 - \xi^*a^2), \quad \xi = |\xi|e^{i\chi}, \qquad (3.205)$$

where ξ is a complex parameter and χ is its (real) phase. It follows now that

$$[S(\xi), a] = -2i\xi a^\dagger, \quad [S(\xi), a^\dagger] = -2i\xi^* a. \qquad (3.206)$$

These, in turn, lead to

$$\underbrace{[S(\xi), \cdots, [S(\xi), [S(\xi), a]]\cdots]}_{2n} = \left((-2i)^2\xi\xi^*\right)^n a = (-4|\xi|^2)^n a,$$

$$\underbrace{[S(\xi), \cdots, [S(\xi), [S(\xi), a]]\cdots]}_{2n+1} = -2i\xi(-4|\xi|^2)^n a^\dagger,$$

$$\underbrace{[S(\xi), \cdots, [S(\xi), [S(\xi), a^\dagger]]\cdots]}_{2n} = (-4|\xi|^2)^n a^\dagger,$$

$$\underbrace{[S(\xi), \cdots, [S(\xi), [S(\xi), a^\dagger]]\cdots]}_{2n+1} = -2i\xi^*(-4|\xi|^2)^n a. \qquad (3.207)$$

If we now define a unitary operator (see also (3.172))

$$U_S(\xi) = e^{-iS(\xi)} = e^{(\xi(a^\dagger)^2 - \xi^* a^2)}, \quad (U_S(\xi))^\dagger = e^{iS(\xi)},$$

$$U_S(\xi)(U_S(\xi))^\dagger = \mathbb{1} = (U_S(\xi))^\dagger U_S(\xi), \tag{3.208}$$

upon using (3.207), it leads to (see also (3.174) and (3.205))

$$(U_S(\xi))^\dagger a\, U_S(\xi) = \sum_{n=0}^{\infty} \Big(\frac{(i)^{2n}}{(2n)!} \underbrace{[S(\xi), \cdots, [S(\xi), [S(\xi), a\,]]\cdots\,]}_{2n}$$

$$+ \frac{(i)^{2n+1}}{(2n+1)!} \underbrace{[S(\xi), \cdots, [S(\xi), [S(\xi), a\,]]\cdots\,]}_{2n+1} \Big)$$

$$= \sum_{n=0}^{\infty} \Big(\frac{(4|\xi|^2)^n}{(2n)!}\, a + \frac{2\xi(4|\xi|^2)^n}{(2n+1)!}\, a^\dagger \Big)$$

$$= \sum_{n=0}^{\infty} \Big(\frac{(2|\xi|)^{2n}}{(2n)!}\, a + \frac{(2|\xi|)^{2n+1}}{(2n+1)!}\, e^{i\chi} a^\dagger \Big)$$

$$= a \cosh 2|\xi| + a^\dagger e^{i\chi} \sinh 2|\xi|, \tag{3.209}$$

where we have used the parameterization of the complex parameter ξ given in (3.205). A similar calculation (or the Hermitian conjugate of (3.209)) gives

$$(U_S(\xi))^\dagger a^\dagger U_S(\xi) = a\, e^{-i\chi} \sinh 2|\xi| + a^\dagger \cosh 2|\xi|. \tag{3.210}$$

Equations (3.209) and (3.210) show that the unitary operator $U_S(\xi)$ does not shift the annihilation/creation operators, rather, it rotates them into each other.

We can now define a new state related to the vacuum state of the oscillator through the unitary operator $U_S(\xi)$ in (3.208) as

$$|\xi\rangle = U_S(\xi)|0\rangle, \tag{3.211}$$

but it is clear that, unlike the coherent state defined in (3.175) (see also (3.176)), the state $|\xi\rangle$ defined in (3.211) is not an eigenstate of either a or $a^\dagger$. However, it satisfies

$$\langle\xi|a|\xi\rangle = \langle0|(U_S(\xi))^\dagger a U_S(\xi)|0\rangle$$
$$= \langle0|(a\cosh 2|\xi| + a^\dagger e^{i\chi}\sinh 2|\xi|)|0\rangle = 0,$$
$$\langle\xi|a^\dagger|\xi\rangle = \langle0|(U_S(\xi))^\dagger a^\dagger U_S(\xi)|0\rangle$$
$$= \langle0|(a e^{-i\chi}\sinh 2|\xi| + a^\dagger\cosh 2|\xi|)|0\rangle = 0. \quad (3.212)$$

It follows from this that (see also (3.179)-(3.180))

$$\langle\xi|X|\xi\rangle = \frac{1}{\sqrt{2}}\langle\xi|(a + a^\dagger)|\xi\rangle = 0,$$

$$\langle\xi|P|\xi\rangle = -\frac{i}{\sqrt{2}}\langle\xi|(a - a^\dagger)|\xi\rangle = 0. \quad (3.213)$$

We can also calculate

$$\langle\xi|X^2|\xi\rangle = \langle0|(U(\xi))^\dagger\frac{1}{2}(a + a^\dagger)^2 U(\xi)|0\rangle$$

$$= \frac{1}{2}\langle0|((U(\xi))^\dagger(a + a^\dagger)U(\xi))^2|0\rangle$$

$$= \frac{1}{2}\langle0|(a(\cosh 2|\xi| + e^{-i\chi}\sinh 2|\xi|)$$

$$+ a^\dagger(\cosh 2|\xi| + e^{i\chi}\sinh 2|\xi|))^2|0\rangle$$

$$= \frac{1}{2}(\cosh 2|\xi| + e^{-i\chi}\sinh 2|\xi|)(\cosh 2|\xi| + e^{i\chi}\sinh 2|\xi|)$$
$$\times \langle0|aa^\dagger|0\rangle$$

$$= \frac{1}{2}(\cosh^2 2|\xi| + \sinh^2 2|\xi| + 2\cosh 2|\xi|\sinh 2|\xi|\cos\chi)$$

$$= \frac{1}{4}(e^{4|\xi|}(1 + \cos\chi) + e^{-4|\xi|}(1 - \cos\chi))$$

$$= \frac{1}{2}(e^{4|\xi|}\cos^2\frac{\chi}{2} + e^{-4|\xi|}\sin^2\frac{\chi}{2}), \quad (3.214)$$

and, similarly, it can be calculated that

$$\langle\xi|P^2|\xi\rangle = -\frac{1}{2}\langle0|(U_S(\xi))^\dagger(a - a^\dagger)^2 U_S(\xi)|0\rangle$$

$$= \frac{1}{2}(e^{4|\xi|}\sin^2\frac{\chi}{2} + e^{-4|\xi|}\cos^2\frac{\chi}{2}). \quad (3.215)$$

Equations (3.213)-(3.215) determine the uncertainties in the position and the momentum to be

$$(\Delta x)^2 = \langle \xi | X^2 | \xi \rangle - \left(\langle \xi | X | \xi \rangle \right)^2$$
$$= \frac{1}{2} \left(e^{4|\xi|} \cos^2 \frac{\chi}{2} + e^{-4|\xi|} \sin^2 \frac{\chi}{2} \right),$$
$$(\Delta p)^2 = \langle \xi | P^2 | \xi \rangle - \left(\langle \xi | P | \xi \rangle \right)^2$$
$$= \frac{1}{2} \left(e^{4|\xi|} \sin^2 \frac{\chi}{2} + e^{-4|\xi|} \cos^2 \frac{\chi}{2} \right). \tag{3.216}$$

As a result, we see that when $\chi = 0, \pi$, namely, when ξ is a real parameter, we have (say, for $\chi = 0$, for $\chi = \pi$, the two factors will be reversed and also remember that $\hbar = 1$)

$$(\Delta x)(\Delta p) = \frac{1}{\sqrt{2}} e^{2|\xi|} \times \frac{1}{\sqrt{2}} e^{-2|\xi|} = \frac{1}{2}, \tag{3.217}$$

so that such a state will also correspond to a minimum uncertainty state just like the coherent state. However, unlike the coherent states for which the uncertainty in the conjugate variables is equal, namely, $\Delta x = \Delta p = \frac{1}{\sqrt{2}}$ (see (3.182)), here the uncertainties in the conjugate variables are unequal and reciprocal,

$$\Delta x = \frac{1}{\sqrt{2}} \begin{cases} e^{2|\xi|}, & \text{for } \chi = 0, \\ e^{-2|\xi|}, & \text{for } \chi = \pi, \end{cases}$$

$$\Delta p = \frac{1}{\sqrt{2}} \begin{cases} e^{-2|\xi|}, & \text{for } \chi = 0, \\ e^{2|\xi|}, & \text{for } \chi = \pi, \end{cases} \tag{3.218}$$

which follows from (3.216). In other words, in this state, the uncertainty in one of the variables is more (less) while that in the conjugate variable is less (more) than the corresponding values for a coherent state. The uncertainty is enhanced in one direction in the phase space while it is squeezed along the conjugate direction. Such a state is correspondingly known as a single mode squeezed state. (This is, however, not the squeezed state that we alluded to earlier in the chapter within the context of the thermal vacuum state and we will come to this next.) The operator $S(\xi)$ which generates the squeezed state is known as the squeezing operator. For completeness, we note here that one can apply the displacement operator $U(\alpha)$ defined in (3.172) to a squeezed state, namely, we can define a state

$$|\alpha, \xi\rangle = U(\alpha)U_S(\xi)|0\rangle, \tag{3.219}$$

which will also produce a state with minimum uncertainty and such a state is known as a coherent-squeezed state.

Let us note from the uncertainties in the coordinate and momentum, given in (3.218), that they coincide with the general results we derived in (3.165) and (3.168) if we identify $\beta = e^{\mp 4|\xi|}$ (for $\chi = 0, \pi$) respectively and $\alpha = 0 = \alpha^*$ (since there is no shift in the annihilation/creation operators). That suggests that, in this case, the squeezed state will correspond to a Gaussian of the form (see (3.160))

$$\psi^{(1)}_{\text{sqeezed}}(x) = \left(\frac{e^{\mp 4|\xi|}}{\pi}\right)^{\frac{1}{4}} e^{-\frac{1}{2}e^{\mp 4|\xi|}x^2}, \tag{3.220}$$

where the superscript in the wave function signifies that this wave function corresponds to the single mode squeezed state. Let us, in fact, derive this directly from the definition of the squeezed state given in (3.211). We note that the unitary operator (3.208) which takes us from the ground state of the oscillator to the squeezed state (see (3.211)) has the form (for $\chi = 0, \pi$)

$$U_S(\xi) = e^{\pm |\xi|((a^\dagger)^2 - a^2)} = e^{\pm \frac{|\xi|}{2}((X - iP)^2 - (X + iP)^2)}$$

$$= e^{\pm |\xi|(-i(XP + PX))} = e^{\mp |\xi|(1 + 2iXP)}$$

$$\rightarrow e^{\mp |\xi|\left(1 + 2x\frac{d}{dx}\right)} = e^{\mp |\xi|} e^{\mp 2|\xi|x\frac{d}{dx}}, \tag{3.221}$$

where, in the last step, we have gone to the coordinate representation. As a result, we can write the single mode squeezed state wave function (for $\chi = 0, \pi$) as

$$\psi^{(1)}_{\text{squeezed}}(x) = \langle x|U_S(\xi)|0\rangle = e^{\mp |\xi|} e^{\mp 2|\xi|x\frac{d}{dx}} \psi_0(x)$$

$$= \left(\frac{e^{\mp 4|\xi|}}{\pi}\right)^{\frac{1}{4}} e^{\mp 2|\xi|x\frac{d}{dx}} e^{-\frac{1}{2}x^2}. \tag{3.222}$$

Let us recall that the operator $e^{ax\frac{d}{dx}}$ scales the coordinate x by e^a so that

$$e^{ax\frac{d}{dx}} x^n = (e^a x)^n, \quad e^{ax\frac{d}{dx}} f(x) = f(e^a x), \tag{3.223}$$

and, using this property of the operator in (3.222), we obtain

$$\psi^{(1)}_{\text{squeezed}}(x) = \left(\frac{e^{\mp 4|\xi|}}{\pi}\right)^{\frac{1}{4}} e^{-\frac{1}{2}\left(e^{\mp 2|\xi|}x\right)^2}$$

$$= \left(\frac{e^{\mp 4|\xi|}}{\pi}\right)^{\frac{1}{4}} e^{-\frac{1}{2}e^{\mp 4|\xi|}x^2}. \tag{3.224}$$

This is indeed the normalized Gaussian we determined from the uncertainty relations in (3.220).

Let us next turn to the two mode squeezed state which is the analog of a thermal state that we have been alluding to. Let us consider a system consisting of two independent and free harmonic oscillators described by the annihilation operators given by a, b which commute with each other (independence). Therefore, the energy eigenstates of such a system would describe a product space of the two oscillators. For example, the normalized ground state wave function for this system will have the form

$$\psi_0(x_a, x_b) = \left(\frac{1}{\pi}\right)^{\frac{1}{4}} e^{-\frac{1}{2}x_a^2} \times \left(\frac{1}{\pi}\right)^{\frac{1}{4}} e^{-\frac{1}{2}x_b^2},$$

$$= \left(\frac{1}{\pi}\right)^{\frac{1}{4}} e^{-\frac{1}{4}(x_a-x_b)^2} \times \left(\frac{1}{\pi}\right)^{\frac{1}{4}} e^{-\frac{1}{4}(x_a+x_b)^2}$$

$$= \left(\frac{1}{\pi}\right)^{\frac{1}{4}} e^{-\frac{1}{2}x^2} \times \left(\frac{1}{\pi}\right)^{\frac{1}{4}} e^{-\frac{1}{2}\tilde{x}^2}, \tag{3.225}$$

where we have written the wave function in an alternate form in the last line for later use with the identification $x = \frac{1}{\sqrt{2}}(x_a - x_b), \tilde{x} = \frac{1}{\sqrt{2}}(x_a + x_b)$.

In this space, let us consider a Hermitian operator of the form

$$S_2(\xi) = i(\xi a^\dagger b^\dagger - \xi^* ba) = (S_2(\xi))^\dagger, \tag{3.226}$$

where the subscript refers to the fact that we are now studying a system with two independent oscillators. As in (3.206)-(3.207), we can determine

$$[S_2(\xi), a] = -i\xi b^\dagger, \qquad [S_2(\xi), a^\dagger] = -i\xi^* b,$$

$$[S_2(\xi), b] = -i\xi a^\dagger, \qquad [S_2(\xi), b^\dagger] = -i\xi^* a, \tag{3.227}$$

which shows that the Hermitian operator $S_2(\xi)$ does not translate the annihilation/creation operators, rather rotates them into each other in a 4-dimensional space (of $a, b, a^\dagger, b^\dagger$). Using these repeatedly we can show, as in (3.207), that

$$[S(\xi), \cdots, \underbrace{[S(\xi), [S(\xi), a]] \cdots]}_{2n} = (-|\xi|^2)^n a,$$

$$[S(\xi), \cdots, \underbrace{[S(\xi), [S(\xi), a^\dagger]] \cdots]}_{2n} = (-|\xi|^2)^n a^\dagger,$$

$$[S(\xi), \cdots, \underbrace{[S(\xi), [S(\xi), b]] \cdots]}_{2n} = (-|\xi|^2)^n b,$$

$$[S(\xi), \cdots, \underbrace{[S(\xi), [S(\xi), b^\dagger]] \cdots]}_{2n} = (-|\xi|^2)^n b^\dagger,$$

$$[S(\xi), \cdots, \underbrace{[S(\xi), [S(\xi), a]] \cdots]}_{2n+1} = -i\xi(-|\xi|^2)^n b^\dagger,$$

$$[S(\xi), \cdots, \underbrace{[S(\xi), [S(\xi), a^\dagger]] \cdots]}_{2n+1} = -i\xi^*(-|\xi|^2)^n b,$$

$$[S(\xi), \cdots, \underbrace{[S(\xi), [S(\xi), b]] \cdots]}_{2n+1} = -i\xi(-|\xi|^2)^n a^\dagger,$$

$$[S(\xi), \cdots, \underbrace{[S(\xi), [S(\xi), b^\dagger]] \cdots]}_{2n+1} = -i\xi^*(-|\xi|^2)^n a. \qquad (3.228)$$

The Hermitian generator $S_2(\xi)$ given in (3.226), then leads to the unitary operator

$$U_{S_2}(\xi) = e^{-iS_2(\xi)} = e^{(\xi a^\dagger b^\dagger - \xi^* ba)}, \qquad (3.229)$$

with the help of which we can define a new state, a squeezed state (to be shown), as

$$|\xi\rangle = U_{S_2}(\xi)|0_a, 0_b\rangle,$$

which, in turn, leads to

$$\langle x_a, x_b|\xi\rangle = \psi_{S_2}(\xi, x_a, x_b) = U_{S_2}(\xi, x_a, x_b)\psi_0(x_a, x_b), \qquad (3.230)$$

where $U_{S_2}(\xi, x_a, x_b)$ denotes the unitary operator in (3.229) in the coordinate representation while $\psi_0(x_a, x_b)$ is the ground state wave function given in (3.225).

If we parameterize the complex parameter of transformation as in (3.205), namely, $\xi = |\xi|e^{i\chi}$, then we can calculate the expectation values of various operators in this state, as in (3.212)-(3.217), using the relations obtained in (3.228) as well as the transformation law for operators, namely, $O(\xi) = U_{S_2}^\dagger(\xi)OU_{S_2}(\xi)$. Without going into details, we simply note here that this leads to

$$(\Delta x_a)^2 = \langle\xi|X_a^2|\xi\rangle - (\langle\xi|X_a|\xi\rangle)^2 = \frac{1}{2}\left(\cosh^2|\xi| + \sinh^2|\xi|\right) - 0$$

$$= \frac{1}{2}\left(\cosh^2|\xi| + \sinh^2|\xi|\right) = \frac{1}{2}\cosh 2|\xi|,$$

$$(\Delta p_a)^2 = \langle\xi|P_a^2|\xi\rangle - (\langle\xi|P_a|\xi\rangle)^2 = \frac{1}{2}\left(\cosh^2|\xi| + \sinh^2|\xi|\right) - 0$$

$$= \frac{1}{2}\left(\cosh^2|\xi| + \sinh^2|\xi|\right) = \frac{1}{2}\cosh 2|\xi|, \tag{3.231}$$

where we have used the standard relations (with $m = \omega = \hbar = 1$) $X_a = \frac{1}{\sqrt{2}}\left(a + a^\dagger\right), P_a = -\frac{i}{\sqrt{2}}\left(a - a^\dagger\right)$ (see, for example, (3.179) or (3.213)). Incidentally, (3.231) holds for any value of the phase χ of the parameter ξ. We can also calculate the same expressions for the "b" degrees of freedom and they lead to identical results. Equation (3.231) leads to

$$(\Delta x_a)(\Delta p_a) = (\Delta x_b)(\Delta p_b) = \frac{1}{2}\cosh 2|\xi|, \tag{3.232}$$

which does not give the minimal uncertainty of $\frac{1}{2}$ unless $|\xi| = 0$ (which would correspond to the original oscillator ground state). Namely, the wave function is no longer a Gaussian in the x_a, x_b variables when $\xi \neq 0$.

On the other hand, it can be checked that if we define a pair of new (canonically) conjugate variables

$$X = \frac{1}{\sqrt{2}}(X_a - X_b) = \frac{1}{2}(a + a^\dagger - (b + b^\dagger)),$$

$$P = \frac{1}{\sqrt{2}}(P_a - P_b) = -\frac{i}{2}(a - a^\dagger - (b - b^\dagger)),$$

$$\widetilde{X} = \frac{1}{\sqrt{2}}(X_a + X_b) = \frac{1}{2}(a + a^\dagger + (b + b^\dagger)),$$

$$\widetilde{P} = \frac{1}{\sqrt{2}}(P_a + P_b) = -\frac{i}{2}(a - a^\dagger + (b - b^\dagger)), \qquad (3.233)$$

then, with a little bit of calculation, we can show that, in the state $|\xi\rangle$ defined in (3.230), we obtain

$$\langle\xi|X|\xi\rangle = 0 = \langle\xi|\widetilde{X}|\xi\rangle = \langle\xi|P|\xi\rangle = \langle\xi|\widetilde{P}|\xi\rangle,$$

$$\langle\xi|X^2|\xi\rangle = \frac{1}{2}\left(\cosh 2|\xi| - \cos\chi \sinh 2|\xi|\right) = \langle\xi|\widetilde{P}^2|\xi\rangle,$$

$$\langle\xi|P^2|\xi\rangle = \frac{1}{2}\left(\cosh 2|\xi| + \cos\chi \sinh 2|\xi|\right) = \langle\xi|\widetilde{X}^2|\xi\rangle. \qquad (3.234)$$

We see that, unlike the case of the single mode squeezed state (see (3.231)), here the expectation values and, therefore, the uncertainties do depend on the value of the phase (in addition to $|\xi|$). And they do not lead to the minimum uncertainty for general values of the phase χ. However, if the phase takes values $\chi = 0, \pi$ respectively, the uncertainties in the coordinates and momenta take the simpler values

$$(\Delta x)^2 = \frac{1}{2}\left(\cosh 2|\xi| \mp \sinh 2|\xi|\right) = \frac{1}{2}e^{\mp 2|\xi|} = (\Delta\widetilde{p})^2,$$

$$(\Delta\widetilde{x})^2 = \frac{1}{2}\left(\cosh 2|\xi| \pm \sinh 2|\xi|\right) = \frac{1}{2}e^{\pm 2|\xi|} = (\Delta p)^2, \qquad (3.235)$$

so that

$$(\Delta x)(\Delta p) = \frac{1}{2} = (\Delta\widetilde{x})(\Delta\widetilde{p}). \qquad (3.236)$$

In other words, we conclude from (3.232) and (3.235) that while the uncertainties in the original canonically conjugate variables, (x_a, p_a) and (x_b, p_b), are not minimal for any value of the phase (χ), the modified canonically conjugate variables $\frac{1}{\sqrt{2}}(x, p)$ and $\frac{1}{\sqrt{2}}(\widetilde{x}, \widetilde{p})$ do lead to a minimum uncertainty for the phase values $\chi = 0, \pi$. Furthermore, for $\chi = 0$, while $(x, \widetilde{p})$ are squeezed, $(p, \widetilde{x})$ are elongated. For $\chi = \pi$, just the opposite happens such that, in either case, the minimum uncertainty of $\frac{1}{2}$ is attained. This is very much like the case of the single mode squeezed state which we have already discussed in (3.216)-(3.217).

Therefore, we expect the squeezed state, defined in (3.230), to be a Gaussian in the variables $(x, \widetilde{x})$ of the form

$$\psi_{S_2}(|\xi|, x_a, x_b) = \psi_{S_2}(|\xi|, x, \widetilde{x})$$

$$= \left(\frac{1}{\pi}\right)^{\frac{1}{4}} e^{-\frac{e^{\pm 2|\xi|}}{2}x^2} \times \left(\frac{1}{\pi}\right)^{\frac{1}{4}} e^{-\frac{e^{\mp 2|\xi|}}{2}\widetilde{x}^2}$$

$$= \left(\frac{e^{\pm 2|\xi|}}{\pi}\right)^{\frac{1}{4}} e^{-\frac{e^{\pm 2|\xi|}}{2}x^2} \times \left(\frac{e^{\mp 2|\xi|}}{\pi}\right)^{\frac{1}{4}} e^{-\frac{e^{\mp 2|\xi|}}{2}\widetilde{x}^2}, \qquad (3.237)$$

which can be compared with (3.160) for $(\alpha = 0, \beta = e^{\pm 2|\xi|})$ and $(\alpha = 0, \beta = e^{\mp 2|\xi|})$ for the two factors respectively. Here, in the last step, we have rewritten the Gaussians in a manner so that each factor is independently normalized (compare with (3.160), for example). Let us see briefly how such a state (wave function) is generated from (3.225) (which also clarifies why we had written the second form of the wave function there) with the operation of the unitary operator in (3.229) on the (oscillator) ground state wave function. First, we note that with $\chi = 0, \pi$, we can write $\xi = \pm|\xi|$ which leads to

$$U_{S_2}(|\xi|) = e^{\pm|\xi|(a^\dagger b^\dagger - ba)} = e^{\frac{\pm i|\xi|}{2}\left((XP+PX)-(\widetilde{X}\widetilde{P}+\widetilde{P}\widetilde{X})\right)}$$

$$\rightarrow e^{\pm\frac{|\xi|}{2}\left((2x\frac{d}{dx}+1)-(2\widetilde{x}\frac{d}{d\widetilde{x}}+1)\right)},$$

$$\text{or,} \;\; U_{S_2}(|\xi|, x, \widetilde{x}) = \left(e^{\pm\frac{|\xi|}{2}} e^{\pm|\xi| x\frac{d}{dx}}\right) \times \left(e^{\mp\frac{|\xi|}{2}} e^{\mp|\xi| \widetilde{x}\frac{d}{d\widetilde{x}}}\right).$$

$$(3.238)$$

Here we have inverted the relations in (3.233) to write the exponential in terms of coordinates and momenta and, in the intermediate step we have written the unitary operator in the coordinate representation since it has to act on the ground state wave function. We also note that, although the constant terms inside the two parenthesis cancel each other, we have kept them explicitly for a reason which will become clear shortly.

It follows now that the transformed state obtained from the ground state of the two decoupled oscillators has the form

$$\psi_{S_2}(|\xi|, x, \widetilde{x}) = U_{S_2}(|\xi|, x, \widetilde{x})\psi_0(x, \widetilde{x})$$

$$
= \left(\frac{1}{\pi}\right)^{\frac{1}{4}} \left(e^{\pm\frac{|\xi|}{2}} e^{\pm|\xi|\, x\frac{\mathrm{d}}{\mathrm{d}x}}\right) e^{-\frac{1}{2}x^2} \times \left(\frac{1}{\pi}\right)^{\frac{1}{4}} \left(e^{\mp\frac{|\xi|}{2}} e^{\mp|\xi|\, \widetilde{x}\frac{\mathrm{d}}{\mathrm{d}\widetilde{x}}}\right) e^{-\frac{1}{2}\widetilde{x}^2}
$$

$$
= \left(\frac{e^{\pm2|\xi|}}{\pi}\right)^{\frac{1}{4}} e^{-\frac{1}{2}\left(e^{\pm|\xi|}x\right)^2} \times \left(\frac{e^{\mp2|\xi|}}{\pi}\right)^{\frac{1}{4}} e^{-\frac{1}{2}\left(e^{\mp|\xi|}\widetilde{x}\right)^2}
$$

$$
= \left(\frac{e^{\pm2|\xi|}}{\pi}\right)^{\frac{1}{4}} e^{-\frac{e^{\pm2|\xi|}}{2}x^2} \times \left(\frac{e^{\mp2|\xi|}}{\pi}\right)^{\frac{1}{4}} e^{-\frac{e^{\mp2|\xi|}}{2}\widetilde{x}^2}, \tag{3.239}
$$

which indeed coincides with (3.237) which we had guessed based on our analysis in (3.160) corresponding to the values $(\alpha = 0, \beta = e^{\pm2|\xi|})$ and $(\alpha = 0, \beta = e^{\mp2|\xi|})$ for the two terms. We note here that, in obtaining the form of the wave function in the intermediate step, we have used the operator relation in (3.223) derived earlier. Furthermore, we see from (3.235) that these are truly a two mode squeezed state, namely, the pair of variables (x, p) and $(\widetilde{x}, \widetilde{p})$ lead to a minimal uncertainty (see (3.236)). However, $(\Delta x) \neq (\Delta p)$ and $(\Delta\widetilde{x}) \neq (\Delta\widetilde{p})$ (unless $|\xi| = 0$ in which case there is no transformation) as we would expect for a coherent state. Rather, while (Δx) is squeezed (elongated), the conjugate variable (Δp) is elongated (squeezed) and conversely, while $(\Delta\widetilde{x})$ is elongated (squeezed), its conjugate variable $(\Delta\widetilde{p})$ is squeezed (elongated) in a mutually compensating manner leading to a minimum uncertainty. In particular, we note that if we identify the doubled degree of freedom with the tilde degree of freedom, $b = \widetilde{a}, b^\dagger = \widetilde{a}^\dagger$, the unitary transformation in (3.229) coincides with the Bogoliubov operator $U(\theta(\beta))$ in (3.67), which takes us from the zero temperature doubled oscillator vacuum to the thermal vacuum, with the identification $\theta(\beta) = \xi = \xi^* = |\xi|$. This is why we have repeatedly emphasized in our discussions that the thermal vacuum is like a (two mode) squeezed state.

3.9.4 Derivation of the propagator in Eq. (3.115).

We noted below (3.115) that the peculiar $i\epsilon$ prescription in the 22 element of the matrix results from the anti-time ordering in that term. This is surprising because in (3.114) we are actually calculating the vacuum expectation value of a time ordered product and the natural question is how can anti-time ordering creep into a time ordered product. The short answer to the change in the sign in the $i\epsilon$ term as well as in the overall sign can be traced to the tilde conjugation rule. Namely, since (see (3.113)-(3.114))

$$\left(iG_{T=0}^{(\mathrm{TFD})}\right)_{11}(x-y) = \langle 0, \widetilde{0}|T(\phi(x)\phi(y))|0, \widetilde{0}\rangle,$$

$$\left(iG_{T=0}^{(\mathrm{TFD})}\right)_{22}(x-y) = \langle 0, \widetilde{0}|T(\widetilde{\phi}(x)\widetilde{\phi}(y))|0, \widetilde{0}\rangle, \tag{3.240}$$

and since, in the momentum space we know that

$$\langle 0, \widetilde{0}|T(\phi(x)\phi(y))|0, \widetilde{0}\rangle \rightarrow \frac{i}{p^2 - m^2 + i\epsilon}, \tag{3.241}$$

it follows from the tilde conjugation relations in (3.88) that

$$\langle 0, \widetilde{0}|T(\widetilde{\phi}(x)\widetilde{\phi}(y))|0, \widetilde{0}\rangle \rightarrow -\frac{i}{p^2 - m^2 - i\epsilon}, \tag{3.242}$$

as we have in (3.115). However, this does not clarify how the tilde conjugation leads to anti-time ordering in the 22 element which we do in some detail in the following.

Let us recall the general field expansion for a Klein-Gordon field

$$\phi(x) = \int \mathrm{d}^3k \left(f_{\mathbf{k}}^{(+)}(x)\, a_{\mathbf{k}} + f_{\mathbf{k}}^{(-)}(x)\, a_{\mathbf{k}}^\dagger \right), \tag{3.243}$$

where

$$f_{\mathbf{k}}^{(\pm)}(x) = \frac{1}{\sqrt{(2\pi)^3 2\omega_{\mathbf{k}}}}\, e^{\mp i\omega_{\mathbf{k}}x^0 \pm i\mathbf{k}\cdot\mathbf{x}}, \quad \omega_{\mathbf{k}} = \sqrt{\mathbf{k}^2 + m^2}, \tag{3.244}$$

denote the positive and negative energy solutions of the Klein-Gordon equation. It follows from the tilde conjugation rule (3.88) that we can write the field expansion for the tilde fields to be Similarly, using the tilde conjugation relation, (3.88), we can write the expansion for the tilde fields to be

$$\widetilde{\phi}(x) = \int \mathrm{d}^3k \left(f_{\mathbf{k}}^{(-)}(x)\, \widetilde{a}_{\mathbf{k}} + f_{\mathbf{k}}^{(+)}(x)\, \widetilde{a}_{\mathbf{k}}^\dagger \right), \tag{3.245}$$

where we have used the fact that, under tilde conjugation, $f_{\mathbf{k}}^{(\pm)}(x) \rightarrow \left(f_{\mathbf{k}}^{(\pm)}(x)\right)^* = f_{\mathbf{k}}^{(\mp)}(x)$. Let us next calculate the expectation value of $\phi(x)\phi(y)$ in the zero temperature doubled vacuum of TFD which gives

$$\langle 0, \tilde{0} | \phi(x) \phi(y) | 0, \tilde{0} \rangle = \int d^3 k d^3 k' \, f_{\mathbf{k}}^{(+)}(x) f_{\mathbf{k}'}^{(-)}(y) \langle 0, \tilde{0} | a_{\mathbf{k}} a_{\mathbf{k}'}^\dagger | 0, \tilde{0} \rangle$$

$$= \int d^3 k d^3 k' \, f_{\mathbf{k}}^{(+)}(x) f_{\mathbf{k}'}^{(-)}(y) \delta^3(k - k') = \int d^3 k \, f_{\mathbf{k}}^{(+)}(x) f_{\mathbf{k}}^{(-)}(y)$$

$$= \int d^3 k \, \frac{e^{-i\omega_{\mathbf{k}}(x^0 - y^0) + i\mathbf{k}\cdot(\mathbf{x}-\mathbf{y})}}{(2\pi)^3 2\omega_{\mathbf{k}}}. \tag{3.246}$$

Similarly, we can calculate the expectation value of $\tilde{\phi}(x)\tilde{\phi}(y)$ in the zero temperature doubled vacuum of TFD and it leads to

$$\langle 0, \tilde{0} | \tilde{\phi}(x) \tilde{\phi}(y) | 0, \tilde{0} \rangle = \int d^3 k \, \frac{e^{i\omega_{\mathbf{k}}(x^0 - y^0) - i\mathbf{k}\cdot(\mathbf{x}-\mathbf{y})}}{(2\pi)^3 2\omega_{\mathbf{k}}}, \tag{3.247}$$

which is consistent with the tilde conjugation rules in (3.88).

We can now calculate the vacuum expectation values of time ordered products. First, we note that the expectation value of the time ordered product can be written as

$$\langle 0, \tilde{0} | T(\phi(x)\phi(y)) | 0, \tilde{0} \rangle$$

$$= \theta(x^0 - y^0) \langle 0, \tilde{0} | \phi(x)\phi(y) | 0, \tilde{0} \rangle + \theta(y^0 - x^0) \langle 0, \tilde{0} | \phi(y)\phi(x) | 0, \tilde{0} \rangle$$

$$= \int \frac{d^3 k}{(2\pi)^3 2\omega_{\mathbf{k}}} \left(\theta(x^0 - y^0) \, e^{-i\omega_{\mathbf{k}}(x^0 - y^0) + i\mathbf{k}\cdot(\mathbf{x}-\mathbf{y})} \right.$$

$$\left. + \theta(y^0 - x^0) \, e^{i\omega_{\mathbf{k}}(x^0 - y^0) - i\mathbf{k}\cdot(\mathbf{x}-\mathbf{y})} \right). \tag{3.248}$$

Furthermore, using the integral representation for the theta function,

$$\theta(x^0 - y^0) = \lim_{\epsilon \to 0+} \frac{i}{2\pi} \int dk^0 \, \frac{e^{-ik^0(x^0 - y^0)}}{k^0 + i\epsilon}, \tag{3.249}$$

equation (3.248) leads to

$$\langle 0, \tilde{0} | T(\phi(x)\phi(y)) | 0, \tilde{0} \rangle$$

$$= \lim_{\epsilon \to 0+} \int \frac{d^4 k}{(2\pi)^4} \frac{i}{2\omega_{\mathbf{k}}} \left(\frac{e^{-i(k^0 + \omega_{\mathbf{k}})(x^0 - y^0) + i\mathbf{k}\cdot(\mathbf{x}-\mathbf{y})}}{k^0 + i\epsilon} \right.$$

$$\left. + \frac{e^{i(k^0 + \omega_{\mathbf{k}})(x^0 - y^0) - i\mathbf{k}\cdot(\mathbf{x}-\mathbf{y})}}{k^0 + i\epsilon} \right)$$

$$= \lim_{\epsilon \to 0+} \int \frac{\mathrm{d}^4 k}{(2\pi)^4} \frac{e^{-ik\cdot(x-y)}}{2\omega_{\mathbf{k}}} \left(\frac{i}{k^0 - \omega_{\mathbf{k}} + i\epsilon} + \frac{i}{-k^0 - \omega_{\mathbf{k}} + i\epsilon} \right)$$

$$= \lim_{\epsilon \to 0+} \int \frac{\mathrm{d}^4 k}{(2\pi)^4} e^{-ik\cdot(x-y)} \frac{i}{k^2 - m^2 + i\epsilon}, \tag{3.250}$$

where we have used the relation $\omega_{\mathbf{k}}^2 = \mathbf{k}^2 + m^2$. This is, of course, the propagator for the free Klein-Gordon field in the coordinate space whose form in the momentum space can be read out from (3.250) to be

$$G(k) = \lim_{\epsilon \to 0+} \frac{i}{k^2 - m^2 + i\epsilon}, \tag{3.251}$$

which is the form given in (3.115) for the 11 element.

On the other hand, comparing (3.246) and (3.247) we see that the two differ in the sign of the exponent (remember, complex conjugation) and, therefore, putting in the integral representations for the theta function given in (3.249), the expectation value of the time ordered tilde fields leads to

$$\langle 0, \widetilde{0} | T(\widetilde{\phi}(x)\widetilde{\phi}(y)) | 0, \widetilde{0} \rangle$$

$$= \lim_{\epsilon \to 0+} \int \frac{\mathrm{d}^4 k}{(2\pi)^4} \frac{i}{2\omega_{\mathbf{k}}} \left(\frac{e^{-i(k^0 - \omega_{\mathbf{k}})(x^0 - y^0) - i\mathbf{k}\cdot(\mathbf{x} - \mathbf{y})}}{k^0 + i\epsilon} \right.$$

$$\left. + \frac{e^{i(k^0 - \omega_{\mathbf{k}})(x^0 - y^0) + i\mathbf{k}\cdot(\mathbf{x} - \mathbf{y})}}{k^0 + i\epsilon} \right)$$

$$= \lim_{\epsilon \to 0+} \int \frac{\mathrm{d}^4 k}{(2\pi)^4} \frac{e^{-ik\cdot(x-y)}}{2\omega_{\mathbf{k}}} \left(\frac{i}{k^0 + \omega_{\mathbf{k}} + i\epsilon} + \frac{i}{-k^0 + \omega_{\mathbf{k}} + i\epsilon} \right)$$

$$= \lim_{\epsilon \to 0+} \int \frac{\mathrm{d}^4 k}{(2\pi)^4} e^{-ik\cdot(x-y)} \frac{(-i)}{k^2 - m^2 - i\epsilon}, \tag{3.252}$$

which can be compared with the 22 element in (3.115). A different way of looking at this result is to note from (3.246) and (3.247) that $\langle 0, \widetilde{0} | \widetilde{\phi}(x)\widetilde{\phi}(y) | 0, \widetilde{0} \rangle$ behaves like $\langle 0, \widetilde{0} | \phi(y)\phi(x) | 0, \widetilde{0} \rangle$. As a result, the expectation value of time ordered tilde fields

$$\langle 0, \widetilde{0} | T(\widetilde{\phi}(x)\widetilde{\phi}(y)) | 0, \widetilde{0} \rangle$$

$$= \theta(x^0 - y^0) \langle 0, \widetilde{0} | \widetilde{\phi}(x)\widetilde{\phi}(y) | 0, \widetilde{0} \rangle + \theta(y^0 - x^0) \langle 0, \widetilde{0} | \widetilde{\phi}(y)\widetilde{\phi}(x) | 0, \widetilde{0} \rangle,$$

behaves like

$$\theta(x^0 - y^0)\langle 0, \widetilde{0}|\phi(y)\phi(x)|0, \widetilde{0}\rangle + \theta(y^0 - x^0)\langle 0, \widetilde{0}|\phi(x)\phi(y)|0, \widetilde{0}\rangle$$
$$= \langle 0, \widetilde{0}|T^*(\phi(x)\phi(y))|0, \widetilde{0}\rangle, \tag{3.253}$$

where T^* denotes the anti-time ordering operator. This brings out the connection between the tilde conjugation and the anti-time ordering in the tilde propagator.

3.10 References

S. N. Biswas and A. Das, Modern Physics Letters **A3**, 549 (1988).

C. M. Caves, Physical Review **D23**, 1693 (1981).

A. Das, Physica **A158**, 1 (1989).

A. Das, *Lectures on Quantum Mechanics* (second edition), Hindustan Book Agency and World Scientific Publishing (2011).

R. J. Glauber, Physical Review Letters **10**, 84 (1963).

R. Kubo, Journal of Physical Society of Japan **12**, 570 (1957).

N . P. Landsman and Ch. G. Van Weert, Physics Reports **145**, 141 (1987).

P. C. Martin and J. Schwinger, Physical Review **115**, 1342 (1959).

A. Niemi and G. Semenoff, Annals of Physics **152**, 105 (1984).

E. C. G. Sudarshan, Physical Review Letters **10**, 277 (1963).

Y. Takahashi and H. Umezawa, Collective Phenomena **2**, 55 (1975).

H. Umezawa, H. Matsumoto and M. Tachiki, *Thermofield Dynamics and Condensed States*, North-Holland (1982).

CHAPTER 4

A general contour in the complex t-plane

4.1 Introduction

In the last three chapters, we have studied three very different formalisms to describe field theories in equilibrium at finite temperature. We have also seen that they can be described by three different time contours in the complex t-plane. For example, the imaginary time formalism is described by a straight path (see Fig. 1.1) along the negative imaginary time axis from $t = (\operatorname{Re} t, \operatorname{Im} t) = (0,0)$ to $t = (0, -\beta)$. On the other hand, in the closed time path formalism the time contour goes from a $t = (T, 0)$ to $(T', 0)$ in a straight path along the real time axis, then comes back in a straight path along the real time axis from $t = (T', 0)$ to $t = (T, 0)$ (which is why it is called the closed time path formalism) and finally goes along a straight path along the negative imaginary time axis from $t = (T, 0)$ to $t = (T, -\beta)$ (see Fig. 2.1). Thermofield dynamics, which primarily started out as an operator description with doubled fields, can also be described with a theory defined on a contour in the complex path originating from $t = (T, 0)$ and going in a straight path along the real time axis to $t = (T', 0)$, then going along the negative imaginary time axis from $t = (T', 0)$ to $t = (T', -\frac{\beta}{2})$, then coming back along the real axis from $t = (T', -\frac{\beta}{2})$ to $t = (T, -\frac{\beta}{2})$ and finally going along a straight path along the negative imaginary time axis from $t = (T, -\frac{\beta}{2})$ to $t = (T, -\beta)$ (see Fig. 3.1). In both closed time path formalism and thermofield dynamics, it is assumed that T is a large negative (real) time while T' is a large positive (real) time tending to $\mp\infty$ respectively and $\beta \, (= \frac{1}{kT})$ represents the inverse temperature of the system in units of the Boltzmann constant k (namely, when $k = 1$).

There are several questions that arise naturally at this point. The first, of course, is whether the paths discussed in Fig. 2.1 and Fig. 3.1 are the only two possible contours in the complex t-plane which can lead to a path integral description of real time thermal field theories or whether other more general contours may also be possible. Indeed it is now known that any general path in the complex t-plane of the form shown in Fig. 4.1 where it is assumed that all the σ_ns are

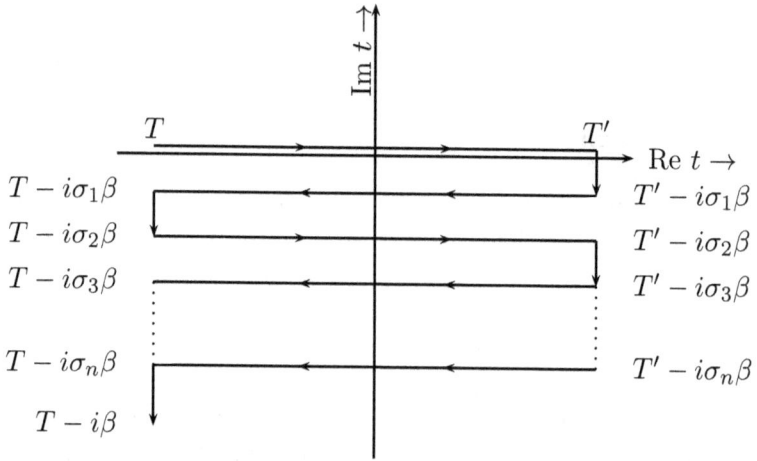

Figure 4.1: General contour in the complex t-plane.

real and positive and satisfy $0 \leq (\sigma_1+\sigma_2+\cdots+\sigma_n) \leq 1$. Namely, the general path consists of a series of real branches criss-crossing along

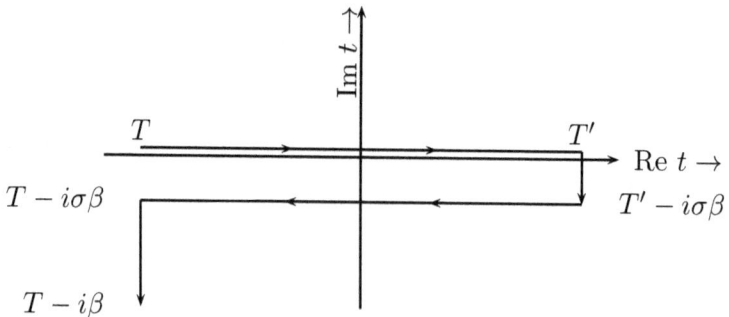

Figure 4.2: A simplified general contour in the complex t-plane.

opposite directions with the paths along the imaginary time axis constantly descending towards more and more negative (imaginary) time values. This is necessary for the convergence of the Boltzmann factor (in the thermal distribution) along such paths. Any such path would lead to a path integral description of thermal field theories. However, for practical purposes, we can simply consider the general graph shown in Fig. 4.2. This simplification can be easily understood from the fact that the thermal expectation value we calculate is that of a physical operator which lies on the first real branch. Furthermore, the group properties of the time evolution operators applied to the neglected branches in Fig. 4.2 compared to Fig. 4.1 ($n - 1$ of them) simply lead to

$$
\begin{aligned}
&U(T - i\beta, T - i\sigma_n\beta)U(T - i\sigma_n\beta, T' - i\sigma_n\beta) \\
&\quad \times U(T' - i\sigma_n\beta, T' - i\sigma_{n-1}\beta)U(T' - i\sigma_{n-1}\beta, T - i\sigma_{n-1}\beta) \\
&\quad \times \cdots \cdots \\
&\quad \times U(T' - i\sigma_2\beta, T - i\sigma_2\beta)U(T - i\sigma_2\beta, T - i\sigma_1\beta) \\
&= U(T - i\beta, T - i\sigma_1\beta).
\end{aligned}
\tag{4.1}
$$

We can now identify $\sigma_1 = \sigma$ which leads to Fig. 4.2. Here it is understood that $0 \leq \sigma \leq 1$. (Note that to go to the point $T - i\beta$ in the complex t-plane starting from T, at least, two real branches are needed in the real time formalism.) This simplified path can also be thought of as resulting from a desire to keep to a two component description of the thermal field theories which would, otherwise, lead to a multicomponent field theory (with multiple real branches), something we would like to avoid for simplicity. Let us note here that when $\sigma = 0$, the path (and the thermal description) coincides with the closed time path (see Fig. 2.1) while when $\sigma = \frac{1}{2}$, the path in Fig. 4.2 reduces to the path of thermofield dynamics (see Fig. 3.1).

Let us note here that, while all such paths in Fig. 4.2 (with $0 \leq \sigma \leq \beta$) lead to a path integral description for thermal field theories, we do know that thermofield dynamics, discussed in detail in the last chapter, also has an operator description which is very useful in studying various fundamental questions. This raises an interesting question, namely, whether thermofield dynamics is unique in that respect or whether thermal field theories associated with all such paths (an infinity of them) also have operator descriptions. It would also be nice to know if thermofield dynamics is special in some respect. This is the question which we will discuss in detail in this chapter. However, before doing that we note that if we define the 2×2 matrix propagator (in the momentum space) as

$$iG^{(\sigma)}(p) = \begin{pmatrix} iG_{11}^{(\sigma)\,\beta}(p) & iG_{12}^{(\sigma)\,\beta}(p) \\ iG_{21}^{(\sigma)\,\beta}(p) & iG_{22}^{(\sigma)\,\beta}(p) \end{pmatrix}, \tag{4.2}$$

where the components have the explicit forms

$$iG_{11}^{(\sigma)\,\beta}(p) = \frac{i}{p^2 - m^2 + i\epsilon} + 2\pi n_B(|p^0|)\delta(p^2 - m^2)$$
$$= iG_{++}(p) = iG_{11}^{(\mathrm{TFD})}(p),$$

$$iG_{12}^{(\sigma)\,\beta}(p) = 2\pi e^{\sigma\beta p^0}(\theta(-p^0) + n_B(|p^0|))\delta(p^2 - m^2)$$
$$= e^{\sigma\beta p^0}iG_{+-}(p) = e^{\lambda\beta p^0}iG_{12}^{(\mathrm{TFD})}(p),$$

$$iG_{21}^{(\sigma)\,\beta}(p) = 2\pi e^{-\sigma\beta p^0}(\theta(p^0) + n_B(|p^0|))\delta(p^2 - m^2)$$
$$= e^{-\sigma\beta p^0}iG_{-+}(p) = e^{-\lambda\beta p^0}iG_{21}^{(\mathrm{TFD})}(p),$$

$$iG_{22}^{(\sigma)\,\beta}(p) = -\frac{i}{p^2 - m^2 - i\epsilon} + 2\pi n_B(|p^0|)\delta(p^2 - m^2)$$
$$= iG_{--}(p) = iG_{22}^{(\mathrm{TFD})}(p), \tag{4.3}$$

where we have defined

$$\lambda = \sigma - \frac{1}{2}, \quad -\frac{1}{2} \le \lambda \le \frac{1}{2}, \tag{4.4}$$

where the last relation follows from the fact that $0 \le \sigma \le 1$. For the arbitrary σ path described in Fig. 4.2, we can now write the thermal propagator explicitly in the matrix form as

$$iG^{(\sigma)}(p) = \begin{pmatrix} \frac{i}{p^2 - m^2 + i\epsilon} & 0 \\ 0 & -\frac{i}{p^2 - m^2 - i\epsilon} \end{pmatrix}$$

$$+ 2\pi n_B(|p^0|)\delta(p^2 - m^2) \begin{pmatrix} 1 & e^{\beta(\frac{|p^0|}{2} + \lambda p^0)} \\ e^{\beta(\frac{1}{2}|p^0| - \lambda p^0)} & 1 \end{pmatrix}, \tag{4.5}$$

where λ is defined in (4.4). We note several things from the structures in (4.3). First, the diagonal components of the propagator, both in the first as well as the second matrix, are independent of the value of σ so that they are the same for the closed time path, thermofield dynamics and for any other path with an arbitrary σ. However, the off-diagonal elements in the second matrix do depend on the value of σ. For $\sigma = 0\,(\lambda = -\frac{1}{2})$, the off-diagonal elements in (4.3) (or (4.5)) reduce to the off-diagonal elements of the closed time path Green's functions in (2.50) (or (2.52)). Furthermore, when $\sigma = \frac{1}{2}\,(\lambda = 0)$, the propagator in (4.3) (or (4.5)) reduces to the propagator for thermofield dynamics given in (3.119) (or (3.127)). Therefore, we see that thermofield dynamics is special in the sense that, in the zero temperature limit, the propagator is diagonal as we would expect for two decoupled independent fields (the thermal doublets). For any other path, off-diagonal elements in the zero temperature limit creep in. With all this preparation, we are now ready to ask whether an operator description exists for an arbitrary path in the real time formalism as shown in Fig. 4.2.

4.2 Bogoliubov transformation for an arbitrary path

As we have already seen in the last chapter, in thermofield dynamics which naturally leads to an operator description of thermal fields, we define a doubled Hilbert space and a thermal vacuum $|0(\beta)\rangle$ in this space. The thermal vacuum is related to the vacuum $|0,\tilde{0}\rangle$ in this doubled space through a Bogoliubov transformation $U(\theta(\beta))$ (see, for example, (3.39)). This Bogoliubov transformation, in turn, rotates the doublet of (doubled) operators in the non tilde-tilde space (see, for example, (3.45)-(3.48) for a fermionic oscillator or (3.70)-(3.74) for a bosonic oscillator or (3.113)-(3.120) for the Klein-Gordon field theory) leading to

$$\Phi(x) \to \overline{U}(-\theta)\Phi(x), \tag{4.6}$$

where, in Fourier components, for example, $\overline{U}(-\theta_{\mathbf{p}})$ is given in (3.118). We also note from (3.116) (or (3.118)) that, in the operator description, the propagator factorizes as shown in (3.116). This is very important because it says that, if an operator description exists for an arbitrary path as in Fig. 4.2, the propagator must factorize as in

(3.116) (with possibly a more complicated matrix with distinct properties). And this matrix may then lead us to a Bogoliubov transformation defining the appropriate thermal vacuum in this case.

Let us note some of the properties of the propagator matrix in (4.5) before proceeding. The diagonal elements are, of course symmetric under reflection of the energy-momentum, namely,

$$iG_{11}^{(\sigma)\,\beta}(-p) = iG_{11}^{(\sigma)\,\beta}(p), \qquad iG_{22}^{(\sigma)\,\beta}(-p) = iG_{22}^{(\sigma)\,\beta}(p). \qquad (4.7)$$

However, for a nontrivial λ, the off-diagonal elements are not symmetric, rather, go into each other, Namely,

$$iG_{12}^{(\sigma)\,\beta}(-p) = iG_{21}^{(\sigma)\,\beta}(p), \qquad iG_{21}^{(\sigma)\,\beta}(-p) = iG_{12}^{(\sigma)\,\beta}(p). \qquad (4.8)$$

The off-diagonal terms are symmetric as well only for $\lambda = 0$ or, eqivalently $\sigma = \frac{1}{2}$ (see (4.4) for the relation between the two). This shows that thermofield dynamics ($\sigma = \frac{1}{2}$) is special in this sense, namely, the propagator matrix is symmetric in energy-momentum which is not true for any other path. Therefore, this should also have some reflection in the Bogoliubov operator in an operator description of such systems. It can be checked that the 2×2 propagator matrix in (4.5) can be factorized as

$$iG^{(\sigma)}(p) = \overline{U}(-\theta_{\mathbf{p}}, \lambda) iG_{(T=0)}^{(\mathrm{TFD})}(p)(\overline{U}(-\theta_{\mathbf{p}}, -\lambda))^T$$

$$= \overline{U}(-\theta_{\mathbf{p}}, \lambda) iG_{(T=0)}^{(\mathrm{TFD})}(p)\overline{U}(-\theta_{\mathbf{p}}, \lambda), \qquad (4.9)$$

where

$$\overline{U}(-\theta_{\mathbf{p}}, \lambda) = \begin{pmatrix} \cosh\theta_{\mathbf{p}} & e^{\lambda\beta p^0}\sinh\theta_{\mathbf{p}} \\ e^{-\lambda\beta p^0}\sinh\theta_{\mathbf{p}} & \cosh\theta_{\mathbf{p}} \end{pmatrix}, \qquad (4.10)$$

and $G_{(T=0)}^{(\mathrm{TFD})}(p)$ is given in (3.115). We note here, for future discussions in chapter **6**, that

$$\lambda \to -\lambda, \quad \text{is also equivalent to} \quad p^0 \to -p^0, \qquad (4.11)$$

(without changing λ) in the matrix in (4.10). Furthermore, $\cosh\theta_{\mathbf{p}}$ and $\sinh\theta_{\mathbf{p}}$ are defined in (3.101). For $\sigma = \frac{1}{2}$ (or $\lambda = 0$), (4.9) and (4.10) indeed reduce to the case of thermofield dynamics (see (3.116)) while, for any other value of σ, the off-diagonal elements of the matrix $\overline{U}(-\theta_{\mathbf{p}})$ are different leading to different off-diagonal components in the thermal Green's function (propagator). It follows, therefore, that, while the matrix $\overline{U}(-\theta_{\mathbf{p}}, \lambda = 0)$ corresponding to the rotation matrix in thermofield dynamics (see (3.116)) is symmetric, for an arbitrary path it is not. Correspondingly, the Bogoliubov transformation for an arbitrary path will be different leading to a different thermal vacuum.

We note from (4.10) that the matrix $\overline{U}(-\theta_{\mathbf{p}}, \lambda)$ satisfies the properties

$$\overline{U}(-\theta_{\mathbf{p}}, \lambda) = (\overline{U}(-\theta_{\mathbf{p}}, -\lambda))^{T},$$

$$\overline{U}(-\theta_{\mathbf{p}}, \lambda)\overline{U}(\theta_{\mathbf{p}}, \lambda) = \mathbb{1}, \quad \overline{U}^{T}(-\theta_{\mathbf{p}}, -\lambda)\sigma_{3}\overline{U}(-\theta_{\mathbf{p}}, \lambda) = \sigma_{3},$$

$$(4.12)$$

the first of which we have already used in (4.9). Let us also note that the matrix in (4.10) can be factorized further as

$$\overline{U}(-\theta_{\mathbf{p}}, \lambda) = \overline{V}(\lambda)\overline{U}(-\theta_{\mathbf{p}})(\overline{V}(\lambda))^{-1}, \qquad (4.13)$$

where $\overline{U}(-\theta_{\mathbf{p}}) = \overline{U}(-\theta_{\mathbf{p}}, \lambda = 0)$ corresponds to the rotation matrix in thermofield dynamics given in (3.116) and the matrix $\overline{V}(\lambda)$ can have one of the two forms, either

$$\overline{V}(\lambda) = \begin{pmatrix} 1 & 0 \\ 0 & e^{-\lambda\beta p^{0}} \end{pmatrix}, \qquad (4.14)$$

or,

$$\overline{V}(\lambda) = \begin{pmatrix} 0 & 1 \\ e^{-\lambda\beta p^{0}} & 0 \end{pmatrix} = \begin{pmatrix} 1 & 0 \\ 0 & e^{-\lambda\beta p^{0}} \end{pmatrix} \begin{pmatrix} 0 & 1 \\ 1 & 0 \end{pmatrix}, \qquad (4.15)$$

with

$$(\overline{V}(\lambda))^{-1} = (\overline{V}(-\lambda))^{T} = (\overline{V}(-\lambda))^{\dagger}, \qquad (4.16)$$

so that

$$(\overline{V}(-\lambda))^{\dagger}\overline{V}(\lambda) = \mathbb{1} = \overline{V}(\lambda)(\overline{V}(-\lambda))^{\dagger}, \qquad (4.17)$$

in either of the two cases. As a result, the factorizing matrix $\overline{V}(\lambda)$, like the original matrix $\overline{U}(-\theta_{\mathbf{p}}, \lambda)$, is not unitary in the Dirac sense.

However, we note that, if we define a matrix (for either of the two matrices in (4.15))

$$\Lambda(\lambda) = \overline{V}(-\lambda), \quad \Lambda^\dagger(\lambda) = (\overline{V}(-\lambda))^\dagger = (\overline{V}(\lambda))^{-1}, \tag{4.18}$$

then, it follows from (4.16)-(4.17) that

$$\Lambda^\dagger(\lambda)\overline{V}(\lambda)\Lambda(\lambda) = \overline{V}(-\lambda), \tag{4.19}$$

namely, $\Lambda(\lambda)$ can be thought of as the matrix (reflecting) changing $\lambda \to -\lambda$ acting on the matrix $\overline{V}(\lambda)$. This allows us to define a modified Hermitian conjugate for $\overline{V}(\lambda)$ as

$$\overline{V}^\ddagger(\lambda) = \Lambda^\dagger(\lambda)\overline{V}^\dagger(\lambda)\Lambda(\lambda) = \overline{V}^{-1}(\lambda), \tag{4.20}$$

making it unitary, namely,

$$\overline{V}^\ddagger(\lambda)\overline{V}(\lambda) = \mathbb{1} = \overline{V}(\lambda)\overline{V}^\ddagger(\lambda). \tag{4.21}$$

Equation (4.13) suggests that the Bogoliubov transformation for the case of an arbitrary path may involve a product of operators. To investigate this issue further, let us recall that the thermal ensemble average of a product of operators is defined as (see (1.13))

$$\langle A_1 A_2 \cdots A_n \rangle_\beta = \mathrm{Tr}\,(\rho(\beta) A_1 A_2 \cdots A_n) = \frac{\mathrm{Tr}\,(e^{-\beta H} A_1 A_2 \cdots A_n)}{\mathrm{Tr}\, e^{-\beta H}}$$

$$= Z^{-1}(\beta)\,\mathrm{Tr}\,(e^{-\beta H} A_1 A_2 \cdots A_n). \tag{4.22}$$

Furthermore, cyclicity under a trace allows us to introduce an arbitrary similarity transformation S (not necessarily unitary) into the ensemble average as

$$\langle A_1 A_2 \cdots A_n \rangle_\beta = Z^{-1}(\beta)\mathrm{Tr}(S^{-1}(\alpha)e^{-\beta H} A_1 A_2 \cdots A_n S(\alpha)). \tag{4.23}$$

If we identify the arbitrary similarity transformation with $S(\alpha) = e^{\alpha\beta H}$, where α is a real arbitrary (constant) dimensionless parameter, we can write (4.23) also as

$$\langle A_1 A_2 \cdots A_n \rangle_\beta = Z^{-1}(\beta)\mathrm{Tr}(e^{-\beta H} e^{-\alpha\beta H} A_1 A_2 \cdots A_n e^{\alpha\beta H})$$

$$= \langle e^{-\alpha\beta H} A_1 A_2 \cdots A_n e^{\alpha\beta H} \rangle_\beta. \tag{4.24}$$

Here we have used the fact that $e^{-\beta H}$ and $e^{-\alpha\beta H}$ commute. Let us note here, for later use, that the similarity transformation $S(\alpha)$ is not unitary in the conventional sense, namely,

$$S^\dagger(\alpha) = e^{\alpha\beta H} = S(\alpha) \neq S^{-1}(\alpha), \tag{4.25}$$

since $H^\dagger = H$, leading to

$$S^\dagger(\alpha)S(\alpha) \neq \mathbb{1}, \quad S(\alpha)S^\dagger(\alpha) \neq \mathbb{1}. \tag{4.26}$$

On the other hand, we also note, for later use, that

$$S^\dagger(-\alpha)S(\alpha) = e^{-\alpha\beta H}e^{\alpha\beta H} = \mathbb{1} = S(\alpha)S^\dagger(-\alpha), \tag{4.27}$$

very much like in (4.16)-(4.18).

In an operator description, the thermal ensemble average can be written as a thermal vacuum expectation value. For example, in thermofield dynamics, we know that we can write the ensemble average of a product of operators as

$$\langle A_1 A_2 \cdots \rangle_\beta = \langle 0, \beta | A_1 A_2 \cdots A_n | 0, \beta \rangle. \tag{4.28}$$

which, upon using the cyclicity of trace (see (4.24)), takes also the form

$$= \langle 0, \beta | e^{-\alpha\beta H} A_1 A_2 \cdots A_n e^{\alpha\beta H} | 0, \beta \rangle, \tag{4.29}$$

$$= \langle 0, \tilde{0} | U^\dagger(\theta) e^{-\alpha\beta H} A_1 A_2 \cdots A_n e^{\alpha\beta H} U(\theta) | 0, \tilde{0} \rangle. \tag{4.30}$$

Here $|0, \beta\rangle$ is the thermal vacuum in the doubled Hilbert space defined, in thermofield dynamics, as $|0, \beta\rangle = U(\theta)|0, \tilde{0}\rangle$ with $U(\theta), \theta = \theta(\beta)$ denoting the Bogoliubov transformation discussed in detail in the last chapter. Let us also recall here from (3.33) as well as (3.64)-(3.65) that the state $|0, \tilde{0}\rangle$ simply corresponds to the zero temperature limit of the thermal vacuum state in thermofield dynamics, namely,

$$|0, \tilde{0}\rangle = |0\rangle \otimes |\tilde{0}\rangle = |0, \beta\rangle|_{T=0}. \tag{4.31}$$

Therefore, we conclude that if an operator description exists for an arbitrary path, we should be able to write the ensemble average in (4.23) also as a thermal vacuum expectation value. It is clear from (4.24) and (4.28)-(4.30) that there are two equivalent ways of obtaining the thermal ensemble average as a thermal vacuum expectation value for an arbitrary path (for an operator description). First, we can look for a new thermal vacuum, different from that of thermofield dynamics which not only carries temperature dependence, but path dependence as well and calculate the expectation value of the doubled zero temperature operators in this (new) thermal vacuum, namely, we can write (4.24) as

$$\langle A_1 A_2 \cdots A_n \rangle_\beta = \langle 0, (\beta, \lambda) | A_1 A_2 \cdots A_n | 0, (\beta, \lambda) \rangle, \qquad (4.32)$$

where

$$|0, (\beta, \lambda)\rangle = U(\theta, \lambda) |0, \widetilde{0}\rangle. \qquad (4.33)$$

Here $U(\theta, \lambda)$ denotes the Bogoliubov transformation (to be determined) which takes the zero temperature doubled vacuum to the new thermal vacuum (for an arbitrary path). This will be more in the spirit of thermofield dynamics. The second equivalent way to proceed is to leave the thermal vacuum of thermofield dynamics, $|0, \beta\rangle$, unchanged and define new (temperature and path dependent) doubled operators through an appropriate transformation whose expectation value can be calculated in the thermal vacuum of thermofield dynamics. This, therefore, would be more in the spirit of defining operators associated with the arbitrary path shown in Fig. 4.2 while the thermal vacuum is that of a fixed path corresponding to $\sigma = \frac{1}{2}$. We note that, while in the first approach, the entire temperature and path dependence is contained in the choice of the thermal vacuum (and the operators are temperature and path independent), in the second approach part of the temperature dependence resides in the thermal vacuum of thermofield dynamics (corresponding to a fixed path) and the rest of temperature and path dependence is transferred to the operators. We will discuss both these procedures in some detail in the following. First, we will describe the first procedure, namely, we will construct the new thermal vacuum which would lead to the propagators we have derived for the arbitrary path and this will bring out some new features that arise in the case of an arbitrary path. Then, we will discuss the second method, namely, we will construct the new thermal operators whose expectation value in the thermal vacuum of thermofield dynamics leads to the thermal ensemble average.

4.2.1 Defining a thermal vacuum for the arbitrary path. As we have noted after (4.15), the factorization of the rotation matrix $\overline{U}(-\theta_{\mathbf{p}}, \lambda)$

in (4.13) suggests that the Bogoliubov transformation, in the operator description for an arbitrary path, may also have a product structure. Furthermore, from (4.13) as well as from (4.23)-(4.24), we suspect that the Bogoliubov transformation can be written as

$$U(\theta, \lambda) = S(\lambda) U(\theta) S^{-1}(\lambda) = e^{\lambda \beta H} U(\theta) e^{-\lambda \beta H}, \qquad (4.34)$$

where $U(\theta)$ denotes the Bogoliubov transformation for thermofield dynamics. In this case, the new thermal vacuum will have the form

$$|0, (\beta, \lambda)\rangle = U(\theta, \lambda)|0, \widetilde{0}\rangle = e^{\lambda \beta H} U(\theta) e^{-\lambda \beta H}|0, \widetilde{0}\rangle$$

$$= e^{\lambda \beta H} U(\theta)|0, \widetilde{0}\rangle = e^{\lambda \beta H}|0, \beta\rangle, \qquad (4.35)$$

where we have used $H|0, \widetilde{0}\rangle = 0$ as well as our earlier result that $U(\theta)|0, \widetilde{0}\rangle = |0, \beta\rangle$, the thermal vacuum of thermofield dynamics. Let us note that, since $\lambda = \sigma - \frac{1}{2}$ is a real parameter and since the Hamiltonian is assumed to be Hermitian, $H^\dagger = H$, it follows from (4.35) that

$$\langle 0, (\beta, \lambda)| = \langle 0, \widetilde{0}|U^\dagger(\theta) e^{\lambda \beta H} = \langle 0, \beta| e^{\lambda \beta H}, \qquad (4.36)$$

so that

$$\langle 0, (\beta, \lambda)|A_1 A_2 \cdots A_n|0, (\beta, \lambda)\rangle$$

$$= \langle 0, \beta| e^{\lambda \beta H} A_1 A_2 \cdots A_n e^{\lambda \beta H}|0, \beta\rangle$$

$$= \langle e^{\lambda \beta H} A_1 A_2 \cdots A_n e^{\lambda \beta H}\rangle_\beta. \qquad (4.37)$$

Clearly, this does not coincide with (4.28)-(4.30) and, therefore, the thermal ensemble average can not be written as a conventional thermal vacuum expectation value in the vacuum state defined in (4.35) when $\lambda \neq 0$.

The root of the problem can be traced to the fact that, while the Bogoliubov transformation for thermofield dynamics (corresponding to $\lambda = 0$) is unitary, $U^\dagger(\theta)U(\theta) = \mathbb{1} = U(\theta)U^\dagger(\theta)$, when $\lambda \neq 0$, the Bogoliubov transformation is not, namely, we see from (4.34) that

$$U^\dagger(\theta, \lambda)U(\theta, \lambda) = e^{-\lambda \beta H} U^\dagger(\theta) e^{\lambda \beta H} e^{\lambda \beta H} U(\theta) e^{-\lambda \beta H} \neq \mathbb{1}, \quad (4.38)$$

and, similarly, $U(\theta, \lambda)U^\dagger(\theta, \lambda) \neq \mathbb{1}$. As a result, we note from (4.38), that the thermal vacuum, defined as in (4.35) is not normalized. On the other hand, we also see, from (4.38), that

$$U^\dagger(\theta, -\lambda)U(\theta, \lambda) = e^{\lambda\beta H}U^\dagger(\theta)e^{-\lambda\beta H}e^{\lambda\beta H}U(\theta)e^{-\lambda\beta H}$$

$$= e^{\lambda\beta H}U^\dagger(\theta)U(\theta)e^{-\lambda\beta H} = e^{\lambda\beta H}e^{-\lambda\beta H}$$

$$= \mathbb{1} = U(\theta, \lambda)U^\dagger(\theta, -\lambda). \tag{4.39}$$

As a result, we conclude that

$$\langle 0, (\beta, -\lambda)|0, (\beta, \lambda)\rangle = \langle 0, \widetilde{0}|U^\dagger(\theta, -\lambda)U(\theta, \lambda)|0, \widetilde{0}\rangle = 1. \tag{4.40}$$

All of this is very similar to what we have seen earlier in (4.17) with the factorizing matrices. Therefore, this suggests that if we have a reflection operator for the parameter λ,

$$\lambda \xrightarrow{\Lambda} -\lambda, \quad \Lambda^2 = \mathbb{1}, \quad \Lambda^\dagger = \Lambda = \Lambda^{-1}, \tag{4.41}$$

then, we can define, as in (4.20)-(4.21), a modified adjoint

$$U^\ddagger(\theta, \lambda) = \Lambda^\dagger U^\dagger(\theta, \lambda)\Lambda = U^\dagger(\theta, -\lambda) = U^{-1}(\theta, \lambda), \tag{4.42}$$

which will lead to

$$U^\ddagger(\theta, \lambda)U(\theta, \lambda) = \mathbb{1} = U(\theta, \lambda)U^\ddagger(\theta, \lambda). \tag{4.43}$$

Let us recall that properties, such as the adjoint of an operator or the unitarity of an operator etc., are defined within the context of the Hilbert space of states and the inner product defined on that space. The standard Dirac inner product leads to the conventional adjoint of operators. However, from our discussions above, (4.38)-(4.43), it appears that, for $\lambda \neq 0$, the inner product on the space of states may be different from the standard Dirac inner product. Let us suppose that the inner product on the Hilbert space of states for $\lambda \neq 0$ is defined by

$$\langle \phi|\psi\rangle_\eta = \langle \phi|\eta|\psi\rangle, \tag{4.44}$$

with the normalization of states given by

$$\langle \psi|\psi\rangle_\eta = \langle \psi|\eta|\psi\rangle = 1, \tag{4.45}$$

where η is known as the (nontrivial) metric operator for the Hilbert space. Therefore, if we identify the metric $\eta = \Lambda = \Lambda^\dagger$, where Λ is the reflection operator defined in (4.41), then it follows that

$$\langle \phi | \psi \rangle_\Lambda = \langle \phi | \Lambda | \psi \rangle = \langle \phi | \Lambda^\dagger | \psi \rangle, \tag{4.46}$$

which leads, in particular to

$$\langle 0, (\beta, \lambda) | 0, (\beta, \lambda) \rangle_\Lambda = \langle 0, (\beta, \lambda) | \Lambda | 0, (\beta, \lambda) \rangle$$
$$= \langle 0, (\beta, -\lambda) | 0, (\beta, \lambda) \rangle = 1. \tag{4.47}$$

This can be compared with (4.40). Furthermore, we note that if we define the adjoint of an operator $\mathcal{O}$ in this space as $\mathcal{O}^\#$, it must satisfy

$$\langle \mathcal{O}^\# \phi | \psi \rangle_\Lambda = \langle \phi | \mathcal{O} \psi \rangle_\Lambda = \langle \phi | \Lambda^\dagger \mathcal{O} \psi \rangle = \langle \phi | \Lambda^\dagger \mathcal{O} \Lambda \Lambda^\dagger \psi \rangle$$
$$= \langle \phi | \Lambda^\dagger \mathcal{O} \Lambda | \Lambda^\dagger \psi \rangle = \langle \phi | (\Lambda^\dagger \mathcal{O}^\dagger \Lambda)^\dagger | \Lambda^\dagger \psi \rangle$$
$$= \langle (\Lambda^\dagger \mathcal{O}^\dagger \Lambda) \phi | \Lambda^\dagger \psi \rangle = \langle (\Lambda^\dagger \mathcal{O}^\dagger \Lambda) \phi | \psi \rangle_\Lambda, \tag{4.48}$$

leading to the adjoint of an operator in this space to be

$$\mathcal{O}^\# = \Lambda^\dagger \mathcal{O} \Lambda. \tag{4.49}$$

It follows now that the adjoint of the Bogoliubov transformation can be written as (Λ simply reflects the parameter λ, namely, $\lambda \to -\lambda$)

$$U^\#(\theta, \lambda) = \Lambda^\dagger U^\dagger(\theta, \lambda) \Lambda = U^\dagger(\theta, -\lambda) = U^\ddagger(\theta, \lambda), \tag{4.50}$$

as defined earlier in (4.42) and satisfying the unitarity condition (4.43)

$$U^\ddagger(\theta, \lambda) U(\theta, \lambda) = \mathbb{1}. \tag{4.51}$$

Therefore, to summarize, we note that when $\lambda \neq 0$, the thermal Hilbert space can be defined and develops a new character in that the inner product is no longer a Dirac space (with a standard inner product), but is a Hilbert space with a nontrivial metric Λ which reduces to identity $\mathbb{1}$ when $\lambda = 0$. The adjoint with respect to this (modified) inner product makes the Bogoliubov transformation unitary. In particular, we note that, for $\lambda \neq 0$, the thermal Hilbert space and the operators acting on it satisfy

$$\langle \phi | \psi \rangle_\Lambda = \langle \phi | \Lambda | \psi \rangle, = \langle \phi | \Lambda^\dagger | \psi \rangle, \qquad (4.52)$$

with the thermal vacuum given by

$$|0, (\beta, \lambda)\rangle = U(\theta, \lambda)|0, \widetilde{0}\rangle, \quad U(\theta, \lambda) = S(\lambda)U(\theta)S^{-1}(\lambda), \quad (4.53)$$

$$(|0, (\beta, \lambda)\rangle)^\ddagger = \langle 0, \widetilde{0}|U^\ddagger(\theta, \lambda) = \langle 0, (\beta, -\lambda)|, \qquad (4.54)$$

$$U^\ddagger(\theta, \lambda) = U^\dagger(\theta, -\lambda), \quad \langle 0, (\beta, -\lambda)|0, (\beta, \lambda)\rangle = 1, \qquad (4.55)$$

$$U^\ddagger(\theta, \lambda)U(\theta, \lambda) = \mathbb{1} = U(\theta, \lambda)U^\ddagger(\theta, \lambda), \qquad (4.56)$$

with $S(\lambda) = e^{\lambda \beta H}$ (defined in (4.34)). Furthermore, with these properties of the thermal vacuum, it is straightforward to show that

$$\begin{aligned}
&\langle 0, (\beta, -\lambda)|A_1 A_2 \cdots A_n|0, (\beta, \lambda)\rangle \\
&= \langle 0, \widetilde{0}|U^\ddagger(\theta, \lambda)A_1 A_2 \cdots A_n U(\theta, \lambda)|0, \widetilde{0}\rangle \\
&= \langle 0, \widetilde{0}|e^{\lambda \beta H}U^\dagger(\theta)e^{-\lambda \beta H}A_1 A_2 \cdots A_n e^{\lambda \beta H}U(\theta)e^{-\lambda \beta H}|0, \widetilde{0}\rangle \\
&= \langle 0, \widetilde{0}|U^\dagger(\theta)e^{-\lambda \beta H}A_1 A_2 \cdots A_n e^{\lambda \beta H}U(\theta)|0, \widetilde{0}\rangle \\
&= \langle 0, \beta|e^{-\lambda \beta H}A_1 A_2 \cdots A_n e^{\lambda \beta H}|0, \beta\rangle \\
&= \langle e^{-\lambda \beta H}A_1 A_2 \cdots A_n e^{\lambda \beta H}\rangle_\beta = \langle A_1 A_2 \cdots A_n\rangle_\beta, \qquad (4.57)
\end{aligned}$$

where we have used the cyclicity under the trace in a thermal ensemble average in the last step as well as $H|0, \widetilde{0}\rangle = 0$ in an intermediate step. This completes the discussion of building a thermal Hilbert space leading to the thermal ensemble average with the zero temperature operators. Of course, it remains to be shown that $|0, (\beta, \lambda)\rangle$, defined as in (4.34)-(4.35), is indeed the thermal vacuum state for an arbitrary path with $U(\theta, \lambda)$ taking the state $|0, \widetilde{0}\rangle$ to this state and we will do it in the next section. In the following subsection we will discuss the second approach to defining a thermal field theory for an arbitrary path shown in Fig. 4.2.

4.2.2 Defining path dependent thermal operators.

In the second approach, we choose (fix) the thermal vacuum to be that of thermofield dynamics, namely, $|0, \beta\rangle$ and define the thermal ensemble average of a product of operators to be the thermal vacuum expectation of a set of path dependent operators. We see, from (4.28)-(4.29), that the thermal ensemble average of a product of operators can be written as

$$\langle A_1 A_2 \cdots A_n \rangle_\beta = \langle 0, \beta | A_1 A_2 \cdots A_n | 0, \beta \rangle,$$
$$= \langle 0, \beta | S^{-1}(\lambda) A_1 A_2 \cdots A_n S(\lambda) | 0, \beta \rangle, \qquad (4.58)$$

where we have identified $S(\lambda) = e^{\lambda \beta H}$.

The thermal Hilbert space of thermofield dynamics, as we have already pointed out earlier, is a Dirac product space, namely, there is no nontrivial metric for this, only when $\lambda \neq 0$ is there a nontrivial inner product space as noted in the last subsection. We can now define path dependent operators through the similarity transformation simply as

$$A(\lambda) = S^{-1}(\lambda) A S(\lambda), \qquad (4.59)$$

so that the thermal ensemble average in (4.58) can be written as

$$\langle A_1 A_2 \cdots A_n \rangle_\beta = \langle 0, \beta | A_1(\lambda) A_2(\lambda) \cdots A_n(\lambda) | 0, \beta \rangle. \qquad (4.60)$$

This, therefore, shows that the thermal ensemble average of a product of operators can indeed be written as the expectation value of a product of path dependent operators in the thermal vacuum of thermofield dynamics. Note that when $\lambda = 0$, the similarity transformation reduces to identity and $A(\lambda = 0) = A$ so that we get back the familiar thermal field description of thermofield dynamics discussed in detail in the last chapter. It is worth noting here that the similarity transformation, $S(\lambda) = e^{\lambda \beta H}$, is not unitary, but that is not a problem since it is not used in defining the thermal vacuum. On the other hand, a similarity transformation preserves commutation relations and, therefore, perfectly respectable as a transformation. So, this defines the second approach. (Alternatively, we note that we can identify $S^{-1}(\lambda) = S^\ddagger(\lambda)$, which satisfies $S(\lambda) S^\ddagger(\lambda) = \mathbb{1} = S^\ddagger(\lambda) S(\lambda)$.) However, what remains to be shown is that both these approaches do indeed lead to the right thermal ensemble average which we will show next.

4.3 Deriving the thermal propagator

Although we have formally introduced the two operatorial approaches to thermal field theory for an arbitrary path (see Fig. 4.2) in the real time formalism, we are yet to show that these descriptions indeed lead to the correct thermal ensemble averages. We will now demonstrate this, for both the approaches, by deriving the thermal propagator in this formalism. To begin with, let us start with the first approach and construct explicitly the Bogoliubov transformation that takes the zero temperature thermofield dynamics vaccuum, $|0, \widetilde{0}\rangle$, to the actual thermal vacuum $|0, (\beta, \lambda)\rangle$. We have seen in (4.34) that this can be formally defined as

$$
U(\theta, \lambda) = S(\lambda)U(\theta)S^{-1}(\lambda) = e^{\lambda \beta H}U(\theta)e^{-\lambda \beta H}
$$
$$
= e^{\lambda \beta H}e^{-Q(\theta)}e^{-\lambda \beta H}, \tag{4.61}
$$

where, in momentum space, the generator of the Bogoliubov transformation for a continuum field theory, like the scalar Klein-Gordon theory, takes the form (see, for example, (3.99), here we have identified $iQ(\theta)$ of (3.99) as $Q(\theta)$)

$$
Q(\theta) = \int d^3p\, \theta_{\mathbf{p}}(\widetilde{a}(\mathbf{p})a(\mathbf{p}) - a^\dagger(\mathbf{p})\widetilde{a}^\dagger(\mathbf{p})), \tag{4.62}
$$

with $\theta_{\mathbf{p}}$ defined in (3.101).

Let us next recall that in a continuum field theory, like the free Klein-Gordon theory, the normal ordered Hamiltonian can be written as

$$
H = \int d^3p\, \omega_{\mathbf{p}} a^\dagger(\mathbf{p})a(\mathbf{p}), \tag{4.63}
$$

where

$$
\omega_{\mathbf{p}} = \sqrt{\mathbf{p}^2 + m^2}, \tag{4.64}
$$

for a particle with momentum $\mathbf{p}$ and mass m. As a result, the first thing that trivially follows from (4.63) is that

$$
[H, \widetilde{a}(\mathbf{p})] = 0 = [H, \widetilde{a}^\dagger(\mathbf{p})]. \tag{4.65}
$$

This is a direct consequence of the fact that tilde and non-tilde operators commute. Furthermore, from the canonical commutation relations between $a(\mathbf{p})$ and $a^\dagger(\mathbf{p})$, it follows that

$$[H, a(\mathbf{p})] = -\omega_\mathbf{p} a(\mathbf{p}), \quad [H, a^\dagger(\mathbf{p})] = \omega_\mathbf{p} a^\dagger(\mathbf{p}). \tag{4.66}$$

We can now calculate, by using nested commutators, as described in chapter **3**, and obtain

$$S(\lambda)\tilde{a}(\mathbf{p})S^{-1}(\lambda) = e^{\lambda\beta H}\tilde{a}(\mathbf{p})e^{-\lambda\beta H} = \tilde{a}(\mathbf{p}),$$

$$S(\lambda)\tilde{a}^\dagger(\mathbf{p})S^{-1}(\lambda) = e^{\lambda\beta H}\tilde{a}^\dagger(\mathbf{p})e^{-\lambda\beta H} = \tilde{a}^\dagger(\mathbf{p}),$$

$$S(\lambda)a(\mathbf{p})S^{-1}(\lambda) = e^{\lambda\beta H}a(\mathbf{p})e^{-\lambda\beta H} = e^{-\lambda\beta\omega_\mathbf{p}}a(\mathbf{p}),$$

$$S(\lambda)a^\dagger(\mathbf{p})S^{-1}(\lambda) = e^{\lambda\beta H}a^\dagger(\mathbf{p})e^{-\lambda\beta H} = e^{\lambda\beta\omega_\mathbf{p}}a^\dagger(\mathbf{p}). \tag{4.67}$$

This shows that, under the similarity transformation $S(\lambda)$, the annihilation and the creation operators scale as

$$a(\mathbf{p}) \to e^{-\lambda\beta\omega_\mathbf{p}}a(\mathbf{p}), \quad a^\dagger(\mathbf{p}) \to e^{\lambda\beta\omega_\mathbf{p}}a^\dagger(\mathbf{p}). \tag{4.68}$$

With these observations, we note that the Bogoliubov transformation in (4.61)-(4.62) can now be written in the explicit form

$$U(\theta, \lambda) = S(\lambda)U(\theta)S^{-1}(\lambda) = e^{-Q(\theta,\lambda)},$$

$$U^\ddagger(\theta, \lambda) = S^{-1}(-\lambda)U^\dagger(\theta)S(-\lambda) = S(\lambda)U^\dagger(\theta)S^{-1}(\lambda)$$

$$= e^{Q(\theta,\lambda)} = U(-\theta, \lambda), \tag{4.69}$$

where we have used $S^{-1}(\lambda) = S(-\lambda), S^\dagger(\lambda) = S(\lambda)$ and where

$$Q(\theta, \lambda) = \int d^3p\, \theta_\mathbf{p}(e^{-\lambda\beta\omega_\mathbf{p}}\tilde{a}(\mathbf{p})a(\mathbf{p}) - e^{\lambda\beta\omega_\mathbf{p}}a^\dagger(\mathbf{p})\tilde{a}^\dagger(\mathbf{p}))$$

$$= -Q^\ddagger(\theta, \lambda). \tag{4.70}$$

We can now calculate the basic commutation relations,

$$[Q(\theta, \lambda), a(\mathbf{p})] = e^{\lambda\beta\omega_\mathbf{p}}\theta_\mathbf{p}\tilde{a}^\dagger(\mathbf{p}),$$

$$[Q(\theta, \lambda), a^\dagger(\mathbf{p})] = e^{-\lambda\beta\omega_\mathbf{p}}\theta_\mathbf{p}\tilde{a}(\mathbf{p}),$$

$$[Q(\theta, \lambda), \tilde{a}(\mathbf{p})] = e^{\lambda\beta\omega_\mathbf{p}}\theta_\mathbf{p}a^\dagger(\mathbf{p}),$$

$$[Q(\theta, \lambda), \tilde{a}^\dagger(\mathbf{p})] = e^{-\lambda\beta\omega_\mathbf{p}}\theta_\mathbf{p}a(\mathbf{p}), \tag{4.71}$$

which upon using (4.69) as well as nested commutators as in (3.41) lead to

$$U^{\ddagger}(\theta,\lambda)a(\mathbf{p})U(\theta,\lambda) = \cosh\theta_{\mathbf{p}}a(\mathbf{p}) + e^{\lambda\beta\omega_{\mathbf{p}}}\sinh\theta_{\mathbf{p}}\tilde{a}^{\dagger}(\mathbf{p}),$$

$$U^{\ddagger}(\theta,\lambda)a^{\dagger}(\mathbf{p})U(\theta,\lambda) = \cosh\theta_{\mathbf{p}}a^{\dagger}(\mathbf{p}) + e^{-\lambda\beta\omega_{\mathbf{p}}}\sinh\theta_{\mathbf{p}}\tilde{a}(\mathbf{p}),$$

$$U^{\ddagger}(\theta,\lambda)\tilde{a}(\mathbf{p})U(\theta,\lambda) = \cosh\theta_{\mathbf{p}}\tilde{a}(\mathbf{p}) + e^{\lambda\beta\omega_{\mathbf{p}}}\sinh\theta_{\mathbf{p}}a^{\dagger}(\mathbf{p}),$$

$$U^{\ddagger}(\theta,\lambda)\tilde{a}^{\dagger}(\mathbf{p})U(\theta,\lambda) = \cosh\theta_{\mathbf{p}}\tilde{a}^{\dagger}(\mathbf{p}) + e^{-\lambda\beta\omega_{\mathbf{p}}}\sinh\theta_{\mathbf{p}}a(\mathbf{p}).$$
$$(4.72)$$

It follows now that if we define two doublets of creation and annihilation operators as

$$A^{(+)}(\mathbf{p}) = \begin{pmatrix} a(\mathbf{p}) \\ \tilde{a}^{\dagger}(\mathbf{p}) \end{pmatrix}, \qquad A^{(-)}(\mathbf{p}) = \begin{pmatrix} a^{\dagger}(\mathbf{p}) \\ \tilde{a}(\mathbf{p}) \end{pmatrix}, \qquad (4.73)$$

then, under the Bogoliubov transformation (4.35), the doublets will transform as

$$A_{\alpha}^{(+)}(\mathbf{p}) \to U^{\ddagger}(\theta,\lambda)A_{\alpha}^{(+)}(\mathbf{p})U(\theta,\lambda) = U(-\theta,\lambda)A_{\alpha}^{(+)}(\mathbf{p})U^{\ddagger}(-\theta,\lambda)$$

$$= \begin{pmatrix} \cosh\theta_{\mathbf{p}} & e^{\lambda\beta\omega_{\mathbf{p}}}\sinh\theta_{\mathbf{p}} \\ e^{-\lambda\beta\omega_{\mathbf{p}}}\sinh\theta_{\mathbf{p}} & \cosh\theta_{\mathbf{p}} \end{pmatrix}_{\alpha\beta} A_{\beta}^{(+)}(\mathbf{p})$$

$$= \overline{U}_{\alpha\beta}(-\theta_{\mathbf{p}},\lambda)A_{\beta}^{(+)}(\mathbf{p}), \qquad (4.74)$$

$$A_{\alpha}^{(-)}(\mathbf{p}) \to U^{\ddagger}(\theta,\lambda)A_{\alpha}^{(-)}(\mathbf{p})U(\theta,\lambda) = U(-\theta,\lambda)A_{\alpha}^{(-)}(\mathbf{p})U^{\ddagger}(-\theta,\lambda)$$

$$= \begin{pmatrix} \cosh\theta_{\mathbf{p}} & e^{-\lambda\beta\omega_{\mathbf{p}}}\sinh\theta_{\mathbf{p}} \\ e^{\lambda\beta\omega_{\mathbf{p}}}\sinh\theta_{\mathbf{p}} & \cosh\theta_{\mathbf{p}} \end{pmatrix}_{\alpha\beta} A_{\beta}^{(-)}(\mathbf{p})$$

$$= \overline{U}_{\alpha\beta}(-\theta_{\mathbf{p}},-\lambda)A_{\beta}^{(-)}(\mathbf{p}), \qquad (4.75)$$

where $\alpha,\beta = 1,2$ (and repeated indices are summed). The matrix

$$\overline{U}(-\theta_{\mathbf{p}},\lambda) = \begin{pmatrix} \cosh\theta_{\mathbf{p}} & e^{\lambda\beta\omega_{\mathbf{p}}}\sinh\theta_{\mathbf{p}} \\ e^{-\lambda\beta\omega_{\mathbf{p}}}\sinh\theta_{\mathbf{p}} & \cosh\theta_{\mathbf{p}} \end{pmatrix}, \qquad (4.76)$$

can be compared with (4.10) and we simply note here for completeness that

$$\overline{U}(-\theta_{\mathbf{p}},-\lambda) = \overline{U}^{T}(-\theta_{\mathbf{p}},\lambda). \qquad (4.77)$$

Defining the doublet of fields, $\Phi(x) = \begin{pmatrix} \phi(x) \\ \widetilde{\phi}(x) \end{pmatrix}$, we note that it has the field decomposition given by

$$\Phi_\alpha(x) = \int d^3p \left(f_{\mathbf{p}}^{(+)}(x) A_\alpha^{(+)}(\mathbf{p}) + f_{\mathbf{p}}^{(-)}(x) A_\alpha^{(-)}(\mathbf{p}) \right), \quad (4.78)$$

where

$$f_{\mathbf{p}}^{(\pm)}(x) = \frac{e^{\mp ip \cdot x}}{\sqrt{(2\pi)^3 2p^0}}, \quad p^0 = \omega_{\mathbf{p}} = \sqrt{\mathbf{p}^2 + m^2} > 0. \quad (4.79)$$

Using (4.74) in the field decomposition above, we see that, under the Bogoliubov transformation, the doublet of fields transform as

$$\Phi_\alpha(x) \rightarrow U^\ddagger(\theta, \lambda) \Phi_\alpha(x) U(\theta, \lambda)$$

$$= \int d^3p \left(f_{\mathbf{p}}^{(+)}(x) \overline{U}_{\alpha\beta}(-\theta_{\mathbf{p}}, \lambda) A_\beta^{(+)}(\mathbf{p}) \right.$$

$$\left. + f_{\mathbf{p}}^{(-)}(x) \overline{U}_{\alpha\beta}(-\theta_{\mathbf{p}}, -\lambda) A_\beta^{(-)}(\mathbf{p}) \right). \quad (4.80)$$

Furthermore, let us note here, for future use, that (these are matrix outer products leading to 2×2 matrices)

$$\langle 0, \widetilde{0} | A_\alpha^{(+)}(\mathbf{p}) A_\beta^{(+)}(\mathbf{p}') | 0, \widetilde{0} \rangle = 0,$$

$$\langle 0, \widetilde{0} | A_\alpha^{(-)}(\mathbf{p}) A_\beta^{(-)}(\mathbf{p}') | 0, \widetilde{0} \rangle = 0,$$

$$\langle 0, \widetilde{0} | A_\alpha^{(+)}(\mathbf{p}) A_\beta^{(-)}(\mathbf{p}') | 0, \widetilde{0} \rangle = \delta^3(\mathbf{p} - \mathbf{p}') \begin{pmatrix} 1 & 0 \\ 0 & 0 \end{pmatrix}_{\alpha\beta},$$

$$\langle 0, \widetilde{0} | A_\alpha^{(-)}(\mathbf{p}) A_\beta^{(+)}(\mathbf{p}') | 0, \widetilde{0} \rangle = \delta^3(\mathbf{p} - \mathbf{p}') \begin{pmatrix} 0 & 0 \\ 0 & 1 \end{pmatrix}_{\alpha\beta}. \quad (4.81)$$

Using all of these results, namely, (4.73)-(4.81), we note that we can write

$$\langle 0, (\beta, \lambda) | \Phi_\alpha(x) \Phi_\beta(y) | 0, (\beta, \lambda) \rangle$$

$$= \langle 0, \widetilde{0} | U^\ddagger(\theta, \lambda) \Phi_\alpha(x) \Phi_\beta(y) U(\theta, \lambda) | 0, \widetilde{0} \rangle$$

$$= \langle 0, \widetilde{0} | U^\ddagger(\theta, \lambda) \Phi_\alpha(x) U(\theta, \lambda) U^\ddagger(\theta, \lambda) \Phi_\beta(y) U(\theta, \lambda) | 0, \widetilde{0} \rangle$$

$$= \int d^3p\, d^3p'\, \delta^3(\mathbf{p} - \mathbf{p}')\, \overline{U}_{\alpha\gamma}(-\theta_{\mathbf{p}}, \lambda)\overline{U}_{\beta\delta}(-\theta_{\mathbf{p}'}, -\lambda)$$

$$\times \begin{pmatrix} f_{\mathbf{p}}^{(+)}(x)f_{\mathbf{p}'}^{(-)}(y) & 0 \\ 0 & f_{\mathbf{p}}^{(-)}(x)f_{\mathbf{p}'}^{(+)}(y) \end{pmatrix}_{\gamma\delta}$$

$$= \int d^3p \left(\overline{U}(-\theta_{\mathbf{p}}, \lambda) \begin{pmatrix} f_{\mathbf{p}}^{(+)}(x)f_{\mathbf{p}}^{(-)}(y) & 0 \\ 0 & f_{\mathbf{p}}^{(-)}(x)f_{\mathbf{p}}^{(+)}(y) \end{pmatrix} \right.$$

$$\left. \times \overline{U}^T(-\theta_{\mathbf{p}}, -\lambda) \right)_{\alpha\beta}$$

$$= \int d^3p \left(\overline{U}(-\theta_{\mathbf{p}}, \lambda) \begin{pmatrix} f_{\mathbf{p}}^{(+)}(x)f_{\mathbf{p}}^{(-)}(y) & 0 \\ 0 & f_{\mathbf{p}}^{(-)}(x)f_{\mathbf{p}}^{(+)}(y) \end{pmatrix} \right.$$

$$\left. \times \overline{U}(-\theta_{\mathbf{p}}, \lambda) \right)_{\alpha\beta}, \tag{4.82}$$

where we have used the symmetry (4.77) in the last factor. Therefore, as a 2×2 matrix, we can write

$$\langle 0, (\beta, \lambda) | \Phi(x)\Phi(y) | 0, (\beta, \lambda) \rangle$$

$$= \int d^3p\, \overline{U}(-\theta_{\mathbf{p}}, \lambda)Q_{\mathbf{p}}(x, y)\overline{U}(-\theta_{\mathbf{p}}, \lambda), \tag{4.83}$$

where we have identified

$$Q_{\mathbf{p}}(x, y) = \begin{pmatrix} f_{\mathbf{p}}^{(+)}(x)f_{\mathbf{p}}^{(-)}(y) & 0 \\ 0 & f_{\mathbf{p}}^{(-)}(x)f_{\mathbf{p}}^{(+)}(y) \end{pmatrix}. \tag{4.84}$$

As a result, we can now obtain the 2×2 matrix propagator for the system for an arbitrary path, as in (3.116),

$$iG^{(\sigma)}(x - y) = \langle 0, (\beta, \lambda) | T(\Phi(x)\Phi(y)) | 0, (\beta, \lambda) \rangle$$

$$= \theta(x^0 - y^0)\langle 0, (\beta, \lambda) | \Phi(x)\Phi(y) | 0, (\beta, \lambda) \rangle$$

$$+ \theta(y^0 - x^0)\langle 0, (\beta, \lambda) | \Phi(y)\Phi(x) | 0, (\beta, \lambda)$$

$$= \int d^3p \left[\overline{U}(-\theta_{\mathbf{p}}, \lambda)\theta(x^0 - y^0)Q_{\mathbf{p}}(x, y)\overline{U}(-\theta_{\mathbf{p}}, \lambda) \right.$$

$$\left. + \overline{U}(-\theta_{\mathbf{p}}, \lambda)\theta(y^0 - x^0)Q_{\mathbf{p}}(y, x)\overline{U}(-\theta_{\mathbf{p}}, \lambda) \right]$$

$$= \int d^3p \left[\overline{U}(-\theta_{\mathbf{p}}, \lambda) \Big(\theta(x^0 - y^0) Q_{\mathbf{p}}(x, y) \right.$$

$$\left. + \theta(y^0 - x^0) Q_{\mathbf{p}}(y, x) \Big) \overline{U}(-\theta_{\mathbf{p}}, \lambda) \right]. \tag{4.85}$$

Let us recall that the theta function has the integral representation given by

$$\theta(x^0 - y^0) = \lim_{\epsilon \to 0^+} \frac{i}{2\pi} \int dp^0 \, \frac{e^{-ip^0(x^0 - y^0)}}{p^0 + i\epsilon}. \tag{4.86}$$

Using this, we obtain (see (4.79) for definitions of $f_{\mathbf{p}}^{(\pm)}(x)$)

$$\theta(x^0 - y^0) f_{\mathbf{p}}^{(\pm)}(x) f_{\mathbf{p}}^{(\mp)}(y)$$

$$= \lim_{\epsilon \to 0^+} \frac{i}{2\pi} \int dp^0 \, \frac{e^{-ip^0(x^0 - y^0)}}{p^0 + i\epsilon} \, \frac{e^{\mp i\omega_{\mathbf{p}}(x^0 - y^0) \pm i\mathbf{p} \cdot (\mathbf{x} - \mathbf{y})}}{(2\pi)^3 \, 2\omega_{\mathbf{p}}}$$

$$= \lim_{\epsilon \to 0^+} \frac{i}{(2\pi)^4 \, 2\omega_{\mathbf{p}}} \int dp^0 \, \frac{e^{-ip \cdot (x - y)}}{p^0 \mp \omega_{\mathbf{p}} + i\epsilon}, \tag{4.87}$$

where we have redefined $p^0 \to p^0 \mp \omega_{\mathbf{p}}, \mathbf{p} \to \pm \mathbf{p}$ in the last line. Similarly, we can show that we can write

$$\theta(y^0 - x^0) f_{\mathbf{p}}^{(\pm)}(y) f_{\mathbf{p}}^{(\mp)}(x)$$

$$= \lim_{\epsilon \to 0^+} -\frac{i}{(2\pi)^4 \, 2\omega_{\mathbf{p}}} \int dp^0 \, \frac{e^{-ip \cdot (x - y)}}{p^0 \pm \omega_{\mathbf{p}} - i\epsilon}, \tag{4.88}$$

which can be obtained from (4.87) by interchanging x and y and then letting $p^\mu \to -p^\mu$. It follows, then, that

$$\theta(x^0 - y^0) f_{\mathbf{p}}^{(+)}(x) f_{\mathbf{p}}^{(-)}(y) + \theta(y^0 - x^0) f_{\mathbf{p}}^{(+)}(y) f_{\mathbf{p}}^{(-)}(x)$$

$$= \lim_{\epsilon \to 0^+} \frac{i}{(2\pi)^4} \int dp^0 \, \frac{e^{-ip \cdot (x - y)}}{(p^0)^2 - \omega_{\mathbf{p}}^2 + i\epsilon}$$

$$= \lim_{\epsilon \to 0^+} \frac{i}{(2\pi)^4} \int dp^0 \, \frac{e^{-ip \cdot (x - y)}}{p^2 - m^2 + i\epsilon}, \tag{4.89}$$

and, similarly,

$$\theta(x^0 - y^0) f_{\mathbf{p}}^{(-)}(x) f_{\mathbf{p}}^{(+)}(y) + \theta(y^0 - x^0) f_{\mathbf{p}}^{(-)}(y) f_{\mathbf{p}}(x)$$

$$= \lim_{\epsilon \to 0^+} -\frac{i}{(2\pi)^4} \int dp^0 \, \frac{e^{-ip \cdot (x - y)}}{p^2 - m^2 - i\epsilon}. \tag{4.90}$$

Substituting these two results into (4.85), we determine the thermal propagator for an arbitrary path to be

$$iG^{(\sigma)}(x-y)$$

$$= \lim_{\epsilon \to 0^+} \int \frac{d^4 p}{(2\pi)^4} e^{-ip \cdot (x-y)} \overline{U}(-\theta_{\mathbf{p}}, \lambda) \begin{pmatrix} \frac{i}{p^2 - m^2 + i\epsilon} & 0 \\ 0 & -\frac{i}{p^2 - m^2 - i\epsilon} \end{pmatrix}$$

$$\times \overline{U}(-\theta_{\mathbf{p}}, \lambda),$$

which can be written in the momentum space as

$$iG^{(\sigma)}(p) = \overline{U}(-\theta_{\mathbf{p}}, \lambda) \left(iG^{(\text{TFD})}_{(T=0)}(p) \right) \overline{U}(-\theta_{\mathbf{p}}, \lambda). \tag{4.91}$$

This can be compared with (4.9) and (4.10). This completes the derivation of the propagator from the time ordered expectation value of (doubled) fields in the vacuum for the arbitrary σ path.

Let us next discuss the calculation of the propagator in the second approach where we keep the vacuum to be that of thermofield dynamics and calculate the expectation value of time ordered products of transformed operators in this vacuum. Namely, we consider the thermal vacuum to correspond to $|0, \beta\rangle$ and the operators to correspond to $O(\lambda) = S^{-1}(\lambda) O S(\lambda)$, $S(\lambda) = e^{\lambda \beta H}$ (see (4.59) and (4.60)). However, since the thermal vacuum state in thermofield dynamics (see (3.77) and (3.78)) satisfies,

$$\widehat{H}|0, \beta\rangle = (H - \widetilde{H})|0, \beta\rangle = 0, \tag{4.92}$$

and $S(\lambda)$ is acting on the thermal vacuum (in thermofield dynamics), we can operationally (and for comparison with earlier results) identify,

$$S(\lambda) = e^{\lambda \beta \widetilde{H}}. \tag{4.93}$$

Then, following (4.67) we can calculate the transformation of the creation and annihilation operators under the similarity transformation to be

$$S^{-1}(\lambda) a S(\lambda) = e^{-\lambda \beta \widetilde{H}} a \, e^{\lambda \beta \widetilde{H}} = a,$$

$$S^{-1}(\lambda) a^\dagger S(\lambda) = e^{-\lambda \beta \widetilde{H}} a^\dagger e^{\lambda \beta \widetilde{H}} = a^\dagger,$$

$$S^{-1}(\lambda) \widetilde{a} S(\lambda) = e^{-\lambda \beta \widetilde{H}} \widetilde{a} \, e^{\lambda \beta \widetilde{H}} = e^{\lambda \beta \omega_{\mathbf{p}}} \, \widetilde{a},$$

$$S^{-1}(\lambda) \widetilde{a}^\dagger S(\lambda) = e^{-\lambda \beta \widetilde{H}} \widetilde{a}^\dagger e^{\lambda \beta \widetilde{H}} = e^{-\lambda \beta \omega_{\mathbf{p}}} \, \widetilde{a}^\dagger. \tag{4.94}$$

This, in turn, leads to the transformation of the doublets in (4.73) under the similarity transformation to be

$$S^{-1}(\lambda)A^{(+)}(\mathbf{p})S(\lambda) = \overline{V}(\lambda)A^{(+)}(\mathbf{p}),$$
$$S^{-1}(\lambda)A^{(-)}(\mathbf{p})S(\lambda) = \overline{V}^{-1}(\lambda)A^{(-)}(\mathbf{p}), \qquad (4.95)$$

where

$$\overline{V}(\lambda) = \begin{pmatrix} 1 & 0 \\ 0 & e^{-\lambda\beta\omega_{\mathbf{p}}} \end{pmatrix}, \quad \overline{V}^{\dagger}(\lambda) = \overline{V}(\lambda),$$
$$\overline{V}^{-1}(\lambda) = \overline{V}(-\lambda) = \overline{V}^{T}(-\lambda) = \overline{V}^{\dagger}(-\lambda). \qquad (4.96)$$

We note here that the 2×2 matrix in (4.96) coincides with the matrix in (4.14).

We can now follow the calculations in (4.73)-(4.91) and show that

$$\langle 0, \beta | S^{-1}(\lambda)A^{(\pm)}(\mathbf{p})A^{(\pm)}(\mathbf{p}')S(\lambda)|0, \beta \rangle$$
$$= \langle 0, \beta | S^{-1}(\lambda)A^{(\pm)}(\mathbf{p})S(\lambda)S^{-1}(\lambda)A^{(\pm)}(\mathbf{p}')S(\lambda)|0, \beta \rangle$$
$$= \overline{V}(\lambda)\langle 0, \beta | A^{(\pm)}(\mathbf{p})A^{(\pm)}(\mathbf{p}')|0, \beta \rangle \overline{V}(-\lambda)$$
$$= \overline{V}(\lambda)\overline{U}(-\theta_{\mathbf{p}})\,\langle 0, \widetilde{0} | A^{(\pm)}(\mathbf{p})A^{(\pm)}(\mathbf{p}')|0, \widetilde{0} \rangle\, \overline{U}(\theta_{\mathbf{p}'})\overline{V}(-\lambda)$$
$$= 0, \qquad (4.97)$$

and, similarly (see (4.81))

$$\langle 0, \beta | S^{-1}(\lambda)A^{(+)}(\mathbf{p})A^{(-)}(\mathbf{p}')S(\lambda)|0, \beta \rangle$$
$$= \delta^{3}(\mathbf{p} - \mathbf{p}')\overline{V}(\lambda)\overline{U}(-\theta_{\mathbf{p}}) \begin{pmatrix} 1 & 0 \\ 0 & 0 \end{pmatrix} \overline{U}(-\theta_{\mathbf{p}})\overline{V}(-\lambda),$$
$$\langle 0, \beta | S^{-1}(\lambda)A^{(-)}(\mathbf{p})A^{(+)}(\mathbf{p}')S(\lambda)|0, \beta \rangle$$
$$= \delta^{3}(\mathbf{p} - \mathbf{p}')\overline{V}(\lambda)\overline{U}(-\theta_{\mathbf{p}}) \begin{pmatrix} 0 & 0 \\ 0 & 1 \end{pmatrix} \overline{U}(-\theta_{\mathbf{p}})\overline{V}(-\lambda). \qquad (4.98)$$

Using these results and following the steps in (4.85)-(4.91), we can show that

$$iG^{(\sigma)}(x-y) = \lim_{\epsilon \to 0^+} \int \frac{\mathrm{d}^4 p}{(2\pi)^4}\, e^{-ip \cdot (x-y)} \overline{V}(\lambda)\overline{U}(-\theta_{\mathbf{p}})$$

$$\times \begin{pmatrix} \frac{i}{p^2-m^2+i\epsilon} & 0 \\ 0 & -\frac{i}{p^2-m^2-i\epsilon} \end{pmatrix} \overline{U}(-\theta_{\mathbf{p}})\overline{V}(-\lambda)$$

$$= \lim_{\epsilon \to 0^+} \int \frac{\mathrm{d}^4 p}{(2\pi)^4}\, e^{-ip \cdot (x-y)} \overline{V}(\lambda)\overline{U}(-\theta_{\mathbf{p}})\overline{V}(-\lambda)$$

$$\times \begin{pmatrix} \frac{i}{p^2-m^2+i\epsilon} & 0 \\ 0 & -\frac{i}{p^2-m^2-i\epsilon} \end{pmatrix} \overline{V}(\lambda)\overline{U}(-\theta_{\mathbf{p}})\overline{V}(-\lambda),$$

where we have used the fact that the diagonal matrices $\overline{V}(\pm\lambda)$ are inverses of each other and that two diagonal matrices commute and this leads to

$$iG^{(\sigma)}(x-y) = \lim_{\epsilon \to 0^+} \int \frac{\mathrm{d}^4 p}{(2\pi)^4}\, e^{-ip \cdot (x-y)}$$

$$\times \overline{U}(-\theta_{\mathbf{p}}, \lambda) \begin{pmatrix} \frac{i}{p^2-m^2+i\epsilon} & 0 \\ 0 & -\frac{i}{p^2-m^2-i\epsilon} \end{pmatrix} \overline{U}(-\theta_{\mathbf{p}}, \lambda).$$

This can be written in the momentum space as

$$iG^{(\sigma)}(p) = \overline{U}(-\theta_{\mathbf{p}}, \lambda)(iG^{(\mathrm{TFD})}_{(T=0)}(p))\overline{U}(-\theta_{\mathbf{p}}, \lambda), \tag{4.99}$$

which can be compared with (4.9)-(4.16). We note here that this also allows us to identify

$$\overline{U}(-\theta_{\mathbf{p}}, \lambda) = \overline{V}(\lambda)\overline{U}(-\theta_{\mathbf{p}})\overline{V}(-\lambda). \tag{4.100}$$

This completes the derivation of the propagator in the second equivalent approach.

4.4 Path independence of physical observables

As we have shown, thermal field theories can be defined on an arbitrary path in the complex t-plane in the real time formalism. This raises an interesting question, namely, whether the ensemble averages of physical observables are independent of the path used in calculating them. After all, if we write, as in (4.29) (with $\alpha = \lambda$),

$$\langle A_1 A_2 \cdots A_n \rangle_\beta = \langle 0, \beta | e^{-\lambda\beta H} A_1 A_2 \cdots A_n e^{\lambda\beta H} | 0, \beta \rangle, \qquad (4.101)$$

the dependence of the ensemble average on the path (remember $\lambda = \sigma - \frac{1}{2}$) is manifest. (Let us recall that the two λ dependent exponential factors were introduced into the thermal vacuum expectation value using the cyclicity of trace in a thermal ensemble average.) Therefore, let us now discuss what we can say about the path independence of thermal vacuum expectation values and ensemble averages of operators, in general, and of physical observables, in particular.

Let us note that physical observables correspond to the operators built out of original field operators and do not involve the fictitious doubled operators. Therefore, when $A_1, A_2, \cdots, A_n$ correspond to such operators, let us first show that the thermal vacuum expectation value and correspondingly the thermal ensemble average will be path independent. Let us recall that the thermal vacuum in thermofield dynamics (see (3.54) and (3.55) as well as from (3.77) and (3.78)) satisfies (both for bosons and fermions)

$$\widehat{H} | 0, \beta \rangle = (H - \widetilde{H}) | 0, \beta \rangle = 0, \qquad (4.102)$$

so that

$$H | 0, \beta \rangle = \widetilde{H} | 0, \beta \rangle. \qquad (4.103)$$

Armed with this relation, let us note that we can write the thermal vacuum expectation value of any product of operators, given in (4.101), equivalently also as

$$\langle A_1 A_2 \cdots A_n \rangle_\beta = \langle 0, \beta | e^{-\lambda\beta\widetilde{H}} A_1 A_2 \cdots A_n e^{\lambda\beta\widetilde{H}} | 0, \beta \rangle. \qquad (4.104)$$

If all the operators, $A_1, A_2, \cdots, A_n$, in the ensemble average are physical (original fields), and we recall that tilde and nontilde operators, in general, commute/anti-commute, we can take the exponential factor of $e^{-\lambda\beta\widetilde{H}}$ in (4.104) through each of the factors $A_1, A_2, \cdots, A_n$ to cancel with the exponential factor $e^{\lambda\beta\widetilde{H}}$ on the right leaving us simply with

$$\langle A_1 A_2 \cdots A_n \rangle_\beta = \langle 0, \beta | A_1 A_2 \cdots A_n | 0, \beta \rangle. \qquad (4.105)$$

The λ (path) dependence has completely cancelled out. This is the simplest way to conclude that the thermal vacuum expectation value and, therefore, the thermal ensemble averages of physical observables are, in fact, path independent, as they should be. As a by-product of this reasoning we can also show that if all the operators $A_1, A_2, \cdots, A_n$ are fictitious (tilde operators), then also the thermal vacuum expectation value is path independent. This follows if we start with (4.101) and assume that all the operators are tilde operators. Then $e^{-\lambda\beta H}$ will commute with each of the tilde operators in the product and the exponential factor can be moved all the way to the right canceling with the inverse exponential factor at the extreme right. This may seem odd at the first sight because there does not seem to be any physical reason for the ensemble average of a product of unphysical fields to be path independent. On the other hand, a little bit of reasoning easily clarifies this in the following way. In a thermal field theory, the ensemble average of a product of operators can be described as an amplitude which is a function. Therefore, the ensemble average of a product of original operators and that of a product of tilde operators can be described by two different functions. However, the two have to be related by the tilde conjugation rules (see, for example, (3.88)) which simply complex conjugates coefficients and functions. Therefore, if the amplitude corresponding to the physical fields is path independent, then, that for the doubled (tilde) fields will also be. This is easily seen in the diagonal elements of the thermal propagator of the scalar field in (4.5) (as also in the derivation in the last chapter). On the other hand, if we have a thermal vacuum expetation of a product of mixed operators, namely, physical as well as fictitious (doubled) operators, neither of the forms in (4.104) or (4.105) would allow us to commute the exponential factor on the left past all the field operators to the right to cancel the inverse exponential factor on the right. As a result, in such thermal vacuum expectation values involving mixed products of operators, the λ dependence would not cancel out and a path dependence would be manifestly present. This is also seen explicitly in the off-diagonal elements of the propagator in (4.5).

There is an alternate way of understanding this result as follows. Let us recall that the thermal vacuum in thermofield dynamics can be written as an expansion of the form (see, for example, (3.10)-(3.18))

$$|0, \beta\rangle = Z^{-\frac{1}{2}}(\beta) \sum_n e^{-\frac{\beta E_n}{2}} |n, \widetilde{n}\rangle$$

$$= Z^{-\frac{1}{2}}(\beta) \sum_n e^{-\frac{\beta E_n}{2}} |n\rangle \otimes |\widetilde{n}\rangle, \qquad (4.106)$$

where $|n\rangle$ and $|\widetilde{n}\rangle$ denote the nth energy state of the original system and the doubled system respectively with energy eigenvalue E_n. As a result, using the above expansion for both the bra and ket thermal vacuum states for a product of (original) physical fields, we can write

$$\langle 0, \beta | e^{-\lambda \beta H} A_1 A_2 \cdots A_n e^{\lambda \beta H} |0, \beta\rangle$$

$$= Z^{-1}(\beta) \sum_{\ell, m} e^{-\lambda \beta E_\ell} e^{-\frac{\beta(E_\ell + E_m)}{2}} \langle \ell, \widetilde{\ell} | A_1 A_2 \cdots A_n | m, \widetilde{m} \rangle e^{\lambda \beta E_m}$$

$$= Z^{-1}(\beta) \sum_{\ell, m} e^{-\lambda \beta (E_\ell - E_m)} e^{-\frac{\beta(E_\ell + E_m)}{2}} \langle \ell | A_1 A_2 \cdots A_n | m \rangle \langle \widetilde{\ell} | \widetilde{m} \rangle$$

$$= Z^{-1}(\beta) \sum_{\ell, m} e^{-\lambda \beta (E_\ell - E_m)} e^{-\frac{\beta(E_\ell + E_m)}{2}} \langle \ell | A_1 A_2 \cdots A_n | m \rangle \delta_{\ell m}$$

$$= Z^{-1}(\beta) \sum_m e^{-\beta E_m} \langle m | A_1 A_2 \cdots A_n | m \rangle$$

$$= Z^{-1}(\beta) \text{Tr} \left(e^{-\beta H} A_1 A_2 \cdots A_n \right) = \langle A_1 A_2 \cdots A_n \rangle_\beta. \qquad (4.107)$$

Similarly, when all the operators are (doubled) tilde operators, using (4.103) we can show that the thermal vacuum expectation value will take the form

$$\langle 0, \beta | e^{-\lambda \beta H} \widetilde{A}_1 \widetilde{A}_2 \cdots \widetilde{A}_n e^{\lambda \beta H} |0, \beta\rangle$$

$$= \langle 0, \beta | e^{-\lambda \beta \widetilde{H}} \widetilde{A}_1 \widetilde{A}_2 \cdots \widetilde{A}_n e^{\lambda \beta \widetilde{H}} |0, \beta\rangle$$

$$= Z^{-1}(\beta) \text{Tr} \left(e^{-\beta \widetilde{H}} \widetilde{A}_1 \widetilde{A}_2 \cdots \widetilde{A}_n \right) = \langle \widetilde{A}_1 \widetilde{A}_2 \cdots \widetilde{A}_n \rangle_\beta. \qquad (4.108)$$

Therefore, for such (diagonal) terms, does indeed represent the thermal ensemble average and, therefore, inside the trace, the λ dependent exponential terms can cancel out and there will not be any path dependence. (Namely, genuine thermal ensemble averages cannot have a path dependence.)

On the other hand, if some of the operators are the (original) physical operators while others are the (doubled) operators, then things are different. We note that, since physical operators and the tilde operators (anti) commute, we can always write such a general thermal vacuum expectation value as

$$\langle 0, \beta | e^{-\lambda \beta H} (A_1 A_2 \cdots A_k)(\widetilde{B}_1 \widetilde{B}_2 \cdots \widetilde{B}_\ell) e^{\lambda \beta H} | 0, \beta \rangle$$

$$= Z^{-1}(\beta) \sum_{m,n} e^{-\lambda \beta (E_m - E_n)} e^{-\frac{\beta(E_m + E_n)}{2}} \langle m | A_1 A_2 \cdots A_k | n \rangle$$

$$\times \langle \widetilde{m} | \widetilde{B}_1 \widetilde{B}_2 \cdots \widetilde{B}_\ell | \widetilde{n} \rangle$$

$$\neq Z^{-1}(\beta) \text{Tr} \left(e^{-\beta H} (A_1 A_2 \cdots A_k)(\widetilde{B}_1 \widetilde{B}_2 \cdots \widetilde{B}_\ell) \right)$$

$$\neq Z^{-1}(\beta) \text{Tr} \left(e^{-\beta \widetilde{H}} (A_1 A_2 \cdots A_k)(\widetilde{B}_1 \widetilde{B}_2 \cdots \widetilde{B}_\ell) \right)$$

and, therefore,

$$\neq \langle (A_1 A_2 \cdots A_k)(\widetilde{B}_1 \widetilde{B}_2 \cdots \widetilde{B}_\ell) \rangle_\beta. \tag{4.109}$$

Since the off-diagonal terms cannot be written as a trace, there is no cyclicity and the two exponential factors $e^{-\lambda \beta H}$ and $e^{\lambda \beta H}$ do not cancel. The manifest λ dependence shows in the second line of the above expression. Another way of saying this is to note that, since this does not represent a thermal ensemble average, there is no physical reason for it to be path independent.

In the next section, which is an appendix for this chapter, we will discuss briefly how and why an arbitrary parameter enters into the theory within the operator formalism and we will illustrate this within the context of the bosonic harmonic oscillator.

4.5 Appendices

In chapter **3**, when we introduced the operator formalism for thermal field theories (see section **3.1**), in general, it led us to thermofield dynamics as the unique thermal field theory we can have. No arbitrary parameter ever entered into our discussion. So, it is natural to go back to our earlier discussion and ask how and where possibly an arbitrary parameter entered into our discussion. We will discuss this interesting question in detail in the following subsections[1] within the context of the bosonic harmonic oscillator which will also lead to a better understanding of everything we have described in this chapter. In some sense, we have already answered this question in section **4.2.1** in general, nonetheless, it is a good idea understand the reason for the arbitrary parameter again within the context of a simple model. In the following subsection, we will derive the thermal propagator of a $0 + 1$ dimensional fermion.

4.5.1 The bosonic harmonic oscillator on the arbitrary path. In section **3.1**, we used the (discrete) energy basis of a quantum mechanical system to write the ensemble average of an operator A as (see (3.5)) as

$$\langle A \rangle_\beta = Z^{-1}(\beta) \sum_n e^{-\beta E_n} \langle n|A|n \rangle, \qquad (4.110)$$

where $Z(\beta)$ denotes the partition function of the system. Here the energy eigenstates $|n\rangle$ satisfy Dirac inner product and define a complete basis. To write the ensemble average as a thermal vacuum expectation of the operator, we recognized that it cannot be written as a superposition of states in the original Hilbert space and we needed to (at least) double the Hilbert space which also required doubling the operators. This led to the introduction of tilde states and tilde operators and we saw that (see (3.10)-(3.18)), in this doubled space ($|n, \tilde{m}\rangle = |n\rangle \otimes |\tilde{m}\rangle$), if we expand the thermal vacuum as

$$|0, \beta\rangle = \sum_n f_n(\beta)|n, \tilde{n}\rangle, \quad f_n(\beta) = \langle n, \tilde{n}|0, \beta\rangle. \qquad (4.111)$$

Requiring the thermal vacuum to be normalized and the expectation value of any operator in this state to lead to the thermal ensemble average determined (see (3.18))

$$f_n(\beta) = f_n^*(\beta) = Z^{-\frac{1}{2}}(\beta) e^{-\frac{\beta E_n}{2}}, \quad Z(\beta) = \sum_n e^{-\beta E_n}. \qquad (4.112)$$

And this led to thermofield dynamics as the unique thermal field theory in the operator language. All of this analysis in section **3.1**,

---

[1]The material in the following three subsections is from unpublished notes by A. Das.

however, had the tacit assumption that the thermal Hilbert space built on the doubled Hilbert space had a Dirac inner product, much like the energy basis states $|n\rangle$ and $|\tilde{n}\rangle$. This is exactly where we missed the possibility for an arbitrary parameter making its way. So, let us redo the analysis next without this particular assumption.

Let us assume that the thermal vacuum depends on a second parameter λ besides the temperature and that dependence leads to a Hilbert space which does not have a Dirac inner product. For example, as we have seen in section **4.2.1**, the Hilbert space built on the thermal vacuum $|0, (\beta, \lambda)\rangle$ has an inner product defined by

$$\langle \mathcal{O} \rangle_\Lambda = \langle 0, (\beta, \lambda)|\Lambda \mathcal{O}(\lambda)|0, (\beta, \lambda)\rangle, \tag{4.113}$$

for any operator $\mathcal{O}(\lambda)$ where the Hermitian operator Λ reflects the parameter λ, namely, $\lambda \xrightarrow{\Lambda} -\lambda$. In particular, when $\mathcal{O} = \mathbb{1}$, this leads to the normalization condition for the state to be

$$\langle \mathbb{1} \rangle_\Lambda = \langle 0, (\beta, \lambda)|\Lambda \mathbb{1}|0, (\beta, \lambda)\rangle = \langle 0, (\beta, -\lambda)|0, (\beta, \lambda)\rangle$$
$$= 1, \tag{4.114}$$

and for any physical operator A, the ensemble average is given by

$$\langle A \rangle_\beta = \langle 0, (\beta, \lambda)|\Lambda A|0, (\beta, \lambda)\rangle = \langle 0, (\beta, -\lambda)|A|0, (\beta, \lambda)\rangle. \tag{4.115}$$

Expanding the thermal vacuum as a linear superposition of spaces in the doubled Hilbert space, we can write, as in (4.111),

$$|0, (\beta, \lambda)\rangle = \sum_n f_n(\beta, \lambda)|n, \tilde{n}\rangle. \tag{4.116}$$

The only difference from (4.111) is that the expansion coefficients now depend on the parameter λ. Substituting (4.116) into (4.114)-(4.115) leads to

$$f_n^*(\beta, -\lambda)f_n(\beta, \lambda) = Z^{-1}(\beta)e^{\beta E_n}, \tag{4.117}$$

where this relation differs from the one in (3.17) in the fact that the coefficients of expansion, $f_n(\beta, \lambda)$, now have a dependence on an arbitrary real parameter λ. This additional dependence, however, now allows a solution to (4.116) of the form

$$f_n(\beta, \lambda) = Z^{-\frac{1}{2}}(\beta)e^{-(\frac{1}{2}-\lambda)\beta E_n},$$

$$f_n^*(\beta, -\lambda) = Z^{-\frac{1}{2}}(\beta)e^{-(\frac{1}{2}+\lambda)\beta E_n}, \tag{4.118}$$

and this is exactly how and where an arbitrary parameter makes its way into the operator formalism. Just for completeness, let us recall that, for the expansion of the thermal vacuum state in (4.116) to be meaningful, the expansion coefficients, $f_n(\beta, \lambda)$ must be well defined in the zero temperature ($\beta \to \infty$) limit. This requires

$$\frac{1}{2} - \lambda \geq 0, \quad \frac{1}{2} + \lambda \geq 0, \tag{4.119}$$

so that

$$-\frac{1}{2} \leq \lambda \leq \frac{1}{2}. \tag{4.120}$$

This is precisely what we have in (4.4) and, therefore, the two (parameters) can be identified.

Let us next note that if the harmonic oscillator has energy levels given by $E_n = n\omega$ (namely, we assume vanishing zero point energy as in a normal ordered theory and also we have set $\hbar = 1$), then, using (4.117) we can write the thermal vacuum in (4.116) as

$$|0, (\beta, \lambda)\rangle = Z^{-\frac{1}{2}}(\beta) \sum_n e^{-(\frac{1}{2} - \lambda)n\beta\omega} |n, \tilde{n}\rangle$$

$$= Z^{-\frac{1}{2}}(\beta) e^{\lambda\beta H} \sum_n e^{-\frac{\beta n\omega}{2}} |n, \tilde{n}\rangle$$

$$= e^{\lambda\beta H} |0, \beta\rangle, \tag{4.121}$$

where we have used the definitions of the thermal vacuum explicitly obtained in the last chapter or given in (4.111)-(4.112). This can be compared with (4.35). Furthermore, it now follows from (4.121) that

$$|0, (\beta, \lambda)\rangle = e^{\lambda\beta H} U(\theta) |0, \tilde{0}\rangle = e^{\lambda\beta H} U(\theta) e^{-\lambda\beta H} |0, \tilde{0}\rangle$$

$$= U(\theta, \lambda) |0, \tilde{0}\rangle, \tag{4.122}$$

where we have used the definition of the Bogoliubov transformation that takes the vacuum state of the doubled Hilbert space to the thermal vacuum in thermofield dynamics given in (3.67)-(3.69) (for the bosonic oscillator) with

$$U(\theta) = e^{-\theta(\beta)(\tilde{a}a - a^\dagger\tilde{a}^\dagger)}, \quad \tanh\theta(\beta) = e^{-\frac{\beta\omega}{2}}. \tag{4.123}$$

Following the steps in (4.65)-(4.68) we can now calculate and show explicitly that

$$U(\theta, \lambda) = e^{-\theta(\beta)(e^{-\lambda\beta\omega}\tilde{a}a - e^{\lambda\beta\omega}a^\dagger\tilde{a}^\dagger)},\tag{4.124}$$

which can be compared with (4.70). We can also explicitly check now that

$$\langle 0, (\beta, \lambda)|0, (\beta, \lambda)\rangle_\Lambda = \langle 0, (\beta, -\lambda)|0, (\beta, \lambda)\rangle$$

$$= Z^{-1}(\beta)\sum_n f_n^*(-\lambda)f_n(\lambda) = Z^{-1}(\beta)\sum_n e^{-\beta E_n} = 1,\tag{4.125}$$

$$U^\ddagger(\theta, \lambda)U(\theta, \lambda) = U^\dagger(\theta, -\lambda)U(\theta, \lambda)$$

$$= e^{\theta(\beta)(e^{-\lambda\beta\omega}\tilde{a}a - e^{\lambda\beta\omega}a^\dagger\tilde{a}^\dagger)}e^{-\theta(\beta)(e^{-\lambda\beta\omega}\tilde{a}a - e^{\lambda\beta\omega}a^\dagger\tilde{a}^\dagger)} = \mathbb{1},\tag{4.126}$$

as well as any other property associated with states and operators on a non-Dirac Hilbert space in this simple quantum mechanical model.

4.5.2 The propagator for a fermionic oscillator on an arbitrary path.
In all our discussions so far, we have talked about bosonic theories and propagators for simplicity. The methods go over to fermions equally well except that the spinor indices of fermions make it a bit cumbersome. In this subsection, we will derive the propagator of a fermion on an arbitrary path (in the real time formalism) in $0 + 1$ dimension where there are no spinor indices which makes the discussion simpler and completely parallel to the bosonic case.

To begin with, we note that the Lagrangian for a free massive fermion (of mass $m = \omega$) in $0 + 1$ dimension is given by ($\overline{\psi}$ simply corresponds to the adjoint $\psi^\dagger$ of ψ in $0 + 1$ dimension)

$$L = \overline{\psi}(i\partial_t - \omega)\psi,\tag{4.127}$$

which leads to the conjugate momenta (for the two fermion field variables) to be (we use left derivatives for fermions)

$$\Pi_\psi = \frac{\partial L}{\partial\dot{\psi}} = -i\overline{\psi}, \qquad \Pi_{\overline{\psi}} = \frac{\partial L}{\partial\dot{\overline{\psi}}} = 0.\tag{4.128}$$

The fermion system is constrained and following the standard methods for a constrained system, we obtain the Hamiltonian for the system to be[2]

$$H = -\Pi_\psi \dot{\psi} + \dot{\overline{\psi}}\Pi_{\overline{\psi}} - L = i\overline{\psi}\dot{\psi} - \overline{\psi}(i\partial_t - \omega)\psi = \omega\overline{\psi}\psi. \quad (4.129)$$

The equation of motion following from (4.127) gives

$$i\dot{\psi} - \omega\psi = 0, \quad \psi(t) = e^{-i\omega t}a_F, \quad \overline{\psi}(t) = e^{i\omega t}a_F^\dagger, \quad (4.130)$$

so that we can write

$$H = \omega a_F^\dagger a_F, \quad (4.131)$$

which can be compared with (3.19). We note here, for completeness, that if we write the Fourier transform of $\psi(t)$ as

$$\psi(t) = \int dp\, e^{-ipt}\, \psi(p), \quad \overline{\psi}(t) = \int dp\, e^{ipt}\, \overline{\psi}(p), \quad (4.132)$$

then, it follows from (4.130) that

$$\psi(p) = \delta(p - \omega)a_F, \quad \overline{\psi}(p) = \delta(p - \omega)a_F^\dagger. \quad (4.133)$$

The fermionic creation and the annihilation operators satisfy the standard anti-commutation relations given in (3.20) and the Hilbert space is two dimensional, there are only two energy states – vacuum state and the one particle state.

For clarity of discussions, we obtain the propagator for the fermion in two steps. First, we develop the system and the propagator in thermofield dynamics and then make the path arbitrary. To introduce temperature into the system, in the thermofield formalism, we recall (from chapter **3**) that we need to double the operators by introducing tilde operators, $\tilde{a}_F, \tilde{a}_F^\dagger$ with identical anti-commutation relations as the original operators. This, in turn, doubles the Hilbert space of states by introducing tilde states and, in this space, we choose the thermal vacuum state as a linear superposition of states of the form

$$|0, \beta\rangle = Z^{-1}(\beta) \sum_{n=0}^{1} f_n(\beta)|n, \tilde{n}\rangle, \quad (4.134)$$

[2]See, for example, section **10.1**, in particular, (10.11) and section **10.5**, in particular, (10.104) in A. Das, *Lectures on Quantum Field Theory* (second edition), World Scientific (2020), Singapore.

where we can determine (see section **3.2**)

$$f_n(\beta) = f_n^*(\beta) = Z^{-\frac{1}{2}}(\beta)e^{-\frac{\beta E_n}{2}},$$

$$Z(\beta) = \sum_{n=0}^{1} e^{-\beta E_n} = (1 + e^{-\beta\omega}), \tag{4.135}$$

by requiring the thermal vacuum state to be normalized and that the ensemble average of a bosonic operator is given by the thermal vacuum expectation value, namely,

$$\langle 0, \beta | 0, \beta \rangle = 1, \qquad \langle A \rangle_\beta = \langle 0, \beta | A | 0, \beta \rangle. \tag{4.136}$$

There are two equivalent ways we can obtain the propagator for the fermion oscillator and we will discuss both in the following. First, we will derive it from the Bogoliubov transformation that takes the zero temperature vacuum to the finite temperature one. In the second approach, we will calculate the fermion propagator directly from the expectation value of fermion operators in the thermal vacuum.

As we have already seen in section **3.2** (see (3.34)-(3.39)), the thermal vacuum in thermofield dynamics, given by (4.134)-(4.135), can be related to the doubled vacuum of the original space through a (unitary) Bogoliubov transformation

$$|0, \beta\rangle = U(\theta)|0, \widetilde{0}\rangle, \quad U(\theta) = e^{-\theta(\beta)(\widetilde{a}_F a_F - a_F^\dagger \widetilde{a}_F^\dagger)}. \tag{4.137}$$

With these, we can define two fermion doublets,

$$\Psi(t) = \begin{pmatrix} \psi(t) \\ \widetilde{\psi}(t) \end{pmatrix} = e^{-i\omega t} \begin{pmatrix} a_F \\ \widetilde{a}_F^\dagger \end{pmatrix},$$

$$\overline{\Psi}(t) = \begin{pmatrix} \overline{\psi}(t) \\ \widetilde{\psi}(t) \end{pmatrix} = e^{i\omega t} \begin{pmatrix} a_F^\dagger \\ \widetilde{a}_F \end{pmatrix}. \tag{4.138}$$

We can now follow the analysis in the last chapter (see (3.41)-(3.47), see also (4.74)-(4.75) with $\lambda = 0$) to obtain that, under a Bogoliubov transformation (4.137),

$$\Psi(t) \to U^\dagger(\theta)\Psi(t)U(\theta) = e^{-i\omega t}\,\overline{U}(-\theta(\beta)) \begin{pmatrix} a_F \\ \tilde{a}_F^\dagger \end{pmatrix},$$

$$\overline{\Psi}(t) \to U^\dagger(\theta)\overline{\Psi}(t)U(\theta) = e^{i\omega t}\,\overline{U}(-\theta(\beta)) \begin{pmatrix} a_F^\dagger \\ \tilde{a}_F \end{pmatrix}, \qquad (4.139)$$

with (see (3.38))

$$\overline{U}(-\theta(\beta)) = \begin{pmatrix} \cos\theta(\beta) & \sin\theta(\beta) \\ -\sin\theta(\beta) & \cos\theta(\beta) \end{pmatrix}, \ \theta(\beta) = \tan^{-1} e^{-\frac{\beta\omega}{2}}.$$
$$(4.140)$$

As a result, we have

$$\langle 0,\beta|\Psi_\alpha(t)\overline{\Psi}_\beta(t')|0,\beta\rangle$$
$$= e^{-i\omega(t-t')}\langle 0,\tilde{0}|(\overline{U}(-\theta(\beta))\Psi(t)(\overline{\Psi}(t'))^T\overline{U}^T(-\theta(\beta)))_{\alpha\beta}|0,\tilde{0}\rangle$$
$$= e^{-i\omega(t-t')}\left(\overline{U}(-\theta(\beta))\langle 0,\tilde{0}|\Psi(t)\overline{\Psi}^T(t')|0,\tilde{0}\rangle\overline{U}^T(-\theta(\beta))\right)_{\alpha\beta}$$
$$= e^{-i\omega(t-t')}\left(\overline{U}(-\theta(\beta)) \begin{pmatrix} 1 & 0 \\ 0 & 0 \end{pmatrix} \overline{U}^T(-\theta(\beta))\right)_{\alpha\beta}, \qquad (4.141)$$

$$\langle 0,\beta|\overline{\Psi}_\beta(t')\Psi_\alpha(t)|0,\beta\rangle$$
$$= e^{-i\omega(t-t')}\langle 0,\tilde{0}|\left(\overline{U}(-\theta(\beta))\overline{\Psi}(t')\Psi^T(t)\overline{U}^T(-\theta(\beta))\right)_{\beta\alpha}|0,\tilde{0}\rangle$$
$$= e^{-i\omega(t-t')}\left(\overline{U}(-\theta(\beta))\langle 0,\tilde{0}|\overline{\Psi}(t')\Psi^T(t)|0,\tilde{0}\rangle\overline{U}^T(-\theta(\beta))\right)_{\beta\alpha}$$
$$= e^{-i\omega(t-t')}\left(\overline{U}(-\theta(\beta)) \begin{pmatrix} 0 & 0 \\ 0 & 1 \end{pmatrix} \overline{U}^T(-\theta(\beta))\right)_{\beta\alpha}. \qquad (4.142)$$

Since the two matrices in (4.141) and (4.142) are symmetric, we obtain the matrix fermion propagator in $0+1$ dimension in thermofield dynamics (at finite temperature) to be (writing $\theta(\beta) = \theta$, for simplicity)

$$iS_{\alpha\beta}^{\text{(TFD)}}(t - t') = \langle 0, \beta | T(\Psi_\alpha(t)\overline{\Psi}_\beta(t')) | 0, \beta \rangle$$

$$= \langle 0, \beta | \theta(t - t')\Psi_\alpha(t)\overline{\Psi}_\beta(t') - \theta(t' - t)\overline{\Psi}_\beta(t')\Psi_\alpha(t) | 0, \beta \rangle$$

$$= e^{-i\omega(t-t')} \left(\overline{U}(-\theta) \begin{pmatrix} \theta(t - t') & 0 \\ 0 & -\theta(t' - t) \end{pmatrix} \overline{U}^T(-\theta) \right)_{\alpha\beta},$$

leading explicitly to

$$= e^{-i\omega(t-t')} \begin{pmatrix} \theta(t - t') - \sin^2\theta & -\frac{1}{2}\sin 2\theta \\ -\frac{1}{2}\sin 2\theta & -\theta(t' - t) + \sin^2\theta \end{pmatrix}_{\alpha\beta}. \quad (4.143)$$

Using (3.38) we can write the 2×2 propagator matrix for the $0 + 1$ dimensional fermion to be

$$iS^{\text{(TFD)}}(t) = \overline{U}(-\theta) \begin{pmatrix} e^{-i\omega t}\theta(t) & 0 \\ 0 & -e^{-i\omega t}\theta(-t) \end{pmatrix} \overline{U}^T(-\theta)$$

$$= e^{-i\omega t} \begin{pmatrix} \theta(t) - n_F(\omega) & -e^{\frac{\beta\omega}{2}} n_F(\omega) \\ -e^{\frac{\beta\omega}{2}} n_F(\omega) & -\theta(-t) + n_F(\omega) \end{pmatrix}$$

$$= e^{-i\omega t} \left[\begin{pmatrix} \theta(t) & 0 \\ 0 & -\theta(-t) \end{pmatrix} - n_F(\omega) \begin{pmatrix} 1 & e^{\frac{\beta\omega}{2}} \\ e^{\frac{\beta\omega}{2}} & -1 \end{pmatrix} \right]. \quad (4.144)$$

The fermion propagator separates into a zero temperature part and a temperature dependent part as we have already seen in the case of a scalar field theory and this form, for the fermion theory, can be compared to (and contrasted with) (3.119). In fact, taking the Fourier transform (and using (4.86)), the fermion propagator in the momentum space can be obtained to be

$$iS^{\text{(TFD)}}(p) = \begin{pmatrix} \frac{i}{p-\omega+i\epsilon} & 0 \\ 0 & -\frac{i}{p-\omega-i\epsilon} \end{pmatrix}$$

$$- 2\pi\delta(p - \omega)n_F(\omega) \begin{pmatrix} 1 & e^{\frac{\beta\omega}{2}} \\ e^{\frac{\beta\omega}{2}} & -1 \end{pmatrix}$$

$$= \overline{U}(-\theta) \begin{pmatrix} \frac{i}{p-\omega+i\epsilon} & 0 \\ 0 & -\frac{i}{p-\omega-i\epsilon} \end{pmatrix} \overline{U}^T(-\theta). \quad (4.145)$$

Therefore, the fermionic propagator, at finite temperature in thermofield dynamics, also factorizes like the bosonic propagator by the Bogoliubov transformation matrices (see (3.118)).

Next, let us calculate the thermal propagator directly and compare with (4.144). We recall that the KMS condition is quite important in the operator formalism (see sections **1.3** and **3.6**) and leads to the periodicity properties. As we have already seen in chapter **3** (for example, see (3.82) and (3.128)), the thermal state conditions lead to the KMS conditions. For example, (3.128) gives the thermal state condition and its adjoint for any operatort $A(t)$, bosonic or fermionic. Identifying $A(t) = \psi(t), \overline{\psi}(t)$ respectively, we obtain

$$\psi(t)|0,\beta\rangle = e^{\frac{\beta\widehat{H}}{2}}\widetilde{\overline{\psi}}(t)|0,\beta\rangle, \qquad \widetilde{\psi}(t)|0,\beta\rangle = -e^{-\frac{\beta\widehat{H}}{2}}\overline{\psi}(t)|0,\beta\rangle,$$

$$\langle 0,\beta|\overline{\psi}(t) = \langle 0,\beta|\widetilde{\psi}(t)e^{\frac{\beta\widehat{H}}{2}}, \qquad \langle 0,\beta|\widetilde{\overline{\psi}}(t) = -\langle 0,\beta|\psi(t)e^{-\frac{\beta\widehat{H}}{2}}.$$

$$\tag{4.146}$$

From these thermal state conditions, we can obtain the KMS conditions for the system, namely,

$$\langle 0,\beta|\psi(t)\overline{\psi}(t')|0,\beta\rangle = \langle 0,\beta|\widetilde{\overline{\psi}}(t)e^{\beta\widehat{H}}\widetilde{\psi}(t')e^{-\beta\widehat{H}}e^{\beta\widehat{H}}|0,\beta\rangle$$

$$= e^{\beta\omega}\langle 0,\beta|\widetilde{\overline{\psi}}(t)\widetilde{\psi}(t')|0,\beta\rangle = e^{-i\omega(t-t')}\,e^{\beta\omega}n_{\mathrm{F}}(\omega)$$

$$= e^{-i\omega(t-t')}\left(1 - n_{\mathrm{F}}(\omega)\right),$$

$$\langle 0,\beta|\psi(t)\widetilde{\psi}(t')|0,\beta\rangle = \langle 0,\beta|\widetilde{\overline{\psi}}(t)\overline{\psi}(t')|0,\beta\rangle$$

$$= -e^{-i\omega(t-t')}\,e^{\frac{\beta\omega}{2}}n_{\mathrm{F}}(\omega),$$

$$\langle 0,\beta|\widetilde{\psi}(t')\psi(t)|0,\beta\rangle = \langle 0,\beta|\overline{\psi}(t')\widetilde{\overline{\psi}}(t)|0,\beta\rangle$$

$$= e^{-i\omega(t-t')}\,e^{\frac{\beta\omega}{2}}n_{\mathrm{F}}(\omega),$$

$$\langle 0,\beta|\widetilde{\psi}(t')\widetilde{\overline{\psi}}(t)|0,\beta\rangle = \langle 0,\beta|\overline{\psi}(t')e^{-\beta\widehat{H}}\psi(t)e^{\beta\widehat{H}}e^{-\beta\widehat{H}}|0,\beta\rangle$$

$$= e^{\beta\omega}\langle 0,\beta|\overline{\psi}(t')\psi(t)|0,\beta\rangle = e^{-i\omega(t-t')}\,e^{\beta\omega}n_{\mathrm{F}}(\omega)$$

$$= e^{-i\omega(t-t')}(1 - n_{\mathrm{F}}(\omega)), \tag{4.147}$$

where we have used $\widehat{H}|0,\beta\rangle = 0$ as well as the facts that

$$e^{\pm\beta\widehat{H}}(a_{\mathrm{F}},\widetilde{a}_{\mathrm{F}}^{\dagger})e^{\mp\beta\widehat{H}} = e^{\mp\beta\omega}(a_{\mathrm{F}},\widetilde{a}_{\mathrm{F}}^{\dagger}),$$

and

$$e^{\pm\beta\hat{H}}(a_{\mathrm{F}}^{\dagger}, \tilde{a}_{\mathrm{F}})e^{\mp\beta\hat{H}} = e^{\pm\beta\omega}(a_{\mathrm{F}}^{\dagger}, \tilde{a}_{\mathrm{F}}). \tag{4.148}$$

If we now define the doublets of fields

$$\Psi(t) = \begin{pmatrix} \psi(t) \\ \tilde{\bar{\psi}}(t) \end{pmatrix}, \quad \overline{\Psi}(t') = \begin{pmatrix} \overline{\psi}(t') \\ \tilde{\psi}(t') \end{pmatrix}, \tag{4.149}$$

then, using (4.149), we can write the 2×2 matrix elements as

$$\langle 0, \beta | \Psi_{\alpha}(t)\overline{\Psi}_{\beta}(t') | 0, \beta \rangle = \langle 0, \beta | \begin{pmatrix} \psi(t)\overline{\psi}(t') & \psi(t)\tilde{\psi}(t') \\ \tilde{\bar{\psi}}(t)\overline{\psi}(t') & \tilde{\bar{\psi}}(t)\tilde{\psi}(t') \end{pmatrix}_{\alpha\beta} | 0, \beta \rangle$$

$$= e^{-i\omega(t-t')} \begin{pmatrix} (1 - n_{\mathrm{F}}(\omega)) & -e^{\frac{\beta\omega}{2}}n_{\mathrm{F}}(\omega) \\ -e^{\frac{\beta\omega}{2}}n_{\mathrm{F}}(\omega) & n_{\mathrm{F}}(\omega) \end{pmatrix}_{\alpha\beta}, \tag{4.150}$$

and, similarly, we obtain

$$\langle 0, \beta | \overline{\Psi}_{\beta}(t')\Psi_{\alpha}(t) | 0, \beta \rangle = \langle 0, \beta | \begin{pmatrix} \overline{\psi}(t')\psi(t) & \overline{\psi}(t')\tilde{\bar{\psi}}(t) \\ \tilde{\psi}(t')\psi(t) & \tilde{\psi}(t')\tilde{\bar{\psi}}(t) \end{pmatrix}_{\beta\alpha} | 0, \beta \rangle$$

$$= e^{-i\omega(t-t')} \begin{pmatrix} n_{\mathrm{F}}(\omega) & e^{\frac{\beta\omega}{2}}n_{\mathrm{F}}(\omega) \\ e^{\frac{\beta\omega}{2}}n_{\mathrm{F}}(\omega) & (1 - n_{\mathrm{F}}(\omega)) \end{pmatrix}_{\beta\alpha}. \tag{4.151}$$

Considering that the two matrices in (4.150) and (4.151) are symmetric, it follows now that the fermion propagator in $0+1$ dimension has the form

$$iS_{\alpha\beta}^{(\mathrm{TFD})}(t - t') = \langle 0, \beta | T(\Psi_{\alpha}(t)\overline{\Psi}_{\beta}(t')) | 0, \beta \rangle$$

$$= \langle 0, \beta | \theta(t - t')\Psi_{\alpha}(t)\overline{\Psi}_{\beta}(t') - \theta(t' - t)\overline{\Psi}_{\beta}(t')\Psi_{\alpha}(t) | 0, \beta \rangle$$

$$= e^{-i\omega(t-t')} \begin{pmatrix} (\theta(t - t') - n_{\mathrm{F}}(\omega)) & -e^{\frac{\beta\omega}{2}}n_{\mathrm{F}}(\omega) \\ -e^{\frac{\beta\omega}{2}}n_{\mathrm{F}}(\omega) & -(\theta(t' - t) - n_{\mathrm{F}}(\omega)) \end{pmatrix}_{\alpha\beta}, \tag{4.152}$$

which coincides with (4.144).

Let us next move on and derive the arbitrary path propagator for the one dimensional fermion. Let us recall from (4.34) and (4.35) that we can define the thermal vacuum for an arbitrary path as

$$|0, (\beta, \lambda)\rangle = U(\theta, \lambda)|0, \widetilde{0}\rangle = e^{\lambda\beta H} U(\theta)|0, \widetilde{0}\rangle$$

$$= e^{\lambda\beta H}|0, \beta\rangle, \tag{4.153}$$

where H denotes the Hamiltonian for the original system and $\lambda = \sigma - \frac{1}{2}$ stands for the real parameter describing the path (see (4.4)). As discussed in detail in section **4.2.1**, such a vacuum leads to a non-Dirac Hilbert space and the adjoint bra vacuum state as well as adjoints of operators are defined as

$$\left(|0, (\beta, \lambda)\rangle\right)^{\ddagger} = \langle 0, \widetilde{0}|U^{\ddagger}(\theta, \lambda) = \langle 0, \widetilde{0}|U^{\dagger}(\theta, -\lambda) = \langle 0, (\beta, -\lambda)|,$$

$$\mathcal{O}^{\ddagger}(\theta, \lambda) = \Lambda^{\dagger}\mathcal{O}^{\dagger}(\theta, \lambda)\Lambda = \mathcal{O}^{\dagger}(\theta, -\lambda), \tag{4.154}$$

where Λ denotes the reflection operator for the parameter λ (namely, takes $\lambda \to -\lambda$).

Let us next recall (see (3.32), (3.37) and (3.38)) that the thermal vacuum for the fermionic oscillator, in thermofield dynamics, can be written as (it is a two level system)

$$|0, \beta\rangle = Z^{-\frac{1}{2}}(\beta)\left(|0, \widetilde{0}\rangle + e^{-\frac{\beta\omega}{2}}|1, \widetilde{1}\rangle\right)$$

$$= \frac{1}{\sqrt{1 + e^{-\beta\omega}}}\left(|0, \widetilde{0}\rangle + e^{-\frac{\beta\omega}{2}}|1, \widetilde{1}\rangle\right)$$

$$= \cos\theta|0, \widetilde{0}\rangle + \sin\theta|1, \widetilde{1}\rangle, \quad \tan\theta = e^{-\frac{\beta\omega}{2}}. \tag{4.155}$$

Furthermore, we note that (see also (4.148))

$$e^{\pm\lambda\beta H} a_{\mathrm{F}} \, e^{\mp\lambda\beta H} = e^{\mp\lambda\beta\omega} a_{\mathrm{F}}, \qquad e^{\pm\lambda\beta H} a_{\mathrm{F}}^{\dagger} \, e^{\mp\lambda\beta H} = e^{\pm\lambda\beta\omega} a_{\mathrm{F}}^{\dagger},$$

$$e^{\lambda\beta H} \widetilde{a}_{\mathrm{F}} \, e^{-\lambda\beta H} = \widetilde{a}_{\mathrm{F}}, \qquad e^{\lambda\beta H} \widetilde{a}_{\mathrm{F}}^{\dagger} \, e^{-\lambda\beta H} = \widetilde{a}_{\mathrm{F}}^{\dagger}, \tag{4.156}$$

so that the thermal vacuum on an arbitrary path (see (4.153)) can be obtained to be

$$|0, (\beta, \lambda)\rangle = e^{\lambda\beta H}|0, \beta\rangle = e^{\lambda\beta H}\left(\cos\theta + \sin\theta a^{\dagger}\widetilde{a}^{\dagger}\right)e^{-\lambda\beta H}|0, \widetilde{0}\rangle$$

$$= \left(\cos\theta + e^{\lambda\beta\omega}\sin\theta a^{\dagger}\widetilde{a}^{\dagger}\right)|0, \widetilde{0}\rangle$$

$$= \cos\theta|0, \widetilde{0}\rangle + e^{\lambda\beta\omega}\sin\theta|1, \widetilde{1}\rangle, \tag{4.157}$$

where we have used $H|0,\widetilde{0}\rangle = 0$. This can be compared with (4.155) and the normalization of the state follows from

$$\langle 0, (\beta, -\lambda)|0, (\beta, \lambda)\rangle = \cos^2\theta \langle 0, \widetilde{0}|0, \widetilde{0}\rangle$$

$$+ e^{-\lambda\beta\omega}\sin\theta \times e^{\lambda\beta\omega}\sin\theta \langle 1, \widetilde{1}|1, \widetilde{1}\rangle$$

$$= \cos^2\theta + \sin^2\theta = 1. \tag{4.158}$$

Therefore, following (4.147)-(4.151) and using (4.156), we can obtain

$$\langle 0, (\beta, -\lambda)|\psi(t)\overline{\psi}(t')|0, (\beta, \lambda)\rangle = \langle 0, \beta|e^{-\lambda\beta H}\psi(t)\overline{\psi}(t')e^{\lambda\beta H}|0, \beta\rangle$$

$$= \langle 0, \beta|e^{-\lambda\beta H}\psi(t)e^{\lambda\beta H}e^{-\lambda\beta H}\,\overline{\psi}(t')e^{\lambda\beta H}|0, \beta\rangle$$

$$= e^{\lambda\beta\omega}e^{-\lambda\beta\omega}\langle 0, \beta|\psi(t)\overline{\psi}(t')|0, \beta\rangle = \langle 0, \beta|\psi(t)\overline{\psi}(t')|0, \beta\rangle,$$

and, similarly,

$$\langle 0, (\beta, -\lambda)|\psi(t)\widetilde{\psi}(t')|0, (\beta, \lambda)\rangle = -\langle 0, (\beta, -\lambda)|\widetilde{\psi}(t')\psi(t)|0, (\beta, \lambda)\rangle$$

$$= e^{\lambda\beta\omega}\langle 0, \beta|\psi(t)\widetilde{\psi}(t')|0, \beta\rangle = -e^{\lambda\beta\omega}\langle 0, \beta|\widetilde{\psi}(t')\psi(t)|0, \beta\rangle,$$

$$\langle 0, (\beta, -\lambda)|\overline{\psi}(t)\widetilde{\overline{\psi}}(t')|0, (\beta, \lambda)\rangle = -\langle 0, (\beta, -\lambda)|\widetilde{\overline{\psi}}(t')\overline{\psi}(t)|0, (\beta, \lambda)\rangle$$

$$= e^{-\lambda\beta\omega}\langle 0, \beta|\overline{\psi}(t)\widetilde{\overline{\psi}}(t')|0, \beta\rangle = -e^{-\lambda\beta\omega}\langle 0, \beta|\widetilde{\overline{\psi}}(t')\overline{\psi}(t)|0, \beta\rangle,$$

$$\langle 0, (\beta, -\lambda)|\widetilde{\overline{\psi}}(t)\widetilde{\psi}(t')|0, (\beta, \lambda)\rangle = \langle 0, \beta|\widetilde{\overline{\psi}}(t)\widetilde{\psi}(t')|0, \beta\rangle,$$

$$\langle 0, (\beta, -\lambda)|\widetilde{\psi}(t)\widetilde{\overline{\psi}}(t')|0, (\beta, \lambda)\rangle = \langle 0, \beta|\widetilde{\psi}(t)\widetilde{\overline{\psi}}(t')|0, \beta\rangle,$$

$$\langle 0, (\beta, -\lambda)|\overline{\psi}(t)\psi(t')|0, (\beta, \lambda)\rangle = \langle 0, \beta|\overline{\psi}(t)\psi(t')|0, \beta\rangle. \tag{4.159}$$

Following the steps in (4.142)-(4.144), we can now show that (we factor out $e^{-i\omega(t-t')}$ for simplicity)

$$\langle 0, (\beta, -\lambda)|\Psi_\alpha(t)\overline{\Psi}_\beta(t')|0, (\beta, \lambda)\rangle$$

$$= \begin{pmatrix} (1 - n_{\mathrm{F}}(\omega)) & -e^{(\lambda+\frac{1}{2})\beta\omega}n_{\mathrm{F}}(\omega) \\ -e^{-(\lambda-\frac{1}{2})\beta\omega}n_{\mathrm{F}}(\omega) & n_{\mathrm{F}}(\omega) \end{pmatrix}_{\alpha\beta}$$

$$= \left[\begin{pmatrix} 1 & 0 \\ 0 & e^{-\lambda\beta\omega} \end{pmatrix}\begin{pmatrix} (1 - n_{\mathrm{F}}(\omega)) & -e^{\frac{\beta\omega}{2}}n_{\mathrm{F}}(\omega) \\ -e^{\frac{\beta\omega}{2}}n_{\mathrm{F}}(\omega) & n_{\mathrm{F}}(\omega) \end{pmatrix}\begin{pmatrix} 1 & 0 \\ 0 & e^{\lambda\beta\omega} \end{pmatrix}\right]_{\alpha\beta}$$

$$= \left(\overline{V}(\lambda)\overline{U}(-\theta) \begin{pmatrix} 1 & 0 \\ 0 & 0 \end{pmatrix} \overline{U}^{-1}(-\theta)\overline{V}^{-1}(\lambda) \right)_{\alpha\beta}$$

$$= \left(\overline{V}(\lambda)\overline{U}(-\theta)\overline{V}^{-1}(\lambda) \begin{pmatrix} 1 & 0 \\ 0 & 0 \end{pmatrix} \overline{V}(\lambda)\overline{U}^{-1}(-\theta)\overline{V}^{-1}(\lambda) \right)_{\alpha\beta},$$

$$(4.160)$$

where $\overline{U}(-\theta)$ is defined in (4.140) and

$$\overline{V}(\lambda) = \begin{pmatrix} 1 & 0 \\ 0 & e^{-\lambda\beta\omega} \end{pmatrix} = \overline{V}^T(\lambda), \quad \overline{V}^{-1}(\lambda) = \overline{V}(-\lambda). \qquad (4.161)$$

We have also used the fact that $\overline{V}(\lambda)$ and $\overline{V}^{-1}(\lambda)$ are diagonal matrices which commute with the diagonal matrix in the middle in (4.160) to give a factor of identity for the additional terms. This also allows us to identify

$$\overline{U}(-\theta, \lambda) = \overline{V}(\lambda)\overline{U}(-\theta)\overline{V}^{-1}(\lambda)$$

$$= \begin{pmatrix} \cos\theta & e^{\lambda\beta\omega}\sin\theta \\ -e^{-\lambda\beta\omega}\sin\theta & \cos\theta \end{pmatrix}, \qquad (4.162)$$

which satisfies

$$\overline{U}^\dagger(-\theta, -\lambda)\overline{U}(-\theta, \lambda) = \overline{U}^\ddagger(-\theta, \lambda)\overline{U}(-\theta, \lambda) = \mathbb{1}. \qquad (4.163)$$

Therefore, we can now write (putting back the exponential factor)

$$\langle 0, (\beta, -\lambda)|\Psi_\alpha(t)\overline{\Psi}_\beta(t')|0, (\beta, \lambda)\rangle$$

$$= e^{-i(\omega(t-t'))} \begin{pmatrix} (1 - n_F(\omega)) & -e^{(\lambda+\frac{1}{2})\beta\omega}n_F(\omega) \\ -e^{-(\lambda-\frac{1}{2})\beta\omega}n_F(\omega) & n_F(\omega) \end{pmatrix}_{\alpha\beta}$$

$$= e^{-i\omega(t-t')} \left(\overline{U}(-\theta, -\lambda) \begin{pmatrix} 1 & 0 \\ 0 & 0 \end{pmatrix} \overline{U}^T(-\theta, \lambda) \right)_{\alpha\beta}. \qquad (4.164)$$

Similarly, we can show that

$$\langle 0, (\beta, \lambda)|\overline{\Psi}_\beta(t')\Psi_\alpha(t)|0, (\beta, \lambda)\rangle$$

$$= e^{-i\omega(t-t')} \begin{pmatrix} n_F(\omega) & -e^{(\lambda+\frac{1}{2})\beta\omega}n_F(\omega) \\ -e^{-(\lambda-\frac{1}{2})\beta\omega}n_F(\omega) & (1 - n_F(\omega)) \end{pmatrix}$$

$$= e^{-i\omega(t-t')} \left(\overline{U}(-\theta, -\lambda) \begin{pmatrix} 0 & 0 \\ 0 & 1 \end{pmatrix} \overline{U}^T(-\theta, \lambda)\right)_{\beta\alpha}. \qquad (4.165)$$

so that the one dimensional fermion propagator on an arbitrary path follows to be (see, for example, (4.152))

$$iS_{\alpha\beta}^{(\lambda)}(t - t') = e^{-i\omega(t-t')}$$

$$\times \left(\overline{U}(-\theta, -\lambda) \begin{pmatrix} \theta(t - t') & 0 \\ 0 & -\theta(t' - t) \end{pmatrix} \overline{U}^T(-\theta, \lambda)\right)_{\alpha\beta}$$

$$= e^{-i\omega(t-t')} \begin{pmatrix} \theta(t - t') - n_F(\omega) & -e^{(\lambda+\frac{1}{2})\beta\omega}n_F(\omega) \\ -e^{-(\lambda-\frac{1}{2})\beta\omega}n_F(\omega) & -\theta(t' - t) + n_F(\omega) \end{pmatrix}_{\alpha\beta}.$$

$$(4.166)$$

For $\lambda = 0$, this indeed reduces to the fermion propagator in thermofield dynamics obtained in (4.152), as it should.

4.5.3 KMS condition on an arbitrary path.

The KMS conditions, for thermofield dynamics, are already discussed in section **3.6**. Here we will derive systematically the corresponding conditions for an arbitrary path. To do that, let us first derive the thermal conditions for an arbitrary path. We note that

$$a|0, \widetilde{0}\rangle = 0 = \widetilde{a}|0, \widetilde{0}\rangle,$$

where $a, \widetilde{a}$ can be bosonic or fermionic annhilation operators and this leads to

$$U(\theta, \lambda)\, a\, U^\dagger(\theta, -\lambda)U(\theta, \lambda)|0, \widetilde{0}\rangle = 0,$$

$$\text{or,} \quad U(\theta, \lambda)\, a\, U^\dagger(\theta, -\lambda)|0, (\beta, \lambda)\rangle = 0, \qquad (4.167)$$

and, similarly,

$$U(\theta, \lambda)\, \widetilde{a}\, U^\dagger(\theta, -\lambda)|0, (\beta, \lambda)\rangle = 0, \qquad (4.168)$$

where

$$U(\theta, \lambda) = e^{\lambda \beta H} U(\theta) e^{-\lambda \beta H} = e^{\lambda \beta H} e^{-\theta(\tilde{a}a - a^\dagger \tilde{a}^\dagger)} e^{-\lambda \beta H}$$

$$= e^{-\theta(e^{-\lambda \beta \omega} \tilde{a}a - e^{\lambda \beta \omega} a^\dagger \tilde{a}^\dagger)}. \tag{4.169}$$

Here we have used (4.156) which holds for both bosonic and fermionic variables. With this, we can calculate and obtain

$$U(\theta, \lambda) \, a \, U^\dagger(\theta, -\lambda) = \cosh \theta \, a - e^{\lambda \beta \omega} \sinh \theta \, \tilde{a}^\dagger, \quad \tanh \theta = e^{-\frac{\beta \omega}{2}},$$

$$U(\theta, \lambda) \, \tilde{a} \, U^\dagger(\theta, -\lambda) = \cosh \theta \, \tilde{a} - e^{\lambda \beta \omega} \sinh \theta \, a^\dagger, \tag{4.170}$$

for bosons, while for fermions, we obtain

$$U(\theta, \lambda) \, a \, U^\dagger(\theta, -\lambda) = \cos \theta \, a - e^{\lambda \beta \omega} \sin \theta \, \tilde{a}^\dagger, \quad \tan \theta = e^{-\frac{\beta \omega}{2}},$$

$$U(\theta, \lambda) \, \tilde{a} \, U^\dagger(\theta, -\lambda) = \cos \theta \, \tilde{a} + e^{\lambda \beta \omega} \sin \theta \, a^\dagger. \tag{4.171}$$

Substituting (4.170) and (4.171) into (4.167) and (4.168) (and factoring out $\cosh \theta$ and $\cos \theta$ respectively), we can write the two relations in a unified manner as

$$a|0, (\beta, \lambda)\rangle = e^{(\lambda - \frac{1}{2})\beta \omega} \, \tilde{a}^\dagger |0, (\beta, \lambda)\rangle$$

$$= e^{-(\lambda - \frac{1}{2})\beta \hat{H}} \, \tilde{a}^\dagger |0, (\beta, \lambda)\rangle, \tag{4.172}$$

$$\tilde{a}|0, (\beta, \lambda)\rangle = (-1)^{|a|} \, e^{(\lambda - \frac{1}{2})\beta \omega} \, a^\dagger |0, (\beta, \lambda)\rangle$$

$$= (-1)^{|a|} \, e^{-(\lambda - \frac{1}{2})\beta \hat{H}} \, a^\dagger |0, (\beta, \lambda)\rangle. \tag{4.173}$$

Here $|a|$ denotes the Grassmann parity of the variable and we have used the fact that $\hat{H}$ is invariant under the Bogoliubov transformation and, therefore, annihilates the vacuum, $\hat{H}|0, (\beta, \lambda)\rangle = 0$. We have also used the relations given in (4.148) in obtaining the final form.

The two relations in (4.172) and (4.173) denote the thermal state conditions for the thermal vacuum on an arbitrary path. We can write these conditions in terms of $\psi(t), \tilde{\psi}(t)$ and they take the forms

$$\psi(t)|0, (\beta, \lambda)\rangle = e^{-(\lambda - \frac{1}{2})\beta \hat{H}} \, \overline{\tilde{\psi}}(t)|0, (\beta, \lambda)\rangle,$$

$$\tilde{\psi}(t)|0, (\beta, \lambda)\rangle = (-1)^{|\psi|} \, e^{-(\lambda - \frac{1}{2})\beta \hat{H}} \, \overline{\psi}(t)|0, (\beta, \lambda)\rangle. \tag{4.174}$$

Taking the adjoints of these two relations (and letting $\lambda \to -\lambda$ since the Hilbert space is not a Dirac space), we obtain

$$\langle 0, (\beta, -\lambda)|\overline{\widetilde{\psi}}(t) = \langle 0, (\beta, -\lambda)|\widetilde{\psi}(t)e^{(\lambda+\frac{1}{2})\beta\widehat{H}},$$

$$\langle 0, (\beta, -\lambda)|\widetilde{\overline{\psi}}(t) = (-1)^{|\psi|} \langle 0, (\beta, -\lambda)|\psi(t)e^{(\lambda+\frac{1}{2})\beta\widehat{H}}. \tag{4.175}$$

Using (4.174) and (4.175), we can now derive KMS conditions for products of operators on an arbitrary path. For example, we can now write

$$\langle 0, (\beta, -\lambda)|\overline{\widetilde{\psi}}(t)\psi(t')|0, (\beta, \lambda)\rangle$$

$$= \langle 0, (\beta, -\lambda)|\widetilde{\psi}(t)e^{(\lambda+\frac{1}{2})\beta\widehat{H}}e^{-(\lambda-\frac{1}{2})\beta\widehat{H}}\,\overline{\widetilde{\psi}}(t')|0, (\beta, \lambda)\rangle$$

$$= \langle 0, (\beta, -\lambda)|\widetilde{\psi}(t)e^{\beta\widehat{H}}\overline{\widetilde{\psi}}(t')e^{-\beta\widehat{H}}|0, (\beta, \lambda)\rangle$$

$$= e^{-\beta\omega}\langle 0, (\beta, -\lambda)|\widetilde{\psi}(t)\overline{\widetilde{\psi}}(t')|0, (\beta, \lambda)\rangle, \tag{4.176}$$

and, similarly,

$$\langle 0, (\beta, -\lambda)|\psi(t)\overline{\widetilde{\psi}}(t')|0, (\beta, \lambda)\rangle$$

$$= e^{\beta\omega}\langle 0, (\beta, -\lambda)|\overline{\widetilde{\psi}}(t)\widetilde{\psi}(t')|0, (\beta, \lambda)\rangle, \tag{4.177}$$

$$\langle 0, (\beta, -\lambda)|\psi(t)\widetilde{\psi}(t')|0, (\beta, \lambda)\rangle$$

$$= \langle 0, (\beta, -\lambda)|\overline{\widetilde{\psi}}(t)e^{-(\lambda+\frac{1}{2})\beta\widehat{H}}e^{-(\lambda-\frac{1}{2})\beta\widehat{H}}\,\overline{\widetilde{\psi}}(t')|0, (\beta, \lambda)\rangle$$

$$= \langle 0, (\beta, -\lambda)|\overline{\widetilde{\psi}}(t)\,e^{-2\lambda\beta\widehat{H}}\,\overline{\widetilde{\psi}}(t')e^{2\lambda\beta\widehat{H}}|0, (\beta, \lambda)\rangle$$

$$= e^{-2\lambda\beta\omega}\langle 0, (\beta, -\lambda)|\overline{\widetilde{\psi}}(t)\overline{\widetilde{\psi}}(t')|0, (\beta, \lambda)\rangle. \tag{4.178}$$

We see the general feature that the diagonal matrix elements in (4.160) (see the second line) and (4.165) are independent of the arbitrary parameter λ while the off-diagonal elements do depend on this parameter. Second, the explicit relation between the two diagonal elements in (4.165) (or the ratio between the two) coincides completely with that given in (4.176) while (4.177) verifies precisely the relation between the two diagonal elements in (4.160). Similarly, the relation between the two off-diagonal elements in (4.160) (or (4.165)) coincides precisely with that given in (4.178). This concludes our discussion of the KMS condition for thermal vacuum expectation values on an arbitrary path.

4.6 References

H. Chu and H. Umezawa, International Journal of Modern Physics **A09**, 2363 (1994).

A. Das (unpublished notes (2021)).

A. Das and L. Greenwood, Physics Letters **B678**, 504 (2009).

A. Das and L. Greenwood, Journal of Mathematical Physics (N. Y.), **51**, 042103 (2010).

A. Das and P. Kalauni, Physical Review **D93**, 125028 (2016).

P. Elmfors and H. Umezawa, Physica **A202**, 577 (1994).

T. S. Evans, I. Hardman, H. Umezawa and Y. Yamanaka, Journal of Mathematical Physics (N. Y.) **33**, 370 (1992).

P. A. Henning and H. Umezawa, Nuclear Physics **B417**, 463 (1994).

P. A. Henning, Physics Reports **253**, 235 (1995).

H. Matsumoto, Y. Nokano, H. Umezawa, F. Mancini and M. Marinaro, Progress of Theoretical Physics **70**, 599 (1983).

H. Matsumoto, Y. Nokano and H. Umezawa, Journal of Mathematical Physics (N. Y.) **25**, 3076 (1984).

CHAPTER 5

Gauge theories

Gauge theories are theories with local invariances. In the simplest form, they might just describe theories invariant under local phase transformations of the field variables, but can also have more complicated symmetry structures. As we understand now, gauge theories are absolutely essential to describe the fundamental forces of nature. It is, therefore, crucial to understand the structure of such theories, both at zero temperature as well as at finite temperature. In the next few sections, we will first try to understand the essential features of these theories at zero temperature before going onto the case of finite temperature.

5.1 Gauge theories at zero temperature

The simplest of the gauge theories, of course, dates back to Maxwell. Maxwell's theory, as a quantum theory, is based on a local $U(1)$ phase invariance of the matter fields. Since the symmetry group $U(1)$ has only one generator, the symmetry algebra is Abelian and correspondingly, Maxwell's theory describes an Abelian gauge theory. This theory describes quite well the interactions between electrons and photons. However, being an Abelian gauge theory, the structure of this theory – Quantum Electrodynamics (QED) – is much simpler compared to the other gauge theories that one deals with. To bring out various interesting structures associated with gauge theories, therefore, we would confine ourselves to non-Abelian gauge theories or Yang-Mills theories in this section.

Let us consider a commonly used group such as $SU(n)$. This is a Lie group with $n^2 - 1$ Hermitian generators, T^a, $a = 1, 2, \cdots, n^2 - 1$, satisfying the Lie algebra (summation over repeated indices is assumed throughout)

$$[T^a, T^b] = if^{abc} T^c, \tag{5.1}$$

where $a, b, c = 1, 2, \cdots, n^2 - 1$ and the real constants, f^{abc}, which are completely antisymmetric in the three indices, are known as the structure constants of the group (algebra). We note that unlike the group $U(1)$, here the generators of the Lie algebra do not commute with one another and, consequently, such an algebra is known as a non-Abelian Lie algebra. The structure constants (including the factor of i), in fact, provide a representation of the generators of the Lie algebra known as the adjoint representation.

The simplest way to introduce the Yang-Mills theory is to look at a free fermion theory where there are n fermions belonging to the fundamental representation of the Lie group $SU(n)$. Thus, we can write the Lagrangian density for such a free theory as

$$\mathcal{L}_0 = \overline{\psi}^i(x)(i\gamma^\mu \partial_\mu - m)\psi^i(x), \tag{5.2}$$

where we assume the fermions to have a mass m and $i = 1, 2, \cdots, n$. It is easy to check that this Lagrangian density is invariant under an infinitesimal global $SU(n)$ transformation of the form

$$\delta\psi^i(x) = -i\epsilon^a (t^a)^{ij} \psi^j(x),$$
$$\delta\overline{\psi}^i(x) = i\epsilon^a \overline{\psi}^j(x)(t^a)^{ji}, \tag{5.3}$$

where the infinitesimal parameters of the transformation, ϵ^a, are real, assumed to be space-time independent and t^a's represent the generators of the $SU(n)$ Lie algebra defined in (5.1) in the fundamental representation to which the fermions belong. (We will reserve T^a to denote the generators in the adjoint representation. We also note here that the global invariance of the Lagrangian density in (5.2) is really $U(n)$. But, for simplicity, we will work with only the $SU(n)$ part of the symmetry of the free theory.) We note here that the conserved current (Nöther current) associated with the $SU(n)$ global invariance under (5.3) of the free theory has the form

$$J^{\mu a}(x) = \overline{\psi}^i(x)\gamma^\mu (t^a)^{ij} \psi^j(x). \tag{5.4}$$

Just like in the case of Quantum Electrodynamics (QED), we can now ask when can this global symmetry be promoted to a local symmetry, namely, when can the parameters of transformation be local. We note that if the parameters of the transformation in (5.3) were local, namely, if $\epsilon^a = \epsilon^a(x)$, then the Lagrangian density in (5.2) will no longer be invariant under the transformations in (5.3). In fact, the change in the Lagrangian density under a local $SU(n)$ transformation of the kind in (5.3) will have the form

$$
\begin{aligned}
\delta\mathcal{L}_0 &= \delta\overline{\psi}^i(x)(i\gamma^\mu\partial_\mu - m)\psi^i(x) + \overline{\psi}^i(x)(i\gamma^\mu\partial_\mu - m)\delta\psi^i(x) \\
&= i\epsilon^a(x)\overline{\psi}^i(x)(t^a)^{ij}(i\gamma^\mu\partial_\mu - m)\psi^j(x) \\
&\quad - i\overline{\psi}^i(x)(i\gamma^\mu\partial_\mu - m)(t^a)^{ij}(\epsilon^a(x)\psi^j(x)) \\
&= (\partial_\mu\epsilon^a(x))\,\overline{\psi}^i(x)\gamma^\mu(t^a)^{ij}\psi^j(x) = (\partial_\mu\epsilon^a(x))J^{\mu a}(x), \quad (5.5)
\end{aligned}
$$

where we have used the fact that the matrices t^a and γ^μ commute since they belong to different spaces (internal symmetry space and Lorentz space respectively). Therefore, if we were to introduce a gauge interaction, completely parallel with QED, it will have the form (we set the gauge coupling to unity for simplicity)

$$
\mathcal{L}_{int} = -J^{\mu a}(x)A^a_\mu(x) = -\overline{\psi}^i(x)\gamma^\mu(t^a)^{ij}\psi^j(x)A^a_\mu(x). \quad (5.6)
$$

We can think of the $A^a_\mu(x)$ as the appropriate gauge fields for the present case and we note that, unlike the case of QED, here there will be $a = 1, 2, \cdots, n^2 - 1$ gauge fields. The change in the interaction Lagrangian density, (5.6), under a local transformation of the kind in (5.3) can also be easily calculated

$$
\begin{aligned}
\delta\mathcal{L}_{int} &= -\delta\overline{\psi}^i(x)\gamma^\mu(t^a)^{ij}\psi^j(x)A^a_\mu(x) - \overline{\psi}^i(x)\gamma^\mu(t^a)^{ij}\delta\psi^j(x)A^a_\mu(x) \\
&\quad - \overline{\psi}^i(x)\gamma^\mu(t^a)^{ij}\psi^j(x)\delta A^a_\mu(x) \\
&= -i\overline{\psi}^i(x)\gamma^\mu(t^c)^{ik}(t^b)^{kj}\psi^j(x)A^b_\mu(x)\epsilon^c(x) \\
&\quad + i\overline{\psi}^i(x)\gamma^\mu(t^b)^{ik}(t^c)^{kj}\psi^j(x)A^b_\mu(x)\epsilon^c(x) \\
&\quad - \overline{\psi}^i(x)\gamma^\mu(t^a)^{ij}\psi^j(x)\delta A^a_\mu(x) \\
&= iA^b_\mu(x)\epsilon^c(x)\overline{\psi}^i(x)\gamma^\mu[t^b, t^c]^{ij}\psi^j(x) - \delta A^a_\mu(x)\overline{\psi}^i\gamma^\mu(t^a)^{ij}\psi^j(x) \\
&= -f^{abc}A^b_\mu(x)\epsilon^c(x)\overline{\psi}^i(x)\gamma^\mu(t^a)^{ij}\psi^j(x)
\end{aligned}
$$

$$- \delta A_\mu^a(x) \overline{\psi}^i(x) \gamma^\mu (t^a)^{ij} \psi^j(x). \qquad (5.7)$$

In deriving the last relation in the above result, we have used the commutation relation for the generators, t^a, given in (5.1) as well as the symmetry properties of the structure constants.

Thus, we see from (5.5) and (5.7) that the total Lagrangian density including the interactions

$$\mathcal{L}_{\mathrm{f}} = \mathcal{L}_0 + \mathcal{L}_{int} = \overline{\psi}^i(x)(i\gamma^\mu(\partial_\mu \psi^i(x) + i(t^a)^{ij} A_\mu^a(x) \psi^j(x)))$$
$$- m\overline{\psi}^i(x)\psi^i(x), \qquad (5.8)$$

changes under the transformations of (5.3) as (when the parameters are local)

$$\delta\mathcal{L}_{\mathrm{f}} = \left[(\partial_\mu \epsilon^a(x) - f^{abc} A_\mu^b(x)\epsilon^c(x)) - \delta A_\mu^a(x)\right]$$
$$\times \overline{\psi}^i(x)\gamma^\mu (t^a)^{ij} \psi^j(x). \qquad (5.9)$$

It is clear now that the total Lagrangian density will be invariant under local transformations of the form (5.3) if we choose the gauge fields also to transform and in the form

$$\delta A_\mu^a(x) = (\partial_\mu \epsilon^a(x) - f^{abc} A_\mu^b(x)\epsilon^c(x)). \qquad (5.10)$$

We note that this is precisely how the gauge field transforms in QED for local gauge invariance where, for $U(1)$, the structure constants vanish. (From now on, we will suppress the coordinate dependence of fields and parameters as much as is possible for simplicity and should be understood.)

For any field, belonging to a given representation of $SU(n)$, we can define a covariant derivative with the generators of the group, in that particular representation, as well as the gauge fields (also called connections). Thus, for example, for the fermions in the fundamental representation in the present example, the covariant derivative is defined to be

$$(D_\mu \psi)^i = (\partial_\mu \psi^i + i(t^a)^{ij} A_\mu^a \psi^j). \qquad (5.11)$$

It can be easily checked that this, indeed, transforms covariantly under the local transformations of the form in (5.3) and (5.10), namely,

$$
\begin{aligned}
\delta(D_\mu \psi)^i &= \partial_\mu \delta \psi^i + i(t^a)^{ij}(\delta A_\mu^a \psi^j + A_\mu^a \delta \psi^j) \\
&= -i(t^a)^{ij}\partial_\mu(\epsilon^a(x)\psi^j) \\
&\quad + i(t^a)^{ij}\left((\partial_\mu \epsilon^a(x) - f^{abc}A_\mu^b \epsilon^c(x))\psi^j - iA_\mu^a \epsilon^b(x)(t^b)^{jk}\psi_k\right) \\
&= -i\epsilon^a(x)(t^a)^{ij}\partial_\mu \psi^j + (t^a t^b)^{ij}A_\mu^b \epsilon^a(x)\psi^j \\
&= -i\epsilon^a(x)(t^a)^{ij}\left(\partial_\mu \psi^j + i(t^b)^{jk}A_\mu^b \psi^k\right) \\
&= -i\epsilon^a(x)(t^a)^{ij}(D_\mu \psi)^j,
\end{aligned}
\tag{5.12}
$$

where we have replaced $if^{abc}(t^a)^{ij}$ by the commutator relation (5.1) in the intermediate step. With this definition of the covariant derivative, we can also write the total fermion Lagrangian density (5.8) more compactly as

$$
\mathcal{L}_{\mathrm{f}} = i\overline{\psi}^i \gamma^\mu (D_\mu \psi)^i - m\overline{\psi}^i \psi^i.
\tag{5.13}
$$

Observing that the covariant derivative transforms covariantly under the local transformations (see (5.12)), the invariance of this Lagrangian density under the local transformation in (5.3) and (5.10) is now manifest.

Let us recall that if^{abc} provides a representation for the generators of $SU(n)$ known as the adjoint representation. In fact, it is easy to check, using Jacobi identity, that the exact identification is given by (the generators, in this case, are $(n^2 - 1) \times (n^2 - 1)$ matrices, see, for example, the discussion in the beginning of the second paragraph of this section)

$$
(T^a)^{bc} = -if^{abc}.
\tag{5.14}
$$

From this as well as from the definition of the covariant derivative in (5.10), we note that the change in the gauge fields under the local transformations can also be written as

$$
\begin{aligned}
\delta A_\mu^a &= \partial_\mu \epsilon^a(x) - f^{abc}A_\mu^b \epsilon^c(x) = \partial_\mu \epsilon^a(x) + f^{bac}A_\mu^b \epsilon^c(x) \\
&= \partial_\mu \epsilon^a(x) + i(-if^{bac})A_\mu^b \epsilon^c(x) = \partial_\mu \epsilon^a(x) + i(T^b)^{ac}A_\mu^b \epsilon^c(x) \\
&= (D_\mu \epsilon(x))^a,
\end{aligned}
\tag{5.15}
$$

where the covariant derivative on the right hand side of (5.15) corresponds to the adjoint representation. This shows that the gauge fields A_μ^a as well as the parameter $\epsilon^a(x)$ belong to the adjoint representation of $SU(n)$. There are, in fact, $n^2 - 1$ components of the gauge fields and, hence, one can also define a matrix valued gauge field as

$$A_\mu = A_\mu^a T^a,$$

where, as we have mentioned earlier, T^a's represent the generators in the adjoint representation. (We can, of course, define a gauge field matrix using the generators in any other representation as well, see, for example, (5.11).)

To give dynamics to the gauge fields, we follow exactly what one does in QED. Namely, we define a field strength tensor associated with the gauge fields (potentials) as

$$F_{\mu\nu}^a = \partial_\mu A_\nu^a - \partial_\nu A_\mu^a - f^{abc} A_\mu^b A_\nu^c. \tag{5.16}$$

This, of course, reduces to the form of the Maxwell field strength tensor if we assume the structure constants to vanish (as is true for Abelian symmetries). We can also scale the gauge fields to bring in a dependence on the coupling constant in the last term in (5.16), which would correspond to the standard form in physical theories. In this form, one would see the essential difference between an Abelian and a non-Abelian field strength tensors in that, because the $SU(n)$ gauge fields carry the internal quantum numbers (generalized charges), they couple to themselves. In contrast, the photon does not carry an electric charge and hence does not have self-couplings. For simplicity, however, we will work with the form of the field strength tensor in (5.16) which can also be equivalently thought of as corresponding to a unit coupling constant. It is easy to check from the definitions in (5.15) and (5.16) that for any quantity, $\alpha^a(x)$, belonging to the adjoint representation, we have

$$([D_\mu, D_\nu]\alpha)^a(x) = D_\mu D_\nu \alpha^a(x) - D_\nu D_\mu \alpha^a(x)$$

$$= \partial_\mu (D_\nu \alpha)^a(x) - f^{abc} A_\mu^b (D_\nu \alpha)^c(x) - (\mu \leftrightarrow \nu)$$

$$= \partial_\mu (\partial_\nu \alpha^a(x) - f^{abc} A_\nu^b \alpha^c(x))$$

$$- f^{abc} A_\mu^b (\partial_\nu \alpha^c(x) - f^{cpq} A_\nu^p \alpha^q(x)) - (\mu \leftrightarrow \nu)$$

$$= -f^{abc}(\partial_\mu A_\nu^b - \partial_\nu A_\mu^b)\alpha^c(x)$$
$$+ (f^{abc}f^{cpq} + f^{apc}f^{cqb})A_\mu^b A_\nu^p \alpha^q(x)$$
$$= -f^{abc}(\partial_\mu A_\nu^b - \partial_\nu A_\mu^b - f^{bpq}A_\mu^p A_\nu^q)\alpha^c(x)$$
$$= -f^{abc} F_{\mu\nu}^b \alpha^c(x), \tag{5.17}$$

where we have used the Jacobi identity for the structure constants in the intermediate step. The definition of the field strength tensor in (5.16) is such that it transforms covariantly under a gauge transformation of the form in (5.15). This, can be checked from the definition in (5.16) with the help of (5.17) as

$$\delta F_{\mu\nu}^a = \partial_\mu(\delta A_\nu^a) - \partial_\nu(\delta A_\mu^a) - f^{abc}(A_\mu^b(\delta A_\nu^c) + (\delta A_\mu^b)A_\nu^c)$$
$$= (D_\mu \delta A_\nu)^a - (D_\nu \delta A_\mu)^a$$
$$= (D_\mu D_\nu \epsilon(x))^a - (D_\nu D_\mu \epsilon(x))^a$$
$$= ([D_\mu, D_\nu]\epsilon(x))^a = -f^{abc} F_{\mu\nu}^b \epsilon^c(x). \tag{5.18}$$

The form of a Lagrangian density for the gauge fields, which will be at most quadratic in the derivatives and will also be gauge invariant, is now easily seen to be (except for the normalization)

$$\mathcal{L}_{\rm G} = -\frac{1}{4} F_{\mu\nu}^a F^{\mu\nu\,a}. \tag{5.19}$$

Gauge invariance is, in fact, obvious, namely,

$$\delta\mathcal{L}_{\rm G} = -\frac{1}{2} F^{\mu\nu\,a}\delta F_{\mu\nu}^a = \frac{1}{2}F^{\mu\nu\,a} f^{abc} F_{\mu\nu}^b \epsilon^c = 0,$$

where the last step follows because of the antisymmetry of f^{abc} in the a, b indices. Thus, the complete Lagrangian density describing the interactions between fermions and the dynamical non-Abelian gauge fields is given by (see (5.13) and (5.19))

$$\mathcal{L} = \mathcal{L}_{\rm G} + \mathcal{L}_{\rm f}$$
$$= -\frac{1}{4}F_{\mu\nu}^a F^{\mu\nu\,a} + i\overline{\psi}^i\gamma^\mu(D_\mu\psi)^i - m\overline{\psi}^i\psi^i. \tag{5.20}$$

While the form of the gauge Lagrangian density in (5.19) is quite similar to that of the Maxwell theory, there are essential differences.

For the non-Abelian gauge fields, $\mathcal{L}_G$ represents a fully interacting theory since the gauge fields couple to themselves and this leads to interesting properties such as asymptotic freedom in such theories. Even the question of quantization of such theories turns out to be nontrivial and that is what we will discuss in the next section.

5.2 BRST invariance

To understand the quantization of non-Abelian gauge theories, the matter part (fermions) is not very relevant since we already know how to quantize fermions and the pure gauge theory represents a fully interacting theory, Therefore, we would consider only the part of the Lagrangian density involving only gauge fields (in other words, set $\psi^i = \overline{\psi}^i = 0$) for the rest of this chapter.

Gauge invariance, as we know from the study of QED, puts a very strong constraint on the structure of the Lagrangian density. In particular, the coefficient matrix of the quadratic terms in the Lagrangian density becomes singular and, therefore, non-invertible. As a result, if we take $\mathcal{L}_G$ in (5.19) as the dynamical theory, we cannot define a propagator for the gauge field and the entire philosophy of doing perturbative calculations with Feynman diagrams breaks down. To circumvent this difficulty, we normally add to the Lagrangian density describing gauge field dynamics a term which breaks gauge invariance and allows us to define the propagator for the gauge fields. Such a term is called a gauge fixing term and any term which makes the coefficient matrix of the quadratic terms non-singular and maintains various global symmetries of the theory is allowed for this purpose. Thus, for example, the standard covariant gauge fixing consists of adding to the Lagrangian density a term of the form

$$\mathcal{L}_{GF} = -\frac{1}{2\xi}(\partial_\mu A^{\mu a})^2. \tag{5.21}$$

Here ξ represents an arbitrary (real) constant parameter known as the gauge fixing parameter. We note here that it is more convenient, for later discussions, to write the gauge fixing Lagrangian density by introducing an auxiliary field (one without any dynamics) as

$$\mathcal{L}_{GF} = \frac{\xi}{2}F^a F^a + \partial^\mu F^a A^a_\mu. \tag{5.22}$$

It is clear from the form of $\mathcal{L}_{GF}$ in (5.22) that the equation of motion for the auxiliary field takes the form

$$\xi F^a = \partial^\mu A_\mu^a, \tag{5.23}$$

and when we eliminate F^a from (5.22) using this equation, we recover the original gauge fixing Lagrangian density, (5.21), if we ignore a total divergence term. Among other things $\mathcal{L}_{GF}$, as written in (5.22), allows us to take such gauge choices as the Landau gauge (corresponding to $\partial_\mu A^\mu = 0$) simply by setting $\xi = 0$ (see (5.23)).

By adding a gauge fixing term to the Lagrangian density for the gauge fields, we have, of course, modified our theory and, therefore, we must compensate for this change. Conventionally, this is done by adding to the Lagrangian density another term known as the ghost Lagrangian density as follows. Suppose $f^a(x) = 0$ corresponds to the gauge condition described by the gauge fixing Lagrangian density, then, we add to our Lagrangian density another term

$$\mathcal{L}_{ghost} = -\int d^4y\, \bar{c}^a(x) \left(\frac{\delta f^a(x)}{\delta A_\mu^b(y)} \right) (D_\mu c(y))^b, \tag{5.24}$$

where the new fields c^a and $\bar{c}^a$ are known as the ghost and the anti-ghost fields respectively. They are spin zero fields which are anti-commuting and are, therefore, unphysical. In general, if the gauge fixing condition involves other fields besides the gauge fields, one can write the ghost action as

$$S_{ghost} = \int d^4x\, \mathcal{L}_{ghost} = -\int d^4x d^4y\, \bar{c}^a(x) \frac{\delta f^{(g)\,a}(x)}{\delta \epsilon^b(y)} c^b(y), \tag{5.25}$$

where $f^{(g)\,a}$ represents the infinitesimal gauge transform of the gauge fixing condition f^a corresponding to the infinitesimal local parameter of transformation $\epsilon^a(x)$. (There is a systematic procedure for introducing all of this in the path integral formulation. Here I am only summarizing what is done in practice in gauge theories.) In the covariant gauge that we are assuming, for example, we see from (5.23) that we can write the gauge fixing condition as

$$f^a = \partial^\mu A_\mu^a - \xi F^a = 0. \tag{5.26}$$

Consequently, the ghost Lagrangian density, for this choice of gauge fixing, follows from (5.24) to be

$$\mathcal{L}_{\text{ghost}} = -\int \mathrm{d}^4 y \, \bar{c}^a(x) \left(\partial_x^\mu \delta^4(x-y)\delta^{ab} \right) (D_\mu c(y))^b$$

$$= -\bar{c}^a(x)\partial^\mu (D_\mu c(x))^a$$

$$= \partial^\mu \bar{c}^a(x)(D_\mu c(x))^a, \tag{5.27}$$

where we have neglected a total derivative term in the last line of the above equation. (We can also derive this result from (5.25) if we assume the auxiliary field to be inert under the gauge transformation.)

With all these modifications, the total Lagrangian density for a gauge theory has the form

$$\mathcal{L}_{\text{TOT}} = \mathcal{L}_{\text{G}} + \mathcal{L}_{\text{GF}} + \mathcal{L}_{\text{ghost}}. \tag{5.28}$$

For the covariant gauge that we are using, the total Lagrangian density is given by

$$\mathcal{L}_{\text{TOT}} = -\frac{1}{4} F_{\mu\nu}^a F^{\mu\nu a} + \frac{\xi}{2} F^a F^a + \partial^\mu F^a A_\mu^a + \partial^\mu \bar{c}^a (D_\mu c)^a. \tag{5.29}$$

As we have mentioned earlier, the gauge fixing and the ghost Lagrangian densities modify the original theory in a compensating manner which allows us to define the Feynman rules of the theory and carry out any perturbative calculation. In a deeper sense, the gauge fixing and the ghost Lagrangians merely correspond to introducing a multiplicative factor of unity in the path integral formulation, which does not change the physical content of the theory. However, without going into technical details, we can see that these additional terms in the Lagrangian density have no physical content in the following way.

The total Lagrangian density of (5.28) or more specifically of (5.29), have been gauge fixed and, therefore, do not have the gauge invariance of the original theory. However, the total Lagrangian density, after gauge fixing, develops a global fermionic symmetry which, in some sense, remembers the gauge invariance of the original theory. It is easy to check that the total Lagrangian density is invariant under the global transformations

$$\delta A_\mu^a = \omega(D_\mu c)^a,$$

$$\delta c^a = \frac{\omega}{2} f^{abc} c^b c^c,$$

$$\delta \bar{c}^a = -\omega F^a,$$

$$\delta F^a = 0. \tag{5.30}$$

Here ω is a constant anticommuting parameter of the global transformation. Before showing the invariance of the total Lagrangian density under these transformations, it is worth noting that these transformations are nilpotent. Namely, for any field variable, it can be easily checked that

$$\delta_2 \delta_1 \phi^a = 0, \tag{5.31}$$

for $\phi^a = A_\mu^a, F^a, c^a, \bar{c}^a$ independent of the parameters of the transformations. The invariance of the Lagrangian density in (5.29) can now be easily checked. First, we note that the transformation of A_μ^a is really a gauge transformation with the parameter $\epsilon^a(x) = \omega c^a(x)$ and, therefore, the gauge Lagrangian density is trivially invariant under these transformation, namely,

$$\delta \mathcal{L}_G = 0.$$

Consequently, we need to worry only about the gauge fixing and the ghost Lagrangian densities. Furthermore, we note from (5.30) that the auxiliary field, F^a, does not transform at all which gives

$$\delta(\mathcal{L}_{GF} + \mathcal{L}_{ghost}) = \partial^\mu F^a \delta A_\mu^a + \partial^\mu \delta \bar{c}^a (D_\mu c)^a + \partial^\mu \bar{c}^a \delta(D_\mu c)^a$$

$$= \omega \partial^\mu F^a (D_\mu c)^a - \omega \partial^\mu F^a (D_\mu c)^a$$

$$= 0. \tag{5.32}$$

In obtaining this result, we have used the fact that $\delta(D_\mu c)^a = 0$ which follows from the nilpotency of the transformations (in particular, for the second variation of the gauge field). We note here, parenthetically, that with the auxiliary fields, the transformations are nilpotent off-shell while without them nilpotency would hold only when equations of motion are used.

These global transformations which define a residual symmetry of the full theory and, in some sense, replace the original local gauge invariance, are known as the BRST transformations and play a fundamental role in the study of gauge theories. There is also a second set of transformations involving the anti-ghost fields of the form

$$\bar{\delta} A_\mu^a = \bar{\omega}(D_\mu \bar{c})^a,$$

$$\bar{\delta} c^a = \bar{\omega}(F^a + f^{abc} c^b \bar{c}^c),$$

$$\bar{\delta} \bar{c}^a = \frac{\bar{\omega}}{2} f^{abc} \bar{c}^b \bar{c}^c,$$

$$\bar{\delta} F^a = -\bar{\omega} f^{abc} F^b \bar{c}^c, \tag{5.33}$$

which leave the full Lagrangian density invariant. These are known as the anti-BRST transformations. However, since they do not lead to any new information beyond what the BRST invariance provides, we will ignore this for the rest of our discussions. In addition to these two anti-commuting symmetries, the total Lagrangian density is also invariant under the bosonic global symmetry transformations

$$\delta c^a = \theta c^a,$$

$$\delta \bar{c}^a = -\theta \bar{c}^a, \tag{5.34}$$

with all other fields remaining unchanged. Here θ represents a constant, bosonic infinitesimal parameter and the generator of the symmetry merely counts the ghost number of the fields. (The fact that these transformations are like scale transformations and not like phase transformations, which normally lead to conserved number operators, has to do with the particular hermiticity properties that the ghost and the anti-ghost fields must satisfy for consistent quantization of the theory. We will, however, not go into the technical details of this here.[1])

The Hilbert space of the full theory, as is clear by now, contains many more states than the physical states of the theory. The physical Hilbert space, therefore, needs to be properly selected out for any discussion of physical questions for the system. Furthermore,

---

[1]See, for example, the discussion around Eqs. (13.34)-(13.37) in A. Das, *Lectures on Quantum Field Theory* (second edition), World Scientific (2020), Singapore.

the physical space must be selected in such a way that it remains invariant under the time evolution of the system. In the covariant gauge in QED, for example, the physical space can be simply selected by requiring that the physical states of the theory satisfy

$$\partial^\mu A_\mu^{(+)}|\text{phys}\rangle = 0, \tag{5.35}$$

where the superscript, $+$ on the left, stands for the positive frequency part of the field. In QED, this works because the operator $\partial^\mu A_\mu$ satisfies the free Klein-Gordon equation in the covariant gauge and hence the physical space so selected remains invariant under time evolution. The corresponding operator in a non-Abelian theory, on the other hand, does not satisfy a free field equation and hence is not suitable for identifying the physical subspace. On the other hand, we note that the generators of the BRST symmetry, Q_{BRST}, and the ghost scaling symmetry, Q_c, are conserved charges (time independent) and hence can be used to define a physical Hilbert space which would remain invariant under the time evolution of the system. (Q_{BRST} and Q_c are the charges constructed from the Nöther current for the appropriate transformations whose explicit forms can be obtained from the Nöther currents given later in (5.53).) Thus, we identify the physical space of the theory as satisfying

$$Q_{\text{BRST}}|\text{phys}\rangle = 0,$$
$$Q_c|\text{phys}\rangle = 0. \tag{5.36}$$

It can be shown in a straightforward manner that the conditions in (5.36) lead to the Gupta-Bleuler condition of (5.35) when restricted to an Abelian theory.

We are now ready to show that the gauge fixing and the ghost Lagrangian densities do not lead to any physical consequence in the following way. We note from (5.29) and (5.30) that these extra terms in the Lagrangian density can, in fact, be written as a BRST variation (without the parameter of transformation), namely,

$$\mathcal{L}_{\text{GF}} + \mathcal{L}_{\text{ghost}} = \delta_{\text{BRST}}(-\partial^\mu \bar{c}^a A_\mu^a - \frac{\xi}{2}\bar{c}^a F^a)$$

$$= [Q_{\text{BRST}}, (-\partial^\mu \bar{c}^a A_\mu^a - \frac{\xi}{2}\bar{c}^a F^a)]_+. \tag{5.37}$$

It follows now from the physical condition in (5.36) that

$$\langle \text{phys} | (\mathcal{L}_{\text{GF}} + \mathcal{L}_{\text{ghost}}) | \text{phys}' \rangle = 0, \tag{5.38}$$

with the use of (5.36), where $|\text{phys}\rangle$ and $|\text{phys}'\rangle$ denote any two physical states of the theory satisfying the physical state conditions given in (5.36). This shows that the terms added to the original Lagrangian density have no contribution to the matrix elements involving the physical states of the theory. Second, this also shows that all the physical matrix elements would be independent of the choice of the gauge and the gauge fixing parameter ξ (since this dependence is entirely in the added terms).

5.3 Ward identities

The BRST invariance of the full theory leads to many relations between various scattering amplitudes of the theory. These are known as the Ward identities (for Abelian theories) or the Slavnov-Taylor identities (for non-Abelian theories) and are quite essential in establishing the renormalizability of gauge field theories. These identities are best described within the context of path integrals which we will do next.

Let us consider an effective Lagrangian density which consists of $\mathcal{L}_{\text{TOT}}$ given in (5.28) in general or (5.29) for the covariant gauge fixing as well as source terms as follows

$$\mathcal{L}_{\text{eff}} = \mathcal{L}_{\text{TOT}} + J^{\mu a} A_\mu^a + J^a F^a + i(\overline{\eta}^a c^a - \overline{c}^a \eta^a)$$

$$+ K^{\mu a}(D_\mu c)^a + K^a(\frac{1}{2} f^{abc} c^b c^c). \tag{5.39}$$

Here, we have not only introduced sources for all the fundamental fields in the theory, but also have added, in addition, sources for the composite variations under the BRST transformation. The usefulness of this will become quite clear shortly. Denoting all the fields and sources generically by A and J respectively, we can write the generating functional for the theory as

$$Z[J] = e^{iW[J]} = \int \mathcal{D}A \, e^{i \int \mathrm{d}^4 x \, \mathcal{L}_{\text{eff}}}. \tag{5.40}$$

The vacuum expectation values of various operators, in the presence of sources, can now be written as

$$\langle A_\mu^a \rangle = A_\mu^{(c)a} = \frac{\delta W}{\delta J^{\mu a}},$$

$$\langle F^a \rangle = F^{(c)a} = \frac{\delta W}{\delta J^a},$$

$$\langle c^a \rangle = c^{(c)a} = -i\frac{\delta W}{\delta \overline{\eta}^a},$$

$$\langle \overline{c}^a \rangle = \overline{c}^{(c)a} = -i\frac{\delta W}{\delta \eta^a},$$

$$\langle (D_\mu c)^a \rangle = \frac{\delta W}{\delta K^{\mu a}},$$

$$\langle \frac{1}{2} f^{abc} c^b c^c \rangle = \frac{\delta W}{\delta K^a}. \tag{5.41}$$

Here, we have assumed a left derivative for the anti-commuting fields. The fields $A^{(c)}$ are known as the classical fields and in what follows, we will ignore the superscript (c) for notational simplicity.

When the external sources are held fixed, the effective Lagrangian density in (5.39) is no longer invariant under the BRST transformations of (5.30). In fact, recalling that $\mathcal{L}_{\text{TOT}}$ is BRST invariant and that the BRST transformations are nilpotent, we obtain the change in $\mathcal{L}_{\text{eff}}$ (without the parameter of transformation and remember that the parameter of transformation is anti-commuting) to be

$$\delta \mathcal{L}_{\text{eff}} = J^{\mu a} \delta A_\mu^a + J^a \delta F^a + i(\overline{\eta}^a \delta c^a - \delta \overline{c}^a \eta^a)$$

$$= J^{\mu a} (D_\mu c)^a - \frac{i}{2} f^{abc} \overline{\eta}^a c^b c^c + iF^a \eta^a. \tag{5.42}$$

On the other hand, the generating functional in (5.40) is defined by integrating over all possible field configurations. Therefore, with a redefinition of the fields under a BRST transformation inside the path integral, the generating functional should be invariant. This immediately leads to

$$\delta Z[J] = 0 = \int \mathcal{D}A \left(i \int d^4 x \, \delta \mathcal{L}_{\text{eff}} \right) e^{i \int d^4 x \mathcal{L}_{\text{eff}}}$$

$$= i \int d^4 x \left(J^{\mu a} \langle (D_\mu c)^a \rangle - i\overline{\eta}^a \langle \frac{1}{2} f^{abc} c^b c^c \rangle + i\eta^a \langle F^a \rangle \right),$$

where we have used (5.42) as well as the definition of the vacuum expectation value in the path integral formalism. This can also be written using (5.41) as

$$\int d^4x \left(J^{\mu a}(x) \frac{\delta W}{\delta K^{\mu a}(x)} - i\overline{\eta}^a(x) \frac{\delta W}{\delta K^a(x)} + i\eta^a(x) \frac{\delta W}{\delta J^a(x)} \right) = 0.$$
(5.43)

This is the Master equation from which one can derive all the identities relating the connected Green's functions of the theory. It is here that the usefulness of the sources for the composite operators becomes obvious.

Most often, however, we are interested in the proper (1PI) vertices of the theory. These can be obtained by passing from the generating functional for the connected Green's functions, $W[J]$, to the generating functional for the proper vertices, $\Gamma[A]$, through a Legendre transformation. Thus, defining a Legendre transformation involving (only) the field variables of the theory (the field variables are really the classical fields and we are dropping the superscript (c) for simplicity), we have

$$\Gamma[A, K] = W[J] - \int d^4x \left(J^{\mu a} A_\mu^a + J^a F^a + i(\overline{\eta}^a c^a - \overline{c}^a \eta^a) \right), \quad (5.44)$$

where K stands generically for the sources for the composite variations. From the definition of the generating functional for the proper vertices in (5.44), it is clear that

$$\frac{\delta \Gamma}{\delta A_\mu^a} = -J^{\mu a},$$

$$\frac{\delta \Gamma}{\delta F^a} = -J^a,$$

$$\frac{\delta \Gamma}{\delta c^a} = i\overline{\eta}^a,$$

$$\frac{\delta \Gamma}{\delta \overline{c}^a} = i\eta^a,$$

$$\frac{\delta \Gamma}{\delta K_\mu^a} = \frac{\delta W}{\delta K_\mu^a},$$

$$\frac{\delta \Gamma}{\delta K^a} = \frac{\delta W}{\delta K^a}.$$
(5.45)

Using the definitions in (5.45), we see that we can rewrite the Master equation of (5.43) in terms of the generating functional of the proper vertices as

$$\int d^4x \left(\frac{\delta \Gamma}{\delta A_\mu^a(x)} \frac{\delta \Gamma}{\delta K^{\mu a}(x)} + \frac{\delta \Gamma}{\delta c^a(x)} \frac{\delta \Gamma}{\delta K^a(x)} - F^a(x) \frac{\delta \Gamma}{\delta \bar{c}^a(x)} \right) = 0.$$

(5.46)

This is the Master equation from which we can derive all the relations between various (1PI) proper vertices resulting from the BRST invariance of the theory. This, in turn, is essential in proving the renormalizability of gauge theories. Thus, for example, let us note that we can write (5.46) in the momentum space as (we have changed the dummy internal index of summation from a to c)

$$\int d^4k \left(\frac{\delta \Gamma}{\delta A_\mu^c(-k)} \frac{\delta \Gamma}{\delta K^{\mu c}(k)} + \frac{\delta \Gamma}{\delta c^c(-k)} \frac{\delta \Gamma}{\delta K^c(k)} - F^c(k) \frac{\delta \Gamma}{\delta \bar{c}^c(k)} \right) = 0.$$

(5.47)

Taking derivative of this with respect to $\frac{\delta^2}{\delta F^a(p) \delta c^b(-p)}$ and setting all the (classical) field variables to zero gives

$$\frac{\delta^2 \Gamma}{\delta F^a(p) \delta A_\mu^c(-p)} \frac{\delta^2 \Gamma}{\delta c^b(-p) \delta K^{\mu c}(p)} - \frac{\delta^2 \Gamma}{\delta c^b(-p) \delta \bar{c}^a(p)} = 0. \quad (5.48)$$

A simple analysis of this relation, then, shows that (5.48) relates the mixed two point function involving F-A_μ with the two point function for the ghost fields and, consequently, the counter terms should satisfy such a relation. The BRST invariance, in this way, is very fundamental in the study of gauge theories as far as renormalizability and gauge independence are concerned. Similarly, taking derivative of (5.47) with respect to $\frac{\delta^2}{\delta A_\mu^a(p) \delta c^b(-p)}$ and setting all the (classical) field variables to zero leads to (we have changed the dummy index of summation in (5.47) from μ to ν)

$$\frac{\delta^2 \Gamma}{\delta A_\mu^a(p) \delta A_\nu^c(-p)} \frac{\delta^2 \Gamma}{\delta c^b(-p) \delta K^{\nu c}(p)} = 0,$$

$$\text{or,} \quad p_\mu \frac{\delta^2 \Gamma}{\delta A_\mu^a(p) \delta A_\nu^b(-p)} = 0, \quad (5.49)$$

which simply says that the two point function (self-energy) of the gauge field is transverse to the external momentum (see (5.39) for the K^c_μ-c^b vertex).

5.4 Unitarity

The BRST invariance of a gauge theory is quite important from yet another consideration. It leads to a formal proof of unitarity of the theory when restricted to the physical subspace of the Hilbert space. We will only outline the proof of this in this section.[2]

From the Lagrangian density in (5.29), we can obtain the canonical momenta conjugate to the various field variables of the theory which are given by

$$\pi^{ia} = \frac{\partial \mathcal{L}}{\partial \dot{A}^a_i} = -F^{0ia},$$

$$\pi^a = \frac{\partial \mathcal{L}}{\partial \dot{F}^a} = A^a_0,$$

$$\pi^a_c = \frac{\partial \mathcal{L}}{\partial \dot{c}^a} = -\dot{\bar{c}}^a,$$

$$\pi^a_{\bar{c}} = \frac{\partial \mathcal{L}}{\partial \dot{\bar{c}}^a} = (D_0 c)^a. \tag{5.50}$$

Here we have used left derivatives for the anti-commuting ghost fields. In particular, we see that A^a_0 plays the role of the momentum conjugate to F^a. The canonical (anti) commutation relations can now be written as ($\hbar = 1$)

$$\left[A^a_i(\mathbf{x}, t), \pi^{jb}(\mathbf{y}, t) \right] = i\delta^{ab} \delta^j_i \delta^3(x - y),$$

$$\left[F^a(\mathbf{x}, t), \pi^b(\mathbf{y}, t) \right] = i\delta^{ab} \delta^3(x - y),$$

$$\left[c^a(\mathbf{x}, t), \pi^b_c(\mathbf{y}, t) \right]_+ = i\delta^{ab} \delta^3(x - y),$$

$$\left[\bar{c}^a(\mathbf{x}, t), \pi^b_{\bar{c}}(\mathbf{y}, t) \right]_+ = i\delta^{ab} \delta^3(x - y). \tag{5.51}$$

---

[2]More details can be found in sections **13.2** and **13.3** in A. Das, *Lectures on Quantum Field Theory* (second edition), World Scientific (2020) Singapore.

The hermiticity conditions for the ghost fields, which arise out of various consistency conditions, are

$$(c^a)^\dagger = c^a,$$
$$(\bar{c}^a)^\dagger = -\bar{c}^a. \tag{5.52}$$

The Nöther current densities associated with the BRST symmetry and the ghost scaling symmetry (see (5.30) and (5.34)) can be obtained in a straightforward manner and have the forms

$$J^\mu_{\text{BRST}} = F^{\mu\nu a}(D_\nu c)^a + F^a(D^\mu c)^a + \frac{1}{2}f^{abc}\,\partial^\mu\bar{c}^a c^b c^c,$$
$$J^\mu_c = \partial^\mu\bar{c}^a c^a - \bar{c}^a(D^\mu c)^a. \tag{5.53}$$

The corresponding conserved charges can also be obtained from these current densities and we can calculate the algebra of charges using the (anti) commutation relations in (5.51). We note, in particular, that

$$[c^a(x), Q_c] = -ic^a(x),$$
$$[\bar{c}^a(x), Q_c] = i\bar{c}^a(x). \tag{5.54}$$

Therefore, we can think of iQ_c as the ghost number operator (with the ghost number for $\bar{c}^a$ being negative).

The algebra of the charges also follows in a straightforward manner.

$$[Q_{\text{BRST}}, Q_{\text{BRST}}]_+ = 0,$$
$$[Q_c, Q_c] = 0,$$
$$[Q_{\text{BRST}}, Q_c] = -iQ_{\text{BRST}}. \tag{5.55}$$

The first equation simply reiterates in an operator language the fact that the BRST transformations are nilpotent. The second equation implies that the ghost scaling symmetry is Abelian. The third merely is a statement of the fact that the BRST charge carries a unit ghost number. An immediate consequence of the third equation in (5.55) is that

$$Q_{\text{BRST}}e^{\pi Q_c} = e^{\pi(Q_c - i)}Q_{\text{BRST}} = -e^{\pi Q_c}Q_{\text{BRST}},$$
$$\text{or,} \quad \left[Q_{\text{BRST}}, e^{\pi Q_c}\right]_+ = 0. \tag{5.56}$$

As we will see later, this relation is quite useful in dealing with gauge theories at finite temperature.

As is clear from our earlier discussions, the total vector space, V, of the complete gauge theory, in addition to the physical states, also contains various unphysical states as well as states with negative norm. Consequently, the metric of this space and, therefore, the inner product becomes indefinite. A probabilistic description of the quantum theory, therefore, is lost unless we can restrict to a suitable subspace with a positive definite inner product. As we have discussed earlier, we can select out a subspace, V_{phys}, by requiring that states in this space are annihilated by Q_{BRST}. Namely, for every $|\Psi\rangle \in V_{\text{phys}}$, we have

$$Q_{\text{BRST}}|\Psi\rangle = 0. \tag{5.57}$$

We note again that such an identification is invariant under the time evolution of the system since Q_{BRST} commutes with the Hamiltonian because the BRST transformations are a symmetry of the full theory, namely,

$$[Q_{\text{BRST}}, H] = 0.$$

Furthermore, there are two possible kinds of states which will satisfy the condition in (5.57). First, if a state Ψ cannot be written as

$$|\Psi\rangle \neq Q_{\text{BRST}}|\widetilde{\Psi}\rangle, \tag{5.58}$$

and still satisfies (5.57), then, it is truly a BRST singlet state. The field operator which creates such a state must necessarily commute with Q_{BRST} and, therefore, must represent a truly gauge invariant field variable, namely, (we assume that the vacuum state is a BRST singlet state, $Q_{\text{BRST}}|0\rangle = 0$)

$$\begin{aligned} Q_{\text{BRST}}|\Psi\rangle &= Q_{\text{BRST}}\Psi|0\rangle \\ &= [Q_{\text{BRST}}, \Psi]\,|0\rangle = 0, \end{aligned} \tag{5.59}$$

where we have assumed that the operator, Ψ, creates the state $|\Psi\rangle$ from vacuum, namely,

$$\Psi\,|0\rangle = |\Psi\rangle. \tag{5.60}$$

Equation (5.59) also implies that

$$[Q_{\text{BRST}}, \Psi] = 0. \tag{5.61}$$

(For, otherwise, if $[Q_{\text{BRST}}, \Psi] = \chi$, (5.59) would imply $Q_{BRST}|\Psi\rangle = |\chi\rangle$ implying that the state $|\Psi\rangle$ transforms under a BRST transformation and is not a singlet state, contradicting the starting assumption.) Such states would, therefore, correspond to truly physical states of the theory and since gauge invariant degrees of freedom have physical (non-negative) commutation relations, such states would have positive norm.

The second class of states which would satisfy (5.57) can be of the form

$$|\Psi\rangle = Q_{\text{BRST}}|\widetilde{\Psi}\rangle. \tag{5.62}$$

We see that the condition (5.57) is trivially satisfied in this case because of the nilpotency of Q_{BRST}, namely, $Q_{\text{BRST}}|\Psi\rangle = Q_{\text{BRST}}^2|\widetilde{\Psi}\rangle = 0$. Such states will not be singlet states and the nilpotency of Q_{BRST} further implies that all such states would have zero norm because (Q_{BRST} is Hermitian with our choice of hermiticity conditions for the ghost fields given in (5.52))

$$\langle\Psi|\Psi'\rangle = \langle\widetilde{\Psi}|Q_{\text{BRST}}\,Q_{\text{BRST}}|\widetilde{\Psi}'\rangle = \langle\widetilde{\Psi}|Q_{\text{BRST}}^2|\widetilde{\Psi}'\rangle = 0.$$

Such a state would also be orthogonal to any physical state of the form in (5.58) and (5.57) because if $|\Psi\rangle$ represents a truly physical state while $|\Psi'\rangle$ a state of the kind in (5.62), then

$$\langle\Psi'|\Psi\rangle = \langle\widetilde{\Psi}'|Q_{\text{BRST}}|\Psi\rangle = 0,$$

which follows because of (5.57). If we denote all the states of V_{phys} of the form as in (5.62) by V_0, then, this would contain all the zero norm states and would be orthogonal to V_{phys} itself. The true physical states, which satisfy (5.57) and (5.58) can, therefore, be identified as belonging to the quotient space $\overline{V}_{\text{phys}} = \frac{V_{\text{phys}}}{V_0}$ and will have positive definite norm.

Every state in V_{phys} can, of course, be projected out according to the number of unphysical particles contained in such a state. Thus, defining $P^{(n)}$ as the projection operator onto the n-unphysical particle sector, we have

$$\sum_{n=0}^{\infty} P^{(n)} = 1,$$

$$P^{(n)} P^{(m)} = \delta_{nm} P^{(n)}. \tag{5.63}$$

It is clear that, by definition, $P^{(0)}$ projects onto the truly physical states of the theory (with no unphysical particles) and from the first equation of (5.63), we note that we can write

$$P^{(0)} = 1 - P', \tag{5.64}$$

where

$$P' = \sum_{n=1}^{\infty} P^{(n)}. \tag{5.65}$$

One can, of course, construct the actual forms of all the projection operators which is not very interesting. What is interesting, however, is the fact that $P^{(0)}$ projects onto the true physical states and, therefore, must be gauge invariant and by (5.61), it will commute with Q_{BRST}. It follows, then, from (5.65) that P' must also commute with Q_{BRST}. Namely, we have

$$\left[Q_{\text{BRST}}, P^{(0)} \right] = 0 = \left[Q_{\text{BRST}}, P' \right]. \tag{5.66}$$

The difference, however, is that P' must necessarily involve unphysical fields since it projects onto unphysical states and, therefore, cannot be truly gauge invariant. Consequently, it must have the form (for some operator, R, whose explicit form is not important for our discussions)

$$P' = [Q_{\text{BRST}}, R]_+, \tag{5.67}$$

which would imply (5.66) because of the nilpotency of Q_{BRST}. In fact, one can show more than this, namely, one can show that for $P^{(n)}$, $n \geq 1$, we can write

$$P^{(n)} = \left[Q_{\text{BRST}}, R^{(n)} \right]_+, \quad \text{for} \quad n \geq 1. \tag{5.68}$$

It is now easy to show that $P^{(n)}$, for $n \geq 1$, and, therefore, P', projects onto the zero norm space, V_0. For example, let $|\Psi\rangle, |\Psi'\rangle \in V_{\text{phys}}$, then,

$$\langle \Psi' | P^{(n)} | \Psi \rangle = \langle \Psi' | \left[Q_{\text{BRST}}, R^{(n)} \right]_+ | \Psi \rangle = 0, \tag{5.69}$$

which follows from the physical state condition in (5.57). Any vector, $|\Psi\rangle \in V_{\text{phys}}$ can now be expanded as

$$|\Psi\rangle = P^{(0)} |\Psi\rangle + P' |\Psi\rangle, \tag{5.70}$$

and the norm of any such state is, then, obtained to be

$$\langle \Psi | \Psi \rangle = \langle \Psi | P^{(0)} | \Psi \rangle \geq 0. \tag{5.71}$$

This, therefore, shows that V_{phys}, defined by (5.57), has a positive semi-definite norm as we should have for a physical vector space and the value of the norm depends on the truly physical component of the state. This also makes it clear that the norm of a state $|\overline{\Psi}\rangle \in \overline{V}_{\text{phys}} = \frac{V_{\text{phys}}}{V_0}$ would have a positive definite norm. This would correspond to the true physical subspace of the total vector space. We also note here that for any two states $|\Psi\rangle, |\Phi\rangle \in V_{\text{phys}}$ and any two operators, A and B, in V_{phys} (namely, any two operators which do not take us out of V_{phys})

$$\langle \Psi | P^{(0)} | \Phi \rangle = \langle \Psi | (\mathbb{1} - P') | \Phi \rangle = \langle \Psi | \Phi \rangle,$$

$$\langle \Psi | A P^{(0)} B | \Phi \rangle = \langle \Psi | A (\mathbb{1} - P') B | \Phi \rangle = \langle \Psi | A B | \Phi \rangle, \tag{5.72}$$

which follow from the conditions given in (5.57) and (5.67).

The formal proof of unitarity now follows. First, we note that, with the hermiticity assignments of (5.52) for the ghost fields, the complete gauge Lagrangian of (5.29) is hermitian. Consequently, the S-matrix of the theory is formally unitary, namely,

$$S^\dagger S = S S^\dagger = \mathbb{1}. \tag{5.73}$$

Furthermore, the S-matrix will be BRST invariant since the full theory is, so that

$$[Q_{\text{BRST}}, S] = 0. \tag{5.74}$$

We note next that given any state $|\Psi\rangle \in V_{\text{phys}}$, we can define a unique state $|\overline{\Psi}\rangle \in \overline{V}_{\text{phys}}$ as follows

$$|\overline{\Psi}\rangle = P^{(0)}|\Psi\rangle. \tag{5.75}$$

In the space $\overline{V}_{\text{phys}}$, of course, $P^{(0)}$ acts as the identity operator, namely,

$$P^{(0)}|\overline{\Psi}\rangle = (P^{(0)})^2|\Psi\rangle = P^{(0)}|\Psi\rangle = |\overline{\Psi}\rangle. \tag{5.76}$$

The two states, $|\Psi\rangle$ and $|\overline{\Psi}\rangle$, differ only by a zero norm state which is orthogonal to every state in V_{phys} so that the inner product of any two such states is the same.

$$\langle\overline{\Psi}|\overline{\Phi}\rangle = \langle\Psi|(P^{(0)})^2|\Phi\rangle = \langle\Psi|P^{(0)}|\Phi\rangle = \langle\Psi|\Phi\rangle,$$

where the last line follows from the first relation in (5.72). Given this, it is clear that we can define the S-matrix which acts on the physical space $\overline{V}_{\text{phys}}$ as satisfying

$$P^{(0)}S = S_{\text{phys}}P^{(0)}, \tag{5.77}$$

so that for $|\Psi\rangle \in V_{\text{phys}}$, $|\overline{\Psi}\rangle \in \overline{V}_{\text{phys}}$ and related through (5.75),

$$S_{\text{phys}}|\overline{\Psi}\rangle = S_{\text{phys}}P^{(0)}|\Psi\rangle = P^0 S|\Psi\rangle = \overline{S|\Psi\rangle}. \tag{5.78}$$

It is clear that since S is BRST invariant ((5.74)), it will leave the space V_{phys} invariant. Furthermore, from the relations in (5.72)-(5.78), we obtain

$$\langle\overline{\Psi}|S_{\text{phys}}^\dagger S_{\text{phys}}|\overline{\Phi}\rangle = \langle\Psi|P^{(0)}S_{\text{phys}}^\dagger S_{\text{phys}}P^{(0)}|\Phi\rangle$$

$$= \langle\Psi|S^\dagger P^{(0)}S|\Phi\rangle$$

$$= \langle\Psi|S^\dagger S|\Phi\rangle = \langle\Psi|\Phi\rangle = \langle\overline{\Psi}|\overline{\Phi}\rangle, \tag{5.79}$$

where we have used the formal unitary nature of S from (5.73). It follows now that

$$S_{\text{phys}}^\dagger S_{\text{phys}} = \mathbb{1}. \tag{5.80}$$

In other words, the physical state condition which naturally follows from the BRST invariance of the theory, also automatically leads to a formal proof of unitarity of the S-matrix in the subspace of the truly physical states of the theory.

5.5 Fermionic boundary condition

As we have noted earlier in chapter **1**, the partition function involving fermionic variables is defined with an anti-periodic boundary condition for the fermionic variables (in the path integral). We will outline in this section how this arises with a simple example of the fermionic quantum mechanical oscillator. Furthermore, we will describe the system in the coherent state representation (see appendix **3.9.2** for a discussion on coherent states) which generalizes naturally to the path integral description for field theories.

Let us consider a fermionic oscillator described by the annihilation and creation operators, a and $a^\dagger$ respectively, satisfying the anti-commutation relations

$$[a, a]_+ = 0 = \left[a^\dagger, a^\dagger\right]_+,$$

$$\left[a, a^\dagger\right]_+ = \mathbb{1}. \tag{5.81}$$

The Hilbert space, in this case, is two dimensional with the basis states $|0\rangle$ and $|1\rangle = a^\dagger |0\rangle$ satisfying the orthonormality and the completeness relations given by

$$\langle n|m\rangle = \delta_{nm}, \qquad n, m = 0, 1,$$

$$|0\rangle\langle 0| + |1\rangle\langle 1| = \mathbb{1}. \tag{5.82}$$

Let us next consider a Grassmann odd parameter ψ and define a state of the theory as

$$|\psi\rangle = e^{-\psi a^\dagger}|0\rangle = |0\rangle - \psi|1\rangle. \tag{5.83}$$

Here, we have used the fact that ψ is Grassmann odd and, therefore, higher powers of ψ vanish ($a^\dagger$ is nilpotent as well, see (5.81)). It also follows from (5.83) that

$$\langle\psi| = \langle 0|e^{-a\psi^*} = \langle 0|e^{\psi^* a} = \langle 0| + \psi^*\langle 1|. \tag{5.84}$$

In deriving (5.84), we have used the fact that ψ (or ψ^*) anti-commutes with the fermionic operators of the theory and that

$$(\psi a^\dagger)^\dagger = a\psi^* = -\psi^* a,$$

which is essential to show that for any two such states

$$\langle\phi|\psi\rangle = (\langle\psi|\phi\rangle)^*.$$

The states defined in (5.83) are known as fermionic coherent states and are eigenstates of the fermionic annihilation operator (compare with (3.176) for bosonic coherent states)

$$a|\psi\rangle = a(|0\rangle - \psi|1\rangle) = \psi|0\rangle = \psi(|0\rangle - \psi|1\rangle) = \psi\,|\psi\rangle.$$

Coherent states are an alternate way of describing the fermion oscillator Hilbert space and as we have mentioned earlier lead naturally to a path integral description of (fermion) field theories. The inner product of two fermionic coherent states can be easily worked out now. For ψ and ϕ Grassmann odd, (5.83) and (5.84) lead to

$$\langle\psi|\phi\rangle = ((\langle 0| + \psi^*\langle 1|)(|0\rangle - \phi|1\rangle))$$

$$= \langle 0|0\rangle + \psi^*\phi\langle 1|1\rangle = 1 + \psi^*\phi = e^{\psi^*\phi}. \qquad (5.85)$$

Once again, in obtaining this, we have used the fact that the Grassmann odd parameters anti-commute with fermionic operators (as well as their nilpotency).

The completeness relation of the coherent states is now easily obtained, using (5.83) and (5.84), namely,

$$|\psi\rangle\langle\psi| = |0\rangle\langle 0| - \psi|1\rangle\langle 0| + \psi^*|0\rangle\langle 1| - \psi^*\psi|1\rangle\langle 1|.$$

Therefore, it now follows that (see (5.82))

$$\int d\psi^* d\psi\, e^{-\psi^*\psi}|\psi\rangle\langle\psi|$$

$$= \int d\psi^* d\psi\, (1 - \psi^*\psi)(|0\rangle\langle 0| - \psi|1\rangle\langle 0| + \psi^*|0\rangle\langle 1| - \psi^*\psi|1\rangle\langle 1|)$$

$$= \int d\psi^* d\psi\, (-\psi^*\psi)(|0\rangle\langle 0| + |1\rangle\langle 1|)$$

$$= |0\rangle\langle 0| + |1\rangle\langle 1| = \mathbb{1}, \qquad (5.86)$$

which can be compared with (3.203) for the bosonic case. In deriving this relation, we have used the nilpotency of the Grassmann odd parameters as well as the Berezin integration rules for Grassmann odd parameters given by

$$\int d\psi\, 1 = 0, \quad \int d\psi\, \psi = 1. \tag{5.87}$$

It can be easily checked now (for example, using (5.85)) that

$$\int d\psi^* d\psi\, e^{-\psi^*\psi}\, |\psi\rangle\langle\psi|\psi'\rangle = \int d\psi^* d\psi\, e^{-\psi^*\psi}\, e^{\psi^*\psi'}\, |\psi\rangle$$

$$= \int d\psi^* d\psi\, e^{-\psi^*(\psi-\psi')}\, |\psi\rangle = \int d\psi\, \delta(\psi - \psi')\, |\psi\rangle = |\psi'\rangle.$$

It is clear now that in the coherent state description, the Gaussian factor $e^{-\psi^*\psi}$ acts like a natural weight factor in defining integrations. With all these, we can now calculate the trace of a bosonic operator A, using (5.86), as follows

$$\text{Tr}\, A = \text{Tr}\, \mathbb{1}\, A = \text{Tr} \int d\psi^* d\psi\, e^{-\psi^*\psi}\, |\psi\rangle\langle\psi|A$$

$$= \int d\psi^* d\psi\, e^{-\psi^*\psi}\, \langle\psi|A| - \psi\rangle. \tag{5.88}$$

The last line in (5.88) is again due to the anti-commuting nature of the Grassmann odd parameter ψ. That (5.88) is, indeed, correct can be verified explicitly using (5.83) and (5.84) as follows.

$$\int d\psi^* d\psi\, e^{-\psi^*\psi}\, \langle\psi|A| - \psi\rangle$$

$$= \int d\psi^* d\psi\, (1 - \psi^*\psi)(\langle 0|A|0\rangle + \psi^*\langle 1|A|0\rangle$$

$$+ \psi\langle 0|A|1\rangle - \psi^*\psi\langle 1|A|1\rangle)$$

$$= \int d\psi^* d\psi\, (-\psi^*\psi)(\langle 0|A|0\rangle + \langle 1|A|1\rangle)$$

$$= \langle 0|A|0\rangle + \langle 1|A|1\rangle = \text{Tr}\, A. \tag{5.89}$$

The relation in (5.88) is quite significant in that it says that the trace of a bosonic operator in a fermionic theory is defined only over anti-periodic states. This is the reason for the anti-periodicity in the definition of the partition function for a fermionic theory (in the path integral).

We note here, for later use, that the number operator for the fermions is idempotent as can be checked from the anti-commutation relations in (5.81), namely,

$$N = a^\dagger a = N^2. \tag{5.90}$$

Consequently, it follows that

$$e^{\pm i\pi N} = 1 + N\left(\pm i\pi + \frac{(\pm i\pi)^2}{2!} + \cdots\right)$$

$$= 1 - N + N\left(1 \pm i\pi + \frac{(\pm i\pi)^2}{2!} + \cdots\right)$$

$$= 1 - N + Ne^{\pm i\pi} = 1 - 2N. \tag{5.91}$$

This, then, leads from the definition of the coherent states in (5.83) to

$$e^{\pm i\pi N}|\psi\rangle = (1 - 2N)(|0\rangle - \psi|1\rangle) = |0\rangle - \psi|1\rangle + 2\psi N|1\rangle$$

$$= |0\rangle + \psi|1\rangle = |-\psi\rangle. \tag{5.92}$$

As a result, for a bosonic operator of the form $e^{-i\pi N}A$, the trace in a fermionic space would have the form

$$\int d\psi^* d\psi \, e^{-\psi^*\psi} \langle\psi|e^{-i\pi N}A| - \psi\rangle$$

$$= \int d\psi^* d\psi \, e^{-\psi^*\psi} \langle-\psi|A| - \psi\rangle. \tag{5.93}$$

In other words, the operator $e^{-i\pi N} = e^{-\mu N}$ corresponds to an imaginary chemical potential of the form

$$\mu = i\pi,$$

and the presence of such a term in the trace of a bosonic operator (over fermionic states) changes the boundary conditions for the fermions leading to periodic boundary conditions much like bosons.

5.6 Partition function

With all this background, we are now ready to define the partition function for a gauge theory. From a practical point of view, if we can give the partition function a path integral representation, it would be useful in carrying out calculations. However, if we try to do that naively, we run into an immediate problem.

As we have seen, the complete theory for a gauge system involves unphysical particles (see (5.29)) and that the BRST invariance plays a very crucial role in the study of gauge theories from various points of view. We note that the ghost fields, c and $\bar{c}$, which effectively behave like the parameters of the BRST transformation, (5.30), are anti-commuting Grassmann fields. They have the statistics of fermions even though they are spin zero fields. Consequently, we expect that at finite temperature they must satisfy anti-periodicity condition like any other fermion. On the other hand, the gauge fields, being bosonic, would naturally obey periodic boundary condition at finite temperature. Thus, we see that there is a mismatch as far as the BRST transformations of (5.30) are concerned. With these periodicity properties, we note that the left hand side and the right hand side of (5.30) do not have a consistent behavior. Therefore, if we accept the periodicity properties as following from the statistics of the various fields, we may have to give up BRST invariance. But this symmetry is too valuable to give up. Consequently, we conclude that if we want to preserve the BRST invariance of the theory, we have to assume a periodic boundary condition for the unphysical ghost fields even though they satisfy anti-commutation relations like fermions. We will see next how this actually arises from the physical state condition of the theory.

In dealing with a gauge theory, we see that it contains unphysical particles which cannot thermalize in a heat bath. Therefore, it is not meaningful to define a partition function as a trace over the complete vector space of the theory. Rather, the partition function leads to physical observables and, therefore, should be defined as a trace over only the physical states of the theory. Thus, it is proper to define the partition function for a gauge theory as

$$Z(\beta) = \overline{\mathrm{Tr}}\, e^{-\beta H}, \tag{5.94}$$

where $\overline{\mathrm{Tr}}$ stands for trace over physical states in $\overline{V}_{\mathrm{phys}}$. Furthermore, recalling that $P^{(0)}$ projects onto the physical subspace of states of the total vector space and that such states have a zero ghost number, we can also rewrite (5.94) by relaxing the sum over the physical states as

$$Z(\beta) = \text{Tr } P^{(0)} \, e^{-\beta H} = \text{Tr } P^{(0)} \, e^{\pi Q_c} \, e^{-\beta H}. \tag{5.95}$$

Here, in the second step, we have used the fact that the physical states have zero ghost number so that the exponential factor involving the ghost number charge Q_c acts like identity in this subspace. This is a very simple fact, but from the point of view of calculations, it is quite useful. To rewrite the partition function in a more familiar form, we use (5.64), (5.65) and (5.67) as well as (5.56) to obtain

$$\begin{aligned} Z(\beta) &= \text{Tr } P^{(0)} \, e^{\pi Q_c} \, e^{-\beta H} \\ &= \text{Tr } \left(\mathbb{1} - P'\right) e^{\pi Q_c} \, e^{-\beta H} \\ &= \text{Tr } \left(\mathbb{1} - [Q_{\text{BRST}}, R]_+\right) e^{\pi Q_c} \, e^{-\beta H} \\ &= \text{Tr } \left(e^{\pi Q_c} \, e^{-\beta H} - R \left[Q_{\text{BRST}}, e^{\pi Q_c}\right]_+ e^{-\beta H}\right) \\ &= \text{Tr } e^{\pi Q_c} \, e^{-\beta H}. \end{aligned} \tag{5.96}$$

In writing the terms in the specific forms in (5.96), we have used the cyclicity of the trace in addition to the fact that the Hamiltonian commutes with Q_{BRST} and Q_c which generate symmetries of the theory as well as (5.56) (which shows that the second term vanishes). Finally, we note from (5.54) that iQ_c really corresponds to N_c – the ghost number operator. Hence it follows from (5.93) that the effect of the first exponential in (5.96) is simply to change the periodicity of the (anti-) ghost fields and, therefore, we can write

$$Z(\beta) = \text{Tr}_{\text{p}} \, e^{-\beta H}, \tag{5.97}$$

where Tr_{p} implies that the (anti-) ghost fields are supposed to satisfy periodic boundary conditions (in the path integral). This is, of course, consistent with our expectation from the BRST invariance point of view. Once, we recognize that we can write the partition function over the unrestricted space as in (5.97), a standard path integral representation follows where the ghost and anti-ghost fields satisfy periodic boundary condition. Following the same steps as in

(5.96), we note here parenthetically that, for any bosonic operator A, we can write the thermal average as

$$\langle A \rangle_\beta = Z^{-1}(\beta) \, \mathrm{Tr} \, P^{(0)} \, e^{\pi Q_c} \, e^{-\beta H} \, A$$

$$= Z^{-1}(\beta) \, \mathrm{Tr} \left(e^{\pi Q_c} \, e^{-\beta H} \, (A - [A, Q_{\mathrm{BRST}}] \, R) \right). \qquad (5.98)$$

For physical, BRST invariant observables (namely, for operators A satisfying $[Q_{\mathrm{BRST}}, A] = 0$), the second term in (5.98) vanishes and this reduces to the more familiar form

$$\langle A \rangle_\beta = Z^{-1}(\beta) \, \mathrm{Tr} \, e^{\pi Q_c} \, e^{-\beta H} \, A Z^{-1}(\beta) \, \mathrm{Tr_p} \, e^{-\beta H} \, A, \qquad (5.99)$$

with periodic ghost fields. However, we also note that for BRST non-invariant operators, the extra term on the right hand side of (5.98) cannot be neglected.

To conclude this section, we note that the presence of ghost fields satisfying periodic boundary conditions in the path integral representation is absolutely crucial for a meaningful, gauge independent definition of the partition function. This is important even in an Abelian gauge theory where the ghosts are noninteracting. To see this, let us note that, if we simplify the full theory of (5.29) by eliminating the auxiliary fields, in the Abelian case, it has the form

$$\mathcal{L}_{\mathrm{TOT}} = -\frac{1}{4} \, F_{\mu\nu} F^{\mu\nu} - \frac{1}{2\xi} (\partial_\mu A^\mu)^2 + \partial^\mu \bar{c} \, \partial_\mu c$$

$$= \frac{1}{2} \, A_\mu \left(\eta^{\mu\nu} \Box - \frac{\xi - 1}{\xi} \partial^\mu \partial^\nu \right) A_\nu + \partial^\mu \bar{c} \, \partial_\mu c. \qquad (5.100)$$

Therefore, if we write the partition function as a path integral assuming that the ghost fields satisfy periodic boundary conditions, we obtain (say in the imaginary time formalism)

$$Z(\beta) = N \int \mathcal{D} A_\mu \, \mathcal{D} \bar{c} \, \mathcal{D} c \, e^{-\int_0^\beta \mathrm{dt} \int \mathrm{d}^3 x \, \mathcal{L}_{\mathrm{TOT}}}$$

$$= N \left(\det(\eta^{\mu\nu} \Box - \frac{\xi - 1}{\xi} \partial^\mu \partial^\nu) \right)^{-1/2} \det(-\Box). \qquad (5.101)$$

Here N is a normalization constant and the first (inverse) determinant in (5.101) arises from the integration of the (bosonic) gauge

fields[3] and the determinant is defined over the finite dimensional
Minkowski space indices ($\mu\nu$) as well as the infinite dimensional coordinate space. The second determinant is from the integration of
the ghost variables and the positive power is because of the anti-
commuting nature of these fields.[4] Because of our choice of periodic
boundary conditions for the ghost fields, both these determinants are
defined over the space of periodic functions.

The first determinant can be evaluated over the finite dimensional Minkowski space and a little bit of work shows that it can be
written as (it has to be a Lorentz scalar) $\det\left(\frac{\Box^4}{\xi}\right)$. The temperature independent constant, ξ, can be absorbed into the normalization
constant so that we can write the partition function in (5.101) as

$$Z(\beta) = N(\det(-\Box)^4)^{-1/2} \det(-\Box) = N(\det(-\Box))^{-1}$$

$$= N\left((\det(-\Box))^{-1/2}\right)^2. \tag{5.102}$$

We note that this partition function is exactly identical to that of
two free scalar fields implying that the gauge theory only has two
physical transverse degrees of freedom as one can explicitly verify
in the unitary gauge. The important thing is that this follows only
because the ghost determinant cancelled out exactly the two unphysical degrees of freedom (the longitudinal and the timelike) from the
gauge field determinant and this was possible only because of the
periodic boundary conditions assigned to the ghost fields (at finite
temperature). Furthermore, the ghost contribution to the partition
function is nontrivial and quite crucial even though the ghost fields
are noninteracting in this theory. Finally, we note that even though
we worked out the partition function in the covariant gauge, the same
conclusion is obtained in any other gauge.

5.7 Ward identities at $T \neq 0$

As is clear from the discussions in this chapter, in defining the partition function of a gauge theory, we have taken care to preserve the

---

[3]See, for example, chapter **3** in A. Das, *Field Theory: A Path Integral Approach*
(third edition), World Scientific (2019).

[4]See, for example, section **5.2** in A. Das, *Field Theory: A Path Integral Approach* (third edition), World Scientific (2019).

BRST symmetry of the zero temperature theory which has many physical consequences. We have also seen that it is the BRST invariance of the complete, gauge fixed theory which leads to the Ward identities of the theory relating various 1PI amplitudes. Therefore, it is clear that the Ward identities will continue to hold even at finite temperature. Let us illustrate this with a simple example. Let us analyze the one loop correction to the polarization tensor of a gauge field, due to the fermion loop, at finite temperature in $1+1$ dimensional massive QED (see Fig. 5.1). At zero temperature, we know that this is completely transverse to the external momentum p^μ (when regularized in a gauge invariant manner using, say, dimensional regularization and this follows from the Ward identities) and what we would like to see is if finite temperature corrections lead to any longitudinal components in the polarization tensor which would violate Ward identities.

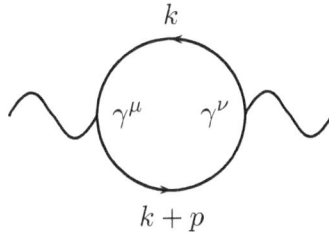

Figure 5.1: Polarization tensor in 1+1 dimensional QED at one loop.

The question of Ward identities is much better studied in a real time formalism and we note from (2.62) that, at finite temperature, the Feynman propagator for the physical fermion fields is given by (we suppress the subscript $++$ for simplicity and note that, at one loop, the other components of the propagator will not contribute to the physical polarization tensor involving the $++$ component)

$$iS(p) = iS^{(0)}(p) + iS^{(\beta)}(p)$$

$$= i(\not{p} + m)\left(\frac{1}{p^2 - m^2 + i\epsilon} + 2i\pi n_F(p)\delta(p^2 - m^2)\right). \quad (5.103)$$

Here $n_F(p)$ represents the fermionic distribution function defined in the covariant notation of chapter **1** (see (1.154)) to be

$$n_F(p) = \frac{1}{e^{\beta|u \cdot p|} + 1},$$

where u^μ, as in chapter **1**, represents the four velocity of the heat bath.

The temperature dependent contribution of the fermion loop to the polarization tensor can now be easily calculated and has the form

$$i\Pi_{\mu\nu}^{(\beta)}(p) = -e^2 \int \frac{d^2k}{(2\pi)^2} \, \text{Tr} \left[\gamma_\mu S^{(0)}(k+p)\gamma_\nu S^{(\beta)}(k) \right.$$

$$\left. + \gamma_\mu S^{(\beta)}(k+p)\gamma_\nu S^{(0)}(k) + \gamma_\mu S^{(\beta)}(k+p)\gamma_\nu S^{(\beta)}(k) \right]$$

$$= -\frac{ie^2}{\pi} \int d^2k$$

$$\times \left[(k+p)_\mu k_\nu + (k+p)_\nu k_\mu - \eta_{\mu\nu}(k \cdot (k+p) - m^2) \right]$$

$$\times \left[\frac{n_F(k)\delta(k^2 - m^2)}{(k+p)^2 - m^2 + i\epsilon} + \frac{n_F(k+p)\delta((k+p)^2 - m^2)}{k^2 - m^2 + i\epsilon} \right.$$

$$\left. + 2i\pi n_F(k)n_F(k+p)\delta(k^2 - m^2)\delta((k+p)^2 - m^2) \right]. \quad (5.104)$$

Here we have used the fact that the temperature dependent terms are finite and, as a result, we have performed the Dirac trace in $1+1$ dimensions.

The simplicity of $1+1$ dimensions allows us to factor the tensor structure out of the integral. In fact, let us define, following the discussion of section **1.8**,

$$\Omega = k \cdot u, \qquad \omega = p \cdot u,$$

$$k^\mu = \Omega u^\mu - \epsilon^{\mu\nu} u_\nu k',$$

$$p^\mu = \omega u^\mu - \epsilon^{\mu\nu} u_\nu p'. \quad (5.105)$$

Here, $\epsilon^{\mu\nu}$ defines the Levi-Civita tensor of $1+1$ dimensions and in our notation, $\epsilon^{01} = 1$. Basically, Ω and k' denote respectively the Lorentz invariant components of the loop momentum k^μ parallel and perpendicular to the four velocity u^μ. We note, for later use, that

(5.105) also gives

$$\epsilon^{\mu\nu} p_\mu u_\nu = p'. \tag{5.106}$$

Substituting these definitions into (5.104), and using the delta function constraints as well as the symmetry properties of the integrals, we find in a straightforward manner that the integral can be written as (the Jacobian of the transformation in going from (k^0, k^1) to (Ω, k') can be easily checked to be unity)

$$i\Pi_{\mu\nu}^{(\beta)}(p) = -e^2 \bar{u}_\mu \bar{u}_\nu (2I + I'), \tag{5.107}$$

where we have defined the velocity transverse to the momentum p^μ as (this can be checked easily with the help of (5.105) and (5.106))

$$\bar{u}_\mu = u_\mu - \frac{\omega}{p'} \epsilon_{\mu\nu} u^\nu, \quad p_\mu \bar{u}^\mu = \omega - \omega = 0, \tag{5.108}$$

and I and I' stand for the integrals

$$I = \frac{ip'}{\omega} \int d\Omega dk' \frac{k'(\omega + \Omega) + \Omega(k' + p')}{\omega(\omega + 2\Omega) - p'(p' + 2k') + i\epsilon} \frac{1}{e^{\beta|\Omega|} + 1}$$
$$\times \delta(\Omega^2 - k'^2 - m^2),$$

$$I' = -\frac{2p'}{\omega} \int d\Omega dk' \frac{k'(\omega + \Omega) + \Omega(k' + p')}{(e^{\beta|\Omega|} + 1)(e^{\beta|\omega + \Omega|} + 1)} \delta(\Omega^2 - k'^2 - m^2)$$
$$\times \delta((\Omega + \omega)^2 - (k' + p')^2 - m^2). \tag{5.109}$$

This shows that independent of the values of the integrals, I and I' (which are finite and temperature dependent), the temperature dependent part of the polarization tensor is manifestly transverse to the external momentum p^μ (see (5.108)), as would be required by the Ward identities. We note that this analysis can be carried over to $3+1$ dimensions as well and the temperature dependent polarization tensor can be shown to be transverse. This is just a simple example to show that the Ward identities are preserved at finite temperature simply because we have taken care to define the partition function carefully preserving the BRST invariance. At zero temperature, however, there are several ways one writes various identities using the inverse of the propagators. At finite temperature, on the other hand, the inverse of the propagator is not unique simply because the temperature dependent part of the propagator is annihilated by, say, $(p^2 - m^2)$ for a scalar. Therefore, one should be careful about what one means by the Ward identities. The correct identities are the ones that follow from the 1PI generating functional as discussed in (5.47).

5.8 References

C. Becchi, A. Rouet and R. Stora, Communications in Mathematical Physics **42**, 127 (1975); Annals of Physics **98**, 287 (1976).

C. Bernard, Physical Review **D9**, 3312 (1974).

A. Das, *Field Theory: A Path Integral Approach* (third edition), World Scientific (2019).

A. Das, *Lectures on Quantum Field Theory* (second edition), World Scientific (2020).

A. Das and M. Hott, Modern Physics Letters **A9**, 3383 (1994).

L. Faddeev and V. N. Popov, Physics Letters **25B**, 29 (1967).

H. Hata and T. Kugo, Physical Review **D21**, 3333 (1980).

G. 't Hooft, Nuclear Physics **B33**, 173 (1971); *ibid* **B35**, 167 (1971).

T. Kugo and I. Ojima, Progress in Theoretical Physics **60**, 1869 (1978); Progress in Theoretical Physics Supplement No. **66** (1979).

I. Ojima, Annals of Physics **137**, 1 (1981).

CHAPTER 6

Thermal operator representation

6.1 Introduction

We have seen, both in imaginary time formalism as well as in real time formalism, that calculations of Feynman diagrams at finite temperature are much more involved than at zero temperature. This, therefore, raises the interesting question as to whether there is any possible simple way a finite temperature Feynman graph may be obtained from the corresponding zero temperature graph. If this is possible, one can possibly obtain the value of the finite temperature Feynman graph in a much simpler manner. More than that, this may also make it possible to carry over various theorems/relations at zero temperature to finite temperature or understand better why certain zero temperature results may not carry over to finite temperature. This is the question which we will discuss in detail in this chapter and derive what is called a "thermal operator" which acting on a zero temperature graph would give the result for the finite temperature graph in a quantum field theory. Clearly, such a "thermal operator" would be temperature dependent such that it will give the entire temperature dependence of a Feynman graph acting on the corresponding zero temperature result.

This result can be shown in both the real time formalism and the imaginary time formalism and the form of the "thermal" operator can be obtained explicitly. However, it is simpler to see this in the closed time formalism of chapter **2** with which we will start and then show the same in the imaginary time formalism. The crucial observation to start with is that a Feynman graph consists of a set of vertices connected to one another through a set of propagators. The vertices do not carry any temperature dependence, rather the propagators do. Therefore, to understand the relation between thermal graphs to the corresponding zero temperature graph, it makes sense to ask if the propagator in a thermal field theory has any relation to its zero temperature counterpart.

6.2 Thermal operator in closed time formalism

The closed time path formalism (as well as thermofield dynamics and the arbitrary path formalism described in chapters **3** and **4** respectively) involve a doubling of field degrees of freedom. As we have done in chapter **2**, we will denote the two field degrees in the closed time formalism by the two subscripts $\pm$. We will study various theories in the closed time path in this section and we start with the simplest case, namely, the scalar field theory.

6.2.1 Real scalar field theory. In this subsection, let us start with a scalar field where the two degrees of freedom are denoted by $\phi_{\pm}$. Correspondingly, the thermal propagator for the scalar field becomes a 2×2 matrix which in the momentum space has the form (see, for example, (2.49) and (2.50))

$$
iG^{(T)}(p) = \begin{pmatrix} iG^{(T)}_{++}(p) & iG^{(T)}_{+-}(p) \\ iG^{(T)}_{-+}(p) & iG^{(T)}_{--}(p) \end{pmatrix},
\tag{6.1}
$$

where (limit $\epsilon \to 0^{+}$ is understood throughout)

$$
iG^{(T)}_{++}(p) = \frac{i}{p^2 - m^2 + i\epsilon} + 2\pi n_B(|p^0|)\delta(p^2 - m^2),
$$

$$
iG^{(T)}_{+-}(p) = 2\pi \left(\theta(-p^0) + n_B(|p^0|) \right) \delta(p^2 - m^2),
$$

$$
iG^{(T)}_{-+}(p) = 2\pi \left(\theta(p^0) + n_B(|p^0|) \right) \delta(p^2 - m^2),
$$

$$
iG^{(T)}_{--}(p) = -\frac{i}{p^2 - m^2 - i\epsilon} + 2\pi n_B(|p^0|)\delta(p^2 - m^2).
\tag{6.2}
$$

Note that the bosonic distribution function $n_B(|p^0|) = \frac{1}{e^{|p^0|/T}-1}$ vanishes at zero temperature $T \to 0$, so that the zero temperature propagator matrix has components given by

$$iG_{++}^{(T=0)}(p) = \frac{i}{p^2 - m^2 + i\epsilon},$$

$$iG_{+-}^{(T=0)}(p) = 2\pi\theta(-p^0)\delta(p^2 - m^2),$$

$$iG_{-+}^{(T=0)}(p) = 2\pi\theta(p^0)\delta(p^2 - m^2),$$

$$iG_{--}^{(T=0)}(p) = -\frac{i}{p^2 - m^2 - i\epsilon}. \tag{6.3}$$

If we Fourier transform the p^0 variable in (6.2)-(6.3) to the t-space, namely, if we use

$$f(t, \mathbf{p}) = \int \frac{dp^0}{2\pi} e^{-ip^0 t} f(p^0, \mathbf{p}),$$

then we obtain (see also (2.42) for $T = 0, \beta \to \infty$)

$$iG_{++}^{(T=0)}(t, \omega) = \frac{1}{2\omega}\left(\theta(t)e^{-i\omega t} + \theta(-t)e^{i\omega t}\right),$$

$$iG_{+-}^{(T=0)}(t, \omega) = \frac{1}{2\omega} e^{i\omega t},$$

$$iG_{-+}^{(T=0)}(t, \omega) = \frac{1}{2\omega} e^{-i\omega t},$$

$$iG_{--}^{(T=0)}(t, \omega) = \frac{1}{2\omega}\left(\theta(-t)e^{-i\omega t} + \theta(t)e^{i\omega t}\right), \tag{6.4}$$

where we have identified

$$\omega = \omega_{\mathbf{p}} = \sqrt{\mathbf{p}^2 + m^2}. \tag{6.5}$$

Similarly, the Fourier transform of (6.2) in the p^0 variable leads to (see (2.42))

$$iG_{++}^{(T)}(t, \omega) = \frac{1}{2\omega}\left((\theta(t) + n_B(\omega))e^{-i\omega t} + (\theta(-t) + n_B(\omega))e^{i\omega t}\right),$$

$$iG_{+-}^{(T)}(t, \omega) = \frac{1}{2\omega}\left(n_B(\omega)e^{-i\omega t} + (1 + n_B(\omega))e^{i\omega t}\right),$$

$$iG_{-+}^{(T)}(t, \omega) = \frac{1}{2\omega}\left((1 + n_B(\omega))e^{-i\omega t} + n_B(\omega)e^{i\omega t}\right), \tag{6.6}$$

$$iG_{--}^{(T)}(t, \omega) = \frac{1}{2\omega}\left((\theta(-t) + n_B(\omega))e^{-i\omega t} + (\theta(t) + n_B(\omega))e^{i\omega t}\right).$$

Let us now introduce a temperature dependent operator

$$\mathcal{O}_B^{(T)}(\omega) = \left(\mathbb{1} + n_B(\omega)(\mathbb{1} - S(\omega))\right), \tag{6.7}$$

where $S(\omega)$ is a reflection operator for ω, namely,

$$S^\dagger(\omega) = S(\omega), \quad (S(\omega))^2 = \mathbb{1},$$
$$S(\omega)f(\omega) = f(-\omega). \tag{6.8}$$

It is clear now that, for each of the components in (6.4)-(6.6), we can write

$$iG_{\pm\pm}^{(T)}(t,\omega) = \mathcal{O}_B^{(T)}(\omega)iG_{\pm\pm}^{(T=0)}(t,\omega),$$

so that we can write the matrix relation

$$iG^{(T)}(t,\omega) = \mathcal{O}_B^{(T)}(\omega)iG^{(T=0)}(t,\omega), \tag{6.9}$$

where we have identified (see also (6.1))

$$iG^{(T)}(t,\omega) = \begin{pmatrix} iG_{++}^{(T)}(t,\omega) & iG_{+-}^{(T)}(t,\omega) \\ iG_{-+}^{(T)}(t,\omega) & iG_{--}^{(T)}(t,\omega) \end{pmatrix},$$

$$iG^{(T=0)}(t,\omega) = \begin{pmatrix} iG_{++}^{(T=0)}(t,\omega) & iG_{+-}^{(T=0)}(t,\omega) \\ iG_{-+}^{(T=0)}(t,\omega) & iG_{--}^{(T=0)}(t,\omega) \end{pmatrix}. \tag{6.10}$$

Relation (6.9) can be easily verified, for each component of the matrices in (6.4)-(6.6), by using the identities

$$S(\omega)n_B(\omega) = n_B(-\omega) = \frac{1}{e^{-\beta\omega}-1} = \frac{e^{\beta\omega}}{1-e^{\beta\omega}}$$
$$= -(1 + n_B(\omega)), \tag{6.11}$$

and, therefore,

$$S(\omega)(1 + n_B(\omega)) = (1 + n_B(-\omega)) = -n_B(\omega), \tag{6.12}$$

which follow from (6.8), namely, $(S(\omega))^2 = \mathbb{1}$.

This shows that the finite temperature matrix propagator in the mixed space $(t, \mathbf{p})$ factorizes as the product of a very simple scalar (multiplicative) operator acting on the zero temperature matrix propagator, namely, each of the matrix components of the finite temperature propagator is related to the corresponding component of the zero temperature propagator by the same (multiplicative) operator. This is very important for various reasons because, in principle, a relation between the two, if it exists, could have involved a non-trivial matrix operator which it is not. We call this operator $\mathcal{O}_B^{(T)}(\omega)$ in (6.7) the "thermal operator". There are various features to note from the structure of this operator. First, as we have already pointed out, it is a scalar operator (and not a matrix). It does not depend on time so that it commutes with the time derivative operator, namely,

$$\partial_t \mathcal{O}_B^{(T)}(\omega) = \mathcal{O}_B^{(T)}(\omega)\partial_t. \tag{6.13}$$

We note here that, since the thermal operator does not depend on the time variable, relation (6.9) can be Fourier transformed back to the p^0 space and it would seem that the relation (6.9) should also hold true in momentum space, namely, we should also have

$$iG^{(T)}(p) = \mathcal{O}_B^{(T)}(\omega)iG^{(T=0)}(p). \tag{6.14}$$

This is indeed true, but to show this one has to be a bit careful. For example, let us note that $(\delta(p^2 - m^2) = \delta((p^0)^2 - \omega^2))$ if we calculate naively we obtain

$$(\mathbb{1} - S(\omega))(\theta(p^0)\delta((p^0)^2 - \omega^2)) = 0, \tag{6.15}$$

since the operator does not act on p^0 (naively) and $\delta((p^0)^2 - \omega^2)$ is an even function of ω. So, it would seem that when the thermal operator acts on, say, $iG_{-+}^{(T=0)}(p)$ in (6.3), all the temperature dependent terms would vanish and we will obtain

$$\mathcal{O}_B^{(T)}(\omega)iG_{-+}^{(T=0)}(p) = 2\pi\theta(p^0)\delta(p^2 - m^2) \neq iG_{-+}^{(T)}(p). \tag{6.16}$$

On the other hand, if we are careful, then, we see that

$$\mathcal{O}_B^{(T)}(\omega)iG_{-+}^{(T=0)}(p)$$
$$= 2\pi\left(\mathbb{1} + n_B(\omega)(\mathbb{1} - S(\omega))\right)\theta(p^0)\delta((p^0)^2 - \omega^2)$$
$$= 2\pi\left(\mathbb{1} + n_B(\omega)(\mathbb{1} - S(\omega))\right)\frac{1}{2\omega}\delta(p^0 - \omega)$$

$$= 2\pi \left(\frac{1}{2\omega} \delta(p^0 - \omega) + \frac{n_B(\omega)}{2\omega} \left(\delta(p^0 - \omega) + \delta(p^0 + \omega) \right) \right)$$

$$= 2\pi \left(\theta(p^0) \delta(p^2 - m^2) + n_B(\omega) \delta(p^2 - m^2) \right)$$

$$= 2\pi \left(\theta(p^0) + n_B(\omega) \right) \delta(p^2 - m^2) = i G_{-+}^{(T)}(p). \tag{6.17}$$

This shows that the thermal operator relation (6.14) indeed holds in momentum space as well, but one has to be very careful in handling how the thermal operator acts. It is for this reason that we restrict our discussions to the mixed space $(t, \mathbf{p})$ description which is much simpler to understand.

Let us next come to a general N-point graph of scalar fields in the mixed space. A Feynman graph, whether at zero temperature or at finite temperature, consists of a set of vertices connected by propagators. In a thermal field theory, for example, in the closed time path formalism, the vertices can be of two types – either all the fields at the vertex are ϕ_+ or they are all ϕ_- (they differ by a negative sign and don't carry temperature). Namely, there is no mixing between the $\phi_\pm$ fields in the tree level Lagrangian density. Let us assume that the N-external time coordinates are labelled $\tau_\alpha = (\tau_1, \tau_2, \cdots, \tau_N)$. In addition, of course, the graph may have n internal vertices with time coordinates $t_a, a = 1, 2, \cdots, n$ which are integrated over all times (from $(-\infty, \infty)$). Let us assume that there are I number of internal lines (propagators) connecting the N external and the n internal vertices with one another. The internal lines (propagators) carry the temperature dependence of the graph and, as we have seen, the temperature dependence is entirely contained in the thermal operators associated with these internal propagators. Therefore, we can represent the complete temperature dependence of a Feynman graph in a scalar thermal field theory as

$$\int \prod_{i=1}^{I} \frac{d^3 k_i}{(2\pi)^3} \prod_{v=1}^{N+n} \delta_v^3(k, p) \gamma_N^{(T)}$$

$$= \int \prod_{i=1}^{I} \frac{d^3 k_i}{(2\pi)^3} \prod_{v=1}^{N+n} \delta_v^3(k, p) \mathcal{O}_B^{(T)} \gamma_N^{(T=0)}. \tag{6.18}$$

Here we have denoted the momenta of the I internal lines (propagators) by k while the N external momenta are denoted by p and

$\delta_v^3(k,p)$ represents the appropriate three dimensional momentum conserving delta function at the vth vertex. The internal three dimensional momenta k are integrated over (while the external momenta are fixed). The external time coordinates are fixed while the n internal time coordinates have been integrated out in the zero temperature amplitude $\gamma_N^{(T=0)}$ as well as in the temperature dependent $\gamma_N^{(T)}$ (the factor of i with possible $\pm$ sign in the amplitude has been factored out). $\mathcal{O}_B^{(T)}$ stands for the product of thermal operators for the I internal lines (propagators) and explicitly, we can write

$$\mathcal{O}_B^{(T)} = \prod_{i=1}^{I} \mathcal{O}_B^{(T)}(\omega_i) = \prod_{i=1}^{I} \left(\mathbb{1} + n_B(\omega_i)(\mathbb{1} - S(\omega_i)) \right), \qquad (6.19)$$

where, as before, we have identified $\omega_i = \sqrt{\mathbf{k}_i^2 + m^2}, i = 1, 2, \cdots, I$. Therefore, we see that every graph, at finite temperature, can be obtained from its zero temperature counterpart through the application of a thermal operator defined in terms of the basic thermal operator for the propagator given in (6.7). Just to make it absolutely clear, let us reiterate that both $\gamma_N^{(T=0)}$ and $\gamma_N^{(T)}$ depend only on the momenta (k,p) as well as the N external times τ_α. The internal time dependence in the graph have been integrated out to obtain γ_N.

We want to add a word of caution here, namely, even when a function may vanish because of conservation laws, the thermal operator acting on it may lead to a nonzero result and, therefore, it would be wrong to set the function itself to zero to begin with and conclude that the thermal operator would also give zero acting on it. A simple example of this error can be seen as follows. Consider the delta function $\delta(\omega_1 + \omega_2)$ which clearly vanishes because both ω_1, ω_2 are positive. Thus, although

$$\delta(\omega_1 + \omega_2) = 0, \qquad (6.20)$$

we note that,

$$\mathcal{O}_B^{(T)}(\omega_1)\mathcal{O}^{(T)}(\omega_2)\delta(\omega_1 + \omega_2)$$
$$= \mathcal{O}_B^{(T)}(\omega_1)\left((1 + n_B(\omega_2))\delta(\omega_1 + \omega_2) - n_B(\omega_2)\delta(\omega_1 - \omega_2)\right)$$
$$= \left((1 + n_B(\omega_1))(1 + n_B(\omega_2)) + n_B(\omega_1)n_B(\omega_2)\right)\delta(\omega_1 + \omega_2)$$

$$- \left((1 + n_B(\omega_1))n_B(\omega_2) + (1 + n_B(\omega_2))n_B(\omega_1)\right)\delta(\omega_1 - \omega_2)$$

$$= -\left((1 + n_B(\omega_1))n_B(\omega_2) + (1 + n_B(\omega_2))n_B(\omega_1)\right)\delta(\omega_1 - \omega_2)$$

$$\neq 0, \tag{6.21}$$

where we have used the fact that the delta function is an even function. On the other hand, if we naively set the delta function to zero to begin with, we would conclude that the thermal operators acting on zero would vanish which is incorrect.

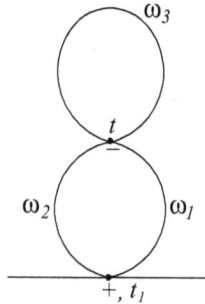

Figure 6.1: A two loop self-energy diagram in the ϕ^4 theory with an internal "$-$" vertex.

The previous example, is not simply academic, but it actually arises in calculations involving higher loop graphs with mixed vertices. For example, consider the two loop graph for the self-energy of a scalar field in the ϕ^4 theory involving a ϕ_- internal vertex as shown in Fig. 6.1. It is easy to see that (we are going to neglect explicit momentum dependences in the graph for simplicity, remember that $\omega_i = \sqrt{\mathbf{k}_i^2 + m^2}$ and we have labelled the external time here as t_1 and not as τ_1 also for simplicity), in this case,

$$-i\gamma_+^{(T=0)}(t_1) = \int_{-\infty}^{\infty} dt \, (-i)iG_{+-}^{(T=0)}(t_1 - t, \omega_1)iG_{+-}^{(T=0)}(t_1 - t, \omega_2)$$

$$\times (i)(iG_{--}^{(T=0)}(0, \omega_3))$$

$$= \frac{1}{(2\omega_1)(2\omega_2)(2\omega_3)}\delta(\omega_1 + \omega_2), \tag{6.22}$$

which vanishes, as we have pointed out in (6.20), since both ω_1, ω_2 are finite positive quantities. So, naively, this would suggest that, at finite temperature, the contribution of this graph should also vanish. However, a direct calculation of the finite temperature graph leads to (we denote $n_B(\omega_i) = n_i, i = 1, 2, 3$)

$$-i\gamma_+^{(T)}(t_1) = \int_{-\infty}^{\infty} dt\, (-i)iG_{+-}^{(T)}(t_1 - t, \omega_1)iG_{+-}^{(T)}(t_1 - t, \omega_2)$$

$$\times (i)(iG_{--}^{(T)}(0, \omega_3))$$

$$= \frac{2\pi(1 + 2n_3)}{(2\omega_1)(2\omega_2)(2\omega_3)} \Big[\big((1 + n_1)(1 + n_2) + n_1 n_2\big)\delta(\omega_1 + \omega_2)$$

$$+ \big(n_1(1 + n_2) + n_2(1 + n_1)\big)\delta(\omega_1 - \omega_2)\Big]. \qquad (6.23)$$

We can, of course, set the $\delta(\omega_1 + \omega_2)$ term to zero and then, we will have the nontrivial contribution to the self-energy graph in Fig. 6.1 coming from the second term in (6.23), namely,

$$-i\gamma_+^{(T)}(t_1) = \frac{2\pi(1 + 2n_3)(n_1 + n_2 + 2n_1 n_2)}{(2\omega_1)(2\omega_2)(2\omega_3)}\delta(\omega_1 - \omega_2). \qquad (6.24)$$

However, we note that if we don't set the first term to zero for the moment, relation (6.23) can be written simply as (see (6.22))

$$-i\gamma_+^{(T)}(t_1) = (1 + n_1(1 - S_1))(1 + n_2(1 - S_2))$$

$$\times (1 + n_3(1 + n_3(1 - S_3)))\frac{2\pi\delta(\omega_1 + \omega_2)}{(2\omega_1)(2\omega_2)(2\omega_3)}$$

$$= \mathcal{O}_1^{(T)}\mathcal{O}_2^{(T)}\mathcal{O}_3^{(T)}\big(-i\gamma_+^{(T=0)}(t_1)\big). \qquad (6.25)$$

Here we have identified $S_i = S(\omega_i), i = 1, 2, 3$ and this exemplifies how the thermal operator representation leads to the correct finite temperature graph even when the zero temperature graph may vanish. We point out here that the overall sign difference between (6.21) and (6.25) arises because the denominators are missing in the previous expression since each of the factors in the denominator changes sign when the reflection operator acts on it.

6.2.2 Origin of the thermal operator. In the last subsection, we have studied in detail the thermal operator representation in the case of the scalar field theory. The two important things which we noted in this connection and which lead to the thermal operator representation of any amplitude are that the vertices in a thermal field theory are temperature independent while the entire temperature dependence is carried by the propagators in the Feynman graph. Furthermore, we saw in detail in (6.1)-(6.17) that the propagator at finite temperature is related to the zero temperature one through the basic thermal operator. This leads to the thermal operator representation of any amplitude graph by graph.

In thermal field theory, the thermal propagator represents the ensemble average of a time ordered product of two field operators and, therefore, an interesting question arises, namely, how can an ensemble average be related to a vacuum expectation value of the time ordered field operators, as the thermal operator representation of the propagator would suggest. Clearly, one way this is possible, is if the (averaged) expectation value of the time ordered product of two operators in every higher energy state is proportional to its vacuum expectation value. That this indeed happens and leads to the thermal operator representation for the propagator can be seen as follows.

The scalar field can be expanded in terms of creation and annihilation operators as[1]

$$\phi(x) = \int \frac{d^3k}{(2\pi)^{\frac{3}{2}}} e^{i\mathbf{k}\cdot\mathbf{x}} \phi(t, \mathbf{k})$$

$$= \int \frac{d^3k}{(2\pi)^{\frac{3}{2}}} e^{i\mathbf{k}\cdot\mathbf{x}} \frac{1}{2\omega} \left(e^{-i\omega t} a(\mathbf{k}) + e^{i\omega t} a^\dagger(-\mathbf{k}) \right), \qquad (6.26)$$

where, as before, $\omega = \omega(\mathbf{k}) = \sqrt{\mathbf{k}^2 + m^2}$. Therefore, the expectation value of a product of two operators in an eigenstate of energy and momentum takes the form

$$\langle n, \mathbf{k} | \phi(t_1, \mathbf{k}_1) \phi(t_2, \mathbf{k}_2) | n, \mathbf{k} \rangle$$

$$= \langle n, \mathbf{k} | \frac{e^{-i\omega_1 t_1 + i\omega_2 t_2}}{2\sqrt{\omega_1 \omega_2}} a(\mathbf{k}_1) a^\dagger(-\mathbf{k}_2)$$

$$+ \frac{e^{i\omega_1 t_1 - i\omega_2 t_2}}{2\sqrt{\omega_1 \omega_2}} a^\dagger(-\mathbf{k}_1) a(\mathbf{k}_2) | n, \mathbf{k} \rangle$$

---

[1] See, for example, section **5.4** in A. Das, *Lectures on Quantum Field Theory* (second edition), World Scientific Publishing, Singapore (2020).

$$= \delta^3(\mathbf{k} + \mathbf{k}_2)\delta^3(\mathbf{k} - \mathbf{k}_1) \frac{e^{-i\omega(t_1 - t_2)}}{2\omega} (1 + n)$$

$$+ \delta^3(\mathbf{k} - \mathbf{k}_2)\delta^3(\mathbf{k} + \mathbf{k}_1) \frac{e^{i\omega(t_1 - t_2)}}{2\omega} n$$

$$= \delta^3(\mathbf{k}_1 + \mathbf{k}_2)\left(\delta^3(\mathbf{k} - \mathbf{k}_1)\frac{e^{-i\omega(t_1 - t_2)}}{2\omega} (1 + n) \right.$$

$$\left. + \delta^3(\mathbf{k} + \mathbf{k}_1) \frac{e^{i\omega(t_1 - t_2)}}{2\omega} n \right), \tag{6.27}$$

and, similarly,

$$\langle n, \mathbf{k} | \phi(t_2, \mathbf{k}_2)\phi(t_1, \mathbf{k}_1) | n, \mathbf{k} \rangle$$

$$= \delta^3(\mathbf{k}_1 + \mathbf{k}_2)\left(\delta^3(\mathbf{k} + \mathbf{k}_1)\frac{e^{i\omega(t_1 - t_2)}}{2\omega} (1 + n) \right.$$

$$\left. + \delta^3(\mathbf{k} - \mathbf{k}_1) \frac{e^{-i\omega(t_1 - t_2)}}{2\omega} n \right). \tag{6.28}$$

There are several things to note here. First, the overall delta function, $\delta^3(\mathbf{k}_1 + \mathbf{k}_2)$, signifies the conservation of momentum for the two point amplitude. Second, n denotes the number of quanta in the state with energy ω and takes values $n = 0, 1, 2, \cdots$. From (6.27) and (6.28) we can write the expectation value of the time ordered product in this energy state as

$$\langle n, \mathbf{k} | T(\phi(t_1, \mathbf{k}_1)\phi(t_2, \mathbf{k}_2)) | n, \mathbf{k} \rangle = \delta^3(\mathbf{k}_1 + \mathbf{k}_2)$$

$$\times \left[\delta^3(\mathbf{k} - \mathbf{k}_1)\left(\theta(t_1 - t_2)(1 + n) + \theta(t_2 - t_1)n \right) \frac{e^{-i\omega(t_1 - t_2)}}{2\omega} \right.$$

$$\left. + \delta^3(\mathbf{k} + \mathbf{k}_1)\left(\theta(t_1 - t_2)n + \theta(t_2 - t_1)(1 + n) \right) \frac{e^{i\omega(t_1 - t_2)}}{2\omega} \right].$$

$$\tag{6.29}$$

This expectation value does not seem like it is proportional to a zero temperature vacuum expectation value. However, let us note that if we were to average this with an even weight function $f(|\mathbf{k}|)$ (as is the case in a thermal averaging), both the delta function terms would lead to nontrivial contributions yielding

$$\int d^3k\, f(|\mathbf{k}|)\, \langle n, \mathbf{k}|T(\phi(t_1, \mathbf{k}_1)\phi(t_2, \mathbf{k}_2))|n, \mathbf{k}\rangle = f(|\mathbf{k}_1|)$$

$$\times\, \delta^3(\mathbf{k}_1 + \mathbf{k}_2)\Big[\theta(t_1 - t_2)\Big((1+n)\frac{e^{-i\omega_1(t_1-t_2)}}{2\omega_1} + n\frac{e^{i\omega_1(t_1-t_2)}}{2\omega_1}\Big)$$

$$+\, \theta(t_2 - t_1)\Big(n\frac{e^{-i\omega_1(t_1-t_2)}}{2\omega_1} + (1+n)\frac{e^{i\omega_1(t_1-t_2)}}{2\omega_1}\Big)\Big]$$

$$= f(|\mathbf{k}_1|)\delta^3(\mathbf{k}_1 + \mathbf{k}_2)\big(\mathbb{1} + n(\mathbb{1} - S(\omega_1))\big)$$

$$\times \Big(\theta(t_1 - t_2)\frac{e^{-i\omega_1(t_1-t_2)}}{2\omega_1} + \theta(t_2 - t_1)\frac{e^{i\omega_1(t_1-t_2)}}{2\omega_1}\Big)$$

$$= f(|\mathbf{k}_1|)\big(\mathbb{1} + n(\mathbb{1} - S(\omega_1))\big)\langle 0|T(\phi(t_1, \mathbf{k}_1)\phi(t_2, \mathbf{k}_2))|0\rangle, \quad (6.30)$$

where $S(\omega_1)$ denotes the reflection operator described earlier. This clearly shows how an averaged expectation value of a time ordered product of two field operators can be related to the vacuum expectation value of the same time ordered product. Now, if we assume that we have a thermal ensemble average, we can identify $f_n(|\mathbf{k}_1|) = e^{-\beta E_{n,\mathbf{k}_1}} = e^{-\beta\omega_1 n}$ (remember that $\hbar = 1$) and, if we sum over all positive integer values of n, we obtain

$$\sum_{n=0}^{\infty} f_n(|\mathbf{k}_1|)\big(\mathbb{1} + n(\mathbb{1} - S(\omega_1))\big) = \sum_{n=0}^{\infty} e^{-\beta\omega_1 n}\big(\mathbb{1} + n(\mathbb{1} - S(\omega_1))\big)$$

$$= \sum_{n=0}^{\infty}\Big[e^{-\beta\omega_1 n}\mathbb{1} - \Big(\frac{\partial}{\partial(\beta\omega_1)}e^{-\beta\omega_1 n}\Big)(\mathbb{1} - S(\omega_1))\Big]$$

$$= \frac{\mathbb{1}}{1 - e^{-\beta\omega_1}} - \Big(\frac{\partial}{\partial(\beta\omega_1)}\frac{1}{1 - e^{-\beta\omega_1}}\Big)(\mathbb{1} - S(\omega_1))$$

$$= (1 + n_B(\omega_1))\big(\mathbb{1} + n_B(\omega_1)(\mathbb{1} - S(\omega_1))\big), \quad (6.31)$$

where $n_B(\omega_1)$ denotes the Bose-Einstein distribution with (energy) frequency ω_1. Furthermore, recalling that the partition function for the bosonic system is obtained from the normalization of the state and is given by

$$Z(\beta) = \sum_{n=0}^{\infty} e^{-\beta\omega_1 n} = \frac{1}{1 - e^{-\beta\omega_1}} = (1 + n_B(\omega_1)), \quad (6.32)$$

we determine the thermal ensemble average of the time ordered two point function (or the thermal propagator) is obtained from (6.30)-(6.32) to be

$$G^{(T)}(t_1 - t_2, \mathbf{k}_1 - \mathbf{k}_2)$$

$$= \frac{1}{Z(\beta)} \sum_{n=0}^{\infty} \int d^3k \, e^{-\beta\omega n} \langle n, \mathbf{k} | T(\phi(t_1, \mathbf{k}_1)\phi(t_2, \mathbf{k}_2)) | n, \mathbf{k} \rangle$$

$$= \left(1 + n_B(\omega_1)(1 - S(\omega_1))\right) G^{(T=0)}(t_1 - t_2, \mathbf{k}_1 - \mathbf{k}_2). \qquad (6.33)$$

This clarifies the origin of the thermal operator as well as the thermal operator representation of the propagator.

6.2.3 Properties of the thermal operator. The thermal operator satisfies a number of interesting properties which we discuss here. First, let us recall, from (6.8) and the two equations in (6.12), the properties satisfied by the reflection operator $S(\omega)$. In particular, we note from the two relations in (6.11) and (6.12) that, as an operator relation, we can write

$$S(\omega)n(\omega) = -(1 + n(\omega))S(\omega), \quad S(\omega)(1 + n(\omega)) = -n(\omega)S(\omega),$$

where we have neglected the subscript B for simplicity. Using these, we can now show that the definition of the thermal operator in (6.7)

$$\mathcal{O}^{(T)}(\omega) = \left(1 + n(\omega)(1 - S(\omega))\right) = (1 - \mathbb{P}(\omega)),$$

$$\mathbb{P}(\omega) = -n(\omega)(1 - S(\omega)), \qquad (6.34)$$

can also be written as

$$= \left((1 + n(\omega))1 - n(\omega)S(\omega)\right), \qquad (6.35)$$

which leads to

$$\left(\mathcal{O}^{(T)}(\omega)\right)^2 = \left((1 + n(\omega))1 - n(\omega)S(\omega)\right)\left((1 + n(\omega))1 - n(\omega)S(\omega)\right)$$

$$= (1 + n(\omega))^2 1 + (1 + n(\omega))(-n(\omega))S(\omega) - n(\omega)(-n(\omega))S(\omega)$$

$$+ n(\omega)(-(1 + n(\omega))S(\omega))S(\omega)$$

$$= (1 + n(\omega))1 - n(\omega)S(\omega) = \mathcal{O}^{(T)}(\omega). \qquad (6.36)$$

Here we have used the property of the reflection operator $(S(\omega))^2 = \mathbb{1}$ given in (6.8) in the last step. This shows that the thermal operator $\mathcal{O}^{(T)}(\omega)$ is idempotent and, therefore, is like a projection operator. As a result, the thermal operator does not have an inverse. Note also, from (6.34), that for $\mathcal{O}^{(T)}(\omega)$ to satisfy (6.36), the operator $\mathbb{P}(\omega)$ must also be a projection operator satisfying

$$\left(\mathbb{P}(\omega)\right)^2 = \mathbb{P}(\omega), \quad \mathbb{P}(\omega) = -n(\omega)(\mathbb{1} - S(\omega)), \tag{6.37}$$

which can be explicitly checked.

Let us also note here that if we were to define a second operator

$$\overline{\mathcal{O}}^{(T)}(\omega) = (\mathbb{1} + n(\omega)(\mathbb{1} + S(\omega))) = (\mathbb{1} - \overline{\mathbb{P}}(\omega)), \tag{6.38}$$

which can also be written as

$$= ((1 + n(\omega))\mathbb{1} + n(\omega)S(\omega)), \tag{6.39}$$

then, it can also be shown in a similar way (as in (6.36)) that

$$\left(\overline{\mathcal{O}}^{(T)}(\omega)\right)^2 = \overline{\mathcal{O}}^{(T)}(\omega), \tag{6.40}$$

$$\left(\overline{\mathbb{P}}(\omega)\right)^2 = \overline{\mathbb{P}}(\omega), \quad \overline{\mathbb{P}}(\omega) = -n(\omega)(\mathbb{1} + S(\omega)), \tag{6.41}$$

namely, these, too, are idempotent and are like projection operators. We note that, with the reflection operator,

$$\begin{aligned}
S(\omega)\mathcal{O}^{(T)}(\omega) &= S(\omega)\big((1 + n(\omega))\mathbb{1} - n(\omega)S(\omega)\big) \\
&= -n(\omega)S(\omega) - (-(1 + n(\omega))\mathbb{1}) \\
&= \mathcal{O}^{(T)}(\omega),
\end{aligned} \tag{6.42}$$

and, similarly,

$$S(\omega)\overline{\mathcal{O}}^{(T)}(\omega) = -\overline{\mathcal{O}}^{(T)}(\omega). \tag{6.43}$$

As a result, we conclude that

$$(\mathbb{1} - S(\omega))\mathcal{O}^{(T)}(\omega) = 0 = (\mathbb{1} + S(\omega))\overline{\mathcal{O}}^{(T)}(\omega), \tag{6.44}$$

while

$$\begin{aligned}
&\frac{1}{2}(\mathbb{1} + S(\omega))\mathcal{O}^{(T)}(\omega) = \mathcal{O}^{(T)}(\omega), \\
&\frac{1}{2}(\mathbb{1} - S(\omega))\overline{\mathcal{O}}^{(T)}(\omega) = \overline{\mathcal{O}}^{(T)}(\omega).
\end{aligned} \tag{6.45}$$

These relations, (6.44) and (6.45), also follow by noting that we can write the two operators in (6.7) and (6.39) alternatively as

$$\mathcal{O}^{(T)}(\omega) = (\mathbb{1} + S(\omega))(1 + n(\omega)),$$
$$\overline{\mathcal{O}}^{(T)}(\omega) = (\mathbb{1} - S(\omega))(1 + n(\omega)), \tag{6.46}$$

and equation (6.44) simply reflects the orthogonality of the projection operators $\frac{1}{2}(\mathbb{1} \pm S(\omega))$.

Finally, to understand what these projection operators mean, let us note that the thermal operators is a sum of two parts

$$iG^{(T)}(t, \omega) = \mathcal{O}^{(T)}(\omega)(iG^{(T=0)}(t, \omega)) = (\mathbb{1} - \mathbb{P}(\omega))iG^{(T=0)}(t, \omega)$$
$$= iG^{(T=0)}(t, \omega) + i\overline{G}(t, \omega), \tag{6.47}$$

where

$$i\overline{G}(t, \omega) = -\mathbb{P}(\omega)iG^{(T=0)}, \tag{6.48}$$

carries all the temperature dependence of the propagator. Furthermore, we note that

$$\mathbb{P}(\omega)i\overline{G}(t, \omega) = -\mathbb{P}^2(\omega)iG^{(T=0)}(t, \omega) = -\mathbb{P}(\omega)iG^{(T=0)}(t, \omega)$$
$$= i\overline{G}(t, \omega), \tag{6.49}$$

where we have used (6.37) as well as (6.48). It now follows from (6.46)-(6.49) that

$$\mathcal{O}^{(T)}i\overline{G}(t, \omega) = (\mathbb{1} - \mathbb{P}(\omega))i\overline{G}(t, \omega) = 0. \tag{6.50}$$

Namely, the thermal operator annihilates the finite temperature part of the thermal propagator. This is also consistent with (6.36) and (6.47), namely,

$$iG^{(T)}(t, \omega) = \mathcal{O}^{(T)}(\omega)iG^{(T=0)}(t, \omega) = iG^{(T=0)}(t, \omega) + i\overline{G}(t, \omega),$$

which leads to

$$\text{or,}\quad \mathcal{O}^{(T)}(\omega)iG^{(T)}(t, \omega) = \mathcal{O}^{(T)}(\omega)iG^{(T=0)}(t, \omega),$$
$$\text{or,}\quad \left(\mathcal{O}^{(T)}(\omega)\right)^2 G^{(T=0)}(t, \omega) = \mathcal{O}^{(T)}(\omega)G^{(T=0)}(t, \omega), \tag{6.51}$$

expressing the idempotent nature of the thermal operator.

6.2.4 Complex scalar field theory. The discussion for the complex scalar field theory is, of course, exactly similar to the real scalar field theory case. However, for a complex scalar field theory, there is the concept of a conserved charge and, therefore, we can discuss about a theory with a chemical potential. The presence of a chemical potential or a conserved charge in a theory changes the thermodynamic ensemble to a grand canonical ensemble as opposed to the canonical ensemble which we were discussing in the earlier subsections.

Let us consider a Lagrangian density of the form

$$\mathcal{L} = \partial_\mu \phi^\dagger \partial^\mu \phi - (m^2 - \mu^2)\phi^\dagger \phi + i\mu \phi^\dagger \overleftrightarrow{\partial_t} \phi, \tag{6.52}$$

where μ denotes the chemical potential satisfying (we assume both m, μ are positive)

$$\mu \leq m. \tag{6.53}$$

We note that

$$i\phi^\dagger \overleftrightarrow{\partial_t} \phi = J^0, \tag{6.54}$$

represents the conserved charge density of the complex scalar field theory so that, in the Hamiltonian, the chemical potential couples to the conserved charge operator, namely,

$$\mathcal{H}^{(\mu)} = -\mu \int d^3x \, J^0 = -\mu Q, \tag{6.55}$$

as it should (see, for example, (1.3)-(1.5)). The chemical potential, in this case, can also be thought of as a constant background electrostatic potential for the complex (charged) scalar field and, in fact, we can rewrite the Lagrangian density in (6.52) also in the more familiar form emphasizing this aspect as

$$\mathcal{L} = (\partial_t + i\mu)\phi^\dagger (\partial_t - i\mu)\phi - \boldsymbol{\nabla}\phi^\dagger \cdot \boldsymbol{\nabla}\phi - m^2\phi^\dagger \phi. \tag{6.56}$$

The presence of the chemical potential clearly only changes quadratic terms in the field variables and, therefore, changes only the propagator, but does not introduce any new (nontrivial) interaction vertices. Therefore, let us determine the propagator for this theory. In the closed time path formalism, the propagator (Green's function), for the theory in (6.56), satisfies the contour ordered differential equation (see chapter **2.6.1**)

$$\left((\partial_t - i\mu)^2 + \omega^2\right) G_c^{(T,\mu)}(t, \omega) = -\delta_c(t),$$ (6.57)

where the subscript c denotes contour ordering, $\omega = \omega_{\mathbf{p}} = (\mathbf{p}^2 + m^2)^{\frac{1}{2}}$ and, for simplicity, we have denoted the time difference $t_1 - t_2$ by t. Following the appendix in **2.6.1**, we write the general solution on the contour as

$$G_c^{(T,\mu)}(t, \omega) = \theta_c(t)\left(\overline{\alpha}\, e^{-i\omega_- t} + \overline{\beta}\, e^{i\omega_+ t}\right)$$
$$+ \theta_c(-t)\left(\overline{\gamma}\, e^{-i\omega_- t} + \overline{\delta}\, e^{i\omega_+ t}\right),$$ (6.58)

where we have identified $\omega_\pm = \omega \pm \mu$ and $\overline{\alpha}, \overline{\beta}, \overline{\gamma}, \overline{\delta}$ are constants to be determined. Acting with the quadratic operator on (6.58) yields (see **2.6.1** for details) leads to

$$\left((\partial_t - i\mu)^2 + \omega^2\right) G_c^{(T,\mu)}(t, \omega)$$
$$= -i\omega(\overline{\alpha} - \overline{\beta} - \overline{\gamma} + \overline{\delta})\,\delta_c(t) + (\overline{\alpha} + \overline{\beta} - \overline{\gamma} - \overline{\delta})\,\delta_c'(t),$$ (6.59)

so that, comparing with the right hand side of (6.57), we can determine

$$\overline{\gamma} = \overline{\alpha} - \frac{1}{2i\omega}, \qquad \overline{\delta} = \overline{\beta} + \frac{1}{2i\omega},$$ (6.60)

so that we can write

$$G_c^{(T,\mu)}(t, \omega) = \theta_c(t)\left(\overline{\alpha}\, e^{-i\omega_- t} + \overline{\beta}\, e^{i\omega_+ t}\right)$$
$$+ \theta_c(-t)\left(\left(\overline{\alpha} - \frac{1}{2i\omega}\right) e^{-i\omega_- t} + \left(\overline{\beta} + \frac{1}{2i\omega}\right) e^{i\omega_+ t}\right).$$ (6.61)

Imposing the periodicity on the closed time contour (see, for example, (2.99))

$$G_c^{(T,\mu)}(T - i\beta - t, \omega) = G_c^{(T,\mu)}(T - t, \omega),$$ (6.62)

where we have identified, as usual, $t_i = T$ and $t_f = T - i\beta$ (not to be confused with the temperature), we determine

$$\overline{\alpha} = \frac{1}{2i\omega}(1 + n_B(\omega_-)), \qquad \overline{\beta} = \frac{1}{2i\omega} n_B(\omega_+),$$ (6.63)

where $n_B(\omega_\pm)$ denote the two bosonic distribution functions which arise corresponding to the energies $\omega_\pm$. Therefore, the tree level propagator is determined to be

$$iG_c^{(T,\mu)}(t,\omega) = \frac{\theta_c(t)}{2\omega}\left((1+n_B(\omega_-))\,e^{-i\omega_- t} + n_B(\omega_+)\,e^{i\omega_+ t}\right)$$

$$+ \frac{\theta_c(-t)}{2\omega}\left(n_B(\omega_-)\,e^{-i\omega_- t} + (1+n_B(\omega_+))\,e^{i\omega_+ t}\right). \quad (6.64)$$

For $\mu = 0$, this can be compared with (2.36).

The four components of the 2×2 propagator matrix (in the closed time path with a chemical potential) can now be obtained from (6.64),

$$iG_{++}^{(T,\mu)}(t,\omega) = \frac{\theta(t)}{2\omega}\left((1+n_B(\omega_-))e^{-i\omega_- t} + n_B(\omega_+)e^{i\omega_+ t}\right)$$

$$+ \frac{\theta(-t)}{2\omega}\left(n_B(\omega_-)e^{-i\omega_- t} + (1+n_B(\omega_+))e^{i\omega_+ t}\right),$$

$$iG_{+-}^{(T,\mu)}(t,\omega) = \frac{1}{2\omega}\left((1+n_B(\omega_-))e^{-i\omega_- t} + n_B(\omega_+)e^{i\omega_+ t}\right), \quad (6.65)$$

$$iG_{-+}^{(T,\mu)}(t,\omega) = \frac{1}{2\omega}\left(n_B(\omega_-)e^{-i\omega_- t} + (1+n_B(\omega_+))e^{i\omega_+ t}\right),$$

$$iG_{--}^{(T,\mu)}(t,\omega) = \frac{\theta(-t)}{2\omega}\left((1+n_B(\omega_-))e^{-i\omega_- t} + n_B(\omega_+)e^{i\omega_+ t}\right)$$

$$+ \frac{\theta(t)}{2\omega}\left(n_B(\omega_-)e^{-i\omega_- t} + (1+n_B(\omega_+))e^{i\omega_+ t}\right).$$

$$(6.66)$$

These components can be compared with the ones in (2.42) for $\mu = 0$. The zero temperature components can be read out from (6.65) and take the forms

$$iG_{++}^{(T=0,\mu)}(t,\omega) = \frac{1}{2\omega}\left(\theta(t)e^{-i\omega_- t} + \theta(-t)e^{i\omega_+ t}\right),$$

$$iG_{+-}^{(T=0,\mu)}(t,\omega) = \frac{1}{2\omega}\,e^{-i\omega_- t},$$

$$iG_{-+}^{(T=0,\mu)}(t,\omega) = \frac{1}{2\omega}\,e^{i\omega_+ t},$$

$$iG_{--}^{(T=0,\mu)}(t,\omega) = \frac{1}{2\omega}\left(\theta(-t)\,e^{-i\omega_- t} + \theta(t)\,e^{i\omega_+ t}\right), \quad (6.67)$$

which can be compared with (6.4) for $\mu = 0$. We note here, for completeness, that the phase factor involving the chemical potential is the same $e^{i\mu t}$ in all the four components of the propagator in (6.65)-(6.67). Therefore, in any loop calculation in a Feynman diagram this phase factor will cancel out. (The distribution functions, however, will continue to carry the dependence on the chemical potential.) Of course, the propagators, in this case, have more structure compared to the case of $\mu = 0$ (see (6.4)-(6.6)). Nonetheless, a nontrivial thermal operator representation can still be obtained if we define

$$\mathcal{O}_B^{(T,\mu)}(\omega) = \left(\mathbb{1} + \widehat{n}_B^{(T,\mu)}(\mathbb{1} - S(\omega))\right), \tag{6.68}$$

where, in addition to the usual reflection operator $S(\omega)$, we have introduced a new operator $\widehat{n}_B^{(T,\mu)}$ whose action on exponential functions is given by

$$\widehat{n}_B^{(T,\mu)} e^{is\omega_s t} = n_B(\omega_s) e^{is\omega_s t}, \quad s = \pm, \ s^2 = 1. \tag{6.69}$$

In fact, such a new operator can be given a simple representation of the form

$$\widehat{n}_B^{(T,\mu)} = n_B(-is\partial_t),$$

so that

$$n_B(-is\partial_t) e^{is\omega_s t} = n_B(\omega_s) e^{is\omega_s t}. \tag{6.70}$$

However, a specific representation of the operator is not necessary for our subsequent discussion. Note also, that when $\mu = 0$, this operator naturally leads to the multiplicative factor $n_B(\omega)$ noted in the thermal operator earlier in (6.7). With this new operator, (6.68), we can check that the finite temperature propagator is related to the zero temperature one, component by component, by the relation

$$iG_{ab}^{(T,\mu)}(t,\omega) = \mathcal{O}_B^{(T,\mu)}(\omega) iG_{ab}^{(T=0,\mu)}(t,\omega), \quad a,b = \pm. \tag{6.71}$$

Since the vertices of the theory do not depend on either temperature or chemical potential, the factorization of any amplitude at finite temperature (and chemical potential) also goes through completely parallel to the earlier discussion in subsection **6.2.1**. To conclude, we want to note here that, in any component of the propagators in (6.65) or (6.67), the phase factor involving the chemical potential is the same $e^{i\mu t}$ and can be factored out as an overall factor. Therefore, in any loop diagram (involving any propagator components), this phase factor would completely cancel out although the chemical potential dependence will still be there in the distribution functions. Of course, this phase factor does not have to cancel out in diagrams which do not involve loops.

6.2.5 Fermion field theory. Let us consider the free Dirac theory of a charged fermion with mass m which is described by the Lagrangian density

$$\mathcal{L} = i\bar{\psi}\slashed{\partial}\psi - m\bar{\psi}\psi, \qquad \slashed{\partial} = \gamma^{\mu}\partial_{\mu}, \tag{6.72}$$

where the index μ, in the last relation, is being summed over its four values $\mu = 0, 1, 2, 3$ (in $3 + 1$ dimensions). We concentrate on the free Lagrangian density simply because the temperature (or the chemical potential) is contained entirely in the propagators which is determined from the free theory, the interaction vertices do not depend on these quantities.

The fermion propagator, at zero temperature, has the familiar form in momentum space given by

$$iS(p) = \frac{i}{\slashed{p} - m + i\epsilon} = \frac{i(\slashed{p} + m)}{p^2 - m^2 + i\epsilon}. \tag{6.73}$$

At finite temperature, however, the degrees of freedom double and in the closed time path formalism, the 2×2 matrix propagator has the components given by

$$iS_{++}^{(T)}(p) = (\slashed{p} + m)\left(\frac{i}{p^2 - m^2 + i\epsilon} - 2\pi n_F(|p^0|)\delta(p^2 - m^2)\right),$$

$$iS_{+-}^{(T)}(p) = 2\pi(\slashed{p} + m)\left(\theta(-p^0) - n_F(|p^0|)\right)\delta(p^2 - m^2),$$

$$iS_{-+}^{(T)}(p) = 2\pi(\slashed{p} + m)\left(\theta(p^0) - n_F(|p^0|)\right)\delta(p^2 - m^2), \tag{6.74}$$

$$iS_{--}^{(T)}(p) = (\slashed{p} + m)\left(-\frac{i}{p^2 - m^2 - i\epsilon} - 2\pi n_F(|p^0|)\delta(p^2 - m^2)\right),$$

where $n_F(|p^0|)$ denotes the fermion distribution function (at finite temperature) and arises from the anti-periodicity condition that the fermion fields have to satisfy. These components can be compared with those for the real scalar field given in (6.2) and it is clear, as is the case there as well, that the thermal parts of the four components in (6.74) are also the same (of course, with a relative negative sign and a different distribution function from (6.2)).

Let us determine the fermion Green's function systematically. In the closed time path formalism it satisfies the equation

$$(i\partial\!\!\!/ - m)S_c(x) = \delta_c(x), \tag{6.75}$$

which in the mixed space (of time and spatial momentum) takes the form

$$(i\gamma^0\partial_t - \boldsymbol{\gamma}\cdot\mathbf{p} - m)S_c(t,\mathbf{p}) = \delta_c(t). \tag{6.76}$$

Since the fermion equation is a first order equation, following **2.6.2**, we can parameterize the propagator as

$$S_c(t,\mathbf{p}) = \theta_c(t)\left(\widetilde{A}(\mathbf{p})e^{-i\omega t} + \widetilde{B}(\mathbf{p})e^{i\omega t}\right)$$
$$- \theta_c(-t)\left(\widetilde{C}(\mathbf{p})e^{-i\omega t} + \widetilde{D}(\mathbf{p})e^{i\omega t}\right), \tag{6.77}$$

where $\omega = \omega_{\mathbf{p}} = \sqrt{\mathbf{p}^2 + m^2}$ and $\widetilde{A}(\mathbf{p}), \widetilde{B}(\mathbf{p}), \widetilde{C}(\mathbf{p}), \widetilde{D}(\mathbf{p})$ are matrices in the Dirac space to be determined. The relative negative sign between the two terms on the right hand side arises from the definition of time ordering for fermions. Substituting this into (6.76), we obtain

$$i\delta_c(t)\gamma^0(\widetilde{A}(\mathbf{p}) + \widetilde{B}(\mathbf{p}) + \widetilde{C}(\mathbf{p}) + \widetilde{D}(\mathbf{p}))$$
$$+ \theta_c(t)\Big((\gamma^0\omega - \boldsymbol{\gamma}\cdot\mathbf{p} - m)\widetilde{A}(\mathbf{p})e^{-i\omega t}$$
$$- (\gamma^0\omega + \boldsymbol{\gamma}\cdot\mathbf{p} + m)\widetilde{B}(\mathbf{p})e^{i\omega t}\Big)$$
$$- \theta_c(-t)\Big((\gamma^0\omega - \boldsymbol{\gamma}\cdot\mathbf{p} - m)\widetilde{C}(\mathbf{p})e^{-i\omega t}$$
$$- (\gamma^0\omega + \boldsymbol{\gamma}\cdot\mathbf{p} + m)\widetilde{D}(\mathbf{p})e^{i\omega t}\Big)$$
$$= \delta_c(t), \tag{6.78}$$

which determines

$$\tilde{A}(\mathbf{p}) + \tilde{B}(\mathbf{p}) + \tilde{C}(\mathbf{p}) + \tilde{D}(\mathbf{p}) = -i\gamma^0,$$

$$(\gamma^0\omega - \boldsymbol{\gamma}\cdot\mathbf{p} - m)\tilde{A}(\mathbf{p}) = 0 = (\gamma^0\omega + \boldsymbol{\gamma}\cdot\mathbf{p} + m)\tilde{B}(\mathbf{p}),$$

$$(\gamma^0\omega - \boldsymbol{\gamma}\cdot\mathbf{p} - m)\tilde{C}(\mathbf{p}) = 0 = (\gamma^0\omega + \boldsymbol{\gamma}\cdot\mathbf{p} + m)\tilde{D}(\mathbf{p}). \quad (6.79)$$

The last two relations of (6.79) are solved by

$$\tilde{A}(\mathbf{p}) = \alpha A(\mathbf{p}) = \alpha(\gamma^0\omega - \boldsymbol{\gamma}\cdot\mathbf{p} + m),$$

$$\tilde{B}(\mathbf{p}) = -\beta B(\mathbf{p}) = -\beta(\gamma^0\omega + \boldsymbol{\gamma}\cdot\mathbf{p} - m),$$

$$\tilde{C}(\mathbf{p}) = \gamma C(\mathbf{p}) = \gamma(\gamma^0\omega - \boldsymbol{\gamma}\cdot\mathbf{p} + m) = \gamma A(\mathbf{p}),$$

$$\tilde{D}(\mathbf{p}) = -\delta D(\mathbf{p}) = -\delta(\gamma^0\omega + \boldsymbol{\gamma}\cdot\mathbf{p} - m) = -\delta B(\mathbf{p}), \quad (6.80)$$

where $\alpha, \beta, \gamma, \delta$ are constant parameters to be determined. Furthermore, the first relation of (6.79) now leads to

$$((\alpha + \gamma) - (\beta + \delta))\gamma^0\omega - ((\alpha + \gamma) + (\beta + \delta))(\boldsymbol{\gamma}\cdot\mathbf{p} - m)$$

$$= -i\gamma^0,$$

which determines

$$\alpha + \gamma = -(\beta + \delta) = -\frac{i}{2\omega} = \frac{1}{2i\omega}. \quad (6.81)$$

As a result, we can write

$$\gamma = -\alpha + \frac{1}{2i\omega}, \quad \delta = -\left(\beta + \frac{1}{2i\omega}\right). \quad (6.82)$$

Relations in (6.80) and (6.82) automatically satisfy the first relation in (6.79).

Therefore, with the two arbitrary constants α, β, we can write the fermion Green's function as

$$S_c(t, \mathbf{p}) = \theta_c(t)\left(\alpha A(\mathbf{p})e^{-i\omega t} - \beta B(\mathbf{p})e^{i\omega t}\right)$$

$$- \theta_c(-t)\left(\gamma A(\mathbf{p})e^{-i\omega t} - \delta B(\mathbf{p})e^{i\omega t}\right)$$

$$= \theta_c(t)\left(\alpha A(\mathbf{p})e^{-i\omega t} - \beta B(\mathbf{p})e^{i\omega t}\right)$$

$$- \theta_c(-t)\left(\left(-\alpha + \frac{1}{2i\omega}\right)A(\mathbf{p})e^{-i\omega t} + \left(\beta + \frac{1}{2i\omega}\right)B(\mathbf{p})e^{i\omega t}\right). \quad (6.83)$$

The parameters α, β can be determined by imposing the periodicity conditions for the fermions. In fact, since fermions satisfy antiperiodic conditions (compare with, say (2.99) or (2.100)), we impose

$$S_c(T - i\beta - t, \mathbf{p}) = -S_c(T - t, \mathbf{p}), \tag{6.84}$$

which determines

$$\alpha = \frac{1}{2i\omega}(1 - n_F(\omega)), \quad \beta = -\frac{1}{2i\omega}n_F(\omega), \tag{6.85}$$

where $n_F(\omega) = \frac{1}{e^{\beta\omega}+1}$ denotes the Fermi distribution function. As a result, the fermion propagator takes the final form

$$iS_c(t, \mathbf{p}) = \frac{1}{2\omega}\Big((\theta_c(t) - n_F(\omega))A(\mathbf{p})\,e^{-i\omega t}$$

$$- (\theta_c(-t) - n_F(\omega))B(\mathbf{p})\,e^{i\omega t}\Big), \tag{6.86}$$

where the matrix functions $A(\mathbf{p})$ and $B(\mathbf{p})$ are defined in (6.80). The four components of the matrix propagator can now be read out from (6.86) and take the forms

$$iS_{++}^{(T)}(t, \mathbf{p}) = \frac{1}{2\omega}\Big((\theta(t) - n_F(\omega))A(\mathbf{p})\,e^{-i\omega t}$$

$$- (\theta(-t) - n_F(\omega))B(\mathbf{p})\,e^{i\omega t}\Big),$$

$$iS_{+-}^{(T)}(t, \omega) = \frac{1}{2\omega}\Big((1 - n_F(\omega))A(\mathbf{p})\,e^{-i\omega t} + n_F(\omega)B(\mathbf{p})\,e^{i\omega t}\Big),$$

$$iS_{-+}^{(T)}(t, \omega) = -\frac{1}{2\omega}\Big(n_F(\omega)A(\mathbf{p})\,e^{-i\omega t} + (1 - n_F(\omega))B(\mathbf{p})\,e^{i\omega t}\Big),$$

$$iS_{--}^{(T)}(t, \mathbf{p}) = \frac{1}{2\omega}\Big((\theta(-t) - n_F(\omega))A(\mathbf{p})\,e^{-i\omega t}$$

$$- (\theta(t) - n_F(\omega))B(\mathbf{p})e^{i\omega t}\Big). \tag{6.87}$$

The corresponding zero temperature propagators have the forms

$$iS_{++}^{(T=0)}(t, \mathbf{p}) = \frac{1}{2\omega}\Big(\theta(t)A(\mathbf{p})\,e^{-i\omega t} - \theta(-t)B(\mathbf{p})\,e^{i\omega t}\Big),$$

$$iS_{+-}^{(T=0)}(t, \omega) = \frac{1}{2\omega}A(\mathbf{p})\,e^{-i\omega t},$$

$$iS_{-+}^{(T=0)}(t,\omega) = \frac{1}{2\omega}B(\mathbf{p})\,e^{i\omega t},$$

$$iS_{--}^{(T=0)}(t,\mathbf{p}) = \frac{1}{2\omega}\Big(\theta(-t)A(\mathbf{p})\,e^{-i\omega t} - \theta(t)B(\mathbf{p})\,e^{i\omega t}\Big). \qquad (6.88)$$

It can now be checked that each of the components in (6.87) and (6.88) are related by the thermal operator as

$$iS_{ab}^{(T)}(t,\mathbf{p}) = \mathcal{O}_F^{(T)}(\omega)iS_{ab}^{(T=0)}(t,\mathbf{p})$$

$$= (1 - n_F(\omega)(\mathbb{1} - S(\omega)))iS_{ab}^{(T=0)}(t,\mathbf{p}), \quad a,b = \pm, \quad (6.89)$$

where $S(\omega)$ represents the reflection operator introduced earlier in (6.8). Furthermore, if we have an interacting field theory of fermions and real scalar fields, since interaction vertices do not carry any temperature dependence and since both the scalar and the fermion propagators factorize with their respective thermal operators, any n-point graph would naturally factorize in a general way, as discussed in section **6.2.1** (see (6.18)).

Let us next note that the fermion fields are complex and, therefore, carry charges. This allows us to introduce a chemical potential to the system. Let us consider the fermion Lagrangian density (see also (6.52) and (6.54))

$$\mathcal{L} = i\overline{\psi}\slashed{\partial}\psi - m\overline{\psi}\psi + \mu J^0 = i\overline{\psi}\slashed{\partial}\psi - m\overline{\psi}\psi + \mu\overline{\psi}\gamma^0\psi, \qquad (6.90)$$

where we have used the fact that, for charged fermions, the conserved current is $J^\mu = \overline{\psi}\gamma^\mu\psi$ so that the charge density has the form $J^0 = \overline{\psi}\gamma^0\psi = \psi^\dagger\psi$. As a result, the equation for the fermion Green's function, on the closed time path contour, takes the form (in the mixed space)

$$\big((\gamma^0(i\partial_t + \mu) - \boldsymbol{\gamma}\cdot\mathbf{p} - m)\big)S_c^{(T,\mu)}(t,\mathbf{p}) = \delta_c(t), \qquad (6.91)$$

which can be compared with (6.76). Therefore, following the steps in (6.76)-(6.86), we can determine the propagator $iS_c^{(T,\mu)}(t,\mathbf{p})$.

Alternatively, we can follow a simpler route as follows. Let us define the fermion Green's function as

$$S_c^{(T,\mu)}(t,\mathbf{p}) = \big(\gamma^0(i\partial_t + \mu) - \boldsymbol{\gamma}\cdot\mathbf{p} + m\big)\mathcal{S}_c^{(T,\mu)}(t,\omega), \qquad (6.92)$$

so that (6.91) takes the form

$$((\partial_t - i\mu)^2 + \omega^2)\mathcal{S}_c^{(T,\mu)}(t,\mathbf{p}) = -\delta_c(t), \tag{6.93}$$

where $\omega = \omega_{\mathbf{p}} = \sqrt{\mathbf{p}^2 + m^2}$, as before. We note that (6.93) coincides with the second order scalar equation given in (6.57) and so we can immediately write down the general solution as in (6.61) as

$$\mathcal{S}_c^{(T,\mu)}(t,\omega) = \theta_c(t)\left(\tilde{\alpha}\,e^{-i\omega_- t} + \tilde{\beta}\,e^{i\omega_+ t}\right)$$

$$+ \theta_c(-t)\left(\left(\tilde{\alpha} - \frac{1}{2i\omega}\right)e^{-i\omega_- t} + \left(\tilde{\beta} + \frac{1}{2i\omega}\right)e^{i\omega_+ t}\right), \tag{6.94}$$

where $\tilde{\alpha}, \tilde{\beta}$ are constants to be determined. The only difference from the earlier case in (6.61) is that we are considering a fermion case now and, therefore, the solution has to satisfy the anti-periodicity condition (see (6.84)) as opposed to the periodicity condition for the bosonic case (see (6.62)). Imposing anti-periodicity for the solution in (6.94), we determine (compare with the two corresponding parameters $\overline{\alpha}, \overline{\beta}$ in (6.63) as well as with α, β in (6.85))

$$\tilde{\alpha} = \frac{1}{2i\omega}(1 - n_F(\omega_-)), \quad \tilde{\beta} = -\frac{1}{2i\omega}n_F(\omega_+), \tag{6.95}$$

where $\omega_\pm = \omega \pm \mu$. With the relations in (6.95), the solution in (6.94) now takes the form

$$\mathcal{S}_c^{(T,\mu)}(t,\omega) = \frac{\theta_c(t)}{2i\omega}\left((1 - n_F(\omega_-))e^{-i\omega_- t} - n_F(\omega_+)e^{i\omega_+ t}\right)$$

$$- \frac{\theta_c(-t)}{2i\omega}\left(n_F(\omega_-)e^{-i\omega_- t} - (1 - n_F(\omega_+))e^{i\omega_+ t}\right). \tag{6.96}$$

Putting (6.96) back into (6.92), the matrix fermion propagator now follows to correspond to

$$iS_c^{(T,\mu)}(t,\mathbf{p}) = i\left(\gamma^0(i\partial_t + \mu) - \boldsymbol{\gamma}\cdot\mathbf{p} + m\right)\mathcal{S}_c^{(T,\mu)}(t,\omega)$$

$$= \frac{1}{2\omega}\left((\theta_c(t) - n_F(\omega_-))(\gamma^0\omega - \boldsymbol{\gamma}\cdot\mathbf{p} + m)e^{-i\omega_- t}\right.$$

$$\left. - (\theta_c(-t) - n_F(\omega_+))(\gamma^0\omega + \boldsymbol{\gamma}\cdot\mathbf{p} - m)e^{i\omega_+ t}\right),$$

$$= \frac{1}{2\omega}\left((\theta_c(t) - n_F(\omega_-))A(\mathbf{p})\,e^{-i\omega_- t}\right.$$

$$\left. - (\theta_c(-t) - n_F(\omega_+))B(\mathbf{p})\,e^{i\omega_+ t}\right), \tag{6.97}$$

which can be compared with (6.86) (with $\mu = 0$). (We note that $A(\mathbf{p}), B(\mathbf{p})$ coincide with the same two (corresponding) matrix functions as in (6.80).)

The components of the 2×2 matrix propagator for the fermion, in the closed time path formalism, can now be read out from (6.97) and take the forms

$$iS_{++}^{(T,\mu)}(t,\mathbf{p}) = \frac{1}{2\omega}\Big((\theta(t) - n_F(\omega_-))A(\mathbf{p})\,e^{-i\omega_- t}$$

$$- (\theta(-t) - n_F(\omega_+))B(\mathbf{p})\,e^{i\omega_+ t}\Big),$$

$$iS_{+-}^{(T,\mu)}(t,\mathbf{p}) = \frac{1}{2\omega}\Big((1 - n_F(\omega_-))A(\mathbf{p})\,e^{-i\omega_- t}$$

$$+ n_F(\omega_+)B(\mathbf{p})\,e^{i\omega_+ t}\Big),$$

$$iS_{-+}^{(T,\mu)}(t,\mathbf{p}) = -\frac{1}{2\omega}\Big(n_F(\omega_-)A(\mathbf{p})\,e^{-i\omega_- t}$$

$$+ (1 - n_F(\omega_+))B(\mathbf{p})\,e^{i\omega_+ t}\Big), \qquad (6.98)$$

$$iS_{--}^{(T,\mu)}(t,\mathbf{p}) = \frac{1}{2\omega}\Big((\theta(-t) - n_F(\omega_-))A(\mathbf{p})\,e^{-i\omega_- t}$$

$$- (\theta(t) - n_F(\omega_+))B(\mathbf{p})\,e^{i\omega_+ t}\Big),$$

where $A(\mathbf{p})$ and $B(\mathbf{p})$ are the matrix functions defined in (6.80) as well as in (6.97). The zero temperature components can be read out from (6.98) and take the forms

$$iS_{++}^{(T=0,\mu)}(t,\mathbf{p}) = \frac{1}{2\omega}\Big(\theta(t)A(\mathbf{p})\,e^{-i\omega_- t} - \theta(-t)B(\mathbf{p})\,e^{i\omega_+ t}\Big),$$

$$iS_{+-}^{(T=0,\mu)}(t,\mathbf{p}) = \frac{1}{2\omega}\,A(\mathbf{p})\,e^{-i\omega_- t},$$

$$iS_{-+}^{(T=0,\mu)}(t,\mathbf{p}) = -\frac{1}{2\omega}\,B(\mathbf{p})\,e^{i\omega_+ t}, \qquad (6.99)$$

$$iS_{--}^{(T=0,\mu)}(t,\mathbf{p}) = \frac{1}{2\omega}\Big(\theta(-t)A(\mathbf{p})\,e^{-i\omega_- t} - \theta(t)B(\mathbf{p})\,e^{i\omega_+ t}\Big).$$

We note here again that the phase factor involving the chemical potential in all the components of the propagators in (6.98)-(6.99) is the same $e^{i\mu t}$ and, therefore, cancels out in a loop calculation of a Feynman diagram. Furthermore, we note that, component by component, the finite temperature propagator is related to the zero temperature propagator as

$$iS_{ab}^{(T,\mu)}(t,\mathbf{p}) = \mathcal{O}_F^{(T,\mu)}(\omega)iS_{ab}^{(T=0,\mu)}, \quad a,b = \pm, \tag{6.100}$$

where the thermal operator, in this case, has the form (see also (6.68))

$$\mathcal{O}_F^{(T,\mu)}(\omega) = \left(\mathbb{1} - \widehat{n}_F^{(T,\mu)}(\mathbb{1} - S(\omega))\right), \tag{6.101}$$

with (see also (6.69))

$$\widehat{n}_F^{(T,\mu)} e^{is\omega_s t} = n_F(\omega_s)e^{is\omega_s t}, \quad s = \pm, \ s^2 = 1. \tag{6.102}$$

In establishing the factorization given above, it is worth noting that

$$\begin{aligned}(S(\omega)A(\mathbf{p})) &= -B(\mathbf{p}), \quad (S(\omega)B(\mathbf{p})) = -A(\mathbf{p}), \\ (S(\omega)\omega_\pm) &= -\omega_\mp.\end{aligned} \tag{6.103}$$

Once again, since the interaction vertices of the theory do not carry any temperature dependence, any Feynman graph at finite temperature can be related to the zero temperature graph through a product of thermal operators as discussed earlier. As in the case of the complex scalar field with a chemical potential (discussed in the last subsection), we again point out here that the phase factor involving the chemical potential in any component of the propagator in (6.98) or (6.99) is the same $e^{i\mu t}$ and can be written as an overall phase factor. As a result, in any loop diagram involving the fermion, this phase factor would completely cancel out although the dependence on the chemical potential would continue to be there in the distribution functions.

6.2.6 Gauge field theory. We have already discussed gauge theories in detail in chapter **5** and, therefore, we introduce the concept of thermal operator for the simplest gauge theory, namely, the Abelian gauge theory, with a covariant gauge fixing. The Lagrangian density for this free theory (with a covariant gauge fixing) is given by

$$\mathcal{L} = -\frac{1}{4}F_{\mu\nu}F^{\mu\nu} - \frac{1}{2\xi}(\partial^\mu A_\mu)^2, \tag{6.104}$$

where ξ is a real constant parameter known as the gauge fixing parameter. The propagator for the photon field at finite temperature,

both in the momentum space as well as the mixed space, can be
derived along the same lines discussed earlier and take the forms

$$iD^{(T)}_{\mu\nu,++}(p) = A_{\mu\nu}(p) \left(\frac{i}{p^2 + i\epsilon} + 2\pi n_B(|p^0|)\delta(p^2)\right),$$

$$iD^{(T)}_{\mu\nu,+-}(p) = A_{\mu\nu}(p) \, 2\pi \left(\theta(-p^0) + n_B(|p^0|)\right)\delta(p^2),$$

$$iD^{(T)}_{\mu\nu,-+}(p) = A_{\mu\nu}(p) \, 2\pi \left(\theta(p^0) + n_B(|p^0|)\right)\delta(p^2),$$

$$iD^{(T)}_{\mu\nu,--}(p) = A_{\mu\nu}(p) \left(-\frac{i}{p^2 - i\epsilon} + 2\pi n_B(|p^0|)\delta(p^2)\right), \qquad (6.105)$$

where we have identified

$$A_{\mu\nu}(p) = -\left(\eta_{\mu\nu} - (1-\xi)\frac{p_\mu p_\nu}{p^2}\right). \qquad (6.106)$$

In the Feynman gauge, which is calculationally much simpler, $\xi = 1$
and the prefactor simplifies to

$$A_{\mu\nu}(p) = -\eta_{\mu\nu}, \qquad (6.107)$$

which is what we will use in the following. The gauge propagator
components for the Yang-Mills theory (or the non-Abelian $SU(N)$
gauge theory) can also be generalized from (6.105) and take the forms

$$iD^{ab\,(T)}_{\mu\nu,++}(p) = -\eta_{\mu\nu}\delta^{ab} \left(\frac{i}{p^2 + i\epsilon} + 2\pi n_B(|p^0|)\delta(p^2)\right),$$

$$iD^{ab\,(T)}_{\mu\nu,+-}(p) = -\eta_{\mu\nu}\delta^{ab} \, 2\pi \left(\theta(-p^0) + n_B(|p^0|)\right)\delta(p^2),$$

$$iD^{ab\,(T)}_{\mu\nu,-+}(p) = -\eta_{\mu\nu}\delta^{ab} \, 2\pi \left(\theta(p^0) + n_B(|p^0|)\right)\delta(p^2),$$

$$iD^{ab\,(T)}_{\mu\nu,--}(p) = -\eta_{\mu\nu}\delta^{ab} \left(-\frac{i}{p^2 - i\epsilon} + 2\pi n_B(|p^0|)\delta(p^2)\right), \qquad (6.108)$$

where $a, b = 1, 2, \cdots, N^2 - 1$ represent the internal $SU(N)$ indices.

In the mixed space, which can be obtained by Fourier transforming the p^0 variable, the components of the photon propagator at finite

temperature take the forms

$$iD^{ab\,(T)}_{\mu\nu,++}(t,\omega) = -\frac{\eta_{\mu\nu}\delta^{ab}}{2\omega}\Big((\theta(t) + n_B(\omega))e^{-i\omega t}$$

$$+ (\theta(-t) + n_B(\omega))e^{i\omega t}\Big),$$

$$iD^{ab\,(T)}_{\mu\nu,+-}(t,\omega) = -\frac{\eta_{\mu\nu}\delta^{ab}}{2\omega}\Big(n_B(\omega)e^{-i\omega t} + (1 + n_B(\omega))e^{i\omega t}\Big),$$

$$iD^{ab\,(T)}_{\mu\nu,-+}(t,\omega) = -\frac{\eta_{\mu\nu}\delta^{ab}}{2\omega}\Big((1 + n_B(\omega))e^{-i\omega t} + n_B(\omega)e^{i\omega t}\Big),$$

$$iD^{ab\,(T)}_{\mu\nu,--}(t,\omega) = -\frac{\eta_{\mu\nu}\delta^{ab}}{2\omega}\Big((\theta(-t) + n_B(\omega))e^{-i\omega t}$$

$$+ (\theta(t) + n_B(\omega))e^{i\omega t}\Big). \tag{6.109}$$

We note that this is exactly the same propagator as in the scalar case given in (6.6) except for the constant tensorial prefactor $-\eta_{\mu\nu}\delta^{ab}$. This shows the usefulness of the Feynman gauge. Furthermore, it now follows as in section **6.2.1** (see (6.9)) that each component of the gauge propagator factorizes as (the prefactor $-\eta_{\mu\nu}\delta^{ab}$ is unaffected by the thermal operator)

$$iD^{ab\,(T)}_{\mu\nu,\alpha\beta}(t,\mathbf{p}) = \mathcal{O}^{(T)}_B(\omega)iD^{ab\,(T=0)}_{\mu\nu,\alpha\beta}(t,\mathbf{p}), \quad \alpha,\beta = \pm, \tag{6.110}$$

where the scalar thermal operator $\mathcal{O}^{(T)}_B(\omega)$ is defined in (6.7).

In addition to the gauge fields, gauge theories also have the ghost fields which arise from gauge fixing the (singular) theory. In an Abelian gauge theory, the ghost fields are noninteracting and, therefore, can be neglected in calculating Feynman graphs. However, they are important in studies of physical quantities at finite temperature such as the partition function, as we have pointed out in section **5.6**. In non-Abelian gauge theories, ghosts are interacting and are important in calculating (physical) amplitudes (besides in physical questions like the partition function). So, we need to calculate the ghost propagator in addition to the gauge propagator. Let us recall (see section **5.6** for details of the argument) that ghost fields are anti-commuting, but since the ghost Lagrangian density arose from gauge fixing the original theory to compensate for the unphysical gauge degrees of freedom, they obey periodic boundary conditions like scalar fields. As a result, the four components of the ghost propagators, in the mixed space, take the forms of the scalar fields given in (6.65) except for the internal indices (even though they are propagators for the anti-commuting ghost fields), namely,

$$iD_{++}^{ab\,(T)}(t,\omega) = \frac{\delta^{ab}}{2\omega}\Big((\theta(t) + n_B(\omega))e^{-i\omega t} + (\theta(-t) + n_B(\omega))e^{i\omega t}\Big),$$

$$iD_{+-}^{ab\,(T)}(t,\omega) = \frac{\delta^{ab}}{2\omega}\Big(n_B(\omega)e^{-i\omega t} + (1 + n_B(\omega))e^{i\omega t}\Big),$$

$$iD_{-+}^{ab\,(T)}(t,\omega) = \frac{\delta^{ab}}{2\omega}\Big((1 + n_B(\omega))e^{-i\omega t} + n_B(\omega)e^{i\omega t}\Big), \qquad (6.111)$$

$$iD_{--}^{ab\,(T)}(t,\omega) = \frac{\delta^{ab}}{2\omega}\Big((\theta(-t) + n_B(\omega))e^{-i\omega t} + (\theta(t) + n_B(\omega))e^{i\omega t}\Big).$$

Like the scalar propagator, the components of the ghost propagator also factorize as (see (6.7) and (6.9))

$$iD_{\alpha\beta}^{ab\,(T)}(t,\omega) = \mathcal{O}_B^{(T)}(\omega)iD_{\alpha\beta}^{ab\,(T=0)}(t,\omega), \qquad (6.112)$$

where $\mathcal{O}_B^{(T)}(\omega)$ denotes the thermal operator (see also (6.110)). This shows that both the gauge and the ghost propagators factorize with the same scalar thermal operator. The vertices also do not carry any temperature dependence like in other theories. However, in the non-Abelian gauge theory, both the gauge fields and ghost fields have derivative coupling. On the other hand, as we have already shown in (6.13), the thermal operator commutes with derivatives, in particular, $\partial_t \mathcal{O}_B^{(T)}(\omega) = \mathcal{O}_B^{(T)}(\omega)\partial_t$. As a result, we can factor out the thermal operator out of any Feynman graph in the non-Abelian gauge theory and it factorizes with a product of thermal operators at finite temperature as in (6.18). The $SU(N)$ gauge fields, A_μ^a are real fields and, therefore, do not carry any physical charge. The ghost field c^a is real while the anti-ghost field is purely imaginary[2] so that they do not carry a physical charge. Rather, they carry a ghost charge which is not physical (the ghost charge operator annihilates the physical vacuum and states). Therefore, we do not have to go into a discussion of chemical potentials for these fields. An interacting $SU(N)$ (or Abelian) gauge theory with fermions and/or scalar fields

---

[2]See Eq. (13.34) in A. Das, *Lectures in Quantum Field Theory* (second edition), World Scientific Publishing (2021).

with/without a chemical potential can now be carried out with all these discussions and it can be shown that any Feynman graph would factorize and would be related to the zero temperature graph through appropriate thermal operators.

6.3 Thermal operator in imaginary time formalism

Let us consider a (charge neutral) massive scalar field in the imaginary time formalism at zero temperature. The Euclidean propagator, in momentum space, takes the form

$$G^{(T=0)}(p) = \frac{1}{p^2 + m^2} = \frac{1}{p_{\mathrm{E}}^2 + \omega^2}, \tag{6.113}$$

where p_{E} denotes the Euclidean energy and $\omega = \sqrt{\mathbf{p}^2 + m^2}$ as before (see, for example, (6.5)). At zero temperature, energy is a continuous variable in the Euclidean space and a Fourier transformation of (6.113) leads to the propagator in the mixed space as

$$G^{(T=0)}(\tau, \omega) = \int \frac{dp_{\mathrm{E}}}{2\pi} \frac{e^{-ip_{\mathrm{E}}\tau}}{p_{\mathrm{E}}^2 + \omega^2}$$

$$= \frac{1}{2\omega}\left(\theta(\tau)e^{-\omega\tau} + \theta(-\tau)e^{\omega\tau}\right), \quad -\infty \le \tau \le \infty. \tag{6.114}$$

At finite temperature, on the other hand, the Euclidean energy, for bosons, takes the discrete values given by (see (1.62) for bosons and (1.64) for fermions with $T = \frac{1}{\beta}$)

$$p_{\mathrm{E},n} = \omega_n = 2n\pi T, \quad n = 0, \pm 1, \pm 2, \cdots. \tag{6.115}$$

Therefore, the momentum integration in (6.114) becomes a sum over the integers and leads to

$$G^{(T)}(\tau, \omega) = T \sum_{n=-\infty}^{\infty} \frac{e^{-2i\pi nT\tau}}{(2\pi nT)^2 + \omega^2}, \quad -\frac{1}{T} \le \tau \le \frac{1}{T},$$

$$= \frac{1}{2\omega}\left((\theta(\tau) + n_B(\omega))e^{-\omega\tau} + (\theta(-\tau) + n_B(\omega))e^{\omega\tau}\right)$$

$$= \left(1 + n_B(\omega)(1 - S(\omega))\right)\frac{1}{2\omega}\left(\theta(\tau)e^{-\omega\tau} + \theta(-\tau)e^{\omega\tau}\right)$$

$$= \mathcal{O}_B^{(T)}(\omega)G^{(T=0)}(\tau, \omega), \tag{6.116}$$

where we recognize that

$$\mathcal{O}_B^{(T)}(\omega) = \left(\mathbb{1} + n_B(\omega)(\mathbb{1} - S(\omega))\right), \tag{6.117}$$

is the same thermal operator which we have already encountered in (6.7). Here, $n_B(\omega)$ represents the Bose-Einstein distribution function and $S(\omega)$ denotes the reflection operator which changes $\omega \to -\omega$.

The simplicity of the imaginary time formalism lies in the fact that there is only one propagator at finite temperature (and not four components as in the closed time formalism). Therefore, factorization of a one loop diagram becomes quite straightforward. For example, in a (real) scalar ϕ^3 theory, if we have an one loop diagram with N external vertices, as shown in Fig. 6.2, where $(\tau_1, \tau_2, \cdots, \tau_N)$,

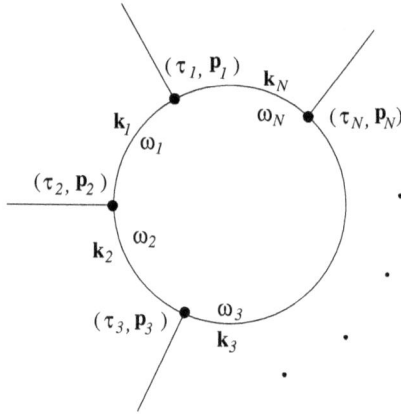

Figure 6.2: A one loop diagram in the ϕ^3 theory with N external vertices.

$(\mathbf{p}_1, \mathbf{p}_2, \cdots, \mathbf{p}_N)$ denote the N external (Euclidean) time coordinates and momenta respectively and $(\mathbf{k}_1, \mathbf{k}_2, \cdots, \mathbf{k}_N)$ denote the N internal momenta, then, we note that the external vertices will not carry any temperature dependence since they are all taken from a Euclidean field theory (there is no temperature dependence in the Lagrangian density). Each of the N internal propagators in the loop, however, will carry temperature dependence and these would give all the temperature dependence to the one loop graph. Therefore, the value of the graph can be immediately written down as

$$\int \prod_{i=1}^{N} d^3k_i \, \delta^3(\mathbf{k}_i - \mathbf{k}_{i+1} + \mathbf{p}_{i+1}) \gamma_N^{(T)}$$

$$= \int \prod_{i=1}^{N} d^3k_i \, \delta^3(\mathbf{k}_i - \mathbf{k}_{i+1} + \mathbf{p}_{i+1}) G^{(T)}(\tau_i - \tau_{i+1}, \omega_i)$$

$$= \int \prod_{i=1}^{N} d^3k_i \, \delta^3(\mathbf{k}_i - \mathbf{k}_{i+1} + \mathbf{p}_{i+1}) \mathcal{O}_B^{(T)}(\omega_i) G^{(T=0)}(\tau_i - \tau_{i+1}, \omega_i)$$

$$= \int \prod_{i=1}^{N} d^3k_i \, \delta^3(\mathbf{k}_i - \mathbf{k}_{i+1} + \mathbf{p}_{i+1}) \mathcal{O}_B^{(T)} \gamma_N^{(T=0)}, \tag{6.118}$$

where, as in (6.19), we have defined

$$\mathcal{O}_B^{(T)} = \prod_{i=1}^{N} \mathcal{O}_B^{(T)}(\omega_i) = \prod_{i=1}^{N} (\mathbb{1} + n_B(\omega_i)(\mathbb{1} - S(\omega_i))). \tag{6.119}$$

Equation (6.118) can be compared to (6.18) and we can conclude that, in the imaginary time formalism, the one loop Feynman graphs factorize through the thermal operator. Similarly, in the ϕ^4 theory, it is straightforward in the imaginary time formalism to show that a factorization through the thermal operator also arises in the two loop graphs shown in Fig. 6.3. This factorization can, in fact, be

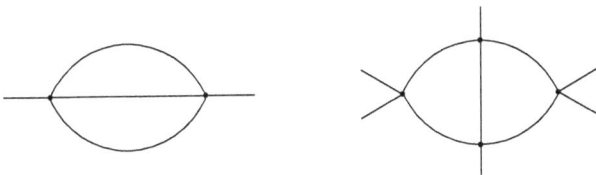

Figure 6.3: Two loop diagrams in the ϕ^4 theory without internal time coordinates.

established for any higher loop graphs where there are no internal time coordinates such as in Fig. 6.4.

The difficulty in establishing factorization, however, arises when we are dealing with higher loop (more than one loop) graphs where

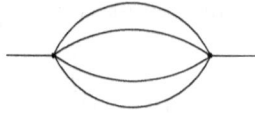

Figure 6.4: A higher loop diagram in the ϕ^4 theory with only external time coordinates.

there are internal time coordinates which need to be integrated over all allowed times. Let us consider, for example, the two two loop self-energy graphs in the ϕ^3 theory shown in Fig. 6.5. Here the two internal time coordinates, τ and $\tilde{\tau}$, have to be integrated over

Figure 6.5: Two loop self-energy diagrams in ϕ^3 theory with internal time coordinates $\tau, \tilde{\tau}$.

all allowed values. We know that, in the imaginary time formalism at finite temperature, the Euclidean time is integrated over a finite range (see, for example, (1.50) or (1.88))

$$\int_0^{\frac{1}{T}} d\tau,$$

while, at zero temperature, the integration is over the entire real line

$$\int_{-\infty}^{\infty} d\tau.$$

(As a result, the imaginary time propagator is defined over the time interval $-\frac{1}{T} \leq \tau \leq \frac{1}{T}$, see, for example, the discussion above (1.58)). Therefore, it seems unlikely that the finite temperature graph can be related to the zero temperature graph (when there are internal time coordinates) simply because of the range of the time integrations in the two cases are so different – unless, of course, the thermal operator (see, for example, (6.118)-(6.119)) gives zero acting on the contribution which arises from the difference in the integration range. This is indeed what happens as we will show in detail next.

Let us consider a general Feynman diagram at an arbitrary loop and focus only on a part of the diagram containing an internal time coordinate τ, connected to N other vertices with time coordinates $\tau_i, i = 1, 2, \cdots, N$ through N propagators. These will represent all the vertices (time coordinates) to which τ is connected. Some of these N time coordinates (vertices) may be external while some others may be internal and we assume that τ is being integrated with the other coordinates (including the internal ones) held fixed. The full diagram, of course, will consist of many such parts combined with all the internal time coordinates integrated over. For example, we note that in the first graph of Fig. 6.5, we can consider a part of the diagram to consist of the internal coordinate τ, connected to three time coordinates, τ_1, τ_2 which are external time coordinates and $\tau_3 = \tilde{\tau}$, an internal time coordinate with τ being integrated over while $\tau_1, \tau_2, \tau_3 = \tilde{\tau}$ held fixed. We can, then, combine this part of the diagram with another part containing the internal vertex $\tau_3 = \tilde{\tau}$ connected to τ_1, τ_2 through two propagators and integrating over $\tau_3 = \tilde{\tau}$ to obtain the value of the complete graph. Therefore, for a general graph, for the first part containing the complete connections of a single internal time coordinate, we have to evaluate an expression of the kind

$$
\begin{aligned}
I_1 &= \int_0^{\frac{1}{T}} d\tau \prod_{i=1}^{N} G^{(T)}(\tau - \tau_i, \omega_i) \\
&= \int_0^{\frac{1}{T}} d\tau \prod_{i=1}^{N} \mathcal{O}_B^{(T)}(\omega_i) G^{(T=0)}(\tau - \tau_i, \omega_i), \\
&= \prod_{i=1}^{N} \mathcal{O}_B^{(T)}(\omega_i) \int_0^{\frac{1}{T}} d\tau\, G^{(T=0)}(\tau - \tau_i, \omega) \\
&= \prod_{i=1}^{N} \mathcal{O}_B^{(T)}(\omega_i) \left(\int_{-\infty}^{\infty} d\tau - \int_{\frac{1}{T}}^{\infty} d\tau - \int_{-\infty}^{0} d\tau \right) G^{(T=0)}(\tau - \tau_i, \omega_i) \\
&= \prod_{i=1}^{N} \mathcal{O}_B^{(T)}(\omega_i) \int_{-\infty}^{\infty} d\tau\, G^{(T=0)}(\tau - \tau_i, \omega_i) - \bar{I}_1, \quad (6.120)
\end{aligned}
$$

where

$$\bar{I}_1 = \prod_{i=1}^{N} \mathcal{O}_B^{(T)}(\omega_i) \left(\int_{\frac{1}{T}}^{\infty} d\tau + \int_{-\infty}^{0} d\tau \right) G^{(T=0)}(\tau - \tau_i, \omega_i). \quad (6.121)$$

Here we could take the thermal operators outside the τ integration only because they are independent of the time coordinate (see (6.13)). We emphasize that, for the purpose of this integration, we assume that all the τ_i are fixed and take values between $(0, \frac{1}{T})$ (whether external or internal). Integration over other internal coordinates will arise as we combine this part of the graph with others to evaluate the complete graph. For this part of the graph, if we can show that $\bar{I}_1 = 0$, we would have shown that the single internal time integration can be extended to the entire real line which is what we discuss next.

Let us recall that the zero temperature Euclidean propagator for the scalar field has the form

$$G^{(T=0)}(\tau - \tau_i, \omega_i) = \frac{1}{2\omega_i} \Big(\theta(\tau - \tau_i) e^{-\omega_i(\tau - \tau_i)}$$

$$+ \theta(\tau_i - \tau) e^{\omega_i(\tau - \tau_i)} \Big), \quad (6.122)$$

and, since $0 \leq \tau_i \leq \frac{1}{T}$,

$$G^{(T=0)}(\tau - \tau_i, \omega_i) = \frac{1}{2\omega_i} e^{-\omega_i(\tau - \tau_i)}, \quad \frac{1}{T} \leq \tau \leq \infty,$$

$$G^{(T=0)}(\tau - \tau_i, \omega_i) = \frac{1}{2\omega_i} e^{\omega_i(\tau - \tau_i)}, \quad -\infty \leq \tau \leq 0. \quad (6.123)$$

Therefore, carrying out the τ integration in (6.121), we obtain

$$\bar{I}_1 = \prod_{i=1}^{N} \mathcal{O}_B^{(T)}(\omega_i) \frac{1}{2\omega_i^2} \left(e^{-\frac{\omega_i}{T} + \omega_i \tau_i} + e^{-\omega_i \tau_i} \right)$$

$$= \prod_{i=1}^{N} \mathcal{O}_B^{(T)}(\omega_i) \frac{1}{2\omega_i^2} \left(\frac{n_B(\omega_i)}{1 + n_B(\omega_i)} e^{\omega_i \tau_i} + e^{-\omega_i \tau_i} \right)$$

$$= 0. \quad (6.124)$$

Equation (6.124) follows simply from the fact that (see (6.35))

$$\mathcal{O}^{(T)}(\omega_i) = \big((1 + n_B(\omega_i)) \mathbb{1} - n_B(\omega_i) S(\omega_i) \big),$$

as well as from the two equations in (6.11) and (6.12) which lead to

$$\left(S(\omega_i) \frac{n_B(\omega_i)}{1 + n_B(\omega_i)} \right) = \frac{1 + n_B(\omega_i)}{n_B(\omega_i)}. \tag{6.125}$$

This shows that, for a single internal time coordinate in a diagram at finite temperature, we can extend the integration range from $(0, \frac{1}{T})$ to the entire real line and the thermal operator representation can be carried over naturally.

Next, let us come to the case where one of the times τ_i, say $\tau_N = \tilde{\tau}$, is an internal time coordinate. (Note that, for a part of the graph to contain an internal time coordinate, it must necessarily contain at least a second internal time coordinate.) Let us enlarge our original part of the graph by joining this second internal time coordinate to all the points in the graph that it is connected with through propagators. Let us label these enlarged coordinate times by $\tilde{\tau}_\alpha, \alpha = 1, 2, \cdots, \tilde{N}$, some of which may be external time coordinates and some internal and we integrate $\tilde{\tau}$ over the allowed values $(0, \frac{1}{T})$ while keeping all the others fixed. (In the example of the first graph in Fig. 6.5 that we alluded to earlier, $\tilde{\tau}$ is connected only to the two external time coordinates τ_1, τ_2. The propagator connecting $\tilde{\tau}$ to τ is already contained in the first part of the graph.) We can again go through the entire procedure as before and conclude that the range of integration can be enlarged to the entire real line and, therefore, a thermal representation for this enlarged part of the diagram also holds. In this way, we can keep on enlarging the original part of the graph one internal coordinate at a time until the part of the graph becomes the entire graph and we would have shown the thermal operator representation for the entire graph at any arbitrary loop, namely, for an N-point graph

$$\int \prod_{i=1}^{I} \frac{d^3 k_i}{(2\pi)^3} \prod_{v=1}^{N+n} \delta_v^3(k, p) \, \gamma_N^{(T)}$$

$$= \int \prod_{i=1}^{I} \frac{d^3 k_i}{(2\pi)^3} \prod_{v=1}^{N+n} \delta_v^3(k, p) \, \mathcal{O}_B^{(T)} \, \gamma_N^{(T=0)}, \tag{6.126}$$

which can be compared with (6.18). Here, as before, N denotes the number of external vertices, n the number of internal vertices (so that the total number of vertices in the graph is given by $V = N + n$) connected through I internal propagators. Furthermore,

$$\mathcal{O}_B^{(T)} = \prod_i^I \mathcal{O}_B^{(T)}(\omega_i) = \prod_{i=1}^I \left(\mathbb{1} + n_B(\omega_i)(\mathbb{1} - S(\omega_i))\right), \qquad (6.127)$$

with $\omega_i = \sqrt{\mathbf{k}_i^2 + m^2}$ (for a fixed i).

The imaginary time formalism has the simplicity that it has only one propagator for a field, unlike real time formalisms where there are four components of the propagator matrix. However, the finite range of integration for the time coordinate is a drawback, but with the discussion of how the range of time integration can be enlarged to the entire real line under the action of the thermal operator and a thermal operator representation for any graph can be obtained, all the discussions of the last section can now be taken over completely. We can study the thermal operator representation for various kinds of field theories in a completely parallel manner. However, we would not go into that in order to avoid repetitions.

6.4 Thermal operator for an arbitrary path

Before we discuss this question in detail, let us briefly recapitulate what we have done so far in this chapter. We found that, both in the imaginary time formalism as well as in the closed time path formalism, the basic propagator in a theory factorizes as a scalar operator (carrying all the temperature dependence) acting on the zero temperature propagator. Since the vertices do not carry any temperature dependence, this leads to the factorization of any Feynman graph in terms of a temperature dependent (thermal) operator acting on the zero temperature graph. We would like to address whether such a factorization also carries over to a theory defined on an arbitrary path in the real time formalism shown in Fig. 6.6 (see also Fig. 3.2). Let us note that we have already discussed and shown, in chapter **4**, that a propagator defined on an arbitrary path in the real time formalism factorizes through the appropriate Bogoliubov transformation. However, that happens to be a matrix factorization and to show the factorization of an arbitrary graph at an arbitrary loop, that would require keeping track of all the matrix indices which would be extremely tedious. In that respect, factorization through a

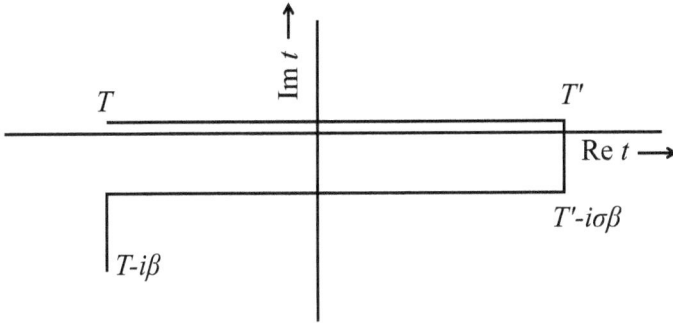

Figure 6.6: An arbitrary path in the complex t-plane.

scalar operator is much simpler to keep track of and calculationally very helpful. A natural question which arises here is if the matrix factorization naturally leads to the scalar (thermal operator) factorization and if so how. If that happens, then, we can carry over the discussions in chapter **4** and possibly find the scalar operator factorization (and the thermal operator) for an arbitrary path in the real time formalism as well. That will help us in establishing a factorization of an arbitrary graph at finite temperature on an arbitrary path (in the real time formalism) in terms of scalar thermal operators.

Therefore, to warm up, let us study this question within the context of the closed time path formalism (corresponding to $\sigma = 0$) where we already know the (scalar) thermal operator factorization of the propagator. The first thing to note is that the entire discussion of factorization in chapter **4** was in momentum space while, the discussion of the scalar thermal operator in this chapter is best carried out in the mixed space. So, let us recapitulate from (6.6) (see also (2.42)) that the matrix components of the propagator at finite temperature take the forms

$$iG^{(T)}_{++}(t, \omega) = \frac{1}{2\omega}\left((\theta(t) + n_B(\omega))e^{-i\omega t} + (\theta(-t) + n_B(\omega))e^{i\omega t}\right),$$

$$iG^{(T)}_{+-}(t, \omega) = \frac{1}{2\omega}\left(n_B(\omega)\, e^{-i\omega t} + (1 + n_B(\omega))e^{i\omega t}\right),$$

$$iG^{(T)}_{-+}(t, \omega) = \frac{1}{2\omega}\left((1 + n_B(\omega))e^{-i\omega t} + n_B(\omega)\, e^{i\omega t}\right), \qquad (6.128)$$

$$iG^{(T)}_{--}(t, \omega) = \frac{1}{2\omega}\left((\theta(-t) + n_B(\omega))e^{-i\omega t} + (\theta(t) + n_B(\omega))e^{i\omega t}\right),$$

where $\omega = \omega_{\mathbf{p}} = \sqrt{\mathbf{p}^2 + m^2}$, as before (see (6.5)). The zero temperature propagator (see (6.4)) follows to be

$$iG_{++}^{(T=0)}(t,\omega) = \frac{1}{2\omega}\left(\theta(t)\,e^{-i\omega t} + \theta(-t)\,e^{i\omega t}\right),$$

$$iG_{+-}^{(T=0)}(t,\omega) = \frac{1}{2\omega}\,e^{i\omega t},$$

$$iG_{-+}^{(T=0)}(t,\omega) = \frac{1}{2\omega}\,e^{-i\omega t},$$

$$iG_{--}^{(T=0)}(t,\omega) = \frac{1}{2\omega}\left(\theta(-t)\,e^{-i\omega t} + \theta(t)\,e^{i\omega t}\right). \tag{6.129}$$

We note here that the theta functions $\theta(t), \theta(-t)$ arise from Fourier transforming the p^0 variable to its conjugate space t and since the two zero temperature propagators respectively have poles at

$$\frac{1}{p^0 \pm (\omega - i\epsilon)} \quad \text{and} \quad \frac{1}{p^0 \pm (\omega + i\epsilon)},$$

the appropriate contour in the upper/lower half of the t-plane needs to be chosen. Let us note from (6.129) the following relations

$$S(\omega)(iG_{\pm\pm}^{(T=0)}(t,\omega)) = -iG_{\mp\mp}^{(T=0)}(t,\omega),$$

$$S(\omega)(iG_{\pm\mp}^{(T=0)}(t,\omega)) = -iG_{\mp\pm}^{(T=0)}(t,\omega), \tag{6.130}$$

as well as

$$iG_{++}(t,\omega) + iG_{--}(t,\omega) = iG_{+-}(t,\omega) + iG_{-+}(t,\omega), \tag{6.131}$$

where the last relation in (6.131) holds both at zero temperature as well as at finite temperature and $S(\omega)$ is the reflection operator defined in (6.8).

The transformation matrix connecting the zero temperature and the finite temperature propagators in (6.128)-(6.129) can now be simply obtained and it takes the form (σ_1 is the Pauli matrix)

$$\overline{U}_{\mathrm{CT}}^{(T)}(t,\omega) = \sqrt{n_B(\omega)}\left(e^{\frac{\beta\omega}{2}}\mathbb{1} + \theta(-t)\,e^{-\frac{\beta\omega}{2}}\sigma_1\right)$$

$$= \sqrt{n_B(\omega)}\begin{pmatrix} e^{\frac{\beta\omega}{2}} & \theta(-t)\,e^{-\frac{\beta\omega}{2}} \\ \theta(-t)\,e^{-\frac{\beta\omega}{2}} & e^{\frac{\beta\omega}{2}} \end{pmatrix}$$

$$= \left(\overline{U}_{\mathrm{CT}}^{(T)}(t,\omega)\right)^{\dagger}, \tag{6.132}$$

so that we can write

$$iG_{\text{CT}}^{(T)}(t,\omega) = \overline{U}_{\text{CT}}^{(T)}(t,\omega) iG_{\text{CT}}^{(T=0)}(t,\omega)(\overline{U}_{\text{CT}}^{(T)}(-t,\omega))^\dagger. \qquad (6.133)$$

Here the change in sign in t in the last factor is a consequence of the nontrivial inner product (which we will discuss in more detail shortly, see also (4.11)), $n_B(\omega) = \frac{1}{e^{\beta\omega}-1}$ denotes the bosonic distribution function and we note that

$$\overline{U}_{\text{CT}}^{(T)}(t,\omega)(\overline{U}_{\text{CT}}^{(T)}(-t,\omega))^\dagger$$

$$= n_B(\omega)\big(e^{\beta\omega}\mathbb{1} + (\theta(t) + \theta(-t))\sigma_1\big) = n_B(\omega)\big(e^{\beta\omega}\mathbb{1} + \sigma_1\big)$$

$$= \mathbb{1} + n_B(\omega)\big(\mathbb{1} + \sigma_1\big). \qquad (6.134)$$

It is to be noted that $\overline{U}_{\text{CT}}^{(T)}(t,\omega)\sigma_3(\overline{U}_{\text{CT}}^{(T)}(-t,\omega))^\dagger \neq \sigma_3$ as we would expect for a bosonic Bogoliubov transformation (see, for example, (3.74)). We will show shortly that the analog of the adjoint satisfies, in this case, is given by

$$(\overline{U}_{\text{CT}}^{(T)}(t,\omega))^\ddagger = (\overline{U}_{\text{CT}}^{(T)}(-t,\omega))^\dagger \qquad (6.135)$$

so that we can, in fact, write (6.134) as

$$\overline{U}_{\text{CT}}^{(T)}(t,\omega)(\overline{U}_{\text{CT}}^{(T)}(t,\omega))^\ddagger = \mathbb{1} + n_B(\omega)\big(\mathbb{1} + \sigma_1\big). \qquad (6.136)$$

Let us next note, using (6.135)-(6.136), that we can write

$$iG_{\text{CT}}^{(T)}(t,\omega) = \overline{U}_{\text{CT}}^{(T)}(t,\omega) iG_{\text{CT}}^{(T=0)}(t,\omega)(\overline{U}_{\text{CT}}^{(T)}(t,\omega))^\ddagger$$

$$= \overline{U}_{\text{CT}}^{(T)}(t,\omega)(\overline{U}_{\text{CT}}^{(T)}(t,\omega))^\ddagger iG_{\text{CT}}^{(T=0)}(t,\omega)$$

$$+ \overline{U}_{\text{CT}}^{(T)}(t,\omega)\big[iG_{\text{CT}}^{(T=0)}(t,\omega), (\overline{U}_{\text{CT}}^{(T)}(t,\omega))^\ddagger\big]$$

$$= \big(\mathbb{1} + n_B(\omega)\big(\mathbb{1} + \sigma_1\big)\big) iG_{\text{CT}}^{(T=0)}(t,\omega)$$

$$+ n_B(\omega)\big(e^{\frac{\beta\omega}{2}}\mathbb{1} + \theta(-t)e^{-\frac{\beta\omega}{2}}\sigma_1\big)$$

$$\times \big[iG_{\text{CT}}^{(T=0)}(t,\omega), \big(e^{\frac{\beta\omega}{2}}\mathbb{1} + \theta(t)e^{-\frac{\beta\omega}{2}}\sigma_1\big)\big]$$

$$= \big(\mathbb{1} + n_B(\omega)\big(\mathbb{1} + \sigma_1\big)\big) iG_{\text{CT}}^{(T=0)}(t,\omega)$$

$$- n_B(\omega)\big(S(\omega) + \sigma_1\big) iG_{\text{CT}}^{(T=0)}(t,\omega)$$

$$= \big(\mathbb{1} + n_B(\omega)(\mathbb{1} - S(\omega))\big) iG_{\text{CT}}^{(T=0)}(t,\omega), \qquad (6.137)$$

where we have used (6.136) as well as the relations in (6.130)-(6.131). $S(\omega)$ is the reflection operator discussed in (6.8). Equation (6.137) can be compared with (6.7) and (6.9). This shows that, in the case of the closed time path formalism, the matrix factorization indeed leads to the scalar thermal operator in a natural way and, therefore, we are now ready to examine the question of a possible scalar thermal operator for the case of an arbitrary path with $\sigma \neq 0$.

For an arbitrary path in the real time formalism, the components of the propagator in the mixed space take the forms

$$iG_{11}^{(\sigma,T)}(t,\omega) = \frac{1}{2\omega}\left((\theta(t) + n_B(\omega))e^{-i\omega t} + (\theta(-t) + n_B(\omega))e^{i\omega t}\right),$$

$$iG_{12}^{(\sigma,T)}(t,\omega) = \frac{1}{2\omega}\left(n_B(\omega)e^{-i\omega(t+i\sigma\beta)} + (1 + n_B(\omega))e^{i\omega(t+i\sigma\beta)}\right),$$

$$iG_{21}^{(\sigma,T)}(t,\omega) = \frac{1}{2\omega}\left((1 + n_B(\omega))e^{-i\omega(t-i\sigma\beta)} + n_B(\omega)e^{i\omega(t-i\sigma\beta)}\right),$$

$$iG_{22}^{(\sigma,T)}(t,\omega) = \frac{1}{2\omega}\left((\theta(-t) + n_B(\omega))e^{-i\omega t} + (\theta(t) + n_B(\omega))e^{i\omega t}\right),$$

$$\tag{6.138}$$

We note here that the diagonal elements of the propagator do not have any σ dependence simply because, along the real branches of the path, the imaginary part of time does not change (recall that $t = t_1 - t_2$). On the other hand, the off-diagonal elements of the propagator represent a transition from one real branch (1 to 2 or *vice versa*) to the other and, consequently, the imaginary part of time does change if $\sigma \neq 0$. For closed time path, on the other hand, $\sigma = 0$ and there is no change in the imaginary part of the time. For $\sigma \neq 0$, it follows from (6.138) that, at zero temperature, $\beta \to \infty$, the components take the forms

$$iG_{11}^{(\sigma,0)}(t,\omega) = \frac{1}{2\omega}\left(\theta(t)e^{-i\omega t} + \theta(-t)e^{i\omega t}\right),$$

$$iG_{12}^{(\sigma,0)}(t,\omega) = 0 = iG_{21}^{(\sigma,0)}(t,\omega),$$

$$iG_{22}^{(\sigma,0)}(t,\omega) = \frac{1}{2\omega}\left(\theta(-t)e^{-i\omega t} + \theta(t)e^{i\omega t}\right).$$

$$\tag{6.139}$$

The striking feature in (6.139) is that, at zero temperature, the off-diagonal components vanish implying that they do not interact. This is, of course, what we expect in an operator description like thermofield dynamics as we have already discussed in chapter **4**. For the closed time path formalism, on the other hand, $\sigma = 0$ and the zero temperature components following from (6.138) do have nonzero off-diagonal values which indeed coincide with (6.129). We have discussed this case in detail earlier and, therefore, let us assume that $\sigma \neq 0$ in the following. It is clear from (6.138) and (6.139) that the propagators (both at zero temperature and at finite temperature) satisfy the relations

$$S(\omega)iG_{11}(t,\omega) = -iG_{22}(t,\omega), \quad S(\omega)iG_{22}(t,\omega) = -iG_{11}(t,\omega),$$

$$S(\omega)iG_{12}(t,\omega) = -iG_{21}(t,\omega), \quad S(\omega)iG_{21}(t,\omega) = -iG_{12}(t,\omega),$$
$$\tag{6.140}$$

as well as

$$iG_{11}(t,\omega) + iG_{22}(t,\omega) = iG_{12}(t,\omega) + iG_{21}(t,\omega), \tag{6.141}$$

which can be compared with (6.130) and (6.131) respectively.

The factorization of the propagator for an arbitrary path (in the operator formalism) has already been discussed in detail in momentum space in sections **4** and **4.2.1** (see, in particular, (4.9)-(4.11)). However, for the discussion of the thermal operator, we need the mixed space description and, therefore, we need to factorize the finite temperature propagator in (6.138) in terms of the zero temperature propagator given in (6.139). Looking at the factorizing matrix for the closed time path formalism in (6.132), we choose a matrix for the arbitrary path of the form

$$\overline{U}^{(\sigma,T)}(t,\omega) = \sqrt{n_B(\omega)}\left(e^{\frac{\beta\omega}{2}}\mathbb{1} + P^{(\sigma,T)}(t,\omega)\sigma_1\right)$$

$$= \sqrt{n_B(\omega)}\begin{pmatrix} e^{\frac{\beta\omega}{2}} & P^{(\sigma,T)}(t,\omega) \\ P^{(\sigma,T)}(t,\omega) & e^{\frac{\beta\omega}{2}} \end{pmatrix}, \tag{6.142}$$

where the function $P^{(\sigma,T)}(t,\omega)$ is yet to be determined and we assume that it satisfies the symmetry property (recall from (4.4) that under $\lambda \to -\lambda, \sigma \to 1-\sigma$)

$$P^{(\sigma,T)}(t,\omega) = P^{(1-\sigma,T)}(-t,\omega). \tag{6.143}$$

With the property (6.143) we note that

$$\left(\overline{U}^{(\sigma,T)}(t,\omega)\right)^{\dagger} = \overline{U}^{(\sigma,T)}(t,\omega),$$

$$\left(\overline{U}^{(\sigma,T)}(t,\omega)\right)^{\ddagger} = \left(\overline{U}^{(1-\sigma,T)}(t,\omega)\right)^{\dagger} = \overline{U}^{(\sigma,T)}(-t,\omega), \qquad (6.144)$$

where $\ddagger$ denotes the modified Hermitian conjugate in the Hilbert space with a nontrivial metric (see, for example, (4.11), (4.21) as well as (4.43)-(4.51)). We note here that $p^0 \to -p^0$ in the momentum space in (4.11) is equivalent to $t \to -t$ in the conjugate mixed space and this explains the change in the sign of t in the (modified) definition of the adjoint matrix in (6.133). Namely, for this matrix to be the factorizing matrix, it has to satisfy

$$iG^{(\sigma,T)}(t,\omega) = \overline{U}^{(\sigma,T)}(t,\omega)\, iG^{(\sigma,0)}(t,\omega) \left(\overline{U}^{(\sigma,T)}(t,\omega)\right)^{\ddagger}$$

$$= \overline{U}^{(\sigma,T)}(t,\omega)\, iG^{(\sigma,0)}(t,\omega)\, \overline{U}^{(\sigma,T)}(-t,\omega). \qquad (6.145)$$

Multiplying out the matrices on the right hand side of (6.145), we obtain the matrix components as

$$iG_{11}^{(\sigma,T)}(t,\omega) = n_B(\omega)\left[e^{\beta\omega}\, iG_{11}^{(\sigma,0)}(t,\omega)\right.$$

$$\left. + P^{(\sigma,T)}(t,\omega)P^{(\sigma,T)}(-t,\omega)\, iG_{22}^{(\sigma,0)}(t,\omega)\right],$$

$$iG_{12}^{(\sigma,T)}(t,\omega) = n_B(\omega)\, e^{\frac{\beta\omega}{2}}\left(P^{(\sigma,T)}(t,\omega)\, iG_{22}^{(\sigma,0)}(t,\omega)\right.$$

$$\left. + P^{(\sigma,T)}(-t,\omega)\, iG_{11}^{(\sigma,0)}(t,\omega)\right),$$

$$iG_{21}^{(\sigma,T)}(t,\omega) = n_B(\omega)\, e^{\frac{\beta\omega}{2}}\left(P^{(\sigma,T)}(t,\omega)\, iG_{11}^{(\sigma,0)}(t,\omega)\right.$$

$$\left. + P^{(\sigma,T)}(-t,\omega)\, iG_{22}^{(\sigma,0)}(t,\omega)\right),$$

$$iG_{22}^{(\sigma,T)}(t,\omega) = n_B(\omega)\left[e^{\beta\omega}\, iG_{22}^{(\sigma,0)}(t,\omega)\right.$$

$$\left. + P^{(\sigma,T)}(t,\omega)P^{(\sigma,T)}(-t,\omega)\, iG_{11}^{(\sigma,0)}(t,\omega)\right]. \qquad (6.146)$$

Putting in the forms of $iG_{11}^{(\sigma,0)}(t,\omega), iG_{22}^{(\sigma,0)}(t,\omega)$ from (6.139) and comparing the results in (6.146) with the actual components of the propagator given in (6.138), we determine that

$$P^{(\sigma,T)}(t,\omega) = \theta(t)\, e^{-(\sigma-\frac{1}{2})\beta\omega} + \theta(-t)\, e^{(\sigma-\frac{1}{2})\beta\omega}$$

$$= P^{(1-\sigma,T)}(-t,\omega), \qquad (6.147)$$

as required (see (6.143)). Furthermore, let us note from (6.147) that

$$P^{(\sigma,T)}(t,\omega)P^{(\sigma,T)}(-t,\omega) = 1. \tag{6.148}$$

This completes the derivation of the factorizing matrix (Bogoliubov transformation in the operator formalism) for the arbitrary path which has the form given in (6.142). It is worth pointing out here that if we set $\sigma = 0$, the factorizing matrix $\overline{U}^{(\sigma=0,T)}(t,\omega)$ does not coincide with the matrix given in (6.132) for the closed time path formalism. This can be understood in the following manner. Here the factorizing matrix connects the finite temperature propagator to a diagonal zero temperature matrix whereas the zero temperature matrix in (6.129) has off-diagonal terms. Nonetheless, it can be checked that, for $\sigma = 0$, the matrix in (6.142) together with (6.147) does take the diagonal zero temperature propagator in (6.139) to the closed time path propagator in (6.128).

One thing is quite clear from the discussion so far. Namely, although we can find a matrix factorization for the four component finite temperature matrix propagator in terms of a diagonal two component zero temperature, it would be impossible to find a well behaved scalar (thermal) operator which acting on the two component zero temperature matrix propagator can yield the four component finite temperature matrix propagator. (In other words, a well behaved operator cannot act on trivial matrix elements to generate a nontrivial element.) So, let us analyze this case a bit more carefully. We note that the factorizing matrix $\overline{U}^{(\sigma,T)}(t,\omega)$ in (6.142) can be further factorized as follows. Let us define two Hermitian matrices

$$\widetilde{U}^{(\sigma,T)}(t,\omega) = \sqrt{n_B(\omega)}\left(e^{\frac{\beta\omega}{2}}\,\mathbb{1} + \theta(-t)e^{(\sigma-\frac{1}{2})\beta\omega}\,\sigma_1\right)$$
$$= \left(\widetilde{U}^{(\sigma,T)}(t,\omega)\right)^{\dagger},$$

$$\overline{V}^{(\sigma,T)}(t,\omega) = \mathbb{1} + \theta(t)\,e^{-\sigma\beta\omega}\,\sigma_1 = \left(\overline{V}^{(\sigma,T)}(t,\omega)\right)^{\dagger}, \tag{6.149}$$

which lead to

$$\widetilde{U}^{(\sigma,T)}(t,\omega)\overline{V}^{(\sigma,T)}(t,\omega) = \sqrt{n_B(\omega)}\left(e^{\frac{\beta\omega}{2}}\,\mathbb{1} + \theta(-t)\,e^{(\sigma-\frac{1}{2})\beta\omega}\,\sigma_1\right)$$
$$\times \left(\mathbb{1} + \theta(t)\,e^{-\sigma\beta\omega}\sigma_1\right)$$
$$= \sqrt{n_B(\omega)}\left(e^{\frac{\beta\omega}{2}}\,\mathbb{1} + (\theta(t)e^{-(\sigma-\frac{1}{2})\beta\omega} + \theta(-t)e^{(\sigma-\frac{1}{2})\beta\omega})\sigma_1\right)$$

$$= \overline{U}^{(\sigma,T)}(t,\omega).\tag{6.150}$$

This is reminiscent of (4.13) although not exactly since the relation there was a conjugation. The reason for considering such a factorization as in (6.149)-(6.150) will become clear shortly. However, let us note at this point that, for $\sigma = 0$, the matrix $\widetilde{U}^{(\sigma,T)}(t,\omega)$ coincides with the factorizing matrix given in (6.132), namely,

$$\widetilde{U}^{(\sigma=0,T)}(t,\omega) = \overline{U}_{\text{CT}}^{(T)}(t,\omega).\tag{6.151}$$

This is already interesting, but let us also note that (see (6.139) for the definition of $iG^{(\sigma,T=0)}(t,\omega)$)

$$i\overline{G}^{(\sigma)}(t,\omega) = \overline{V}^{(\sigma,T)}(t,\omega)iG^{(\sigma,T=0)}(t,\omega)\big(\overline{V}^{(\sigma,T)}(-t,\omega)\big)^{\dagger}$$

$$= \frac{1}{2\omega}\begin{pmatrix} \theta(t)e^{-i\omega t} + \theta(-t)e^{i\omega t} & e^{i\omega(t+i\sigma\beta)} \\ e^{-i\omega(t-i\sigma\beta)} & \theta(-t)e^{-i\omega t} + \theta(t)e^{i\omega t} \end{pmatrix},\tag{6.152}$$

which, for $\sigma = 0$, coincides with $G_{\text{CT}}^{(T=0)}(t,\omega)$ given in (6.129), namely,

$$iG_{\text{CT}}^{(T=0)}(t,\omega) = i\overline{G}^{(\sigma=0)}(t,\omega).\tag{6.153}$$

This also clarifies the meaning of the matrix $\overline{V}^{(\sigma,T)}(t,\omega)$. Namely, the diagonal matrix, $iG^{(\sigma,T=0)}(t,\omega)$ in (6.139) is the natural zero temperature propagator in an operator description where the tilde fields are assumed to decouple at zero temperature. In the diagrammatic (path integral) description (see Fig. 6.6), however, the four components of the propagator matrix are all nontrivial (namely, the off-diagonal components $iG_{12}^{(\sigma,T)}(t,\omega), iG_{21}^{(\sigma,T)}(t,\omega)$) are nonzero and do carry temperature dependence unless $\sigma = 0$. The matrix $\overline{V}^{(\sigma,T)}(t,\omega)$ simply connects the basic zero temperature propagator in the operator formalism to the basic propagator in an arbitrary path calculation. Furthermore, we note that we can now write (see (6.145))

$$iG^{(\sigma,T)}(t,\omega) = \overline{U}^{(\sigma,T)}(t,\omega)iG^{(\sigma,0)}(t,\omega)\big(\overline{U}^{(\sigma,T)}(t,\omega)\big)^{\ddagger}$$

$$= \widetilde{U}^{(\sigma,T)}(t,\omega)\overline{V}^{(\sigma,T)}(t,\omega)iG^{(\sigma,0)}(t,\omega)\overline{V}^{(\sigma,T)}(-t,\omega)\widetilde{U}^{(\sigma,T)}(-t,\omega)$$

$$= \widetilde{U}^{(\sigma,T)}(t,\omega)i\overline{G}^{(\sigma)}(t,\omega)\widetilde{U}^{(\sigma,T)}(-t,\omega).\tag{6.154}$$

From the definition in (6.149), we note that the matrix $\widetilde{U}^{(\sigma,T)}(t,\omega)$ satisfies

$$\widetilde{U}^{(\sigma,T)}(t,\omega)\big(\widetilde{U}^{(\sigma,T)}(t,\omega)\big)^{\ddagger} = \widetilde{U}^{(\sigma,T)}(t,\omega)\widetilde{U}^{(\sigma,T)}(-t,\omega)$$

$$= n_B(\omega)\big(e^{\beta\omega}\mathbb{1} + e^{\sigma\beta\omega}\,\sigma_1\big) = \mathbb{1} + n_B(\omega)\big(1 + e^{\sigma\beta\omega}\,\sigma_1\big), \quad (6.155)$$

which can be compared with (6.136) for the closed time case ($\sigma = 0$). Therefore, as in (6.137), we can proceed from (6.154) and write

$$iG^{(\sigma,T)}(t,\omega) = \widetilde{U}^{(\sigma,T)}(t,\omega)i\overline{G}^{(\sigma)}(t,\omega)\widetilde{U}^{(\sigma,T)}(-t,\omega)$$

$$= \widetilde{U}^{(\sigma,T)}(t,\omega)\widetilde{U}^{(\sigma,T)}(-t,\omega)i\overline{G}^{((\sigma))}(t,\omega)$$

$$+ \widetilde{U}^{(\sigma,T)}(t,\omega)\big[i\overline{G}^{(\sigma)}(t,\omega),\widetilde{U}^{(\sigma,T)}(-t,\omega)\big]$$

$$= \big(\mathbb{1} + n_B(\omega)(1 + e^{\sigma\beta\omega}\sigma_1)\big)i\overline{G}^{(\sigma)}(t,\omega)$$

$$+ \widetilde{U}^{(\sigma,T)}(t,\omega)\,\sqrt{n_B(\omega)}\big[i\overline{G}^{(\sigma)}(t,\omega),\theta(t)e^{(\sigma-\frac{1}{2})\beta\omega}\sigma_1\big],$$

where we have used the form of $\widetilde{U}^{(\sigma,T)}(-t,\omega)$ from (6.150) as well as the fact that the identity matrix commutes with everything. Furthermore, using the form of $i\overline{G}^{(\sigma)}(t,\omega)$ from (6.152), this leads to

$$= \big(\mathbb{1} + n_B(\omega)(1 + e^{\sigma\beta\omega}\sigma_1)\big)i\overline{G}^{(\sigma)}(t,\omega)$$

$$- n_B(\omega)\big(S(\omega) + e^{\sigma\beta\omega}\sigma_1\big)i\overline{G}^{(\sigma)}(t,\omega)$$

$$= \big(\mathbb{1} + n_B(\omega)(1 - S(\omega))\big)i\overline{G}^{(\sigma)}(t,\omega)$$

$$= \mathcal{O}^{(T)}(t,\omega)\,i\overline{G}^{(\sigma)}(t,\omega). \quad (6.156)$$

This determines the scalar thermal operator for an arbitrary path in the real time formalism and we note that the scalar thermal operator coincides with that of the closed time path (see (6.7), (6.9)). However, it is worth noting that it acts on the four component basic propagator $i\overline{G}^{(\sigma)}(t,\omega)$ (and not on the diagonal two component zero temperature propagator of the operator formalism) which is relevant in a path integral calculation (which happens to be the zero temperature propagator for closed time path). As a result of this scalar factorization of the propagator (given in (6.156)), any finite temperature graph can now be factorized in terms of the basic graph of the path integral formalism. This analysis can be extended to any other field theory (fermions, gauge fields etc.) in a straightforward manner. All of this discussion can be carried over to complex scalar fields and fermions (with or without chemical potential) as well as to gauge fields in a straightforward manner.

6.5 Some applications

There are several things that can be beneficially done with the thermal operator representation. Finite temperature calculations of Feynman graphs are generally notoriously difficult compared to the zero temperature ones. However, the thermal operator method greatly simplifies that because once a graph is calculated at zero temperature, one can simply apply the appropriate thermal operator (involving products of operators) to obtain the value of the graph at finite temperature. Similarly, thermal operator representation also allows us to generalize results (theorems) to finite temperature if they hold at zero temperature and if the thermal operators do not violate any of the conditions associated with the theorems. A case in point is the cutting rules in obtaining the imaginary part of a Feynman graph. At zero temperature these are known as the Cutkosky rules and generalization of this result to finite temperature (which we will discuss in detail in chapter **8**) is highly nontrivial. However, *a posteriori*, this extension can be intuitively seen through the action of the thermal operators. Namely, if the cutting rules for the imaginary part of a Feynman graph holds at zero temperature, it must also hold at finite temperature since the thermal operator is real and, therefore, the proof should easily go through, namely, if

$$\text{Im } \gamma_N^{(T=0)} = \sum (\text{cut diagrams})^{(T=0)},$$

$$\text{Im } \gamma_N^{(T)} = \text{Im } \mathcal{O}^{(T)} \gamma^{(T=0)} = \mathcal{O}^{(T)} \text{Im } \gamma_N^{(T=0)}$$

$$= \sum \mathcal{O}^{(T)} (\text{cut diagrams})^{(T=0)}$$

$$= \sum (\text{cut diagrams})^{(T)}, \tag{6.157}$$

where γ_N denotes an arbitrary N-point Feynman graph of the theory. There are many such results that can be extended to finite temperature through the thermal operator representation. In this section, we will discuss only one calculation in some detail and point out where else it may be useful in later chapters as we go along.

6.5.1 0 + 1 dimensional Chern-Simons QED. In $0 + 1$ dimension, the Lagrangian for Chern-Simons QED is given by (for the fermion part of the Lagrangian, see, for example, (5.11) and (5.13) with $t^a = \mathbb{1}$ and without the internal i, j indices)

$$L = \overline{\psi}(i\partial_t - m - A)\psi - \kappa A, \tag{6.158}$$

where κ is known as the Chern-Simons coefficient and the last term in (6.158) is known as the Chern-Simons term. We note here that, even if $\kappa = 0$ in the tree level Lagrangian, a Chern-Simons term is generated through radiative corrections arising from the interaction between the photon and the electron (fermion). Under a Euclidean rotation of the time coordinate (see (1.94) with $\mu = 0$ as well as section **1.9.1**),

$$t \to -i\tau, \quad \partial_t \to i\partial_\tau, \quad A \to iA, \quad L \to -L_E, \tag{6.159}$$

so that, in the Euclidean space the Lagrangian has the form

$$L_E = \overline{\psi}(\partial_\tau + m + iA)\psi + i\kappa A. \tag{6.160}$$

There are several things to point out here. For example, the mass m can be thought of as a chemical potential since it couples to a conserved charge (in $0 + 1$ dimension, $\overline{\psi}$ really stands for $\psi^\dagger$ and, therefore, $\overline{\psi}\psi = N$ is the conserved number operator). Furthermore, in $0 + 1$ dimension, a vector can only have one component, the time component, which is neglected (and should be understood) in the gauge field terms involving A. As a result of this, there is no kinetic energy term for the photon since $F_{\mu\nu} = 0$. We can only have an Abelian Chern-Simons term in $0 + 1$ dimension since, otherwise, the non-Abelian index cannot be saturated. Furthermore, the theory is invariant under an Abelian gauge transformation, the quadratic part involving the fermions is because of minimal coupling and the Chern-Simons term changes by a total derivative (under a gauge transformation) so that the action is invariant. In fact, this theory is invariant under (Abelian) large gauge transformations (and not just infinitesimal gauge transformations). Since the photon has no kinetic energy term, the photon does not propagate (it does not have a propagator) in this theory. Consequently, any nontrivial graph in this theory would involve only external photon lines and internal fermion lines. Furthermore, there cannot be any nontrivial Feynman graphs beyond one loop because that would necessarily involve photon lines (and the photon propagator is trivial). Therefore, the only one loop graph in this theory can have N external photon lines attached to a single fermion loop as shown in Fig. 6.7.

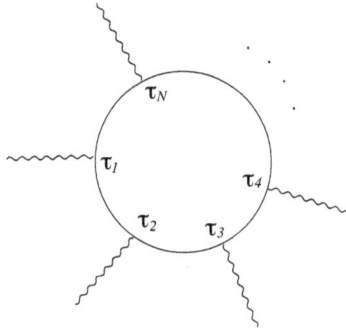

Figure 6.7: A fermion one loop graph with N external photons.

We have already derived the fermion propagator at finite temperature in chapter **2** (see section **2.6.2**, in particular (2.116)). In imaginary time formalism $t \to -i\tau$, it has the form (we are suppressing the subscript F signifying "Feynman" for simplicity)

$$S^{(T,m)}(\tau) = e^{-m\tau}(\theta(\tau) - n_F(m)),$$
$$S^{(T=0,m)}(\tau) = e^{-m\tau}\,\theta(\tau), \quad S^{(T=0,m=0)}(\tau) = \theta(\tau), \tag{6.161}$$

so that we can write

$$S^{(T,m)}(\tau) = e^{-m\tau}\left(\mathbb{1} - n_F(m)(1 + \mathcal{S}(\tau))\right)S^{(T=0,m=0)}(\tau)$$
$$= e^{-m\tau}\,\mathcal{O}_F^{(T,m)}(\tau)\,S^{(T=0,m=0)}(\tau), \tag{6.162}$$

where $\mathcal{S}(\tau)$ reflects the τ coordinate. Even though the thermal operator in (6.162) is explicitly time (τ) dependent, it does not cause any problem since the only meaningful diagram in this theory is the one loop diagram which does not involve integration over time. For the one loop N-point Feynman graph shown in Fig. 6.7, we note that the exponential term cancels out in the closed loop both at zero temperature as well as at finite temperature and, therefore, the value of the graph at zero temperature is simply given by (note that we have chosen a unit coupling constant for the coupling of the photon to fermions in (6.158))

$$-\gamma_N^{(T=0)}(\tau_1, \tau_2, \cdots, \tau_N) = -\frac{(-i)^N}{N!} \prod_{i=1}^{N} \theta(\tau_i - \tau_{i+1}), \qquad (6.163)$$

with the identification $\tau_{N+1} = \tau_1$. The overall negative sign is because of the fermion loop (the vertex $(-i)$ arises because in Euclidean space, the vertices are obtained from $-S_E$) and we note that the product of the theta functions in (6.163) vanishes for $N \geq 2$ because of opposing theta functions. Only for the 1-point function, the graph has a nontrivial value of $-\frac{i}{2}$ (if we assume the average value $\theta(0) = \frac{1}{2}$). On the other hand, it follows from the form of the propagator at finite temperature in (6.161), that the graph in Fig. 6.7 would have a nontrivial value for any N. (The graph will have explicit temperature dependence with the zero temperature contribution vanishing for $N \geq 2$.) Therefore, it is not a priori clear how the thermal operator acting on trivial zero temperature values can lead to nontrivial values for the graph at finite temperature when $N \geq 2$. The way out of this apparent problem is to write the value of the graph at finite temperature in Fig. 6.7 as

$$-\gamma_N^{(T)}(\tau_1, \tau_2, \cdots, \tau_N) = -\frac{(-i)^N}{N!} \prod_{i=1}^{N} S^{(T,m)}(\tau_i - \tau_{i+1})$$

$$= -\frac{(-i)^N}{N!} \prod_{i=1}^{N} e^{-m(\tau_i - \tau_{i+1})} (\theta(\tau_i - \tau_{i+1}) - n_F(m))$$

$$= -\frac{(-i)^N}{N!} e^{-m(\tau_1 - \tau_{N+1})} \prod_{i=1}^{N} (\theta(\tau_i - \tau_{i+1}) - n_F(m)), \qquad (6.164)$$

and take the limit $\tau_{N+1} = \tau_1$ only at the end. In this case, we see that the exponential will become unity in the limit at the end independent of the other factors. The term depending only on temperature independent product of theta functions will also vanish in this limit at the end for $N \geq 2$, signifying that there is no zero temperature contribution for the higher point graphs. There would, however, remain other terms which will involve products of theta functions as well as factors of $n_F(m)$. Furthermore, using (6.162) as well as all these observations, we can write the value of the graph now as

$$-\gamma_N^{(T)}(\tau_1, \tau_2, \cdots, \tau_N)$$

$$= -\frac{(-i)^N}{N!} \prod_i^N \mathcal{O}_F^{(T,m)}(\tau_i - \tau_{i+1}) S^{(T=0,m=0)}(\tau_i - \tau_{i+1}), \quad (6.165)$$

with the identification $\tau_{N+1} = \tau_1$ only after the action of the thermal operator on the zero temperature propagator. For the complete N-point amplitude, $\Gamma_N^{(T)}$, however, we have to add all graphs where we keep one point fixed, say τ_1, and permute all the other points $\tau_2, \tau_3, \cdots, \tau_N$ in all possible ways (remember that $\tau_{N+1} = \tau_1$). This would lead to the theta function terms would add up to unity in each of the terms involving a given power of $n_F(m)$. For example, we list below the values of the first few amplitudes

$$-\Gamma_1^{(T)}(\tau_1) = \lim_{\tau_2 \to \tau_1}$$

$$(-) - i\mathcal{O}_F^{(T,m)}(\tau_1 - \tau_2)\theta(\tau_1 - \tau_2) = i(\theta(0) - n_F(m))$$

$$= \frac{i}{2}(1 - 2n_F(m)) = \frac{i}{2}\tanh\frac{\beta m}{2},$$

$$-\Gamma_2^{(T)}(\tau_1, \tau_2) = \lim_{\tau_3 \to \tau_1}$$

$$- \frac{(-i)^2}{2!} \mathcal{O}_F^{(T,m)}(\tau_1 - \tau_2)\mathcal{O}_F^{(T,m)}(\tau_2 - \tau_3)$$

$$\times \theta(\tau_1 - \tau_2)\theta(\tau_2 - \tau_3)$$

$$= \frac{1}{2!} n_F(m)(1 - n_F(m)) = \frac{1}{2!}\frac{1}{4}\operatorname{sech}^2\frac{\beta m}{2},$$

$$-\Gamma_3^{(T)}(\tau_1, \tau_2, \tau_3) = \lim_{\tau_4 \to \tau_1}$$

$$- \frac{(-i)^3}{3!} \left[\mathcal{O}_F^{(T,m)}(\tau_1 - \tau_2)\mathcal{O}_F^{(T,m)}(\tau_2 - \tau_3)\mathcal{O}_F^{(T,m)}(\tau_3 - \tau_4) \right.$$

$$\left. \theta(\tau_1 - \tau_2)\theta(\tau_2 - \tau_3)\theta(\tau_3 - \tau_4) + \tau_2 \leftrightarrow \tau_3 \right]$$

$$= \frac{i}{3!} n_F(m)(1 - n_F(m))(1 - 2n_F(m))$$

$$= \frac{i}{3!} \frac{1}{4} \tanh \frac{\beta m}{2} \operatorname{sech}^2 \frac{\beta m}{2}, \tag{6.166}$$

and so on. There are several things to note from these results. First, the amplitudes seem to be time independent (in general, in the mixed space the amplitudes do depend on time) which can also be seen by taking time derivative of the products of fermion propagator in the amplitude. This signifies that the effective action, which is obtained by multiplying the external photon fields with external time coordinates and integrating over the corresponding time coordinates, would simply involve powers of $d\tau\, A(\tau)$. This is, of course, the Chern-Simons action (see (6.160)). Therefore, the radiative corrections simply renormalize the Chern-Simons coefficient κ. Second, each higher point amplitude, say $\Gamma_N^{(T,m)}$, seems to be related to the previous one, $\Gamma_{N-1}^{(T,m)}$, through a derivative with respect to $-\frac{i}{N} \frac{\partial}{\partial \beta m}$. This, in fact, is the manifestation of the Ward identity in $0 + 1$ dimension. Furthermore, this also implies that the N-point amplitude is simply related recursively to the 1-point (tadpole) amplitude and the effective action can be easily summed in a closed form.[3]

6.6 References

P. F. Bedaque and A. Das, Modern Physics Letters **A8**, 3151 (1993).

F. T. Brandt, A. Das, O. Espinosa, J. Frenkel and S. Perez, Physical Review **D72**, 085006 (2005).

F. T. Brandt, A. Das, O. Espinosa, J. Frenkel and S. Perez, Physical Review **D73**, 065010 (2006).

F. T. Brandt, A. Das, O. Espinosa, J. Frenkel and S. Perez, Physical Review **D73**, 067702 (2006).

F. T. Brandt, A. Das, O. Espinosa, J. Frenkel and S. Perez, Physical Review **D74**, 085006 (2006).

---

[3] For more details, see, A. Das and G. V. Dunne, Physical Review **D57**, 5023 (1998).

F. T. Brandt, A. Das, O. Espinosa, J. Frenkel and S. Perez, Physical Review **D74**, 125005 (2006).

A. Das and G. V. Dunne, Physical Review **D57**, 5023 (1998).

A. Das and J. Frenkel, Physical Review **D76**, 087701 (2007).

A. Das, A. Deshmukhya, P. Kalauni, S. Panda, Physical Review **D97**, 045015 (2018).

A. Das and P. Kalauni, Physical Review **D93**, 125028 (2016).

G. V. Dunne, K. Lee and C. Lu, Physical Review **78**, 3434 (1997).

O. Espinosa and E. Stockmeyer, Physical Review **D69**, 065004 (2004).

O. Espinosa, Physical Review **D71**, 065009 (2005).

M. Inui, H. Kohyama and A. Niegawa, Physical Review **D73**, 047702 (2006).

CHAPTER 7

Light-front field theories at finite temperature

So far, we have considered only conventional field theories, namely, field theories quantized on a space-like surface such as the equal-time surface. However, there are other alternatives to the conventional quantization which are useful in the study of various questions of physical interest. Light-front quantization is one such alternative where one quantizes a field theory on a light-like surface and interest in such a quantization dates back to Dirac. There are several distinctive characteristics that arise in such a quantization that are very different from the conventional equal time quantization. For example, among the many advantages which naturally develop in such a quantization, the most interesting is that the theory becomes first order in the "time" derivative which renders several of the Poincaré generators to be simply kinematical leading to a trivial structure for the vacuum state of the theory. This, in principle, allows us to carry out nonperturbative studies in a simple manner. As a result, it becomes an interesting approach to studying difficult physical questions in branches such as QCD, string theories and membrane theories among others. Quite independently, within the context of studies in current algebra, it was known that, if one studies field theories in the infinite momentum frame (frames which travel at the speed of light), various simplifications arise. In particular, the structure of the vacuum becomes very simple. In this infinite momentum frame, a residual two dimensional Galilean subgroup plays an important role. So, this seems very much like the light-front frame (which should also move with the speed of light) and, in fact, the two can be shown to be the same.

7.1 Conventional light-front coordinates

The idea behind light-front quantization is very simple. Let us consider an n-dimensional Minkowski space with coordinates $x^\alpha = (x^0, x^i, x^{n-1})$, $\alpha = 0, i\, (= 1, 2, \cdots, n-2), n-1$, where $x^0 = t$ is identified with the conventional Minkowski time coordinate (we have also set $c = 1$ for simplicity). The equal-time surface of conventional quantization can be identified with $x^0 = x'^0$ (or $t = t'$). Covariantly, the equal time surface can be written as

$$n^{(\text{ET})} \cdot x = n^{(\text{ET})} \cdot x', \quad \text{where} \quad n_\mu^{(\text{ET})} = (1, 0, 0, \cdots, 0). \qquad (7.1)$$

In this Minkowski space-time, let us define new coordinates, the light-front coordinates, as $\overline{x}^\mu = (x^+, x^i, x^-)$, $\mu = 0, i\, (= 1, 2, \cdots, n-2), n-1$, where

$$\overline{x}^0 = x^+ = x^0 + x^{n-1}, \qquad \overline{x}^{n-1} = x^- = x^0 - x^{n-1}. \qquad (7.2)$$

x^+ is known as the light-front time and, in light-front quantization, one quantizes fields on a light-front surface where $x^+ = x'^+$ which can be written covariantly as

$$n^{(\text{LF})} \cdot x = n^{(\text{LF})} \cdot x', \quad n_\mu^{(\text{LF})} = (1, 0, 0, \cdots, 1). \qquad (7.3)$$

The canonically conjugate momentum variable in Minkowski space $p^\alpha = (p^0, p^i, p^{n-1})$, then, goes over to the light-front momentum variable $\overline{p}^\mu = (p^+, p^i, p^-)$, where

$$\overline{p}^0 = p^+ = p^0 + p^{n-1}, \qquad \overline{p}^{n-1} = p^- = p^0 - p^{n-1}. \qquad (7.4)$$

In $3 + 1$ dimensional Minkowski space, for example, the light-front coordinates are defined as

$$\overline{x}^\mu = (x^+, x^1, x^2, x^-),$$

with

$$\overline{x}^0 = x^+ = x^0 + x^3, \qquad \overline{x}^3 = x^- = x^0 - x^3. \qquad (7.5)$$

Similarly, the conjugate momenta are defined as

$$\overline{p}^\mu = (p^+, p^1, p^2, p^-),$$

with

$$\bar{p}^0 = p^+ = p^0 + p^3, \qquad \bar{p}^3 = p^- = p^0 - p^3. \tag{7.6}$$

This describes the conventional light-front coordinates.

It is clear, from (7.5) and (7.6), that the Minkowski and the light-front coordinates (and momenta) are related through the linear transformation matrix

$$\bar{x}^\mu = \Lambda^\mu{}_\alpha x^\alpha, \qquad \bar{p}^\mu = \Lambda^\mu{}_\alpha p^\alpha, \quad \mu, \alpha = 0, 1, 2, 3,$$

with

$$\Lambda^\mu{}_\alpha = \begin{pmatrix} 1 & 0 & 0 & 1 \\ 0 & 1 & 0 & 0 \\ 0 & 0 & 1 & 0 \\ 1 & 0 & 0 & -1 \end{pmatrix}. \tag{7.7}$$

Let us denote the inverse transformation of (7.7) as

$$x^\alpha = \tilde{\Lambda}^\alpha{}_\mu \bar{x}^\mu, \tag{7.8}$$

Together with (7.7) (see the first relation), (7.8) determines

$$\Lambda^\mu{}_\alpha \tilde{\Lambda}^\alpha{}_\nu = \delta^\mu{}_\nu, \qquad \tilde{\Lambda}^\alpha{}_\mu \Lambda^\mu{}_\beta = \delta^\alpha{}_\beta, \tag{7.9}$$

and, therefore, using the explicit matrix form given in (7.7), we determine,

$$\tilde{\Lambda}^\alpha{}_\mu = \begin{pmatrix} \frac{1}{2} & 0 & 0 & \frac{1}{2} \\ 0 & 1 & 0 & 0 \\ 0 & 0 & 1 & 0 \\ \frac{1}{2} & 0 & 0 & -\frac{1}{2} \end{pmatrix}. \tag{7.10}$$

The metric in the transformed frame can now be determined from the contravariant transformation rules in (7.7) (see the first relation) to correspond to

$$\bar{\eta}^{\mu\nu} = \Lambda^\mu{}_\alpha \eta^{\alpha\beta} \Lambda^\nu{}_\beta = (\Lambda \eta \Lambda^T)^{\mu\nu}$$

$$= \begin{pmatrix} 0 & 0 & 0 & 2 \\ 0 & -1 & 0 & 0 \\ 0 & 0 & -1 & 0 \\ 2 & 0 & 0 & 0 \end{pmatrix} = \bar{\eta}^{\nu\mu}, \tag{7.11}$$

where $\eta^{\alpha\beta}$ is the standard Minkowski metric and, similarly,

$$\overline{\eta}_{\mu\nu} = \left(\widetilde{\Lambda}^T \eta \widetilde{\Lambda}\right)_{\mu\nu} = \begin{pmatrix} 0 & 0 & 0 & \frac{1}{2} \\ 0 & -1 & 0 & 0 \\ 0 & 0 & -1 & 0 \\ \frac{1}{2} & 0 & 0 & 0 \end{pmatrix} = \overline{\eta}_{\nu\mu}, \qquad (7.12)$$

where $\overline{\eta}_{\mu\nu}$ and $\overline{\eta}^{\mu\nu}$ are known as the covariant and contravariant conventional light-front metrics and they satisfy, as they should,

$$\overline{\eta}^{\mu\lambda}\,\overline{\eta}_{\lambda\nu} = \delta^{\mu}{}_{\nu}. \qquad (7.13)$$

From (7.12), let us note here, for later use, that

$$\overline{\eta} = \det\overline{\eta}_{\mu\nu} = -\frac{1}{2} \times 1 \times \frac{1}{2} = -\frac{1}{4}, \quad \text{so that} \quad \sqrt{-\overline{\eta}} = \frac{1}{2}. \quad (7.14)$$

It follows now that the invariant volumes in the transformed (coordinate and momentum) spaces are given by (recall that, in Minkowski space, $\eta = \det\eta_{\mu\nu} = -1$)

$$\int \sqrt{-\overline{\eta}}\,\mathrm{d}^4\overline{x} = \frac{1}{2}\int \mathrm{d}^4\overline{x} = \int \sqrt{-\eta}\,\mathrm{d}^4 x = \int \mathrm{d}^4 x,$$

$$\int \left(\sqrt{-\overline{\eta}}\right)^{-1}\mathrm{d}^4\overline{p} = 2\int \mathrm{d}^4\overline{p} = \int \left(\sqrt{-\eta}\right)^{-1}\mathrm{d}^4 p = \int \mathrm{d}^4 p. \quad (7.15)$$

We can now easily determine the transformation laws for the covariant vectors from

$$\overline{x}_\mu = \overline{\eta}_{\mu\nu}\overline{x}^\nu = \overline{\eta}_{\mu\nu}\Lambda^\nu{}_\alpha x^\alpha = \Lambda_{\mu\alpha}x^\alpha = \Lambda_\mu{}^\alpha x_\alpha, \qquad (7.16)$$

$$x_\alpha = \eta_{\alpha\beta}x^\beta = \eta_{\alpha\beta}\widetilde{\Lambda}^\beta{}_\mu \overline{x}^\mu = \widetilde{\Lambda}_\alpha{}^\mu \overline{x}_\mu. \qquad (7.17)$$

The explicit forms of these matrices, $\Lambda_\mu{}^\alpha, \widetilde{\Lambda}_\alpha{}^\mu$ can now be calculated from (7.7), (7.10)-(7.12) and have the forms

$$\Lambda_\mu{}^\alpha = \begin{pmatrix} \frac{1}{2} & 0 & 0 & \frac{1}{2} \\ 0 & 1 & 0 & 0 \\ 0 & 0 & 1 & 0 \\ \frac{1}{2} & 0 & 0 & -\frac{1}{2} \end{pmatrix} = \widetilde{\Lambda}^\alpha{}_\mu, \qquad (7.18)$$

$$\tilde{\Lambda}_\alpha{}^\mu = \begin{pmatrix} 1 & 0 & 0 & 1 \\ 0 & 1 & 0 & 0 \\ 0 & 0 & 1 & 0 \\ 1 & 0 & 0 & -1 \end{pmatrix} = \Lambda^\mu{}_\alpha, \tag{7.19}$$

which determines (remember that in Minkowski space, space components change sign in raising and lowering)

$$\bar{x}_\mu = (\frac{1}{2}x^-, -x^1, -x^2, \frac{1}{2}x^+),$$

$$\bar{p}_\mu = (\frac{1}{2}p^-, -p^1, -p^2, \frac{1}{2}p^+), \tag{7.20}$$

where we have used the definitions given in (7.5)-(7.6). Furthermore, let us note that, using (7.11) and (7.12), we obtain

$$\bar{x}^2 = \bar{x}^\mu \bar{\eta}_{\mu\nu} \bar{x}^\nu = \bar{x}^0 \bar{x}^3 - (\bar{x}^1)^2 - (\bar{x}^2)^2 = x^+ x^- - (x^1)^2 - (x^2)^2$$

$$= (x^0)^2 - (x^1)^2 - (x^2)^2 - (x^3)^2 = \eta_{\alpha\beta} x^\alpha x^\beta = x^2. \tag{7.21}$$

Similarly,

$$\bar{p}^2 = \bar{p}_\mu \bar{\eta}^{\mu\nu} \bar{p}_\nu = 4\bar{p}_0 \bar{p}_3 - (\bar{p}_1)^2 - (\bar{p}_2)^2 = p^+ p^- - (p^1)^2 - (p^2)^2$$

$$= (p^0)^2 - (p^1)^2 - (p^2)^2 - (p^3)^2 = \eta^{\alpha\beta} p_\alpha p_\beta = p^2. \tag{7.22}$$

As a result, we conclude that the coordinate transformation between Minkowski coordinates and the light-front coordinates given in (7.7) and (7.8) are length preserving linear transformations. In fact, for any two arbitrary four vectors, A^μ, B^μ, it is now straightforward to show that

$$\bar{A} \cdot \bar{B} = \bar{A}_\mu \bar{B}^\mu = \Lambda_\mu{}^\alpha A_\alpha \Lambda^\mu{}_\beta B^\beta = \Lambda_\mu{}^\alpha \Lambda^\mu{}_\beta A_\alpha B^\beta$$

$$= \tilde{\Lambda}^\alpha{}_\mu \Lambda^\mu{}_\beta A_\alpha B^\beta = \delta^\alpha{}_\beta A_\alpha B^\beta = A_\alpha B^\alpha = A \cdot B, \tag{7.23}$$

where we have used (7.9) and (7.18). However, we note that both (7.21) and (7.22) show that, in spite of the length being the same, the light-front coordinates introduce a one parameter family of uncertainty in the components themselves, namely,

$$\bar{x}^0 \to a\bar{x}^0, \quad \bar{x}^3 \to \frac{1}{a}\bar{x}^3, \quad \text{and,} \quad \bar{p}_0 \to \frac{1}{a}\bar{p}_0, \quad \bar{p}_3 \to a\bar{p}_3, \tag{7.24}$$

solve (7.21) and (7.22) respectively for an arbitrary constant parameter a. (The inverse scaling of the momentum variables is to preserve commutation relations.)

Although the light-front field theories have been studied widely at zero temperature, the thermal properties of such theories had not drawn as much attention as those for theories quantized on equal time surfaces. In the next section, we will describe how statistical mechanics can be systematically developed for field theories quantized on the light-front.

7.1.1 Generalization to finite temperature.

Let us recall from the earlier discussions, in particular (7.5)-(7.6) as well as (7.20), that the canonically conjugate coordinate and momentum operators satisfy the commutation relation (on the light-front)

$$[\bar{x}^{\mu}, \bar{p}_{\nu}]| = i\hbar \delta^{\mu}{}_{\nu} \mathbb{1}, \qquad (7.25)$$

where $\hbar$ denotes the Planck's constant and the restriction on the commutator implies that it is evaluated on the light-front. As a result, we conclude that the conjugate momentum to the light-front time coordinate $\bar{x}^0 = x^+ = x^0 + x^3$ is the momentum variable $\bar{p}_0 = \frac{1}{2}p^- = \frac{1}{2}(p^0 - p^3)$ (see (7.5) and (7.20)). This shows that the Hamiltonian, in a theory conventionally quantized on the light-front, can be identified with $\bar{p}_0$, namely,

$$\overline{H} = \bar{p}_0 = \frac{1}{2}p^- = \frac{1}{2}(p^0 - p^3). \qquad (7.26)$$

Therefore, from our discussions in the earlier chapters of the book so far, it will seem logical to define the thermal ensemble average of an arbitrary operator (observable) $\mathcal{O}$ in the light-front quantization as (see, for example, (1.11))

$$\langle \mathcal{O} \rangle_{\beta} = \mathrm{Tr}\, \rho(\beta) \mathcal{O} = \frac{1}{Z(\beta)} \mathrm{Tr}\, e^{-\beta \overline{H}} \mathcal{O} = \frac{1}{Z(\beta)} \mathrm{Tr}\, e^{-\beta \bar{p}_0} \mathcal{O},$$

$$Z(\beta) = \mathrm{Tr}\, \rho(\beta) = \mathrm{Tr}\, e^{-\beta \overline{H}} = \mathrm{Tr}\, e^{-\beta \bar{p}_0} = \mathrm{Tr}\, e^{-\frac{1}{2}\beta p^-}, \qquad (7.27)$$

where we have used (7.26) and $\rho(\beta)$ denotes the density matrix. We note here that $Z(\beta)$ denotes the partition function, as before, with $\beta = \frac{1}{kT}$ where k stands for the Boltzmann constant. As we will show below, this generalization of the density matrix (partition function), however, leads to serious problems.

Real time formalism

Let us start with the real time formalism of the closed time path discussed in chapter **2**. First, we note from (7.22) that

$$\bar{p}^2 = \bar{p}_\mu \bar{\eta}^{\mu\nu} \bar{p}_\nu = 4\bar{p}_0 \bar{p}_3 - (\bar{p}_1)^2 - (\bar{p}_2)^2 = 4\bar{p}_0 \bar{p}_3 - \mathbf{p}^2, \qquad (7.28)$$

where we have denoted the square of the two dimensional momentum vector by

$$\mathbf{p}^2 = \bar{p}_i^2 = (\bar{p}_1)^2 + (\bar{p}_2)^2 = (p_1)^2 + (p_2)^2 = p_i^2 = \mathbf{p}^2, \ i = 1, 2. \tag{7.29}$$

Therefore, the Einstein relation (see (7.28)), in this case, leads to

$$\bar{p}^2 = 4\bar{p}_0 \bar{p}_3 - \bar{p}^2 = m^2,$$

$$\text{or,} \quad \bar{p}_0 = \frac{\mathbf{p}^2 + m^2}{4\bar{p}_3} = \frac{\omega_p^2}{2p^+}, \qquad (7.30)$$

where we have defined

$$\omega_p^2 = \mathbf{p}^2 + m^2 = \omega_{\bar{p}}^2, \tag{7.31}$$

with $\mathbf{p}^2 = \bar{\mathbf{p}}^2$ defined in (7.29). We note from (7.30) that the Einstein relation becomes linear in the energy variable in the light-front theory and this is a distinctive difference from equal-time quantization.

With these preliminaries, let us consider the self-interacting ϕ^4 theory discussed in (2.81)-(2.82) with the Lagrangian density

$$\mathcal{L}(\phi) = \frac{1}{2} \bar{\eta}^{\mu\nu} \bar{\partial}_\mu \phi \bar{\partial}_\nu \phi - \frac{m^2}{2} \phi^2 - \frac{\lambda}{4!} \phi^4, \qquad (7.32)$$

with $\bar{\partial}_\mu = \frac{\partial}{\partial \bar{x}^\mu}$. The propagator, on the ++ branch, is already given in (2.50) for a theory quantized on the equal-time surface and transforms to the light-front theory as (remember $\bar{\mathbf{p}} = \mathbf{p}$)

$$i\bar{G}_{++}(\bar{p}) = \frac{i}{4\bar{p}_0 \bar{p}_3 - \bar{\mathbf{p}}^2 - m^2 + i\epsilon} + 2\pi n_B(|\bar{p}_0|)\delta(4\bar{p}_0 \bar{p}_3 - \bar{\mathbf{p}}^2 - m^2),$$

$$= \frac{i}{4\bar{p}_0 \bar{p}_3 - \omega_p^2 + i\epsilon} + 2\pi n_B(|\bar{p}_0|)\delta(4\bar{p}_0 \bar{p}_3 - \omega_p^2), \qquad (7.33)$$

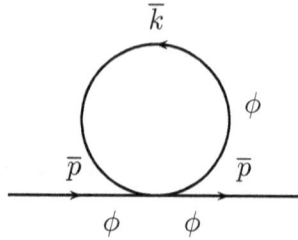

Figure 7.1: One loop correction to the self-energy in the ϕ^4 theory.

where $\overline{\mathbf{p}}^2$ and ω_p are defined in (7.29) and (7.31) respectively and we recall from (7.26) that $\overline{p}_0$ is the energy variable in light-front theories. Here $n_B(|\overline{p}_0|)$ denotes the boson distribution function with (light-front) energy $|\overline{p}_0|$. The interaction vertices do not change in the light-front theories and, therefore, the correction to the scalar self-energy at one loop (on the $++$ branch), shown in Fig. 7.1, can be calculated following (2.84)-(2.85) leading to the mass correction ($\frac{1}{2}$ represents the symmetry factor of the diagram)

$$i\Pi_{++}(\overline{p}) = -i\Delta m^2$$

$$= -\frac{i\lambda}{2} \int (\sqrt{-\overline{\eta}})^{-1} \frac{d\overline{k}_0 d\overline{k}_3 d^2\overline{k}}{(2\pi)^4} iG_{++}(\overline{k}), \qquad (7.34)$$

where we have written out explicitly the four dimensional invariant volume element in the light-front momentum space $(\sqrt{-\overline{\eta}})^{-1}d^4\overline{k}$. Keeping only the temperature dependent (second) term in the propagator in (7.33), we obtain the temperature dependent correction to the mass term to be

$$\Delta m_T^2 = \frac{\lambda}{2} \int 2\frac{d\overline{k}_0 d\overline{k}_3 d^2\overline{k}}{(2\pi)^3} n_B(|\overline{k}_0|)\delta(4\overline{k}_0\overline{k}_3 - \omega_k^2)$$

$$= \frac{\lambda}{(2\pi)^3} \int d^2\overline{k} \times 2\int_0^\infty \frac{d\overline{k}_3}{4\overline{k}_3} n_B\left(\frac{\omega_k^2}{4\overline{k}_3}\right), \qquad (7.35)$$

where we have used the value of $\sqrt{-\overline{\eta}}$ from (7.14) (see also (7.15)).

We note that the integral over $\overline{k}_3$ is divergent at the ultraviolet limit, $\overline{k}_3 \to \infty$, since the bosonic distribution function does not provide any damping there (at the lower limit, $\overline{k}_3 \to 0$, there is exponential damping). This is, in fact, quite surprising since we do not expect divergences to arise at finite temperature. Thermal interactions lead to on-shell contributions coming from possible additional channels of physical reactions and this is why it is believed that zero temperature counterterms are sufficient to renormalize the theory (without any need for temperature dependent counterterms). The divergence in (7.35), therefore, indicates a marked difference from the case in the equal-time quantization and we handle it by simply regularizing the factor $\frac{1}{\overline{k}_3}$ as

$$\Delta m_T^2 = \lim_{\epsilon \to 0} \frac{\lambda}{4\pi^3} \int d^2\overline{k} \int_0^\infty \frac{d\overline{k}_3}{4(\overline{k}_3)^{1-\epsilon}}\, n_B\left(\frac{\omega_k^2}{4\overline{k}_3}\right). \tag{7.36}$$

Let us further define a new variable

$$x = \frac{\omega_k^2}{4\overline{k}_3}, \quad dx = -\frac{\omega_k^2}{4\overline{k}_3^2}\, d\overline{k}_3, \quad \frac{d\overline{k}_3}{4\overline{k}_3} = -\frac{dx}{4x}, \tag{7.37}$$

so that we can write the mass relation in (7.35) as

$$\begin{aligned}
\Delta m_T^2 &= \lim_{\epsilon \to 0} \frac{\lambda}{16\pi^3} \int d^2\overline{k} \left(\frac{4}{\omega_k^2}\right)^\epsilon \int_0^\infty dx\, x^{\epsilon-1}\, n_B(x) \\
&= \lim_{\epsilon \to 0} \frac{\lambda}{16\pi^3} \int d^2\overline{k} \left(\frac{4}{\omega_k^2}\right)^\epsilon \int_0^\infty dx\, \frac{x^{\epsilon-1}}{e^{\beta x} - 1} \\
&= \lim_{\epsilon \to 0} \frac{\lambda}{16\pi^3} \int d^2\overline{k} \left(\frac{4}{\beta\omega_k^2}\right)^\epsilon \int_0^\infty dx\, \frac{x^{\epsilon-1}}{e^x - 1},
\end{aligned} \tag{7.38}$$

where we have scaled $x \to \frac{x}{\beta}$ in the last step. We note that the negative sign in (7.37) is absorbed into flipping the limits of the integral in (7.38) and, furthermore, the divergence at $\overline{k}_3 \to \infty$ now manifests at $x \to 0$. The x integral can be exactly evaluated using the standard tables[1] and leads to

$$\Delta m_T^2 = \lim_{\epsilon \to 0} \frac{\lambda}{16\pi^3} \int d^2\overline{k} \left(\frac{4}{\beta\omega_k^2}\right)^\epsilon \Gamma(\epsilon)\zeta(\epsilon),$$

which can be expanded in powers of ϵ to give

$$= \lim_{\epsilon \to 0} \frac{\lambda}{32\pi^3} \int d^2\overline{k} \left(-\frac{1}{\epsilon} + \mathbf{C} - \ln\frac{8\pi}{\beta\omega_k^2} + O(\epsilon)\right), \tag{7.39}$$

---

[1] See, for example, formula 3.411.1 in Gradshteyn and Ryzhik.

where $\mathbf{C}$ denotes the Euler's constant and we have used the relations

$$\Gamma(\epsilon) \approx \frac{1}{\epsilon} - \mathbf{C} + O(\epsilon), \quad \zeta(\epsilon) \approx -\frac{1}{2} - \frac{\epsilon}{2}\ln 2\pi + O(\epsilon). \quad (7.40)$$

This brings out another interesting puzzle, namely, even though this is supposed to give only the finite temperature correction to mass, it has divergent terms which are independent of temperature. Therefore, it would seem that we would need (zero temperature) counterterms beyond what is already needed to renormalize the theory at zero temperature. Furthermore, such divergences will arise at every loop. Besides this, the temperature dependent term in (7.39) (for example, $\ln\beta$) is also divergent unless we are in $1+1$ dimensions (where there are no transverse directions, $\mathbf{k}$, to integrate over). All of this is very much against the general properties of finite temperature field theories and, therefore, we need to analyze what may have gone wrong in the naive generalization of light-front field theories to finite temperature and how it can be fixed. In retrospect, the sign of all these problems is already present in the form of the propagator in (7.33) (which we have completely missed), namely, the finite temperature part of the propagator (the second term) does not really vanish in the zero temperature limit ($\beta \to \infty$) unlike in equal-time theories. Here, this nonvanishing arises because of the linear nature of the Einstein relation (7.30), namely, it allows for $|\overline{p}_0| \to 0, \overline{p}_3 \to \infty$ such that that the delta function constraint in (7.33) can be satisfied with a nonvanishing boson distribution function. In fact, as we have noted earlier, the Einstein relation in (7.30) does allow for a one parameter family of solutions. Namely, $\overline{p}_0 \to \frac{1}{\alpha}\overline{p}_0$, $\overline{p}_3 \to \alpha\overline{p}_3$ also solves the relation (7.30) for an arbitrary constant parameter α (see also (7.24)). Before analyzing the source for all these problems, let us carry out the same calculation in the imaginary time formalism to make sure that this is not an issue only with the real time (closed time) formalism.

Imaginary time formalism

In the imaginary time formalism, described in detail in chapter **1**, the theory is rotated to the Euclidean space so that, in the light-front coordinates, $\overline{p}_0 \to i\overline{p}_4$ (see the discussion just below (1.78)) and the scalar propagator, in the Euclidean space, is obtained from (7.30) or

(7.33) to have the zero temperature form (see (1.80) or (1.82))

$$\mathcal{G}(\bar{p}_4, \bar{p}_i, \bar{p}_3) = \frac{1}{4i\bar{p}_4\bar{p}_3 - \mathbf{p}^2 - m^2}, \quad i = 1, 2,$$

$$= \frac{1}{4i\bar{p}_4\bar{p}_3 - \omega_p^2}, \tag{7.41}$$

which can be compared with (1.80) or (1.82). The first thing to note here is that the form of the zero temperature Euclidean propagator (7.41), in the light-front coordinates, is invariant under the scaling $\bar{p}_4 \to \frac{1}{\alpha}\bar{p}_4$, $\bar{p}_3 \to \alpha\bar{p}_3$, as was also the case in the real time description in section **7.1.1**. At finite temperature, the energy becomes discrete in the imaginary time formalism and the propagator is obtained by letting $\bar{p}_4 \to \frac{2n\pi}{\beta}$ (see (1.73)) with $\beta = \frac{1}{kT}$ and the integral over energy replaced by a discrete sum as given in (1.75) so that we can write

$$\mathcal{G}_\beta(n, \omega_p, \bar{p}_3) = \frac{1}{\frac{8in\pi\bar{p}_3}{\beta} - \omega_p^2}. \tag{7.42}$$

With this finite temperature scalar propagator, in the light-front quantization, the one loop self-energy diagram in the imaginary time formalism gives (see (1.75) and recall that the quartic interaction vertex in Euclidean space is given by $-\lambda$)

$$-\Pi(\bar{p}) = -\Delta m^2 = -\frac{\lambda}{2}\int(\sqrt{-\bar{\eta}})^{-1}\frac{d^2\bar{k}d\bar{k}_3}{(2\pi)^3}\frac{1}{\beta}\sum_{n=-\infty}^{\infty}\mathcal{G}_\beta(n, \omega_k, \bar{k}_3),$$

which leads to (remember that $(\sqrt{-\bar{\eta}})^{-1} = 2$)

$$\Delta m^2 = \frac{\lambda}{(2\pi)^3}\int d^2\bar{k}d\bar{k}_3 \frac{1}{\beta}\sum_{n=-\infty}^{\infty}\frac{1}{\frac{8in\pi\bar{k}_3}{\beta} - \omega_k^2}$$

$$= \frac{\lambda}{(2\pi)^3}\int d^2\bar{k}d\bar{k}_3 \frac{1}{8i\pi\bar{k}_3}\sum_{n=-\infty}^{\infty}\frac{1}{n + i\frac{\beta\omega_k^2}{8\pi\bar{k}_3}}$$

$$= \frac{\lambda}{(2\pi)^3}\int d^2\bar{k} \times 2\int_0^\infty \frac{d\bar{k}_3}{\bar{k}_3}\frac{1}{8i\pi}(-i\pi)\coth\frac{\beta\omega_k^2}{8\bar{k}_3},$$

where we have used (1.104) as well as the symmetry of the $\bar{k}_3$ integral leading to

$$= -\frac{\lambda}{(2\pi)^3} \int d^2\bar{k} \times 2 \int_0^\infty \frac{d\bar{k}_3}{8\bar{k}_3} \left(1 + 2n_B\left(\frac{\beta\omega_k^2}{4\bar{k}_3}\right)\right), \quad (7.43)$$

where we have used (1.106). Subtracting out the zero temperature part (1) in (7.43), we obtain

$$\Delta m_T^2 = -\frac{\lambda}{4\pi^3} \int d^2\bar{k} \int_0^\infty \frac{d\bar{k}_3}{4\bar{k}_3} n_B\left(\frac{\beta\omega_k^2}{4\bar{k}_3}\right), \quad (7.44)$$

which is exactly the same result obtained in the real time formalism in (7.35). This shows that the divergence encountered in the thermal mass arises both in the real time as well as the imaginary time formalism. Therefore, we have to understand the origin of this trouble and construct a unique and meaningful passage of light-front field theories to finite temperature.

7.2 General light-front field theories

Let us recall the basic idea behind statistical mechanics in usual equal-time quantization. We assume that a very large heat bath is at rest with which the quantum particles are interacting, exchanging heat and particles etc. The statistical distribution of particles in such a set up (in the rest frame of the heat bath) is given by a generalized Boltzmann distribution $e^{-\beta H} = e^{-\beta p_0}$ which leads to a statistical average for any observable $\mathcal{O}$ expressed as

$$\langle \mathcal{O} \rangle_\beta = \text{Tr}\,\rho(\beta)\mathcal{O} = \frac{1}{Z(\beta)} \text{Tr}\,e^{-\beta H}\mathcal{O} = \frac{1}{Z(\beta)} \text{Tr}\,e^{-\beta p_0}\,\mathcal{O}. \quad (7.45)$$

Here $\rho(\beta)$ stands for the density matrix while $\beta = \frac{1}{T}$ denotes the inverse temperature in units of the Boltzmann constant k and $Z(\beta)$ denotes the partition function for the system. Basically, it is this idea that we have carried over to light-front quantization in (7.27). There are various ways to see that such a generalization is bound to fail in the conventional light-front theories. For example, a rest frame of the heat bath is not a Lorentz covariant concept and is certainly inconsistent with the concept of a light-front frame (infinite momentum frame). Let us explain here, in some detail, why the rest frame (of the heat bath) is not appropriate in the light-front quantization. If we naively define a rest frame in the light-front coordinates, the four velocity vector will have the form

$$\overline{u}^{\mu}_{\text{rest}} = (1, 0, 0, 0), \tag{7.46}$$

leading to

$$\overline{u}^{\mu}_{\text{rest}} \overline{u}_{\text{rest } \mu} = \overline{u}^{\mu}_{\text{rest}} \overline{\eta}_{\mu\nu} \overline{u}^{\nu}_{\text{rest}},$$

which, upon using the covariant metric in (7.12) yields

$$\overline{u}^{\mu}_{\text{rest}} \overline{u}_{\text{rest } \mu} = 0. \tag{7.47}$$

Therefore, the "naive" four velocity for the rest frame is not normalizable (see, for example, (7.48) in the following) and is not a meaningful concept in the light-front quantization. There are other reasons also for (7.27) to fail in the conventional light-front systems as we will discuss shortly (for example, see (7.54)-(7.59)). So, we will have to reanalyze the light-front calculations keeping these ideas in mind. Of course, we have briefly described covariant statistical mechanics in section **1.8**, but we will describe it here with a bit more detail.

Let us assume that the heat bath is moving along the x^3 axis in the Minkowski space with a speed v (alternatively, we can look at the system from a Lorentz frame boosted along the x^3 axis with a speed v). Let us assume that the coordinates of the heat bath (HB), in the equal-time description, are given by $(x^0_{\text{HB}}, x^1_{\text{HB}}, x^2_{\text{HB}}, x^3_{\text{HB}})$. The normalized four velocity of the heat bath is easily obtained from its worldline to be (remember that $x^0 = ct$ and we are choosing $c = 1$)

$$u^{\alpha} = \frac{\mathrm{d} x^{\alpha}_{\text{HB}}}{\mathrm{d}\tau} = \left(\frac{\mathrm{d} t_{\text{HB}}}{\mathrm{d}\tau}, 0, 0, \frac{\mathrm{d} t_{\text{HB}}}{\mathrm{d}\tau} \frac{\mathrm{d} x^3_{\text{HB}}}{\mathrm{d} t_{\text{HB}}} \right) = \frac{\mathrm{d} t_{\text{HB}}}{\mathrm{d}\tau} \left(1, 0, 0, \frac{\mathrm{d} x^3_{\text{HB}}}{\mathrm{d} t_{\text{HB}}} \right)$$

$$= \gamma(1, 0, 0, v), \tag{7.48}$$

where we have identified, as usual, the (three) velocity, v, as well as the Lorentz factor, γ, associated with the heat bath as

$$v = \frac{\mathrm{d} x^3_{\text{HB}}}{\mathrm{d} t_{\text{HB}}}, \quad \gamma = \frac{1}{\sqrt{1 - v^2}}. \tag{7.49}$$

It follows now that

$$u^\alpha \eta_{\alpha\beta} u^\beta = \gamma \begin{pmatrix} 1 & 0 & 0 & v \end{pmatrix} \begin{pmatrix} 1 & 0 & 0 & 0 \\ 0 & -1 & 0 & 0 \\ 0 & 0 & -1 & 0 \\ 0 & 0 & 0 & -1 \end{pmatrix} \gamma \begin{pmatrix} 1 \\ 0 \\ 0 \\ v \end{pmatrix}$$

$$= \gamma^2 (1 - v^2) = 1. \tag{7.50}$$

Namely, the four velocity of the (moving) heat bath is normalized. (This is normalized even when $v = 0$, namely, a rest frame of the heat bath can be satisfactorily defined in the Minkowski space or equal-time quantization.) Using (7.7) and (7.13), we can now calculate the four velocity of the heat bath in the light-front coordinates as

$$\bar{u}^\mu = \Lambda^\mu{}_\alpha u^\alpha = \begin{pmatrix} 1 & 0 & 0 & 1 \\ 0 & 1 & 0 & 0 \\ 0 & 0 & 1 & 0 \\ 1 & 0 & 0 & -1 \end{pmatrix} \gamma \begin{pmatrix} 1 \\ 0 \\ 0 \\ v \end{pmatrix} = \gamma \begin{pmatrix} 1 + v \\ 0 \\ 0 \\ 1 - v \end{pmatrix},$$

leading to

$$\bar{u}^\mu \bar{u}_\mu = \bar{u}^\mu \bar{\eta}_{\mu\nu} \bar{u}^\nu = \gamma^2 (1 - v^2) = 1. \tag{7.51}$$

Namely, the four velocity continues to be normalized in the light-front coordinates as well which is expected since scalars do not change under a coordinate redefinition. Note that this vector is normalized even when $v = 0$, but the four vector, in this case, has the form

$$\bar{u}^\mu_{\text{``rest''}} = \begin{pmatrix} 1 \\ 0 \\ 0 \\ 1 \end{pmatrix}, \tag{7.52}$$

which can be compared with (7.46).

From (7.20), (7.48) and (7.49), we note that

$$\bar{u} \cdot \bar{p} = u \cdot p = u^\alpha p_\alpha = (u^0 p_0 + u^3 p_3)$$

$$= \gamma(p_0 + v p_3) = \gamma(p^0 - v p^3). \tag{7.53}$$

Therefore, following the discussion in section **1.8** (see, in particular, (1.153)), we note from (7.50) that we can write the density matrix for such a system as

$$\rho(\beta) = \frac{1}{Z(\beta)}\, e^{-\beta u \cdot p} = \frac{1}{Z(\beta)}\, e^{-\beta \bar{u} \cdot \bar{p}}$$

$$= \frac{1}{Z(\beta)}\, e^{-\beta\gamma(p^0 - vp^3)}, \tag{7.54}$$

where we have used (7.23) which says that the scalar product of two four vectors remains invariant under a transformation to the light-front coordinates. There are several things to note from (7.54). First, it is clear from the first line of (7.54) that, in order for the density matrix to define a meaningful (Boltzmann) distribution, we must have

$$\bar{u} \cdot \bar{p} = u \cdot p > 0. \tag{7.55}$$

Second, comparing the final result in (7.54) with (7.45), it is clear that it has the form of a statistical (thermal) density matrix with the governing Hamiltonian and a rescaled temperature given by

$$\rho(\beta) = \frac{1}{Z(\beta)}\, e^{-\bar{\beta}\bar{p}_0}, \quad \bar{p}_0 = p^0 - vp^3, \tag{7.56}$$

with

$$\bar{\beta} = \frac{1}{k\bar{T}} = \beta\gamma = \frac{1}{kT\sqrt{1 - v^2}}. \tag{7.57}$$

In particular, we note from (7.56)-(7.57) that, for $v = 0$ (the rest frame), we will have

$$\bar{p}_0 = p^0 \neq p^0 - p^3 = \overline{H}, \quad \text{and} \quad \bar{\beta} = \beta. \tag{7.58}$$

Furthermore, for nonzero v, (7.56) would coincide with the density matrix for the light-front quantization used in (7.27) (namely, $\bar{p}_0 = \overline{H} = p^0 - p^3$) only if $v = 1$, namely, if the heat bath is moving with the speed of light (remember that we are using $c = 1$). However, in that case, it also follows that the rescaled temperature, for the system, given in (7.57) would correspond to

$$\overline{T} = T\sqrt{1 - v^2} \to 0, \quad \bar{\beta} \to \infty, \quad \text{as} \quad v \to 1. \tag{7.59}$$

Namely, the conventional light-front quantization does not lead to a satisfactory thermal (statistical) description. Therefore, we need to look for general light-front coordinates which will not only allow theories to be quantized on the light-front, but also will permit a satisfactory statistical (thermal) description.

7.2.1 General light-front coordinates.

Let us define a new set of co-ordinates $\bar{x}^\mu, \mu = 0, 1, 2, 3$ which is related to the Minkowski coordinates $x^\alpha, \alpha = 0, 1, 2, 3$ as

$$\bar{x}^0 = x^0 + x^3, \quad \bar{x}^3 = Ax^0 + Bx^3, \quad \bar{x}^i = x^i, \quad i = 1, 2, \quad (7.60)$$

where A, B are real constants with the only restriction that $A \neq B$ (or $A - B \neq 0$). This restriction arises because otherwise, we will have $\bar{x}^3 = A\bar{x}^0$ which would correspond to a singular, non-invertible coordinate redefinition. We will assume this restriction throughout our discussion in this chapter. These coordinates can be compared with (7.5). We note that the time coordinate $\bar{x}^0$, given in (7.60), coincides with the conventional light-front time coordinate (see (7.5)) and is defined keeping in mind that we will like to quantize the theory on the light-front, namely, we can still define the surface of quantization to be the light-front surface as in (7.3) with $n^{(L)} \cdot x = n^{(L)} \cdot x'$. In fact, for $A = 1 = -B$, $\bar{x}^3$ also coincides with the corresponding coordinate in the conventional light-front coordinates given in (7.5), but, in general, (namely, for arbitrary A, B) it is a different coordinate system which we call the general light-front coordinates. The two coordinate systems are related to each other through the linear transformations

$$\bar{x}^\mu = L^\mu_{\ \alpha} x^\alpha, \quad x^\alpha = \tilde{L}^\alpha_{\ \mu} \bar{x}^\mu, \quad (7.61)$$

where

$$L^\mu_{\ \alpha} = \begin{pmatrix} 1 & 0 & 0 & 1 \\ 0 & 1 & 0 & 0 \\ 0 & 0 & 1 & 0 \\ A & 0 & 0 & B \end{pmatrix}, \quad \tilde{L}^\alpha_{\ \mu} = \begin{pmatrix} \frac{B}{B-A} & 0 & 0 & \frac{1}{A-B} \\ 0 & 1 & 0 & 0 \\ 0 & 0 & 1 & 0 \\ \frac{A}{A-B} & 0 & 0 & \frac{1}{B-A} \end{pmatrix}, \quad (7.62)$$

so that

$$L^\mu_{\ \alpha} \tilde{L}^\alpha_{\ \nu} = \delta^\mu_{\ \nu}, \quad \tilde{L}^\alpha_{\ \mu} L^\mu_{\ \beta} = \delta^\alpha_{\ \beta}. \quad (7.63)$$

For $A = 1, B = -1$, we note that these matrices reduce respectively to $\Lambda^\mu_{\ \alpha}$ and $\Lambda^\alpha_{\ \mu}$ given in (7.7) and (7.10). On the other hand, for $A = B$, the singular nature of the transformation is reflected in the fact that

$$\det L^\mu_{\ \alpha} = 0, \tag{7.64}$$

so that the transformation matrix is non-invertible. We note from (7.61)-(7.62) that the four velocity of the heat bath, in these general coordinate system, has the form

$$\bar{u}^\mu = L^\mu_{\ \alpha} u^\alpha = \gamma \begin{pmatrix} 1 + v \\ 0 \\ 0 \\ A + vB \end{pmatrix}. \tag{7.65}$$

Similarly, the matrix describing the transformation of covariant vectors is given by

$$L_\mu^{\ \alpha} = \begin{pmatrix} -\frac{B}{A-B} & 0 & 0 & \frac{A}{A-B} \\ 0 & 1 & 0 & 0 \\ 0 & 0 & 1 & 0 \\ \frac{1}{A-B} & 0 & 0 & -\frac{1}{A-B} \end{pmatrix}, \tag{7.66}$$

so that, for any arbitrary contravariant vector V_α, we can write

$$\bar{V}_\mu = L_\mu^{\ \alpha} V_\alpha. \tag{7.67}$$

It leads, in particular, to the covariant four velocity and momentum vectors to correspond to

$$\bar{u}_\mu = L_\mu^{\ \alpha} u_\alpha = \frac{1}{A-B} \begin{pmatrix} -Bu_0 + Au_3 \\ u_1 \\ u_2 \\ u_0 - u_3 \end{pmatrix}$$

$$= \frac{\gamma}{A-B} \begin{pmatrix} -(B + vA) \\ 0 \\ 0 \\ 1 + v \end{pmatrix}, \tag{7.68}$$

and

$$\bar{p}_\mu = L_\mu{}^\alpha p_\alpha = \frac{1}{A-B} \begin{pmatrix} -Bp_0 + Ap_3 \\ p_1 \\ p_2 \\ p_0 - p_3 \end{pmatrix}$$

$$= \frac{1}{A-B} \begin{pmatrix} -(Bp^0 + Ap^3) \\ -p^1 \\ -p^2 \\ p^0 + p^3 \end{pmatrix}. \tag{7.69}$$

The transformation matrix in (7.66) and the covariant momentum in (7.69) reduce respectively to (7.18) and (7.20) for $A = 1 = -B$. Furthermore, we note from (7.65) and (7.68) that

$$\bar{u}^\mu \bar{u}_\mu = \frac{1}{A-B} \times (A-B)\gamma^2(1-v^2) = 1, \tag{7.70}$$

namely, the four velocity in the general light-front coordinates is normalized (even when $v = 0$, namely, in the rest frame).

The metric, in the transformed space, can now be constructed as in (7.11) and (7.12) (we call it $\bar{g}^{\mu\nu}$ and $\bar{g}_{\mu\nu}$ to distinguish from the earlier metric $\bar{\eta}^{\mu\nu}$ and $\bar{\eta}_{\mu\nu}$ in the conventional light-front coordinates) and take the forms

$$\bar{g}^{\mu\nu} = L^\mu{}_\alpha \eta^{\alpha\beta} L^\nu{}_\beta = \left(L\eta L^T\right)^{\mu\nu}$$

$$= \begin{pmatrix} 0 & 0 & 0 & (A-B) \\ 0 & -1 & 0 & 0 \\ 0 & 0 & -1 & 0 \\ (A-B) & 0 & 0 & (A^2-B^2) \end{pmatrix} = \bar{g}^{\nu\mu}, \tag{7.71}$$

and, similarly,

$$\bar{g}_{\mu\nu} = \tilde{L}^\alpha{}_\mu \eta_{\alpha\beta} \tilde{L}^\beta{}_\nu = \left(\tilde{L}^T \eta \tilde{L}\right)_{\mu\nu}$$

$$= \begin{pmatrix} -\frac{A+B}{A-B} & 0 & 0 & \frac{1}{A-B} \\ 0 & -1 & 0 & 0 \\ 0 & 0 & -1 & 0 \\ \frac{1}{A-B} & 0 & 0 & 0 \end{pmatrix} = \bar{g}_{\nu\mu}. \tag{7.72}$$

For $\overline{x}^0$ to represent the time coordinate, we must have

$$\overline{g}_{00} = \frac{A+B}{B-A} \geq 0, \quad \text{which leads to} \quad |B| \geq |A|, \tag{7.73}$$

and obviously holds for the conventional light-front coordinates for which $A = 1, B = -1$ where $\overline{g}_{00} = 0$. However, for a consistent thermal description we must have

$$\overline{g}_{00} > 0, \tag{7.74}$$

which fails for the conventional light-front coordinates and the statistical description fails, as we have already discussed in (7.46)-(7.47) as well as in (7.57)-(7.59). On the other hand, for $A = 0$ and $B = 1$, the coordinate system known as the oblique light-front coordinates and used earlier in connection with studies in light-front field theories at finite temperature,[2] it follows from (7.72) that $\overline{g}_{00} = 1$ which is consistent with (7.74). In this case, a statistical description does exist and has been studied in detail for various theories. We also note that, for $A = 1 = -B$, the transformed metrics in (7.71) and (7.72) reduce respectively to (7.11) and (7.12) as they should. Note also that the scalar product of two arbitrary vectors $\overline{V}^\mu$ and $\overline{W}^\nu$ is obtained from (7.71)-(7.72) to be

$$\overline{V} \cdot \overline{W} = \overline{g}^{\mu\nu} \overline{V}_\mu \overline{W}_\nu = \overline{g}_{\mu\nu} \overline{V}^\mu \overline{W}^\nu$$

$$= (A-B)(\overline{V}_0 \overline{W}_3 + \overline{V}_3 \overline{W}_0) - \overline{V}_1 \overline{W}_1 - \overline{V}_2 \overline{W}_2$$

$$- (B^2 - A^2)\overline{V}_3 \overline{W}_3. \tag{7.75}$$

Furthermore, we note here (for later use) that the metrics give us the invariant volumes, namely,

$$\det(-\overline{g}_{\mu\nu}) = (-\overline{g}) = \frac{1}{(A-B)^2} > 0,$$

$$\sqrt{-\overline{g}} = \frac{1}{|A-B|}, \quad \sqrt{-\overline{g}}\,(A-B) = \text{sgn}\,(A-B), \tag{7.76}$$

so that we can write

$$\int \sqrt{-\overline{g}}\,\mathrm{d}^4\overline{x} = \int \frac{1}{|A-B|}\,\mathrm{d}^4\overline{x} = \int \mathrm{d}^4 x,$$

$$\int \left(-\sqrt{\overline{g}}\right)^{-1}\mathrm{d}^4\overline{p} = \int |A-B|\,\mathrm{d}^4\overline{p} = \int \mathrm{d}^4 p. \tag{7.77}$$

---

[2] See, V. S. Alves, A. Das and S. Perez, Physical Review **D66**, 125008 (2002).

For $A = 1 = -B$, these are precisely the relations we have used in (7.35).

Let us also note that, using (7.66), the (covariant) momenta can be obtained in the transformed system from

$$\bar{p}_\mu = L_\mu{}^\alpha p_\alpha,$$

which, upon using (7.66)-(7.67), yield

$$\bar{p}_0 = \frac{1}{A - B}(-Bp_0 + Ap_3), \quad \bar{p}_i = p_i, \quad i = 1, 2,$$

$$\bar{p}_3 = \frac{1}{A - B}(p_0 - p_3). \tag{7.78}$$

As a result, we note that even in the generalized light-front coordinates we continue to have (which is expected because it has to equal $u \cdot p$)

$$\bar{u} \cdot \bar{p} = u \cdot p = \gamma(p^0 - vp^3), \tag{7.79}$$

as in (7.50). But, more importantly, we note that a rest frame given by (7.46) is meaningful in the generalized light-front coordinate as long as $A + B \neq 0$, namely,

$$\bar{u}^\mu_{\text{rest}} \bar{g}_{\mu\nu} \bar{u}^\nu_{\text{rest}} = 1, \quad \bar{u}^\mu_{\text{rest}} = \frac{1}{\sqrt{\bar{g}_{00}}} \begin{pmatrix} 1 \\ 0 \\ 0 \\ 0 \end{pmatrix} = \sqrt{\frac{B - A}{A + B}} \begin{pmatrix} 1 \\ 0 \\ 0 \\ 0 \end{pmatrix}, \tag{7.80}$$

and we note that a meaningful normalized rest frame exists only for $A + B \neq 0$. This shows that the conventional light-front coordinates are quite singular from the point of view of a statistical description and so a general light-front description is preferable which allows for a rest frame description of the heat bath.

There is one other ingredient that needs to be discussed. Namely, even though the temperature T or $\beta = \frac{1}{kT}$ appears to be a scalar under a coordinate redefinition and has been treated as such, it is not. We have already seen in chapter **1** that β can be associated with the Euclidean time interval (see section **1.6**, in particular Fig. 1.1). In fact, with a couple of simple assumptions, it has been shown by Tolman and Ehrenfest (and, consequently, it is known as the

Tolman-Ehrenfest effect) that, in the presence of a static time-like Killing vector, the temperature scales under a coordinate redefinition with the square root of $\overline{g}_{00}(x)$ leading to a local temperature at every space-time coordinate. In a flat space-time, on the other hand, this simplifies to a global scaling of the temperature for the general light-front coordinates as

$$\overline{T} = \sqrt{\overline{g}_{00}}\, T = \sqrt{\frac{A+B}{B-A}}\, T, \quad \overline{\beta} = \frac{\beta}{\sqrt{\overline{g}_{00}}} = \sqrt{\frac{B-A}{A+B}}\, \beta, \quad (7.81)$$

where the unbarred quantities refer to Minkowski space quantities.[3] In our discussions in the following, therefore, we will choose the density matrix defined with a heat bath at rest (in the transformed coordinates) and the thermal average given by

$$\rho(\beta) = \frac{e^{-\overline{\beta}\,\overline{u}_{\text{rest}}\cdot\overline{p}}}{Z(\beta)} = \frac{e^{-\frac{\overline{\beta}}{\sqrt{\overline{g}_{00}}}\overline{P}_0}}{Z(\beta)} = \frac{e^{-\frac{\beta}{\overline{g}_{00}}\overline{P}_0}}{Z(\beta)}, \quad (7.82)$$

where we have used (7.80)-(7.81) and

$$Z(\beta) = \operatorname{Tr} e^{-\frac{\beta}{\overline{g}_{00}}\overline{P}_0},$$
$$\langle \mathcal{O} \rangle_\beta = \operatorname{Tr} \rho(\beta)\mathcal{O} = \frac{1}{Z(\beta)}\operatorname{Tr} e^{-\frac{\beta}{\overline{g}_{00}}\overline{P}_0}\, \mathcal{O}, \quad (7.83)$$

and it is worth recalling that $\overline{g}_{00} > 0$ which is essential for a statistical description to hold. We note here from (7.80)-(7.83) that a statistical description is not possible if $A + B = 0$ which corresponds to the conventional light-front coordinates. On the other hand, we note that, for $A = 0, B = 1$, we have $\overline{\beta} = \beta$ or $\overline{T} = T$ (and $g_{00} = 1 > 0$) which is the case of the oblique light-front coordinates used earlier.[4]

The Einstein relation, in the generalized light-front coordinates, becomes

$$\overline{p}^2 = \overline{g}^{\mu\nu}\overline{p}_\mu\overline{p}_\nu = m^2,$$

---

[3]We point out here that, if the assumptions of Tolman-Ehrenfest are violated in a (quantum) theory, the temperature redefinition may be modified, H. A. Weldon, private communication.

[4]See, for example, V. S. Alves, A. Das and S. Perez, Physical Review **D66**, 125008 (2002).

which, upon using (7.75), yields

$$2(A - B)\bar{p}_0 \bar{p}_3 - \bar{p}_i^2 - (B^2 - A^2)\bar{p}_3^2 = m^2, \ i = 1, 2, \quad (7.84)$$

$$\text{or,} \quad \bar{p}_0 = \frac{\omega_{\bar{p}}^2 + (B^2 - A^2)\bar{p}_3^2}{2(A - B)\bar{p}_3}, \quad (7.85)$$

where, in order to avoid any confusion, we have continued to define (as before)

$$\omega_{\bar{p}}^2 = \bar{\mathbf{p}}^2 + m^2 = \bar{p}_i^2 + m^2, \quad i = 1, 2, \quad (7.86)$$

which reduce to conventional light-front results given in (7.30)-(7.31) for $A = 1 = -B$.

For completeness, we end this section by noting that, in a boosted frame along the z-axis, we have

$$x'^0 = \gamma(x^0 - vx^3) = \cosh\phi \, x^0 - \sinh\phi \, x^3,$$

$$x'^3 = \gamma(-vx^0 + x^3) = -\sinh\phi \, x^0 + \cosh\phi \, x^3,$$

where we have denoted

$$\gamma = \cosh\phi, \quad \gamma v = \sinh\phi,$$

and this leads to

$$\bar{x}'^0 = x'^0 + x'^3 = e^{-\phi} (x^0 + x^3) = e^{-\phi} \bar{x}^0,$$

$$\bar{x}'^3 = Ax'^0 + Bx'^3 = e^{\phi} \bar{x}^3 - (A + B)\sinh\phi \, \bar{x}^0. \quad (7.87)$$

This shows that the boost acts as a scale transformation for $\bar{x}^0$, but not for $\bar{x}^3$, in general, unless $A + B = 0$ which, of course, holds for the conventional light-front coordinates. We note that the quantization surface $\bar{x}^0 = 0$ remains invariant under such a transformation.

7.2.2 Calculation of the one loop self-energy for the scalar field.

As we have pointed out earlier, the general light-front coordinates do allow us to quantize a theory on the light-front surface $\bar{x}^0 = \bar{x}'^0$ and the field quantization of scalar, fermion and gauge theories have been discussed in detail in the literature.[5] Therefore, we will redo only

---

[5] A. Das and S. Perez, Physical Review **D70**, 065006 (2004).

the calculations of the one loop self-energy in the scalar ϕ^4 theory (see section **7.1.1**) here in these new coordinates. Let us consider the real time calculation first and we will comment on the imaginary time calculation briefly at the end. The Lagrangian density of the scalar ϕ^4 theory, in the general light-front frame, is given by (see (7.32) for the Lagrangian density in the conventional light-front quantization)

$$\mathcal{L}(\phi) = \frac{1}{2}\bar{g}^{\mu\nu}\bar{\partial}_\mu\phi\,\bar{\partial}_\nu\phi - \frac{m^2}{2}\phi^2 - \frac{\lambda}{4!}\phi^4, \tag{7.88}$$

where $\bar{\partial}_\mu = \frac{\partial}{\partial\bar{x}^\mu}$ and the transformed contravariant metric $\bar{g}^{\mu\nu}$ is given in (7.71).

The first thing we note from (7.79)-(7.86) that the thermal propagator for the scalar field on the $++$ branch in the closed time path, following from (7.88), can be written as (compare with (7.33) and note that the rest frame of the heat bath is given in (7.65) with $v = 0$ and it is normalized as shown in (7.70))

$$i\bar{G}_{++}(\bar{p}) = \frac{i}{\bar{p}^2 - m^2 + i\epsilon} + 2\pi n_B(|\bar{p}_0|)\delta(\bar{p}^2 - m^2)$$

$$= \frac{i}{2(A - B)\bar{p}_0\bar{p}_3 - \omega_{\bar{p}}^2 - (B^2 - A^2)\bar{p}_3^2 + i\epsilon}$$

$$+ 2\pi n_B(|\bar{p}_0|)\delta(2(A - B)\bar{p}_0\bar{p}_3 - \omega_{\bar{p}}^2 - (B^2 - A^2)\bar{p}_3^2), \tag{7.89}$$

where we have identified $\bar{p}^2 = \bar{g}^{\mu\nu}\bar{p}_\mu\bar{p}_\nu$ (see, for example, (7.84)) and $\omega_{\bar{p}}^2$ is defined, as in the case of conventional light-front, in (7.86). The bosonic distribution function is given by

$$n_B(|\bar{p}_0|) = \frac{1}{e^{\frac{\beta}{\bar{g}_{00}}|\bar{p}_0|} - 1}, \tag{7.90}$$

where β denotes the inverse Minkowski temperature in units of the Boltzmann constant. It is clear now from the Einstein relation (7.85)-(7.86) (imposed by the delta function in (7.89)) that

$$|\bar{p}_0| \to \infty, \quad \text{both as} \quad \bar{p}_3 \to 0, \quad \text{as well as when} \quad \bar{p}_3 \to \infty, \tag{7.91}$$

and the bosonic distribution function $n_B(|\bar{p}_0|) = n_B(|\frac{\omega_{\bar{p}}^2 + (B^2 - A^2)\bar{p}_3^2}{2(A - B)\bar{p}_3}|)$, unlike in (7.35), is well behaved (damped) for both large and small $\bar{p}_3$ as long as $A + B \neq 0$ (when $A + B = 0$, the $\bar{p}_3^2$ term is not present in the Einstein relation) and we do not expect any divergence coming from the finite temperature part of the propagator in contrast with the conventional light-front quantization. The improved behavior comes primarily from the form of the Einstein relation in (7.85)-(7.86) where we note that the numerator for the energy does depend on $\bar{p}_3^2$ in addition to $\omega_{\bar{p}}^2$ when $A + B \neq 0$ and this $\bar{p}_3^2$ term dominates when $\bar{p}_3 \to \infty$. As $\bar{p}_3 \to 0$, this term vanishes, but the denominator in (7.85) also vanishes so that the argument of the distribution function diverge leading to exponential damping.

With this observation, let us redo the calculation in (7.34)-(7.35) which leads to

$$\Delta m^2 = \frac{\lambda}{2} \int |A - B| \frac{d\bar{k}_0 d\bar{k}_3 d^2\bar{k}}{(2\pi)^4} \, i\bar{G}_{++}(\bar{k}), \tag{7.92}$$

where we have used (7.77). The temperature dependent mass correction follows from this to be

$$\Delta m_T^2 = \frac{\lambda |A - B|}{2(2\pi)^3} \int d\bar{k}_0 d\bar{k}_3 d^2\bar{k} \, n_B(|\bar{k}_0|)$$
$$\times \delta\big(2(A - B)\bar{k}_0\bar{k}_3 - \omega_{\bar{k}}^2 - (B^2 - A^2)\bar{k}_3^2\big)$$
$$= \frac{\lambda |A - B|}{2(2\pi)^3} \int d^2\bar{k} \times 2 \int_0^\infty \frac{d\bar{k}_3}{2|A - B|\bar{k}_3} \, n_B\Big(\frac{\omega_{\bar{k}}^2 + (B^2 - A^2)\bar{k}_3^2}{2|A - B|\bar{k}_3}\Big)$$
$$= \frac{\lambda}{2(2\pi)^3} \int d^2\bar{k} \int_0^\infty \frac{d\bar{k}_3}{\bar{k}_3} \, n_B\Big(\frac{\omega_{\bar{k}}^2}{2|A - B|\bar{k}_3} + \frac{|A + B|}{2}\bar{k}_3\Big). \tag{7.93}$$

Using the fact that the distribution function is given by $n_B(x) = \frac{1}{e^{\frac{\beta}{g_{00}}x} - 1}$ (see (7.90)), the integral over $\bar{k}_3$ can be done using the standard tables of integrals and the result yields[6]

$$\Delta m_T^2 = \frac{\lambda}{2(2\pi)^3} \int d^2\bar{k} \times 2 \sum_{n=1}^{\infty} K_0\Big(n\sqrt{\frac{|A + B|}{|A - B|}} \frac{\beta}{g_{00}} \omega_{\bar{k}}\Big). \tag{7.94}$$

The integral in (7.93), like most finite temperature integrals, cannot be evaluated in a closed form. However, in the limit of high temperature, $\beta m \ll 1$, its leading behavior can be calculated. We note that

---

[6]See, for example, formula 3.471.9 in I. S. Gradshteyn and I. M. Ryzhik.

at high temperature, we can approximate

$$\beta\omega_{\overline{k}} = \beta\sqrt{\overline{k}^2 + m^2} \approx \beta\overline{k}. \tag{7.95}$$

Substituting this into (7.94), we obtain

$$\Delta m_T^2 \approx \frac{\lambda}{(2\pi)^3} \sum_{n=1}^{\infty}(2\pi) \int_0^{\infty} d\overline{k}\,\overline{k}\, K_0(n\sqrt{\left|\frac{A+B}{A-B}\right|}\frac{\beta}{\overline{g}_{00}}\overline{k}),$$

$$= \frac{\lambda}{(2\pi)^2}\frac{|A-B|}{|A+B|}\frac{(\overline{g}_{00})^2}{\beta^2} \sum_{n=1}^{\infty}\frac{1}{n^2} = \frac{\lambda}{(2\pi)^2}\frac{|A-B|}{|A+B|}\frac{(\overline{g}_{00})^2}{\beta^2}\zeta(2)$$

$$= \frac{\lambda}{(2\pi)^2}\frac{|A-B|}{|A+B|}\frac{(\overline{g}_{00})^2}{\beta^2} \times \frac{\pi^2}{6}$$

$$= \frac{|A-B|}{|A+B|} \times \frac{(A+B)^2}{(B-A)^2}\frac{\lambda}{24\beta^2} = \frac{A+B}{B-A}\frac{\lambda}{24\beta^2} > 0. \tag{7.96}$$

where we have used the table of integrals[7] in the second line as well
as the value of $\overline{g}_{00}$ given in (7.73) in the last step. (As we have repeatedly emphasized, all the discussions assume $A + B \neq 0$ which
describes the conventional light-front quantization. (Also, tacitly we
are assuming, as pointed out earlier, that $A - B \neq 0$ which will lead to
a singular coordinate transformation.) The temperature dependent
mass correction is positive, as in the case of equal-time quantization,
which is crucial for restoration of spontaneously broken symmetries
at finite temperature, but it now depends on the parameters A and
B and hence is different from the equal-time calculation, in general,
unless $A = 0$ (see, for example, (1.113) and (2.87)). This is, of course,
the case for the oblique light-front for which $A = 0, B = 1$ discussed
earlier,[8] but this agreement holds also for the entire family of coordinate redefinitions for which $A = 0, B =$ any real constant (it simply
corresponds to a scaling of the x^3 coordinate, namely, $\overline{x}^3 = Bx^3$).
There is no divergence for $A + B \neq 0$ in the temperature dependent
term so that there is no need for thermal counter terms and the zero
temperature counter terms suffice to renormalize the theory. Only
for $A + B = 0$, which corresponds to the conventional light-front
quantization does the one loop correction diverge as we have noted

---

[7]See, for example, formula 6.561.16 in I. S. Gradshteyn and I. M. Ryzhik.
[8]See the reference for Alves, Das and Perez.

earlier in (7.35)-(7.39). Without describing the calculations here we simply note here that this calculation can also be carried out in the imaginary time formalism, as in section **7.1.2** (taking into account the appropriate periodicity), and because $\dfrac{\omega_{\bar{k}}^2}{2(A-B)\bar{k}_3}$ is now well behaved at both the ultraviolet and the infrared limits (for $A+B \neq 0$), there will be no divergence and the result will coincide with (7.96).

Finally, we end this section by looking at the zero temperature part of the self-energy graph in Fig. 7.1. From (7.89) and (7.92), we obtain the zero temperature contribution to the graph to be

$$\Delta m_0^2 = \frac{i\lambda|A-B|}{2(2\pi)^4} \int d\bar{k}_0 d\bar{k}_3 d^2\bar{k}_i \; \frac{1}{\bar{k}^2 - m^2 + i\epsilon}. \tag{7.97}$$

Let us next change the variables of integration to (see (7.78))

$$\bar{k}_0 = \frac{1}{A-B}\left(-Bk_0 + Ak_3\right), \quad \bar{k}_i = k_i, \quad i = 1, 2,$$

$$\bar{k}_3 = \frac{1}{A-B}\left(k_0 - k_3\right), \tag{7.98}$$

with a Jacobian of transformation

$$|J| = \left|\frac{\partial(\bar{k}_\mu)}{\partial(k_\nu)}\right| = \frac{1}{|A-B|}. \tag{7.99}$$

Furthermore, this change of variables leads to

$$\bar{k}^2 - m^2 = k_0^2 - (k_1^2 + k_2^2 + k_3^2 + m^2) = k_0^2 - \omega_k^2. \tag{7.100}$$

Using (7.98)-(7.100), we can write (7.97) as ($d^3k = dk_1 dk_2 dk_3$)

$$\Delta m_0^2 = \frac{i\lambda|A-B|}{2(2\pi)^4} \int \frac{1}{|A-B|} dk_0 d^3k \; \frac{1}{k_0^2 - \omega_k^2 + i\epsilon},$$

$$= \frac{i\lambda}{2(2\pi)^4} \int d^3k \int_{-\infty}^{\infty} dk_0 \; \frac{1}{k_0^2 - \omega_k^2 + i\epsilon},$$

and the k_0 integral can be evaluated using the method of residues to yield

$$\Delta m_0^2 = \frac{i\lambda}{2(2\pi)^4} \int d^3k \, (-2\pi i)\frac{1}{2\omega_k} = \frac{\lambda}{4} \int \frac{d^3k}{(2\pi)^3} \frac{1}{\omega_k}. \tag{7.101}$$

This is precisely the value of the zero temperature mass correction in the equal-time quantization (both in the imaginary time and the real time formalisms) as seen in (1.108) and (2.86) respectively. As pointed out there, only the zero temperature part of the mass correction contains a divergence which can be taken care of by a (renormalization) counterterm whereas the finite temperature part is completely finite.

The results in (7.96) and (7.101) bring out a very interesting fact. Namely, the zero temperature part of the self-energy correction, in the light-front quantization, coincides with that in the equal-time quantization requiring the same divergent counterterms, while the finite temperature parts in the two quantizations are, in general different (depending on the arbitrary constants A, B in the definition of the general light-front frame). As long as $A + B \neq 0$, the finite temperature parts are finite without any divergence. (However, for the conventional light-front quantization, corresponding to $A + B = 0$, the finite temperature results are singular which, as we have discussed in detail in section **7.1.1**, can be traced to the fact that conventional light-front quantization does not allow for a statistical description.) The reason behind this interesting general conclusion lies in the fact that, inside the integral, the zero temperature propagators for bosons and fermions in the general light-front quantization go into that the equal-time quantization under a change of integration variables (which are the same as the transformation which connects the two coordinate systems), making the (zero temperature) Feynman amplitudes to lead to identical values. On the other hand, while parts of the finite temperature propagators, such as the delta functions enforcing the on-shell nature of the thermal interactions (Einstein relation) are also related by the same change of (integration) variables, the distribution function $n_B(|\bar{k}_0|)$ and $n_F(|\bar{k}_0|)$, in the general light-front coordinates do not go over to $n_B(|k_0|)$ and $n_F(|k_0|)$ respectively by the same redefinition. Furthermore, under a change of variables of integration, temperature does not change (it is not a function of the momentum variables). This is where the difference comes in, leading to differences in the finite temperature thermal contributions in the two quantizations. We will see this in the following example as well involving fermions.

7.2.3 Chiral anomaly in the Schwinger model. Schwinger model is the theory of massless (charged) fermions interacting with a photon field in $1+1$ dimensions and, in the general light-front coordinates, the Lagrangian density has the form

$$\mathcal{L} = -\frac{1}{4}\bar{g}^{\mu\lambda}\bar{g}^{\nu\rho}\overline{F}_{\mu\nu}\overline{F}_{\lambda\rho} + i\overline{\psi}\bar{g}^{\mu\nu}\overline{\gamma}_{\mu}\overline{D}_{\nu}\psi, \quad \mu,\nu = 0,1, \quad (7.102)$$

where (see also (7.71)-(7.72))

$$\bar{g}^{\mu\nu} = (A-B)\begin{pmatrix} 0 & 1 \\ 1 & A+B \end{pmatrix},$$

$$\bar{g}_{\mu\nu} = \frac{1}{A-B}\begin{pmatrix} -(A+B) & 1 \\ 1 & 0 \end{pmatrix}. \quad (7.103)$$

We see that the transformed matrices are symmetric, $\bar{g}^{\mu\nu} = \bar{g}^{\nu\mu}$, $\bar{g}_{\mu\nu} = \bar{g}_{\nu\mu}$, but they are no longer diagonal as in the Minkowski space. Furthermore, the Dirac matrices are defined as

$$\overline{\gamma}^{\mu} = L^{\mu}{}_{\alpha}\gamma^{\alpha}, \quad \overline{\gamma}_{\mu} = L_{\mu}{}^{\alpha}\gamma_{\alpha}, \quad \mu,\alpha = 0,1, \quad (7.104)$$

where $\gamma^{\alpha}, \gamma_{\alpha}$ denote respectively the contravariant and the covariant Dirac matrices in the Minkowski space, with the forms for the contravariant matrices given by

$$\gamma^0 = \begin{pmatrix} 1 & 0 \\ 0 & -1 \end{pmatrix} = \gamma_0, \quad \gamma^1 = \begin{pmatrix} 0 & 1 \\ -1 & 0 \end{pmatrix} = -\gamma_1, \quad (7.105)$$

and the transformation matrices for the contravariant and covariant Dirac matrices can be obtained from (7.63) and (7.66) respectively and have the forms

$$L^{\mu}{}_{\alpha} = \begin{pmatrix} 1 & 1 \\ A & B \end{pmatrix}, \quad L_{\mu}{}^{\alpha} = \frac{1}{A-B}\begin{pmatrix} -B & A \\ 1 & -1 \end{pmatrix}. \quad (7.106)$$

The transformed gamma matrices can be seen, using (7.104)-(7.106), to have the explicit forms

$$\overline{\gamma}^0 = \gamma^0 + \gamma^1 = \begin{pmatrix} 1 & 1 \\ -1 & -1 \end{pmatrix},$$

$$\overline{\gamma}_0 = \frac{1}{A - B}\left(-B\gamma_0 + A\gamma_1\right) = \frac{1}{A - B}\begin{pmatrix} -B & -A \\ A & B \end{pmatrix},$$

$$\overline{\gamma}^1 = A\gamma^0 + B\gamma^1 = \begin{pmatrix} A & B \\ -B & -A \end{pmatrix},$$

$$\overline{\gamma}_1 = \frac{1}{A - B}\left(\gamma_0 - \gamma_1\right) = \frac{1}{A - B}\begin{pmatrix} 1 & 1 \\ -1 & -1 \end{pmatrix}, \tag{7.107}$$

which lead to (as in the Minkowski space-time)

$$\overline{\gamma}^\mu \overline{\gamma}_\mu = \overline{\gamma}^0 \overline{\gamma}_0 + \overline{\gamma}^1 \overline{\gamma}_1 = 2\mathbb{1}. \tag{7.108}$$

We note from (7.107) that the transformed gamma (Dirac) matrices do not have simple Hermiticity properties as in the Minkowski space. Instead, they satisfy

$$(\overline{\gamma}^0)^\dagger = \frac{1}{A - B}\left(-(A + B)\overline{\gamma}^0 + 2\overline{\gamma}^1\right) = \begin{pmatrix} 1 & -1 \\ 1 & -1 \end{pmatrix},$$

$$(\overline{\gamma}^1)^\dagger = \frac{1}{A - B}\left(-2AB\overline{\gamma}^0 + (A + B)\overline{\gamma}^1\right) = \begin{pmatrix} A & -B \\ B & -A \end{pmatrix}. \tag{7.109}$$

The gamma matrices satisfy the Dirac algebra

$$[\overline{\gamma}^\mu, \overline{\gamma}^\nu]_+ = 2\overline{g}^{\mu\nu}\mathbb{1}, \quad \overline{\gamma}^\mu = \overline{g}^{\mu\nu}\overline{\gamma}_\nu, \quad \mu, \nu = 0, 1. \tag{7.110}$$

The interesting fact to note from the above is that (see (7.103))

$$(\overline{\gamma}^0)^2 = \overline{g}^{00}\mathbb{1} = 0, \tag{7.111}$$

which can be contrasted with the Minkowski space result $(\gamma^0)^2 = \mathbb{1}$. It also follows from (7.103) and (7.110) that (with $|B| \geq |A|$, see (7.73))

$$(\overline{\gamma}^1)^2 = \overline{g}^{11}\mathbb{1} = (A^2 - B^2)\mathbb{1} \leq 0, \tag{7.112}$$

which can also be obtained from (7.107) and is similar in spirit to the case of the Minkowski space (unless $A = \pm B$). We also note from (7.110) that $\bar{\gamma}^0$ and $\bar{\gamma}^1$ do not anti-commute, since (see (7.103) and (7.110))

$$[\bar{\gamma}^0, \bar{\gamma}^1]_+ = 2\bar{g}^{01}\mathbb{1} = 2\bar{g}^{10}\mathbb{1} = 2(A - B)\mathbb{1}, \qquad (7.113)$$

rather lead to

$$\bar{\gamma}^0\bar{\gamma}^1 = (A-B)\begin{pmatrix} 1 & -1 \\ -1 & 1 \end{pmatrix}, \quad \bar{\gamma}^1\bar{\gamma}^0 = (A-B)\begin{pmatrix} 1 & 1 \\ 1 & 1 \end{pmatrix}. \qquad (7.114)$$

Since $\bar{\gamma}^0$ and $\bar{\gamma}^1$ do not anti-commute, we can no longer define $\bar{\gamma}_5 = \bar{\gamma}^0\bar{\gamma}^1$ or $\bar{\gamma}_5 = \bar{\gamma}^1\bar{\gamma}^0$. Rather, we have to define it as

$$\bar{\gamma}_5 = -\frac{1}{2}\bar{\epsilon}_{\mu\nu}\bar{\gamma}^\mu\bar{\gamma}^\nu, \qquad (7.115)$$

where the two dimensional transformed Levi-Civita tensor takes the form (see (7.71) and (7.72) for the definition of $\bar{g}$ in $3+1$ dimensions)

$$\bar{\epsilon}^{\mu\nu} = L^\mu{}_\alpha \epsilon^{\alpha\beta} L^\nu{}_\beta = (B - A)\begin{pmatrix} 0 & 1 \\ -1 & 0 \end{pmatrix}^{\mu\nu} = (B - A)\epsilon^{\mu\nu}$$

$$= \left(\sqrt{-\bar{g}}\right)^{-1}\epsilon^{\mu\nu}, \quad \mu, \nu, \alpha, \beta = 0, 1, \qquad (7.116)$$

$$\bar{\epsilon}_{\mu\nu} = L_\mu{}^\alpha \epsilon_{\alpha\beta} L_\nu{}^\beta = \frac{1}{B - A}\begin{pmatrix} 0 & -1 \\ 1 & 0 \end{pmatrix}_{\mu\nu} = \frac{1}{B - A}\epsilon_{\mu\nu}$$

$$= \sqrt{-\bar{g}}\,\epsilon_{\mu\nu}. \qquad (7.117)$$

As a result, using (7.114), we obtain from (7.115)

$$\bar{\gamma}_5 = -\frac{1}{2}\bar{\epsilon}_{\mu\nu}\bar{\gamma}^\mu\bar{\gamma}^\nu = -\bar{\epsilon}_{01}\frac{(A - B)}{2}\left[\begin{pmatrix} 1 & -1 \\ -1 & 1 \end{pmatrix} - \begin{pmatrix} 1 & 1 \\ 1 & 1 \end{pmatrix}\right]$$

$$= \frac{1}{B - A} \times \frac{(A - B)}{2} \times (-2)\begin{pmatrix} 0 & 1 \\ 1 & 0 \end{pmatrix}$$

$$= \begin{pmatrix} 0 & 1 \\ 1 & 0 \end{pmatrix} = \gamma^0\gamma^1 = \gamma_5 = \bar{\gamma}_5^\dagger. \qquad (7.118)$$

Namely, the γ_5 matrix (defined properly), is invariant under the transformation to the general light-front coordinates and remains Hermitian unlike $\overline{\gamma}^0$ and $\overline{\gamma}^1$ (see (7.109)).

It also follows from (7.107) and (7.118) that

$$\left(\overline{\gamma}_5\right)^2 = \mathbb{1}, \quad \mathrm{Tr}\,\overline{\gamma}_5 = 0, \quad [\overline{\gamma}_5, \overline{\gamma}^\mu]_+ = 0. \tag{7.119}$$

We note that the usual decomposition of the product of two gamma matrices in $1+1$ dimensions continues to hold in the general light-front coordinates, namely,

$$\overline{\gamma}^\mu \overline{\gamma}^\nu = \overline{g}^{\mu\nu} + \overline{\epsilon}^{\mu\nu} \overline{\gamma}_5, \tag{7.120}$$

which is compatible with (7.118) and leads to the relation (using (7.108), for example)

$$\overline{\gamma}_5 \overline{\gamma}^\mu = \overline{\epsilon}^{\mu\nu} \overline{\gamma}_\nu. \tag{7.121}$$

Let us note from (7.114) as well as (7.118) that we can define the two projection operators

$$P^{(+)} = \frac{1}{2(A-B)} \overline{\gamma}^1 \overline{\gamma}^0 = \frac{1}{2} \begin{pmatrix} 1 & 1 \\ 1 & 1 \end{pmatrix} = \frac{1}{2}(\mathbb{1} + \overline{\gamma}_5),$$

$$P^{(-)} = \frac{1}{2(A-B)} \overline{\gamma}^0 \overline{\gamma}^1 = \frac{1}{2} \begin{pmatrix} 1 & -1 \\ -1 & 1 \end{pmatrix} = \frac{1}{2}(\mathbb{1} - \overline{\gamma}_5). \tag{7.122}$$

They satisfy the usual properties of projection operators, namely,

$$P^{(+)} + P^{(-)} = \mathbb{1}, \quad P^{(+)} P^{(-)} = 0 = P^{(-)} P^{(+)},$$

$$(P^{(+)})^2 = P^{(+)}, \quad (P^{(-)})^2 = P^{(-)}, \tag{7.123}$$

and we recognize them to be the usual chirality projection operators which can be used to define chiral fermions as

$$\psi^{(+)} = P^{(+)}\psi, \quad \psi^{(-)} = P^{(-)}\psi, \tag{7.124}$$

and, when used in the Lagrangian density (7.102), the component $\psi^{(-)}$ can be seen to become nondynamical because of the nilpotent property of the $\overline{\gamma}^0$ matrix given in (7.111). However, we would work with the full fermion ψ without projecting on to chiral components.

The covariant derivative for the fermions is defined as

$$\overline{D}_\mu \psi = (\overline{\partial}_\mu + ie\overline{A}_\mu)\psi, \tag{7.125}$$

where e denotes the electric charge and the anti-symmetric field strength tensor in (7.102) is given, as usual, by

$$\overline{F}_{\mu\nu} = \overline{\partial}_\mu \overline{A}_\nu - \overline{\partial}_\nu \overline{A}_\mu = -\overline{F}_{\nu\mu}. \tag{7.126}$$

This theory, like QED, has an Abelian gauge invariance[9] and we need to add a gauge fixing term to the Lagrangian density which we choose to be the covariant gauge, for simplicity, given by

$$\mathcal{L}_{\mathrm{GF}} = -\frac{1}{2\xi}\left(\overline{\partial}^\mu \overline{A}_\mu\right)^2 = -\frac{1}{2\xi}\left(\overline{g}^{\mu\nu}\overline{\partial}_\mu \overline{A}_\nu\right)^2. \tag{7.127}$$

The gauge fixing term, in turn, would induce a ghost Lagrangian density. However, since the ghosts are free in an Abelian gauge theory, we neglect them in our discussions. Thus, with all these details, we can now write down the propagator for the massless fermion at

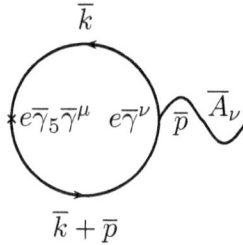

Figure 7.2: Chiral anomaly in the Schwinger model.

finite temperature which takes the form (the gauge field propagator is not relevant for the study of the one loop chiral anomaly as is clear from Fig. 7.2)

$$i\overline{S}(\overline{p}) = \overline{\not{p}}\left(\frac{i}{\overline{p}^2 + i\epsilon} - 2\pi n_F(|\overline{p}_0|)\delta(\overline{p}^2)\right), \tag{7.128}$$

which can be compared with the $++$ component of the fermion propagator in (6.74) with $m = 0$ and $\overline{\not{p}} = \overline{\gamma}_\mu \overline{p}^\mu$. We also note here that

---

[9]See section **12.3** and, in particular, Eq. (12.165) for the covariant gauge, in A. Das, *Lectures on Quantum Field Theory* (second edition), World Scientific (2020).

since we are considering a massless fermion, the Einstein relation (see (7.79)-(7.86) with $m = 0 = \overline{p}_i$ and with the identification $\overline{p}_3 \to \overline{p}_1$) gives

$$\overline{p}^2 = (A - B)(2\overline{p}_0\overline{p}_1 + (A + B)\overline{p}_1^2). \tag{7.129}$$

The graph in Fig. 7.2 describes a massless fermion loop with an axial vector current insertion $(e\overline{\gamma}_5\overline{\gamma}^\mu)$ at the left vertex and a photon coupling at the right and represents the thermal average of the amplitude $\langle \overline{j}_5^\mu \overline{j}^\nu \rangle_\beta \overline{A}_\nu$ at one loop. Here we have assumed that

$$\overline{j}_5^\mu = e\overline{\psi}\,\overline{\gamma}_5\overline{\gamma}^\mu\psi, \quad \overline{j}^\nu = e\overline{\psi}\,\overline{\gamma}^\nu\psi. \tag{7.130}$$

The amplitude in Fig. 7.2 involves a fermion loop which leads to a trace over the Dirac matrix indices with a negative sign (since fermions anti-commute) and, therefore, we can write the amplitude (neglecting the external photon field) in the momentum space as

$$\overline{I}_5^{\mu\nu}(\overline{p}) = -e^2 \int (\sqrt{-\overline{g}})^{-1} \frac{\mathrm{d}^2\overline{k}}{(2\pi)^2} \,\mathrm{Tr}\,\overline{\gamma}_5\overline{\gamma}^\mu iS(\overline{k})\overline{\gamma}^\nu iS(\overline{k} + \overline{p})$$

$$= -\frac{e^2|A - B|}{(2\pi)^2} \int \mathrm{d}^2\overline{k}\,\mathrm{Tr}\,\left(\overline{\gamma}_5\overline{\gamma}^\mu \overline{k}\overline{\gamma}^\nu(\overline{k} + \overline{p})\right)$$

$$\times \left(\frac{i}{\overline{k}^2 + i\epsilon} - 2\pi n_F(|\overline{k}_0|)\delta(\overline{k}^2)\right)$$

$$\times \left(\frac{i}{(\overline{k} + \overline{p})^2 + i\epsilon} - 2\pi n_F(|\overline{k}_0 + \overline{p}_0|)\delta((\overline{k} + \overline{p})^2)\right). \tag{7.131}$$

If we contract this amplitude with the external momentum $\overline{p}_\mu$, then (if it is non-zero) this leads to an anomaly in the chiral current. Namely, since the theory described by the Lagrangian density in (7.102) involves massless fermions, it is invariant under a global chiral transformation,

$$\psi \to e^{i\alpha\overline{\gamma}_5}\psi, \quad \overline{\psi} \to \overline{\psi}e^{i\alpha\overline{\gamma}_5}, \quad \text{with } \alpha \text{ arbitrary constant}, \tag{7.132}$$

leading to the conserved current

$$\overline{\partial}_\mu \overline{j}_5^\mu = 0. \tag{7.133}$$

This conservation, on the other hand, is violated when an anomaly is present (the current is no longer conserved) and the current becomes anomalous. Therefore, if we are looking for the anomaly in the axial (chiral) current, we have (in momentum space $\overline{\partial}_\mu \to -i\overline{p}_\mu$)

$$\overline{p}_\mu \overline{I}_5^{\mu\nu}(\overline{p}) = \frac{ie^2|A-B|}{(2\pi)^2} \int d^2\overline{k} \, \mathrm{Tr} \left(\overline{\gamma}_5 \overline{\slashed{p}} \overline{\slashed{k}} \overline{\gamma}^\nu (\overline{\slashed{k}} + \overline{\slashed{p}}) \right)$$

$$\times \left(\frac{i}{\overline{k}^2 + i\epsilon} - 2\pi n_F(|\overline{k}_0|)\delta(\overline{k}^2) \right)$$

$$\times \left(\frac{i}{(\overline{k}+\overline{p})^2 + i\epsilon} - 2\pi n_F(|\overline{k}_0 + \overline{p}_0|)\delta((\overline{k}+\overline{p})^2) \right). \quad (7.134)$$

We note that there are three temperature dependent terms on the right hand side of (7.134), two with a single delta function and one with two delta functions. The two terms with a single delta function go into each other under $\overline{k} \to -(\overline{k}+\overline{p})$, particularly since

$$\mathrm{Tr} \left(\overline{\gamma}_5 \overline{\slashed{p}} \overline{\slashed{k}} \overline{\gamma}^\nu (\overline{\slashed{k}}+\overline{\slashed{p}}) \right) = \mathrm{Tr} \left(\overline{\gamma}_5 [\overline{\slashed{p}}(\overline{\slashed{k}}+\overline{\slashed{p}})\overline{\gamma}^\nu \overline{\slashed{k}} - \overline{p}^2 \overline{\gamma}^\nu \overline{\slashed{k}} - \overline{p}^2 \overline{\slashed{k}} \overline{\gamma}^\nu] \right)$$

$$= \mathrm{Tr} \left(\overline{\gamma}_5 \overline{\slashed{p}}(\overline{\slashed{k}}+\overline{\slashed{p}})\overline{\gamma}^\nu \overline{\slashed{k}} \right) - 2\overline{p}^2 \overline{k}^\nu \, \mathrm{Tr} \, \overline{\gamma}_5$$

$$= \mathrm{Tr} \left(\overline{\gamma}_5 \overline{\slashed{p}}(\overline{\slashed{k}}+\overline{\slashed{p}})\overline{\gamma}^\nu \overline{\slashed{k}} \right), \quad (7.135)$$

where, in the first line, we have used cyclicity of trace to bring $\overline{\slashed{p}}$ to the front and taken it through $\overline{\gamma}_5$ giving a negative sign. In the second line, we have used the familiar property, (7.119), that the trace of $\overline{\gamma}_5$ vanishes. Equation (7.135) essentially establishes that the two single delta terms in (7.134) contribute equally to the anomaly. Furthermore, since these terms are finite, we can use the two dimensional identity in (7.121) to write the trace as

$$\mathrm{Tr} \left(\overline{\gamma}_5 \overline{\slashed{p}} \overline{\slashed{k}} \overline{\gamma}^\nu (\overline{\slashed{k}}+\overline{\slashed{p}}) \right) = \overline{\epsilon}^{\nu\sigma} \, \mathrm{Tr} \left(\overline{\slashed{p}} \overline{\slashed{k}} \overline{\gamma}_\sigma (\overline{\slashed{k}}+\overline{\slashed{p}}) \right)$$

$$= \overline{\epsilon}^{\nu\sigma} \left[\mathrm{Tr} \left(\overline{\slashed{p}} \overline{\slashed{k}} \overline{\gamma}_\sigma \overline{\slashed{k}} \right) + \overline{p}^2 \, \mathrm{Tr} \, \overline{\slashed{k}} \overline{\gamma}_\sigma \right]$$

$$= 2\overline{\epsilon}^{\nu\sigma} \left((\overline{p}\cdot\overline{k})\overline{k}_\sigma - \overline{k}^2 \overline{p}_\sigma + (\overline{p}\cdot\overline{k})\overline{k}_\sigma \right) + \overline{p}^2 \overline{k}_\sigma \right)$$

$$= 2\overline{\epsilon}^{\nu\sigma} \left(\overline{k}_\sigma(\overline{k}+\overline{p})^2 - (\overline{k}+\overline{p})_\sigma \overline{k}^2 \right). \quad (7.136)$$

Substituting this into the integrand of one of the single delta terms in (7.134), we obtain

$$\int d^2\overline{k}\, 2\overline{\epsilon}^{\nu\sigma}\left(\overline{k}_\sigma(\overline{k}+\overline{p})^2 - (\overline{k}+\overline{p})_\sigma\,\overline{k}^2\right)\frac{(-2i\pi n_F(|\overline{k}_0|))}{(\overline{k}+\overline{p})^2}\,\delta(\overline{k}^2)$$

$$= -4i\pi\overline{\epsilon}^{\nu\sigma}\int d^2\overline{k}\,\overline{k}_\sigma n_F(|\overline{k}_0|)\,\delta(\overline{k}^2) = 0, \qquad (7.137)$$

where the vanishing of the integral in the last step follows from the anti-symmetry of the integrand under $\overline{k}_\mu \to -\overline{k}_\mu$. Therefore, the single delta terms do not contribute to the anomaly. The double delta terms also do not contribute to the anomaly simply from the fact that each of the two terms in the "Tr" in (7.136) has a square which vanishes because the integrand involves a product of two delta functions with exactly those squares as arguments. This shows that there is no temperature dependent contribution to the anomaly which is also the conclusion obtained in the equal-time quantization of the theory. However, that does not mean that the equal-time and the general light-front quantizations give identical results for the temperature dependent part of the chiral anomaly, the coefficients (of zero) may be different. This is easily noted, for example, from the fact that the argument $|\overline{k}_0|$ of the distribution function $n_F(|\overline{k}_0|)$ depends on the parameters A, B when the Einstein relation is used (see (7.129)). Similarly, as we have seen in the case of the scalar self-energy in the last section, the temperature factor is also multiplied by $\overline{g}_{00}$ which involves the constants A, B.

The zero temperature contribution to the anomaly in (7.134) comes from the term without the delta functions and can be evaluated in the following way. Let us redefine the variables of integration as (see, for example, (7.78) for this $1+1$ dimensional case)

$$\overline{k}_0 = \frac{1}{A-B}\left(-Bk_0 + Ak_1\right), \quad \overline{k}_1 = \frac{1}{A-B}(k_0 - k_1). \qquad (7.138)$$

The Jacobian for this change of variables is given by

$$|J| = \left|\frac{\partial(\overline{k}_0, \overline{k}_1)}{\partial(k_0, k_1)}\right| = \frac{1}{|A-B|}. \qquad (7.139)$$

Furthermore, under this change of variables, we have (see (7.129))

$$\overline{k}^2 = (A-B)(2\overline{k}_0\overline{k}_1 - (A+B)\overline{k}_1^2) = (k_0^2 - k_1^2) = k^2,$$

and, similarly,

$$(\overline{k} + \overline{p})^2 = ((k_0 + p_0)^2 - (k_1 + p_1)^2) = (k + p)^2, \qquad (7.140)$$

so that we can apply dimensional regularization. In n-dimensions, the zero temperature anomaly in (7.134) would take the form

$$\overline{p}_\mu \overline{I}_5^{\mu\nu}(\overline{p}) = -ie^2 |A - B| \operatorname{Tr}\left(\overline{\gamma}_5 \overline{\not{p}} \overline{\gamma}^\lambda \overline{\gamma}^\nu \overline{\gamma}^\rho\right)$$

$$\times \int \frac{d^2 \overline{k}}{(2\pi)^2} \frac{\overline{k}_\lambda (\overline{k} + \overline{p})_\rho}{(\overline{k}^2 + i\epsilon)((\overline{k} + \overline{p})^2 + i\epsilon)}$$

$$= -ie^2 |A - B| \operatorname{Tr}\left(\overline{\gamma}_5 \overline{\not{p}} \overline{\gamma}^\lambda \overline{\gamma}^\nu \overline{\gamma}^\rho\right)$$

$$\times \int \frac{1}{|A - B|} \frac{d^2 k}{(2\pi)^2} \frac{L_\lambda^\alpha L_\rho^\beta k_\alpha (k + p)_\beta}{(k^2 + i\epsilon)((k + p)^2 + i\epsilon)}$$

$$= -ie^2 \operatorname{Tr}\left(\overline{\gamma}_5 \overline{\not{p}} \overline{\gamma}^\lambda \overline{\gamma}^\nu \overline{\gamma}^\rho\right)$$

$$\times L_\lambda^\alpha L_\rho^\beta \lim_{\epsilon \to 0^+} \int \frac{d^n k}{(2\pi)^n} \frac{k_\alpha (k + p)_\beta}{k^2 (k + p)^2}, \qquad (7.141)$$

where we have extended the integration to $n = 2 - \epsilon$ dimensions. Let us look at the second factor of (7.141) which involves the momentum integration. Use of standard results in dimensional regularization determines the value of the integral to be

$$L_\lambda^\alpha L_\rho^\beta \lim_{\epsilon \to 0^+} \int \frac{d^n k}{(2\pi)^n} \frac{k_\alpha (k + p)_\beta}{k^2 (k + p)^2}$$

$$= -\lim_{\epsilon \to 0^+} \frac{ie^2}{(4\pi)^{1 - \frac{\epsilon}{2}}} \int_0^1 dx \, L_\lambda^\alpha L_\rho^\beta \left[-\frac{1}{2} \frac{\Gamma(\frac{\epsilon}{2}) \eta_{\alpha\beta}}{(-x(1 - x)p^2)^{\frac{\epsilon}{2}}} \right.$$

$$\left. - \Gamma(1 + \frac{\epsilon}{2}) \frac{x(1 - x)p_\alpha p_\beta}{(-x(1 - x)p^2)^{1 + \frac{\epsilon}{2}}} \right]$$

$$= -\lim_{\epsilon \to 0^+} \frac{ie^2}{(4\pi)^{1 - \frac{\epsilon}{2}}} \int_0^1 dx \left[-\frac{1}{2} \frac{\Gamma(\frac{\epsilon}{2}) \overline{g}_{\lambda\rho}}{(-x(1 - x)p^2)^{\frac{\epsilon}{2}}} \right.$$

$$\left. - \Gamma(1 + \frac{\epsilon}{2}) \frac{x(1 - x)\overline{p}_\lambda \overline{p}_\rho}{(-x(1 - x)p^2)^{1 + \frac{\epsilon}{2}}} \right]. \qquad (7.142)$$

We note that the first term inside the bracket in (7.142) is divergent in the limit, but the second term is finite. So, the Dirac trace of the gamma matrices (in the first factor of (7.141)) can be done in 2 dimensions in the second term. On the other hand, we also note that doing the Dirac trace in n dimensions in the first term in (7.142) leads to,

$$\text{Tr}\left(\overline{\gamma}_5\, \slashed{\overline{p}}\,\overline{\gamma}^\lambda\,\overline{\gamma}^\nu\,\overline{\gamma}^\rho\right)\overline{g}_{\lambda\rho} = (2-n)\text{Tr}\left(\overline{\gamma}_5\,\slashed{\overline{p}}\,\overline{\gamma}^\nu\right)$$

$$= \epsilon\,\text{Tr}\left(\overline{\gamma}_5\slashed{\overline{p}}\,\overline{\gamma}^\nu\right). \tag{7.143}$$

This makes even the first term in (7.142) finite so that the Dirac traces can now be evaluated in 2 dimensions and the limit $\epsilon \to 0^+$ can be taken using the identities

$$\text{Tr}\left(\overline{\gamma}_5\overline{\gamma}^\mu\overline{\gamma}^\nu\right) = 2\overline{\epsilon}^{\mu\nu},$$

$$\text{Tr}\left(\overline{\gamma}_5\overline{\gamma}^\mu\overline{\gamma}^\lambda\overline{\gamma}^\nu\overline{\gamma}^\rho\right) = 2(\overline{\epsilon}^{\mu\lambda}\overline{g}^{\nu\rho} - \overline{\epsilon}^{\mu\nu}\overline{g}^{\lambda\rho} + \overline{\epsilon}^{\mu\rho}\overline{g}^{\lambda\nu}). \tag{7.144}$$

This makes the zero temperature anomaly to be (x integral is trivial)

$$\overline{p}_\mu \overline{I}_5^{\mu\nu} = \frac{ie^2}{4\pi} \times 4\overline{\epsilon}^{\mu\nu}\overline{p}_\mu \int_0^1 \mathrm{d}x = \frac{ie^2}{\pi} \overline{\epsilon}^{\mu\nu}\overline{p}_\mu. \tag{7.145}$$

In the coordinate space, then, this leads to the anomaly (with $\overline{A}_\nu$ put back, see Fig. 7.2 and remember $\overline{p}_\mu \to i\overline{\partial}_\mu$)

$$\overline{\partial}_\mu \overline{j}_5^\mu(x) = \frac{e^2}{\pi} \overline{\epsilon}^{\mu\nu}\, \overline{\partial}_\mu \overline{A}_\nu = \frac{e^2}{2\pi} \overline{\epsilon}^{\mu\nu}\, \overline{F}_{\mu\nu}, \tag{7.146}$$

which is also the zero temperature anomaly in the equal-time quantization. This shows again that the zero temperature amplitude has the same value in the general light-front quantization as in equal time quantization. However, the temperature dependent parts are different.

7.3 Unruh effect in the general light-front

The Unruh effect can be described as saying that an observer in a uniformly accelerated frame in Minkowski space-time sees the vacuum of the equal time quantum field theory as a thermal vacuum with temperature

$$T_\text{M} = \frac{\alpha}{2\pi}, \tag{7.147}$$

where α denotes the constant proper acceleration of the observer. So, let us discuss, in some detail, the accelerated trajectory in the Minkowski space.

7.3.1 Accelerated motion in Minkowski space. We note that the line element for the Minkowski space is given by (the metric has the diagonal form $\eta_{\mu\nu} = (+, -, -, -)$),

$$d\tau^2 = \eta_{\mu\nu}dx^\mu dx^\nu = dt^2 - dx^2 - dy^2 - dz^2, \tag{7.148}$$

where τ denotes the proper time and we have made the familiar identifications $x^0 = t, x^1 = x, x^2 = y, x^3 = z$. It reflects the isotropy of the Minkowski space in the three spatial directions x, y, z. As a result of this isotropy, (7.147) holds independent of the direction of the acceleration. Therefore, let us assume that the particle is moving along the x (or y or z)-axis with velocity v so that the Lorentz factor is given by

$$\gamma = \frac{dt}{d\tau} = \frac{1}{\sqrt{1-v^2}}, \quad v = \frac{dx^1}{dt} = \frac{dx}{dt}, \tag{7.149}$$

which leads to

$$\gamma^2 v^2 + 1 = \gamma^2. \tag{7.150}$$

The four velocity of the particle can now be obtained to be

$$u^\mu(\tau) = \frac{dx^\mu}{d\tau} = \frac{dt}{d\tau}\frac{dx^\mu}{dt} = \gamma(1, v, 0, 0) = (\gamma, \gamma v, 0, 0), \tag{7.151}$$

which satisfies

$$\eta_{\mu\nu}u^\mu(\tau)u^\nu(\tau) = 1. \tag{7.152}$$

If the moving particle is also being uniformly accelerated in the x direction, then we can define the proper acceleration four vector as

$$a^\mu(\tau) = \frac{du^\mu(\tau)}{d\tau} = \frac{dt}{d\tau}\frac{du^\mu}{dt} = \gamma\left(\frac{d\gamma}{dt}, \frac{d(\gamma v)}{dt}, 0, 0\right). \tag{7.153}$$

Using (7.150) and identifying,

$$\frac{d\gamma}{dt} = \frac{d(\gamma v)}{dt}v = \alpha v, \quad \alpha = \frac{d(\gamma v)}{dt}, \tag{7.154}$$

we can obtain the proper acceleration from (7.151)

$$a^\mu(\tau) = \gamma\left(\frac{d\gamma}{dt}, \frac{d(\gamma v)}{dt}, 0, 0\right) = \alpha(\gamma v, \gamma, 0, 0). \tag{7.155}$$

We see from (7.154) that, in the instantaneous rest frame ($v = 0$) of the particle, we have $\gamma = 1$ which leads to

$$\alpha = \frac{dv}{dt}. \tag{7.156}$$

Therefore, we can think of α as the constant instantaneous acceleration (along the x-axis) of the particle (along the x-axis). The proper four acceleration a^μ clearly satisfies

$$\eta_{\mu\nu} a^\mu(\tau) a^\nu(\tau) = -\alpha^2, \quad \eta_{\mu\nu} u^\mu a^\nu = 0, \tag{7.157}$$

showing that the proper acceleration is space-like and is orthogonal to the four velocity of the particle. Therefore, the four velocity continues to satisfy (7.152) even in the presence of an acceleration. It follows from (7.151) and (7.155) that

$$\frac{d^2 x^\mu(\tau)}{d\tau^2} = \frac{du^\mu}{d\tau} = a^\mu = \alpha(\gamma v, \gamma, 0, 0), \tag{7.158}$$

where we have used (7.155).

The solution for the trajectory can now be obtained easily. From the deinition of α in (7.154) we obtain

$$\gamma v = \alpha t, \tag{7.159}$$

where we have used the fact that α is constant and a constant of integration has been absorbed into the initial condition $v(t = 0) = 0$. With a little bit of algebra, this determines

$$v(t) = \frac{\alpha t}{\sqrt{1 + \alpha^2 t^2}}. \tag{7.160}$$

Furthermore using (7.160), the definition of the velocity in (7.149) can now be integrated with the initial condition $x(t = 0) = \frac{1}{\alpha}$ to lead to the complete solution of the accelerated motion of the particle

$$t(\tau) = \frac{\sinh \alpha \tau}{\alpha}, \qquad x(\tau) = \frac{\cosh \alpha \tau}{\alpha},$$

$$\gamma(\tau) = \cosh \alpha \tau, \qquad v(\tau) = \tanh \alpha \tau. \tag{7.161}$$

The solutions clearly satisfies the equation of a hyperbola

$$\eta_{\mu\nu} x^\mu(\tau) x^\nu(\tau) = t^2(\tau) - x^2(\tau) = -\frac{1}{\alpha^2}, \tag{7.162}$$

as opposed to a parabolic motion in the nonrelativistic case.

7.3.2 Accelerated motion in general light-front coordinates. On the other hand, if we are in the general light-front frame, the line element has the form

$$d\tau^2 = \bar{g}_{\mu\nu}d\bar{x}^\mu d\bar{x}^\nu$$

$$= \frac{A+B}{B-A}d\bar{t}^2 - d\bar{x}^2 - d\bar{y}^2 - \frac{2}{B-A}d\bar{t}d\bar{z}, \qquad (7.163)$$

where we have used the form of the metric given in (7.72) and have identified $\bar{x}^0 = \bar{t}$, $\bar{x}^1 = \bar{x}$, $\bar{x}^2 = \bar{y}$, $\bar{x}^3 = \bar{z}$. The Lorentz factor, in this case, is easily worked out to be

$$\bar{\gamma} = \frac{d\bar{t}}{d\tau} = \frac{1}{\sqrt{\frac{A+B}{B-A} - \bar{v}_{\bar{x}}^2 - \bar{v}_{\bar{y}}^2 - \frac{2}{B-A}\bar{v}_{\bar{z}}}}, \qquad (7.164)$$

with

$$\bar{v}_{\bar{x}} = \frac{d\bar{x}}{d\bar{t}}, \quad \bar{v}_{\bar{y}} = \frac{d\bar{y}}{d\bar{t}}, \quad \bar{v}_{\bar{z}} = \frac{d\bar{z}}{d\bar{t}}. \qquad (7.165)$$

We note from (7.163) that, in the general light-front coordinates, there is only a two dimensional (spatial) isotropy in the $\bar{x} - \bar{y}$ plane. Consequently, there are two cases to consider, namely, motion along the $\bar{x}$ (or $\bar{y}$) direction and motion along the $\bar{z}$ direction. Let us begin with the analysis of motion along the $\bar{x}$ direction.

Motion along $\bar{x}$

If the particle is moving along the $\bar{x}$ axis with velocity $\bar{v}_{\bar{x}}$, then we have (see (7.164))

$$\bar{v}_{\bar{y}} = 0 = \bar{v}_{\bar{z}}, \quad \bar{\gamma} = \frac{d\bar{t}}{d\tau} = \frac{1}{\sqrt{\frac{A+B}{B-A} - \bar{v}_{\bar{x}}^2}}. \qquad (7.166)$$

The proper (four) velocity, therefore, has the form

$$\bar{u}^\mu(\tau) = \frac{d\bar{x}^\mu}{d\tau} = \frac{d\bar{t}}{d\tau}\frac{d\bar{x}^\mu}{d\bar{t}} = \bar{\gamma}(1, \bar{v}_{\bar{x}}, 0, 0), \qquad (7.167)$$

and it satisfies (as it should, see, for example, (7.152) and (7.166))

$$\bar{g}_{\mu\nu}\bar{u}^\mu(\tau)\bar{u}^\nu(\tau) = \bar{\gamma}^2\left(\frac{A+B}{B-A} - \bar{v}_{\bar{x}}^2\right) = 1. \qquad (7.168)$$

This relation also implies (compare with (7.150)) that

$$\overline{\gamma}^2 \overline{v}_{\overline{x}}^2 + 1 = \frac{A+B}{B-A}\,\overline{\gamma}^2. \tag{7.169}$$

This, in turn, leads to

$$\frac{A+B}{B-A}\frac{\mathrm{d}\overline{\gamma}}{\mathrm{d}\overline{t}} = \frac{\mathrm{d}(\overline{\gamma}\,\overline{v}_{\overline{x}})}{\mathrm{d}\overline{t}}\,\overline{v}_{\overline{x}} = \overline{\alpha}\,\overline{v}_{\overline{x}}, \quad \overline{\alpha} = \sqrt{\frac{B-A}{A+B}}\frac{\mathrm{d}(\overline{\gamma}\,\overline{v}_{\overline{x}})}{\mathrm{d}\overline{t}}, \tag{7.170}$$

which can be compared with (7.154). As before, we note that, in the instantaneous rest frame ($\overline{v}_{\overline{x}} = 0, \overline{\gamma} = 1$),

$$\overline{\alpha} = \sqrt{\frac{B-A}{A+B}}\frac{\mathrm{d}\overline{v}_{\overline{x}}}{\mathrm{d}\overline{t}}, \tag{7.171}$$

can be thought of as the constant, instantaneous acceleration along $\overline{x}$. The proper acceleration can now be obtained (as in (7.155))

$$\begin{aligned}
\overline{a}^\mu(\tau) &= \frac{\mathrm{d}\overline{u}^\mu}{\mathrm{d}\tau} = \frac{\mathrm{d}\overline{t}}{\mathrm{d}\tau}\left(\frac{\mathrm{d}\overline{\gamma}}{\mathrm{d}\overline{t}}, \frac{\mathrm{d}(\overline{\gamma}\,\overline{v}_{\overline{x}})}{\mathrm{d}\overline{t}}, 0, 0\right) \\
&= \overline{\gamma}\left(\sqrt{\frac{B-A}{A+B}}\,\overline{\alpha}\,\overline{v}_{\overline{x}}, \sqrt{\frac{A+B}{B-A}}\,\overline{\alpha}, 0, 0\right) \\
&= \overline{\alpha}\left(\sqrt{\frac{B-A}{A+B}}\,\overline{\gamma}\,\overline{v}_{\overline{x}}, \sqrt{\frac{A+B}{B-A}}\,\overline{\gamma}, 0, 0\right),
\end{aligned} \tag{7.172}$$

where we have used (7.170). Once again, using the metric in (7.72), we see that the proper acceleration satisfies

$$\begin{aligned}
\overline{a}^2 &= \overline{g}_{\mu\nu}\overline{a}^\mu \overline{a}^\nu = \overline{\alpha}^2\overline{\gamma}^2\left(\frac{A+B}{B-A}\times\frac{B-A}{A+B}\,\overline{v}_{\overline{x}}^2 - \frac{A+B}{B-A}\right) \\
&= \overline{\alpha}^2\overline{\gamma}^2\left(\overline{v}_{\overline{x}}^2 - \frac{A+B}{B-A}\right) = -\overline{\alpha}^2,
\end{aligned} \tag{7.173}$$

where we have used (7.169) and

$$\begin{aligned}
\overline{g}_{\mu\nu}\overline{a}^\mu \overline{u}^\nu &= \overline{\alpha}\,\overline{\gamma}^2\left(\frac{A+B}{B-A}\sqrt{\frac{B-A}{A+B}}\,\overline{v}_{\overline{x}} - \sqrt{\frac{A+B}{B-A}}\,\overline{v}_{\overline{x}}\right) \\
&= 0.
\end{aligned} \tag{7.174}$$

These two relations can be compared with the Minkowski space ones given in (7.157). For uniform (constant) acceleration, we have (see, for example, (7.158)-(7.161)

$$\frac{d^2\bar{x}^\mu}{d\tau^2} = \frac{d\bar{u}^\mu}{d\tau} = \bar{a}^\mu = \bar{\alpha}\bar{\gamma}\left(\sqrt{\frac{B-A}{A+B}}\,\bar{v}_{\bar{x}}, \sqrt{\frac{A+B}{B-A}}, 0, 0\right), \quad (7.175)$$

which can be integrated, subject to the initial conditions $\bar{x}(\tau = 0) = \frac{1}{\bar{\alpha}}$, $\bar{t}(\tau = 0) = 0 = \bar{v}_{\bar{x}}(\tau = 0)$, to yield

$$\bar{t}(\tau) = \sqrt{\frac{B-A}{A+B}}\,\frac{\sinh\bar{\alpha}\tau}{\bar{\alpha}}, \quad \bar{x}(\tau) = \frac{\cosh\bar{\alpha}\tau}{\bar{\alpha}},$$

$$\bar{\gamma}(\tau) = \sqrt{\frac{B-A}{A+B}}\,\cosh\bar{\alpha}\tau, \quad \bar{v}_{\bar{x}}(\tau) = \sqrt{\frac{A+B}{B-A}}\,\tanh\bar{\alpha}\tau. \quad (7.176)$$

Here $\bar{\alpha}$ is the constant acceleration given in (7.171) and these solutions can be compared with those in Minkowski coordinates given in (7.161). The motion, in this case, is along a hyperbola given by

$$\bar{g}_{\mu\nu}\bar{x}^\mu\bar{x}^\nu = \frac{A+B}{B-A}\,\bar{t}^2 - \bar{x}^2 = -\frac{1}{\bar{\alpha}^2}, \quad (7.177)$$

as in (7.162).

Motion along $\bar{z}$

A similar derivation can be done for uniform accelerated motion along the $\bar{z}$ axis. We will give only the essential results here. In this case, we have $\bar{v}_{\bar{x}} = 0 = \bar{v}_{\bar{y}}$ so that we have (see (7.164))

$$\bar{\gamma} = \frac{d\bar{t}}{d\tau} = \frac{1}{\sqrt{\frac{A+B}{B-A} - \frac{2}{B-A}\bar{v}_{\bar{z}}}} = \sqrt{\frac{B-A}{A+B-2\bar{v}_{\bar{z}}}}, \quad (7.178)$$

and the proper velocity has the form

$$\bar{u}^\mu = \frac{d\bar{x}^\mu}{d\tau} = \bar{\gamma}(1, 0, 0, \bar{v}_{\bar{z}}), \quad (7.179)$$

leading to (see (7.178))

$$\bar{g}_{\mu\nu}\bar{u}^\mu\bar{u}^\nu = \bar{\gamma}^2\left(\frac{A+B}{B-A} - \frac{2}{B-A}\bar{v}_{\bar{z}}\right) = 1. \quad (7.180)$$

It can be checked from (7.178) that

$$\frac{d\bar{\gamma}}{d\bar{t}} = \frac{\bar{\gamma}}{A + B - 2\bar{v}_{\bar{z}}} \frac{d\bar{v}_{\bar{z}}}{d\bar{t}},$$

$$\frac{d(\bar{\gamma}\,\bar{v}_{\bar{z}})}{d\tau} = (A + B - \bar{v}_{\bar{z}}) \frac{d\bar{\gamma}}{d\bar{t}}. \tag{7.181}$$

The proper acceleration follows from (7.179) and (7.181) to be

$$\bar{a}^{\mu} = \frac{d\bar{u}^{\mu}}{d\tau} = (1, 0, 0, (A + B - \bar{v}_{\bar{z}}))\,\bar{\gamma}\frac{d\bar{\gamma}}{d\bar{t}}, \tag{7.182}$$

which leads to (upon using (7.178))

$$\bar{a}^2 = \bar{g}_{\mu\nu}\bar{a}^{\mu}\bar{a}^{\nu} = -\left(\frac{d\bar{\gamma}}{d\bar{t}}\right)^2 = -\bar{\alpha}^2, \tag{7.183}$$

with $\bar{\alpha}$ representing the constant (proper) acceleration and, in the instantaneous rest frame, is given by

$$\bar{\alpha} = \frac{1}{A + B} \frac{d\bar{v}_{\bar{z}}}{d\bar{t}}. \tag{7.184}$$

The trajectory of the particle can now be derived from the dynamical equation

$$\frac{d^2\bar{x}^{\mu}}{d\tau^2} = \frac{d\bar{u}^{\mu}}{d\tau} = \bar{a}^{\mu} = (1, 0, 0, (A + B - \bar{v}_{\bar{z}}))\bar{\gamma}\,\bar{\alpha}. \tag{7.185}$$

For initial conditions $t(\tau = 0) = \sqrt{\frac{B-A}{A+B}}\frac{1}{\bar{\alpha}}, \bar{z}(\tau = 0) = \text{sgn}(B - A)\frac{\sqrt{B^2-A^2}}{\bar{\alpha}}$ and $\bar{v}_{\bar{z}}(\tau = 0) = 0$, we obtain

$$\bar{t}(\tau) = \sqrt{\frac{B - A}{A + B}}\frac{1}{\bar{\alpha}}\,e^{\bar{\alpha}\tau}, \quad \bar{z}(\tau) = \text{sgn}(B - A)\frac{\sqrt{B^2 - A^2}}{\bar{\alpha}}\cosh\bar{\alpha}\tau,$$

$$\bar{\gamma}(\tau) = \sqrt{\frac{B - A}{A + B}}\,e^{\bar{\alpha}\tau}, \quad \bar{v}_{\bar{z}}(\tau) = \frac{A + B}{2}(1 - e^{-2\bar{\alpha}\tau}). \tag{7.186}$$

It can now be checked that the trajectory lies on a hyperbola given by

$$\bar{x}^2 = \bar{g}_{\mu\nu}\bar{x}^{\mu}\bar{x}^{\nu} = \frac{A + B}{B - A}\bar{t}^2(\tau) - \frac{2}{B - A}\bar{t}(\tau)\bar{z}(\tau) = -\frac{1}{\bar{\alpha}^2}. \tag{7.187}$$

Therefore, we see, from (7.162), (7.177) and (7.187), that for the same constant accelerations $\alpha = \bar{\alpha}$, the motion is along the same hyperbola, be it on the $t - x$ $(\bar{t} - \bar{x})$ plane or the $t - y$ $(\bar{t} - \bar{y})$ plane or the $t - z$ $(\bar{t} - \bar{z})$ plane.

7.3.3 Coordinate Green's function. Let us present here a simpler discussion of the Unruh effect in terms of Green's functions. A more rigorous Hamiltonian analysis can be found in the literature,[10] which we do not get into here (for space reasons). We will discuss the Green's function both in the Minkowski space as well as in the general light-front space.

Minkowski space

Let us first determine the Green's function in the Minkowski space. For simplicity, let us consider a massless scalar field whose time ordered propagator (Feynman propagator), in the momentum space (at zero temperature), has the form

$$iG^{(T_M=0)}(p) = \lim_{\epsilon \to 0^+} \frac{i}{p^2 + i\epsilon} = \lim_{\epsilon \to 0^+} \frac{i}{p_0^2 - \mathbf{p}^2 + i\epsilon}, \qquad (7.188)$$

where $G(p)$ denotes the Green's function. The coordinate space Feynman propagator, at zero temperature, is obtained by taking the Fourier transformation and leads to

$$iG^{(T_M=0)}(x_1 - x_2) = \lim_{\epsilon \to 0^+} \int \frac{\mathrm{d}^4 p}{(2\pi)^4} e^{-ip \cdot (x_1 - x_2)} \frac{i}{p^2 + i\epsilon}$$

$$= \lim_{\epsilon \to 0^+} \frac{i}{(2\pi)^4} \int \mathrm{d}^3 p \, e^{i\mathbf{p} \cdot (\mathbf{x}_1 - \mathbf{x}_2)} \int \mathrm{d}p_0 \frac{e^{-ip_0(x_1^0 - x_2^0)}}{p_0^2 - \mathbf{p}^2 + i\epsilon}$$

$$= \lim_{\epsilon \to 0^+} \frac{i}{(2\pi)^4} \int_0^\infty p^2 \mathrm{d}p \int_{-1}^1 \mathrm{d}\cos\theta \int_0^{2\pi} \mathrm{d}\phi \, e^{i|\mathbf{p}||\mathbf{x}_1 - \mathbf{x}_2|\cos\theta}$$

$$\times \int_{-\infty}^\infty \mathrm{d}p_0 \frac{e^{-ip_0(x_1^0 - x_2^0)}}{(p_0 + |\mathbf{p}| - i\epsilon)(p_0 - |\mathbf{p}| + i\epsilon)},$$

where we have identified $p = |\mathbf{p}|$. If we assume $x_1^0 - x_2^0 > 0$, then, the contour has to be closed in the lower half plane in the clockwise direction (for damping of the exponential), picking up the residue of the pole in the lower half plane and leading to

$$= \frac{i}{(2\pi)^4} \int_0^\infty p^2 \mathrm{d}p \int_{-1}^1 \mathrm{d}\cos\theta \, (2\pi) \, e^{ip|\mathbf{x}_1 - \mathbf{x}_2|\cos\theta} \, (-2\pi i) \frac{e^{-ip(x_1^0 - x_2^0)}}{2p}$$

---

[10]See, for example, references by P. C. W. Davies, by W. G. Unruh as well as by A. Das, J. Frenkel and S. Perez at the end of this chapter.

$$= \frac{1}{2(2\pi)^2} \int_0^\infty p dp \, \frac{e^{-ip|\mathbf{x}_1 - \mathbf{x}_2|} - e^{ip|\mathbf{x}_1 - \mathbf{x}_2|}}{ip|\mathbf{x}_1 - \mathbf{x}_2|} e^{-ip_0(x_1^0 - x_2^0)}$$

$$= \frac{1}{i(2\pi)^2} \int_0^\infty dp \left(\frac{e^{-ip(x_1^0 - x_2^0 - i\epsilon - |\mathbf{x}_1 - \mathbf{x}_2|)} - e^{-ip(x_1^0 - x_2^0 - i\epsilon + |\mathbf{x}_1 - \mathbf{x}_2|)}}{2|\mathbf{x}_1 - \mathbf{x}_2|} \right)$$

$$= \lim_{\epsilon \to 0^+} -\frac{1}{(2\pi)^2} \frac{1}{(x_1 - x_2)^2 - i\epsilon} = -\frac{1}{(2\pi)^2} \frac{1}{(x_1 - x_2)^2}. \quad (7.189)$$

The zero temperature propagator can be thought of as the vacuum expectation value of the time ordered product of field operators,

$$iG^{(T_M=0)}(x_1 - x_2) = \theta(x_1^0 - x_2^0)\langle 0|\phi(x_1)\phi(x_2)|0\rangle$$
$$+ \theta(x_2^0 - x_1^0)\langle 0|\phi(x_2)\phi(x_1^0)|0\rangle = \langle 0|\phi(x_1)\phi(x_2)|0\rangle, \quad (7.190)$$

since we have already assumed that $x_1^0 - x_2^0 > 0$.

At finite temperature, the Feynman propagator for this theory, in momentum space, is given by

$$iG^{T_M}(p) = \frac{i}{p^2 + i\epsilon} + 2\pi n_B(|p_0|)\delta(p^2)$$

$$= iG^{(T_M=0)}(p) + i\widetilde{G}^{T_M}(p), \quad (7.191)$$

where $n_B(|p_0|)$ denotes the Bose-Einstein distribution function and we have assumed the heat bath to be at rest. The coordinate space propagator is obtained by taking the Fourier transformation as in (7.189), where we have already evaluated the zero temperature part. Therefore, let us focus on the finite temperature part which leads to

$$i\widetilde{G}^{T_M}(x_1 - x_2) = \int \frac{d^4 p}{(2\pi)^4} e^{-ip \cdot (x_1 - x_2)} (2\pi) n_B(|p_0|)\,\delta(p^2)$$

$$= \frac{1}{(2\pi)^3} \int d^3 p \, e^{ip|\mathbf{x}_1 - \mathbf{x}_2|\cos\theta} \int dp_0 \, e^{-ip_0(x_1^0 - x_2^0)} n_B(|p_0|)\delta(p_0^2 - p^2)$$

$$= \frac{1}{(2\pi)^2|\mathbf{x}_1 - \mathbf{x}_2|} \int_0^\infty dp \, \frac{1}{e^{\beta_M p} - 1}$$

$$\times \left((\sin p(x_1^0 - x_2^0 + |\mathbf{x}_1 - \mathbf{x}_2|)) - \sin p(x_1^0 - x_2^0 - |\mathbf{x}_1 - \mathbf{x}_2|) \right),$$

where we have identified $p = |\mathbf{p}|$, $\beta_M = \frac{1}{T_M}$ (in units of the Boltzmann constant) and the θ, ϕ integrals have been carried out as in (7.189). The integral over p can also be done using the standard tables.[11] leading to

$$= \frac{1}{(2\pi)^2} \left[\frac{\pi}{2\beta_M |\mathbf{x}_1 - \mathbf{x}_2|} \left(\coth \frac{\pi(x_1^0 - x_2^0 + |\mathbf{x}_1 - \mathbf{x}_2|)}{\beta_M} \right. \right.$$

$$\left. - \coth \frac{\pi(x_1^0 - x_2^0 - |\mathbf{x}_1 - \mathbf{x}_2|)}{\beta_M} \right)$$

$$\left. + \frac{1}{(x_1^0 - x_2^0)^2 - |\mathbf{x}_1 - \mathbf{x}_2|^2} \right]. \qquad (7.192)$$

It is surprising to find a temperature independent term (the last term) in the finite temperature part of the propagator, but this arises specifically because of the infrared divergence present in the Bose-Einstein distribution function. Furthermore, this zero temperature term has the same value as in (7.189), but with an opposite sign. Therefore, in the complete thermal propagator, (7.191), this term cancels out. Using standard trigonometric identities and in the limit $\mathbf{x}_1 = \mathbf{x}_2$, we obtain (for $t_1 - t_2 > 0$)

$$iG^{T_M}(t_1 - t_2, \mathbf{0}) = \langle \phi(t_1)\phi(t_2) \rangle_\beta$$

$$= \lim_{\mathbf{x}_1 \to \mathbf{x}_2} \int \frac{d^4 p}{(2\pi)^4} e^{-ip \cdot (x_1 - x_2)} iG^{T_M}(p)$$

$$= \lim_{\mathbf{x}_1 \to \mathbf{x}_2} -\frac{1}{(2\pi)^2} \frac{\pi}{2\beta_M} \frac{\sinh \frac{2\pi|\mathbf{x}_1 - \mathbf{x}_2|}{\beta_M}}{|\mathbf{x}_1 - \mathbf{x}_2|} \frac{1}{\sinh^2 \frac{\pi(t_1 - t_2)}{\beta_M}}$$

$$= -\frac{1}{(2\pi)^2} \frac{\pi}{2\beta_M} \times \frac{2\pi}{\beta_M} \operatorname{cosech}^2 \frac{\pi(t_1 - t_2)}{\beta_M}$$

$$= -\frac{1}{(2\pi)^2} \left(\frac{\pi}{\beta_M} \right)^2 \operatorname{cosech}^2 \frac{\pi(t_1 - t_2)}{\beta_M}$$

$$= -\frac{1}{(2\pi)^2} (\pi T_M)^2 \operatorname{cosech}^2 \pi T_M \tau, \qquad (7.193)$$

where we have identified $\beta_M = \frac{1}{T_M}$ (in units of the Boltzmann constant). We note here that setting $\mathbf{x}_1 = \mathbf{x}_2$ is equivalent to assuming that the system is at rest. Furthermore, we have also used the fact that, in Minkowski space, for $\mathbf{x}_1 = \mathbf{x}_2$ (system at rest), the proper line equation (7.148) leads to

$$t_1 - t_2 = \frac{\tau_1 - \tau_2}{\sqrt{\eta_{00}}} = \tau_1 - \tau_2 = \tau. \tag{7.194}$$

Let us next look at the zero temperature propagator given in (7.189) and analyze its form as seen by an observer in a frame accelerating along the x-axis. Clearly, for an observer at rest in such an accelerated frame, the system would appear to be moving in an accelerated trajectory along the x axis as given in (7.161) (described in detail in **7.3.1**) even though the system may be at rest (as discussed above). Using the solution for $t(\tau), x(\tau)$ in (7.161), we obtain

$$\begin{aligned}
(x_1 - x_2)^2 &= (t(\tau_1) - t(\tau_2))^2 - (x(\tau_1) - x(\tau_2))^2 \\
&= \frac{1}{\alpha^2}\left((\sinh \alpha\tau_1 - \sinh \alpha\tau_2)^2 - (\cosh \alpha\tau_1 - \cosh \alpha\tau_2)^2\right) \\
&= \frac{4}{\alpha^2} \sinh^2 \frac{\alpha(\tau_1 - \tau_2)}{2}(\cosh^2 \frac{\alpha(\tau_1 + \tau_2)}{2} - \sinh^2 \frac{\alpha(\tau_1 + \tau_2)}{2}) \\
&= \frac{4}{\alpha^2} \sinh^2 \frac{\alpha\tau}{2}, \tag{7.195}
\end{aligned}$$

which can be substituted into (7.189) which leads to

$$iG^{(T_{\mathrm{M}}=0)}(x_1 - x_2) = -\frac{1}{(2\pi)^2} \left(\frac{\alpha}{2}\right)^2 \operatorname{cosech}^2 \frac{\alpha\tau}{2}. \tag{7.196}$$

Comparing this with (7.193) we conclude that an observer in an accelerated frame with a constant acceleration α will see the zero temperature Green's function of a scalar field as if it is in a heat bath with temperature

$$\pi T_{\mathrm{M}} = \frac{\alpha}{2}, \quad \text{or,} \quad T_{\mathrm{M}} = \frac{\alpha}{2\pi}, \tag{7.197}$$

which coincides with (7.147) and says that the temperature which an accelerated observer sees for the system is proportional to the acceleration of the observer's frame.

General light-front space

We can do similar analyses for the system in general light-front coordinates. In this case, the general form of the zero temperature

Feynman propagator for a scalar field, in the coordinate representation, can be calculated as before for $\overline{x}_1^0 - \overline{x}_2^0 > 0$ and leads to

$$iG^{(T_{\mathrm{GLF}}=0)}(\overline{x}_1 - \overline{x}_2) = \langle 0|\phi(\overline{x}_1)\phi(\overline{x}_2)|0\rangle = \int \frac{\mathrm{d}^4\overline{p}}{(2\pi)^4} \frac{ie^{-i\overline{p}\cdot(\overline{x}_1 - \overline{x}_2)}}{\overline{p}^2 + i\epsilon}$$

$$= -\frac{1}{(2\pi)^2} \frac{1}{(\overline{x}_1 - \overline{x}_2)^2 - i\epsilon} = -\frac{1}{(2\pi)^2} \frac{1}{(\overline{x}_1 - \overline{x}_2)^2}$$

$$= -\frac{1}{(2\pi)^2} \frac{1}{\overline{g}_{\mu\nu}(\overline{x}_1 - \overline{x}_2)^\mu(\overline{x}_1 - \overline{x}_2)^\nu}, \tag{7.198}$$

where $\overline{g}_{\mu\nu}$ denotes the metric tensor for the general light-front coordinates given in (7.163). As we have argued earlier (see, for example, (7.97)-(7.101)), we can alternatively make a change of variables of integration in (7.198) back to the Minkowski space variables and the same result would follow. On the other hand, if we calculate, as before, the complete thermal propagator, in the rest frame of the heat bath, we note that

$$iG^{T_{\mathrm{GLF}}}(p) = \frac{i}{\overline{p}^2 + i\epsilon} + 2\pi n_B(|\overline{p}_0|)\delta(\overline{p}^2), \tag{7.199}$$

which is exactly like (7.191) except that it is in the general light-front coordinates and, consequently, involves a nontrivial metric. Furthermore, in the rest frame of the heat bath, the distribution function has the form

$$n_B(|\overline{p}_0|) = \frac{1}{e^{\beta_{\mathrm{GLF}}|\overline{p}_0|} - 1}, \tag{7.200}$$

with $\beta_{\mathrm{GLF}} = \frac{1}{T_{\mathrm{GLF}}}$ in units of the Boltzmann constant and T_{GLF} denoting the temperature of the system in the general light-front coordinates. If we assume further that $\overline{\mathbf{x}}_1 = \overline{\mathbf{x}}_2$ (namely, the system is at rest), the Fourier transformation would lead to (after using tables of integrals)

$$iG^{T_{\mathrm{GLF}}}(\overline{t}_1 - \overline{t}_2, \mathbf{0}) = \lim_{\mathbf{x}_1 \to \mathbf{x}_2} \int \frac{\mathrm{d}^4\overline{p}}{(2\pi)^4} e^{-i\overline{p}\cdot(\overline{x}_1 - \overline{x}_2)} iG^{T_{\mathrm{GLF}}}(p)$$

$$= -\frac{1}{(2\pi)^2}\left(\pi T_{\mathrm{GLF}}\sqrt{\frac{B-A}{A+B}}\right)^2 \mathrm{cosech}^2\pi T_{\mathrm{GLF}}(\overline{t}_1 - \overline{t}_2)$$

$$= -\frac{1}{(2\pi)^2}\left(\frac{\pi T_{\mathrm{GLF}}}{\sqrt{\overline{g}_{00}}}\right)^2 \mathrm{cosech}^2\,\pi T_{\mathrm{GLF}}(\overline{t}_1 - \overline{t}_2). \tag{7.201}$$

If we now use the fact that, for $\overline{\mathbf{x}}_1 = \overline{\mathbf{x}}_2$, the line element in (7.163) gives

$$\overline{t}_1 - \overline{t}_2 = \frac{1}{\sqrt{\overline{g}_{00}}}(\tau_1 - \tau_2) = \sqrt{\frac{B-A}{A+B}}\,\tau, \quad \tau = \tau_1 - \tau_2, \quad (7.202)$$

we can write (7.201) as

$$i\overline{G}^{T_{\mathrm{GLF}}}(\overline{t}_1 - \overline{t}_2, \mathbf{0}) = -\frac{1}{(2\pi)^2}\left(\frac{\pi T_{\mathrm{GLF}}}{\sqrt{\overline{g}_{00}}}\right)^2 \operatorname{cosech}^2 \frac{\pi T_{\mathrm{GLF}}}{\sqrt{\overline{g}_{00}}}\tau, \quad (7.203)$$

which can be compared with the result in Minkowski space given in (7.193).

Let us next go back to the zero temperature propagator for the system at rest and examine how it would appear to an observer in a frame accelerating uniformly. As we already know, there can be two distinct cases here, the observer may be accelerating along the $\overline{x}$ (or $\overline{y}$) axis or it may be accelerating along the $\overline{z}$ axis. Suppose the constant acceleration is $\overline{\alpha}_x$ along the $\overline{x}$ axis, then the system would appear to the observer (at rest on the accelerating frame) as accelerating in the opposite direction. Solutions for such a trajectory are already worked out in (7.176) and they satisfy

$$(\overline{x}_1 - \overline{x}_2)^2 = \frac{A+B}{B-A}(\overline{t}(\tau_1) - \overline{t}(\tau_2))^2 - (\overline{x}(\tau_1) - \overline{x}(\tau_2))^2$$

$$= \frac{1}{\overline{\alpha}_x^2}\left(\frac{A+B}{B-A} \times \frac{B-A}{A+B}(\sinh\overline{\alpha}_x\tau_1 - \sinh\overline{\alpha}_x\tau_2)^2\right.$$

$$\left. - (\cosh\overline{\alpha}_x\tau_1 - \cosh\overline{\alpha}_x\tau_2)^2\right)$$

$$= \left(\frac{2}{\overline{\alpha}_x}\right)^2 \sinh^2\frac{\overline{\alpha}_x\tau}{2}, \quad \tau = \tau_1 - \tau_2, \quad (7.204)$$

where we have used (7.195) in the last step. Using this, we obtain from (7.198),

$$i\overline{G}^{(T_{\mathrm{GLF}}=0)}(\overline{x}_1 - \overline{x}_2) = -\frac{1}{(2\pi)^2}\left(\frac{\overline{\alpha}_x}{2}\right)^2 \operatorname{cosech}^2 \frac{\overline{\alpha}_x\tau}{2}. \quad (7.205)$$

Comparing this with (7.203), we conclude that the observer in an accelerating frame along the $\overline{x}$ axis would see the zero temperature system (at rest) to be at a temperature given by

$$\frac{\overline{\alpha}_x}{2} = \frac{\pi T_{\text{GLF}}}{\sqrt{\overline{g}_{00}}}, \quad \text{or}, T_{\text{GLF}} = \frac{\sqrt{\overline{g}_{00}}\,\overline{\alpha}_x}{2\pi} = \frac{\sqrt{\frac{A+B}{B-A}}\,\overline{\alpha}_x}{2\pi}. \tag{7.206}$$

On the other hand, if the observer was accelerating along the $\overline{z}$ axis with a constant acceleration $\overline{\alpha}_z$, a similar calculation would lead to the observer finding the system at a temperature given by

$$T_{\text{GLF}} = \frac{\sqrt{\overline{g}_{00}}\,\overline{\alpha}_z}{2\pi} = \frac{\sqrt{\frac{A+B}{B-A}}\,\overline{\alpha}_z}{2\pi}. \tag{7.207}$$

It is clear from (7.206) and (7.207) that, for the same constant acceleration $\overline{\alpha}_x = \overline{\alpha}_z$, the observer would find the system at the same temperature, independent of the direction of the acceleration, even though the line element, in (7.163) for general light-front coordinates does not show any isotropy in the $\overline{x}\,(\overline{y}) - \overline{z}$ coordinates. Furthermore, comparing (7.206)-(7.207) with the Minkowski space results in (7.197), we see that, for the same constant acceleration $\alpha = \overline{\alpha}_x = \overline{\alpha}_z$, the temperatures which the observer would see are related by

$$\frac{T_{\text{GLF}}}{\sqrt{\overline{g}_{00}}} = \frac{T_{\text{M}}}{\sqrt{\eta_{00}}} = \frac{\alpha}{2\pi}, \tag{7.208}$$

and this is also consistent with the Tolman-Ehrenfest effect discussed in (7.81).

7.4 References

V. S. Alves, A. Das and S. Perez, Physical Review **D66**, 125008 (2002).

S. Brodsky, H. C. Pauli and S. S. Pinsky, Physics Reports **301**, 299 (1998).

M. Burkhadt, Advances in Nuclear Physics **23**, 1 (1996).

A. Das, *Lectures on Quantum Field Theory* (second edition), World Scientific (2020).

A. Das, J. Frenkel and S. Perez, Physical Review **D71**, 105018 (2005).

A. Das and S. Perez, Physical Review **D70**, 065006 (2004).

A. Das and X. Zhou, Physical Review **D68**, 065017 (2003).

P. C. W. Davies, Journal of Physics **A8**, 609 (1975).

P. A. M. Dirac, Reviews of Modern Physics **21**, 392 (1949).

P. A. M. Dirac, *Lectures in Quantum Mechanics*, Benjamin, New York (1964).

I. S. Gradshteyn and I. M. Ryzhik, *Table of Integrals, Series and Products*, Academic Press, San Diego (1980).

A. J. Hanson, T. Regge and C. Teitelboim, *Constrained Hamiltonian Systems*, Academia Nazionale dei Lincei, Rome (1976).

T. Heinzl, *Lecture Notes in Physics* **572**, 55 (2001).

W. Israel, Annals of Physics (N. Y.) **100**, 310 (1976); Physica **A106**, 204 (1981).

J. B. Kogut and D. E. Soper, Physical Review **D1**, 2901 (1970).

C. Rovelli and M. Smerlak, Classical and Quantum Gravity **28**, 75007 (2011).

L. Susskind, Physical Review **165**, 1535 (1968).

R. C. Tolman, Physical Review **35**, 904 (1930).

R. C. Tolman, *Relativity, Thermodynamics, and Cosmology*, Oxford University Press, Oxford, England (1934).

R. C. Tolman and P. Ehrenfest, Physical Review **36**, 1791, (1930).

W. G. Unruh, Physical Review **D14**, 870 (1976).

S. Weinberg, Physical Review **150**, 1313 (1966).

H. A. Weldon, Physical Review **D26**, 1394 (1982); *ibid* **D67**, 085027 (2003).

K. Yamawaki, *QCD, Light Cone Physics and Hadron Phenomenology (NuSS 97)*, 116-199 (1998) (hep-th/9802037).

CHAPTER 8

Cutting rules at finite temperature

In dealing with quantum field theories, often we are interested in calculating the imaginary part of a Feynman amplitude. For example, at zero temperature, the imaginary part of a scattering amplitude is related to the scattering cross section through the unitarity relation. Namely, let

$$S = \mathbb{1} + iT, \tag{8.1}$$

where we have expressed the S-matrix in terms of the pure scattering matrix, T. Formal unitarity of the S-matrix, then, implies that

$$SS^\dagger = \mathbb{1},$$

$$\text{or,} \quad -i(T - T^\dagger) = TT^\dagger,$$

$$\text{or,} \quad -i\langle f|(T - T^\dagger)|i\rangle = \langle f|TT^\dagger|i\rangle = \sum_p \langle f|T|p\rangle\langle p|T^\dagger|i\rangle, \tag{8.2}$$

where $|p\rangle$ denotes an intermediate state. The left hand side of this equation simply gives the imaginary part of the scattering matrix taken between an initial state $|i\rangle$ and a final state $|f\rangle$ while the right hand side expresses it as a product of transition amplitudes summed over all possible intermediate states. Similarly, at finite temperature, the imaginary parts of various correlation functions can be related to different transport coefficients which are physically meaningful.

Thus, we see that the calculation of imaginary parts of amplitudes is quite important both at zero temperature as well as at finite temperature. But, as is clear from any direct computation, even at zero temperature, such a calculation is quite nontrivial. The cutting rules or Cutkosky rules as they are called, give a simple and systematic way for calculating the imaginary part of a Feynman amplitude which further makes the connection with unitarity manifest. They express the imaginary part of an n-loop amplitude in terms of physical amplitudes of lower order (namely, $(n-1)$-loop or lower). It is in this sense that they are quite useful. In the next section, we will discuss the cutting rules at zero temperature which we will subsequently generalize to finite temperature.

8.1 Zero temperature

The discussion of cutting rules is quite straightforward in the diagrammatic language and that is what we will follow in this chapter. In calculating the imaginary part of a scattering amplitude, it is quite clear that the imaginary parts can arise from two distinct sources. First, the vertices are defined from $i\mathcal{L}$ where $\mathcal{L}$ represents the Lagrangian density of the theory. We see that if the interactions are Hermitian, then, the vertices are purely imaginary. Second, the Feynman propagators are complex because of the $i\epsilon$ (Feynman) prescription (as well as the multiplicative factor of "i" with which a propagator is defined) and both of these will contribute to the imaginary part of an amplitude. It is clear that since the imaginary part of an amplitude can be obtained from the amplitude and its complex conjugate, we can give it a diagrammatic representation if we enlarge our theory to include also the complex conjugate vertices as well as the complex conjugate propagators of our original theory. We recognize that, for Hermitian interactions, complex conjugating the vertex corresponds to simply changing the sign of the vertex. However, the complex conjugate of the propagator obviously is not so simple.

For simplicity, let us consider a scalar field theory although the discussion can be carried over to any other theory. The Feynman propagator is the time ordered two point Green's function defined to be

$$iG_F(x) = \langle 0|T(\phi(x)\phi(0))|0\rangle. \tag{8.3}$$

(The propagator is normally denoted by iG. However, for simplicity of notation, we will represent the propagators (*only in this chapter*) by G. The formulae of chapters **2** and **3** can, therefore, be compared with those of the present chapter keeping in mind the identification

$iG \rightarrow G$.) In $3+1$ dimensions, the (scalar) Feynman propagator has the explicit representation

$$G_F(x) = \theta(x^0)G^{(+)}(x) + \theta(-x^0)G^{(-)}(x)$$

$$= \int \frac{d^4p}{(2\pi)^4} \frac{i}{p^2 - m^2 + i\epsilon} e^{-ip\cdot x}, \tag{8.4}$$

where we are assuming that the scalar particles have a mass m. Here $G^{(\pm)}(x)$ are the forward and backward moving components of the time ordered propagator in terms of which all other Green's functions can be expressed and $G^{(\pm)}(x)$ have the explicit representations

$$G^{(\pm)}(x) = \int \frac{d^4p}{(2\pi)^4} 2\pi\theta(\pm p_0)\delta(p^2 - m^2) e^{-ip\cdot x}. \tag{8.5}$$

We note, from (8.5), that, at zero temperature, the forward moving component of the time ordered propagator depends only on positive frequency while the backward moving component depends only on negative frequency. As we will see in the next section, this will not be true at finite temperature. We also see from (8.5) that the positive and the negative frequency components have support only on the mass shell (which will also be true at finite temperature) while the Feynman propagator is an off shell Green's function. Parenthetically, we note here that the Feynman propagator of (8.4) also has the Källen-Lehmann spectral representation given by

$$G_F(x) = i \int_0^\infty ds \int \frac{d^4p}{(2\pi)^4} \frac{\rho(s,p)}{p^2 - s + i\epsilon} e^{-ip\cdot x}, \tag{8.6}$$

where the spectral function, $\rho(s,p)$, at zero temperature, has the explicit form

$$\rho(s,p) = \rho(s) = \delta(s - m^2). \tag{8.7}$$

The coefficients of $\theta(\pm x^0)$ of a Green's function can be easily read out once we know its spectral decomposition.

We note from (8.5) that the forward and backward moving components of the propagator, $G^{(\pm)}(x)$, are not even functions. Rather, they satisfy the following symmetry properties.

$$G^{(\pm)}(-x) = G^{(\mp)}(x) = (G^{(\pm)}(x))^*. \tag{8.8}$$

They are, in fact, complex conjugates of each other. On the other hand, we see from (8.4) that, while the Feynman propagator is not a real function, it is an even function, namely,

$$G_F(-x) = G_F(x). \tag{8.9}$$

We also note that if we were to define an anti-time ordered propagator, it will have the form

$$\widetilde{G}(x) = \theta(x^0)G^{(-)}(x) + \theta(-x^0)G^{(+)}(x)$$
$$= \int \frac{d^4p}{(2\pi)^4} \frac{(-i)}{p^2 - m^2 - i\epsilon} e^{-ip\cdot x}. \tag{8.10}$$

It is immediately clear from (8.4), (8.8), (8.10) that the anti-time ordered propagator is the complex conjugate of the Feynman propagator, namely, (with $p \to -p$ in the integrand, say, in (8.10))

$$\widetilde{G}(x) = (G_F(x))^*. \tag{8.11}$$

We also note here that not all of these propagators are independent. In fact, from the definitions in (8.4) and (8.10), as well as the properties of the θ-functions, we see that

$$G_F(x) + \widetilde{G}(x)$$
$$= (\theta(x^0) + \theta(-x^0))G^{(+)}(x) + (\theta(x^0) + \theta(-x^0))G^{(-)}(x)$$
$$= G^{(+)}(x) + G^{(-)}(x). \tag{8.12}$$

Given a theory, we can define the Feynman rules associated with the theory, namely, the Feynman propagator as well as the interaction vertices of the theory. Any Feynman diagram, then, simply corresponds to a number of vertices and propagators connected to one another in a given way and integrated over internal vertices. In the coordinate representation, therefore, the integrand of the diagram can be written as a function of all the coordinates of the vertices (that is, before integrating over the internal coordinates). Thus, for example, for the scalar theory with ϕ^3 interactions, we can write the

Feynman rules of the theory to be

$$x_1 \quad\rule{2cm}{0.4pt}\quad x_2 = G_F(x_1 - x_2),$$

$$= -ig, \tag{8.13}$$

where g denotes the coupling constant (or the strength of the interaction) and from these Feynman rules, we can obtain a representative set of Feynman diagrams to correspond to (the diagrams are to be understood without the external propagators)

$$= F(x_1, x_2)$$
$$= (-ig)^2 G_F(x_1 - x_2) G_F(x_2 - x_1), \tag{8.14}$$

$$= F(x_1, x_2, x_3)$$
$$= (-ig)^3 G_F(x_1 - x_2) G_F(x_2 - x_3) G_F(x_3 - x_1). \tag{8.15}$$

Let us, next, enlarge our theory such that the Feynman rules of the enlarged theory are given by

$$x_1 \quad\rule{2cm}{0.4pt}\quad x_2 = G_F(x_1 - x_2),$$
$$x_1 \quad\rule{2cm}{0.4pt}\!\!\bullet\!\!\rule{0.5cm}{0.4pt} x_2 = G^{(-)}(x_1 - x_2) = G^{(+)}(x_2 - x_1),$$

$$x_1 \!\!\! \circ\!\!-\!\!\!- x_2 \;=\; G^{(+)}(x_1 - x_2) = G^{(-)}(x_2 - x_1),$$

$$x_1 \!\!\! \circ\!\!-\!\!\!- \!\!\! \circ x_2 \;=\; G_F^*(x_1 - x_2) = \widetilde{G}_F(x_1 - x_2), \qquad (8.16)$$

$$\begin{aligned} x_1 \\ \\ x_2 \end{aligned} \!\!\!\!\!\!\! \Big\rangle\!\!-\!\!\!\!-\!\!\!- x_3 \;=\; -ig,$$

$$\begin{aligned} x_1 \\ \\ x_2 \end{aligned} \!\!\!\!\!\!\! \Big\rangle\!\!\odot\!\!\!-\!\!\!- x_3 \;=\; ig. \qquad (8.17)$$

In the momentum space, the propagators have the following forms

$$\underset{p}{\overline{}} \;=\; \frac{i}{p^2 - m^2 + i\epsilon},$$

$$\underset{p}{-\!\!\!\!\longrightarrow\!\!\circ} \;=\; 2\pi\theta(-p_0)\delta(p^2 - m^2),$$

$$\underset{p}{\circ\!\!\longrightarrow\!\!\!-} \;=\; 2\pi\theta(p_0)\delta(p^2 - m^2),$$

$$\underset{p}{\circ\!\!-\!\!\!-\!\!\circ} \;=\; -\frac{i}{p^2 - m^2 - i\epsilon}, \qquad (8.18)$$

where we note that the direction of the momentum flow is quite important when one of the ends of a propagator is circled because, as we have observed earlier in (8.8), such propagators correspond to the negative/positive frequency propagators which are not even functions.

The important thing to note from the enlarged set of Feynman rules given in (8.16)-(8.18) is the fact that the circled vertex is merely the complex conjugate of the original vertex and, similarly, the doubly circled propagator (or the anti-time ordered propagator) is the complex conjugate of the Feynman propagator of the ϕ^3 theory. It is clear, therefore, that the new Feynman rules of the enlarged theory also contain the complex conjugates of the interaction vertex as well as that of the propagator of the original theory. Furthermore, with these new Feynman rules, we can now draw new kinds of Feynman diagrams where some of the vertices may be circled. The integrand of a Feynman diagram, in this new theory, can still be represented as a function of the coordinates of the vertices (external as well as internal) and we will identify a circled vertex merely by underlying the corresponding coordinate. With this convention, then, a small representative of the new Feynman diagrams, for example, would correspond to

$$= F(x_1, \underline{x_2}),$$
$$= (-ig)(ig)G^{(-)}(x_1 - x_2)G^{(-)}(x_1 - x_2), \qquad (8.19)$$

$$= F(\underline{x_1}, \underline{x_2})$$
$$= (ig)^2 G_F^*(x_1 - x_2)G_F^*(x_1 - x_2)$$

$$\qquad\qquad (8.20)$$

$$= F(x_1, x_2, \underline{x_3})$$

$$= (-ig)^2(ig)G_F(x_1 - x_2)G^{(-)}(x_2 - x_3)G^{(-)}(x_1 - x_3). \qquad (8.21)$$

It is now clear that a Feynman diagram where all the vertices are circled is nothing other than the complex conjugate of the diagram

without any circling (namely, of the original theory). This is because
in such a case, all the vertices as well as the propagators are the
complex conjugates of the original diagram. Algebraically, we can
write

$$F(\underline{x_1}, \underline{x_2}, \cdots, \underline{x_n}) = (F(x_1, x_2, \cdots, x_n))^*. \qquad (8.22)$$

This, therefore, gives us a diagrammatic method for constructing the
complex conjugate and, therefore, the imaginary part of a Feynman
amplitude of the original theory. However, it is not yet in a simple
form which is convenient for computations.

The graphs with circled vertices satisfy many identities simply
because the propagators satisfy various identities. Let us note here
one which will be useful in the description of the cutting rules. Let
us first consider a two point Feynman diagram with coordinates x_1
and x_2 where we assume that $x_1^0 > x_2^0$. Then, it follows that (the
dots represent vertices and we are neglecting the coupling constants
for simplicity, just the signs of the vertices are relevant here, the first
two diagrams can be thought of as being internal parts of a larger
Feynman diagram as is clear in the examples following them)

$$x_1 \; \bullet\!\!-\!\!-\!\!-\!\!\bullet \; x_2 + x_1 \; \circledcirc\!\!-\!\!-\!\!-\!\!\bullet \; x_2$$

$$= \; G^{(+)}(x_1 - x_2) + (-)G^{(+)}(x_1 - x_2)$$

$$= 0. \qquad (8.23)$$

$$x_1 \; \bullet\!\!-\!\!-\!\!\circledcirc \, x_2 + x_1 \; \circledcirc\!\!-\!\!-\!\!\circledcirc \, x_2$$

$$= (-)G^{(-)}(x_1 - x_2) + (-)^2 G^{(-)}(x_1 - x_2)$$

$$= 0. \qquad (8.24)$$

$$= (-ig)^2 (G^{(+)}(x_1 - x_2))^2 + (-ig)(ig)(G^{(+)}(x_1 - x_2))^2$$

$$= 0. \qquad (8.25)$$

$$= (-ig)(ig)(G^{(-)}(x_1 - x_2))^2 + (ig)^2(G^{(-)}(x_1 - x_2))^2$$
$$= 0. \tag{8.26}$$

In deriving these relations, we have used the fact that a circled vertex gives an extra $(-)$ sign and that, for $x_1^0 > x_2^0$, $G_F(x_1 - x_2) = G^{(+)}(x_1 - x_2)$ and $\tilde{G}(x_1 - x_2) = G^{(-)}(x_1 - x_2)$. Next, let us consider a three point function with coordinates x_1, x_2 and x_3 and let us assume that x_1^0 represents the largest of the three time coordinates. Then, it is easy to check that

$$= (-ig)^3 \left[G^{(+)}(x_1 - x_3) \, G_F(x_2 - x_3) \, G^{(+)}(x_1 - x_2) \right.$$
$$\left. + (-) \, G^{(+)}(x_1 - x_3) \, G_F(x_2 - x_3) \, G^{(+)}(x_1 - x_2) \right] = 0. \tag{8.27}$$

$$= (-ig)^3 \left[(-) \, G^{(-)}(x_1 - x_3) \, G^{(+)}(x_3 - x_2) \, G^{(+)}(x_1 - x_2) \right.$$
$$\left. + (-)^2 \, G^{(-)}(x_1 - x_3) G^{(+)}(x_3 - x_2) G^{(+)}(x_1 - x_2) \right] = 0. \tag{8.28}$$

In this way, one derives an interesting relation. Namely, in this enlarged theory, if we take an amplitude with the largest time vertex uncircled and add to it the corresponding amplitude with the largest time vertex circled, then, the sum vanishes. This is known as the largest time equation. (Similarly, there is a corresponding smallest time equation also.) An immediate consequence of the largest time equation (or the smallest time equation) is that for a particular amplitude, the sum of graphs with all possible circlings vanish. This can be easily checked, for example, from

$$= (-ig)^2 \big[(G_F(x_1 - x_2))^2 + (-)(G^{(-)}(x_1 - x_2))^2$$

$$+ (-)(G^{(+)}(x_1 - x_2))^2 + (-)^2(\widetilde{G}(x_1 - x_2))^2 \big]$$

$$= (-ig)^2 \big[(\theta(x_1^0 - x_2^0)(G^{(+)}(x_1 - x_2))^2$$

$$+ \theta(x_2^0 - x_1^0)(G^{(-)}(x_1 - x_2))^2)$$

$$- (G^{(-)}(x_1 - x_2))^2 - (G^{(+)}(x_1 - x_2))^2$$

$$+ (\theta(x_1^0 - x_2^0)(G^{(-)}(x_1 - x_2))^2 + \theta(x_2^0 - x_1^0)(G^{(+)}(x_1 - x_2))^2) \big]$$

$$= 0. \tag{8.29}$$

This can be intuitively expected because for a given distribution of the time arguments of a graph, diagrams with the largest time circled and uncircled would cancel pairwise because of the largest time equation. (We note that the number of graphs with all possible circlings corresponding to any given amplitude is even.) It is also worth emphasizing that all these identities hold both in the coordinate space as well as in the momentum space.

Algebraically, therefore, we see that for any Feynman diagram, we can write

$$\sum_{\text{all underlinings}} F(x_1, x_2, \cdots, \underline{x_i}, \cdots, x_n) = 0. \tag{8.30}$$

This also implies that

$$F(x_1, x_2, \cdot, x_n) + F(\underline{x_1}, \underline{x_2}, \cdots, \underline{x_n})$$

$$= -\sum{}' F(x_1, \cdots, \underline{x_i}, \cdots, x_n). \tag{8.31}$$

Here the prime on the right hand side over the sum stands for all possible circlings except the cases without any and with all circlings. Furthermore, recognizing that a Feynman diagram really gives "i" times a T-matrix element, namely,

$$F \sim iT, \tag{8.32}$$

we see that we can relate the left hand side of (8.31) to the imaginary part of a given Feynman amplitude and it can be diagrammatically expressed in terms of graphs with all possible circlings (except no circling and all circlings). The right hand side of (8.31) still involves a large number of graphs and hence is not in a simple form for calculations. Besides, it is not yet in a form where the unitarity relation of (8.2) is manifest.

Further simplification of the right hand side of (8.31) comes from the following observations. Let us consider the following self-energy graph with an isolated internal vertex which is circled. In this case, we see quite easily, from the definition of the propagators in (8.18) that

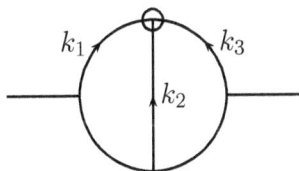

$$\sim G^{(-)}(k_1) G^{(-)}(k_2) G^{(-)}(k_3)$$

$$\sim \theta(-(k_1)_0) \theta(-(k_2)_0) \theta(-(k_3)_0) = 0, \tag{8.33}$$

where the last equality follows because, by energy-momentum conservation, $k_1 + k_2 + k_3 = 0$. Therefore, at least one of the k_0s has to be positive which makes, at least, one of the theta functions vanish. This simple analysis shows that any Feynman diagram with an isolated circled internal vertex vanishes. This reduces the number of diagrams to be considered on the right hand side of (8.31). In a completely parallel manner, we can show that energy-momentum conservation makes Feynman diagrams with an isolated uncircled internal vertex also to vanish. (In fact, this follows trivially because a diagram with an isolated uncircled vertex is the complex conjugate of a diagram with an isolated circled vertex.) Furthermore, if we assume that particles are incoming from the left and are going out at the right – that is, if we assume that energy is flowing from left to right – then, it is straightforward to show that we cannot have any uncircled vertex attached to the external lines on the left surrounded by circled vertices. Thus, for example, the following graph would vanish (see, (8.18)), namely,

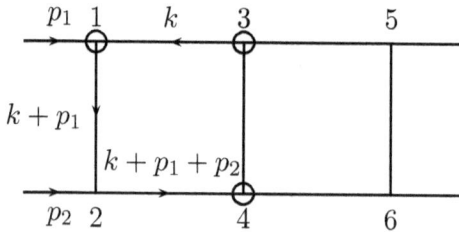

$$\sim \theta(k_0 + (p_1)_0)\theta(-k_0 - (p_1)_0 - (p_2)_0) = 0, \tag{8.34}$$

since, by assumption, $(p_2)_0$ is positive (incoming energy) and $k_0 + (p_1)_0$ has to be positive (because of the first theta function). In fact, a systematic analysis shows that the only diagrams that can give nonvanishing contribution to the right hand side of (8.31) are the ones where the circled and the uncircled vertices form connected regions, connected with the external lines with no isolated islands (of either circled or uncircled type). Furthermore, with our convention of representing the incoming particles as coming in from the left and leaving at the right, diagrams with only circled vertices attached to the external lines at the left and uncircled vertices attached to the external lines at the right can give nonvanishing contribution. (With the opposite convention, namely, particles coming in from the right and going out at the left, the nonvanishing diagrams would merely correspond to the transposed ones.)

With these properties, we realize that a large number of graphs on the right hand side of (8.31) vanish and that the ones that are nontrivial can be given a cutting description since the circled and the uncircled vertices form connected regions. Thus, for example, we can represent

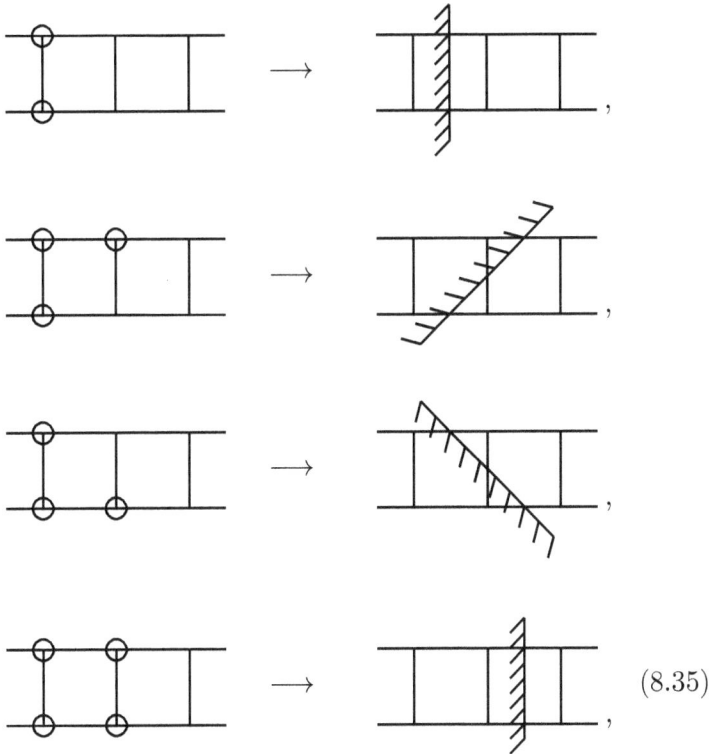

$$(8.35)$$

where any vertex in the shadow region is understood to be a circled vertex. Since all the vertices and the propagators in the shadow region correspond to the complex conjugates of vertices and propagators of the original theory, the part of the diagram in the shadow can be thought of as the complex conjugate of a matrix element of the original theory. As a consequence, we can write the imaginary part of any Feynman amplitude in terms of cut diagrams which will also lead to a natural interpretation in terms of unitarity (see (8.2)).

Thus, for example, we note that

$$\begin{aligned} 2\,\mathrm{Im} \quad & \\ &= \qquad\qquad = \qquad\qquad , \end{aligned} \tag{8.36}$$

$$\begin{aligned} 2\,\mathrm{Im} \quad & \\ &= \qquad\qquad + \qquad\qquad \\ &+ \qquad\qquad + \qquad\qquad \\ &= \qquad\qquad + \qquad\qquad \\ &+ \qquad\qquad + \qquad\qquad . \end{aligned} \tag{8.37}$$

In general, we can write (for a two point function, in this case)

$$\mathrm{Im} \quad = \frac{1}{2} \qquad . \tag{8.38}$$

These are the cutting rules at zero temperature which say that the imaginary part of a given Feynman diagram is given by the sum of all possible cuttings of the diagram which separate the circled and the uncircled vertices as well as the external lines such that there are no isolated islands of circled/uncircled internal vertices not connected to the external lines. We note that a cut propagator is an on-shell propagator (since it connects a circled and an uncircled vertex) and, therefore, the cutting rules express the imaginary part of a Feynman amplitude in terms of physical, on-shell amplitudes of lower order. We also note here that, crucial to the cutting description of the imaginary part of a diagram, is the fact that a diagram with an isolated circled/uncircled internal vertex vanishes. This follows because, at zero temperature, a propagator with one end circled and the other uncircled has contribution only from positive/negative energy (see, (8.18)), namely,

$$\underset{p}{\xrightarrow{\hspace{2cm}}}\!\!\!\bigcirc \quad = 2\pi\theta(-p_0)\delta(p^2 - m^2),$$

$$\bigcirc\!\!\!\underset{p}{\xrightarrow{\hspace{2cm}}} \quad = 2\pi\theta(p_0)\delta(p^2 - m^2). \tag{8.39}$$

As we will see later, this is no longer true at finite temperature. In fact, let us note here that if such diagrams, with an isolated circled/uncircled internal vertex, did not vanish, then, the imaginary part of an amplitude cannot have a cutting description in terms of physical amplitudes which is evident from the following simple self-energy diagram

$$\tag{8.40}$$

which does not allow a cutting description in the sense described above.

8.2 Finite temperature

We have already seen, in chapter **1**, in the imaginary time formalism, that the one-loop self-energy at finite temperature exhibits a cutting structure as long as we allow for additional channels of reaction to be present in a medium (see (1.135) as well as the discussion following that). It is natural, therefore, to ask whether the notion of cutting generalizes to all orders at finite temperature. As we have noted earlier, if this were true, calculation of imaginary parts of amplitudes which are related to transport coefficients would simplify enormously.

The question of cutting rules is best described in the real time formalism and we will use the closed time path formalism (see, chapter **2**) for our discussions. (Our arguments can be extended to thermofield dynamics as well as to arbitrary paths in the real time formalism with appropriate modifications.) The real time formalism at finite temperature differs from the zero temperature field theory mainly in two respects. First, there is a (thermal) doubling of the degrees of freedom (fields) which leads to a 2×2 matrix structure for the propagator and also leads to two kinds of vertices (on the two branches of the real time path) with opposite sign. Second, the fields obey a periodic/anti-periodic boundary condition (for bosons/fermions) with the Green's functions satisfying the appropriate KMS conditions (see (1.36) or (3.129) or (4.176)-(4.178)). Thus, for example, for a ϕ^3 theory, the 2×2 propagator matrix, for the scalar field (satisfying periodic boundary condition), as we have seen in (2.50), has four components in the momentum space, given by (recall that, in this chapter, we are using $iG \to G$)

$$G_{++}(p) = \frac{i}{p^2 - m^2 + i\epsilon} + 2\pi n(|p_0|)\delta(p^2 - m^2),$$

$$G_{+-}(p) = 2\pi(\theta(-p_0) + n(|p_0|))\delta(p^2 - m^2),$$

$$G_{-+}(p) = 2\pi(\theta(p_0) + n(|p_0|))\delta(p^2 - m^2),$$

$$G_{--}(p) = -\frac{i}{p^2 - m^2 - i\epsilon} + 2\pi n(|p_0|)\delta(p^2 - m^2), \tag{8.41}$$

where the boson distribution function is given by

$$n(|p_0|) = n_B(|p_0|) = \frac{1}{e^{\beta|p_0|} - 1}.$$

The two kinds of vertices are, similarly, given by

$$= -ig,$$

$$= ig. \tag{8.42}$$

It is worth noting here, from the structure of the components of the propagator given in (8.41), that

$$G_{--}(p) = (G_{++}(p))^*,$$

$$G_{++}(p) + G_{--}(p) = G_{+-}(p) + G_{-+}(p). \tag{8.43}$$

Namely, these identities involving the thermal propagators are very similar to the zero temperature relations given in (8.11) and (8.12) respectively although the meanings are quite different. Here the matrix components refer to propagation between the two different real branches and, as a result, the off-diagonal elements no longer correspond to having only positive/negative frequency (energy) components (see, for example, (8.18) or (8.39)).

From the forms of the momentum space propagators given in (8.41), we can obtain the (Källen-Lehmann) spectral representation for the matrix components of the propagator. Because of the δ-functions in (8.41), the components, (unlike the zero temperature case), have contributions from the entire complex energy plane and consequently, the spectral representation takes the form (compare with (8.5)-(8.6))

$$G_{ab}(x) = i \int_0^\infty ds \int \frac{d^4p}{(2\pi)^4} \left[\frac{\rho_{ab}(s,p)}{p^2 - s + i\epsilon} + \frac{\widetilde{\rho}_{ab}(s,p)}{p^2 - s - i\epsilon} \right] e^{-ip \cdot x}, \tag{8.44}$$

where the spectral functions ρ_{ab} and $\widetilde{\rho}_{ab}$ can be read out from (8.41) to be

$$\rho_{++}(s,p) = (1 + n(|p_0|))\delta(s - m^2) = -\widetilde{\rho}_{--}(s,p),$$

$$\rho_{--}(s,p) = n(|p_0|)\delta(s - m^2) = -\widetilde{\rho}_{++}(s,p),$$

$$\rho_{+-}(s,p) = (\theta(-p_0) + n(|p_0|))\delta(s - m^2) = -\widetilde{\rho}_{+-}(s,p),$$

$$\rho_{-+}(s,p) = (\theta(p_0) + n(|p_0|))\delta(s - m^2) = -\widetilde{\rho}_{-+}(s,p). \tag{8.45}$$

These can be compared with the zero temperature case described in (8.5)-(8.7). Given the spectral representations in (8.44) and (8.45), we can now express the decomposition of the propagators as

$$G_{ab}(x) = \theta(x^0)G_{ab}^{(+)}(x) + \theta(-x^0)G_{ab}^{(-)}(x), \tag{8.46}$$

where $G_{ab}^{(\pm)}(x)$ continue to be the forward/backward moving components of the propagator, but now have the explicit forms

$$G_{ab}^{(\pm)}(x) = \int_0^\infty ds \int \frac{d^4p}{(2\pi)^4}\, 2\pi\delta(p^2 - m^2)\, e^{-ip\cdot x}$$

$$\times \left[\theta(\pm p_0)\rho_{ab}(s,p) - \theta(\mp p_0)\tilde\rho_{ab}(s,p)\right], \quad (8.47)$$

satisfying, as in (8.8),

$$G_{ab}^{(\pm)}(-x) = G_{ab}^{(\mp)}(x) = \left(G_{ab}^{(\pm)}(x)\right)^*. \quad (8.48)$$

As a result, we see that the forward and backward moving components of the propagator no longer coincide with the positive and negative frequency components of the propagator, as was the case at zero temperature (see (8.5)) and, therefore, lot of the arguments used in obtaining the cutting rules at zero temperature have to be now scrutinized and modified at finite temperature, if cutting rules have to hold.

Following the discussion of the zero temperature case, we can now introduce the circled propagators for our enlarged theory at finite temperature (recall that the circled vertices involve complex conjugation). They are given by $(a, b = \pm)$

$$\underset{a,x \qquad\qquad b,y}{\rule{3cm}{0.4pt}} = G_{ab}(x-y),$$

$$\underset{a,x \qquad\qquad b,y}{\rule{3cm}{0.4pt}\circ} = G_{ab}^{(-)}(x-y),$$

$$\circ\underset{a,x \qquad\qquad b,y}{\rule{3cm}{0.4pt}} = G_{ab}^{(+)}(x-y),$$

$$\circ\underset{a,x \qquad\qquad b,y}{\rule{3cm}{0.4pt}}\circ = \widetilde{G}_{ab}(x-y) = (G_{ab}(x-y))^*, \quad (8.49)$$

where the superscripts $(\pm)$ denote respectively the forward and backward moving components of the propagator. For the propagators where only one end is circled, the momentum flow is assumed to be from the first thermal index to the second, namely, $a \to b$ (see (8.39)). The important point to note here is that, like at zero temperature, all the propagators can be expressed in terms of the two fundamental functions $G_{ab}^{(\pm)}(x-y)$. In fact, from the structure of the spectral functions in (8.45), we can easily show that (8.47) gives

$$G_{+-}^{(+)}(x) = G_{+-}^{(-)}(x) = G_{+-}(x),$$

$$G_{-+}^{(+)}(x) = G_{-+}^{(-)}(x) = G_{-+}(x),$$

$$G_{++}^{(+)}(x) = G_{--}^{(-)}(x) = G_{-+}(x),$$

$$G_{++}^{(-)}(x) = G_{--}^{(+)}(x) = G_{+-}(x). \tag{8.50}$$

For later use, let us note here from (8.47) and (8.49) that

$$\tilde{G}_{ab}^{(\pm)}(x) = \left(G_{ab}^{(\pm)}(x)\right)^* = \int_0^\infty \mathrm{d}s \int \frac{\mathrm{d}^4 p}{(2\pi)^4}\, 2\pi\delta(p^2 - m^2)\, e^{ip\cdot x}$$

$$\times\, [\theta(\pm p_0)\rho_{ab}(s,p) - \theta(\mp p_0)\tilde{\rho}_{ab}(s,p)]$$

$$= \int_0^\infty \mathrm{d}s \int \frac{\mathrm{d}^4 p}{(2\pi)^4}\, 2\pi\delta(p^2 - m^2)\, e^{-ip\cdot x}$$

$$\times\, [\theta(\mp p_0)\rho_{ab}(s,p) - \theta(\pm p_0)\tilde{\rho}_{ab}(s,p)]$$

$$= G_{ab}^{(\mp)}(x), \tag{8.51}$$

so that we can write

$$\tilde{G}_{ab}(x) = \theta(x^0)\tilde{G}_{ab}^{(+)}(x) + \theta(-x^0)\tilde{G}_{ab}^{(-)}(x)$$

$$= \theta(x^0)G_{ab}^{(-)}(x) + \theta(-x^0)G_{ab}^{(+)}(x). \tag{8.52}$$

This can be compared with (8.10) and (8.50)-(8.52) further allow us to write explicitly the propagators where both the ends are circled as

$$\tilde{G}_{+-}^{(+)}(x) = G_{+-}^{(-)}(x) = G_{+-}^{(+)}(x) = \tilde{G}_{+-}^{(-)}(x) = G_{+-}(x),$$

$$\tilde{G}_{-+}^{(+)}(x) = G_{-+}^{(-)}(x) = G_{-+}^{(+)}(x) = \tilde{G}_{-+}^{(-)}(x) = G_{-+}(x),$$

$$\tilde{G}_{++}^{(+)}(x) = G_{++}^{(-)}(x) = G_{--}^{(+)}(x) = \tilde{G}_{--}^{(-)}(x) = G_{+-}(x),$$

$$\tilde{G}_{++}^{(-)}(x) = G_{++}^{(+)}(x) = G_{--}^{(-)}(x) = \tilde{G}_{--}^{(+)}(x) = G_{-+}(x). \tag{8.53}$$

This shows that the two fundamental (independent) components of the propagators are $G_{+-}(x)$ and $G_{-+}(x)$ in terms of which all the components of the propagators can be described and from the explicit forms of these Green's functions in momentum space, given in (8.41), we note that

$$G_{+-}(p) = 2\pi(\theta(-p_0) + n(|p_0|))\delta(p^2 - m^2),$$

$$G_{-+}(p) = 2\pi(\theta(p_0) + n(|p_0|))\delta(p^2 - m^2), \qquad (8.54)$$

where the momentum flow is from the first thermal index to the second. Therefore, we see that while a propagator with one end circled and the other uncircled (described by (8.54)) is still on-shell, it contains both positive and negative energy contributions unlike at zero temperature. This has the immediate consequence that, unlike at zero temperature, at finite temperature isolated circled vertices will no longer vanish (see (8.39) and the discussion there). As we had argued earlier, vanishing of such diagrams was quite essential for the cutting description to hold at zero temperature and, therefore, this raises a major challenge for generalizing the cutting rules to finite temperature.

To see how exactly the cutting rules generalize to finite temperature, we note that, in this enlarged theory at finite temperature, we would have four possible vertices (unlike two at zero temperature given in (8.17))

$$= -ig,$$

$$= ig,$$

$$= ig,$$

$$= -ig. \qquad (8.55)$$

With these, one can derive, once again like at zero temperature, the largest time (smallest time) equation. This would again lead to the result that, for a given Feynman diagram with a given distribution of thermal vertices, the sum over all possible circlings must vanish, namely,

$$\sum_{\text{all underlinings}} F_{a_1 a_2 \cdots a_n}(x_1, x_2, \cdots, \underline{x_i}, \cdots, x_n) = 0, \tag{8.56}$$

with $a_1, a_2, \cdots, a_n$ fixed with the indices taking values $(\pm)$. This, in turn, will yield the imaginary part of a diagram with a given distribution of thermal vertices (remember $F \sim iT$) as in (8.31), namely,

$$F_{a_1 \cdots a_n}(x_1, \cdots, x_n) + F_{a_1 \cdots a_n}(\underline{x_1}, \cdots, \underline{x_n})$$

$$= - \sum_{\text{all underlinings}}' F_{a_1 \cdots a_n}(x_1, \cdots, \underline{x_i}, \cdots, x_n). \tag{8.57}$$

As we have argued before, imaginary parts of these individual diagrams cannot have a cutting description at finite temperature simply because isolated circled vertices no longer vanish (compare with (8.33)-(8.34)) at finite temperature since the forward and backward moving components of the propagator (one end uncircled and the other circled) have both positive and negative frequency (energy) contributions (see (8.47)-(8.49)). (Incidentally, such nontrivial graphs arise only at two loops or higher order.) However, at finite temperature, a physical amplitude is obtained by summing over all possible thermal indices of the intermediate internal vertices (mainly because of the matrix structure of the theory). Therefore, although at finite temperature, non-cuttable diagrams occur graph by graph, it is possible that when summed over the thermal indices of the internal vertices, all such diagrams would cancel giving a cutting description to the complete, physical amplitude. This is exactly what happens as we will show next.

To show the cancellation of the non-cuttable graphs, let us start with the following interesting observations. First, let us explicitly write down the forms of the propagators where one end is circled (which follows from (8.49) and (8.50))

$$
\begin{aligned}
\underset{+,x \qquad\qquad +,y}{\rule{4cm}{0.4pt}\!\!\circ} &= G_{+-}(x-y), \\[1ex]
\underset{+,x \qquad\qquad -,y}{\rule{4cm}{0.4pt}\!\!\circ} &= G_{+-}(x-y), \\[1ex]
\underset{-,x \qquad\qquad +,y}{\rule{4cm}{0.4pt}\!\!\circ} &= G_{-+}(x-y), \\[1ex]
\underset{-,x \qquad\qquad -,y}{\rule{4cm}{0.4pt}\!\!\circ} &= G_{-+}(x-y).
\end{aligned}
\tag{8.58}
$$

It is immediately clear from (8.58) that the form of the propagators, where one end is circled, is independent of the thermal index of the circled vertex and is completely determined by the thermal index of the uncircled vertex (in fact, it is the opposite of the uncircled vertex). This, as we will see, is crucial to proving various cancellations. The other fact that is quite important is that, at finite temperature, Green's functions obey the KMS condition (see (1.36) or (3.129)) which implies (which can also be explicitly checked from (2.42)) that, for boson propagators,

$$
G_{+-}(t,\mathbf{x}) = G_{-+}(t - i\beta, \mathbf{x}).
\tag{8.59}
$$

In the momentum space, this translates to

$$
G_{+-}(p) = e^{-\beta p_0}\, G_{-+}(p),
$$

$$
\text{or,}\quad G_{-+}(p) = e^{\beta p_0}\, G_{+-}(p).
\tag{8.60}
$$

This can, of course, be checked explicitly from the forms of the propagators given in (8.54) and the properties of the distribution function, for example,

$$
\begin{aligned}
G_{+-}(p) &= (\theta(-p_0) + n(|p_0|)) \\
&= (\theta(p_0)n(|p_0|) + \theta(-p_0)(1 + n(|p_0|))) \\
&= (\theta(p_0) + \theta(-p_0)e^{-\beta p_0})n(|p_0|) \\
&= e^{-\beta p_0}(\theta(p_0)e^{\beta p_0} + \theta(-p_0))n(|p_0|) \\
&= e^{-\beta p_0}(\theta(p_0) + n(|p_0|)) = e^{-\beta p_0}\, G_{-+}(p),
\end{aligned}
\tag{8.61}
$$

where the last line results from using the identity $e^{\beta|p_0|} n(|p_0|) = (1 + n(|p_0|))$ in the first term. In general, we can write the KMS condition to give

$$G_{a(-a)}(p) = e^{-a\beta p_0} \, G_{(-a)a}(p). \tag{8.62}$$

This relation is self-consistent in the sense that

$$G_{a(-a)}(p) = e^{-a\beta p_0} \, G_{(-a)a}(p) = e^{-a\beta p_0} \times e^{a\beta p_0} \, G_{a(-a)}(p)$$
$$\equiv G_{a(-a)}(p). \tag{8.63}$$

The next important thing to recognize is that any diagram in this enlarged theory at finite temperatures is just a collection of circled and uncircled vertices (of a given distribution of thermal indices) connected to one another through appropriate propagators and can be represented in one of the following three topologically distinct categories. (There can also be class I diagrams with the blobs Γ_1 and Γ_2 interchanged.)

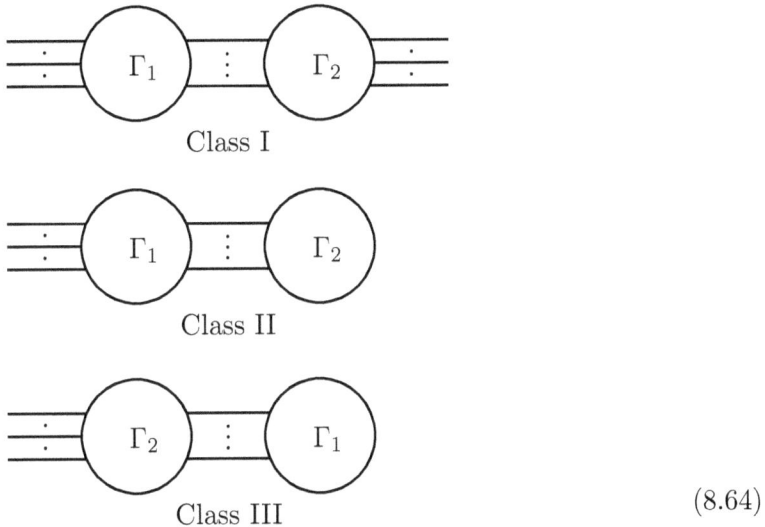

Class I

Class II

Class III

$$\tag{8.64}$$

Here we assume that the blob, Γ_1, contains only circled vertices whereas the blob, Γ_2, contains only uncircled vertices. (Namely, we have grouped all the circled and uncircled vertices into two distinct blobs.) The vertices between the blobs may be connected in a highly nontrivial manner and that not all vertices in a (single) blob may be connected to one another within the same blob. Thus, for example, we can represent

$$(8.65)$$

In the second and the third topological classes of diagrams in (8.64), we note that vertices in Γ_2/Γ_1, respectively, are not connected to external lines. Therefore, these classes of diagrams cannot be given a cutting description because cutting all the (intermediate) propagators connecting the two blobs would give rise to an island of isolated uncircled/circled vertices (respectively) not connected to external lines. Only graphs of class I can, possibly, be given a cutting description by cutting the intermediate propagators (see, for example, (8.38)). However, parts of the first topological class of diagrams where cuttings can produce isolated islands of disconnected circled/uncircled vertices cannot also be given a meaningful cutting description. For example, this can happen if there is an island of vertices within a blob, say Γ_1, connected among themselves and only to the blob Γ_2, but not connected to the external lines. Therefore, such diagrams have to vanish when summed over the internal thermal indices if cutting rules have to generalize to finite temperature. The only class of graphs which can be given a cutting description will have to come from graphs of the type class I, where vertices in each of the two blobs Γ_1 and Γ_2 are connected to the external lines, among themselves (without any isolated islands) and with vertices in the other blob so that when the intermediate lines are cut, they separate into a set of circled/uncircled vertices connected to external lines (as well as to cut internal lines) and do not produce an island of vertices (circled/uncircled) not connected to external lines (see (8.38)). This will generalize the cutting rule to finite temperature as well, for the whole amplitude – not graph by graph.

Let us now consider the class of diagrams where there is an isolated set of circled internal vertices, not connected to the external lines. This could be, for example, diagrams of the class III or those of class I where there exists a disconnected set of internal circled vertices. Clearly, since these circled vertices are all internal, their thermal indices have to be summed over $(\pm)$ in the calculation of any amplitude. Let us label the coordinate of the vertex among these circled ones with the smallest time to be x. This vertex, of course, may be connected to uncircled vertices of the type $+$ at coordinates $y_1, y_2, \cdots, y_n$ as well as to uncircled vertices of the type $-$ at coordinates $z_1, z_2, \cdots, z_m$. In addition, the circled vertex at x may also be connected to the circled vertices of the cluster of the type $+$ at $r_1, r_2, \cdots, r_p$ and of the type $-$ at $s_1, s_2, \cdots, s_q$. The vertex x may, in general, also be connected to itself l-times. Let us next sum over the thermal index of the circled coordinate at x, for such a general graph, with all other thermal indices fixed.

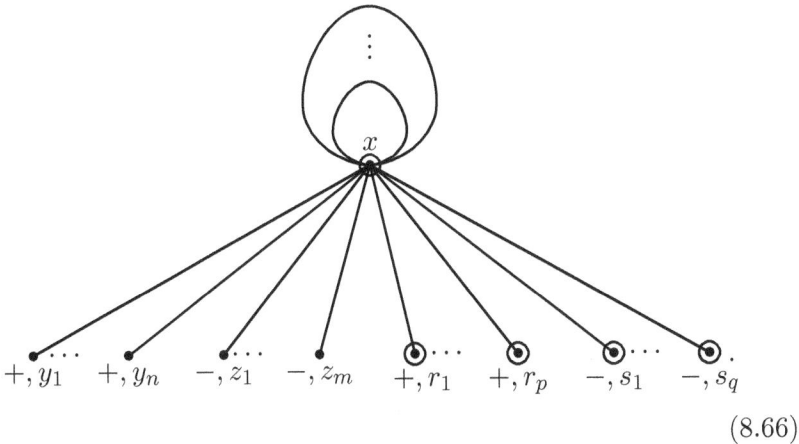

$$(8.66)$$

In evaluating this sum, the crucial thing to note is the observation we made earlier, namely, the propagator connecting a circled and an uncircled vertex does not depend on the thermal index of the circled vertex (see, (8.58)) and, in fact, the propagator behaves as if the thermal index of the circled vertex is the opposite of that of the uncircled vertex. As a result, in the sum over the thermal index of the internal vertex x in (8.66), the propagators connecting x to the uncircled vertices (at $y_1, \cdots, y_n$ and $z_1, \cdots, z_m$) factorize (since we have fixed the thermal indices of the uncircled vertices) and we obtain (remember that x has the smallest time coordinate)

$$
\begin{aligned}
\sum_{a=\pm} \text{diagram} &= R G^{(-)}_{-+}(x - y_1) \cdots G^{(-)}_{-+}(x - y_n) \\
&\quad \times G^{(-)}_{+-}(x - z_1) \cdots G^{(-)}_{+-}(x - z_m) \\
&\quad \times \Big[(\widetilde{G}_{--}(0))^l \big(\widetilde{G}^{(-)}_{-+}(x - r_1) \cdots \widetilde{G}^{(-)}_{-+}(x - r_p) \\
&\qquad\qquad \times \widetilde{G}^{(-)}_{--}(x - s_1) \cdots \widetilde{G}^{(-)}_{--}(x - s_q) \big) \\
&\quad - (\widetilde{G}_{++}(0))^l \big(\widetilde{G}^{(-)}_{++}(x - r_1) \cdots \widetilde{G}^{(-)}_{++}(x - r_p) \\
&\qquad\qquad \times \widetilde{G}^{(-)}_{+-}(x - s_1) \cdots \widetilde{G}^{(-)}_{+-}(x - s_q) \big) \Big]. \quad (8.67)
\end{aligned}
$$

Here R stands for the rest of the diagram whose actual form is not very important except for simply noting that it remains the same for the two values of the thermal index at x, $a = \pm$. The relative sign difference between the two terms in the bracket comes from the change in the vertex type, namely, $(- \to +)$. Furthermore, using the relations given in (8.46), (8.50), (8.52) and (8.53), we obtain

$$
\widetilde{G}_{++}(0) = G_{++}(0) = G_{--}(0) = \widetilde{G}_{--}(0), \quad (8.68)
$$

as well as

$$
\begin{aligned}
\widetilde{G}^{(-)}_{\mp\pm}(x) &= G_{\mp\pm}(x), \\
\widetilde{G}^{(-)}_{++}(x - s_i) &= G_{-+}(x - s_i), \quad i = 1, 2, \cdots, q, \\
\widetilde{G}^{(-)}_{--}(x - r_j) &= G_{+-}(x - r_j), \quad j = 1, 2, \cdots, p, \quad (8.69)
\end{aligned}
$$

which lead to the result that

$$
\begin{aligned}
\sum_{a=\pm} \text{diagram} &= R(G_{++}(0))^l G_{-+}(x - y_1) \cdots G_{-+}(x - y_n) \\
&\quad \times G_{+-}(x - z_1) \cdots G_{+-}(x - z_m) \\
&\quad \times [G_{-+}(x - r_1) \cdots G_{-+}(x - r_p) G_{+-}(x - s_1) \cdots G_{+-}(x - s_q) \\
&\quad - G_{-+}(x - r_1) \cdots G_{-+}(x - r_p) G_{+-}(x - s_1) \cdots G_{+-}(x - s_q)] \\
&= 0. \quad (8.70)
\end{aligned}
$$

This result is quite important, for it says that diagrams of class III as well as those of class I with an isolated internal island of circled vertices (see (8.64)) vanish when summed over the internal thermal indices. It is worth emphasizing here that there were two essential ingredients in proving this result at finite temperature. First, as we have noted before (see (8.52) and the discussion following the equation), the propagators where one end is circled and the other is not, does not depend on the thermal index of the circled vertex and, therefore, such propagators remain the same when we sum over the thermal index of the circled vertex and factorize. The second crucial point is that we are able to choose a vertex with the smallest time among the circled vertices whose thermal index is being summed. We note that, since the circled vertices are internal vertices, we have to integrate over their coordinates as well as sum over their thermal indices in evaluating the amplitude and, consequently, among the circled vertices, we can always choose one coordinate with the smallest time (for a given distribution of coordinates before integration) whose thermal index is (also) being summed. In contrast, if any of these vertices were connected to the external lines, then, for some distribution of the time variables of the circled vertices, the time coordinate of the external circled vertex will be the smallest. But being an external vertex, such a vertex is not summed over the two values of the thermal index and hence our result in (8.70) would not follow and such diagrams will not vanish.

From the result that we have just proved, it may seem that we can also prove the complementary result trivially, namely, that diagrams with an isolated island of uncircled internal vertices (not connected to external lines) also vanish when summed over the thermal indices of the internal vertices. But, in fact, the previous result does not automatically generalize if we interchange the circled and uncircled (internal) vertices. This is because, in the second case, the internal vertices are uncircled and we are summing over the thermal indices of these uncircled vertices. Let us recollect (as we have already noted earlier also) that propagators connecting a circled and an uncircled vertex are completely independent of the thermal index of the circled vertex and, therefore, remain the same and factorize when we sum over the thermal index of the circled vertex. In contrast, when we sum over the thermal index of an uncircled vertex, such propagators do not remain the same and do not factorize, unlike in the earlier case, and the analogous proof (see (8.70)) does not go through. To understand better the structure of such diagrams and to show that they, too, vanish when summed over the internal thermal indices, we somehow need to factor out the propagators connecting a circled and an uncircled vertex (as in the proof in (8.70)).

The factorization of these propagators can be more easily seen in the momentum space as follows. Let us consider a class of diagrams where there is an isolated island of internal uncircled vertices not connected to external lines. This could, for example, be diagrams of class II or those of class I where there is a disconnected island of uncircled internal vertices in Γ_1. Let $k_1, k_2, \cdots, k_n$ represent the momenta flowing along propagators connecting Γ_1 to the isolated island of uncircled vertices in Λ not connected to external lines. Λ, as we have noted earlier, can represent Γ_2 in the class II of diagrams in (8.64) or an isolated island of vertices in Γ_2 not connected to the external lines in the class I of diagrams. Since these propagators necessarily have one end circled (in Γ_1) and the other end uncircled (in Λ which belongs to Γ_2), their forms are completely determined by the thermal indices of the uncircled vertices in Λ (see, for example, (8.58)). In fact, such propagators have opposite thermal indices at the two ends, completely determined by the indices of the uncircled vertices in Λ. Let us consider a fixed distribution of thermal indices for the circled internal vertices in Γ_1 from which propagators originate and end on internal vertices in Λ and we sum over the thermal indices of these internal vertices in Λ. In such a case, we can write the self-energy $\Pi(p)$ (for fermions, the self-energy is denoted as $\Sigma(p)$) as

$$\Pi(p) = \Gamma_n(p, k_1, \cdots, k_n) \sum_{\alpha_1, \cdots, \alpha_n = \pm} \Lambda_{\alpha_1 \cdots \alpha_n}(k_1, \cdots, k_n)$$

$$\times G_{-\alpha_1 \alpha_1}(k_1) \cdots G_{-\alpha_n \alpha_n}(k_n), \quad (8.71)$$

where the internal momenta need to be integrated as well which we are neglecting for simplicity. Note that $\alpha_1, \alpha_2, \cdots, \alpha_n$ are the internal vertices in Λ whose thermal indices are being summed and the thermal indices of the other ends of the propagators in Γ_1 are completely determined by these to be $-\alpha_i, i = 1, 2, \cdots, n$. Furthermore, p represents the total momentum flowing between Γ_1 and the isolated island Λ (not connected to external lines) through these propagators such that

$$p = k_1 + k_2 + \cdots + k_n. \tag{8.72}$$

Furthermore, $\Gamma_n(p, k_1, \cdots, k_n)$ denotes the part of the amplitude in Γ_1 which contains only doubly circled propagators (as well as vertices connected to external lines) while $\Lambda_{\alpha_1 \cdots \alpha_n}(k_1, \cdots, k_n)$ represents the part of the amplitude in Λ which contains only uncircled propagators and vertices not connected to external lines. Let us also assume, for simplicity, that all the factors of i coming from the vertices are contained in $\Gamma_n(p, k_1, \cdots, k_n)$.

We concentrate next on the sum over only the i-th thermal index on the right side of (8.71) (namely, we write out the sum over only $\alpha_i = \pm$ explicitly). This would give

$$\Pi(p) = \Gamma_n(p, k_1, \cdots, k_n)$$

$$\times \sum{}' \left[G_{-+}(k_i) \Lambda_{\alpha_1 \cdots + \cdots \alpha_n} + G_{+-}(k_i) \Lambda_{\alpha_1 \cdots - \cdots \alpha_n} \right]$$

$$\times G_{-\alpha_1 \alpha_1}(k_1) \cdots G_{-\alpha_n \alpha_n}(k_n). \tag{8.73}$$

Here the prime on the summation sign, on the right hand side, signifies that the sum is over the thermal indices of all vertices except the i-th one. This makes it explicitly clear now that when we sum over the thermal index of an uncircled vertex of a propagator connected to a circled vertex, the two corresponding multiplicative propagators in the sum are not the same (so that they cannot be simply factored out). On the other hand, we can use the KMS relation of (8.60) to relate the two and, thereby, factorize the propagator out as

$$\Pi(p) = \Gamma_n(p, k_1, \cdots, k_n) G_{+-}(k_i) \sum{}' \left[e^{\beta(k_i)_0} \Lambda_{\alpha_1 \cdots + \cdots \alpha_n} \right.$$

$$\left. + \Lambda_{\alpha_1 \cdots - \cdots \alpha_n} \right] \times G_{-\alpha_1 \alpha_1}(k_1) \cdots G_{-\alpha_n \alpha_n}(k_n). \tag{8.74}$$

The sum over the rest of the thermal indices can now be done pairwise in a similar manner leading to factorization of all the propagators and yielding

$$\Pi(p) = \Gamma_n(p, k_1, \cdots, k_n) G_{+-}(k_1) \cdots G_{+-}(k_n)$$

$$\times \sum_{\alpha_1, \cdots, \alpha_n = \pm} \Lambda_{\alpha_1 \cdots \alpha_n} \, e^{\sum_{j=1}^{n} \frac{\alpha_j + 1}{2} \beta(k_j)_0}, \tag{8.75}$$

where we have identified

$$
a_j = \begin{cases} +1 & \text{for} \quad \alpha_j = +, \\ -1 & \text{for} \quad \alpha_j = -, \end{cases}
\tag{8.76}
$$

so that $\frac{a_j+1}{2} = 1, 0$, for $\alpha_j = \pm$ respectively. The important thing to note here is that all the propagators connecting a circled and an uncircled vertex in the isolated island have now been factored out and, in the sum that remains, there is an appropriate nontrivial exponential factor for every positive (thermal) index α_j. Therefore, for every term in the sum (8.75), we can think of the exponential as $e^{\beta p_0}$ where p_0 represents the total energy entering the Feynman diagram $\Lambda_{\alpha_1 \cdots \alpha_n}(k_1, \cdots, k_n)$ through its vertices with $+$ thermal index. (Recall that the exponent becomes trivial for every negative α_j (or a_j) so that the nontrivial exponential, in a given term of the sum in (8.75), is determined completely by the number of positive thermal indices in Λ.)

A general term, $\Lambda_{\alpha_1 \cdots \alpha_n}$, in the sum (8.75), contains only uncircled vertices, p of $+$ type and q of $-$ type such that $p + q = n$. It is clear that $p, q = 0, 1, 2, \cdots, n$. Each such term can be represented as a diagram of the following form

$$
\tag{8.77}
$$

Here the cluster Λ_+^p contains only vertices of "$+$" type (p of them) and the cluster Λ_-^q contains only vertices of "$-$" type (q of them).

The proper way to visualize the case under discussion is to imagine the propagators at the two ends in (8.77) to be connected to Γ_1, some coming into Λ through vertices with positive thermal index and others through vertices with negative thermal index (so that when we cut all of these propagators Λ becomes disconnected). In addition, there may be propagators with momenta $l_1, \cdots, l_t$ connecting the positive and the negative clusters Λ_+^p and Λ_-^q and the total momentum flowing in from Γ_1 to Λ_+^p exits through these intermediate propagators to the cluster Λ_-^q. These will necessarily be propagators of the type G_{+-}/G_{-+}. In the clusters Λ_+^p and Λ_-^q, there can also be propagators of the kind G_{++} and G_{--} respectively. Let us recall some of the properties of these propagators (in the Λ island) in the momentum space (see (8.41) and (8.60))

$$G_{++}(p) = (G_{--}(p))^*,$$
$$G_{+-}(p) = (G_{+-}(p))^* = \left(e^{-\beta p_0}(G_{-+}(p))\right)^* = e^{-\beta p_0}(G_{-+}(p))^*$$
$$= e^{-\beta p_0}G_{-+}(p). \tag{8.78}$$

Let us next consider a term in the sum in (8.75) of the form $\Lambda_{\underbrace{+ + \cdots +}_{p}\underbrace{- - \cdots -}_{q}}$ which can be characterized by the diagram in (8.77). Therefore, we note that if we were to complex conjugate a graph representing a term $\Lambda_{\alpha_1 \cdots \alpha_n}(k_1, \cdots, k_n)$ in (8.75), then, according to (8.78),

$$\left(G_{++}(k)\right)^* \leftrightarrow G_{--}(k),$$
$$\left(G_{+-}(k)\right)^* \rightarrow e^{-\beta k_0} G_{-+}(k). \tag{8.79}$$

Consequently, we can think of complex conjugating any diagram corresponding to a term in the sum (8.75) as giving a diagram where the positive and the negative vertices and the propagators connecting them are interchanged within the two clusters. The propagators connecting the vertices of the two clusters are unchanged. However, we can interchange their indices as well (namely, $G_{+-} \rightarrow G_{-+}$) if we use the KMS condition which gives rise to an exponential factor (see (8.79)) for each of the intermediate G_{+-} propagators. Adding up all such exponents for momenta flowing from the positive to the negative cluster of the original diagram yields the exponential $e^{-\beta p_0}$

where p_0 is the total energy flowing from the positive to the negative cluster through the intermediate propagators (which is also the same as the total energy entring Λ from Γ_1 through the positive vertices) of the original diagram. On the other hand, complex conjugation interchanges the positive and the negative clusters (see, for example, (8.80)) and, therefore the total energy entering Λ from Γ_1 through the positive vertices changes the sign and, as a result, the exponential factor becomes $e^{\beta p_0}$. This is diagrammatically represented as

$$\left(\begin{array}{c} \cdots \\ \Lambda_-^q \\ \cdots \\ \Lambda_+^p \\ \cdots \end{array} \right)^* = e^{\beta p_0} \begin{array}{c} \cdots \\ \Lambda_+^q \\ \cdots \\ \Lambda_-^p \\ \cdots \end{array} \tag{8.80}$$

where P represents the total momentum entering the diagram on the right hand side of (8.80) through the positive vertices. Algebraically, this is equivalent to saying that

$$\Lambda^*_{\underbrace{+ + \cdots +}_{p} \underbrace{- - \cdots -}_{q}} = e^{\beta p_0} \, \Lambda_{\underbrace{- - \cdots -}_{p} \underbrace{+ + \cdots +}_{q}}, \tag{8.81}$$

with $p + q = n$. Equation (8.81) is quite crucial for it says that we can eliminate all the exponential factors in the sum in (8.75) to write

$$\Pi(p) = \Gamma(p, k_1, \cdots, k_n) G_{+-}(k_1) \cdots G_{+-}(k_n)$$

$$\times \sum_{\alpha_1, \cdots, \alpha_n = \pm} \Lambda^*_{\alpha_1 \cdots \alpha_n}(k_1, \cdots, k_n)$$

$$= \Gamma(p, k_1, \cdots, k_n) G_{+-}(k_1) \cdots G_{+-}(k_n)$$

$$\times \left(\sum_{\alpha_1, \cdots, \alpha_n = \pm} \Lambda_{\alpha_1 \cdots \alpha_n}(k_1, \cdots, k_n) \right)^* . \tag{8.82}$$

The sum on the right hand side, inside the parenthesis (namely, before complex conjugation), can now be shown to vanish through the largest time equation in the coordinate space. Each term in the sum in (8.82) represents a graph with only uncircled vertices of $+$ and $-$ type that are internal vertices and, therefore, the thermal indices are summed over. Let x represent the coordinate of one such vertex with the largest time and we will sum over the thermal index of this vertex $\alpha = \pm$ holding the thermal indices of all other vertices fixed. This vertex will be connected to v vertices of $+$ type located at, say, $y_1, \cdots, y_v$ and w vertices of $-$ type located at $z_1, \cdots, z_w$ such that $v + w = n - 1$. Then, if x^0 is the largest time coordinate, then summing over its thermal index, the sum in (8.82) would lead to (in the coordinate space before taking the complex conjugation)

$$\sum_{\alpha = \pm} \Lambda_{\alpha + \cdots + - \cdots -}(x, y_1, \cdots, y_v, z_1, \cdots, z_w)$$

$$= R\left[G_{++}(x - y_1) \cdots G_{++}(x - y_v) G_{+-}(x - z_1) \cdots G_{+-}(x - z_w) \right.$$

$$\left. - G_{-+}(x - y_1) \cdots G_{-+}(x - y_v) G_{--}(x - z_1) \cdots G_{--}(x - z_w) \right]$$

$$= R\left[G_{-+}(x - y_1) \cdots G_{-+}(x - y_v) G_{+-}(x - z_1) \cdots G_{+-}(x - z_w) \right.$$

$$\left. - G_{-+}(x - y_1) \cdots G_{-+}(x - y_v) G_{+-}(x - z_1) \cdots G_{+-}(x - z_w) \right]$$

$$= 0, \tag{8.83}$$

where R denotes the rest of the diagram whose form is not very relevant for the conclusion. Here we have used (8.46) as well as (8.50) and this, then, proves that $\Pi(p) = 0$ (see (8.82)). In other words, a diagram with an isolated island of uncircled internal vertices must vanish when summed over the internal thermal indices. This result, then, shows that diagrams of type class II as well as parts of those in class I where there occur isolated islands of internal uncircled vertices must vanish. The two results together (namely, (8.70) and (8.83)) then prove that when we sum over the thermal indices of all the internal vertices (circled and uncircled), the only diagrams that give nonvanishing contribution are of type class I where there are no isolated islands of internal circled/uncircled vertices.

This would, therefore, seem to imply that the cutting rules of zero temperature generalize to finite temperature such that the imaginary part of an amplitude can be given by all possible cuttings when the internal thermal indices are summed (see (8.31) and the subsequent discussions). However, there is one difference between this result and the one at zero temperature. To make things absolutely clear, let us look at the imaginary part of the self-energy which we have also discussed at zero temperature in (8.38). We note that, at zero temperature, because of the properties of the cut propagators (namely, they depend either on positive or negative energies, see, (8.39)), the only diagrams that contribute to the imaginary part, by conservation of energy and momentum, are the ones where the shadow of the cut is to the left (with our convention of incoming and outgoing particles). At finite temperature, however, the cut propagators have contributions from both positive and negative energies (see (8.46)-(8.50)). As a result, energy-momentum conservation allows cut diagrams to have shadows to the left as well as to the right. Therefore, to make our discussion completely parallel to the zero temperature result, we need one final ingredient. We note that if we are looking at the imaginary part of the retarded propagator, which is defined as (see (2.56)),

$$G_R = G_{++} - G_{+-} = G_{-+} - G_{--}, \tag{8.84}$$

then, the thermal index on the second vertex is effectively being summed with the first index held fixed. (Note that there is an extra $(-)$ factor arising between the two terms from a $-$ type vertex and, hence, this really represents a sum of two amplitudes.) For such an amplitude, therefore, diagrams where the second (right) vertex is circled give vanishing contribution because of the first theorem that we have proved in (8.70). Consequently, the only graphs that will contribute nontrivially to the imaginary part of the retarded self-energy are the ones where the circled vertices are at the left. This is quite interesting because, as we have seen in (2.78), the imaginary part of the retarded Green's function is related to the Feynman Green's function through a multiplicative constant. And, what we find so far from our discussions is that, as long as we are looking at the imaginary part of the retarded function, the generalization of the cutting rules is completely parallel to what we had discussed at zero temperature in (8.38), namely, we must take all possible cuttings with the shadow on the left (with our convention of incoming and outgoing particles).

Consequently, we see that the cutting rules of zero temperature do generalize to finite temperature as well, not graph by graph, but in the whole amplitude when we sum over the thermal indices of the internal vertices for a retarded amplitude. We would like to emphasize here that it is only a coincidence that at zero temperature a cutting description holds for the imaginary part, graph by graph, but unitarity only requires that such a description hold for the complete amplitude when summed over all intermediate states (channels) (see, for example, (8.2)) and that is exactly what we find at finite temperature. Diagrammatically, we can, therefore, write that, at finite temperature, when we sum over all the internal thermal indices, the imaginary part of the retarded self energy is given by (we suppress the label "retarded" in the diagram only for simplicity)

$$\text{Im} \; \multimap\!\!\bigcirc\!\!\multimap \quad = \quad \frac{1}{2} \; \multimap\!\!\oslash\!\!\multimap \; , \tag{8.85}$$

which can be compared with (8.38).

8.3 Example

We will end this discussion with a simple example of the cutting rules at finite temperature and at one loop for the theory of two interacting scalar fields ϕ and B) described by the Lagrangian density in (1.114),

$$\mathcal{L} = \frac{1}{2} \partial_\mu \phi \partial^\mu \phi - \frac{m^2}{2} \phi^2 + \frac{1}{2} \partial_\mu B \partial^\mu B - \frac{M^2}{2} B^2 - \frac{g}{2} \phi B^2. \tag{8.86}$$

We can try to obtain the imaginary part for the retarded self-energy for the ϕ field (arising from the loop correction due to the B field) from the cutting rules that we have described so far. Let us recall, from (8.49), (8.50) and (8.54), that (with momentum flowing from the circled to the uncircled vertex) the relevant components of the propagator for the B field are given by

$$G_{\oplus+}(k) = 2\pi(\theta(k_0) + n(|k_0|))\delta(k^2 - M^2),$$

$$G_{\oplus-}(k) = 2\pi(\theta(-k_0) + n(|k_0|))\delta(k^2 - M^2), \tag{8.87}$$

where the subscript $\oplus$ signifies that the end with thermal index $+$ is circled.

We note that the cut graph with $++$ component of the ϕ self energy due to the B-loop in (8.85) (which gives $\operatorname{Im}\Pi_{++}(p)$) has the form (remember the symmetry factor $\frac{1}{2}$ for the self-energy diagram in Fig. 1.3, see (1.115))

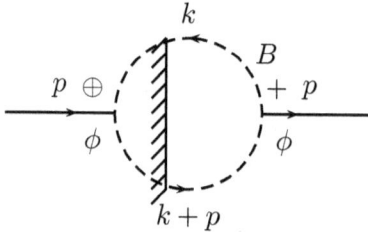

$$\operatorname{Im}\Pi_{++}(p)$$

$$= \frac{g^2}{2}\int \frac{d^4k}{(2\pi)^4}\, G_{\oplus+}(-k)G_{\oplus+}(k+p)$$

$$= \frac{g^2}{2}\int \frac{d^4k}{(2\pi)^4}\,(2\pi)\big(\theta(-k_0) + n(|k_0|)\big)$$

$$\times (2\pi)\big(\theta(k_0 + p_0) + n(|k_0 + p_0|)\big)$$

$$\times \delta(k^2 - M^2)\delta((k+p)^2 - M^2)$$

$$= \frac{g^2}{2}\int \frac{d^3k}{(2\pi)^2}\frac{1}{4\omega_k\omega_{k+p}}\big[n(\omega_k)(1 + n(\omega_{k+p}))\delta(p_0 + \omega_k - \omega_{k+p})$$

$$+ (1 + n(\omega_k))n(\omega_{k+p})\delta(p_0 - \omega_k + \omega_{k+p})$$

$$+ n(\omega_k)n(\omega_{k+p})\delta(p_0 + \omega_k + \omega_{k+p})$$

$$+ (1 + n(\omega_k))(1 + n(\omega_{k+p}))\delta(p_0 - \omega_k - \omega_{k+p})\big], \tag{8.88}$$

where we have used the first of the two cut propagators defined in (8.87) as well as identified ω_k and ω_{k+p} as in (1.116).

In a similar manner, we can evaluate the cut diagram for the $+-$ component of the self energy due to the B-loop (which gives $\mathrm{Im}\,\Pi_{+-}(p)$ and it has the form (the overall negative sign is because of the "$-$" vertex)

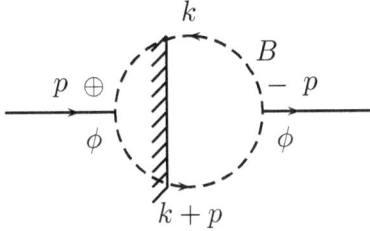

$\mathrm{Im}\,\Pi_{+-}(p)$

$$= -\frac{g^2}{2} \int \frac{d^4 k}{(2\pi)^4} \, G_{\oplus -}(-k) G_{\oplus -}(k+p)$$

$$= -\frac{g^2}{2} \int \frac{d^4 k}{(2\pi)^4} \, (2\pi)\big(\theta(k_0) + n(|k_0|)\big)$$

$$\times (2\pi)\big(\theta(-k_0 - p_0) + n(|k_0 + p_0|)\big)$$

$$\times \delta(k^2 - M^2)\delta((k+p)^2 - M^2)$$

$$= -\frac{g^2}{2} \int \frac{d^3 k}{(2\pi)^2} \frac{1}{4\omega_k \omega_{k+p}} \Big[(1 + n(\omega_k)) n(\omega_{k+p}) \delta(p_0 + \omega_k - \omega_{k+p})$$

$$+ n(\omega_k)(1 + n(\omega_{k+p})) \delta(p_0 - \omega_k + \omega_{k+p})$$

$$+ (1 + n(\omega_k))(1 + n(\omega_{k+p})) \delta(p_0 + \omega_k + \omega_{k+p})$$

$$+ n(\omega_k) n(\omega_{k+p}) \delta(p_0 - \omega_k - \omega_{k+p}) \Big]. \tag{8.89}$$

If we add the two contributions in (8.88) and (8.89), we obtain the cut diagram for the retarded self energy and, from (8.85), we obtain the imaginary part of the retarded self energy to be given by

$$\mathrm{Im}\,\Pi_R(p) = \frac{1}{2}\,\mathrm{Im}\,\big(\Pi_{++}(p) + \Pi_{+-}(p)\big)$$

$$= -\frac{g^2}{16} \int \frac{d^3 k}{(2\pi)^2} \frac{1}{\omega_k \omega_{k+p}}$$

$$\times \Big[\big((1 + n(\omega_k) + n(\omega_{k+p}))\big)$$

$$\times \left(\delta(p_0 + \omega_k + \omega_{k+p}) - \delta(p_0 - \omega_k - \omega_{k+p}) \right)$$
$$+ \left(n(\omega_k) - n(\omega_{k+p}) \right)$$
$$\times \left(\delta(p_0 - \omega_k + \omega_{k+p}) - \delta(p_0 + \omega_k - \omega_{k+p}) \right)]. \quad (8.90)$$

This can be compared with (1.135) keeping in mind the identification that $p_0 = \omega$. The on-shell nature (delta function dependences) of the imaginary part in (8.90) shows that it is directly related to various new channels of physical processes $(n(\omega_k), n(\omega_{k+p})$ dependent terms) at finite temperature. This derivation shows how the generalization of the cutting rules work at finite temperature and, even though we have only worked out a simple one loop example, it can be carried out explicitly to higher loops where more interesting structures (physical processes) arise.

We have, of course, already commented, in chapter **6** in (6.157), on the relevance of the thermal operator representation to cutting rules at finite temperature. Here, let us see how equation (8.90) can also be obtained from the thermal operator representation. For that, let us note, from (8.87), that at zero temperature $(n(\omega_k) = n(\omega_{k+p}) = 0)$ the relevant cut propagators for the two diagrams given in (8.88) and (8.89) coincide with the ones given in(8.49), (8.50) and (8.54) in the zero temperature limit, namely,

$$G_{\oplus +}^{(T=0)}(k) = 2\pi\theta(k_0)\delta(k^2 - M^2),$$
$$G_{\oplus -}^{(T=0)}(k) = 2\pi\theta(-k_0)\delta(k^2 - M^2), \quad (8.91)$$

Therefore, a direct calculation following (8.88) and (8.89) leads to the imaginary part of the retarded self-energy at zero temperature to be

$$\mathrm{Im}\,\Pi_R^{(T=0)}(p) = -\frac{g^2}{16} \int \frac{d^3k}{(2\pi)^2} \frac{1}{\omega_k \omega_{k+p}}$$
$$\times \left(\delta(p_0 + \omega_k + \omega_{k+p}) - \delta(p_0 - \omega_k - \omega_{k+p}) \right), \quad (8.92)$$

which coincides with the zero temperature limit of (8.90). We note, following our discussions in (6.14)-(6.17) that we can write the finite temperature propagators in (8.87) can be written as

$$G_{\oplus+}(k) = \mathcal{O}_B^{(T)}(\omega_k) G_{\oplus+}^{(T=0)}(k)$$
$$= 2\pi(\theta(k_0) + n(|k_0|))\delta(k^2 - M^2),$$

$$G_{\oplus-}(k) = \mathcal{O}_B^{(T)}(\omega_k) G_{\oplus-}^{(T=0)}(k)$$
$$= 2\pi(\theta(-k_0) + n(|k_0|))\delta(k^2 - M^2), \tag{8.93}$$

where

$$\mathcal{O}_B^{(T)}(\omega) = (\mathbb{1} + n(\omega)(\mathbb{1} - S(\omega))), \tag{8.94}$$

denotes the thermal operator defined in (6.7).

Since the retarded self-energy in (8.92) involves two propagators (at zero temperature), applying the relevant two thermal operators to the integrand in (8.92), we obtain

$$\mathrm{Im}\,\Pi_R^{(T)}(p) = -\frac{g^2}{16} \int \frac{\mathrm{d}^3 k}{(2\pi)^2} \, \mathcal{O}_B^{(T)}(\omega_k)\mathcal{O}_B^{(T)}(\omega_{k+p}) \frac{1}{\omega_k \omega_{k+p}}$$

$$\times \left(\delta(p_0 + \omega_k + \omega_{k+p}) - \delta(p_0 - \omega_k - \omega_{k+p})\right)$$

$$= -\frac{g^2}{16} \int \frac{\mathrm{d}^3 k}{(2\pi)^2} \, \mathcal{O}_B^{(T)}(\omega_k) \left[\frac{1}{\omega_k \omega_{k+p}}\right.$$

$$\times \left\{(1 + n(\omega_{k+p}))\left(\delta(p_0 + \omega_k + \omega_{k+p}) - \delta(p_0 - \omega_k - \omega_{k+p})\right)\right.$$

$$\left.\left. + n(\omega_{k+p})\left(\delta(p_0 + \omega_k - \omega_{k+p}) - \delta(p_0 - \omega_k + \omega_{k+p})\right)\right\}\right]$$

$$= -\frac{g^2}{16} \int \frac{\mathrm{d}^3 k}{(2\pi)^2} \frac{1}{\omega_k \omega_{k+p}}$$

$$\times \left[\left((1 + n(\omega_k) + n(\omega_{k+p}))\right)\right.$$

$$\times \left(\delta(p_0 + \omega_k + \omega_{k+p}) - \delta(p_0 - \omega_k - \omega_{k+p})\right)$$

$$+ \left(n(\omega_k) - n(\omega_{k+p})\right)$$

$$\left.\times \left(\delta(p_0 - \omega_k + \omega_{k+p}) - \delta(p_0 + \omega_k - \omega_{k+p})\right)\right], \tag{8.95}$$

which coincides with the direct calculation given in (8.90).

8.4 Cutting rules in the presence of a chemical potential

The cutting rules or the Cutcosky rules can also be generalized to the case when a chemical potential is present in addition to temperature. This can be *a posteriori* understood because, as we have mentioned in the beginning of section **6.5**, the thermal operator is a real operator (does not have any imaginary part) and, therefore, should not cause any complication in the study of the imaginary part of an amplitude and, if the cutting rules hold at zero temperature, they should also hold at finite temperature with or without a chemical potential. This can, in fact, be proved explicitly as in the earlier discussions in this chapter (without a chemical potential).[1] Here we would demonstrate this through the use of the thermal operator in the calculation of the self-energy of a scalar field when a chemical potential is present. Let us next consider a generalized Lagrangian density (see (6.52) or (6.56) as well as (8.86)) for a complex B field of the form

$$
\mathcal{L} = \frac{1}{2}\partial_\mu \phi \partial^\mu \phi - \frac{m^2}{2}\phi^2 + ((\partial_t - i\mu)B)^*(\partial_t - i\mu)B
$$
$$
- (\boldsymbol{\nabla} B)^* \cdot (\boldsymbol{\nabla} B) - M^2 B^* B - g\phi B^* B, \tag{8.96}
$$

where the real constant $\mu < M$ (see (6.53)) denotes the chemical potential for the complex B field. We would like to calculate the imaginary part of the retarded self-energy for the B-field in this theory with a chemical potential so that we can compare it with our calculation in the last section. Such a calculation is certainly more involved to carry out in the momentum space than in the mixed space.[2] Nonetheless, we do it in the momentum space here to bring out the technical details as well as to be consistent with the rest of the discussions in this chapter.

The propagators for complex fields with a chemical potential are discussed in detail in section **6.2.4** and taking the Fourier transform of (6.65), we note that the four components of the B-field propagator can be written as

$$
G_{++}^{(T,\mu)}(k) = \frac{i}{(k_0 + \mu)^2 - \omega_k^2 + i\epsilon} + 2\pi n(|k_0|)\delta((k_0 + \mu)^2 - \omega_k^2),
$$

---

[1]See, F. T. Brandt, A. Das, O. Espinosa, J. Frenkel and S. Perez, Physical Review **D74**, 085006 (2006).

[2]See, the earlier footnote.

$$G_{+-}^{(T,\mu)}(k) = 2\pi(\theta(-k_0 - \mu) + n(|k_0|))\delta((k_0 + \mu)^2 - \omega_k^2),$$

$$G_{-+}^{(T,\mu)}(k) = 2\pi(\theta(k_0 + \mu) + n(|k_0|))\delta((k_0 + \mu)^2 - \omega_k^2), \qquad (8.97)$$

$$G_{--}^{(T,\mu)}(k) = \frac{(-i)}{(k_0 + \mu)^2 - \omega_k^2 - i\epsilon} + 2\pi n(|k_0|)\delta((k_0 + \mu)^2 - \omega_k^2),$$

where (see (6.53))

$$\omega_k = \sqrt{\mathbf{k}^2 + M^2} > \mu. \qquad (8.98)$$

The zero temperature propagator components can be obtained from (8.97) in a straightforward manner and they take the forms

$$G_{++}^{(T=0,\mu)}(k) = \frac{i}{(k_0 + \mu)^2 - \omega_k^2 + i\epsilon},$$

$$G_{+-}^{(T=0,\mu)}(k) = 2\pi\theta(-k_0 - \mu)\delta((k_0 + \mu)^2 - \omega_k^2),$$

$$G_{-+}^{(T=0,\mu)}(k) = 2\pi\theta(k_0 + \mu)\delta((k_0 + \mu)^2 - \omega_k^2),$$

$$G_{--}^{(T=0,\mu)}(k) = \frac{(-i)}{(k_0 + \mu)^2 - \omega_k^2 - i\epsilon}. \qquad (8.99)$$

Following the discussions in section **6.2.4**, in particular, (6.68) we note that the thermal operator, in this case, can be written as

$$\mathcal{O}_B^{(T,\mu)}(\omega) = \left(\mathbb{1} + \widehat{n}(\omega)(\mathbb{1} - S(\omega))\right), \qquad (8.100)$$

where $\widehat{n}(\omega)$ acts only on functions of $\omega^{(\pm)} = \omega \pm \mu$ (and not on ω) as

$$\widehat{n}(\omega)f(\omega^{(\pm)}) = n(\omega^{(\pm)})f(\omega^{(\pm)}), \quad \omega^{(\pm)} = \omega \pm \mu, \qquad (8.101)$$

and $S(\omega)$ is the reflection operator defined in (6.8), namely,

$$S(\omega)f(\omega) = f(-\omega), \quad \text{so that} \quad S(\omega)\omega_\pm = -\omega_\mp. \qquad (8.102)$$

As a result, we can relate the components in (8.98) and (8.99) as

$$G_{ab}^{(T,\mu)}(p) = \mathcal{O}_B^{(T,\mu)}(\omega)G_{ab}^{(0,\mu)}(p), \quad \text{for } a, b = \pm. \qquad (8.103)$$

Of course, as in (6.17), the thermal operator in momentum space needs to be applied carefully. For example, we note that

$$\delta((k_0 + \mu)^2 - \omega_k^2) = \frac{1}{2|k_0 + \mu|}\left(\delta(k_0 + \mu + \omega_k) + \delta(k_0 + \mu - \omega_k)\right)$$

$$= \frac{1}{2\omega_k}\left(\delta(k_0 + \omega_k^{(+)}) + \delta(k_0 - \omega_k^{(-)})\right), \tag{8.104}$$

which leads to

$$\theta(-k_0)\delta((k_0 + \mu)^2 - \omega_k^2) = \frac{1}{2\omega_k}\delta(k_0 + \omega_k^{(+)}),$$

$$\theta(k_0)\delta((k_0 + \mu)^2 - \omega_k^2) = \frac{1}{2\omega_k}\delta(k_0 - \omega_k^{(-)}), \tag{8.105}$$

so that we have

$$\widehat{n}(\omega_k)(\mathbb{1} - S(\omega_k))\theta(-k_0)\delta((k_0 + \mu)^2 - \omega_k^2)$$

$$= \widehat{n}(\omega_k)\frac{1}{2\omega_k}\left(\delta(k_0 + \omega_k^{(+)}) + \delta(k_0 - \omega_k^{(-)})\right),$$

$$= \frac{1}{2\omega_k}\left(n(\omega_k^{(+)})\delta(k_0 + \omega_k^{(+)}) + n(\omega_k^{(-)})\delta(k_0 - \omega_k^{(-)})\right)$$

$$= \frac{n(|k_0|)}{2\omega_k}\left(\delta(k_0 + \omega_k^{(+)}) + \delta(k_0 - \omega_k^{(-)})\right)$$

$$= n(|k_0|)\delta((k_0 + \mu)^2 - \omega_k^2), \tag{8.106}$$

and, similarly,

$$\widehat{n}(\omega_k)(\mathbb{1} - S(\omega_k))\theta(k_0)\delta((k_0 + \mu)^2 - \omega_k^2)$$

$$= n(|k_0|)\delta((k_0 + \mu)^2 - \omega_k^2). \tag{8.107}$$

Equation (8.104)-(8.107) now lead to, as in (6.68),

$$G_{+-}^{(T,\mu)}(k) = \left(\mathbb{1} + \widehat{n}(\omega_k)(\mathbb{1} - S(\omega_k))\right)G_{+-}^{(T=0,\mu)}(k),$$

$$G_{-+}^{(T,\mu)}(k) = \left(\mathbb{1} + \widehat{n}(\omega_k)(\mathbb{1} - S(\omega_k))\right)G_{-+}^{(T=0,\mu)}(k),$$

$$\tag{8.108}$$

which prove (8.103) for the +− and −+ components of the propagator. The thermal operator relation (8.103) can be shown to hold for other components in a similar manner.

We can now determine the relevant propagators with one end circled necessary for calculation of the retarded self-energy of the B field in this theory (8.96) and, at $T = 0$, they take the simple forms (see (8.92) with momentum flowing from the circled vertex)

$$G_{\oplus +}^{(T=0,\mu)}(k) = 2\pi\theta(k_0 + \mu)\delta((k_0 + \mu)^2 - \omega_k^2),$$

$$G_{\oplus -}^{(T=0,\mu)}(k) = 2\pi\theta(-k_0 - \mu)\delta((k_0 + \mu)^2 - \omega_k^2), \tag{8.109}$$

which can be compared with (8.91). We can now calculate the imaginary parts of the amplitudes $\Pi_{++}(p)$ and $\Pi_{+-}(p)$ (see (8.88) and (8.89) at zero temperature and they lead to

$$\operatorname{Im}\Pi_{++}^{(T=0,\mu)}(p) = \frac{g^2}{2} \int \frac{d^4k}{(2\pi)^4}\, G_{\oplus +}^{(T=0,\mu)}(-k)G_{\oplus +}^{(T=0,\mu)}(k+p)$$

$$= \frac{g^2}{2} \int \frac{d^4k}{(2\pi)^2}\, \theta(-k_0 - \mu)\delta((k_0 + \mu)^2 - \omega_k^2)$$

$$\times \theta(k_0 + p_0 + \mu)\delta((k_0 + p_0 + \mu)^2 - \omega_{k+p}^2)$$

$$= \frac{g^2}{8} \int \frac{d^3k}{(2\pi)^2} \frac{1}{\omega_k\omega_{k+p}}\, \delta(p_0 - \omega_k^{(+)} - \omega_{k+p}^{(-)}), \tag{8.110}$$

which can be compared with the zero temperature limit of (8.88). Note that the μ dependence inside the delta function cancels completely. This is a consequence of the fact that, at every vertex, not only is the energy-momentum conserved, but the μ dependence in the propagators must also add up to zero since the vertex does not know anything about the chemical potential. This dictates the form of $G_{\oplus +}^{(T=0,\mu)}(-k)$ in (8.110) (namely, $k_0 \to -k_0$ accompanied by $\mu \to -\mu$). Note that the delta functions basically express the conservation of energy-momentum as well as the μ independence at a vertex.

In a similar manner, we can calculate the imaginary part of $\Pi_{+-}(p)$ and it leads to

$$\operatorname{Im}\Pi_{+-}^{(T=0,\mu)}(p) = -\frac{g^2}{2} \int \frac{d^4k}{(2\pi)^4}\, G_{\oplus -}^{(T=0,\mu)}(-k)G_{\oplus -}^{(T=0,\mu)}(k+p)$$

$$= -\frac{g^2}{8} \int \frac{d^3k}{(2\pi)^2} \frac{1}{\omega_k\omega_{k+p}}\, \delta(p_0 + \omega_k^{(-)} + \omega_{k+p}^{(+)}), \tag{8.111}$$

which can also be compared with the zero temperature limit of (8.89) and it is worth noting that all the μ dependence inside the delta function cancels out as it should. It follows, therefore, that the imaginary part of the retarded self-energy at zero temperature is given by

$$
\mathrm{Im}\,\Pi_{\mathrm{R}}^{(T=0,\mu)}(p) = \frac{1}{2}\left(\mathrm{Im}\,\Pi_{++}^{(T=0,\mu)}(p) + \mathrm{Im}\,\Pi_{+-}^{(T=0,\mu)}(p)\right)
$$

$$
= -\frac{g^2}{16}\int \frac{d^3k}{(2\pi)^2}\,\frac{1}{\omega_k\omega_{k+p}}
$$

$$
\times\left(\delta(p_0 + \omega_k^{(-)} + \omega_{k+p}^{(+)}) - \delta(p_0 - \omega_k^{(+)} - \omega_{k+p}^{(-)})\right). \quad (8.112)
$$

We note that this coincides with (8.92) for $\mu = 0$.

To obtain the imaginary part of the retarded self-energy at finite temperature (and chemical potential), we can apply the thermal operator defined in (8.100)-(8.102). In fact, noticing that the one loop self-energy involves two propagators with energies ω_k and ω_{k+p}, we can write

$$
\mathrm{Im}\,\Pi_{\mathrm{R}}^{(T,\mu)}(p) = \frac{1}{2}\left(\mathrm{Im}\,\Pi_{++}^{(T,\mu)}(p) + \mathrm{Im}\,\Pi_{+-}^{(T,\mu)}(p)\right)
$$

$$
= -\frac{g^2}{16}\int \frac{d^3k}{(2\pi)^2}\,\mathcal{O}_B^{(T,\mu)}(\omega_{k+p})\mathcal{O}_B^{(T,\mu)}(\omega_k)\left(\frac{1}{\omega_k\omega_{k+p}}\right.
$$

$$
\left.\times\left(\delta(p_0 + \omega_k^{(-)} + \omega_{k+p}^{(+)}) - \delta(p_0 - \omega_k^{(+)} - \omega_{k+p}^{(-)})\right)\right). \quad (8.113)
$$

Using the relations from (8.104)-(8.107), we note that

$$
\mathcal{O}_B^{(T,\mu)}(\omega_k)\left(\frac{1}{\omega_k\omega_{k+p}}\left(\delta(p_0 + \omega_k^{(-)} + \omega_{k+p}^{(+)})\right.\right.
$$

$$
\left.\left. - \delta(p_0 - \omega^{(+)} - \omega_{k+p}^{(-)})\right)\right)
$$

$$
= \frac{1}{\omega_k\omega_{k+p}}\left(\delta(p_0 + \omega_k^{(-)} + \omega_{k+p}^{(+)}) - \delta(p_0 - \omega_k^{(+)} - \omega_{k+p}^{(-)})\right)
$$

$$
+ n(\omega_k^{(-)})\left(\delta(p_0 + \omega_k^{(-)} + \omega_{k+p}^{(+)}) - \delta(p_0 + \omega_k^{(-)} - \omega_{k+p}^{(-)})\right)
$$

$$
+ n(\omega_k^{(+)})\left(\delta(p_0 - \omega_k^{(+)} + \omega_{k+p}^{(+)}) - \delta(p_0 - \omega_k^{(+)} - \omega_{k+p}^{(-)})\right).
$$

$$
(8.114)
$$

Applying $\mathcal{O}_B^{(T,\mu)}(\omega_{k+p})$ to (8.114) in a similar manner and simplifying, we obtain the imaginary part of the retarded self-energy from (8.113) to be

$$\operatorname{Im}\Pi_R^{(T,\mu)}(p) = -\frac{g^2}{16} \int \frac{d^3k}{(2\pi)^2} \frac{1}{\omega_k \omega_{k+p}}$$
$$\times \Big[\big(1 + n(\omega_k^{(-)}) + n(\omega_{k+p}^{(+)})\big)\delta(p_0 + \omega_k^{(-)} + \omega_{k+p}^{(+)})$$
$$- \big(1 + n(\omega_k^{(+)}) + n(\omega_{k+p}^{(-)})\big)\delta(p_0 - \omega_k^{(+)} - \omega_{k+p}^{(-)})$$
$$- \big(n(\omega_k^{(-)}) - n(\omega_{k+p}^{(-)})\big)\delta(p_0 + \omega_k^{(-)} - \omega_{k+p}^{(-)})$$
$$+ \big(n(\omega_k^{(+)}) - n(\omega_{k+p}^{(+)})\big)\delta(p_0 - \omega_k^{(+)} + \omega_{k+p}^{(+)})\Big]. \tag{8.115}$$

Once again, the μ dependence cancels out in the delta functions, but we have kept them so that following the calculations will be easier. This result reduces in the zero temperature limit (when $n(\omega_k) = 0 = n(\omega_{k+p})$) agrees with (8.112) and the result in (8.115) can also be obtained from a direct calculation of the imaginary part of the retarded self-energy as done earlier in section **8.3** for the case $\mu = 0$.

8.5 References

P. F. Bedaque, A. Das and S. Naik, Modern Physics Letters **A12**, 2623 (1997).

F. T. Brandt, A. Das, O. Espinosa, J. Frenkel and S. Perez, Physical Review **D74**, 085006 (2006).

R. Cutkosky, Journal of Mathematical Physics **1**, 429 (1960).

G. 't Hooft and M. Veltman, *Diagrammar*, CERN 73-9.

P. V. Landshoff, Physics Letters **B386**, 291 (1996).

H. A. Weldon, Physical Review **D28**, 2007 (1983).

CHAPTER 9

Spontaneous symmetry breaking

Symmetry plays a fundamental role in the study of physical phenomena. Global symmetry imposes various constraints on the structure of the theory and local symmetry which leads to physical forces is even more restrictive. Sometimes, a given symmetry in a theory may be spontaneously broken by the choice of a vacuum (ground) state and this has profound consequences. Examples of symmetry breaking that may only be too familiar are the existence of phase transitions in various physical systems where the different phases may exhibit different symmetries. We have alluded to some of the consequences of a spontaneous breakdown of a continuous global symmetry briefly in chapter **3**. In such a case, the physical system develops long range correlations (in the language of relativistic quantum field theories, there arise massless quanta in the theory). The consequences are even more interesting when a local symmetry is involved. We also know from the study of physical systems such as magnets – where rotational symmetry is spontaneously broken (the magnet has a preferred direction in space) – that when the magnet is placed in a heat bath, magnetization is lost for temperatures higher than a certain critical temperature. Above this critical temperature, the random thermal motions overtake any tendency in the elementary spins to align. As a result, the rotational symmetry which is spontaneously broken at low temperatures, is restored. In this chapter, we would study such phenomena within the context of relativistic quantum field theories.

9.1 Global symmetry

As a starting example, let us consider a simple scalar field theory which involves a single complex scalar field ϕ and is described by the Lagrangian density

$$\mathcal{L} = \partial_\mu \phi^\dagger \partial^\mu \phi + m^2 \phi^\dagger \phi - \frac{\lambda}{2}(\phi^\dagger \phi)^2, \qquad \lambda, m^2 > 0. \qquad (9.1)$$

The complex fields can also be equivalently expressed in terms of two Hermitian (real) fields σ and χ which may sometimes be useful. Thus, let us define

$$\phi(x) = \frac{1}{\sqrt{2}}(\sigma(x) + i\chi(x)),$$

$$\phi^\dagger(x) = \frac{1}{\sqrt{2}}(\sigma(x) - i\chi(x)). \qquad (9.2)$$

In terms of these fields, the Lagrangian density of (9.1) takes the form

$$\mathcal{L} = \frac{1}{2}\partial_\mu \sigma \partial^\mu \sigma + \frac{1}{2}\partial_\mu \chi \partial^\mu \chi + \frac{m^2}{2}(\sigma^2 + \chi^2) - \frac{\lambda}{8}(\sigma^2 + \chi^2)^2. \qquad (9.3)$$

The Lagrangian density of (9.1) can be easily checked to be invariant under the infinitesimal global phase transformations

$$\delta\phi = -i\epsilon\phi, \quad \delta\phi^\dagger = i\epsilon\phi^\dagger, \qquad (9.4)$$

where ϵ is a constant, infinitesimal, real parameter of the phase transformation (invariance under a finite transformation $\phi \to e^{-i\theta}\phi$, $\phi^\dagger \to e^{i\theta}\phi^\dagger$, with θ a real constant parameter of transformation, holds as well). Equivalently, we can show that the Lagrangian density of (9.3) is invariant under the infinitesimal global transformations

$$\delta\sigma = \epsilon\chi, \quad \delta\chi = -\epsilon\sigma. \qquad (9.5)$$

The Nöther current density associated with the symmetry transformations, (9.4) or (9.5), can be derived in a straightforward manner and has the form

$$J^\mu = i(\phi^\dagger \partial^\mu \phi - (\partial^\mu \phi^\dagger)\phi) = (\chi\partial^\mu \sigma - \sigma\partial^\mu \chi), \qquad (9.6)$$

from which the conserved charge is obtained to be

$$Q = \int d^3x \, J^0 = \int d^3x \, i(\phi^\dagger \dot{\phi} - \dot{\phi}^\dagger \phi) = \int d^3x \, (\chi \dot{\sigma} - \sigma \dot{\chi}). \quad (9.7)$$

The Nöther charge Q is the generator of the infinitesimal transformations in (9.4) or (9.5) and since these define a symmetry of the theory (leave the Hamiltonian invariant), we can check (with the canonical commutation relations of the theory) that Q commutes with the Hamiltonian, namely,

$$[Q, H] = 0, \quad (9.8)$$

with H representing the Hamiltonian which governs the dynamics of the theory and has the form (we are going to be slightly sloppy – the $\dot{\phi}$ and $\dot{\phi}^\dagger$ etc should really be replaced by the appropriate canonical conjugate momenta)

$$H = \int d^3x \left[\dot{\phi}^\dagger \dot{\phi} + \boldsymbol{\nabla}\phi^\dagger \cdot \boldsymbol{\nabla}\phi - m^2 \phi^\dagger \phi + \frac{\lambda}{2}(\phi^\dagger \phi)^2 \right]$$

$$= \int d^3x \left[\frac{1}{2}(\dot{\sigma}^2 + \dot{\chi}^2) + \frac{1}{2}(\boldsymbol{\nabla}\sigma \cdot \boldsymbol{\nabla}\sigma + \boldsymbol{\nabla}\chi \cdot \boldsymbol{\nabla}\chi) \right.$$

$$\left. - \frac{m^2}{2}(\sigma^2 + \chi^2) + \frac{\lambda}{8}(\sigma^2 + \chi^2)^2 \right]. \quad (9.9)$$

Furthermore, being the generator of the symmetry transformations, the commutator of Q with the field variables gives the infinitesimal transformations of the corresponding field variables (with a factor of i and without the parameter of transformation ϵ), namely,

$$[Q, \phi(x)] = i\delta\phi(x), \qquad [Q, \phi^\dagger(x)] = i\delta\phi^\dagger(x),$$

or, alternatively,

$$[Q, \sigma(x)] = i\delta\sigma(x), \qquad [Q, \chi(x)] = i\delta\chi(x). \quad (9.10)$$

While the Lagrangian density (or the Hamiltonian) of the theory is invariant under the transformations in (9.4) or (9.5), the ground state of the theory, in this case, is not. This is more easily seen in terms of the fields σ and χ as follows. First, we note that the theory in (9.1) or (9.3) is different from the standard theory for a complex scalar field in that the sign of the mass term is reversed (it is not $-m^2\phi^\dagger\phi$). Second, we note that the equation of motion following from the Lagrangian density in (9.3) (the Euler-Lagrange equations) are given by

$$\left(\Box - m^2 + \frac{\lambda}{2}(\sigma^2 + \chi^2)\right)\sigma = 0,$$

$$\left(\Box - m^2 + \frac{\lambda}{2}(\sigma^2 + \chi^2)\right)\chi = 0. \tag{9.11}$$

It is clear from the form of H in (9.9) that constant field configurations $(\sigma(x) = \sigma = \text{constant}, \chi(x) = \chi = \text{constant})$ would lead to a minimum of energy (the kinetic energy terms, which normally contribute positively to energy, will vanish in this case), provided

$$\frac{\partial V}{\partial \sigma} = 0 = \frac{\partial V}{\partial \chi}, \tag{9.12}$$

and such constant field configurations would also be a solution of the equations of motion (9.11). We note, from (9.9), that for constant field configurations,

$$V(\sigma, \chi) = -\frac{m^2}{2}(\sigma^2 + \chi^2) + \frac{\lambda}{8}(\sigma^2 + \chi^2)^2, \tag{9.13}$$

which is known as the tree level potential of the theory. For constant σ, χ, it is a simple function and that is the reason why we do not have functional derivatives in (9.12).

From (9.12) and (9.13), we see that for $\lambda, m^2 > 0$, there are two extrema of the potential which occur at

$$\sigma = 0 = \chi, \quad \text{or}, \quad (\sigma^2 + \chi^2) = \frac{2m^2}{\lambda}. \tag{9.14}$$

It can be easily checked that the first solution corresponding to the origin in the (constant) field configuration space is a local maximum whereas it is the second solution which represents the true minimum of the potential. This is the difference between this theory and the standard complex scalar field theory where the mass term has a negative sign (in the Lagrangian density) which leads to the (unique) minimum of the potential at the origin of the field configuration space (see Fig. 9.1). On the other hand, it is also clear that the second so-

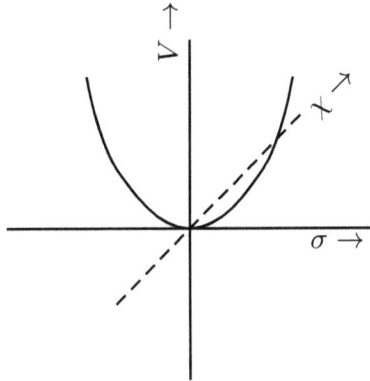

Figure 9.1: The form of the potential in the σ-χ field space for $m^2 <$ 0.

lution in (9.14), giving the minimum of the potential, defines a circle in the field space (see Fig. 9.2) which is invariant under the transformations in (9.5). Every point on this circle defines a minimum of energy and, consequently, we have an infinitely degenerate set of minima. However, once we choose a particular point on this circle as representing the minimum of the potential of our theory, the global invariance of (9.5) would be broken.

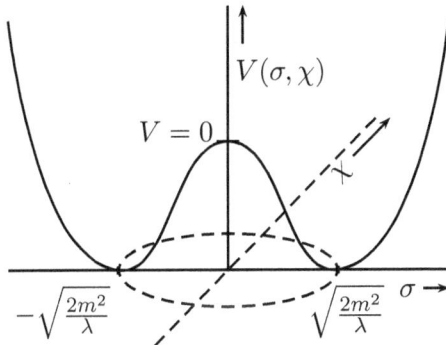

Figure 9.2: The form of the potential in the σ-χ field space for $m^2 >$ 0.

For simplicity, let us choose

$$\sigma = v = \sqrt{\frac{2m^2}{\lambda}}, \qquad \chi = 0, \tag{9.15}$$

as the minimum of the potential (see Fig. 9.2) of our theory. In the quantum theory, as we will see later, this corresponds to choosing a ground state (vacuum state) such that the vacuum expectation values of the two fields are given by

$$\langle 0|\sigma|0\rangle = \langle \sigma \rangle = v, \qquad \langle 0|\chi|0\rangle = \langle \chi \rangle = 0. \tag{9.16}$$

Consequently, $v = \sqrt{\frac{2m^2}{\lambda}}$ is also known as the vacuum expectation value of the σ field. It is now easy to see that the global symmetry of (9.5) is spontaneously broken in the following way. Taking the vacuum expectation values of (9.5), (9.10) and using (9.16), we obtain

$$\langle 0| [Q, \chi] |0\rangle = i\delta\langle 0|\chi|0\rangle = -i\langle 0|\sigma|0\rangle = -iv \neq 0. \tag{9.17}$$

For this to be true, the generator of infinitesimal transformations, Q, cannot annihilate the ground state (otherwise, the left hand side of (9.17) would vanish), namely,

$$Q|0\rangle \neq 0. \tag{9.18}$$

Equivalently, the ground state or the vacuum of the theory will not be invariant under the group of global transformations. When the Hamiltonian of the theory is invariant under a symmetry transformation, but the ground state is not, we say that the symmetry is spontaneously broken (by the choice of a ground state). (More intuitively, every point on the circle of minima in (9.14) defines an inequivalent vacuum of the quantized theory. Under the action of the global transformations, each point on this circle maps to another point and correspondingly, a given vacuum of the quantized theory under the action of the group of transformations goes to another inequivalent ground state.)

The field whose transformation picks up a nonzero vacuum expectation value can be shown to become massless (there will be no mass term for this field in the theory as we will see shortly) and is known as the Goldstone boson. Thus, in the present case, we see from (9.17), that χ would represent the Goldstone boson field. To see that this field, indeed, becomes massless, we note that we can define our quantum theory by redefining the field variables as

$$\sigma \to \sigma + v, \qquad \chi \to \chi, \qquad v = \sqrt{\frac{2m^2}{\lambda}}, \qquad (9.19)$$

so that the new fields would have zero vacuum expectation value and this is what we would expect for a normal field theory. With this redefinition of fields, the Lagrangian density of (9.3) becomes

$$\mathcal{L} = \frac{1}{2}\partial_\mu\sigma\partial^\mu\sigma + \frac{1}{2}\partial_\mu\chi\partial^\mu\chi - m^2\sigma^2 + \frac{m^4}{2\lambda}$$
$$- m\sqrt{\frac{\lambda}{2}}\,\sigma(\sigma^2 + \chi^2) - \frac{\lambda}{8}(\sigma^2 + \chi^2)^2. \qquad (9.20)$$

This clearly shows that the σ-field has become massive with a mass $\sqrt{2}m$ while the χ-field has become massless. This is the Goldstone boson. According to Goldstone's theorem, whenever a continuous global symmetry is spontaneously broken in a manifestly Lorentz invariant theory with a positive metric Hilbert space, massless particles must arise as a consequence of symmetry breaking. Here, we have explicitly seen this to be the case. More intuitively, Goldstone's theorem can be understood from the structure of the potential of the theory as follows. For constant field configurations, we have seen that the potential, in the case of spontaneous symmetry breaking, has the structure of a double well, rotated around the vertical axis as shown in Fig. 9.2. (This is also commonly known as the Mexican hat potential.) The true minimum of the potential, which occurs for a finite value of the field variables, allows for two kinds of motion. The small oscillations in the well, obviously, are harmonic and would correspond to massive quanta. On the other hand, the motion along the valley of the minimum would cost no energy and, therefore, can be associated with massless quanta. That is physically the origin of Goldstone bosons (massless particles) and the existence of massless quanta in the theory would imply long range correlations in the theory. However, in nature we do not see such massless particles with spin 0. As we will see in the next section, in the presence of gauge interactions, such massless particles are naturally removed.

9.2 Local symmetry

Let us consider, once again, the complex scalar field theory of the previous section, but with the complex scalar fields now interacting with an Abelian gauge field. (This is only for simplicity of discussions, the interaction can be with non-Abelian gauge fields as well.) The Lagrangian density now has the form (see the discussion in chapter **5**, particularly in section **5.1**)

$$\mathcal{L} = -\frac{1}{4}F_{\mu\nu}F^{\mu\nu} + (D_\mu\phi)^\dagger D^\mu\phi + m^2\phi^\dagger\phi - \frac{\lambda}{2}(\phi^\dagger\phi)^2. \tag{9.21}$$

Here, the covariant derivative is defined as (see (5.11))

$$D_\mu\phi = \partial_\mu\phi + ieA_\mu\phi, \tag{9.22}$$

where A_μ represents the Abelian gauge field and e the coupling constant or the interaction strength. This model is commonly known as the Abelian Higgs model.

The Lagrangian density, (9.21), can also be written in the σ-χ basis and takes the form

$$\mathcal{L} = -\frac{1}{4}F_{\mu\nu}F^{\mu\nu} + \frac{1}{2}\partial_\mu\sigma\partial^\mu\sigma + \frac{1}{2}\partial_\mu\chi\partial^\mu\chi + \frac{m^2}{2}(\sigma^2 + \chi^2)$$

$$- e(\chi\partial^\mu\sigma - \sigma\partial^\mu\chi)A_\mu + \frac{e^2}{2}(\sigma^2 + \chi^2)A_\mu A^\mu - \frac{\lambda}{8}(\sigma^2 + \chi^2)^2. \tag{9.23}$$

The Lagrangian densities in (9.21) or (9.23) can be easily seen to be invariant under the transformations in (9.4) or (9.5) respectively with the parameter of transformation a local function of space-time with the gauge fields also transforming as in (5.15) (note that $f^{abc} = 0$ for an Abelian theory), namely,

$$\delta\phi = -ie\epsilon(x)\phi,$$
$$\delta\phi^\dagger = ie\epsilon(x)\phi^\dagger,$$
$$\delta A_\mu = \partial_\mu\epsilon(x). \tag{9.24}$$

Here we have emphasized the space-time coordinate dependence of the infinitesimal parameter $\epsilon(x)$ while suppressing the coordinate dependence of the fields for simplicity. Equivalently, we can write the transformations in (9.24) in terms of the σ and the χ fields as

$$\delta\sigma = e\epsilon(x)\chi,$$
$$\delta\chi = -e\epsilon(x)\sigma,$$
$$\delta A_\mu = \partial_\mu \epsilon(x). \tag{9.25}$$

The global invariance of the theory, therefore, has been promoted to a local invariance in this theory because of the presence of the Abelian gauge field.

On the other hand, the scalar potential of the theory remains unchanged from that of the previous section. Consequently, following our discussions of the last section, we note that (see (9.15)) the minimum of the potential can be chosen to occur at nontrivial values of the field configuration

$$\langle 0|\sigma|0\rangle = \langle\sigma\rangle = v, \quad \langle 0|\chi|0\rangle = \langle\chi\rangle = 0, \quad v = \sqrt{\frac{2m^2}{\lambda}}.$$

The continuous phase symmetry is spontaneously broken and again we can expand our quantum field theory around this nontrivial vacuum (see (9.19)) and the shifted Lagrangian density (9.23) takes the form

$$\mathcal{L} = -\frac{1}{4}F_{\mu\nu}F^{\mu\nu} + \frac{1}{2}\partial_\mu\sigma\partial^\mu\sigma + \frac{1}{2}\partial_\mu\chi\partial^\mu\chi - m^2\sigma^2 + \frac{m^4}{2\lambda}$$
$$+ \frac{e^2v^2}{2}A_\mu A^\mu + ev\partial_\mu\chi A^\mu - e(\chi\partial_\mu\sigma - \sigma\partial_\mu\chi)A^\mu + e^2v\sigma A_\mu A^\mu$$
$$+ \frac{e^2}{2}(\sigma^2 + \chi^2)A_\mu A^\mu - \frac{\lambda v}{2}\sigma(\sigma^2 + \chi^2) - \frac{\lambda}{8}(\sigma^2 + \chi^2)^2. \tag{9.26}$$

Equation (9.26) is quite interesting in that it shows that the Abelian gauge field and the "Goldstone" field χ mix because of the symmetry breaking. To see the spectrum of the theory, we need to diagonalize only the quadratic part of the Lagrangian density in (9.26) which has the form

$$\mathcal{L}_q = -\frac{1}{4}F_{\mu\nu}F^{\mu\nu} + \frac{1}{2}\partial_\mu\sigma\partial^\mu\sigma + \frac{1}{2}\partial_\mu\chi\partial^\mu\chi - m^2\sigma^2$$
$$+ ev\partial_\mu\chi A^\mu + \frac{e^2v^2}{2}A_\mu A^\mu. \tag{9.27}$$

This has the interesting feature that if we define a new gauge field as

$$B_\mu = A_\mu + \frac{1}{ev} \partial_\mu \chi, \quad F_{\mu\nu} = \partial_\mu A_\nu - \partial_\nu A_\mu = \partial_\mu B_\nu - \partial_\nu B_\mu, \quad (9.28)$$

then, the quadratic Lagrangian density of (9.27) can be written as

$$\mathcal{L}_q = -\frac{1}{4}(\partial_\mu B_\nu - \partial_\nu B_\mu)(\partial^\mu B^\nu - \partial^\nu B^\mu) + \frac{e^2 v^2}{2} B_\mu B^\mu$$

$$+ \frac{1}{2} \partial_\mu \sigma \partial^\mu \sigma - m^2 \sigma^2. \quad (9.29)$$

We see that the quadratic Lagrangian density has diagonalized and, with the redefinition in (9.28), the "Goldstone" field (massless χ field) has disappeared completely from the quadratic part of the theory. The redefinition in (9.28), of course, has precisely the same form as a gauge transformation (see (9.25)) and so it is clear that the χ field has been absorbed into the gauge degrees of freedom of the A_μ field. Furthermore, we also see from (9.29) that the (redefined) Abelian gauge field (photon) has now become massive with a mass

$$m_{\rm g} = ev = e\sqrt{\frac{2m^2}{\lambda}}. \quad (9.30)$$

Thus, we see that, in this case, we have a massive gauge field with mass $m_{\rm g}$ and a scalar field (σ) with mass $\sqrt{2}m$, but there is no massless scalar field in the quadratic part of the Lagrangian density any more. The χ degree of freedom has gone into the massive degree of freedom of the gauge field and so the number of physical degrees of freedom is maintained (there is no violation of unitarity). Conventionally, one says that the gauge field has eaten up the "Goldstone" field to become massive.

The "Goldstone" field has, of course, disappeared from the part of the Lagrangian density which is quadratic in fields, but we can also show that it does not have any physical effect in the full theory. This can be seen most easily by noting that we have a gauge theory with a local gauge invariance. As we have discussed in chapter **5**, in such a case, we need to choose a gauge for any physical, perturbative calculation. (We point out here that in a lattice calculation, in contrast, there is no need for a choice of gauge.) In particular, if we choose the gauge

$$\chi(x) = 0, \tag{9.31}$$

then, in this gauge, the Lagrangian density of (9.26) including interactions would become

$$\mathcal{L} = -\frac{1}{4} F_{\mu\nu} F^{\mu\nu} + \frac{1}{2} \partial_\mu \sigma \partial^\mu \sigma - m^2 \sigma^2 + \frac{e^2 v^2}{2} A_\mu A^\mu + \frac{m^4}{2\lambda}$$
$$+ e^2 v \sigma A_\mu A^\mu + \frac{e^2}{2} \sigma^2 A_\mu A^\mu - \frac{\lambda v}{2} \sigma^3 - \frac{\lambda}{8} \sigma^4. \tag{9.32}$$

In this gauge, the "Goldstone" field has disappeared and the complete Lagrangian density is written in terms of only the physical field variables. As a result such a gauge choice is also known as the unitary gauge. However, this is not a particularly convenient gauge choice from the point of view of doing calculations.[1] A more commonly used covariant gauge which is convenient for calculations in such a theory is known as the R_ξ-gauge. Here one adds to the Lagrangian density of (9.26) a gauge fixing as well as a ghost Lagrangian density of the form (without the auxiliary fields of chapter 5, see, for example, (5.25) and (9.25))

$$\mathcal{L}_{\text{gf}} + \mathcal{L}_{\text{ghost}}$$
$$= -\frac{1}{2\xi} (\partial_\mu A^\mu - \xi ev\chi)^2 + \partial_\mu \bar{c} \, \partial^\mu c - \xi (ev)^2 \bar{c}c - \xi e^2 v\sigma \bar{c}c$$
$$= -\frac{1}{2\xi} (\partial_\mu A^\mu)^2 + ev\chi \partial_\mu A^\mu - \frac{\xi(ev)^2}{2} \chi^2 + \partial_\mu \bar{c} \, \partial^\mu c$$
$$- \xi(ev)^2 \bar{c}c - \xi e^2 v\sigma \bar{c}c. \tag{9.33}$$

Comparing this with (9.26), we see that the mixing between χ and the A_μ fields disappears in this gauge. However, both the χ field and the ghost fields pick up an identical, gauge parameter dependent mass (see also (9.30))

$$m_\chi^2 = m_{\text{ghost}}^2 = \xi (ev)^2 = \xi \, m_{\text{g}}^2. \tag{9.34}$$

---

[1]See A. Das, *Lectures on Quantum Field Theory* (second edition), World Scientific (2020).

We would end this section with a brief discussion of why the "Goldstone" theorem fails in this case. We recall that the "Goldstone" theorem necessarily predicts the existence of massless particles in a theory when a continuous symmetry is spontaneously broken. The assumption, in the theorem, however, is that the theory should be manifestly Lorentz invariant with a positive metric for the Hilbert space. These assumptions are clearly violated in the case of a gauge theory. As we have seen in chapter **5**, if we want to describe a gauge theory in a manifestly Lorentz invariant manner, we must necessarily have unphysical particles (degrees of freedom) with negative norm and consequently, we cannot have a positive metric for the vector space. Conversely, if we choose to describe the theory only with the physical degrees of freedom – the transverse components – then, the Hilbert space will have a positive definite inner product, but manifest Lorentz invariance will be lost. Thus, we see that when there is a local invariance, the basic assumptions of the "Goldstone" theorem fail (become incompatible). Consequently, we avoid the problem of having massless (spin zero) fields (particles) which were not observed experimentally.

9.3 Effective potential

It is clear from the discussion of the last two sections that the classical potential of a quantum field theory plays a crucial role in studying the question of symmetry breaking. However, sometimes the radiative corrections in a quantum theory may change the behavior of the classical potential. The minimum of the classical potential may be unstable under radiative corrections. A symmetry that is spontaneously broken at the classical level may be restored or even a symmetry that is unbroken at the classical level may be spontaneously broken because of quantum effects. The important quantity to consider, therefore, is not the tree level (classical) potential, but the effective potential of the theory which takes into account all the quantum corrections in the theory. The effective potential, unfortunately, does not have a closed form expression unlike the classical potential and, for this reason, one has to analyze this quantity order by order in perturbation theory. In our discussions, we will restrict ourselves to the effective potential taking into account only the one loop corrections. There are, however, systematic procedures to go beyond one loop.

Let us, for simplicity, consider a self-interacting scalar ϕ^4-theory in the presence of a source described by the action (the source allows us to define Green's functions of the theory)

$$
\begin{aligned}
S^J &= \int d^4x \left[\frac{1}{2} \partial_\mu \phi \partial^\mu \phi - \frac{m^2}{2} \phi^2 - \frac{\lambda}{4!} \phi^4 + J\phi \right] \\
&= S + \int d^4x \, J\phi.
\end{aligned}
\tag{9.35}
$$

Here we have assumed the tree level mass term to have the right sign (as opposed to the last two sections) and the classical or the tree level potential is clearly given by (with $J = 0$, the superscript on the potential denotes the loop order)

$$
V^{(0)} = \frac{m^2}{2} \phi^2 + \frac{\lambda}{4!} \phi^4.
\tag{9.36}
$$

The generating functional, in this case, is defined to be

$$
Z[J] = e^{\frac{i}{\hbar} W[J]} = \int \mathcal{D}\phi \, e^{\frac{i}{\hbar} S^J}.
\tag{9.37}
$$

Here, $W[J]$ is known as the generating functional for the connected Green's functions and gives (the superscript "c" signifies connected Green's functions)

$$
\langle 0|T(\phi(x_1) \cdots \phi(x_n))|0\rangle^c = (-i\hbar)^{n-1} \frac{\delta^n W}{\delta J(x_1) \cdots \delta J(x_n)} \bigg|_{J=0}.
\tag{9.38}
$$

From equation (9.38), we note, in particular, that the one point function (in the presence of the source) is given by

$$
\phi_c(x) = \frac{\delta W}{\delta J(x)} = \langle 0|\phi(x)|0\rangle_J,
\tag{9.39}
$$

which is known as the classical field (note the subscript "c") and (9.39) shows that the classical field, ϕ_c, which represents the vacuum expectation value of the field, ϕ, would be a constant independent of space-time when the source is turned off simply because of the translation invariance of the vacuum. In general, however, it will be a functional of the source J and is known to generate the tree diagrams of the theory.

The Euler-Lagrange equation following from the form of the action in (9.35) is given by (note that $\frac{\delta S}{\delta \phi} = \frac{\partial \mathcal{L}}{\partial \phi} - \partial_\mu \frac{\partial \mathcal{L}}{\partial \partial_\mu \phi}$ for the present theory)

$$\frac{\delta S^J}{\delta \phi(x)} = \frac{\delta S}{\delta \phi(x)} + J(x) = 0,$$

$$\text{or,} \quad (\Box + m^2)\phi + \frac{\lambda}{3!}\phi^3 - J = 0. \tag{9.40}$$

For simplicity, let us denote this as

$$F(\phi) = J, \tag{9.41}$$

where $F(\phi)$ represents the Euler operator

$$F(\phi) = (\Box + m^2)\phi + \frac{\lambda}{3!}\phi^3. \tag{9.42}$$

Next, we note from the definition of the generating functional in (9.37) that it has no dependence on the field variable (namely, all possible field configurations are being integrated over in the path integral). Consequently, if we were to redefine

$$\phi \to \phi + \delta\phi,$$

inside the path integral, with $\delta\phi$ infinitesimal and arbitrary, the generating functional would not change. This, therefore, leads to the identity (under such a field redefinition)

$$\delta Z[J] = 0, \tag{9.43}$$

which, in turn, would imply that (see (9.37) and (9.40)-(9.42))

$$\int \mathcal{D}\phi \, \frac{\delta S^J}{\delta \phi} \, e^{\frac{i}{\hbar} S^J} = 0,$$

$$\text{or,} \quad \int \mathcal{D}\phi \, (F(\phi) - J) \, e^{\frac{i}{\hbar} S^J} = 0,$$

$$\text{or,} \quad \int \mathcal{D}\phi \, \left(F\!\left(-i\hbar\frac{\delta}{\delta J} \right) - J \right) e^{\frac{i}{\hbar} S^J} = 0,$$

$$\text{or,} \quad \left(F\!\left(-i\hbar\frac{\delta}{\delta J} \right) - J \right) Z[J] = 0.$$

In terms of $W[J]$ (see (9.37)) this, then, gives

$$e^{-\frac{i}{\hbar}W[J]}F\left(-i\hbar\frac{\delta}{\delta J(x)}\right)e^{\frac{i}{\hbar}W[J]} = J(x),$$

$$\text{or,}\quad F\left(\frac{\delta W}{\delta J(x)} - i\hbar\frac{\delta}{\delta J(x)}\right) = J(x). \tag{9.44}$$

(Here, we are assuming that the functional derivative, after simplification, acts on the identity on the right.) In the limit of vanishing $\hbar$, this equation reduces to (see (9.39))

$$F\left(\frac{\delta W}{\delta J(x)}\right) = J(x),$$

$$\text{or,}\quad F(\phi_{\mathrm{c}}) = J(x). \tag{9.45}$$

Thus, we see that the classical field satisfies the same equation as the Euler-Lagrange equation of the original theory (see (9.41)) when $\hbar \to 0$. If we now switch off the source, then, as we have noted earlier, the classical field would have a constant value. Consequently, in that limit, equation (9.45) would lead to

$$\frac{\partial V^{(0)}(\phi_{\mathrm{c}})}{\partial \phi_{\mathrm{c}}} = 0. \tag{9.46}$$

This is the minimum equation for the potential in terms of ϕ_{c} and as we have seen earlier, a nontrivial minimum of the tree level potential would correspond to a nontrivial solution for (9.46). On the other hand, a nontrivial solution for (9.46) would imply

$$\phi_{\mathrm{c}} = \langle 0|\phi|0\rangle \neq 0. \tag{9.47}$$

This, therefore, ties in with our earlier assertion (see (9.15), (9.16) and the discussion there) that a nontrivial minimum for the classical potential would imply that the quantum fields would have nonzero vacuum expectation value in the theory. We can think of $\phi_{\mathrm{c}} = \langle 0|\phi|0\rangle$ as an order parameter of symmetry breaking and functionally, we would recognize spontaneous symmetry breaking if (see (9.39) and (9.47))

$$\left.\frac{\delta W}{\delta J}\right|_{J=0} \neq 0. \tag{9.48}$$

In a quantum field theory, however, we are more interested in studying the effective action or the generating functional for the 1PI (one particle irreducible) graphs which is the Legendre transformation of $W[J]$ defined as follows

$$\Gamma[\phi_c] = W[J] - \int d^4x\, J(x)\phi_c(x). \tag{9.49}$$

This definition immediately leads to

$$\frac{\delta\Gamma}{\delta\phi_c(x)} = -J(x). \tag{9.50}$$

The symmetry breaking condition of (9.48) can now be written as

$$\left.\frac{\delta\Gamma}{\delta\phi_c}\right|_{\phi_c=v\neq0} = 0. \tag{9.51}$$

As we will see shortly, this condition is more like the minimum condition of (9.46).

The effective action, Γ, is a functional of the classical field and there are two equivalent ways of expanding it. First, we can expand it as a power series (Taylor series) in ϕ_c to write

$$\Gamma[\phi_c] = \sum_n \frac{1}{n!} \int d^4x_1 \cdots d^4x_n\, \Gamma^{(n)}(x_1, \cdots, x_n)$$

$$\times \phi_c(x_1) \cdots \phi_c(x_n), \tag{9.52}$$

where the coefficients of expansion

$$\Gamma^{(n)}(x_1, \cdots, x_n) = \left.\frac{\delta^n\Gamma}{\delta\phi_c(x_1)\cdots\delta\phi_c(x_n)}\right|_{\phi_c=0}, \tag{9.53}$$

represent the 1PI n-point vertex functions. This expansion assumes that there is no symmetry breaking. On the other hand, if we had a spontaneous breakdown of symmetry, the corresponding formulae would take the forms

$$\Gamma[\phi_c] = \sum_n \frac{1}{n!} \int d^4x_1 \cdots d^4x_n\, \Gamma^{(n)}(x_1, \cdots, x_n)$$

$$\times (\phi_c(x_1) - v) \cdots (\phi_c(x_n) - v), \tag{9.54}$$

where the 1PI vertex functions, in the symmetry broken case, are defined as

$$\Gamma^{(n)}(x_1, \cdots, x_n) = \left. \frac{\delta^n \Gamma}{\delta \phi_c(x_1) \cdots \delta \phi_c(x_n)} \right|_{\phi_c = v}, \tag{9.55}$$

with v denoting the solution of the minimum equation (9.51).

An alternate expansion of the effective action is in terms of powers of momentum (or derivatives) as

$$\Gamma[\phi_c] = \int d^4x \left[-V(\phi_c) + \frac{1}{2} \partial_\mu \phi_c \partial^\mu \phi_c + \cdots \right]. \tag{9.56}$$

Here the "$\cdots$" represent terms which involve higher derivatives of ϕ_c. It is easy to see that such an expansion is possible simply because $\Gamma[\phi_c]$ represents an effective action and must necessarily contain the structure of the classical action in (9.35). This form of the expansion of the effective action, however, is more convenient from the point of view of studying symmetry breaking. Thus, for example, we have seen that when $J \to 0$, $\phi_c(x) = \phi$ is a constant (which may or may not be zero). In this limit, then, all the derivative terms in the expansion in (9.56) vanish and we have

$$\Gamma(\phi) = -\Omega V_{\text{eff}}(\phi) = -(2\pi)^4 \delta(0) V_{\text{eff}}(\phi), \tag{9.57}$$

where $V_{\text{eff}}(\phi)$ is known as the effective potential (including quantum corrections). The multiplicative factor, Ω, simply represents the space-time volume and because ϕ is a constant, we only have a function (and not a functional) of this variable. Consequently, the minimum equation of (9.51) becomes

$$\frac{\partial V_{\text{eff}}(\phi)}{\partial \phi} = 0, \tag{9.58}$$

which is more like (9.46). However, V_{eff} now contains quantum corrections to the original classical potential $V^{(0)}$ of (9.36).

9.4 One loop potential

As we have noted earlier, V_{eff} represents the effective potential including quantum corrections to all orders. However, unlike the classical (tree level) potential, it cannot be expressed explicitly in a closed form. Consequently, its structure has to be studied order by order in perturbation theory. A perturbative expansion, which is particularly convenient in connection with symmetry breaking, is in terms of the number of loops of a Feynman diagram which also coincides with an expansion in powers of $\hbar$ (Planck's constant). In our discussions, we will restrict ourselves only to one loop for simplicity, although the calculations can be carried out systematically to higher loops as well. Thus, we write

$$V_{\text{eff}} = V^{(0)} + V^{(1)} + O(\hbar^2), \tag{9.59}$$

where we know the tree level potential, $V^{(0)}$, from (9.36), and we are interested in calculating the one loop correction, $V^{(1)}$.

There are several ways of calculating the one loop correction to the effective potential and we will use the one that naturally generalizes to the case of finite temperature as well. Let us assume that $\phi_0(x)$ represents a solution of the tree level Euler-Lagrange equation

$$\left. \frac{\delta S^J[\phi]}{\delta \phi(x)} \right|_{\phi=\phi_0(x)} = 0. \tag{9.60}$$

In such a case, we can expand the field variables around this classical solution as

$$\phi(x) = \phi_0(x) + \sqrt{\hbar}\chi(x). \tag{9.61}$$

Correspondingly, the classical action would have the expansion

$$\begin{aligned}
S^J[\phi] &= S^J[\phi_0 + \sqrt{\hbar}\chi] \\
&= S^J[\phi_0] + \sqrt{\hbar} \int \mathrm{d}^4x\, \chi(x) \left. \frac{\delta S^J}{\delta \phi(x)} \right|_{\phi=\phi_0} \\
&\quad + \frac{\hbar}{2} \int \mathrm{d}^4x \mathrm{d}^4y\, \chi(x) \left. \frac{\delta^2 S^J}{\delta \phi(x) \delta \phi(y)} \right|_{\phi=\phi_0} \chi(y) + \cdots \\
&= S^J[\phi_0] - \frac{\hbar}{2} \int \mathrm{d}^4x \mathrm{d}^4y\, \chi(x) G^{-1}(x,y;\phi_0)\chi(y) + \cdots. \tag{9.62}
\end{aligned}$$

There are several things to note about the expansion in (9.62). First, the linear term in χ vanishes because of (9.60), namely, because ϕ_0 represents a solution of the classical Euler-Lagrange equation.

Second, the terms represented by "$\cdots$" are higher order in the fields χ and, therefore, from the expansion in (9.61), we see that they would contribute with a higher power of $\hbar$ (namely, $\hbar^{\frac{3}{2}}$ and higher). Consequently, we can neglect them. Furthermore, since the source term in S^J is linear in the field variable, it does not contribute to the second derivative terms and we have identified

$$G^{-1}(x, y; \phi_0) = -\left.\frac{\delta^2 S^J}{\delta\phi(x)\delta\phi(y)}\right|_{\phi_0} = -\left.\frac{\delta^2 S}{\delta\phi(x)\delta\phi(y)}\right|_{\phi_0}. \qquad (9.63)$$

This is simply the inverse of the propagator (up to factors of i and can be related to the two point function) in the presence of a background field ϕ_0 and, as we will see shortly, would correspond to a propagator (Green's function) with an effective mass in this calculation of the effective potential.

Substituting the expansion in (9.62) (keeping only up to quadratic terms in χ) into the expression for the generating functional in (9.37), we have

$$Z[J] = e^{\frac{i}{\hbar}W[J]} = \int \mathcal{D}\chi \, e^{\frac{i}{\hbar}S^J[\phi_0 + \sqrt{\hbar}\chi]}$$

$$\approx e^{\frac{i}{\hbar}S^J[\phi_0]} \int \mathcal{D}\chi \, e^{-\frac{i}{2}\int d^4x d^4y \, \chi(x) \, G^{-1}(x,y;\phi_0)\chi(y)}$$

$$= e^{\frac{i}{\hbar}S^J[\phi_0]} \left[\det(G^{-1})\right]^{-\frac{1}{2}}. \qquad (9.64)$$

It follows, therefore, that (recall that $\det A = e^{\operatorname{Tr} \ln A}$)

$$W[J] = S^J[\phi_0] - i\hbar \, \ln \left[\det(G^{-1})\right]^{-\frac{1}{2}}$$

$$= S^J[\phi_0] + \frac{i\hbar}{2} \operatorname{Tr} \ln G^{-1}. \qquad (9.65)$$

The determinant in the above equation, of course, needs to be evaluated carefully by properly rotating to the Euclidean space, but we will come to that shortly.

To go to the effective action, we note that ϕ_c, the classical field, can also be expanded in powers of $\hbar$. However, since it satisfies the classical solution in the limit $\hbar \to 0$ (see (9.60)-(9.61)), we can write its expansion as

$$\phi_c(x) = \phi_0(x) + \phi_1(x) + \cdots, \qquad (9.66)$$

where by definition, ϕ_1 is of order $\hbar$ while the "$\cdots$" represent higher order terms in the expansion. With this, we note now that

$$
\begin{aligned}
S[\phi_c] &\sim S[\phi_0 + \phi_1] \\
&= S[\phi_0] + \int \mathrm{d}^4x\, \phi_1(x) \left.\frac{\delta S}{\delta \phi(x)}\right|_{\phi_0} + O(\hbar^2) \\
&\approx S[\phi_0] - \int \mathrm{d}^4x\, \phi_1(x) J(x),
\end{aligned}
\tag{9.67}
$$

where we have used the classical Euler lagrange equation (9.40) and have neglected the higher order terms.

It now follows from Eqs. (9.49) and (9.65) that the effective action, to this order, has the form

$$
\begin{aligned}
\Gamma[\phi_c] &= W[J] - \int \mathrm{d}^4x\, \phi_c(x) J(x) \\
&= S^J[\phi_0] + \frac{i\hbar}{2} \operatorname{Tr}\ln G^{-1} - \int \mathrm{d}^4x\, (\phi_0 + \phi_1) J \\
&= S[\phi_0] + \int \mathrm{d}^4x\, \phi_0 J + \frac{i\hbar}{2} \operatorname{Tr}\ln G^{-1} - \int \mathrm{d}^4x\, (\phi_0 + \phi_1) J \\
&= S[\phi_0] - \int \mathrm{d}^4x\, \phi_1 J + \frac{i\hbar}{2} \operatorname{Tr}\ln G^{-1} \\
&= S[\phi_c] + \frac{i\hbar}{2} \operatorname{Tr}\ln G^{-1}.
\end{aligned}
\tag{9.68}
$$

Here we used the identification in (9.67) in the last step. For constant field configurations ($\phi_c = $ constant), we determine the effective potential from (9.68) up to this order to be ($V(\phi)$ denotes the tree level scalar potential)

$$
\Gamma(\phi_c) = -\Omega\, V_{\text{eff}}(\phi_c) = -\Omega\, V(\phi_c) + \frac{i\hbar}{2} \operatorname{Tr}\ln G^{-1},
$$

$$
\text{or,} \quad V_{\text{eff}}(\phi_c) = V(\phi_c) - \frac{i\hbar}{2} \Omega^{-1} \operatorname{Tr}\ln G^{-1}.
\tag{9.69}
$$

With the identification in (9.59), we now see that the one loop contribution to the effective potential is given by

$$
V^{(1)}(\phi_c) = -\frac{i\hbar}{2} \Omega^{-1} \operatorname{Tr}\ln G^{-1}.
\tag{9.70}
$$

To explicitly evaluate the one loop correction to the classical potential, we first note, from the definition in (9.63) and the form of S in (9.35), that

$$\langle x|G^{-1}|y\rangle = G^{-1}(x,y) = \left(\Box + m^2 + \frac{\lambda}{2}\phi_0^2\right)\delta^4(x-y)$$

$$= \left(\Box + m^2 + \frac{\lambda}{2}\phi_c^2\right)\delta^4(x-y)$$

$$= \left(\Box + m_{\text{eff}}^2\right)\delta^4(x-y), \tag{9.71}$$

where we have used the fact that to the order of our calculation, we can use $\phi_c \sim \phi_0$ (remember that the expression in (9.70) has an overall factor of $\hbar$). This also shows, as we had noted earlier, that for the calculation of the effective potential, G behaves like a propagator (Green's function) with an effective mass

$$m_{\text{eff}}^2 = m^2 + \frac{\lambda}{2}\phi_c^2. \tag{9.72}$$

Taking the Fourier transform of (9.71), we determine

$$\langle k|G^{-1}|k'\rangle = G^{-1}(k,k') = (-k^2 + m_{\text{eff}}^2)\,\delta^4(k-k'). \tag{9.73}$$

Consequently, using the fact that for a diagonal matrix we have,

$$\langle i|\ln A_D|j\rangle = \ln\langle i|A_D|j\rangle, \tag{9.74}$$

we can now obtain

$$\text{Tr}\ln G^{-1} = \int d^4x\,\langle x|\ln G^{-1}|x\rangle$$

$$= \int d^4x\,d^4k\,d^4k'\,\langle x|k\rangle\langle k|\ln G^{-1}|k'\rangle\langle k'|x\rangle$$

$$= \int d^4x\,d^4k\,d^4k'\,\frac{e^{-i(k-k')\cdot x}}{(2\pi)^4}\,\ln(-k^2 + m_{\text{eff}}^2)\,\delta^4(k-k')$$

$$= \int d^4x\,\frac{d^4k}{(2\pi)^4}\,\ln(-k^2 + m_{\text{eff}}^2)$$

$$= \Omega\int\frac{d^4k}{(2\pi)^4}\,\ln(-k^2 + m_{\text{eff}}^2), \tag{9.75}$$

where Ω stands for the space-time volume.

Thus, from (9.70), we see that we can identify

$$V^{(1)}(\phi_c) = -\frac{i\hbar}{2} \int \frac{d^4k}{(2\pi)^4} \ln(-k^2 + m^2_{\text{eff}}). \tag{9.76}$$

This integral can be evaluated by rotating to the Euclidean space with

$$k^0 \to ik_4, \quad k^2 \to k^2_E = -k^2,$$

so that the one loop correction to the potential in (9.76) becomes

$$\begin{aligned}
V^{(1)}(\phi_c) &= -\frac{i\hbar}{2} \int \frac{i d^4k_E}{(2\pi)^4} \ln(k^2_E + m^2_{\text{eff}}) \\
&= \frac{\hbar}{2} \frac{2\pi^2}{(2\pi)^4} \int_0^\infty dk_E \, k^3_E \, \ln(k^2_E + m^2_{\text{eff}}).
\end{aligned} \tag{9.77}$$

This integral diverges at the upper limit and hence needs to be regularized. Using a large cut-off, Λ, we can evaluate the integral to be

$$\begin{aligned}
V^{(1)}(\phi_c) &= \frac{\hbar}{32\pi^2} \int_0^{\Lambda^2} dx \, x \, \ln(x + m^2_{\text{eff}}) \\
&= \frac{\hbar}{32\pi^2} \Big[\frac{\Lambda^4}{2}(\ln \Lambda^2 - \frac{1}{2}) + m^2_{\text{eff}}\Lambda^2 \\
&\quad + \frac{m^4_{\text{eff}}}{2}(\ln \frac{m^2_{\text{eff}}}{\Lambda^2} - \frac{1}{2}) + O(\frac{1}{\Lambda^2}) \Big].
\end{aligned} \tag{9.78}$$

The important point to note is that the one loop correction is explicitly divergent as $\Lambda \to \infty$. Consequently, we must add to our theory counterterms which would take care of these divergences. This is the standard renormalization procedure in a quantum field theory. If we add counterterms and renormalize our theory by demanding that the renormalized mass and the coupling constants of the theory up to one loop are given by

$$\left. \frac{d^2V_{\text{eff}}(\phi_c)}{d\phi_c^2} \right|_{\phi_c=0} = m^2,$$

$$\left. \frac{d^4V_{\text{eff}}(\phi_c)}{d\phi_c^4} \right|_{\phi_c=0} = \lambda, \tag{9.79}$$

then, the effective potential, up to one loop, can be rewritten (with an unimportant constant) to be

$$V_{\text{eff}}(\phi_c) = V^{(0)}(\phi_c) + \frac{\hbar}{64\pi^2} \, (m_{\text{eff}}^4 \ln \frac{m_{\text{eff}}^2}{m^2} - \frac{3}{2}(m_{\text{eff}}^2 - \frac{2}{3}m^2)^2).$$
(9.80)

It is worth noting here that we are considering a massive scalar theory and, consequently, the renormalization conditions in (9.79) can be used without running into infrared divergences. In a massless theory, on the other hand, one cannot use these conditions. In such a case, one defines the renormalized parameters at an arbitrary renormalization scale μ and the constants appearing in the expression for the effective potential would be different. Physical observables, however, can be shown to be independent of such an arbitrary energy scale.

We note here, for use in the next section, that the one loop effective potential in (9.77) can also be written alternatively by integrating out the fourth component of the Euclidean momentum, k_4, as (we also add an unimportant constant factor)

$$\begin{aligned}
V^{(1)}(\phi_c) &= \frac{\hbar}{2} \int \frac{d^3k}{(2\pi)^3} \int \frac{dk_4}{2\pi} \, \ln(k_4^2 + \omega^2) \\
&= \frac{\hbar}{2} \int \frac{d^3k}{(2\pi)^3} \int \frac{dk_4}{2\pi} \int_0^1 d\alpha \, \frac{\omega^2}{k_4^2 + \alpha\omega^2},
\end{aligned}$$
(9.81)

where we have defined

$$\omega^2 = \mathbf{k}^2 + m_{\text{eff}}^2.$$
(9.82)

The integrand in (9.81) has poles at $k_4 = \pm i\sqrt{\alpha}\omega$. Consequently, if we interchange the order of the α and the k_4 integrations, we can evaluate the k_4 integral using the residue theorem leading to

$$\begin{aligned}
\int_0^1 d\alpha \int \frac{dk_4}{2\pi} \frac{\omega^2}{k_4^2 + \alpha\omega^2} \\
= \int_0^1 d\alpha \, \frac{1}{2\pi} \, (2\pi i) \frac{\omega^2}{2i\sqrt{\alpha}\omega} = \frac{\omega}{2} \int_0^1 \frac{d\alpha}{\sqrt{\alpha}} = \omega.
\end{aligned}$$
(9.83)

Substituting this into (9.81), the one loop effective potential takes the form

$$V^{(1)}(\phi_c) = \frac{\hbar}{2} \int \frac{d^3k}{(2\pi)^3}\, \omega. \tag{9.84}$$

Finally, we conclude this section by noting that we have calculated the one loop effective potential in a theory involving only a single scalar field. If the scalar field carries an internal index or if there are fermions and gauge fields present, the one loop contribution of these fields to the scalar effective potential can also be calculated in a parallel manner and can be shown to have the leading general structure (see (9.80))

$$V^{(1)} = \frac{\hbar}{64\pi^2} \text{Tr}\left(M_S^4 \ln M_S^2 - 4(M_F^\dagger M_F)^2 \ln(M_F^\dagger M_F) + 3M_V^4 \ln M_V^2 \right), \tag{9.85}$$

where we have defined the scalar, fermion and the gauge mass matrices from the (tree level) classical action S with zero momentum as (with ϕ_c constant and all other fields vanishing)

$$M_S^{2\,ij} = -\left.\frac{\partial^2 S}{\partial \phi^i \partial \phi^j}\right|_{\phi=\phi_c},$$

$$M_F^{ab} = \left.\frac{\partial^2 S}{\partial \bar{\psi}^a \partial \psi^b}\right|_{\phi=\phi_c},$$

$$\eta^{\mu\nu} M_V^{2\,\alpha\beta} = \left.\frac{\partial^2 S}{\partial A_\mu^\alpha \partial A_\nu^\beta}\right|_{\phi=\phi_c}. \tag{9.86}$$

There are several things to note here. First, the fermion derivatives are defined to be left derivatives. Second, the fermion mass matrix will, in general, be complex depending on its coupling to the scalar fields and, consequently, the contribution of the fermions comes in the proper Hermitian combination (and with a negative sign because the integration over the fermion variables gives a determinant, and not the inverse as is true for bosonic fields, see for example, (5.102)). Furthermore, in general, when we shift the scalar fields by ϕ_c, there would be mixing terms between the scalar and the gauge fields. However, such terms give vanishing contribution in the Landau gauge and the expression in (9.86) corresponds to this choice of the gauge. The effective potential, as we will discuss in more detail in the next chapter, is a gauge dependent quantity although the physical observables derived from it are gauge independent if there are no (infrared) singularities.

9.5 Symmetry restoration

As we have noted in the beginning of this chapter, in physical systems such as a magnet where rotational symmetry is spontaneously broken, systems go through a phase transition when the temperature is raised and above a critical temperature the symmetry is restored. It is meaningful to ask whether a similar thing also happens in the case of a quantum field theory with a spontaneously broken symmetry. Let us, for simplicity, consider the theory of (9.1) or (9.3) with a spontaneous breakdown of a global symmetry. We have seen from (9.12)-(9.14) that a nontrivial vacuum exists leading to a spontaneous breaking of the symmetry only because the mass term has a negative sign. On the other hand, on physical grounds, we expect that if a particle moves in a hot medium, it must pick up a temperature dependent effective mass. In fact, we have already seen (see (1.113)) that at one loop, the mass term in a quantum field theory becomes temperature dependent and that the temperature dependent corrections come with a positive sign. So, intuitively, it is clear that above a certain temperature, this temperature dependent correction will take over the tree level negative mass squared term and symmetry will be restored. Let us now study this phenomenon a bit more quantitatively. For simplicity and ease of comparison with the results of chapter 1, we will now set $\hbar = 1$. Second, since the effective potential, by definition, is a static quantity, we will use the imaginary time formalism for its evaluation although any of the real time formalisms can also be used with equal ease.

Let us consider the Lagrangian density of (9.3). We can, of course, evaluate the one loop effective potential in full generality following the discussions of the previous section. However, let us simplify our calculation with the following observation. First, if a nontrivial vacuum does exist, then the solution will have the form of the second relation in (9.14) and we can always choose the minimum of the potential to be at $\chi = 0$ (as in (9.16)). Therefore, it is meaningful and calculationally simple, for the purpose of evaluating the effective potential, to shift the fields as (σ_c = constant)

$$\sigma \to \sigma + \sigma_c \quad \chi \to \chi. \tag{9.87}$$

With this shifting, the quadratic Lagrangian density in the quantum fields becomes (see (9.3))

$$\mathcal{L}_q = \frac{1}{2}\partial_\mu \sigma \partial^\mu \sigma - \frac{M_1^2}{2}\sigma^2 + \frac{1}{2}\partial_\mu \chi \partial^\mu \chi - \frac{M_2^2}{2}\chi^2, \tag{9.88}$$

where we have defined

$$M_1^2 = (-m^2 + \frac{3\lambda}{2}\sigma_c^2), \quad M_2^2 = (-m^2 + \frac{\lambda}{2}\sigma_c^2). \tag{9.89}$$

It is now clear from the discussion of the last section (see (9.76)) that the one loop correction is given by (remember $\hbar = 1$)

$$V^{(1)}(\sigma_c) = -\frac{i}{2}\int \frac{d^4k}{(2\pi)^4}\left[\ln(-k^2 + M_1^2) + \ln(-k^2 + M_2^2)\right]. \tag{9.90}$$

The only thing to remember in the case of the imaginary time formalism is that we have to rotate to Euclidean space with the energies taking discrete values (see (1.73)) and the integration over the energy being replaced by a sum (see (1.99)). Thus, the effective potential, in this case, takes the form (with n taking both positive as well as negative integer values)

$$V^{(1)}(\sigma_c) = \frac{1}{2\beta}\sum_n \int \frac{d^3k}{(2\pi)^3}\left[\ln\left(\left(\frac{2n\pi}{\beta}\right)^2 + \omega_1^2\right) + \ln\left(\left(\frac{2n\pi}{\beta}\right)^2 + \omega_2^2\right)\right], \tag{9.91}$$

where we have defined

$$\omega_{1,2}^2 = \mathbf{k}^2 + M_{1,2}^2. \tag{9.92}$$

Let us, therefore, evaluate a generic expression of the form

$$I(M) = \frac{1}{2\beta}\sum_n \int \frac{d^3k}{(2\pi)^3}\ln\left(\left(\frac{2n\pi}{\beta}\right)^2 + \mathbf{k}^2 + M^2\right)$$

$$= \frac{1}{2\beta}\sum_n \int \frac{d^3k}{(2\pi)^3}\ln\left(\left(\frac{2n\pi}{\beta}\right)^2 + \omega^2\right)$$

$$= \int \frac{d^3k}{(2\pi)^3}\tilde{I}(\omega), \tag{9.93}$$

where

$$\tilde{I}(\omega) = \frac{1}{2\beta} \sum_n \ln \left((\frac{2n\pi}{\beta})^2 + \omega^2 \right), \tag{9.94}$$

with $\omega^2 = \mathbf{k}^2 + M^2$. Taking the derivative of $\tilde{I}(\omega)$ with respect to ω, we obtain

$$\frac{\partial \tilde{I}(\omega)}{\partial \omega} = \frac{1}{\beta} \sum_n \frac{\omega}{\omega^2 + (\frac{2n\pi}{\beta})^2}. \tag{9.95}$$

This sum can be easily evaluated using (1.102) and gives

$$\frac{\partial \tilde{I}(\omega)}{\partial \omega} = \left(\frac{1}{2} + \frac{1}{e^{\beta\omega} - 1} \right),$$

or, $\tilde{I}(\omega) = \frac{\omega}{2} + \frac{1}{\beta} \ln(1 - e^{-\beta\omega}).$ \tag{9.96}

Consequently, we can now write

$$I(M) = \int \frac{\mathrm{d}^3 k}{(2\pi)^3} \left(\frac{\omega}{2} + \frac{1}{\beta} \ln(1 - e^{-\beta\omega}) \right). \tag{9.97}$$

There are several things to note here. First, we see that for vanishing temperature (or $\beta \to \infty$), only the first term contributes and the expression reduces to (9.84) (with $\hbar = 1$) as it should. The second term in the integrand, therefore, represents the true temperature dependent corrections and we will denote this by $I^{(\beta)}$. Second, the temperature dependent part of the integral in (9.97), $I^{(\beta)}$, cannot, in general, be evaluated in closed form, as is normally the case for most integrals in statistical mechanics. Even a high temperature expansion (β small) needs to be done carefully because some of the terms in the expansion become complex (since it corresponds to expanding in powers of M and M^2 and can become negative when there is spontaneous symmetry breaking) signalling a breakdown of the approximation. The first few dominant terms in such a high temperature expansion can, however, be unambiguously calculated and is enough to give an approximate critical temperature for symmetry restoration and that is what we will calculate next. Let us note that the angular integration gives a factor of 4π and with $x = \beta|\mathbf{k}|$, we obtain

$$I^{(\beta)}(M)\Big|_{M^2=0} = \frac{1}{2\pi^2\beta^4} \int_0^\infty dx\, x^2 \ln(1 - e^{-x}) = -\frac{\pi^2}{90\beta^4},$$

$$\frac{\partial I^{(\beta)}(M)}{\partial M^2}\Big|_{M^2=0} = \frac{1}{4\pi^2\beta^2} \int_0^\infty dx\, \frac{x}{e^x - 1} = \frac{1}{24\beta^2}, \qquad (9.98)$$

where both these integrals are standard integrals appearing in statistical mechanics and we have simply used their values. This now shows that, to order M^2, we can write

$$I^{(\beta)}(M) = -\frac{\pi^2}{90\beta^4} + \frac{M^2}{24\beta^2} + \cdots. \qquad (9.99)$$

Here "$\cdots$" represent higher order terms that cannot be calculated reliably at this order of perturbation. We note here that the second term in (9.99) has exactly the same structure as what we had calculated earlier in (1.113).

Putting all of these into (9.91), we see that the complete potential at one loop can be written as ($V^{(0)}$ represents the zero temperature potential including the one loop corrections.)

$$V(\sigma_c) = V^{(T=0)}(\sigma_c) + V^{(1)\,(\beta)}(\sigma_c)$$

$$= V^{(T=0)}(\sigma_c) - \frac{\pi^2}{45\beta^4} + \frac{M_1^2 + M_2^2}{24\beta^2} + \cdots. \qquad (9.100)$$

As we have seen earlier (see (9.79)), the second derivative of the zero temperature effective potential, $V^{(T=0)}$, gives the renormalized mass of the theory (remember that this mass square is negative in a spontaneously (symmetry) broken theory). Using this we obtain the temperature dependent effective mass of the theory to be

$$m^2(\beta) = \frac{d^2 V^{(T=0)}}{d\sigma_c d\sigma_c}\Big|_{\sigma_c=0} + \frac{d^2 V^{(1)\,(\beta)}}{d\sigma_c d\sigma_c}\Big|_{\sigma_c=0}$$

$$= -m^2 + \frac{\lambda}{6\beta^2}. \qquad (9.101)$$

As we have discussed earlier, the nontrivial vacuum and the symmetry breaking disappear when the mass squared becomes nonnegative.

This, therefore, defines a natural critical temperature for such a phase transition to be

$$m^2(\beta_c) = 0, \quad \text{or,} \quad \beta_c^2 = \frac{\lambda}{6m^2}. \tag{9.102}$$

For temperatures larger than this critical temperature, the mass would be positive and there will be no symmetry breaking. The spontaneously broken global symmetry of zero temperature is, therefore, restored above this critical temperature. Exactly similar analysis goes through for the case of a spontaneously broken local symmetry although the calculations are a bit more involved. The qualitative conclusions, however, remain the same.

9.6 Dynamical symmetry breaking

So far, we have discussed the case of symmetry breaking where a fundamental scalar field picks up a nonzero vacuum expectation value. Conventionally, such a symmetry breaking is known as spontaneous breaking of symmetry. There are situations, however, where a composite operator picks up a nonzero vacuum expectation value, instead of a fundamental scalar, also leading to the breaking of a symmetry. In such a case, one says that the symmetry is dynamically broken. The most popular of the composite operators whose vacuum expectation value one considers as breaking a symmetry is that of a pair of fermions. The classic examples are the pairing of fermions in superconductivity or the fermion condensation in the case of dynamical chiral symmetry breaking and so on. Dynamical symmetry breaking has a characteristic difference from spontaneous symmetry breaking in the following sense. In most cases of dynamical symmetry breaking, the nonzero value of the order parameter (vacuum expectation value of the composite operator) is associated with an anomaly of the theory such as the chiral anomaly. Anomalies, on the other hand, arise because of our inability to regularize the ultraviolet divergences of a quantum field theory maintaining certain symmetries. They really represent the ultraviolet behavior of a theory after regularization. However, as we have argued in chapter **2**, the finite temperature effects do not change the ultraviolet behavior of a theory. So, intuitively, we expect temperature effects not to change the anomaly in a theory. Consequently, we would expect the dynamical breaking of a symmetry, associated with an anomaly, to be uninfluenced by temperature effects. In particular, unlike the case of the spontaneous breakdown of a symmetry, we do not expect a dynamically broken symmetry to be restored at high temperatures. (It is worth noting here that there are some models of dynamical symmetry breaking such as the Nambu-Jona-Lasinio model and QCD where the dynamical symmetry breaking is not associated with an anomaly. The behavior of these theories, as far as symmetry restoration is concerned, is more like the theories with spontaneous symmetry breaking and our discussion of this section does not apply to such theories.) In this section, we will show that this is indeed true with the simple example of the Schwinger model in $1+1$ dimensions.

The Schwinger model describes massless QED in $1+1$ dimensions where the Lagrangian density is given by (see also section **7.2.3**

$$\mathcal{L} = -\frac{1}{4}F_{\mu\nu}F^{\mu\nu} + i\bar{\psi}\gamma^{\mu}(\partial_{\mu} + ieA_{\mu})\psi, \quad \mu, \nu = 0, 1. \tag{9.103}$$

This is a gauge invariant Lagrangian density with a massless fermion and, in $1+1$ dimensions, the electromagnetic coupling e has the dimensions of a mass. In addition to the local gauge invariance, the theory is also invariant under a global chiral transformation since the fermions are massless. However, as is well known, the chiral symmetry is anomalous in the quantum theory which leads to the photon becoming massive in the full theory. In this section, we will show that the chiral anomaly is independent of temperature and, consequently, the dynamical mass generation of photon is also temperature independent.

Let us note that the anomaly equation in this theory can be written as

$$\partial_{\mu}\langle T(j_5^{\mu}(x)j^{\nu}(y))\rangle = G^{\nu}(x - y)$$

with

$$j_5^{\mu}(x) = e\bar{\psi}(x)\gamma_5\gamma^{\mu}\psi(x), \quad j^{\nu}(x) = e\bar{\psi}(x)\gamma^{\nu}\psi(x), \tag{9.104}$$

where the explicit form of $G^{\nu}(x - y)$ is not important for our discussions, except for the fact that it is related to the chiral anomaly of the theory. It is clear that the question of temperature dependence of the anomaly is best studied within the framework of the real time formalism, for example, the closed time path formalism, where propagators naturally separate into a temperature independent part and a temperature dependent part (see (2.50) and (2.62)). Diagrammatically, the anomaly term in $1+1$ dimensions arises from a self-energy graph at one loop with a $\gamma_5\gamma^\mu$ vertex as shown in Fig. 9.3. The γ^ν vertex is assumed to be coupled to a photon (since the interaction in the Lagrangian density (9.103) has the form $-j^\nu(y)A_\nu(y)$). The

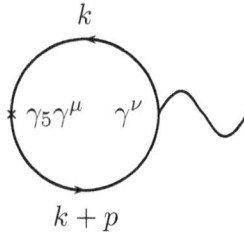

Figure 9.3: Chiral anomaly graph in the Schwinger model.

fermion propagator at finite temperature, in the closed time path formalism, has the momentum space form given by (see (2.62) for the $++$ component and also remember that we have a massless fermion in the present case)

$$iS(p) = i\not{p}\left(\frac{1}{p^2 + i\epsilon} + 2i\pi n_{\mathrm{F}}(|p^0|)\delta(p^2)\right). \tag{9.105}$$

Thus, we see that at finite temperature, the anomaly in (9.104) (given by ∂_μ acting on Fig. 9.3) is obtained to be

$$(-)ip_\mu \int \frac{\mathrm{d}^2k}{(2\pi)^2} \,\mathrm{Tr}\,\left((-e\gamma_5\gamma^\mu)(iS(k))(-e\gamma^\nu)(iS(k+p))\right)$$

$$= ie^2 \int \frac{\mathrm{d}^2k}{(2\pi)^2}\,\left(\mathrm{Tr}\,(\gamma_5\not{p}S(k)\gamma^\nu S(k+p))\right)$$

$$= ie^2 \int \frac{\mathrm{d}^2k}{(2\pi)^2}\,\mathrm{Tr}\left(\gamma_5\not{p}\not{k}\gamma^\nu(\not{k}+\not{p})\right)\left[\frac{1}{k^2 + i\epsilon} + 2i\pi n_{\mathrm{F}}(|k^0|)\delta(k^2)\right]$$

$$\times \left[\frac{1}{(k+p)^2 + i\epsilon} + 2i\pi n_{\mathrm{F}}(|(k^0 + p^0)|)\delta((k+p)^2)\right]. \tag{9.106}$$

Let us point out here that the overall negative sign in the first line of the equation above is because we have a fermionic loop. The negative sign in the two vertices arises from our convention of left derivatives for the fermions.

It is clear now that the amplitude has four contributions, namely, one zero temperature part and three temperature dependent ones given by

$$I_0 = ie^2 \int \frac{d^2k}{(2\pi)^2} \text{ Tr } \gamma_5 \not{p} \not{k} \gamma^\nu (\not{k} + \not{p}) \frac{1}{k^2(k+p)^2},$$

$$I_1 = -2\pi e^2 \int \frac{d^2k}{(2\pi)^2} \text{ Tr } \gamma_5 \not{p} \not{k} \gamma^\nu (\not{k} + \not{p}) \frac{1}{(k+p)^2} \frac{1}{e^{\beta |k^0|} + 1} \delta(k^2),$$

$$I_2 = -2\pi e^2 \int \frac{d^2k}{(2\pi)^2} \text{ Tr } \gamma_5 \not{p} \not{k} \gamma^\nu (\not{k} + \not{p}) \frac{1}{k^2} \frac{1}{e^{\beta |k^0 + p^0|} + 1} \delta((k+p)^2),$$

$$I_3 = -i(2\pi)^2 e^2 \int \frac{d^2k}{(2\pi)^2} \text{ Tr } \gamma_5 \not{p} \not{k} \gamma^\nu (\not{k} + \not{p})$$

$$\times \frac{1}{e^{|k^0 + p^0|} + 1} \frac{1}{e^{\beta |k^0|} + 1} \delta(k^2) \delta((k+p)^2). \quad (9.107)$$

It is easy to show that I_1 and I_2 give the same contribution. This follows from the gamma matrix trace identity. First, we note that if we let $k \leftrightarrow -(k+p)$ in I_2 (the temperature dependent integrals are finite, since they are on-shell, and hence there is no problem with the change of variables), it coincides with I_1 except for the trace part. However,

$$\text{Tr } \gamma_5 \not{p} (\not{k} + \not{p}) \gamma^\nu \not{k} = \text{Tr } \gamma_5 \not{p} (\not{k} \gamma^\nu \not{k} + \not{p} \gamma^\nu \not{k})$$

$$= \text{Tr } \gamma_5 \not{p} [\not{k} \gamma^\nu (\not{k} + \not{p}) - \not{k} \gamma^\nu \not{p} + \not{p} \gamma^\nu \not{k}]$$

$$= \text{Tr } [\gamma_5 \not{p} \not{k} \gamma^\nu (\not{k} + \not{p}) + p^2 \gamma_5 (\not{k} \gamma^\nu + \gamma^\nu \not{k})]$$

$$= \text{Tr } [\gamma_5 \not{p} \not{k} \gamma^\nu (\not{k} + \not{p}) + 2p^2 k^\nu \gamma_5]$$

$$= \text{Tr } \gamma_5 \not{p} \not{k} \gamma^\nu (\not{k} + \not{p}), \quad (9.108)$$

which is identical to the trace in I_1. Here we have used the cyclicity of trace in the intermediate step as well as the standard Dirac algebra. We note here that the temperature dependent contributions in the anomaly diagram come from I_1, I_2 and I_3 in (9.107). If we now use the two dimensional identity

$$\gamma_5 \gamma^\nu = \epsilon^{\nu\lambda} \gamma_\lambda,$$

we can write

$I_1 + I_2$

$$= -4\pi e^2 \epsilon^{\nu\lambda} \int \frac{d^2k}{(2\pi)^2} \, \text{Tr} \, (\not{p}\not{k}\gamma_\lambda(\not{k}+\not{p})) \frac{1}{(k+p)^2} \frac{1}{e^{\beta|k^0|}+1} \delta(k^2)$$

$$= -4\pi e^2 \epsilon^{\nu\lambda} \int \frac{d^2k}{(2\pi)^2} \, \text{Tr} \, (((\not{k}+\not{p})-\not{k})\not{k}\gamma_\lambda(\not{k}+\not{p}))$$

$$\times \frac{1}{(k+p)^2} \frac{1}{e^{\beta|k^0|}+1} \delta(k^2)$$

$$= -8\pi e^2 \epsilon^{\nu\lambda} \int \frac{d^2k}{(2\pi)^2} \, (k_\lambda(k+p)^2) \frac{1}{(k+p)^2} \frac{1}{e^{\beta|k^0|}+1} \delta(k^2)$$

$$= -8\pi e^2 \epsilon^{\nu\lambda} \int \frac{d^2k}{(2\pi)^2} \, k_\lambda \frac{1}{e^{\beta|k^0|}+1} \delta(k^2) = 0. \qquad (9.109)$$

Here we have used the cyclicity of the trace as well as the fact that the k^2 term vanishes because of the delta function in the intermediate steps and in the last step we have used the fact that the integral vanishes because the integrand is odd under $k \to -k$. It now follows trivially from the above calculation that (because the integrand in I_3 in (9.107) involves an additional $\delta((k+p)^2)$ term in addition to $\delta(k^2)$)

$$I_3 = 0. \qquad (9.110)$$

This shows that the temperature dependent terms in the anomaly identically vanish and what is left is I_0 which is the zero temperature anomaly. Consequently, this shows that the anomaly is independent of temperature and so is the dynamical mass generated in this model. We conclude with a cautionary note that, while temperature dependent parts of an amplitude are ultraviolet finite in general because they receive only on-shell contributions (from delta functions imposing the on-shell conditions), the infrared behavior may be much more severe at finite temperature and needs to be studied carefully.

9.7 References

S. Coleman and E. Weinberg, Physical Review **D7**, 1883 (1973).

A. Das and A. Karev, Physical Review **D36**, 623 (1987).

A. Das, *Field Theory: A Path Integral Approach* (third edition), World Scientific Publishing (2019).

A. Das, *Lectures on Quantum Field Theory* (second edition), World Scientific Publishing (2020).

L. Dolan and R. Jackiw, Physical Review **D9**, 3320 (1974).

J. Goldstone, Nuovo Cimento **19**, 154 (1961).

G. Guralnik *et al*, in *Advances in Particle Physics* (volume 2), Ed. R. L. Cool and R. E. Marshak, John Wiley and Sons (1968).

R. Jackiw, Physical Review **D9**, 1686 (1974).

G. Jona-Lasinio, Nuovo Cimento **34**, 1790 (1964).

Y. Nambu, Physics Letters **26B**, 626 (1966).

Y. Nambu and G. Jona-Lasinio, Physical Review **122**, 345 (1961).

J. Schwinger, Physical Review **128**, 2425 (1962).

S. Weinberg, Physical Review **D9**, 3357 (1974).

CHAPTER 10

Nielsen identities

As we have seen in the last chapter, the effective potential plays a crucial role in studying symmetry breaking and restoration in a quantum field theory. A theory may already display spontaneous breakdown of a symmetry at the tree level and yet it is useful to examine the behavior of the effective potential at higher loops for various reasons. First, of course, we would like to know whether the tree level symmetry breaking is stable under radiative corrections. Second, sometimes in a gauge theory, the Higgs potential may have a larger symmetry than is present in the full theory. In such a case, the tree level breaking of this larger symmetry of the Higgs potential may lead to the appearance of massless "pseudo-Goldstone" bosons in the theory which may become massive due to radiative corrections. More importantly, sometimes a symmetry may not be broken at the tree level and yet a breaking may be induced when radiative corrections are included. For all these reasons, as well as to determine if there is a restoration of symmetry, it is useful to study the behavior of the effective potential in a quantum field theory including radiative corrections. However, a serious conceptual problem arises in this endeavor in that, in a gauge theory, the effective potential is not a gauge independent quantity. And, therefore, it is not clear *a priori* how much of the analysis resulting from a study of the effective potential is physically relevant or can be taken seriously. In this chapter, we will discuss this issue in some detail.

10.1 An obvious gauge problem

To begin with, let us recall that the effective potential is obtained from the effective action of the theory (see (9.56) and (9.57)) when the scalar fields take a constant value (with all other fields vanishing). Therefore, it is a generator of Green's functions with zero external momentum which are by definition off-shell. Off-shell Green's functions, on the other hand, need not be gauge independent and that shows that, in general, the effective potential in a gauge theory is not gauge independent. However, we also recognize (from the argument following (9.11)) that the value of the effective potential at its minimum represents the ground state energy of a system which is a physical quantity and, therefore, we expect this to be gauge independent. Thus, we already see from this that the gauge dependence of the effective potential and various related questions need to be studied more carefully.

To see that the effective potential can, in fact, become gauge dependent, even at the tree level, it is best to look at the Abelian Higgs model in the R_ξ gauge (see (9.33)). The Lagrangian density, before shifting the fields, becomes in this gauge (it is the same as the sum of (9.23) and (9.33) only if we remember that the fields are not yet shifted),

$$
\begin{aligned}
\mathcal{L} = &-\frac{1}{4} F_{\mu\nu} F^{\mu\nu} + \frac{1}{2} \partial_\mu \sigma \partial^\mu \sigma + \frac{1}{2} \partial_\mu \chi \partial^\mu \chi + \frac{m^2}{2}(\sigma^2 + \chi^2) \\
&- e(\chi \partial^\mu \sigma - \sigma \partial^\mu \chi) A_\mu + \frac{e^2}{2}(\sigma^2 + \chi^2) A_\mu A^\mu - \frac{\lambda}{8}(\sigma^2 + \chi^2)^2 \\
&- \frac{1}{2\xi}(\partial_\mu A^\mu - \xi e v \chi)^2 + \partial_\mu \bar{c} \partial^\mu c - \xi e^2 v \sigma \bar{c} c.
\end{aligned} \tag{10.1}
$$

Here, we have identified $v = \langle \sigma \rangle$. It is clear now that for constant values of the σ and the χ fields, with all other fields set equal to zero, the tree level potential becomes (which we normally denote by $V(\phi)$, as in the last chapter)

$$
V^{(0)}(\sigma, \chi) = -\frac{m^2}{2}(\sigma^2 + \chi^2) + \frac{\lambda}{8}(\sigma^2 + \chi^2)^2 - \frac{\xi(ev)^2}{2}\chi^2. \tag{10.2}
$$

This shows explicitly that even the tree level potential is gauge dependent in this simple model. (We have, of course, already noted earlier, in (9.34), that the tree level mass for the χ field in the R_ξ gauge is gauge dependent.)

The argument above holds for any generalized R_ξ gauge as well and has led to a lot of confusion in the literature. Therefore, let us understand the structure of the tree level potential first. There are several objections to the structure of the theory in (10.1) and the potential in (10.2). First, there is a conceptual question. We have included in the gauge fixing term (and, therefore, in the tree level potential) the vacuum expectation value of the scalar field even before minimizing the potential and so, naturally, the question arises as to whether the whole procedure is consistent. In fact, if one tries to minimize the potential in the R_ξ gauge (generalized or not), one finds spurious gauge dependent solutions. Second, for various reasons, it has been argued in the literature that the R_ξ gauge may not be a good gauge for the study of effective potential. The answer to both these questions, as we will see below, is that the choice of the gauge fixing term does not influence the minimization of the potential and that the R_ξ gauge is perfectly fine as far as the study of the effective potential is concerned. This is best seen if we use an auxiliary field to write the gauge fixing Lagrangian density (see (5.21) and (5.22)). Thus, to make the discussion clear, we will deviate from the standard description of the subject in the literature and use the auxiliary field formulation. We note that with the auxiliary field, $\mathcal{L}_{\mathrm{GF}}$ takes the form (see (5.22))

$$\mathcal{L}_{\mathrm{GF}} = \frac{\xi}{2}F^2 + \partial_\mu F A^\mu + \xi e v F \chi, \tag{10.3}$$

which upon using the equation for the auxiliary field

$$F = \frac{1}{\xi}(\partial_\mu A^\mu - \xi e v \chi), \tag{10.4}$$

would lead to the gauge fixing Lagrangian density in the R_ξ gauge of (10.1) if we ignore a total divergence term.

The important point to recall here is the discussion of chapter **5**, namely, (5.30), which says that the BRST variation of the anti-ghost field is the auxiliary field and since the BRST transformations are nilpotent, the auxiliary field remains invariant under a BRST transformation. This implies that

$$[Q_{\mathrm{BRST}}, \bar{c}]_+ = -F,$$
$$[Q_{\mathrm{BRST}}, F] = 0. \tag{10.5}$$

Combining this with the physical state condition of (5.36), we obtain now (from the first relation in (10.5)) that in the true vacuum of the theory

$$\langle 0|F|0\rangle = 0. \tag{10.6}$$

This shows that the gauge fixing Lagrangian density gives a vanishing contribution to the energy of the true vacuum and hence cannot influence the minimization of the potential (see (10.4) and (10.1)). Incidentally, we note here that a bad choice of the gauge fixing Lagrangian density may force that the gauge field takes a nonlocal, space-time dependent value at the minimum for (10.6) to be true, but the gauge fixing Lagrangian density would still contribute trivially to the minimum of the potential. What we have in mind here is as follows. Suppose, we want to choose as a minimum $\langle \sigma \rangle \neq 0$ and $\langle \chi \rangle = 0$. In such a case, if we had chosen a gauge fixing Lagrangian of the form

$$\mathcal{L}_{\text{GF}} = \frac{\xi}{2}F^2 + \partial_\mu F A^\mu + \beta F \sigma,$$

with β an arbitrary constant parameter, then the equation for the auxiliary field would yield

$$\xi F = \partial_\mu A^\mu - \beta \sigma.$$

It is clear now that, at the minimum, for $\langle F \rangle$ to vanish, we must have

$$\langle \partial_\mu A^\mu \rangle = \beta \langle \sigma \rangle \neq 0.$$

And, for consistency, that would require that the vacuum expectation value of the gauge field cannot be set equal to zero.

In fact, from (10.5) and the physical state condition (5.36), it is easy to see that any correlation function of the F field must vanish in vacuum

$$\langle 0|F(x_1)F(x_2)\cdots F(x_n)|0\rangle = 0. \tag{10.7}$$

This, in particular, implies that the two point correlation involving the auxiliary fields and, therefore, the propagator for the F field vanishes

$$\langle 0|F(x_1)F(x_2)|0\rangle = 0. \tag{10.8}$$

This can, in fact, be explicitly checked. We see from (10.1) and (10.3) that there is mixing between the A_μ and F fields. The two point function for the

$$\begin{pmatrix} F \\ A_\mu \end{pmatrix},$$

system has the form (in momentum space)

$$\begin{pmatrix} \xi & ik^\mu \\ -ik^\nu & -(\eta^{\mu\nu}k^2 - k^\mu k^\nu) \end{pmatrix}. \tag{10.9}$$

The inverse of this matrix can be worked out in a straightforward manner and has the form

$$\begin{pmatrix} 0 & \frac{ik_\nu}{k^2} \\ -\frac{ik_\mu}{k^2} & -\frac{1}{k^2}(\eta_{\mu\nu} - (1-\xi)\frac{k_\mu k_\nu}{k^2}) \end{pmatrix}. \tag{10.10}$$

This shows explicitly that the F propagator indeed vanishes. This is consistent with (10.7) or (10.8).

We note from (10.2) that while the tree level potential is explicitly gauge dependent, its value at the minimum (see (9.15)), namely, for

$$\langle \sigma \rangle = v, \qquad \langle \chi \rangle = 0,$$

is independent of the gauge fixing parameter. It can be seen more generally as follows. From (10.1) and (10.3), we see that

$$\langle 0|\frac{\partial S}{\partial \xi}|0\rangle = \int d^4x \, \langle 0|(\frac{1}{2}F^2 + evF\chi - e^2v\sigma\bar{c}c)|0\rangle$$

$$= \int d^4x \, \langle 0| - \delta_{\text{BRST}}[\bar{c}(\frac{1}{2}F + ev\chi)]|0\rangle$$

$$= -\int d^4x \, \langle 0| \left[Q_{\text{BRST}}, \bar{c}(\frac{1}{2}F + ev\chi)\right]_+ |0\rangle$$

$$= 0, \tag{10.11}$$

where S represents the complete action of the tree level theory and where we have used the physical state condition of (5.36). (The BRST variation is without the parameter of transformation and in taking out the parameter of transformation, we have to remember that it anti-commutes with the anti-ghost field, $\bar{c}$. The explicit BRST transformations are given later in (10.13).) This shows that the value of the tree level potential at its minimum is gauge independent as it should be since it represents the ground state energy of the system which is an observable quantity.

The discussion so far makes it very clear that since the effective potential is a gauge dependent quantity, we should determine systematically which of the quantities derived from it will be gauge independent. It is worth noting here that with the use of the auxiliary fields, one can carry out the discussion of Abelian as well as non-Abelian gauge theories with equal ease. For simplicity, however, we will restrict ourselves to the example of the Abelian Higgs model for the rest of this chapter.

10.1.1 Nielsen identities at $T = 0$. Nielsen identities are identities which describe the gauge dependence of the effective action and, therefore, of the effective potential. As we have seen in chapter **5**, amplitudes which are not gauge invariant satisfy Ward identities following from the BRST invariance of the theory. Ward identities, basically, express how non-gauge invariant amplitudes would change under a gauge transformation or a change of gauge. Therefore, we see that since the effective potential is not gauge independent, the relations (identities) that it must satisfy are best described within the framework of BRST invariance. And in trying to derive these, we will not follow the conventional description, rather, we will use the auxiliary field formalism which is much simpler.

Let us, for simplicity, consider the Abelian Higgs model in the R_ξ gauge (before shifting) although, as we have tried to emphasize earlier, with the auxiliary fields non-Abelian theories are described with equal ease as well. The Lagrangian density is given by (see (10.1))

$$\mathcal{L} = -\frac{1}{4}F_{\mu\nu}F^{\mu\nu} + \frac{1}{2}\partial_\mu\sigma\partial^\mu\sigma + \frac{1}{2}\partial_\mu\chi\partial^\mu\chi + \frac{m^2}{2}(\sigma^2 + \chi^2)$$

$$- e(\chi\partial^\mu\sigma - \sigma\partial^\mu\chi)A_\mu + \frac{e^2}{2}(\sigma^2 + \chi^2)A_\mu A^\mu - \frac{\lambda}{8}(\sigma^2 + \chi^2)^2$$

$$+ \frac{\xi}{2}F^2 + \partial_\mu F A^\mu + \xi ev F\chi + \partial_\mu \bar{c}\partial^\mu c - \xi e^2 v\sigma\bar{c}c. \tag{10.12}$$

Following our discussion of the derivation of Ward identities in chapter **5**, we now introduce sources for the various fields of our theory as well as their BRST variations. However, since this is a simple Abelian theory, the BRST variations are much simpler (see (5.30) and (9.25))

$$\delta_{\text{BRST}} A_\mu = \omega \partial_\mu c,$$

$$\delta_{\text{BRST}} \sigma = \omega e \chi c,$$

$$\delta_{\text{BRST}} \chi = -\omega e \sigma c,$$

$$\delta_{\text{BRST}} c = 0,$$

$$\delta_{\text{BRST}} \bar{c} = -\omega F,$$

$$\delta_{\text{BRST}} F = 0. \tag{10.13}$$

Consequently, the only two composite variations for which we have to add sources are $\delta_{\text{BRST}} \sigma$ and $\delta_{\text{BRST}} \chi$.

Thus, the source Lagrangian density takes the form

$$\mathcal{L}_{\text{source}} = J_\mu A^\mu + JF + K\sigma + L\chi + i(\bar{\eta}c - \bar{c}\eta)$$

$$+ M(e\chi c) + N(-e\sigma c) + H[\bar{c}\,(\tfrac{1}{2}F + ev\chi)], \tag{10.14}$$

where, in addition to the usual sources used in the derivation of the Ward identities (see (5.39)), we have added an extra source, H, in the last term. Comparing this last term with (10.11), it is clear that the variation of this last term under a BRST transformation would yield a term which would give the gauge dependence of the effective potential. (We note here, parenthetically, that, with the auxiliary field, the gauge dependence is obtained exactly from the BRST variation of this term and we do not have to worry about a divergence or the ghost equations of motion etc., as is normally the case in the literature.) Thus, with the two Lagrangian densities in (10.12) and (10.14) together, we can write the generating functional for the system as

$$Z[J] = e^{iW[J]} = \int \mathcal{D}\phi \, e^{i \int d^4x \, (\mathcal{L} + \mathcal{L}_{\text{source}})}, \tag{10.15}$$

where we have denoted all the sources and fields generically by J and ϕ respectively. Let us next shift all the fields under the integral by

their BRST variations holding the sources fixed. In such a case, as
we have noted earlier in (5.42)-(5.43),

$$
\int \mathcal{D}\phi \int d^4x \left[J_\mu \delta_{\text{BRST}} A^\mu + K \delta_{\text{BRST}}\sigma + L\delta_{\text{BRST}}\chi - i(\delta_{\text{BRST}}\bar{c})\eta \right.
$$

$$
\left. + H\delta_{\text{BRST}}(\bar{c}(\frac{1}{2}F + ev\chi)) \right] \times e^{i \int d^4x \, (\mathcal{L} + \mathcal{L}_{\text{source}})} = 0. \qquad (10.16)
$$

We can rearrange the terms and note from the BRST variations in
(10.13) that we can write this also as

$$
e^{-iW} \int \mathcal{D}\phi \int d^4x \, H(x) \frac{\partial \mathcal{L}}{\partial \xi} \, e^{i \int d^4x \, (\mathcal{L} + \mathcal{L}_{\text{source}})}
$$

$$
= \int d^4x \left(iJ_\mu \partial^\mu \frac{\delta W}{\delta \bar{\eta}} - K \frac{\delta W}{\delta M} - L \frac{\delta W}{\delta N} + i\frac{\delta W}{\delta J}\eta \right), \qquad (10.17)
$$

where we have also used our earlier observation in (10.11).

We see from (10.17) that if we take the functional derivative of
the left hand side with respect to $H(x)$ and integrate over x (and
set $H = 0$), we will obtain the gauge dependence of the generating
functional, namely, (10.17), in such a case, would give

$$
\frac{\partial W}{\partial \xi} = \int d^4x \, d^4y \left(iJ_\mu(y)\partial_y^\mu \frac{\delta^2 W}{\delta H(x)\delta\bar{\eta}(y)} - K(y)\frac{\delta^2 W}{\delta H(x)\delta M(y)} \right.
$$

$$
\left. -L(y)\frac{\delta^2 W}{\delta H(x)\delta N(y)} + i\frac{\delta^2 W}{\delta H(x)\delta J(y)}\eta(y) \right). \qquad (10.18)
$$

Here we have assumed that all the sources are linearly independent.
Furthermore, we note that if we now set all the sources equal to
zero, we will obtain a relation describing the gauge dependence of the
effective potential. (Setting all the sources equal to zero is equivalent
to putting all the classical fields except the scalar fields to zero.)

A more direct way to obtain a relation for the effective potential
is to go to the effective action. Thus, as in chapter **5**, we define a
Legendre transformation with respect to the sources J_μ, J, K, L, η
and $\bar{\eta}$ as

$$
\Gamma = W - \int d^4x \left(J_\mu A^\mu + JF + K\sigma + L\chi + i(\bar{\eta}c - \bar{c}\eta) \right), \qquad (10.19)
$$

where, as before, the fields appearing on the right hand side are the
classical fields although we do not write them explicitly. Since the

Legendre transformation does not involve the sources M, N and H (as well as the gauge fixing parameter ξ), it follows now that

$$\frac{\delta\Gamma}{\delta M} = \frac{\delta W}{\delta M}, \qquad \frac{\delta\Gamma}{\delta N} = \frac{\delta W}{\delta N},$$

$$\frac{\delta\Gamma}{\delta H} = \frac{\delta W}{\delta H}, \qquad \frac{\partial\Gamma}{\partial\xi} = \frac{\partial W}{\partial\xi}, \qquad (10.20)$$

while for the derivatives with respect to the classical fields, we have

$$\frac{\delta\Gamma}{\delta A^\mu} = -J_\mu, \qquad \frac{\delta\Gamma}{\delta F} = -J,$$

$$\frac{\delta\Gamma}{\delta\sigma} = -K, \qquad \frac{\delta\Gamma}{\delta\chi} = -L,$$

$$\frac{\delta\Gamma}{\delta c} = i\bar{\eta}, \qquad \frac{\delta\Gamma}{\delta\bar{c}} = i\eta. \qquad (10.21)$$

With these relations, (10.18) can now be written as (with $H = 0$)

$$\frac{\partial\Gamma}{\partial\xi} = \int d^4x\, d^4y \left(\frac{\delta\Gamma}{\delta A^\mu(y)} \partial_y^\mu \frac{\delta c(y)}{\delta H(x)} + \frac{\delta\Gamma}{\delta\sigma(y)} \frac{\delta^2\Gamma}{\delta H(x)\delta M(y)} \right.$$

$$\left. + \frac{\delta\Gamma}{\delta\chi(y)} \frac{\delta^2\Gamma}{\delta H(x)\delta N(y)} + \frac{\delta F(y)}{\delta H(x)} \frac{\delta\Gamma}{\delta\bar{c}(y)} \right). \qquad (10.22)$$

This describes how the effective action changes as the gauge fixing parameter is changed. If we now set all the classical fields to zero except σ which we assume to be a constant (*i.e.*, we are assuming a particular form of the minimum of the potential consistent with our choice of the gauge fixing Lagrangian, but the arguments do not depend on the choice of the minimum at all as we have emphasized before), then, the left hand side of (10.22) represents the derivative of the effective potential with respect to the gauge fixing parameter up to a proportionality factor. On the right hand side, the first term vanishes because of Lorentz invariance (namely, there is no one point function for the gauge field) and the last term vanishes because of ghost number conservation (there is no one point function for the ghost field). The other two terms, in principle, survive. But, if we look at the structure of $\mathcal{L}_{\text{source}}$ in (10.14), it is clear that the radiative corrections can lead to a two point function involving H and M, but not for H and N. Therefore, in this limit, (10.22) becomes

$$\frac{\partial V_{\text{eff}}}{\partial \xi} = -C(\sigma, \xi)\frac{\partial V_{\text{eff}}}{\partial \sigma}, \qquad (10.23)$$

where we have used the translational invariance of the two point function to define

$$C(\sigma, \xi) = -\int d^4x \, \frac{\delta^2\Gamma}{\delta H(x)\delta M(0)}, \qquad (10.24)$$

with all fields set to zero except σ. This is the essence of the Master identity due to Nielsen which can be thought of as a Ward Identity for the gauge (parameter) variation of the effective potential. Note that since the two point function in (10.24) has contributions at one loop and beyond, (10.23) implies that the gauge parameter dependence of the effective potential also arises because of radiative corrections.

To understand the meaning of (10.23), we note that even though the effective potential is gauge dependent, at the minimum where

$$\frac{\partial V_{\text{eff}}}{\partial \sigma} = 0,$$

we have from (10.23)

$$\left.\frac{\partial V_{\text{eff}}}{\partial \xi}\right|_{\text{min}} = 0 \qquad (10.25)$$

This shows that the value of the effective potential at its minimum is gauge (parameter) independent as it should be considering that it represents the ground state energy of the system. (Incidentally, we note here that not only is the value of the potential gauge (parameter) independent at its minimum, it is also gauge independent at all its extrema. This, then, would imply that the difference in the values of the potential at the false vacuum and the true one would also be gauge independent. This height is quite relevant in the study of various physical phenomena such as tunnelling, rate of decay of the false vacuum or some questions related to phase transitions and it is reassuring to know that this does not depend on the gauge fixing parameter.) Furthermore, as a partial differential equation, (10.23) expresses the fact that the effective potential V_{eff} is stationary (with respect to variations in ξ) along the characteristics given by

$$\frac{\partial \sigma}{\partial \xi} = C(\sigma, \xi), \qquad (10.26)$$

and that (10.23) can be expressed, in such a case, as

$$\frac{dV_{\text{eff}}}{d\xi} = \frac{\partial V_{\text{eff}}}{\partial \xi} + \frac{\partial \sigma}{\partial \xi}\frac{\partial V_{\text{eff}}}{\partial \sigma} = 0. \tag{10.27}$$

Namely, we see from (10.26) that the expectation value of the scalar field is gauge dependent leading to both an implicit and an explicit dependence of the effective potential on the gauge fixing parameter. However, the two contributions cancel each other along the characteristics leading to the fact that the value of the effective potential is truly independent of the gauge fixing parameter along the characteristics (*i.e.*, when we consider the total gauge dependence both explicit and implicit). This is true even if we are not at a minimum of the effective potential.

It is interesting to note that the expectation value of the scalar field, although gauge independent at the tree level, can become gauge dependent at higher orders. We could have derived directly the gauge dependence of this from (10.18). In fact, noting that (see, for example, (5.41) or (9.39))

$$\frac{\delta W}{\delta K(x)} = \sigma(x),$$

we see that if we take the functional derivative of (10.18) with respect to $K(y)$ and set all the sources equal to zero (which is also equivalent to setting all the fields to zero except $\sigma(x) = \sigma$, a constant), we obtain

$$\frac{\partial \sigma}{\partial \xi} = -\int d^4x\, \frac{\delta^2 W}{\delta H(x)\delta M(0)}$$

$$= -\int d^4x\, \frac{\delta^2 \Gamma}{\delta H(x)\delta M(0)}$$

$$= C(\sigma, \xi), \tag{10.28}$$

where we have used translation invariance of the two point function as well as the identification in (10.24) in the last step. This result is quite interesting in the following sense. We have argued before that we can think of the vacuum expectation value of the scalar field as an order parameter of symmetry breaking. However, we find that this is a gauge dependent quantity. It is not clear, therefore, whether we can use this quantity to study phenomena such as symmetry restoration and the critical temperature associated with the phase transition. Furthermore, it is not clear whether physical quantities, such as the mass, derived from the effective potential will be independent of the gauge fixing parameter (we expect that it should be). All such questions can, in fact, be studied starting from the Master equation in (10.23).

Let us look at the gauge independence of the Higgs mass (mass of the σ field) for simplicity. The discussion for other physical quantities can be carried out along the same lines. We note that there are two kinds of masses that we can define in a field theory. First, there is the mass parameter obtained from the second derivative of the effective potential. We can think of this as the two point function at vanishing momentum. The physical mass, on the other hand, is obtained from the pole of the propagator which we can also think of as the value of p^2 (four momentum squared) for which the two point function vanishes. The two masses agree only at the tree level. However, to obtain the physical mass, we should really look at the zero of the inverse propagator or the two point function. Keeping this in mind, we take the second functional derivative of (10.22) with respect to $\sigma(x)$ and $\sigma(y)$ and then set all the classical fields to zero except σ which takes a constant value. In this case, we obtain, from (10.12),

$$
\frac{\partial}{\partial \xi} \frac{\delta^2 \Gamma}{\delta\sigma(x)\delta\sigma(y)}\bigg|_{\min} = \int \mathrm{d}^4 w \mathrm{d}^4 z \left[\frac{\delta^2 \Gamma}{\delta\sigma(z)\delta\sigma(x)} \frac{\delta^3 \Gamma}{\delta H(w)\delta M(z)\delta\sigma(y)} \right.
$$
$$
+ \frac{\delta^2 \Gamma}{\delta\sigma(z)\delta\sigma(y)} \frac{\delta^3 \Gamma}{\delta H(w)\delta M(z)\delta\sigma(x)}
$$
$$
\left. + \frac{\delta^3 \Gamma}{\delta\sigma(z)\delta\sigma(x)\delta\sigma(y)} \frac{\delta^2 \Gamma}{\delta H(w)\delta M(z)} \right]_{\min}.
$$
$$(10.29)$$

Here the "min" stands for setting all fields to zero except σ which takes a constant value.

We note here that, by definition,

$$
\frac{\delta^2 \Gamma}{\delta\sigma(x)\delta\sigma(y)}\bigg|_{\min} = G^{-1}(x - y; \sigma), \qquad (10.30)
$$

where G represents the renormalized propagator for the σ field. Furthermore, we are interested in the form of the equation (10.29) at points where the inverse propagators vanish. It is clear that, at such points, the first two terms on the right hand side of (10.29) would vanish. Therefore, in the momentum space, (10.29) would take the form (for vanishing two point function)

$$\frac{\partial G^{-1}(p^2;\sigma)}{\partial \xi} = -C(\sigma,\xi)\frac{\partial G^{-1}(p^2;\sigma)}{\partial \sigma}, \tag{10.31}$$

where we have used the definition in (10.24). It is clear now that if we represent (up to a multiplicative constant near the zero of the two point function)

$$G^{-1}(p^2;\sigma) = (p^2 - m_P^2(\sigma,\xi)), \tag{10.32}$$

where m_P represents the physical mass corresponding to the pole of the propagator, (10.31) would take the form

$$\frac{\partial m_P^2}{\partial \xi} = -C(\sigma,\xi)\frac{\partial m_P^2}{\partial \sigma},$$

$$\text{or,} \quad \frac{dm_P^2}{d\xi} = 0, \tag{10.33}$$

along the characteristics (see (10.26)). Once again, following our earlier discussion, this shows that the physical Higgs mass is independent of the gauge fixing parameter along any characteristics. The gauge independence of other physical quantities can also be derived following the same line of reasoning.

Next, we come to the question of symmetry restoration. We have seen, in chapter **9**, that a spontaneously broken symmetry is restored when the scalar vacuum expectation value vanishes. We have also seen that this happens when the mass parameter (the second derivative of the effective potential with respect to the scalar field) vanishes. But what we have seen so far, from the discussions in this chapter, is that the vacuum expectation value of the scalar field (the "order" parameter) is gauge dependent. A natural question that arises, then, is whether such an analysis is meaningful at all. To examine this question, we go to the Master equation (10.23) and take the second derivative with respect to σ which gives, at the minimum of the potential,

$$\frac{\partial}{\partial \xi}\frac{\partial^2 V_{\text{eff}}}{\partial \sigma^2} + C(\sigma,\xi)\frac{\partial}{\partial \sigma}\frac{\partial^2 V_{\text{eff}}}{\partial \sigma^2} + 2\frac{\partial C}{\partial \sigma}\frac{\partial^2 V_{\text{eff}}}{\partial \sigma^2} = 0. \tag{10.34}$$

This, as we can clearly see, does not have a gauge independent structure because of the last term (see (10.27) or (10.33)). But we note that if the second derivative of the effective potential vanishes (*i.e.* when the mass parameter in the theory vanishes) the equation takes a gauge independent form

$$\frac{\partial}{\partial\xi}\frac{\partial^2 V_{\text{eff}}}{\partial\sigma^2} + C(\sigma,\xi)\frac{\partial}{\partial\sigma}\frac{\partial^2 V_{\text{eff}}}{\partial\sigma^2} = 0. \tag{10.35}$$

This shows that even if the scalar vacuum expectation value is gauge dependent, we can still study symmetry restoration and calculate a critical temperature associated with such a phase transition in a gauge independent fashion by analyzing when the mass parameter of the theory vanishes.

Finally, we note here without going into details that the gauge fixing Lagrangian density, as we have seen, does not lead to any observable properties. From (10.12), we see that the gauge fixing Lagrangian density depends on another parameter v and so we conclude that the physical quantities cannot depend on this parameter either. In fact, following the line of argument that led to identities involving ξ dependence, we can also derive identities involving v dependence which would also show that physical observables will be independent of the choice of this parameter.

10.1.2 Verification of identities. The interesting thing about the Nielsen identities (10.23) is that the function $C(\sigma,\xi)$ has a closed form expression given in (10.24). Therefore, the identities can be verified order by order in perturbation theory. Thus, for example, we have already seen that at the tree level, there is no gauge dependence of the potential, namely,

$$\frac{\partial V^{(0)}}{\partial\xi} = 0. \tag{10.36}$$

Furthermore, looking at the source Lagrangian in (10.14), we note that at the tree level, there is no mixing between the sources H and M and consequently, from (10.24) we conclude that

$$C^{(0)}(\sigma,\xi) = 0, \tag{10.37}$$

where the superscript denotes the order (of loop) at which various quantities are evaluated. Therefore, the Nielsen identity trivially holds at the tree level,

$$\frac{\partial V^{(0)}}{\partial \xi} + C^{(0)} \frac{\partial V^{(0)}}{\partial \sigma} = 0. \tag{10.38}$$

To order one loop, we can write the relation in (10.23) as

$$\frac{\partial V^{(1)}}{\partial \xi} + C^{(0)} \frac{\partial V^{(1)}}{\partial \sigma} + C^{(1)} \frac{\partial V^{(0)}}{\partial \sigma} = 0,$$

or, $\quad \dfrac{\partial V^{(1)}}{\partial \xi} + C^{(1)} \dfrac{\partial V^{(0)}}{\partial \sigma} = 0. \tag{10.39}$

We will show that this relation holds true through an explicit calculation.

To evaluate the effective potential at one loop, we follow the discussion of chapter **9**. In this simple case, with our choice of symmetry breaking, we only need to shift

$$\sigma \longrightarrow \sigma + \sigma_{\mathrm{c}}.$$

For simplicity, we will also identify v in the (gauge fixing) Lagrangian density (10.12) with σ_{c} although, in general, one can leave this as an arbitrary parameter. Thus, we obtain from the Lagrangian density (10.12) (see also (10.2))

$$V^{(0)}(\sigma_{\mathrm{c}}) = -\frac{m^2}{2}\sigma_{\mathrm{c}}^2 + \frac{\lambda}{8}\sigma_{\mathrm{c}}^4,$$

so that, we have

$$\frac{\partial V^{(0)}}{\partial \sigma_{\mathrm{c}}} = -\sigma_{\mathrm{c}}\left(m^2 - \frac{\lambda}{2}\sigma_{\mathrm{c}}^2\right). \tag{10.40}$$

For the evaluation of the one loop effective potential, we need the part of the Lagrangian density in (10.12) which is quadratic in the quantum fields after shifting and this has the form

$$\mathcal{L}_{\mathrm{Q}} = \frac{1}{2}A_\mu\left(\eta^{\mu\nu}(\Box + e^2\sigma_{\mathrm{c}}^2) - (1 - \frac{1}{\xi})\partial^\mu\partial^\nu\right)A_\nu$$

$$+ \frac{1}{2}\sigma(-\Box + m^2 - \frac{3\lambda}{2}\sigma_{\mathrm{c}}^2)\sigma + \frac{1}{2}\chi(-\Box + m^2 - (\frac{\lambda}{2} + \xi e^2)\sigma_{\mathrm{c}}^2)\chi$$

$$+ \frac{\xi}{2}F^2 - \bar{c}(\Box + \xi e^2\sigma_{\mathrm{c}}^2)c. \tag{10.41}$$

In writing the above Lagrangian, we have diagonalised terms involving mixing between the F, A_μ and the χ fields by redefining

$$F \longrightarrow F + \frac{1}{\xi}(\partial_\mu A^\mu - \xi e \sigma_c \chi), \tag{10.42}$$

which makes the calculation simpler. Denoting the two point functions for the gauge field, the scalar fields and the ghost fields by $G^{-1\,\mu\nu}$, G_σ^{-1}, G_χ^{-1}, G_F^{-1} and S^{-1} respectively, we note from our discussions in chapter **6** that the one loop effective potential can be written as ($\hbar = 1$)

$$V^{(1)} = -\frac{i}{2}\Omega^{-1}\mathrm{Tr}\left[\ln G^{-1\,\mu\nu} + \ln G_\sigma^{-1} + \ln G_\chi^{-1} + \ln G_F^{-1}\right.$$
$$\left. -2\ln S^{-1}\right]. \tag{10.43}$$

Here the ghost contribution is different from the others because, as we have discussed earlier, the ghost fields are anti-commuting and are more like fermions leading to the negative sign and the factor of two arises because they are complex. We note here that G_σ^{-1} does not involve the gauge fixing parameter and, therefore, would not contribute to $\frac{\partial V^{(1)}}{\partial \xi}$. Consequently, we have

$$\frac{\partial V^{(1)}}{\partial \xi} = -\frac{i}{2}\Omega^{-1}\mathrm{Tr}\left[G_{\mu\nu}\frac{\partial G^{-1\,\nu\mu}}{\partial \xi} + G_\chi\frac{\partial G_\chi^{-1}}{\partial \xi} + G_F\frac{\partial G_F^{-1}}{\partial \xi}\right.$$
$$\left. -2S\frac{\partial S^{-1}}{\partial \xi}\right]. \tag{10.44}$$

The relevant functions can be read out from (10.41) to have the form, in momentum space,

$$G^{-1\,\mu\nu} = \left(\eta^{\mu\nu}(-p^2 + e^2\sigma_c^2) + (1 - \frac{1}{\xi})p^\mu p^\nu\right),$$

$$G_{\mu\nu} = \frac{1}{-p^2 + e^2\sigma_c^2}\left(\eta_{\mu\nu} + (\xi - 1)\frac{p_\mu p_\nu}{p^2 - \xi e^2\sigma_c^2}\right),$$

$$G_\chi^{-1} = (p^2 + m^2 - (\frac{\lambda}{2} + \xi e^2)\sigma_c^2),$$

$$G_F^{-1} = \xi,$$

$$S^{-1} = (p^2 - \xi e^2\sigma_c^2). \tag{10.45}$$

Using these relations and following the discussion in chapter **6**, we can now simplify (10.44)) to obtain

$$\frac{\partial V^{(1)}}{\partial \xi} = -\frac{i}{2} \int \frac{d^4 p}{(2\pi)^4} \frac{e^2 \sigma_c^2 (m^2 - \frac{\lambda}{2} \sigma_c^2)}{(p^2 - \xi e^2 \sigma_c^2)(p^2 + m^2 - (\frac{\lambda}{2} + \xi e^2) \sigma_c^2)}.$$

$$(10.46)$$

The only thing that remains to be evaluated is $C^{(1)}$. We note that, at one loop, the Feynman diagram which leads to $C^{(1)}$ (*i.e.* the two point function involving the sources H and M) has the form shown in Fig. 10.1.

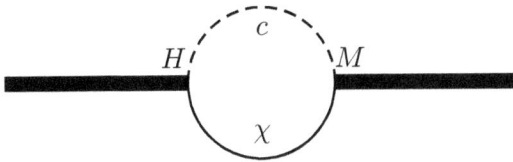

Figure 10.1: Feynman diagram at one loop contributing to the mixing of H and M.

We already know the propagators for the χ field and the ghost field from (10.45)). We also see that under the shifting, Eq. (10.42), the source Lagrangian density in (10.14)) takes the form

$$\mathcal{L}_{\text{source}} \to J_\mu A^\mu + JF + K\sigma + L\chi + i(\bar{\eta}c - \bar{c}\eta) + M(e\chi c)$$

$$+ N(-e\sigma c) + \frac{1}{2} H[\bar{c}(F + \frac{1}{\xi} \partial_\mu A^\mu + e\sigma_c \chi)]. \quad (10.47)$$

Thus, the calculation of $C^{(1)}$ is now straightforward (taking into account the factors of i coming from the vertices, the propagators and the definition of the amplitude which corresponds to $(-i)$ times the Feynman diagram because of the definition (10.24)

$$C^{(1)} = -\frac{i}{2} e^2 \sigma_c \int \frac{d^4 p}{(2\pi)^4} S(p) G_\chi(-p)$$

$$= -\frac{i}{2} e^2 \sigma_c \int \frac{d^4 p}{(2\pi)^4} \frac{1}{(p^2 - \xi e^2 \sigma_c^2)(p^2 + m^2 - (\frac{\lambda}{2} + \xi e^2) \sigma_c^2)}.$$

$$(10.48)$$

From (10.40), (10.46) and (10.48), we now see that the identity
(10.39) holds, verifying the Nielsen identity at one loop. The identity
involving the mass, (10.33), can also be verified in a similar manner.
However, since this is slightly more involved, we will not go into it
here.

10.1.3 Nielsen identities at $T \neq 0$.

As we have seen, in chapter **5**, the
partition function in the presence of gauge interactions is defined in
a BRST invariant manner. As a result, the Ward identities of the
theory continue to hold even at finite temperature. Nielsen identities
expressing the dependence of the effective potential on the gauge fixing parameter can also be thought of as a sort of Ward identity and
consequently, we expect that these identities should also continue to
hold at nonzero temperature. The derivation of the identities goes
through without any modification. Verifying the identities is also
equally straightforward provided we remember that at finite temperature (in the imaginary time formalism) energy takes discrete values
(the description in the real time formalism is also equally straightforward). Thus, for the gauge field, the σ and the χ fields, we have
from (1.73)

$$p^0 \to i\omega_n, \qquad \omega_n = \frac{2n\pi}{\beta}, \quad n = \text{integer.} \tag{10.49}$$

The ghost fields, on the other hand, satisfy anti-commutation relations like fermions and we would expect their energy values to be
quantized like fermions (see (1.73)). However, as we have discussed
in detail in chapter **5** (see the discussion in section **5.6**), in spite of
their Grassmann nature, the ghost fields do satisfy periodic boundary conditions like any bosonic field so that for the ghost fields, too,
we can write

$$p^0 \to i\omega_n, \qquad \omega_n = \frac{2n\pi}{\beta}, \quad n = \text{integer.} \tag{10.50}$$

With these relations (and the discussion on evaluations in the
imaginary time formalism in chapter **1**), we note that (10.46) takes

the form (see also (9.91))

$$\frac{\partial V^{(1)}}{\partial \xi} = \frac{1}{2\beta} \sum_n \int \frac{d^3 p}{(2\pi)^3}$$

$$\times \frac{e^2 \sigma_c^2 (m^2 - \frac{\lambda}{2}\sigma_c^2)}{((\frac{2n\pi}{\beta})^2 + \mathbf{p}^2 + \xi e^2 \sigma_c^2)((\frac{2n\pi}{\beta})^2 + \mathbf{p}^2 - m^2 + (\frac{\lambda}{2} + \xi e^2)\sigma_c^2)}.$$

$$(10.51)$$

Similarly, (10.48) takes the form

$$C^{(1)} = \frac{e^2 \sigma_c}{2\beta} \sum_n \int \frac{d^3 p}{(2\pi)^3}$$

$$\times \frac{1}{((\frac{2n\pi}{\beta})^2 + \mathbf{p}^2 + \xi e^2 \sigma_c^2)((\frac{2n\pi}{\beta})^2 + \mathbf{p}^2 - m^2 + (\frac{\lambda}{2} + \xi e^2)\sigma_c^2)}.$$

$$(10.52)$$

On the other hand, at finite temperature, the classical Lagrangian density remains the same so that (10.40) remains unchanged. From (10.40), (10.51) and (10.52), we see that the one loop Nielsen identity, (10.39), continues to hold at finite temperature.

10.2 A broader gauge problem for physical parameters

In the last section, we discussed the issue of gauge parameter dependence of the effective scalar potential within the context of covariant gauge choices. We described in detail how Nielsen identities help us show that, in spite of the effective scalar potential being gauge parameter dependent, its value at the minimum (or any extremum) is gauge parameter independent (see (10.25)). Furthermore, the identities also lead us to the fact that the physical mass of the scalar field (particle) is also independent of the gauge fixing parameter in the covariant gauge (see (10.33)).

However, that raises an interesting question like how can one show the gauge independence of physical parameters which are not derived from the scalar potential. For example, the physical mass of a fermion or any other particle (like the gauge particle) is not associated with the scalar potential and, therefore, the discussions of the last section does not help. Furthermore, the discussion in the last section was completely within the framework of the covariant gauge and, therefore, the gauge independence simply corresponded to showing the lack of dependence of the minimum of the potential or the physical mass of the scalar field only on the gauge fixing parameter ξ of the covariant gauge. However, there are also other possible classes of gauge choices which are noncovariant, such as the Coulomb gauge or the axial gauge, which are quite useful for different kinds of studies. A physical parameter such as the physical mass of a fermion has to be independent, not only of the gauge fixing parameter within a class of gauge choice, but also independent of the particular class of gauge choice made. This would then correspond to a complete gauge independence of the physical parameter. In this section, we will discuss in detail how such a question of complete gauge independence can be addressed with the example of the physical mass of a fermion, interacting with an Abelian gauge field (for simplicity). We will also point out, with an example, circumstances under which such a complete gauge independence may not hold.

10.2.1 Covariant and noncovariant gauges. We have already discussed covariant gauges in a non-Abelian gauge theory in (5.21), which, for an Abelian theory, takes the form ($\xi = \frac{1}{\alpha^2}$)

$$\mathcal{L}_{\mathrm{GF}} = -\frac{1}{2}(\Lambda_\mu(\partial)A^\mu)^2 = -\frac{\alpha^2}{2}(\partial_\mu A^\mu)^2, \ \Lambda_\mu(\partial) = \alpha\partial_\mu, \quad (10.53)$$

with α representing the gauge fixing parameter. With this gauge fixing Lagrangian, the propagator for the gauge field in the momentum space, $iD_{\mu\nu}^{(\alpha)}(p)$, takes the form (the $i\epsilon$ prescription is understood)

$$D_{\mu\nu}^{(\alpha)}(p) = -\frac{1}{p^2}\left[\eta_{\mu\nu} - \frac{\Lambda_\mu(p)p_\nu + \Lambda_\nu(p)p_\mu}{\Lambda(p)\cdot p}\right.$$
$$\left. + \frac{p_\mu p_\nu (\Lambda(p))^2}{(\Lambda(p)\cdot p)^2}\left(1 + \frac{p^2}{(\Lambda(p))^2}\right)\right] \quad (10.54)$$
$$= -\frac{1}{p^2}\left(\eta_{\mu\nu} - (1 - \frac{1}{\alpha^2})\frac{p_\mu p_\nu}{p^2}\right), \quad (10.55)$$

which, in the Feynman gauge $\alpha = 1$ ($\xi = 1$), takes the simple form

$$D_{\mu\nu}^{(\alpha=1)}(p) = -\frac{\eta_{\mu\nu}}{p^2}. \quad (10.56)$$

On the other hand, in noncovariant gauges, there is a constant four vector n^μ which selects out a fixed direction in space-time. Every vector (tensor) can be decomposed into components along this direction or perpendicular to it. For example, given any four vector V_μ we can define its longitudinal and transverse components with respect to this vector as

$$V_\mu^L = \frac{(n \cdot V)}{n^2} n_\mu, \quad V_\mu^T = V_\mu - V_\mu^L = \left(\eta_{\mu\nu} - \frac{n_\mu n_\nu}{n^2}\right) V^\nu, \quad (10.57)$$

where we are assuming that $n^2 = n \cdot n = n_\mu n^\mu \neq 0$. It is clear from the definitions in (10.57) that $n_\mu^L = n_\mu$ and $n_\mu^T = n_\mu - n_\mu^L = 0$ (unless we are in $1 + 1$ dimensions where we can define $n_\mu^T = \epsilon_{\mu\nu} n^\nu$). We also note that the gradient four vector can aslo be decomposed in the presence of n_μ as

$$\partial_\mu^L = \frac{n \cdot \partial}{n^2} n_\mu, \quad \partial_\mu^T = \partial_\mu - \partial_\mu^L. \quad (10.58)$$

If one uses the longitudinal component of the gradient for gauge fixing, for example,

$$\mathcal{L}_{\text{GF}} = -\frac{1}{2}(\Lambda^\mu(\partial) A_\mu)^2 = -\frac{\beta^2}{2}(\partial^L \cdot A)^2, \quad \Lambda_\mu(\partial) = \beta \partial_\mu^L, \quad (10.59)$$

where β is the constant gauge fixing parameter, then such a gauge choices are known as generalized axial gauge while if one uses the transverse component of the gradient for gauge fixing and n_μ is time-like, for example,

$$\mathcal{L}_{\text{GF}} = -\frac{1}{2}(\Lambda^\mu(\partial) A_\mu)^2 = -\frac{\beta^2}{2}(\partial^T \cdot A)^2, \quad \Lambda_\mu(\partial) = \beta \partial_\mu^T, \quad (10.60)$$

then, such a gauge choice is known as the generalized Coulomb gauge. In the generalized axial gauge of (10.59), for example, using (10.54) and (10.58), we obtain the propagator to have the form

$$D_{\mu\nu}^{(\beta)}(p) = -\frac{1}{p^2}\left[\eta_{\mu\nu} - \frac{n_\mu p_\nu + n_\nu p_\mu}{n \cdot p} + \frac{p_\mu p_\nu n^2}{(n \cdot p)^2}\left(1 + \frac{n^2 p^2}{\beta^2 (n \cdot p)^2}\right)\right].$$

$$(10.61)$$

We can, similarly derive the propagator in the generalized Coulomb gauge of (10.60).

10.2.2 Interpolating gauge. To show the gauge independence of any physical parameter, we need to have a gauge choice which can interpolate between the covariant as well as the two types of generalized noncovariant gauge. Keeping that in mind, we introduce a gauge fixing Lagrangian density

$$\mathcal{L}_{\text{GF}} = -\frac{1}{2}\left(\Lambda^{\mu}_{\phi^{(a)}}(\partial)A_{\mu}\right)^2 = \frac{1}{2}F^2 + \left(\Lambda^{\mu}_{\phi^{(a)}}F\right)A_{\mu}, \tag{10.62}$$

with

$$\Lambda^{\mu}_{\phi^{(a)}}(\partial) = \alpha\partial^{\mu} + \beta\partial^{\mu,L}. \tag{10.63}$$

This gauge fixing Lagrangian density consists of three arbitrary parameters,

$$\phi^{(a)} = (\alpha, \beta, n^{\mu}), \tag{10.64}$$

and we note that when $\beta = 0$, the Lagrangian density in (10.62) reduces to the covariant gauge fixing given in (10.53). On the other hand, when $\alpha = 0$, the Lagrangian density in (10.62) reduces to (10.59) corresponding to the generalized axial gauge fixing, while, when $\alpha = -\beta$ and n^{μ} is time-like ($n^2 > 0$), we recover the gauge fixing Lagrangian density of (10.60) corresponding to the generalized Coulomb gauge. The propagator, in this interpolating gauge, has the general form given in (10.54) with $\Lambda^{\mu}(p) \equiv \Lambda^{\mu}_{(\alpha,\beta,n)}(p)$ given in (10.62).

10.2.3 Physical mass of a fermion. The physical mass of a particle is defined from the Einstein relation $p^2 = m^2$ where m denotes the rest mass of the particle (at the tree level). This, on the other hand, translates to one of the Casimir operators of the Poincaré algebra, $P^2 = P_{\mu}P^{\mu}$, of a relativistic theory acting on a single particle (energy-momentum) state to give

$$P^2|p\rangle = p^2|p\rangle = M_{\text{P}}^2|p\rangle, \quad \text{or,} \quad (P^2 - M_{\text{P}}^2\mathbb{1})|p\rangle = 0. \tag{10.65}$$

Therefore, the physical mass of a particle can be determined from the zeros of the two point function or from the poles of the propagator since the two are inversely related. Namely, from the dynamical equations of the theory, we can symbolically write the equation for the Green's function of the theory (for a scalar field theory) as

$$\Gamma(p)G(p) = \mathbb{1}, \quad \Gamma(p) = p^2 - m^2 - \Sigma(p,m) = G^{-1}(p), \quad (10.66)$$

where $\Gamma(p)$ denotes the complete two point function including quantum corrections and correspondingly $G(p)$ represents the complete propagator (Green's function) of the theory. Furthermore, $\Sigma(p,m)$ stands for the self-energy which contains all the quantum corrections (beyond the tree level) to the two point function.

For a fermion, on the other hand, the dynamical equation has a matrix structure (because of the Dirac matrices) and, consequently, the equation for the Green's function (compare with (10.66)) takes the form

$$S^{-1}(p)S(p) = (\not{p} - m - \Sigma(p,m))S(p) = \mathbb{1}, \qquad (10.67)$$

where the self-energy $\Sigma(p,m)$, containing the quantum corrections, has a matrix structure. As a result, extracting the physical mass or the zeros of the two point function is a lot more involved. Let us recall that the Dirac matrices, γ^μ, allow us to have other matrix structures (in $3+1$ dimensions) such as $\mathbb{1}, \gamma^\mu, \gamma^{[\mu}\gamma^{\nu]}, \gamma^{[\mu}\gamma^\nu\gamma^{\lambda]}, \gamma^{[\mu}\gamma^\nu\gamma^\lambda\gamma^{\rho]}$. Here the square brackets imply anti-symmetrization in the respective indices. Since $S^{-1}(p)$ does not have any vector index, if only one external vector, p_μ, is available, the most general self-energy matrix we can construct has the form

$$\Sigma(p,m) = \Sigma^{(\mathrm{c})}(p,m) = mA(p^2,m) + B(p^2,m)\not{p}, \qquad (10.68)$$

where $A(p^2,m), B(p^2,m)$ are dimensionless functions. This would be the case if we were working with a covariant gauge choice (which is the reason for the superscript "(c)"). On the other hand, the structure of the self-energy clearly would have more structures if, in addition to p_μ, we also had a second vector structure such as n_μ in a noncovariant gauge. However, some of these structures can be eliminated by imposing the charge conjugation invariance of the theory which leads to the final structure

$$\Sigma^{(\mathrm{nc})}(p,m) = mA + B\not{p} + C\not{p}_L, \qquad (10.69)$$

where the scalar coefficients A, B, C are dimensionless (Lorentz) scalar functions of (p^2, p_L^2, n^2, m) and the longitudinal component of the momentum, $p_\mu^L = \frac{(n \cdot p)}{n^2} n_\mu$, follows from (10.57). It follows from (10.69) that if there is no second structure, namely, if $n_\mu = 0$, the last term in $\Sigma^{(\mathrm{nc})}(p, m)$ vanishes and (10.69) reduces to $\Sigma^{(\mathrm{c})}(p, m)$ in (10.68).

10.2.4 Determining the fermion physical mass.

As we have pointed out already, the physical mass corresponds to the location of the pole of the propagator. The two point function is the inverse of the Green's function and, therefore, the physical mass can also be thought of as the location of the zero of the two point function. This is very useful in scalar field theories. Things become a bit more complicated when fermions are involved since the two point function becomes a matrix and, therefore, extracting the physical mass of a fermion from the zero of the two point function becomes nontrivial. Let us discuss here, in some detail, what kind of problems arise and how the gauge independence of the fermion mass is demonstrated.

As we have seen in (10.67)-(10.68), with the covariant gauge choice, the Dirac equation

$$S^{-1}(p) = \slashed{p} - m - \Sigma^{(\mathrm{c})},$$

so that the zero of the Dirac equation for the fermion leads to

$$S^{-1}(p)u(p)\big| = 0 = (\slashed{p} - m - \Sigma^{(\mathrm{c})})u(p)\big|,$$

$$\text{or,} \quad \bar{u}(p)S^{-1}(p)u(p)\big| = 0 = \bar{u}(p)(\slashed{p} - m - \Sigma^{(\mathrm{c})})u(p)\big|,$$

$$\text{or,} \quad M_{\mathrm{P}} = m + \bar{u}(p)\Sigma^{(\mathrm{c})}u(p)\big|, \tag{10.70}$$

where we have used the fact that the Dirac spinor is normalized in the first two terms and the restriction "$|$" indicates that the quantities have to be evaluated at $p^2 = M_{\mathrm{P}}^2$. The self-energy can be calculated order by order in perturbation theory and, therefore, M_{P} can be determined order by order. Furthermore, this mass is also known to coincide with the pole (mass) of the fermion propagator and is gauge independent (independent of the parameter α in (10.62)).

The situation, however, is quite different in noncovariant gauges. If we follow the steps given in (10.70) with a noncovariant self-energy (see (10.69)), we have

$$\bar{u}(p)S^{-1}(p)u(p)\big| = 0 = \bar{u}(p)(\slashed{p} - m - \Sigma^{(\mathrm{nc})})u(p)\big|,$$

$$\text{or,} \quad \widetilde{M} = m + \bar{u}(p)\Sigma^{(\mathrm{nc})}u(p)\big|, \tag{10.71}$$

where the restriction now stands for quantities to be evaluated at $p^2 = \widetilde{M}^2$. In this case, also the mass $\widetilde{M}$ can be evaluated order by order in perturbation theory. However, the explicit evaluation shows that the mass determined in this way is not gauge parameter independent and, second, it does not coincide with the pole mass (where the pole of the Green's function lies which is gauge parameter independent) beginning at two loops. The first behavior is rather easy to understand. We know that $\Sigma^{(\mathrm{nc})}(p, m)$ and the parameters A, B, C in (10.69) depend on both p^2 as well as p_L^2. On the other hand, in determining the mass in (10.71), the expectation of $\Sigma^{(\mathrm{nc})}$ is evaluated at $p^2 = \widetilde{M}^2$. That leaves the dependence on p_L^2 arbitrary (unless these terms cancel out exactly) which is explicitly dependent on the gauge parameter n^μ. The second behavior can also be understood in the following way. We note from (10.70) that $S^{-1}(p)u(p)\| = 0$, does imply $\bar{u}(p)S^{-1}(p)u(p)\| = 0$. However, the converse is not true, namely $\bar{u}(p)S^{-1}(p)u(p)\| = 0$ does not imply $S^{-1}(p)u(p)\| = 0$, which is what we have used in (10.71). As a result, in the case of a noncovariant gauge fixing, we have to analyze the pole of the propagator (Green's function) to derive any conclusion.

10.2.5 Fermion propagator. When there are two four vectors available, such as p_μ and n_μ, as we have shown in (10.69), the fermion two point function has a more general structure, namely,

$$S^{-1}(p) = \not{p} - m - \Sigma^{(\mathrm{nc})} = \mathcal{A}(p) - \mathcal{B}(p)\mathbb{1}, \qquad (10.72)$$

where (see (10.69))

$$\mathcal{A}(p) = (1 - B)\not{p} - C\not{p}_L, \quad \mathcal{B}(p) = m(1 + A), \qquad (10.73)$$

so that, while $\mathcal{A}$ is a matrix in the Dirac space, $\mathcal{B}$ is a scalar function. This form of the fermion two point function holds not only in a noncovariant gauge fixing, but also in the interpolating gauge as well. Furthermore, it can be checked, using the algebra of Dirac matrices, that $\mathcal{A}^2$ is proportional to the identity matrix in the Dirac space and, explicitly, it has the form

$$\mathcal{A}^2(p) = \left((1 - B)^2 p^2 + (C^2 - 2C(1 - B))p_L^2\right)\mathbb{1}. \qquad (10.74)$$

Therefore, the Green's function (the propagator is i times the Green's function) will have the form

$$S(p) = \frac{\mathcal{A} + \mathcal{B}\mathbb{1}}{(\mathcal{A} + \mathcal{B}\mathbb{1})(\mathcal{A} - \mathcal{B}\mathbb{1})} = \frac{\mathcal{N}}{\mathcal{D}\mathbb{1}}, \tag{10.75}$$

where (remember that A, B, C are functions of p^2, p_L^2 and, therefore, are invariant under $p \to -pf$)

$$\mathcal{N} = \mathcal{A}(p) + \mathcal{B}(p)\mathbb{1} = -(\mathcal{A}(-p) - \mathcal{B}(-p)\mathbb{1})$$
$$= -S^{-1}(-p) = -\mathcal{C}(S^{-1}(p))^T \mathcal{C}^{-1}, \tag{10.76}$$

$$\mathcal{D} = \mathcal{A}^2 - \mathcal{B}^2$$
$$= (1 - B)^2 p^2 + (C^2 - 2C(1 - B))p_L^2 - m^2(1 + A)^2. \tag{10.77}$$

Here $\mathcal{C}$ denotes the charge conjugation matrix satisfying

$$\mathcal{C}^{-1}\gamma^\mu\mathcal{C} = -(\gamma^\mu)^T, \quad \mathcal{C}(\gamma^\mu)^T\mathcal{C}^{-1} = -\gamma^\mu. \tag{10.78}$$

The pole of the Green's function (propagator) corresponds to the zeros of $\mathcal{D}$. If we assume that the pole occurs at $p^2 = M_P^2$, then we determine from (10.77) that

$$M_P^2 = \frac{m^2(1 + A)^2 - (C^2 - 2C(1 - B))p_L^2}{(1 - B)^2}\bigg|_{p^2 = M_P^2}. \tag{10.79}$$

Therefore, we see that a mass cannot even be defined unless all the dependence on p_L^2 canceled out. As we will see a little later that this is indeed the case which follows from the Nielsen identities and that the physical mass, indeed, is independent of the gauge fixing parameters (α, β, n^μ) (in the general interpolating gauge given in (10.62)). Let us note here, for later use, that the relations in (10.75)-(10.78) allow us to write $(\mathcal{D}(-p) = \mathcal{D}(p))$

$$\mathcal{D}\mathbb{1} = -S^{-1}(-p)S^{-1}(p) = -S^{-1}(p)S^{-1}(-p),$$
$$= -S^{-1}(p)\mathcal{C}(S^{-1}(p))^T\mathcal{C}^{-1},$$

so that, taking the trace over the Dirac indices on both sides, we can write (in n dimensions)

$$\mathcal{D} = -\frac{1}{2^{\lfloor \frac{n}{2} \rfloor}} \operatorname{Tr}\left(S^{-1}(p)\mathcal{C}(S^{-1}(p))^T\mathcal{C}^{-1}\right), \tag{10.80}$$

where n represents the dimensionality of space-time, $[\frac{n}{2}]$ is the integer part of the fraction and we have also used the fact that the Dirac matrices in n dimensions are $2^{[\frac{n}{2}]} \times 2^{[\frac{n}{2}]}$ dimensional.

10.2.6 Nielsen identities. Following the discussions in the previous section (section **10.1**), we can derive the Nielsen identity for the case of QED and since we are interested in studying the general gauge independence of the mass of the fermion, we can take as the starting theory

$$\mathcal{L}_{\text{QED}} = -\frac{1}{4}F_{\mu\nu}F^{\mu\nu} + \overline{\psi}(i\gamma^{\mu}D_{\mu} - m)\psi, \tag{10.81}$$

where the covariant derivative is defined as

$$D_{\mu}\psi = (\partial_{\mu} + ieA_{\mu})\psi. \tag{10.82}$$

We can add to it the Lagrangian density describing the interpolating gauge fixing, given in (10.62)-(10.64) with auxiliary fields which is useful in deriving Nielsen identities (see, for example, (10.3)), namely,

$$\mathcal{L}_{\text{GF}} = \frac{1}{2}\,F^2 + \left(\Lambda^{\mu}_{\phi(a)}(\partial)F\right)A_{\mu}. \tag{10.83}$$

The gauge fixing Lagrangian density in (10.83), in turn, induces the ghost Lagrangian density given by

$$\mathcal{L}_{\text{ghost}} = \left(\Lambda^{\mu}_{\phi(a)}(\partial)\overline{c}\right)\partial_{\mu}c. \tag{10.84}$$

The sum of these three Lagrangian densities

$$\mathcal{L} = \mathcal{L}_{\text{QED}} + \mathcal{L}_{\text{GF}} + \mathcal{L}_{\text{ghost}}, \tag{10.85}$$

would be invariant under the simple Abelian BRST transformations involving the ghost fields, namely,

$$\delta_{\text{BRST}}A_{\mu} = \omega\partial_{\mu}c, \qquad \delta_{\text{BRST}}F = 0, \qquad \delta_{\text{BRST}}\psi = -ie\omega c\psi,$$

$$\delta_{\text{BRST}}\overline{\psi} = -ie\omega\overline{\psi}c, \qquad \delta_{\text{BRST}}c = 0, \qquad \delta_{\text{BRST}}\overline{c} = -\omega F, \tag{10.86}$$

where ω denotes an arbitrary constant Grassmann parameter, namely,

$$\delta_{\text{BRST}}\mathcal{L} = 0. \tag{10.87}$$

In addition to these standard Lagrangian densities, we need to introduce the Lagrangian density containing the sources for the fields $(A_\mu, \overline{\psi}, \psi, F, \overline{c}, c)$ if we want to study their Green's functions. Furthermore, we need to introduce sources for the composite BRST variations of fields (see (10.86)) which, in this simple Abelian theory, would correspond to the BRST variations of $(\overline{\psi}, \psi)$. The source Lagrangian density takes the form

$$\mathcal{L}_{\text{source}} = J^\mu A_\mu + JF + i(\overline{\chi}\psi - \overline{\psi}\chi) + i(\overline{\eta}c - \overline{c}\eta)$$

$$+ ie(\overline{M}c\psi - \overline{\psi}cM) + \left(H_{(\alpha)}(\partial^\mu \overline{c}) + H_{(\beta)}(\partial_L^\mu \overline{c})\right)A_\mu$$

$$+ \beta H_{(n)\mu}(N^{\mu\nu}\overline{c})A_\nu, \tag{10.88}$$

where

$$N^{\mu\nu} = \frac{\partial \partial_L^\nu}{\partial n_\mu} = \frac{\partial}{\partial n_\mu}\frac{(n \cdot \partial)n^\nu}{n^2}$$

$$= \frac{(n \cdot \partial)}{n^2}\left(\eta^{\mu\nu} + \frac{\partial^\mu n^\nu}{n \cdot \partial} - \frac{2n^\mu n^\nu}{n^2}\right). \tag{10.89}$$

Here $(J^\mu, J, \overline{\chi}, \chi, \overline{\eta}, \eta)$ are the sources for the fields $(A_\mu, F, \psi, \overline{\psi}, c, \overline{c})$ respectively, $(\overline{M}, M)$ are the sources for the composite BRST variations $(\delta\psi, \delta\overline{\psi})$ respectively (without the BRST transformation parameter ω, see (10.86)). We also note here that, in addition to the sources $(\overline{\chi}, \chi), (\overline{\eta}, \eta)$, the sources $(H_{(\alpha)}, H_{(\beta)}, H_{(n)\mu}$ are also fermionic (anti-commuting) in nature. Finally, the BRST variation of the three terms with sources $H_{\phi(a)} = (H_{(\alpha)}, H_{(\beta)}, H_{(n)\mu})$ generate the variation of $\mathcal{L}$ in (10.85) with the three parameters $\phi^{(a)} = (\alpha, \beta, n_\mu)$. Adding all the terms in (10.85) and (10.88), we have the complete Lagrangian density given by

$$\mathcal{L}_{\text{TOT}} = \mathcal{L} + \mathcal{L}_{\text{source}}, \tag{10.90}$$

and the complete action in n dimensions is

$$S_{\text{TOT}} = \int d^n x\, \mathcal{L}_{\text{TOT}}. \tag{10.91}$$

We note here that, unlike $\mathcal{L}$ (see (10.87)), the source Lagrangian density $\mathcal{L}_{\text{source}}$ (see (10.88)) is not invariant under the BRST transformations of (10.86). In fact, the variation leads to (without the parameter of transformation ω)

$$
\begin{aligned}
\delta_{\text{BRST}} \mathcal{L}_{\text{TOT}} &= J^\mu \partial_\mu c - e(\bar{\chi} c \psi + \overline{\psi} c \chi) + i F \eta \\
&\quad + H_{(\alpha)} \big((\partial^\mu F) A_\mu + (\partial^\mu \bar{c}) \partial_\mu c \big) + H_{(\beta)} \big((\partial_L^\mu F) A_\mu + (\partial^\mu \bar{c}) \partial_\mu c \big) \\
&\quad + \beta H_{(n)\mu} \big((N^{\mu\nu} F) A_\nu + (N^{\mu\nu} \bar{c}) \partial_\nu c \big) \\
&= J^\mu \partial_\mu c - e(\bar{\chi} c \psi + \overline{\psi} c \chi) + i F \eta \\
&\quad + H_{(\alpha)} \frac{\partial \mathcal{L}_{\text{TOT}}}{\partial \alpha} + H_{(\beta)} \Big(\frac{\partial \mathcal{L}_{\text{TOT}}}{\partial \beta} - H_{(n)\mu}(N^{\mu\nu} \bar{c}) \partial_\nu c \Big) \\
&\quad + H_{(n)\mu} \Big(\frac{\partial \mathcal{L}_{\text{TOT}}}{\partial n_\mu} - H_{(\beta)}(N^{\mu\nu} \bar{c}) A_\nu - \beta H_{(n)\nu} \Big(\frac{\partial N^{\nu\lambda}}{\partial n_\mu} \bar{c} \Big) A_\lambda \Big) \\
&= J^\mu \partial_\mu c - e(\bar{\chi} c \psi + \overline{\psi} c \chi) + i F \eta \\
&\quad + H_{(a)} \frac{\partial \mathcal{L}_{\text{TOT}}}{\partial \phi_{(a)}} + O(\text{quadratic in } H_{(a)}).
\end{aligned} \tag{10.92}
$$

As a result, we see that neither $\mathcal{L}_{\text{TOT}}$ nor S_{TOT}, given above in (10.90) or in (10.91) respectively, is invariant under the BRST transformation of (10.86).

The generating functional for this gauge fixed theory is now given by ($\hbar = 1$)

$$
Z[\mathcal{J}] = e^{iW[\mathcal{J}]} = N \int \mathcal{D}\varphi \, e^{iS_{\text{TOT}}[\varphi]}, \tag{10.93}
$$

where N is a normalization constant and we have denoted collectively all the fields and sources respectively by φ and $\mathcal{J}$. Since the path integral is over all field configurations, it should not change if we make a BRST shift of the field variables, namely, under

$$
\varphi \to \varphi + \delta_{\text{BRST}} \varphi, \tag{10.94}
$$

the change in the generating functional should vanish (another way of saying this is to note that $Z[\mathcal{J}]$ does not depend on φ and, therefore, should not change),

$$
\delta Z[\mathcal{J}] = iN \int \mathcal{D}\varphi \, (\delta_{\text{BRST}} S_{\text{TOT}}) \, e^{iS_{\text{TOT}}} = 0. \tag{10.95}
$$

Using (10.92) for the variation of the total Lagrangian density in (10.95), we obtain an identity (see (10.17) as well)

$$
Ne^{-iW[\mathcal{J}]} \int \mathcal{D}\varphi \int \mathrm{d}^n x \left(H_{(a)} \frac{\partial \mathcal{L}_{\mathrm{TOT}}}{\partial \phi_{(a)}} + O((H_{(a)})^2) \right)
$$
$$
= i \int \mathrm{d}^n x \left(J^\mu \partial_\mu \frac{\delta W}{\delta \bar{\eta}} - \bar{\chi} \frac{\delta W}{\delta \overline{M}} + \frac{\delta W}{\delta M} \chi - \frac{\delta W}{\delta J} \eta \right). \tag{10.96}
$$

If we now take derivative of (10.96) with respect to $H_{(a)}(y)$ and set the sources $H_{(a)} = 0$ and then integrate over $\int \mathrm{d}^n y$, we obtain (remember that $H_{(a)}$ are anti-commuting)

$$
\frac{\partial W[\mathcal{J}]}{\partial \phi_{(a)}} = i \int \mathrm{d}^n x \mathrm{d}^n y \left(J^\mu(x) \partial_\mu^x \frac{\delta^2 W[\mathcal{J}]}{\delta H_{(a)}(y) \delta \bar{\eta}(x)} \right.
$$
$$
+ \bar{\chi}(x) \frac{\delta^2 W[\mathcal{J}]}{\delta H_{(a)}(y) \delta \overline{M}(x)} + \frac{\delta^2 W[\mathcal{J}]}{\delta H_{(a)}(y) \delta M(x)} \chi(x)
$$
$$
\left. - \frac{\delta^2 W[\mathcal{J}]}{\delta H_{(a)}(y) \delta J(x)} \eta(x) \right). \tag{10.97}
$$

$W[\mathcal{J}]$ generates connected Green's functions and (10.97) gives the master identity for the gauge parameter dependence of the connected Green's functions. Therefore, by taking derivatives with respect to sources (for the field variables) and setting these sources to zero, we can derive the gauge parameter dependence of various connected Green's functions of the theory.

To determine the gauge parameter dependence of the one particle irreducible (1PI) amplitudes of the theory, we simply look at the Legendre transformation of $W[\mathcal{J}]$ with respect to the sources of the (classical) dynamical fields of the theory (see, for example, (5.41)), namely,

$$
\Gamma = W - \int \mathrm{d}^n x \left(J^\mu A_\mu + JF + i(\bar{\chi}\psi - \overline{\psi}\chi) + i(\bar{\eta}c - \bar{c}\eta) \right), \tag{10.98}
$$

where the fields $(A_\mu, F, \psi, \overline{\psi}, c, \bar{c})$ inside the integral correspond to the classical fields. We can derive from (10.97) the master equation for the 1PI amplitudes and it takes the form

$$\frac{\partial \Gamma}{\partial \phi_{(a)}} = \int d^n z\, d^n w \left(\frac{\delta \Gamma}{\delta \psi_\gamma(w)} \frac{\delta^2 \Gamma}{\delta \overline{M}_\gamma(w) \delta H_{(a)}(z)} \right.$$

$$\left. + \frac{\delta^2 \Gamma}{\delta H_{(a)}(z) \delta M_\gamma(w)} \frac{\delta \Gamma}{\delta \overline{\psi}_\gamma(w)} \right), \qquad (10.99)$$

where the sources $H_{(a)}$ have been set to zero. This is the master identity (Nielsen identity) from which the gauge parameter dependence of various 1PI amplitudes can be obtained by taking derivatives with respect to appropriate field variables and setting all fields (and M and $\overline{M}$) to zero. For example, if we take the derivative of both sides in (10.99) with respect to $\frac{\delta^2}{\delta \psi_\beta(y) \delta \overline{\psi}_\alpha(x)}$ and set all fields to zero, we obtain

$$\frac{\partial S^{-1}_{\alpha\beta}(x - y)}{\partial \phi_{(a)}} = \int d^n z\, d^n w \left(\mathcal{F}^{(a)}_{\alpha\gamma}(x, z, w) S^{-1}_{\gamma\beta}(w - y) \right.$$

$$\left. + S^{-1}_{\alpha\gamma}(x - w) \mathcal{G}^{(a)}_{\gamma\beta}(w, z, y) \right), \qquad (10.100)$$

where we have defined

$$S^{-1}_{\alpha\beta}(x - y) = \frac{\delta^2 \Gamma}{\delta \psi_\beta(y) \delta \overline{\psi}_\alpha(x)},$$

$$\mathcal{F}^{(a)}_{\alpha\beta}(x, z, w) = \frac{\delta^3 \Gamma}{\delta \overline{\psi}_\alpha(x) \delta H_{(a)}(z) \delta M_\gamma(w)},$$

$$\mathcal{G}^{(a)}_{\gamma\beta}(w, z, y) = \frac{\delta^3 \Gamma}{\delta \overline{M}_\gamma(w) \delta H_{(a)}(z) \delta \psi_\beta(y)}, \qquad (10.101)$$

with all field variables (including $(M, \overline{M})$) to zero. In the momentum space, (10.100) takes the simpler form

$$\frac{\partial S^{-1}_{\alpha\beta}(p)}{\partial \phi_{(a)}} = \mathcal{F}^{(a)}_{\alpha\gamma}(p) S^{-1}_{\gamma\beta}(p) + S^{-1}_{\alpha\gamma}(p) \mathcal{G}^{(a)}_{\gamma\beta}(p), \qquad (10.102)$$

where we have identified

$$\mathcal{F}^{(a)}_{\alpha\gamma}(p) = \mathcal{F}^{(a)}_{\alpha\gamma}(-p, 0, p),$$

$$\mathcal{G}^{(a)}_{\gamma\beta}(p) = \mathcal{G}^{(a)}_{\gamma\beta}(-p, 0, p). \qquad (10.103)$$

Equation (10.102) is the Nielsen identity for the fermion two point function which can be calculated order by order in perturbation theory. With these we are now ready to study the gauge parameter dependence of the pole of the fermion propagator.

10.2.7 Gauge parameter dependence of the pole. The pole of the fermion propagator is given by the zero of the denominator (see, for example, (10.75)). Furthermore, we see, from (10.80), that it is related to a product of the fermion two point function whose gauge parameter dependence, in momentum space, is given in (10.102). Therefore, using these relations, we can calculate the gauge parameter dependence of the denominator of the fermion propagator in a straightforward manner which leads to

$$
\frac{\partial \mathcal{D}(p)}{\partial \phi_{(a)}}
$$

$$
= -\frac{1}{2^{[\frac{n}{2}]}} \mathrm{Tr}\left(\frac{\partial S^{-1}(p)}{\partial \phi_{(a)}} \mathcal{C}(S^{-1}(p))^T \mathcal{C}^{-1} + S^{-1}(p)\mathcal{C}\frac{\partial (S^{-1}(p))^T}{\partial \phi_{(a)}} \mathcal{C}^{-1} \right)
$$

$$
= -\frac{1}{2^{[\frac{n}{2}]}} \mathrm{Tr}\left((\mathcal{F}^{(a)}(p)S^{-1}(p) + S^{-1}(p)\mathcal{G}^{(a)}(p))\mathcal{C}(S^{-1}(p))^T \mathcal{C}^{-1} \right.
$$

$$
\left. + S^{-1}(p)\mathcal{C}\left((S^{-1}(p))^T (\mathcal{F}^{(a)}(p))^T + (\mathcal{G}^{(a)}(p))^T (S^{-1}(p))^T \right)\mathcal{C}^{-1} \right)
$$

$$
= \mathcal{D}(p)\, \mathrm{Tr}\left(\mathcal{F}^{(a)}(p) + \mathcal{G}^{(a)}(p) + \mathcal{C}(\mathcal{F}^{(a)}(p) + \mathcal{G}^{(a)}(p))^T \mathcal{C}^{-1} \right)
$$

$$
= \mathcal{D}(p)\, \mathrm{Tr}\left((\mathcal{F}^{(a)} + \mathcal{G}^{(a)}(p)) + (\mathcal{F}^{(a)}(p) + \mathcal{G}^{(a)}(p))^T \right)
$$

$$
= 2\mathcal{D}(p)\, \mathrm{Tr}\left(\mathcal{F}^{(a)}(p) + \mathcal{G}^{(a)}(p) \right). \tag{10.104}
$$

Here we have used (10.80) as well as the fact that the trace of the transpose of a matrix is the same as the trace of the matrix itself and also the cyclicity of the trace.

Equation (10.104) is interesting in the sense that the denominator of the propagator, in general, depends on the three gauge parameters. However, since the variation of the denominator with respect to the gauge parameters is proportional to the denominator itself, as long as the multiplying factors $\mathcal{F}^{(a)}(p), \mathcal{G}^{(a)}(p)$ are well behaved, the zero of the denominator (or the pole of the propagator) is independent of the gauge parameters, namely,

$$\frac{\partial \mathcal{D}(p)}{\partial \phi_{(a)}}\bigg|_{\mathcal{D}(p)=0} = 0. \tag{10.105}$$

Furthermore, near the zero of the denominator (or the pole of the propagator), we can write

$$\mathcal{D}(p) \xrightarrow{p^2 \to M_{\mathrm{P}}^2} Z(p^2 - M_{\mathrm{P}}^2), \tag{10.106}$$

where Z is a constant related to the wave function normalization Z_2^{-1} of the fermion and (10.106) leads to

$$\frac{\partial M_{\mathrm{P}}}{\partial \phi_{(a)}} = 0. \tag{10.107}$$

In other words, the physical fermion mass is independent of all the three gauge fixing parameters (α, β, n^μ). This is more general than the discussion in section **10.1** where we only discussed the gauge parameter independence within the class of covariant gauge fixing. Here, on the other hand, the physical mass being independent of all the three parameters shows that it has the same value in the covariant gauge ($\beta = 0$) or in the generalized axial gauges ($\alpha = 0$) or in the generalized Coulomb gauges ($\alpha = -\beta$) with n^μ time-like. Namely, the physical mass has the same value across all gauges as a physical quantity should be. As we have mentioned above, this holds only if $\mathcal{F}^{(a)}, \mathcal{G}^{(a)}$ are well behaved and $\mathcal{D}$ itself does not have any singularity, namely, there are no infrared divergences or mass shell singularities. This is generally true in QED or QCD as well as other massive theories in four space-time dimensions. However, in dimensions $n < 4$, the infrared divergences and mass-shell singularities ($p^2 \to m_{\mathrm{P}}^2$) become severe and, in that case, (10.105) and (10.107) may not hold and the pole of the propagator may become gauge parameter dependent. In the next section, we will give an explicit example of this phenomenon with the Schwinger model (both massive and massless) in $1 + 1$ dimensions.

10.3 Example: Schwinger model

Schwinger model is also known as the Abelian QED in $1 + 1$ dimensions. Here the fermion can have a mass (massive Schwinger model)

or may be massless (massless Schwinger model which is normally referred to as the Schwinger model) at the tree level. The analysis of
this model would clarify some of the points we have tried to make towards the end of the previous section. Let us consider massive QED
in D dimensions. The basic interacting Lagrangian density continues
to be given as in (10.81) and we can add to it the gauge fixing and
ghost Lagrangian densities as described in (10.81)-(10.85). Let us
analyze the one loop fermion self-energy graph shown in Fig. 10.2.
Here the photon is massless while we assume that the fermion to have
a mass m which can be set to zero if we are interested in a massless
fermion theory. Let us look at the case of a general covariant gauge
fixing where the photon propagator has the form (see (10.55) with
$\xi = \frac{1}{\alpha^2}$)

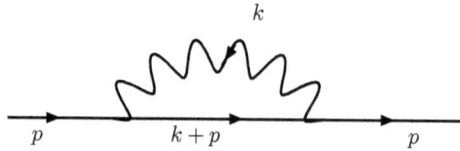

Figure 10.2: One loop fermion self-energy in QED.

$$D_{\mu\nu}(p) = -\frac{1}{p^2}\left(\eta_{\mu\nu} - (1-\xi)\frac{p_\mu p_\nu}{p^2}\right). \tag{10.108}$$

Therefore, the one loop fermion self-energy can be written as

$$\Sigma_{(c)}^{(1)}(p) = e^2 \int \frac{\mathrm{d}^D k}{(2\pi)^D}\, \gamma^\mu S(k+p)\gamma^\nu D_{\mu\nu}(p)$$

$$= -e^2 \int \frac{\mathrm{d}^D k}{(2\pi)^D}\left[(2-D)(\slashed{k}+\slashed{p}) + Dm + (1-\xi)(\slashed{p}-m)\right.$$

$$\left. + (1-\xi)(p^2 - m^2)\frac{\slashed{k}}{k^2}\right]\frac{1}{k^2((k+p)^2 - m^2)}. \tag{10.109}$$

This integral can be done in D dimensions and has the behavior

$$\Sigma_{(c)}^{(1)}(p) = -\frac{1}{2^D \pi^{\frac{D}{2}}}\left(\frac{e}{m^{2-\frac{D}{2}}}\right)^2 \Gamma(3-D)$$

$$\times \left[-m(D-1+\xi)\left(1-\frac{p^2}{m^2}\right)^{D-3} \Gamma\left(\frac{D}{2}-1\right)\right.$$

$$+ 2\xi\not p\left(1-\frac{p^2}{m^2}\right)^{D-3} \Gamma\left(\frac{D}{2}\right)$$

$$+ \left((-2+(D+1)\xi)-(D-1+\xi)m\right)$$

$$\times \left(1-\frac{p^2}{m^2}\right)^{D-2} \Gamma\left(\frac{D}{2}\right)$$

$$\left. - (\xi\not p - (D-1+\xi)m)\frac{\Gamma(2-\frac{D}{2})}{\Gamma(4-D)}\right]. \tag{10.110}$$

We see, from (10.110), that for $D > 3$, the self-energy has the usual ultraviolet divergences (for example, $\Gamma(3-D), \Gamma(2-\frac{D}{2}), \Gamma(4-D)$ terms) which can be handled through renormalization counterterms. However, there are no mass-shell singularities (in this one loop order, $p^2 \to m^2$). On the other hand, in $D \le 3$ dimensions, there are not only infrared divergences (gamma function singularities which can be identified with infrared divergences), but also mass-shell singularities (for example, $\left(1-\frac{p^2}{m^2}\right)^{D-3}, \left(1-\frac{p^2}{m^2}\right)^{D-2}$ terms). As a result, in dimensions, $D \le 3$, the fermion two point function at one loop

$$S^{-1}(p) = \not p - m - \Sigma^{(1)}_{(c)}(p), \tag{10.111}$$

would have infrared and/or mass-shell singularities which would manifest in $\mathcal{D}(p)$ (see (10.80)) as well. In such a case, a Taylor expansion of the self-energy to determine the mass is not possible. In fact, let us note from (10.110) that, for $D = 2+2\epsilon$, the leading term in the self-energy comes from the first term and leads to

$$\Sigma^{(1)}_{(c)\,\text{L}}(p) = \frac{e^2}{4\pi m}(1+\xi)\left(1-\frac{p^2}{m^2}\right)^{-1}\frac{1}{\epsilon}$$

$$= -\frac{e^2 m}{4\pi}(1+\xi)\frac{1}{p^2-m^2}\frac{1}{\epsilon}. \tag{10.112}$$

This makes it clear that the leading term in the one loop self-energy has an infrared divergence ($\frac{1}{\epsilon}$) as well as a mass-shell singularity ($\frac{1}{p^2-m^2}$). Furthermore, it is explicitly gauge parameter dependent. Although the leading divergent term can be set to zero at one loop order by choosing $\xi = -1$, it will reappear in higher orders and there is no value of ξ which can get rid of this leading behavior at all orders.

We recall that the physical mass is obtained from the self-energy order by order in perturbation theory. For example, in the covariant gauge, the general form of the self-energy can be written as (see also (10.68))

$$\Sigma_{(c)}(p) = A(p^2)m + B(p^2)\not{p},$$

so that

$$S^{-1}(p) = (1 - B(p^2))\not{p} - m(1 + A(p^2)). \tag{10.113}$$

If $m_{\rm P}$, denotes the physical mass, it would correspond to the zero of $S^{-1}(p)$ or the pole of the propagator $S(p)$. Therefore, Taylor expanding the two point function around $\not{p} = m_{\rm P}$ ($p^2 = \not{p}^2 = m_{\rm P}^2$), we obtain

$$S^{-1}(p) = m_{\rm P}(1 - B(m_{\rm P}^2)) - m(1 + A(m_{\rm P}^2))$$
$$+ (\not{p} - m_{\rm P})\big((1 - B(m_{\rm P}^2)) - 2m_{\rm P}(mA'(m_{\rm P}^2) + m_{\rm P}B'(m_{\rm P}^2))\big)$$
$$+ O(p^2 - m_{\rm P}^2). \tag{10.114}$$

As a result, $S^{-1}(\not{p} = m_{\rm P}) = 0$ determines

$$m_{\rm P} = m(1 + A(m_{\rm P}^2)) + m_{\rm P}B(m_{\rm P}^2). \tag{10.115}$$

This equation has to be solved order by order consistently to determine the physical mass. For example, at one loop, since the coefficients A, B are already of one loop order (they have a factor of $\hbar$ which we have set to unity), the solution gives

$$m_{\rm P}^{(1)} = m(1 + A(m^2)) + mB(m^2) = m + m(A(m^2) + B(m^2))$$
$$= m_{\rm P}^{(0)} + m(A(m^2) + B(m^2)), \tag{10.116}$$

where we have used the fact that, at the tree level, the fermion mass is $m_{\rm P}^{(0)} = m$ and, therefore, the second term is the genuine one loop contribution to the fermion mass. (The primes on $(A(m_{\rm P}^2), B(m_{\rm P}^2))$ simply stand for a single derivative of $(A(m_{\rm P}^2), B(m_{\rm P}^2))$ with respect to $m_{\rm P}^2$.) On the other hand, we see from (10.112) that, at one loop, $A(p^2) = -\frac{e^2}{4\pi}(1 + \xi)\frac{1}{p^2 - m^2}\frac{1}{\epsilon}$ while $B(p^2) = 0$. As a result, $A(p^2 = m^2) \to \infty$ and $m_{\rm P}^{(1)}$ diverges as well in a gauge parameter dependent manner. Namely, the mass-shell singularity may or may not allow an order by order determination of the physical mass (independent of the problem of infrared divergence) and gauge parameter dependence is manifest.

It is known that the (massless) Schwinger model does not have any infrared problem. Therefore, it is a bit surprising that the massive Schwinger model will have an infrared problem. The origin of this is easy to see from (10.110) in the Feynman gauge ($\xi = 1$). We see that the integrand, in $D = 2 + 2\epsilon$ dimensions, has the form

$$\frac{-2\epsilon(\not{k} + \not{p}) + (2 + 2\epsilon)m}{k^2(k + p)^2}. \tag{10.117}$$

Therefore, when $m \neq 0$, the numerator of the integrand is finite in the limit $\epsilon \to 0$ and does not moderate the infrared divergence arising from the denominator in the integral. On the other hand, if $m = 0$ (in the massless Schwinger model), the numerator is proportional to ϵ and moderates the infrared divergence coming from the denominator making the integral finite. In fact, in this case ($m = 0$), the fermion self-energy at one loop has the value (in the covariant gauge, emphasized by the subscript (c) below)

$$\Sigma_{(c)}^{(1)}(p) = \frac{\xi e^2}{\pi p^2} \not{p}. \tag{10.118}$$

The one loop self-energy does not have any infrared divergence, but does show a mass-shell singularity at $p^2 = 0$ (remember that we have a massless fermion). Nonetheless, in this case, a mass can be defined, namely,

$$S_{(c)}^{-1\,(1)}(p) = \not{p} - \Sigma_{(c)}^{(1)}(p) = \not{p} - \frac{\xi e^2}{\pi p^2} \not{p},$$

leading to

$$S_{(c)}^{(1)}(p) = \frac{\not{p}}{p^2 - \frac{\xi e^2}{\pi}}, \tag{10.119}$$

which shows that the fermion pole mass, in the (massless) Schwinger model, has a manifestly gauge parameter dependent value in the covariant gauge

$$m_{\text{pole}}^2 = \frac{\xi e^2}{\pi}. \tag{10.120}$$

This simple example makes it clear that, in the covariant gauge, while the mass-shell singularity in the massive Schwinger model does not allow a mass to be defined, it is possible to define a pole mass in the (massless) Schwinger model in spite of the mass-shell singularity, which, however, is manifestly gauge parameter dependent.

Figure 10.3: Quenched rainbow approximation.

Coming back to the massive Schwinger model, it is quite natural to ask whether it is possible to rearrange the perturbation theory (some form of resummation) which will allow a definition of mass in this theory. A very common resummation which is used is known as the rainbow approximation is shown in Fig. 10.3. Normally, the overlapping graphs can be neglected in this approximation if we are interested in the leading divergent terms which helps a lot. Second, in the rainbow approximation, we generally allow the photon self-energy graphs which can help improve the infrared behavior. However, in this two dimensional model, we note that the one loop photon polarization tensor has the explicit form

$$\Pi^{\mu\nu}(p) = \left(\eta^{\mu\nu} - \frac{p^\mu p^\nu}{p^2}\right)\Pi_T(p^2),$$

which is manifestly transverse, as it should be, and where

$$\Pi_T(p^2) = -\frac{e^2}{\pi}\left[1 - \int_0^1 \mathrm{d}x \, \frac{m^2}{m^2 - x(1-x)p^2}\right]. \tag{10.121}$$

When $m = 0$ (massless Schwinger model), the second term vanishes and equation (10.121) clearly leads to

$$\Pi_T(p^2) = -\frac{e^2}{\pi}. \tag{10.122}$$

Namely, the photon becomes massive in the (massless) Schwinger model and this Higgs phenomenon is well known. On the other hand, if the fermion is massive, $\Pi_T(p^2)$ in (10.121) becomes non-analytic. In fact, from (10.121), we see that, for $0 \le p^2 \le 4m^2$,

$$\Pi_T(p^2) = -\frac{e^2}{\pi}\left[1 - \frac{\frac{4m^2}{p^2}}{\sqrt{\frac{4m^2}{p^2} - 1}} \arctan \frac{1}{\sqrt{\frac{4m^2}{p^2} - 1}}\right], \qquad (10.123)$$

whereas, for $p^2 \ge 4m^2$, we have

$$\Pi_T(p^2) = -\frac{e^2}{\pi}\left[1 - \frac{\frac{2m^2}{p^2}}{\sqrt{1 - \frac{4m^2}{p^2}}} \ln \frac{1 - \sqrt{1 - \frac{4m^2}{p^2}}}{1 + \sqrt{1 - \frac{4m^2}{p^2}}}\right]. \qquad (10.124)$$

Equation (10.124) allows us to take the massless limit leading to (10.122). On the other hand, (10.123) allows us to take the limit $p^2 \to 0$ for the massive fermion yielding

$$\Pi_T(p^2) \xrightarrow{p^2 \to 0} -\frac{e^2}{\pi}\left[1 - \frac{4m^2}{\sqrt{p^2(4m^2 - p^2)}} \times \sqrt{\frac{p^2}{(4m^2 - p^2)}}\right]$$

$$= -\frac{e^2}{\pi}(1 - 1) = 0, \qquad (10.125)$$

showing that there is no mass generated for the photon in the massive Schwinger model (in contrast to the massless Schwinger model, see (10.122)). This absence of a Higgs phenomenon in the massive Schwinger model is known from a different analysis as well.[1] As a result, it does not help to include the photon self-energy in the rainbow approximation. When the photon self-energy is not included, the approximation is known as the quenched rainbow approximation shown in Fig. 10.3.

In the quenched rainbow approximation, it is clear from the diagram in Fig. 10.3 that the leading term in the fermion self-energy at the $(n+1)$ the loop is related to that at the n th loop as

$$\Sigma_{(c)L}^{(n+1)}(p) = \Sigma_{(c)L}^{(n)}(p)\Sigma_{(c)L}^{(1)}(p)\frac{1}{\not{p} - m}\bigg|_{\not{p} \to m}. \qquad (10.126)$$

[1]See, for example, S. Coleman, R. Jackiw and L. Susskind, Annals of Physics (N. Y.) **93**, 267 (1974) and S. Coleman, Annalas of Physics (N. Y.) **101**, 239 (1976).

Since the recursion relation in (10.126) is simple, it allows us to sum the leading behavior of the self-energy iteratively to all orders and we obtain

$$\Sigma_{(c)\,L}(p) = \sum_{n=0}^{\infty} \Sigma_{(c)\,L}^{(n+1)} = \Sigma_{(c)\,L}^{(1)}(p) \sum_{n=0}^{\infty} \left(\Sigma_{(c)\,L}^{(1)}(p) \frac{1}{\not{p} - m} \right)^{n} \Bigg|_{\not{p} \to m}$$

$$= \Sigma_{(c)\,L}^{(1)}(p) \frac{1}{\not{p} - m - \Sigma_{(c)\,L}^{(1)}(p)} (\not{p} - m) \Bigg|_{\not{p} \to m} \to 0.$$

$$(10.127)$$

This makes it clear that when we sum over all loops in the quenched rainbow approximation, the leading divergence in the fermion self-energy vanishes for any value of the gauge parameter ξ. This behavior can also be seen in an alternative manner which we do not go into here.[2] However, we point out that this derivation does not say anything about the subleading behavior which is, in general, much harder to calculate because they do include the overlapping graphs. However, this result suggests that the fermion self-energy may be better behaved in physical gauges.

10.3.1 Self-energy in physical gauges.

Let us note here that, in a covariant gauge, the theory contains both physical as well as unphysical degrees of freedom and, as a result, a resummation (such as the quenched rainbow approximation) opens up the possibility for cancellation of contributions among these degrees of freedom leading to an improved behavior. However, in a physical gauge, such as the axial gauge or the Coulomb gauge, there are no unphysical degrees of freedom so that there is no question of cancellation among the degrees of freedom. As a result, if the result has to be well behaved in the infrared, it must be so without any resummation and the fermion self-energy must be well behaved already at the one loop level. The physical gauges are known to soften infrared divergences and we will discuss this in some detail in the following.

Let us consider the homogeneous axial gauge fixing for the photon field, namely,

$$n \cdot A = n^{\mu} A_{\mu} = 0. \tag{10.128}$$

---

[2]See, for example, A. Das, J. Frenkel and C. Schubert, Physics Letters **B720**, 414 (2013).

In this case, the photon Green's function has the form (see (10.61) with $\beta \to \infty$)

$$D_{\mu\nu}(p) = -\frac{1}{p^2}\left(\eta_{\mu\nu} - \frac{n_\mu p_\nu + n_\nu p_\mu}{n \cdot p} + \frac{n^2 p_\mu p_\nu}{(n \cdot p)^2}\right), \qquad (10.129)$$

which clearly satisfies

$$n^\mu D_{\mu\nu}(p) = 0 = D_{\mu\nu}(p)n^\nu. \qquad (10.130)$$

The transversality of the photon Green's function (with respect to n^μ) is expected, in this homogeneous gauge choice of (10.128), since the propagator is defined as

$$iD_{\mu\nu}(p) = \langle 0|T(A_\mu(p)A_\nu(0))|0\rangle, \quad \text{and} \quad n^\mu A_\mu = 0. \qquad (10.131)$$

We note that there are two kinds of poles in the Green's function (propagator) in (10.129). The pole at $p^2 = 0$ is handled by the standard Feynman prescription while the poles at $n \cdot p = 0$ use the PV (principal value) prescription.[3] The PV prescription, however, is known to be problematic for time-like $(n^2 > 0)$ as well as light-like $(n^2 = 0)$ axial gauges. Therefore, we restrict ourselves to space-like $(n^2 < 0)$ axial gauges. It is worth noting here that, in $1 + 1$ dimensions, space-like axial gauges also include the Coulomb gauge. For example, in $1 + 1$ dimensions,

$$\nabla \cdot \mathbf{A} = \partial^1 A_1 = 0, \qquad (10.132)$$

implies that $A_1 = A_1(t)$. However, this does not uniquely determine $A_1(t)$, namely, we can always define a new field

$$A_1(t) \to A_1(t) + \alpha(t), \qquad (10.133)$$

where $\alpha(t)$ is arbitrary and the new field will also satisfy the Coulomb gauge condition, (10.132). Taking advantage of this arbitrariness, we can make the solution of the Coulomb gauge condition in (10.132) to vanish (for example, by choosing $\alpha(t) = -A_1(t)$) so that we can write

$$A_1(t) = 0, \quad \text{which satisfies} \quad n^\mu A_\mu = 0, \qquad (10.134)$$

---

[3]W. Kummer, Acta Physica Austriaca **41**, 16 (1975) and J. Frenkel, Physical Review **D13**, 2325 (1976).

with a space-like $n^\mu = (0, 1)$ $(n^2 = -1)$. Coulomb gauge is notoriously singular in $1 + 1$ dimensions. However, viewed as a family of the class of axial gauges, it inherits the PV prescription giving it an improved behavior.

We can now evaluate the one loop fermion self-energy shown in Fig. 10.2 as in (10.109), but with the photon Green's function given in (10.129). (Remember also the decompositions discussed in (10.57)-(10.58) which allows for a manifestly covariant form of the result.) With the PV prescription for the poles at $n \cdot p = 0$, the one loop self-energy takes the form ($\not{p}_L = \frac{(n \cdot p)}{n^2} \not{n}$, see (10.57) and (10.69))

$$\Sigma^{(1)}_{(nc)}(p) = \frac{e^2}{2\pi m} G - \frac{e^2}{2\pi} (1 + G) \frac{\not{n}}{n \cdot p}$$

$$= \frac{e^2}{2\pi m} G - \frac{e^2}{2\pi} (1 + G) \frac{1}{p_L^2} \not{p}_L, \tag{10.135}$$

where

$$G = (\Omega^2 - 1)\left(1 - \frac{\Omega}{2} \ln \frac{1 + \Omega}{1 - \Omega}\right),$$

$$\Omega = \frac{n \cdot p}{\sqrt{(n \cdot p)^2 - m^2 n^2}}. \tag{10.136}$$

It is clear that G is a function (not a matrix) and comparing (10.135) with (10.69), we can identify that, at one loop, we have

$$A = \frac{e^2}{2\pi m^2} G, \quad B = 0, \quad C = -\frac{e^2}{2\pi p_L^2} (1 + G). \tag{10.137}$$

We can now write the fermion two point function, at one loop, as

$$S^{-1(1)}_{(nc)}(p) = \not{p} - m - \Sigma^{(1)}_{(nc)} = \not{p} + C\not{p}_L - m(1 + A). \tag{10.138}$$

Using (10.137), the denominator of the fermion propagator is now easily obtained to be (see (10.72), (10.73) and (10.77))

$$\mathcal{D} = p^2 + (C^2 - 2C)p_L^2 - m^2(1 + A)^2,$$

which can be compared with (10.77) and leads to the pole mass (zero of the denominator) at one loop to be (see also (10.79))

$$p^2 = m_{\text{pole}}^2 = m^2(1+A)^2 + (2C - C^2)p_L^2\Big|_{p^2=m_{\text{pole}}^2}$$

$$= m^2\left(1 + \frac{e^2}{2\pi m^2}G\right)^2 + \left(-\frac{e^2}{\pi}(1+G) - \left(\frac{e^2}{2\pi}\right)^2\frac{1}{p_L^2}(1+G)^2\right)$$

$$= \left(m^2 + \frac{e^2}{\pi}G + O(e^4)\right) - \left(\frac{e^2}{\pi}(1+G) + O(e^4)\right)$$

$$= m^2 - \frac{e^2}{\pi} + O(e^4). \tag{10.139}$$

This is the "pole" mass of the fermion, in the massive Schwinger model, in the physical(axial/Coulomb) gauge up to one loop (order e^2) and is independent of the gauge parameter. However, that should not mislead us into thinking that it is a gauge independent physical parameter because we cannot even define a fermion mass in the covariant gauge because of infrared divergences and mass singularities. Therefore, it is not the value of the mass across all possible gauge classes. In fact, this becomes even more clear in the case of the (massless) Schwinger model ($m = 0$) where the physical gauge calculation (10.139) leads to

$$m_{\text{Schwinger}}^2 = -\frac{e^2}{\pi}, \tag{10.140}$$

whereas, in the cavariant gauge, as we have seen in (10.120)

$$m_{\text{Schwinger}}^2 = \frac{\xi e^2}{\pi}. \tag{10.141}$$

Not only the values of the "pole" masses different in the two classes of gauges, the mass, in the physical gauge, given in (10.140) is not even real (it is imaginary).

Let us conclude this section by simply saying that, through the simple example of the $1+1$ dimensional QED (massless/massive), we have shown that the gauge independence of the physical parameters such as mass of a particle can be studied, in the broader sense, through Nielsen identities. When there are no infrared divergences or mass-shell singularities, a gauge independent physical mass can always be defined. When infrared divergences and mass-shell singularities are present in a covariant gauge, it may or may not be possible to define a consistent "pole" mass. Even when the "pole" mass" can be defined, it will not be gauge parameter independent. On the other hand, in the physical gauges, when a "pole" mass can be defined, it will be independent of the gauge parameter (in that class of gauges). However, it will not be gauge independent across the different possible classes of gauges.

10.4 References

I. J. R. Aitchison and C. M. Fraser, Annals of Physics **156**, 1 (1984).

J. C. Breckenridge, M. J. Lavelle and T. G. Steele, Zeitschrift für Physik **C65**, 155 (1995).

S. Coleman, Annalas of Physics (N. Y.) **101**, 239 (1976).

S. Coleman, R. Jackiw and L. Susskind, Annals of Physics (N. Y.) **93**, 267 (1974).

A. Das, unpublished notes.

A. Das, J. Frenkel and C. Schubert, Physics Letters **B720**, 414 (2013).

A. Das and J. Frenkel, Physics Letters **B726**, 493 (2013).

A. Das, R. Francisco and J. Frenkel, Physical Review **D88**, 085012 (2013).

L. Dolan and R. Jackiw, Physical Review **D9**, 2904 (1974).

J. Frenkel, Physical Review **D13**, 2325 (1976).

R. Fukuda and T. Kugo, Physical Review **D13**, 3469 (1976).

D. A. Johnston, Nuclear Physics **B253**, 687 (1985).

D. Johnston, preprint LPTHE Orsay 86/49 (unpublished).

W. Kummer, Acta Physica Austriaca **41**, 16 (1975).

N. K. Nielsen, Nuclear Physics **101**, 173 (1975).

S. Ramaswamy, Nuclear Physics **B453**, 240 (1995).

B. de Wit, Physical Review **D12**, 1843 (1975).

CHAPTER 11

Subtleties at $T \neq 0$

In chapter **1**, we calculated the self-energy diagram for a bosonic field at finite temperature in the imaginary time formalism (see section **1.7.2**) and showed that this amplitude is nonanalytic at the origin in the energy-momentum space (see (1.123)-(1.124)). We can try to calculate the same quantity in the real time formalism and, of course, the answer turns out to be identical. (In fact, we have already calculated the imaginary part of the same amplitude, in the real time formalism, using the cutting rules in (8.90) and found that it is identical to the imaginary time calculation.) However, in the process, we learn of subtleties that one should be careful about, in the real time formalism, at finite temperature. In the next section, we will discuss one such subtlety, which arises with the simple example of the degenerate electron gas at zero temperature and will take up a second example in the following section. We could have taken up the same example of the bosonic self-energy as in chapter **1**, but this is a much simpler problem which clarifies the point quite nicely.

Before we go into the discussion, however, a bit of history would be helpful to fully appreciate the issue. The self-energy of the bosonic theory was first calculated in the imaginary time formalism and the nonanalyticity was already noted quite early on. With the development of the real time formalism, then, people became interested in checking the imaginary time calculation because such results are unexpected from our experience with calculations at zero temperature. The real time calculations are, however, a lot trickier since they involve products of ill-defined quantities (like delta functions). So, while a nonrigorous calculation in the real time formalism showed that the amplitude is nonanalytic as suggested by the imaginary time calculations, there was not sufficient confidence in the manner in which the result was obtained. At the same time, many proponents of real time formalism did not quite accept a nonanalytic behavior of the amplitude and even proposed additional *ad hoc* finite temperature Feynman rules to give the amplitude an analytic structure. Furthermore, to add to the confusion, when a careful, real time calculation was carried out with a proper regularization for the ill-defined terms of the propagator, it led to an analytic expression for the self-energy which was in disagreement with the imaginary time result. Therefore, if the two formalisms were to coincide, either the regularization procedure was in error or something else in the calculation which needed to be looked at carefully. It turned out that it was something else - in fact, something that we are least likely to suspect, namely, it is the Feynman combination formula that breaks down (gets modified) at finite temperature.

11.1 Degenerate electron gas

Let us consider a system of nonrelativistic degenerate electron gas at zero temperature. This is a system that is widely studied and its properties are quite well measured experimentally. Even though we are considering a system at zero temperature, surprisingly enough, it has all the characteristics of a system at finite temperature, as we will see shortly. The propagator for a nonrelativistic, degenerate electron gas at zero temperature is well known and has the form

$$
\begin{aligned}
iS(p) &= \frac{i}{p^0 - \omega_p + \mu + i\epsilon \, \mathrm{sgn} \, (\omega_p - \mu)} \\
&= \frac{i}{p^0 - \omega_p + \mu + i\epsilon \, \mathrm{sgn} \, (p^0)}.
\end{aligned} \tag{11.1}
$$

Here μ represents the chemical potential which can be identified with the Fermi energy and

$$
\omega_p = \frac{\mathbf{p}^2}{2m}. \tag{11.2}
$$

We also note that while the two forms of the propagator in (11.1) are completely equivalent, it is the second form which is calculationally more convenient and, hence, we will use this. Note here that the propagator in (11.1) does not have a well behaved analytic behavior in the sense that for positive p^0, it has a pole in the lower half plane while for negative values of p^0 the location of the pole is in the upper half of the complex p^0-plane. This is reminiscent of the structure of the propagators at finite temperature (see, for example, (2.53)) even though the system under study is at zero temperature.

We can calculate the self-energy, in the present case, in a direct manner. For simplicity, let us concentrate only on the real part of the self-energy (shown in Fig. 11.1) which takes the form

Figure 11.1: Self-energy diagram at one loop. The two dots represent the points where the amputated external (scalar) legs are connected.

$$\operatorname{Re}\Pi(p) = \operatorname{Re}\left((-2i)\int \frac{d^4k}{(2\pi)^4} S(k)S(k-p)\right)$$

$$= \operatorname{Re}\left((-2i)\int \frac{d^4k}{(2\pi)^4} \frac{1}{k^0 - \omega_k + \mu + i\epsilon\operatorname{sgn}(k^0)}\right.$$

$$\left.\times \frac{1}{k^0 - p^0 - \omega_{k-p} + \mu + i\epsilon\operatorname{sgn}(k^0 - p^0)}\right), \quad (11.3)$$

where we have normalized the coupling strength to unity and we note that the factor of 2 comes from a trace over the spin index. (The negative sign is for the fermion loop and the factor of "i" comes because the diagram is normally identified with $-i\Pi$.) The propagators can be expanded using the standard representation

$$\lim_{\epsilon \to 0^+} \frac{1}{x + i\alpha\epsilon} = \frac{1}{x} - i\pi\operatorname{sgn}(\alpha)\delta(x), \quad (11.4)$$

where the first term denotes the "principal value" (PV). With this,

(11.3) takes the form

$$\operatorname{Re}\Pi(p) = -\int \frac{\mathrm{d}^4 k}{(2\pi)^3}$$

$$\times \left(\frac{1}{k^0 - \omega_k + \mu} \operatorname{sgn}(k^0 - p^0) \, \delta(k^0 - p^0 - \omega_{k-p} + \mu) \right.$$

$$\left. + \frac{1}{k^0 - p^0 - \omega_{k-p} + \mu} \operatorname{sgn}(k^0) \, \delta(k^0 - \omega_k + \mu) \right)$$

$$= -\int \frac{\mathrm{d}^4 k}{(2\pi)^3} \operatorname{sgn}(k^0) \delta(k^0 - \omega_k + \mu)$$

$$\times \left(\frac{1}{k^0 + p^0 - \omega_{k+p} + \mu} + (p^\mu \to -p^\mu) \right). \qquad (11.5)$$

Here we have let $k^\mu \to k^\mu + p^\mu$ in the first term of the integrand. The k^0 integral can now be simply done, using the delta function, and we obtain

$$\operatorname{Re}\Pi(p) = -\int \frac{\mathrm{d}^3 k}{(2\pi)^3} \operatorname{sgn}(\omega_k - \mu)$$

$$\times \left(\frac{1}{p^0 + \omega_k - \omega_{k+p}} + (p^\mu \to -p^\mu) \right). \qquad (11.6)$$

This is, indeed, the standard expression for the self-energy as can be seen from any traditional text. In fact, if we let

$$\mathbf{k} \to -\mathbf{k},$$

in the second term in (11.6), we can write the above expression also as

$$\operatorname{Re}\Pi(p)$$

$$= -\int \frac{\mathrm{d}^3 k}{(2\pi)^3} \operatorname{sgn}(\omega_k - \mu) \left(\frac{1}{p^0 + \omega_k - \omega_{k+p}} - \frac{1}{p^0 + \omega_{k+p} - \omega_k} \right)$$

$$= 2 \int \frac{\mathrm{d}^3 k}{(2\pi)^3} \theta(\mu - \omega_k) \left(\frac{1}{p^0 + \omega_k - \omega_{k+p}} - \frac{1}{p^0 + \omega_{k+p} - \omega_k} \right). \qquad (11.7)$$

We have used here the standard representation of the sign function

$$\text{sgn}(x) = \theta(x) - \theta(-x) = 1 - 2\theta(-x),$$

as well as the fact that the integral of the constant first term (1) vanishes, since the integrand in the parenthesis is odd under the redefinition

$$\mathbf{k} \leftrightarrow -(\mathbf{k} + \mathbf{p}).$$

Furthermore, if we write the chemical potential in terms of a Fermi momentum as (note that k_F is the Fermi momentum of the degenerate gas and has nothing to do with the integration variable k)

$$\mu = \omega_F = \frac{k_F^2}{2m}, \tag{11.8}$$

and scale all the momenta by the Fermi momentum (we measure all momenta with respect to the Fermi momentum k_F), namely, let us define new variables

$$\overline{\mathbf{k}} = \frac{\mathbf{k}}{k_F}, \qquad \overline{\mathbf{p}} = \frac{\mathbf{p}}{k_F}, \tag{11.9}$$

then, we have,

$$d^3k = k_F^3 \, d\overline{k}^3, \qquad \omega_k - \omega_{k+p} = \frac{k_F^2}{m}\left(-\overline{\mathbf{k}} \cdot \overline{\mathbf{p}} - \frac{\overline{\mathbf{p}}^2}{2}\right),$$

and

$$\theta(\mu - \omega_k) = \theta(\omega_F - \omega_k) = \theta\left(\frac{k_F^2}{2m}(1 - \overline{k}^2)\right) = \theta(1 - \overline{k}), \tag{11.10}$$

where we have used (11.2) as well as the fact that $k = |\mathbf{k}| > 0$. Let us also define a dimensionless energy as

$$\nu = \frac{p^0}{2\mu} = \frac{p^0}{2\omega_F} = \frac{m}{k_F^2} p^0, \quad \text{so that} \quad p^0 = \frac{k_F^2}{m} \nu, \tag{11.11}$$

then, the self-energy in (11.7)) can be written as

$$\text{Re}\,\Pi(\nu, \overline{p})$$

$$= 2mk_F \int \frac{d^3\overline{k}}{(2\pi)^3} \theta(1 - \overline{k}) \left(\frac{1}{\nu - \overline{\mathbf{k}} \cdot \overline{\mathbf{p}} - \frac{\overline{p}^2}{2}} - \frac{1}{\nu + \overline{\mathbf{k}} \cdot \overline{\mathbf{p}} + \frac{\overline{p}^2}{2}}\right)$$

$$= 2mk_F \int \frac{d^3\overline{k}}{(2\pi)^3} \theta(1 - \overline{k}) \left(\frac{1}{\nu - \overline{k}\overline{p}\cos\theta - \frac{\overline{p}^2}{2}} - \frac{1}{\nu + \overline{k}\overline{p}\cos\theta + \frac{\overline{p}^2}{2}}\right).$$

$$\tag{11.12}$$

Here $\bar{k}$ and $\bar{p}$ represent the magnitudes of the three vectors $\mathbf{k}$ and $\mathbf{p}$ respectively.

Expression (11.12) is identical to what is known in the standard literature.[1] The integral can be easily evaluated (or directly taken from any standard table) which leads to a self-energy of the form

$$
\mathrm{Re}\,\Pi(\nu,\bar{p}) = \frac{mk_F}{2\pi^2}\left\{ -1 \right.
$$

$$
+ \frac{1}{2\bar{p}}\left(1 - \left(\frac{\nu}{\bar{p}} - \frac{\bar{p}}{2}\right)^2\right)\ln\left|\frac{1 + (\frac{\nu}{\bar{p}} - \frac{\bar{p}}{2})}{1 - (\frac{\nu}{\bar{p}} - \frac{\bar{p}}{2})}\right|
$$

$$
\left. - \frac{1}{2\bar{p}}\left(1 - \left(\frac{\nu}{\bar{p}} + \frac{\bar{p}}{2}\right)^2\right)\ln\left|\frac{1 + (\frac{\nu}{\bar{p}} + \frac{\bar{p}}{2})}{1 - (\frac{\nu}{\bar{p}} + \frac{\bar{p}}{2})}\right| \right\}. \tag{11.13}
$$

This self-energy does not have a covariant dependence on energy and momentum (as we have often emphasized) and is nonanalytic at the origin in energy-momentum space. The simplest way to see this is to define

$$
\nu = \alpha\bar{p}, \tag{11.14}
$$

and take the limit $\bar{p} \to 0$ (this also implies $p \to 0$) with α fixed. In this limit, the self-energy in (11.13) is easily seen to have the form

$$
\lim_{\bar{p}\to 0}\mathrm{Re}\,\Pi(\alpha\bar{p},\bar{p}) = \lim_{\bar{p}\to 0}\frac{mk_F}{2\pi^2}\left[-1 \right.
$$

$$
+ \left(\left(\frac{(1-\alpha^2)}{2\bar{p}} + \frac{\alpha}{2}\right)\ln\left|\frac{1+\alpha}{1-\alpha}\right| - \frac{1}{2}\right)
$$

$$
\left. - \left(\left(\frac{(1-\alpha^2)}{2\bar{p}} - \frac{\alpha}{2}\right)\ln\left|\frac{1+\alpha}{1-\alpha}\right| + \frac{1}{2}\right)\right]
$$

$$
= \frac{mk_F}{2\pi^2}\left(-2 + \alpha\ln\left|\frac{1+\alpha}{1-\alpha}\right|\right). \tag{11.15}
$$

The α dependence in the real part of the self-energy shows clearly that the value at the origin depends on how the limits, $p \to 0$ and $\nu \to 0$ are taken.

---

[1] See, for example, the unnumbered equation following (12.35) as well as (12.36) for the following relation (with the identifications $q = \bar{p}$ and $k = \bar{k}$) in A. L. Fetter and J. D. Walecka, *Quantum Theory of Many Particle Systems*, McGraw-Hill Book company (1971).

11.2 Calculation with ε-regularization

In the real time formalism, calculations often involve products of singular functions. The simplest way to see this is to note from (11.5) that if we had set $p^0 = 0$ and $\mathbf{p} = 0$ in the integrand itself, we would have products of terms like

$$\frac{1}{x}\,\delta(x),$$

where the first factor corresponds to the principal value of $\frac{1}{x}$. Such products are highly singular and need to be regularized properly. A standard regularization that is used at finite temperature is known as the ε–regularization which simply corresponds to the prescription that we use the representations

$$\frac{1}{x} = P\frac{1}{x} = \lim_{\epsilon\to 0^+} \frac{x}{x^2 + \epsilon^2} = \lim_{\epsilon\to 0^+} \frac{1}{2}\left(\frac{1}{x + i\epsilon} + \frac{1}{x - i\epsilon}\right),$$

$$\delta(x) = \lim_{\epsilon\to 0^+} \frac{1}{\pi}\frac{\epsilon}{x^2 + \epsilon^2} = \lim_{\epsilon\to 0^+} \frac{i}{2\pi}\left(\frac{1}{x + i\epsilon} - \frac{1}{x - i\epsilon}\right), \qquad (11.16)$$

where it is understood that the limit is to be taken only at the end of the calculations. It is worth checking, in this example of the degenerate electron gas, whether this regularization is, indeed, meaningful or whether it can lead to subtleties. To that end, we will evaluate the self-energy for the degenerate electron gas through this regularization as well.

We note that if we use the representations for the delta function as well as the principal value, given in (11.16), in the integrand of (11.5), we can write the real part of the self-energy also as (we will always understand the limiting procedure of (11.16) to be taken at the end although we will not write it explicitly for simplicity)

$$\operatorname{Re}\Pi(p) = -\frac{i}{2}\int \frac{\mathrm{d}^4k}{(2\pi)^4}\left(\frac{1}{k^0 + p^0 - \omega_{k+p} + \mu + i\epsilon}\right.$$

$$+ \frac{1}{k^0 + p^0 - \omega_{k+p} + \mu - i\epsilon} + (p^\mu \to -p^\mu)\right)$$

$$\times \operatorname{sgn}(k^0)\left(\frac{1}{k^0 - \omega_k + \mu + i\epsilon} - \frac{1}{k^0 - \omega_k + \mu - i\epsilon}\right)$$

$$= -\frac{i}{2} \int \frac{\mathrm{d}^4 k}{(2\pi)^4} \, \mathrm{sgn}(k^0)$$

$$\Bigg(\frac{1}{k^0 + p^0 - \omega_{k+p} + \mu + i\epsilon} \frac{1}{k^0 - \omega_k + \mu + i\epsilon}$$

$$- \frac{1}{k^0 + p^0 - \omega_{k+p} + \mu - i\epsilon} \frac{1}{k^0 - \omega_k + \mu - i\epsilon}$$

$$- \frac{1}{k^0 + p^0 - \omega_{k+p} + \mu + i\epsilon} \frac{1}{k^0 - \omega_k + \mu - i\epsilon}$$

$$+ \frac{1}{k^0 + p^0 - \omega_{k+p} + \mu - i\epsilon} \frac{1}{k^0 - \omega_k + \mu + i\epsilon}$$

$$+ (p^\mu \to -p^\mu) \Bigg)$$

$$= I_1 + I_2, \tag{11.17}$$

where we have grouped terms in the integrand with similar analytic behavior (namely, $+i\epsilon$ or $-i\epsilon$) for the two factors into I_1, while I_2 contains terms where the two factors have opposite analytic behavior.

The integrands in (11.17) for both I_1 and I_2 are nonanalytic because of the overall $\mathrm{sgn}(k^0)$ term. Therefore, to evaluate the k^0 integral, we first rewrite the integrals as follows

$$I_1 =$$

$$-\frac{i}{2} \int \frac{\mathrm{d}^3 k}{(2\pi)^4} \int_0^\infty \mathrm{d}k^0 \Bigg[\Bigg(\frac{1}{k^0 + p^0 - \omega_{k+p} + \mu + i\epsilon} \frac{1}{k^0 - \omega_k + \mu + i\epsilon}$$

$$- \frac{1}{k^0 - p^0 + \omega_{k+p} - \mu - i\epsilon} \frac{1}{k^0 + \omega_k - \mu - i\epsilon}$$

$$- \frac{1}{k^0 + p^0 - \omega_{k+p} + \mu - i\epsilon} \frac{1}{k^0 - \omega_k + \mu - i\epsilon}$$

$$+ \frac{1}{k^0 - p^0 + \omega_{k+p} - \mu + i\epsilon} \frac{1}{k^0 + \omega_k - \mu + i\epsilon} \Bigg)$$

$$+ (p^\mu \to -p^\mu) \Bigg]. \tag{11.18}$$

The k^0 integral in I_1 in (11.17) can now be done in a straightforward manner using the method of residues with a contour in the upper right quadrant and we obtain

$$
I_1 = -\frac{i}{2} \int \frac{d^3 k}{(2\pi)^4} \, 2i\pi \left(-\frac{\theta(\mu + p^0 - \omega_{k+p})}{p^0 + \omega_k - \omega_{k+p}} - \frac{\theta(\mu - \omega_k)}{\omega_{k+p} - p^0 - \omega_k} \right.
$$

$$
\left. -\frac{\theta(\omega_{k+p} - p^0 - \mu)}{\omega_{k+p} - p^0 - \omega_k} - \frac{\theta(\omega_k - \mu)}{\omega_k - \omega_{k+p} + p^0} + (p^\mu \to -p^\mu) \right)
$$

$$
= -\frac{1}{2} \int \frac{d^3 k}{(2\pi)^3} \left(\frac{(\mathrm{sgn}(\omega_k - \mu) - \mathrm{sgn}(\omega_{k+p} - p^0 - \mu))}{p^0 + \omega_k - \omega_{k+p}} \right.
$$

$$
\left. + (p^\mu \to -p^\mu) \right). \tag{11.19}
$$

The k^0 integral in I_2 in (11.17) can again be performed exactly in the same manner and yields

$$
I_2 = -\frac{1}{2} \int \frac{d^3 k}{(2\pi)^3} \left(\frac{(\mathrm{sgn}(\omega_k - \mu) + \mathrm{sgn}(\omega_{k+p} - p^0 - \mu))}{p^0 + \omega_k - \omega_{k+p}} \right.
$$

$$
\left. + (p^\mu \to -p^\mu) \right). \tag{11.20}
$$

From (11.17), (11.19) and (11.20), then, we obtain

$$
\mathrm{Re}\,\Pi(p) = I_1 + I_2 = - \int \frac{d^3 k}{(2\pi)^3} \mathrm{sgn}(\omega_k - \mu)
$$

$$
\times \left(\frac{1}{p^0 + \omega_k - \omega_{k+p}} + (p^\mu \to -p^\mu) \right). \tag{11.21}
$$

This is, of course, exactly the same result that is obtained by direct integration in (11.6). This shows that the ϵ–regularization does not lead to any particular subtlety if used properly.

11.3 Feynman parameterization

Next, let us see what would the result be if we used the naive Feynman combination formula with ϵ–regularization. We note that the naive combination formula gives[2]

$$\frac{1}{A + i\alpha\epsilon}\frac{1}{B + i\beta\epsilon}$$

$$= \int\limits_0^1 \frac{dx}{[x(A + i\alpha\epsilon) + (1 - x)(B + i\beta\epsilon)]^2}. \quad (11.22)$$

If we use this formula in (11.18), then we obtain

$$I_1 = -\frac{i}{2}\int\frac{d^3k}{(2\pi)^4}\int\limits_0^\infty\int\limits_0^1 dx\left(\frac{1}{[k^0 + x(p^0 + \omega_k - \omega_{k+p}) - \omega_k + \mu + i\epsilon]^2}\right.$$

$$-\frac{1}{[k^0 - x(p^0 + \omega_k - \omega_{k+p}) + \omega_k - \mu - i\epsilon]^2}$$

$$-\frac{1}{[k^0 + x(p^0 + \omega_k - \omega_{k+p}) - \omega_k + \mu - i\epsilon]^2}$$

$$\left.+\frac{1}{[k^0 - x(p^0 + \omega_k - \omega_{k+p}) + \omega_k + i\epsilon]^2} + (p^\mu \to -p^\mu)\right).$$

$$(11.23)$$

The integral in (11.23) can be done trivially over k^0 and yields

$$I_1 = -\frac{i}{2}\int\frac{d^3k}{(2\pi)^4}\int\limits_0^1 dx\left(\frac{1}{x(p^0 + \omega_k - \omega_{k+p}) - \omega_k + \mu + i\epsilon}\right.$$

$$+\frac{1}{x(p^0 + \omega_k - \omega_{k+p}) - \omega_k + \mu + i\epsilon}$$

$$-\frac{1}{x(p^0 + \omega_k - \omega_{k+p}) - \omega_k + \mu - i\epsilon}$$

$$\left.-\frac{1}{x(p^0 + \omega_k - \omega_{k+p}) - \omega_k + \mu - i\epsilon} + (p^\mu \to -p^\mu)\right)$$

---

[2]See any standard text on quantum field theory, for example, Eq. (15.22) in A. Das, *Lectures on Quantum Field Theory* (second edition), World Scientific Publishing (2021).

$$= -i \int \frac{\mathrm{d}^3 k}{(2\pi)^4} \int_0^1 \mathrm{d}x \, (-2i\pi) \left(\delta(x(p^0 + \omega_k - \omega_{k+p}) - \omega_k + \mu) \right.$$

$$\left. + (p^\mu \to -p^\mu) \right)$$

$$= -\int \frac{\mathrm{d}^3 k}{(2\pi)^3} \int_{-\omega_k + \mu}^{p^0 + \mu - \omega_{k+p}} \mathrm{d}s \left(\frac{\delta(s)}{p^0 + \omega_k - \omega_{k+p}} + (p^\mu \to -p^\mu) \right)$$

$$= -\frac{1}{2} \int \frac{\mathrm{d}^3 k}{(2\pi)^3} \left[\left\{ \left(\mathrm{sgn}(\omega_k - \mu) - \mathrm{sgn}(\omega_{k+p} - p^0 - \mu) \right) \right. \right.$$

$$\left. \left. \times \frac{1}{p^0 + \omega_k - \omega_{k+p}} \right\} + (p^\mu \to -p^\mu) \right]. \qquad (11.24)$$

Here we have used formula (11.4) in the intermediate steps. We note that this agrees with (11.19) completely.

We can, similarly, evaluate I_2 using the Feynman combination formula. Without going into details, let us simply note here that, in this case, we obtain

$$I_2 = -\frac{1}{2} \int \frac{\mathrm{d}^3 k}{(2\pi)^3} \left[\left\{ 2 \, \mathrm{sgn} \left(\frac{p^0 - \omega_k - \omega_{k+p}}{2} + \mu \right) \right. \right.$$

$$\left. + \mathrm{sgn}(\omega_k - \mu) + \mathrm{sgn}(\omega_{k+p} - p^0 - \mu) \right\}$$

$$\left. \times \frac{1}{p^0 + \omega_k - \omega_{k+p}} + (p^\mu \to -p^\mu) \right]. \qquad (11.25)$$

This is, however, different from (11.20) and, therefore, it is clear that if we evaluate the self-energy using the naive Feynman combination formula, the result will be inaccurate. It is interesting to note also that while the combination formula in (11.22) seems to work when the factors have the same analyticity properties, it breaks down for factors with opposite analyticity properties. We will next see how the Feynman formula modifies in such cases.

11.4 Modification of Feynman formula

Let us look at (11.22)) and note that if the integrand, on the right hand side, does not have any poles on the real x–axis, then, we have

$$\int_0^1 \frac{\mathrm{d}x}{[x(A+i\alpha\epsilon)+(1-x)(B+i\beta\epsilon)]^2}$$

$$= -\frac{1}{A-B+i(\alpha-\beta)\epsilon}\, \frac{1}{x(A-B+i(\alpha-\beta)\epsilon)+B+i\beta\epsilon}\bigg|_0^1$$

$$= -\frac{1}{A-B+i(\alpha-\beta)\epsilon}\left(\frac{1}{A+i\alpha\epsilon}-\frac{1}{B+i\beta\epsilon}\right)$$

$$= \frac{1}{A+i\alpha\epsilon}\, \frac{1}{B+i\beta\epsilon}. \tag{11.26}$$

This is, of course, the standard combination formula. However, let us assume that the parameters α and β are such that the integrand on the right hand side of (11.22) has a pole on the real x–axis between 0 and 1. (Note that without loss of generality, we can choose $|\alpha| = |\beta| = 1$.) We see immediately that the integrand can have a pole on the real x–axis at x_0 if

$$(x_0 A + (1-x_0)B) + i\epsilon(x_0\alpha + (1-x_0)\beta) = 0,$$

$$\text{or,} \quad x_0(A-B)+B = 0 = (\beta + x_0(\alpha-\beta)). \tag{11.27}$$

The solution to these two equations gives

$$x_0 = \frac{\beta}{\beta-\alpha}, \quad \beta A = \alpha B. \tag{11.28}$$

It is clear from (11.28) that such a pole will lie between 0 and 1 on the real axis only if α and β have opposite sign. That is, only when the two factors have opposite analytic behavior will the pole on the real axis contribute nontrivially to the integration. Let us assume that $0 \leq x_0 = \frac{\beta}{\beta-\alpha} \leq 1$.

In this case, we can evaluate the right hand side of (11.22) using the principal value since the pole is on the real axis and we can write

$$
\int_0^1 \frac{\mathrm{d}x}{[x(A + i\alpha\epsilon) + (1 - x)(B + i\beta\epsilon)]^2}
$$

$$
= \lim_{\eta \to 0^+} \left(\int_0^{\frac{\beta}{\beta-\alpha}-\eta} \frac{\mathrm{d}x}{[x(A - B + i\epsilon(\alpha - \beta)) + B + i\beta\epsilon]^2} \right.
$$

$$
\left. + \int_{\frac{\beta}{\beta-\alpha}+\eta}^1 \frac{\mathrm{d}x}{[x(A - B + i\epsilon(\alpha - \beta)) + B + i\beta\epsilon]^2} \right)
$$

$$
= -\frac{1}{A - B + i(\alpha - \beta)\epsilon} \times
$$

$$
\lim_{\eta \to 0^+} \left[\frac{1}{x(A - B + i\alpha\epsilon(\alpha - \beta)) + B + i\beta\epsilon} \Big|_0^{\frac{\beta}{\beta-\alpha}-\eta} \right.
$$

$$
\left. + \frac{1}{x(A - B + i\alpha\epsilon(\alpha - \beta)) + B + i\beta\epsilon} \Big|_{\frac{\beta}{\beta-\alpha}+\eta}^1 \right]
$$

$$
= \frac{1}{A - B + i(\alpha - \beta)\epsilon} \times
$$

$$
\lim_{\eta \to 0^+} \left[\left\{ \frac{1}{B + i\beta\epsilon} - \frac{(\beta - \alpha)}{\beta A - \alpha B + i\eta\epsilon} \right\} \right.
$$

$$
\left. - \left\{ \frac{1}{A + i\alpha\epsilon} + \frac{(\beta - \alpha)}{\beta A - \alpha B - i\eta\epsilon} \right\} \right]
$$

$$
= \frac{1}{A + i\alpha\epsilon} \frac{1}{B + i\beta\epsilon} - 2i\pi \frac{(\alpha - \beta)\delta(\beta A - \alpha B)}{A - B + i(\alpha - \beta)\epsilon}. \tag{11.29}
$$

In other words, when the two factors have opposite analytic properties, the combination formula (11.22)) modifies and, in general, we can write

$$\frac{1}{A + i\alpha\epsilon}\frac{1}{B + i\beta\epsilon} = \int_0^1 \frac{\mathrm{d}x}{[x(A + i\alpha\epsilon) + (1 - x)(B + i\beta\epsilon)]^2}$$

$$+ 2i\pi \frac{(\alpha - \beta)\delta(\beta A - \alpha B)}{A - B + i(\alpha - \beta)\epsilon}. \qquad (11.30)$$

Since I_2 (see (11.17)) involves factors with opposite analyticity, this extra term in the combination formula is likely to contribute there. In fact, we can simply calculate the correction that the δ–function term would give rise to in I_2 and it has the form

$$I_2' = \int \frac{\mathrm{d}^3 k}{(2\pi)^3}\left[\mathrm{sgn}\left(\frac{p^0 - \omega_k - \omega_{k+p}}{2} + \mu\right)\frac{1}{p^0 + \omega_k - \omega_{k+p}}\right.$$

$$\left. + (p^\mu \to -p^\mu)\right]. \qquad (11.31)$$

It is now clear that the sum of (11.25) and (11.31) coincide exactly with (11.20). This simple example shows that since, at finite temperature, the propagator contains terms with opposite analytic behavior, in calculating, one should be careful and use the modified Feynman combination formula whenever necessary. It was, in fact, this particular error that led to contradicting results early on for the self-energy at finite temperature, when calculated in the real time formalism.

11.5 Catalysis at $T \neq 0$

The nonanalyticity at $p^\mu = 0$ for nonzero temperatures is only one of many such nonanalyticities that show up at finite temperature. In this section, we will study in some detail another kind of nonanalyticity which is related to the vanishing mass limit.

To begin with, let us consider the $2 + 1$ dimensional massive fermion theory described by the Lagrangian density

$$\mathcal{L} = i\overline{\psi}\gamma^\mu \partial_\mu \psi - m\overline{\psi}\psi, \quad \mu = 0, 1, 2. \qquad (11.32)$$

In $2 + 1$ dimensions, the spinor representation is 2-dimensional and so, we can think of ψ as a 2-component spinor while we can identify

$$\gamma^0 = \sigma_2, \quad \gamma^1 = i\sigma_1, \quad \gamma^2 = -i\sigma_3. \qquad (11.33)$$

Here the three σ_i's represent the three Pauli matrices and γ^0 is Hermitian while γ^1, γ^2 are anti-Hermitian. As usual, we have defined

$$\overline{\psi} = \psi^\dagger \gamma^0,$$

in (11.32). In this two dimensional representation, it is clear that there cannot be a γ_5 matrix since

$$\gamma^0 \gamma^1 \gamma^2 = \sigma_2 \sigma_1 \sigma_3 = -i\mathbb{1},$$

where $\mathbb{1}$ represents the 2×2 identity matrix and, in fact, there is no other linearly independent 2×2 matrix beyond these four basic matrices. As a result, there is no notion of a chiral transformation in this theory.

On the other hand, in $2 + 1$ dimensions, one can define a parity operation as reflecting through only one of the two axes, namely,

$$(x, y) \to (-x, y), \quad \text{or,} \quad (x, y) \to (x, -y), \tag{11.34}$$

since a reflection about both the axes is equivalent to a rotation by an angle π (around the z-axis). Therefore, choosing the first of (11.34) as defining a parity transformation (one can choose the second as well without altering any of the conclusions), we note that, under parity (η is a phase factor corresponding to the intrinsic parity of the field),

$$\psi \to \eta \gamma^1 \psi, \qquad \overline{\psi} \to -\eta^* \overline{\psi} \gamma^1, \tag{11.35}$$

so that the fermion bilinears transform, under parity, as

$$\overline{\psi} \gamma^0 \psi \to \overline{\psi} \gamma^0 \psi,$$
$$\overline{\psi} \gamma^1 \psi \to -\overline{\psi} \gamma^1 \psi,$$
$$\overline{\psi} \gamma^2 \psi \to \overline{\psi} \gamma^2 \psi. \tag{11.36}$$

It is clear now from the definitions of the fermion transformations in (11.35) that, under parity,

$$\overline{\psi} \psi \to -\overline{\psi} \psi, \tag{11.37}$$

which shows that the massive 2-component spinor theory described by (11.32), in $2 + 1$ dimensions, is not invariant under parity since the mass term would break the symmetry.

On the other hand, we can consider a fermion theory where the fermion field is a 4-component spinor corresponding to a reducible representation of the Lorentz group. Thus, writing the fermion field in terms of two 2-component spinor fields, ψ_1 and ψ_2 as

$$\psi = \begin{pmatrix} \psi_1 \\ \psi_2 \end{pmatrix}, \tag{11.38}$$

where each of ψ_1 and ψ_2 is a two component spinor, we see that we can choose the three "gamma" matrices (of $2+1$ dimension) to be the block diagonal 4×4 matrices given by

$$\gamma^0 = \begin{pmatrix} \sigma_3 & 0 \\ 0 & -\sigma_3 \end{pmatrix},$$

$$\gamma^1 = \begin{pmatrix} i\sigma_1 & 0 \\ 0 & -i\sigma_1 \end{pmatrix},$$

$$\gamma^2 = \begin{pmatrix} i\sigma_2 & 0 \\ 0 & -i\sigma_2 \end{pmatrix}. \tag{11.39}$$

In this reducible representation, it is interesting to note that, we can define two "γ_5" matrices which anti-commute with all three of the "gamma" matrices, γ^0, γ^1 and γ^2. Denoting these two matrices by ($\mathbb{1}$ denotes the 2×2 identity matrix)

$$\gamma^3 = i \begin{pmatrix} 0 & \mathbb{1} \\ \mathbb{1} & 0 \end{pmatrix} = -(\gamma^3)^\dagger,$$

$$\gamma_5 = i \begin{pmatrix} 0 & \mathbb{1} \\ -\mathbb{1} & 0 \end{pmatrix} = \gamma_5^\dagger, \tag{11.40}$$

we recognize that

$$\gamma_5 = i\gamma^0\gamma^1\gamma^2\gamma^3.$$

Furthermore, it is easy to see that, in this fermion theory of (11.32) (with 4-component fields), we can now define two (unitary) global "chiral" transformations, generated respectively by γ^3 and γ_5, namely,

$$\psi \to e^{\alpha\gamma^3}\psi, \qquad\qquad \overline{\psi} \to \overline{\psi}e^{\alpha\gamma^3},$$

$$\psi \to e^{i\tilde{\alpha}\gamma_5}\psi, \qquad\qquad \overline{\psi} \to \overline{\psi}e^{-i\tilde{\alpha}\gamma_5}, \tag{11.41}$$

where α and $\tilde{\alpha}$ denote two arbitrary real constant parameters. It is easily checked that the kinetic energy term is invariant under a $U(2)$ global transformation where the generators of the transformation are $\mathbb{1}$, γ_5, $-i\gamma^3$ and $\gamma^3\gamma_5$. That is, in such a case, there are two flavors of fermionic fields and $U(2)$ simply represents the flavor symmetry of the massless theory. Furthermore, it is obvious from (11.41) that the mass term, on the other hand, breaks the global flavor symmetry down to the product of two symmetries (corresponding to simple phase transformations, generated by the identity matrix), namely, in the presence of the mass term,

$$U(2) \to U(1) \times U(1).$$

The mass can, of course, be thought of as a source for the bilinear $\overline{\psi}\psi$ and, therefore, we see that if in the fermion theory, a condensate develops (when the mass vanishes), then, the flavor symmetry would have broken down to $U(1) \times U(1)$.

It is also interesting to note that, we can now define a parity transformation for the fermion fields as (similar to (11.35)))

$$\psi \to -\eta\gamma^1\gamma_5\psi, \qquad \overline{\psi} \to \eta^*\overline{\psi}\gamma_5\gamma^1, \qquad (11.42)$$

so that, under parity, the bilinears transform as in (11.36). However, it is interesting to note that under the parity transformation, (11.42), we now have

$$\overline{\psi}\psi \to \overline{\psi}\psi. \qquad (11.43)$$

In other words, the massive fermion theory with doubled degrees of freedom is not only invariant under "chiral" transformations (corresponding to the two $U(1)$ symmetries), but is also parity invariant.

Let us next study, in some detail, this 4-component theory interacting with a constant external magnetic field along the negative z-axis. First, let us obtain the first quantized spinor wave functions for such a system. The Lagrangian density, in this case, has the form (see (5.11), (5.13) and (5.20))

$$\mathcal{L} = i\overline{\psi}\gamma^\mu(\partial_\mu - ieA_\mu)\psi - m\overline{\psi}\psi. \qquad (11.44)$$

The equations of motion for the two 2-component spinors can now be obtained from the Euler-Lagrange equations following from $\mathcal{L}$ in

(11.44) and take the forms

$$\left(i\sigma_3(\partial_0 - ieA_0) - \sigma_1(\partial_1 - ieA_1) - \sigma_2(\partial_2 - ieA_2) - m\right)\psi_1 = 0,$$

$$\left(i\sigma_3(\partial_0 - ieA_0) - \sigma_1(\partial_1 - ieA_1) - \sigma_2(\partial_2 - ieA_2) + m\right)\psi_2 = 0.$$
$$\tag{11.45}$$

Thus, we see that the forms of the two equations satisfied by ψ_1 and ψ_2 are the same except for the sign of the mass term. Therefore, let us analyze only one of them and the other can be obtained from the solutions of this.

For a constant magnetic field along the z-axis, we can choose the vector potentials to be of the form (the choice of the potential is, of course, not unique and can be changed with a gauge transformation, but the physical results would be independent of any such choice)

$$A_0 = 0 = A_1, \qquad A_2 = Bx, \tag{11.46}$$

where B represents a constant magnetic field. Writing the 2-component spinor field, ψ_1, as

$$\psi_1 = \begin{pmatrix} \psi_1^+ \\ \psi_1^- \end{pmatrix}, \tag{11.47}$$

we see that the equation for ψ_1 from (11.45) takes the form

$$(i\frac{\partial}{\partial t} - m)\psi_1^+ + (-\frac{\partial}{\partial x} + i\frac{\partial}{\partial y} + eBx)\psi_1^- = 0,$$

$$(-\frac{\partial}{\partial x} - i\frac{\partial}{\partial y} - eBx)\psi_1^+ - (i\frac{\partial}{\partial t} + m)\psi_1^- = 0. \tag{11.48}$$

We note from (11.48) that the solution will have a plane wave form in t and y and, therefore, factoring out

$$e^{-iEt+ipy},$$

where p denotes the momentum along the y-axis, the two coupled equations in (11.48) take the form

$$(E - m)\psi_1^+ - (\frac{\partial}{\partial \bar{x}} - eB\bar{x})\psi_1^- = 0,$$

$$(\frac{\partial}{\partial \bar{x}} + eB\bar{x})\psi_1^+ + (E + m)\psi_1^- = 0, \tag{11.49}$$

where we have defined

$$\bar{x} = x - \frac{p}{eB}. \tag{11.50}$$

The two first order coupled equations, in (11.49), can be decoupled and written in terms of second order equations as

$$\left(\frac{\partial^2}{\partial \bar{x}^2} - e^2 B^2 \bar{x}^2 + E^2 - m^2 + eB \right) \psi_1^+ = 0,$$

$$\left(\frac{\partial^2}{\partial \bar{x}^2} - e^2 B^2 \bar{x}^2 + E^2 - m^2 - eB \right) \psi_1^- = 0. \tag{11.51}$$

These are (shifted) harmonic oscillator equations and, if we assume that $eB > 0$, the energy eigenvalues can be obtained to be[3]

$$E_n^+ = \pm\sqrt{2neB + m^2},$$

$$E_n^- = \pm\sqrt{2(n+1)eB + m^2}, \qquad n = 0, 1, 2, \cdots. \tag{11.52}$$

These are basically the Landau levels of the theory and we note that ψ_2 would also have identical energy levels since the energy eigenvalues are invariant under $m \to -m$.

With a little bit of more analysis, we can obtain the normalized eigenfunctions for ψ_1 and ψ_2 which are related to the harmonic oscillator solutions (Hermite polynomials). If we now put together both these solutions, then, the 4-component solutions will have the form (assuming $m > 0$)

$$\psi_1^+ = N_n e^{-i(|E_n|t - py)} \begin{pmatrix} (|E_n| + m)I(n, p, x) \\ -\sqrt{2eBn}I(n-1, p, x) \\ 0 \\ 0 \end{pmatrix},$$

$$\psi_2^+ = N_n e^{-i(|E_n|t - py)} \begin{pmatrix} 0 \\ 0 \\ -\sqrt{2eBn}I(n, p, x) \\ (|E_n| + m)I(n-1, p, x) \end{pmatrix},$$

[3]See, for example, M. Kobayashi and M. Sakamoto, Progress in Theoretical Physics **70**, 1375 (1983).

$$\psi_1^- = N_n e^{i(|E_n|t - py)} \begin{pmatrix} \sqrt{2eBn} I(n, -p, x) \\ (|E_n| + m)I(n-1, -p, x) \\ 0 \\ 0 \end{pmatrix},$$

$$\psi_2^- = N_n e^{i(|E_n|t - py)} \begin{pmatrix} 0 \\ 0 \\ (|E_n| + m)I(n, -p, x) \\ \sqrt{2eBn} I(n-1, -p, x) \end{pmatrix}, \qquad (11.53)$$

where we have defined

$$|E_n| = \sqrt{2eBn + m^2},$$

$$N_n = \frac{1}{\sqrt{2|E_n|(|E_n| + m)}},$$

$$I(n, p, x) = \left(\frac{eB}{\pi} \right)^{\frac{1}{4}} \frac{1}{\sqrt{2^n n!}} H_n \left(\sqrt{eB}(x - \frac{p}{eB}) \right) e^{-\frac{eB}{2}(x - \frac{p}{eB})^2},$$

$$I(n = -1, p, x) = 0. \qquad (11.54)$$

Here H_n denotes the Hermite polynomials of order n. It is interesting to note from the solutions in (11.53) that, for $n \neq 0$, the energy levels are doubly degenerate whereas for $n = 0$, the two states are nondegenerate and have the forms

$$\psi^+ = e^{-i(|E_0|t - py)} \begin{pmatrix} I(0, p, x) \\ 0 \\ 0 \\ 0 \end{pmatrix},$$

$$\psi^- = e^{i(|E_0|t - py)} \begin{pmatrix} 0 \\ 0 \\ I(0, -p, x) \\ 0 \end{pmatrix}. \qquad (11.55)$$

This is, of course, beyond the degeneracy in the momentum variable $p_y = p$.

Let us next note that we can expand the second quantized spinor field operator in the complete basis provided by the solutions in (11.53)-(11.55) as

$$\psi(\mathbf{x}, t) = \sum_n \sum_{i=1,2}{}' \int \frac{dp}{\sqrt{2\pi}} \left(a_i(n, p)\psi_i^+ + b_i^\dagger(n, p)\psi_i^- \right), \quad (11.56)$$

where the prime stands for the fact that the second sum is only for the two degenerate $n \neq 0$ terms. The fermionic creation and annihilation operators satisfy the standard anti-commutation relations and the nontrivial ones are given by

$$\left[a_i(n, p), a_j^\dagger(n', p') \right]_+ = \delta_{ij}\delta_{nn'}\delta(p - p') = \left[b_i(n, p), b_j^\dagger(n', p') \right]_+.$$
$$(11.57)$$

Using the field expansion in (11.56) and the anti-commutation relations in (11.57), we note that the only bilinears which would give rise to nontrivial vacuum expectation values are

$$\langle 0|a_i(n, p)a_j^\dagger(n', p')|0\rangle = \delta_{ij}\delta_{nn'}\delta(p - p') = \langle 0|b_i(n, p)b_j^\dagger(n', p')|0\rangle.$$
$$(11.58)$$

As a result, separating the sum into $n = 0$ and $n \neq 0$, we can now calculate the vacuum expectation value of the fermion condensate in the presence of a constant, external magnetic field which takes the form

$$\langle 0|\overline{\psi}(\mathbf{x}, t)\psi(\mathbf{x}, t)|0\rangle = -\frac{m}{|m|}\frac{eB}{2\pi} - \frac{meB}{\pi}\sum_{n=1}^{\infty}\frac{1}{|E_n|}. \quad (11.59)$$

Here, the sum in the second term leads to the Hurwitz ζ-function and, as is clear from the form of $|E_n|$ in (11.54), this reduces to the Riemann ζ-function, $\zeta_R(\frac{1}{2})$, when $m \to 0$. The Riemann zeta function,[4] $\zeta_R(s)$ is an analytic function in the entire complex s-plane except for a simple pole at $s = 1$. In fact, for $s = \frac{1}{2}$, the value is known to be $\zeta_R(\frac{1}{2}) = -1.46$ which is finite. Thus, we see that in the

---

[4]See, for example, chapter **17** in A. Das, *Field Theory: A Path Integral Approach* (third edition), World Scientific Publishing (2020).

limit $m \rightarrow 0$, the second term in (11.59) would vanish because of the factor m in the numerator. As a result, we see that

$$\lim_{m \rightarrow 0} \langle 0|\overline{\psi}(\mathbf{x}, t)\psi(\mathbf{x}, t)|0\rangle = -\text{sgn}(m)\frac{eB}{2\pi}. \qquad (11.60)$$

This analysis shows that, in the limit when $m \rightarrow 0$, the theory develops a nonzero fermion condensate in the presence of a constant, external magnetic field. As we have discussed earlier, such a nonzero condensate would break the $U(2)$ flavor symmetry of the massless theory down to $U(1) \times U(1)$. A little bit more analysis also would show that even though the value of the fermion condensate is nonzero, it does not imply a mass for the fermions. The phenomenon, nonetheless, is interesting for the following reason. Dynamical mass generation in fermionic theories has been of great interest for a variety of reasons. A particular model which displays such a phenomenon in $3 + 1$ dimensions is known as the Nambu-Jona-Lasinio model (we will discuss this model in detail in section **14.3**) and involves quartic fermion interactions. This is a nonrenormalizable model in $3 + 1$ dimensions, but can be thought of as a low energy limit of a renormalizable theory and it is known that when the strength of the four-fermion interaction exceeds a critical value, the theory has dynamical generation of mass for the fermions. However, as we have shown above, if, in $2 + 1$ dimensions, the fermions were to interact with a constant, external magnetic field in addition to the four-fermion interaction, then there would be a nonzero fermion condensate independent of the strength of the four-fermion interaction. This, in turn, would lead to a mass for the fermions (namely, when the four-fermion terms are (expanded) shifted around the nonzero condensate, they would give rise to a fermion mass term) for any value of the four fermion interaction strength. One says that the presence of an external magnetic field catalyses the dynamical generation of mass in this theory.

All of the above analysis is, of course, at zero temperature and it is interesting to ask what effect would temperature have on such a phenomenon. This question is easily analyzed using the operator formalism of thermofield dynamics of chapter **3**, although other approaches also lead to equivalent results. Let us recall that in thermofield dynamics, for every field, we have to introduce a tilde field and following (3.45) or (3.103), we define a thermal doublet field operator $\Psi(\mathbf{x}, t)$, for the present case, as

$$\Psi(\mathbf{x}, t) = \begin{pmatrix} \psi(\mathbf{x}, t) \\ \widetilde{\psi}^\dagger(\mathbf{x}, t) \end{pmatrix}. \tag{11.61}$$

We can now expand this thermal doublet operator in terms of the complete basis of fermionic wave functions given in (11.53)-(11.54) (see also (11.56)) as

$$\Psi(\mathbf{x}, t) = \sum_n \sum_{i=1,2}{}' \int \frac{dp}{\sqrt{2\pi}} \left[\begin{pmatrix} a_i(n, p) \\ \widetilde{a}_i^\dagger(n, p) \end{pmatrix} \psi_i^+ + \begin{pmatrix} b_i^\dagger(n, p) \\ \widetilde{b}_i(n, p) \end{pmatrix} \psi_i^- \right]. \tag{11.62}$$

To construct the thermal vacuum and the temperature dependent field operators, we can introduce a thermal Bogoliubov transformation, following the discussions of chapter **3**, as

$$U(\theta) = e^{iQ(\theta)}, \tag{11.63}$$

where the form of $Q(\theta)$ can be determined (see (3.35)-(3.36)) to be

$$Q(\theta) = i \sum_n \sum_{i=1,2}{}' \int dp \, [\theta_i^+(n)(\widetilde{a}_i a_i - a_i^\dagger \widetilde{a}_i^\dagger) + \theta_i^-(n)(\widetilde{b}_i b_i - b_i^\dagger \widetilde{b}_i^\dagger)],$$

with

$$\sin^2 \theta_i^+(n) = n_F(|E_n|),$$

$$\sin^2 \theta_i^-(n) = 1 - n_F(-|E_n|),$$

$$n_F(|E_n|) = \frac{1}{e^{\beta(|E_n| - \mu)} + 1}, \tag{11.64}$$

where we have introduced a chemical potential for generality.

Following the discussion in (3.37) or (3.68), we can now define the thermal vacuum as

$$|0, \beta\rangle = U(\theta)|0, \widetilde{0}\rangle. \tag{11.65}$$

Similarly, we can obtain the thermal creation and annihilation operators through the Bogoliubov transformation as (see (3.46))

$$\begin{pmatrix} a_i^\beta(n, p) \\ \widetilde{a}_i^{\beta\dagger}(n, p) \end{pmatrix} = \begin{pmatrix} \cos \theta_i^+(n) & -\sin \theta_i^+(n) \\ \sin \theta_i^+(n) & \cos \theta_i^+(n) \end{pmatrix} \begin{pmatrix} a_i(n, p) \\ \widetilde{a}_i^\dagger(n, p) \end{pmatrix},$$

and, conversely,

$$\begin{pmatrix} a_i(n,p) \\ \tilde{a}_i^\dagger(n,p) \end{pmatrix} = \begin{pmatrix} \cos\theta_i^+(n) & \sin\theta_i^+(n) \\ -\sin\theta_i^+(n) & \cos\theta_i^+(n) \end{pmatrix} \begin{pmatrix} a_i^\beta(n,p) \\ \tilde{a}_i^{\beta\dagger}(n,p) \end{pmatrix}, \quad (11.66)$$

where $\sin\theta_i^+$ (and, therefore, $\cos\theta_i^+$) is defined in (11.64). Similarly, the relations for the operators b_i and $\tilde{b}_i$ can be obtained by letting $\theta_i^+ \to \theta_i^-$.

With all these as well as the action of the thermal creation and annihilation operators on the thermal vacuum (discussed in chapter **3**, see, in particular (3.52)), we can now calculate the value of the fermion condensate in the thermal vacuum to be

$$\langle 0, \beta | \overline{\psi}(\mathbf{x},t)\psi(\mathbf{x},t) | 0, \beta \rangle$$

$$= \sum_n \sum_{i=1,2}{}' \int \frac{dp}{2\pi} \left(\sin^2\theta_i^+(n)\overline{\psi}_i^+ \psi_i^+ + \cos^2\theta_i^-(n)\overline{\psi}_i^- \psi_i^- \right)$$

$$= \frac{m}{|m|} \frac{eB}{2\pi} \left(1 - \frac{1}{e^{\beta(|m|-\mu)}+1} - \frac{1}{e^{\beta(|m|+\mu)}+1} \right)$$

$$- \frac{meB}{\pi} \sum_{n=1}^{\infty} \frac{1}{|E_n|} \left(1 - \frac{1}{e^{\beta(|E_n|-\mu)}+1} - \frac{1}{e^{\beta(|E_n|+\mu)}+1} \right).$$

$$(11.67)$$

It is interesting to note from this equation that for any finite temperature (finite β),

$$\lim_{m\to 0} \langle 0, \beta | \overline{\psi}(\mathbf{x},t)\psi(\mathbf{x},t) | 0, \beta \rangle = 0, \quad (11.68)$$

which follows from the algebraic identity

$$\frac{1}{e^{-\beta\mu}+1} + \frac{1}{e^{\beta\mu}+1} = \frac{e^{\beta\mu}}{e^{\beta\mu}+1} + \frac{1}{e^{\beta\mu}+1} = 1.$$

In other words, as soon as the system is put in contact with a heat bath, the phase transition breaking the $U(2)$ flavor symmetry disappears. Namely, the dynamical breaking of the symmetry in the presence of an external magnetic field is extremely unstable to the introduction of temperature.

It is more interesting to note from (11.67) that the condensate develops a nonanalytic structure with the introduction of temperature. In fact, let us, for simplicity, set $\mu = 0$ and take the limits

$$m \to 0, \quad \beta \to \infty, \quad |m|\beta = \alpha. \tag{11.69}$$

In this zero temperature limit, the value of the condensate in (11.67) becomes

$$\langle 0, \beta|\overline{\psi}(\mathbf{x}, t)\psi(\mathbf{x}, t)|0, \beta\rangle = -\mathrm{sgn}(m)\frac{eB}{2\pi}\left(1 - \frac{2}{e^\alpha + 1}\right). \tag{11.70}$$

This explicitly shows that the limits $m \to 0$ and $\beta \to \infty$ (or, $T \to 0$) are not commutative, the value of the condensate depends on how we approach the origin in the (m, T)-plane. Namely, the value of the condensate is not analytic at the origin in the (m, T)-plane. This is again similar to the nonanalyticity at the origin in the $(p^0, \mathbf{p})$ plane which we have discussed earlier. To understand this nonanalyticity a little better, let us again set $\mu = 0$. In this case, we note from (11.64) that

$$\theta_i^+(n) = \theta_i^-(n). \tag{11.71}$$

In fact, there is only one $\theta(n)$, in this case, satisfying (see, for example, (3.52))

$$\tan\theta(n) = e^{-\frac{\beta|E_n|}{2}}, \tag{11.72}$$

and we can write the generator of the Bogoliubov transformation, in this case, as

$$Q(\theta) = i\sum_n \sum_{i=1,2}{}'\int \mathrm{d}p\,\theta(n)(\tilde{a}_i a_i - a_i^\dagger \tilde{a}_i^\dagger + \tilde{b}_i b_i - b_i^\dagger \tilde{b}_i^\dagger). \tag{11.73}$$

For $m \neq 0$, this, indeed, has the right zero temperature behavior, namely, we see from (11.72) that when $\beta \to \infty$ ($T \to 0$),

$$\tan\theta(n) = 0, \implies \theta(n) = 0, \tag{11.74}$$

so that $U(\theta) = \mathbb{1}$ (see, for example, (11.66)). On the other hand, if we set $m = 0$ and then take the limit $\beta \to \infty$, we note from (11.72) that all the $\theta(n)$'s vanish except

$$\tan\theta(n = 0) = 1, \implies \theta(n = 0) = \frac{\pi}{4}. \tag{11.75}$$

Namely, it is the ground state that contributes to the condensate and we see that, for this state, θ has the unique value of $\frac{\pi}{4}$. The Bogoliubov transformation, of course, produces a squeezed state which, for this particular value of the parameter, behaves like a crossed polarizer leading to the vanishing of the condensate in the zero mass limit (see section **3.9.3** for a detailed discussion on the squeezed states).

Let us also note that in the general limit given in (11.69), all the $\theta(n)$'s can be seen to vanish from (11.72) except the one corresponding to the ground state which takes the value

$$\tan \theta(n = 0) = e^{-\frac{\alpha}{2}}, \quad \Longrightarrow \theta(n = 0) = \theta(\alpha). \tag{11.76}$$

Namely, we see that the parameter θ itself becomes nonanalytic and, consequently, the generator of the Bogoliubov transformation as well as the transformation itself become nonanalytic at the origin of the (m, T) plane. The Bogoliubov transformation does not necessarily reduce to the Identity operator in the zero temperature limit – rather, the structure depends on how this limit is taken. It is clear now that, because the Bogoliubov transformation develops a nonanalytic structure, not only the condensate, but most other observables are also likely to inherit this nonanalytic structure.

11.6 References

T. Appelquist *et al*, Physical Review **D33**, 3704 (1986).

P. F. Bedaque and A. Das, Physical Review **D45**, 2906 (1992).

P. F. Bedaque and A. Das, Physical Review **D47**, 601 (1993).

A. Das, *Field Theory: A Path Integral Approach* (third edition), World Scientific Publishing (2020).

A. Das, *Lectures in Quantum Field Theory* (second edition), World Scientific Publishing (2021).

A. Das and M. Hott, Physical Review **D53**, 2252 (1996).

A. L. Fetter and J. D. Walecka, *Quantum Theory of Many Particle Systems*, McGraw-Hill Book company (1971).

Y. Fujimoto and H. Yamada, Zeit für Physik **C37**, 265 (1988).

P. S. Gribosky and B. R. Holstein, Zeit für Physik **C47**, 205 (1990).

V. P. Gusynin *et al*, Physical Review Letters **73**, 3499 (1994).

K. G. Klimenko, Zeit für Physik **C54**, 323 (1992).

M. Kobayashi and M. Sakamoto, Progress in Theoretical Physics **70**, 1375 (1983).

H. Weldon, Physical Review **D47**, 594 (1993).

CHAPTER 12

Supersymmetry at $T \neq 0$

12.1 Supersymmetry

As we have seen in chapter **9**, most symmetries which are spontaneously broken at zero temperature are restored at some finite critical temperature. That is, field theories with spontaneously broken symmetries, behave like magnets where the magnetization vanishes beyond the Curie temperature restoring rotational invariance. There are, of course, some exceptions and one such case that we discussed earlier is a system with a dynamically broken symmetry. There is another fundamental symmetry, known as supersymmetry, that is not restored at finite temperature either – in fact, it is spontaneously broken at finite temperature. Since supersymmetry is quite fundamental in the study of various grand unified theories as well as in the study of various cosmological questions, we will discuss supersymmetry and its behavior at finite temperature in some detail in this chapter.

Most ordinary symmetries are based on Lie Groups and the multiplets of these symmetry groups consist of particles of the same kind, namely, either bosons or fermions. Therefore, most ordinary symmetry transformations map either bosons into bosons or fermions into fermions. Supersymmetry, on the other hand, is a symmetry which transforms bosons into fermions and *vice versa*. This is the ultimate form of symmetry that one can imagine in the sense that it would imply that bosons and fermions are different manifestations of a single entity.

Since supersymmetry transformations take bosons into fermions and *vice versa*, it is clear that, unlike ordinary symmetries, the generators of supersymmetry will be fermionic in character. The algebra of these generators, which would involve anti-commutators, would clearly give rise to a generator which would have a bosonic character and, consequently, would satisfy commutation relations. As a result, we see that the complete algebra of supersymmetry would correspond to a graded algebra which would involve both commutators and anti-commutators. To see the simplest of such algebras, let us assume that Q and $\overline{Q}$ are two fermionic conserved charges (both adjoints of each other) in a quantum mechanical theory described by a quantum mechanical Hamiltonian, H. Then, the simplest graded algebra that we can write down, involving these three generators, has the form

$$[H, H] = 0,$$
$$[Q, Q]_{+} = 0 = \left[\overline{Q}, \overline{Q}\right]_{+},$$
$$[Q, \overline{Q}]_{+} = H. \tag{12.1}$$

As we will see, this is the algebra satisfied by the conserved charges in the supersymmetric, non-relativistic quantum mechanics. We can also easily check in this simple algebra that the super-Jacobi identity defined by

$$(-1)^{|A||C|} [[A, B], C] + (-1)^{|B||C|} [[C, A], B]$$
$$+ (-1)^{|A||B|} [[B, C], A] = 0, \tag{12.2}$$

is satisfied. Here $|\phi|$ ($\phi = (A, B, C)$) represents the Grassmann parity of the variable ϕ (which takes the value 0 for bosons and 1 for fermions) and the bracket for any two variables is defined to be

$$[A, B] = AB - (-1)^{|A||B|} BA. \tag{12.3}$$

The relativistic supersymmetric field theories, on the other hand, are realizations of the graded Poincaré algebra. Let P_μ and $M_{\mu\nu}$ represent the usual translation and Lorentz generators of the Poincaré algebra. In addition, if we have two fermionic generators represented by the Majorana spinors, Q_α and $\overline{Q}_\alpha$ (both are adjoints of each other), $\alpha = 1, 2, 3, 4$, then, the graded Poincaré algebra has the additional relations (*i.e.*, in addition to the usual Poincaré algebra relations)

$$[P_\mu, Q_\alpha] = 0 = [P_\mu, \overline{Q}_\alpha],$$

$$[M_{\mu\nu}, Q_\alpha] = -\frac{1}{2}(\sigma_{\mu\nu})_{\alpha\beta}Q_\beta,$$

$$[M_{\mu\nu}, \overline{Q}_\alpha] = -\frac{1}{2}(\sigma_{\mu\nu})_{\alpha\beta}\overline{Q}_\beta,$$

$$[Q_\alpha, \overline{Q}_\beta]_+ = (\gamma^\mu)_{\alpha\beta}P_\mu, \tag{12.4}$$

where γ^μ's represent the Dirac matrices and

$$\sigma_{\mu\nu} = \frac{i}{2}\,[\gamma_\mu, \gamma_\nu]\,. \tag{12.5}$$

The relations, in (12.4), can be easily understood as follows. The first relation merely says that supersymmetry transformations commute with translations so that supersymmetric theories will be translation invariant. The second and the third give the transformation properties of Q_α and $\overline{Q}_\alpha$ under Lorentz transformations, namely, they transform as spinors. The last, on the other hand, represents the fact that two supersymmetry transformations are equivalent to a translation. (One can also assume the super charges, Q_α and $\overline{Q}_\alpha$, carry an internal symmetry index which would, then, lead to an extended graded algebra. However, we will not go into this.) The representations of the supersymmetry algebra can be easily worked out much the same way as is done for ordinary Lie algebras and consist of multiplets containing bosons and fermions of degenerate mass (or energy in the non-relativistic case). It is in this sense that supersymmetry unifies the two different kinds of particles known in nature. The supersymmetry generators acting on a bosonic (fermionic) state take it to a fermionic (bosonic) state of same mass (energy). This is easily seen from (12.4) as follows (it can also be seen from (12.1)). Let $|\psi\rangle$ represent a state with energy-momentum p^μ. Thus,

$$P^\mu|\psi\rangle = p^\mu|\psi\rangle,$$

$$P^\mu(Q_\alpha|\psi\rangle) = Q_\alpha P^\mu|\psi\rangle = p^\mu(Q_\alpha|\psi\rangle). \tag{12.6}$$

The equality of masses for the superpartner states, $|\psi\rangle$ and $(Q_\alpha|\psi\rangle)$, then, follows from $p^2 = p^\mu p_\mu = m^2$. (An alternative way of seeing this is to note that the only Casimir operator of the graded Poincaré algebra is $P^2 = P^\mu P_\mu$ and, hence, all representations must consist of multiplets with degenerate mass.)

The supersymmetric theories are of interest for various reasons. The supersymmetric field theories have extremely well behaved ultraviolet properties. The bosonic and fermionic loop contributions in many physical processes cancel leading to a much milder divergence structure than in ordinary field theories. Supersymmetry also gives a physical reason for the existence of scalar fields – they can simply be thought of as the superpartners of fermionic fields. Furthermore, supersymmetry implies a vanishing ground state energy which is quite significant in connection with the question of the cosmological constant. The vanishing of the ground state energy can be easily seen from (12.1) (or equally well from (12.4)) as follows. If $|0\rangle$ represents the ground state, then, from (12.1), we have

$$\langle 0|Q\overline{Q} + \overline{Q}Q|0\rangle = \langle 0|H|0\rangle = E_0. \tag{12.7}$$

If supersymmetry is unbroken, then, we have

$$Q|0\rangle = 0 = \overline{Q}|0\rangle. \tag{12.8}$$

It follows from this, as well as from (12.7), that, if supersymmetry is a true symmetry of the theory, then the ground state energy must vanish, namely,

$$E_0 = 0. \tag{12.9}$$

Incidentally, we can generalize the argument in the above equation to show that the energy of any state in a supersymmetric theory is necessarily positive semi-definite.

While supersymmetry is a symmetry with a rich structure, experimentally, supersymmetry is unobserved. In fact, the spectrum of known elementary particles does not display any degeneracy in the masses of bosons and fermions. Therefore, it is clear that if supersymmetry is a symmetry of our physical theory, it must be spontaneously broken so as to maintain all the good features following from the symmetry while avoiding degenerate bosonic and fermionic states. Furthermore, we are interested in the question of whether thermal effects lead to a spontaneous breakdown of supersymmetry.

Intuitively, it is clear that since bosons and fermions obey different statistics, any symmetry that transforms bosons into fermions and *vice versa* would be broken at finite temperature. However, we would like to study the structure of this systematically. To study the phenomenon of supersymmetry breaking, we again have to look for order parameters associated with this symmetry. It is true that when supersymmetry is spontaneously broken, as in ordinary symmetries, the charges do not annihilate the vacuum

$$Q|0\rangle \neq 0, \qquad \overline{Q}|0\rangle \neq 0. \tag{12.10}$$

This, also implies, as in the case of spontaneous breaking of an ordinary symmetry (see, for example, (9.17)), that the expectation value of the transformation of a dynamical variable under supersymmetry must be nonzero. The only difference is that in this case, the dynamical variable whose transformation picks up a nonzero value must be fermionic. However, it may not always be easy to check whether the charges annihilate the ground state or not. On the other hand, if the charges do not annihilate the vacuum, then, it follows, from ((12.7) and (12.10)) that when supersymmetry is spontaneously broken, the ground state energy will be nonzero (and positive), namely,

$$E_0 > 0. \tag{12.11}$$

Thus, we can think of the ground state energy in a supersymmetric theory as an order parameter of supersymmetry breaking.

There is also an alternative order parameter of supersymmetry breaking known as the Witten index. Let us define

$$(-1)^{N_F} = e^{i\pi N_F} = 1 - 2N_F, \tag{12.12}$$

where N_F represents the fermion number operator and takes the values $0, 1$ in any state. For the bosonic states, the fermion number takes the value 0 while the value 1 holds for fermionic states. Clearly, therefore, if we define the Witten index as

$$\Delta = \langle 0|(-1)^{N_F}|0\rangle = \sum_n \langle n|(-1)^{N_F}|n\rangle = \text{Tr}(-1)^{N_F}, \tag{12.13}$$

then, this measures the difference in the number of zero energy bosonic and fermionic states in the theory. The second equality follows from the fact that all the higher energy bosonic and fermionic

states are paired giving zero to the sum. If the Witten index in
(12.13) takes a nonzero value, then, one can show that supersymmetry is unbroken. In fact, it is clear that the ground state, when
supersymmetry is unbroken, is bosonic and has no superpartner state
giving a value 1 to the index. However, when the value of the index
vanishes, supersymmetry may be spontaneously broken. For example, if the number of zero energy bosonic as well as fermionic states
vanishes, then, the Witten index will identically vanish. In this case,
however, there will be no zero energy state and, consequently, it follows from the supersymmetry algebra that supersymmetry must be
spontaneously broken. Thus, the Witten index can also be taken
as an order parameter in the study of spontaneous breakdown of
supersymmetry.

In the next section, we will study some simple quantum mechanical systems to bring out the various features in connection with
supersymmetry breaking.

12.2 Supersymmetric oscillator

Let us begin with the simplest of supersymmetric theories in this section. The Lagrangian for a general one-dimensional supersymmetric
quantum mechanical theory is given by

$$L = \frac{1}{2}\dot{q}^2 - \frac{1}{2}(f(q))^2 + i\overline{\psi}\dot{\psi} - \frac{1}{2}f'(q)\left[\overline{\psi}, \psi\right], \qquad (12.14)$$

where $f(q)$ is an arbitrary polynomial of the coordinate q and may
involve various coupling constants and other parameters of the theory. ψ and $\overline{\psi}$ are treated as completely independent anti-commuting
quantities describing the fermionic degrees of freedom of the theory.
(Supersymmetric theories normally involve some auxiliary fields for
the closure of the algebra. In writing the Lagrangian completely in
terms of dynamical variables as in (12.14), we have eliminated the
auxiliary fields. Furthermore, for certain classes of $f(q)$, it is known
that instanton solutions are present leading to a breaking of supersymmetry. In all our discussions, we will assume that $f(q)$ has the
form such that instantons are not present.) It can be easily checked
that the action obtained from the Lagrangian in (12.14) is invariant

under the following two sets of supersymmetry transformations

$$
\begin{array}{ll}
\delta q = \frac{1}{\sqrt{2}}\overline{\psi}\epsilon, & \overline{\delta} q = \frac{1}{\sqrt{2}}\overline{\epsilon}\psi, \\
\delta\psi = -\frac{i}{\sqrt{2}}\dot{q}\epsilon - \frac{1}{\sqrt{2}}f(q)\epsilon, & \overline{\delta}\psi = 0, \\
\delta\overline{\psi} = 0, & \overline{\delta}\,\overline{\psi} = \frac{i}{\sqrt{2}}\dot{q}\overline{\epsilon} - \frac{1}{\sqrt{2}}f(q)\overline{\epsilon}.
\end{array}
\tag{12.15}
$$

Here ϵ and $\overline{\epsilon}$ are two constant anti-commuting parameters of the transformations and the structure of the transformations in (12.15) shows clearly that under a supersymmetry transformation, $q \leftrightarrow \psi\,(\overline{\psi})$, namely, the bosonic and the fermionic degrees of freedom are transformed into each other.

The canonical momenta can be obtained from the Lagrangian in Eq. ((12.14)) and the Hamiltonian can be constructed to be

$$
H = \frac{1}{2}\left(p^2 + (f(q))^2 + f'(q)\left[\overline{\psi}, \psi\right]\right),
\tag{12.16}
$$

and the fundamental nontrivial commutation and anti-commutation relations of the theory are given by ($\hbar = 1$)

$$
[q, p] = i, \qquad \left[\psi, \overline{\psi}\right]_+ = 1.
\tag{12.17}
$$

The supersymmetry charges can now be constructed to be

$$
Q = \frac{1}{\sqrt{2}}(p + if(q))\overline{\psi}, \qquad \overline{Q} = \frac{1}{\sqrt{2}}(p - if(q))\psi,
\tag{12.18}
$$

and it can be easily checked with the (anti) commutation relations of Eq. ((12.17)) that these charges, indeed, generate the supersymmetry transformations of (12.15). Furthermore, Q, $\overline{Q}$ and H can also be easily seen through the (anti) commutation relations of (12.17) to satisfy the supersymmetry algebra of Eq. ((12.1)).

Let us next specialize to the case where $f(q)$ is a linear function of the bosonic coordinate q, namely,

$$
f(q) = \omega q.
\tag{12.19}
$$

In this case, the Hamiltonian of the system given in (12.16) becomes

$$
H = \frac{1}{2}\left(p^2 + \omega^2 q^2 + \omega\left[\overline{\psi}, \psi\right]\right),
\tag{12.20}
$$

and corresponds to the Hamiltonian for the supersymmetric oscillator (with $m = 1$). To see this, let us define, as usual, the bosonic and the fermionic creation and annihilation operators as

$$a_B = \frac{1}{\sqrt{2\omega}}(p - i\omega q), \quad a_B^\dagger = \frac{1}{\sqrt{2\omega}}(p + i\omega q),$$
$$a_F = \overline{\psi}, \qquad\qquad a_F^\dagger = \psi,$$

(12.21)

and we can derive the nontrivial canonical (anti-) commutation relations between these operators to be

$$\left[a_B, a_B^\dagger\right] = 1 = \left[a_F, a_F^\dagger\right]_+ .$$

(12.22)

The Hamiltonian of (12.20) can then be written in terms of the creation and the annihilation operators as

$$H = \omega(a_B^\dagger a_B + a_F^\dagger a_F),$$

(12.23)

which is, indeed, the Hamiltonian for the supersymmetric oscillator. The supersymmetry charges, in this case, take the form (see (12.18), (12.19) and (12.21))

$$Q = \sqrt{\omega}\, a_B^\dagger a_F, \qquad \overline{Q} = \sqrt{\omega}\, a_F^\dagger a_B.$$

(12.24)

Once again, we can check using the (anti-) commutation relations of (12.22) that Q, $\overline{Q}$ and H of ((12.24)) and ((12.23)) satisfy the supersymmetry algebra of (12.1).

The Hilbert space of the supersymmetric oscillator can be worked out in a straightforward manner. Let us define the bosonic and the fermionic number operator as usual as

$$N_B = a_B^\dagger a_B, \qquad N_F = a_F^\dagger a_F,$$

(12.25)

so that the Hamiltonian of (12.23) can also be written as

$$H = \omega(N_B + N_F).$$

(12.26)

The eigenvalues of the number operators can be determined in the standard manner to take the values

$$n_B = 0, 1, 2, \cdots , \qquad n_F = 0, 1.$$

(12.27)

The eigenstates of the Hamiltonian, $|n_B, n_F\rangle$, can now be seen to satisfy

$$H|n_B, n_F\rangle = \omega(n_B + n_F)|n_B, n_F\rangle, \tag{12.28}$$

with n_B and n_F taking the values of (12.27). The ground state of the theory, as usual, satisfies

$$a_B|0\rangle = 0 = a_F|0\rangle. \tag{12.29}$$

The higher energy states are, then, explicitly determined to have the form

$$|n_B, n_F\rangle = \frac{(a_B^\dagger)^{n_B}}{\sqrt{n_B!}}(a_F^\dagger)^{n_F}|0\rangle. \tag{12.30}$$

Let us now note from (12.27) and (12.28) that, for any state $|n_B, n_F\rangle$,

$$\langle n_B, n_F|H|n_B, n_F\rangle = \omega(n_B + n_F) \geq 0, \tag{12.31}$$

as it should be for a supersymmetric theory. In particular, we note that

$$\langle 0|H|0\rangle = 0. \tag{12.32}$$

That is, the ground state energy, unlike the case of a bosonic or a fermionic oscillator, vanishes. This would further emphasize that supersymmetry is an unbroken symmetry in this theory. In fact, we can check explicitly from the definition of (12.24) as well as the ground state condition of (12.29) that

$$Q|0\rangle = 0 = \overline{Q}|0\rangle. \tag{12.33}$$

We note from (12.27) and (12.28) that, except for the ground state, all the energy eigenstates are doubly degenerate. In fact, we see that the effect of the supercharges acting on the energy eigenstates are given by (for $n_B \neq 0$)

$$Q|n_B, 1\rangle = \sqrt{\omega(n_B + 1)}|n_B + 1, 0\rangle,$$

$$Q|n_B, 0\rangle = 0,$$

$$\overline{Q}|n_B, 0\rangle = \sqrt{\frac{\omega}{n_B}}|n_B - 1, 1\rangle,$$

$$\overline{Q}|n_B, 1\rangle = 0. \tag{12.34}$$

It is now clear from (12.31) that the superpartner states $|n_B, 0\rangle$ and $|n_B - 1, 1\rangle$ (for $n_B \neq 0$) are degenerate in energy as we would expect in a supersymmetric theory. The Witten index can also be calculated for this theory and has the form

$$\Delta = \langle 0|(1 - 2N_F)|0\rangle = 1 = \text{Tr}(1 - 2N_F). \tag{12.35}$$

Thus, we see that all the order parameters, in the case of the supersymmetric oscillator, indicate that supersymmetry is a true symmetry of this theory at zero temperature.

Let us next analyze this model at finite temperature. Since we are interested in the structure of the Hilbert space and the action of various operators on the space of states, in this case, it is more convenient to use the formalism of thermofield dynamics discussed in chapter **3**. As we have seen there, in this formalism one doubles the degrees of freedom by introducing "tilde" variables (see, for example, ((3.22)) and ((3.54))). The combined Hamiltonian which governs the dynamics of the "tilde" and the "non-tilde" systems, in this case, can be written as

$$\hat{H} = H - \widetilde{H}$$

$$= m(a_B^\dagger a_B + a_F^\dagger a_F) - m(\widetilde{a}_B^\dagger \widetilde{a}_B + \widetilde{a}_F^\dagger \widetilde{a}_F). \tag{12.36}$$

We assume, as was discussed in chapter **3**, that the "tilde" operators satisfy the same commutation (anti-commutation) relations as the original operators and that the "tilde" and the "non-tilde" operators commute (anti-commute) among each other.

The states, in this doubled Hilbert space can be labelled as

$$|0\rangle = |0\rangle \otimes |\widetilde{0}\rangle,$$

$$|n_B, n_F; \widetilde{n}_B, \widetilde{n}_F\rangle = |n_B, n_F\rangle \otimes |\widetilde{n}_B, \widetilde{n}_F\rangle. \tag{12.37}$$

We can now construct the thermal vacuum following the discussion of chapter **3** (see (3.35)-(3.37) as well as (3.68)) as

$$|0, \beta\rangle = U(\theta)|0\rangle = e^{-iG(\theta)}|0\rangle, \tag{12.38}$$

where the generator of the Bogoliubov transformation is given by (see (3.38) as well as (3.69))

$$G(\theta) = -i\theta_F(\beta)(\widetilde{a}_F a_F - a_F^\dagger \widetilde{a}_F^\dagger) - i\theta_B(\beta)(\widetilde{a}_B a_B - a_B^\dagger \widetilde{a}_B^\dagger),$$

where

$$\tan \theta_F(\beta) = e^{-\beta\omega/2} = \tanh \theta_B(\beta). \tag{12.39}$$

(Note that we are denoting, in this chapter, the generator of the Bogoliubov transformation by $G(\theta)$, which is different from that in chapter **3**, simply because conventionally, the supersymmetric charges are represented by Q.)

Once the Bogoliubov transformation defining the thermal vacuum has been determined, we can obtain the thermal operators corresponding to any given zero temperature operator A simply as

$$A(\beta) = U(\theta)AU^\dagger(\theta).$$

In particular, this determines the thermal creation and annihilation operators as the linear combinations

$$a_B(\beta) = a_B \cosh \theta_B - \widetilde{a}_B^\dagger \sinh \theta_B,$$

$$\widetilde{a}_B(\beta) = \widetilde{a}_B \cosh \theta_B - a_B^\dagger \sinh \theta_B,$$

$$a_F(\beta) = a_F \cos \theta_F - \widetilde{a}_F^\dagger \sin \theta_F,$$

$$\widetilde{a}_F(\beta) = \widetilde{a}_F \cos \theta_F + a_F^\dagger \sin \theta_F, \tag{12.40}$$

and their hermitian conjugates. Of course, these relations can be inverted and the zero temperature operators can also be expressed in terms of the thermal creation and annihilation operators as

$$a_B = a_B(\beta) \cosh \theta_B - \widetilde{a}_B^\dagger(\beta) \sinh \theta_B,$$

$$\widetilde{a}_B = \widetilde{a}_B(\beta) \cosh \theta_B - a_B^\dagger(\beta) \sinh \theta_B,$$

$$a_F = a_F(\beta) \cos \theta_F + \widetilde{a}_F^\dagger(\beta) \sin \theta_F,$$

$$\widetilde{a}_F = \widetilde{a}_F(\beta) \cos \theta_F - a_F^\dagger(\beta) \sin \theta_F. \tag{12.41}$$

The thermal operators also satisfy the same commutation (anticommutation) relations as the zero temperature ones since the transformation connecting the two is unitary which can be explicitly checked. Furthermore, the thermal Hilbert space can be constructed from the thermal vacuum by the operation of the thermal creation operators. We note, in particular, that the thermal vacuum is normalized to unity since the Bogoliubov transformation is a unitary transformation (in general, at least formally). Furthermore, the thermal vacuum is annihilated by all the thermal annihilation operators, namely,

$$a(\beta)|0, \beta\rangle = 0, \tag{12.42}$$

where $a(\beta)$ can be any one from the set $(a_B(\beta), \tilde{a}_B(\beta), a_F(\beta), \tilde{a}_F(\beta))$. We are now ready to calculate various statistical averages of interest in this theory. First, we note that the energy of the thermal vacuum is given by

$$
\begin{aligned}
E_0(\beta) &= \langle 0, \beta|H|0, \beta\rangle = \langle 0, \beta|\omega(a_B^\dagger a_B + a_F^\dagger a_F)|0, \beta\rangle \\
&= \omega(\sinh^2 \theta_B + \sin^2 \theta_F) \\
&= \omega\left(\frac{e^{-\beta\omega}}{1 - e^{-\beta\omega}} + \frac{e^{-\beta\omega}}{1 + e^{-\beta\omega}}\right),
\end{aligned} \tag{12.43}
$$

where we have used ((12.41)) and ((12.42)) as well as the commutation (anti-commutation) relations of the thermal creation and annihilation operators.

There are several things to note from (12.43). First, for $T \to 0$ or equivalently, $\beta \to \infty$, $E_0 \to 0$ as we would expect at zero temperature. At any finite temperature, however, the thermal energy is nonzero and positive. In fact, it is simply the sum of the thermal energies for a bosonic and a fermionic oscillator system of the same frequency. This is the first indication that supersymmetry may be spontaneously broken at finite temperature. In fact, since there is no state with zero energy, from the discussion following (12.13), we expect the Witten index to vanish corresponding to the supersymmetry broken phase. However, we also point out that in this theory, the value of the Witten index can be explicitly calculated and leads to the following result

$$
\begin{aligned}
\Delta(\beta) &= \langle 0, \beta|(1 - 2N_F)|0, \beta\rangle = \langle 0, \beta|(1 - 2a_F^\dagger a_F)|0, \beta\rangle \\
&= (1 - 2\sin^2 \theta_F) \\
&= \frac{1 - e^{-\beta\omega}}{1 + e^{-\beta\omega}}.
\end{aligned} \tag{12.44}
$$

We note here that for $T \to 0$ ($\beta \to \infty$), $\Delta(\beta) \to 1$ consistent with (12.35). Furthermore, for $T \to \infty$ ($\beta \to 0$), $\Delta(\beta) \to 0$ consistent with the fact that supersymmetry is spontaneously broken at very high temperatures. At any finite temperature, however, (12.44) gives a fractional value for the Witten index. This shows that at finite temperatures, $(1 - 2N_F)$ does not give a good representation for the Witten index and one should calculate this quantity more carefully. (By definition, (12.13), Witten index measures the difference in the number of zero energy bosonic and fermionic states and, therefore, must be an integer.)

In this simple model, we can, of course, explicitly check whether supersymmetry is spontaneously broken from the action of the supersymmetry charges on the thermal vacuum. In fact, we note from (12.24) that

$$Q|0, \beta\rangle = \sqrt{\omega} a_B^\dagger a_F |0, \beta\rangle$$

$$= \sqrt{\omega} \cosh \theta_B \sin \theta_F$$

$$\times |n_B(\beta) = 1, n_F(\beta) = 0; \tilde{n}_B(\beta) = 0, \tilde{n}_F(\beta) = 1\rangle$$

$$= \frac{\sqrt{\omega} e^{-\beta\omega/2}}{[(1 - e^{-\beta\omega})(1 + e^{-\beta\omega})]^{1/2}} |\chi_1(\beta)\rangle,$$

$$\overline{Q}|0, \beta\rangle = \sqrt{\omega} a_F^\dagger a_B |0, \beta\rangle$$

$$= \sqrt{\omega} \sinh \theta_B \cos \theta_F$$

$$\times |n_B(\beta) = 0, n_F(\beta) = 1; \tilde{n}_B(\beta) = 1, \tilde{n}_F(\beta) = 0\rangle$$

$$= \frac{\sqrt{\omega} e^{-\beta\omega/2}}{[(1 - e^{-\beta\omega})(1 + e^{-\beta\omega})]^{1/2}} |\chi_2(\beta)\rangle. \tag{12.45}$$

This shows explicitly that as $T \to 0$ ($\beta \to \infty$), the supercharges annihilate the vacuum consistent with our earlier discussion that at zero temperature supersymmetry is unbroken. However, at any finite temperature, we see that the supercharges do not annihilate the thermal vacuum indicating that supersymmetry is spontaneously broken at finite temperature. Intuitively, it is clear that since supersymmetry transforms bosons into fermions and since bosons and fermions respond differently to temperature (their statistics are quite distinct), supersymmetry will be broken with the introduction of a heat bath or for that matter any boundary condition which will distinguish between the two. What is interesting, however, is the nature of the breaking, namely, the fact that supersymmetry is spontaneously broken. In fact, following our discussion in chapter **6**, we recognize that the states $|\chi_1(\beta)\rangle$ and $|\chi_2(\beta)\rangle$ represent the Goldstino states at finite temperature. (The Goldstone states, in this case, are fermionic since the generators of symmetry, Q and $\overline{Q}$ are and, conventionally, such states are known as Goldstino states.)

12.3 Interacting theory

The supersymmetric oscillator is a free theory. Therefore, while the discussion of the last section clearly shows that supersymmetry is spontaneously broken in this theory at finite temperature, a question naturally arises as to whether interactions can change this result. More specifically, it will be nice to know whether in an interacting theory, the ground state energy would continue to be nonzero at finite temperature. This question is particularly significant because, as we know, in a supersymmetric theory the bosonic and fermionic contributions tend to cancel leading to a theory which is much better behaved than theories without supersymmetry. Intuitively, however, we know that we are looking for thermal contributions and they generally have a tendency to add rather than cancel. It is, therefore, of interest to analyze this question in some detail.

Let us, for simplicity, consider a one-dimensional, interacting supersymmetric quantum mechanical theory described by the Lagrangian (see (12.14))

$$L = \frac{1}{2}(\dot{q})^2 - \frac{1}{2}(\xi + \omega q + g q^2)^2 + i\overline{\psi}\dot{\psi} - \frac{1}{2}(\omega + 2gq)\left[\overline{\psi}, \psi\right]. \quad (12.46)$$

Here ξ is a constant parameter known as the Fayet-Illiopoulous parameter which can lead to supersymmetry breaking in more realistic field theories. For $\xi > 0$, supersymmetry is normally spontaneously broken in realistic field theories while when $\xi < 0$, supersymmetry is unbroken. The sign of this term, therefore, is quite crucial in realistic field theories to determine whether supersymmetry is spontaneously broken. Furthermore, g represents the strength of interaction which controls both the Yukawa as well as the quartic interactions which is a standard feature of supersymmetric theories.

We can now obtain the Hamiltonian for the interacting theory in (12.46) in a straightforward manner which takes the form

$$
\begin{aligned}
H &= \frac{p^2}{2} + \frac{1}{2}(\xi + \omega q + gq^2)^2 + \frac{1}{2}(\omega + 2gq)\left[\bar{\psi}, \psi\right] \\
&= \frac{p^2}{2} + \frac{\omega^2 q^2}{2} + \frac{\omega}{2}\left[\bar{\psi}, \psi\right] \\
&\quad + \frac{1}{2}[(\xi + gq^2)^2 + 2\omega q(\xi + gq^2)] + gq\left[\bar{\psi}, \psi\right] \\
&= H_0 + H_I.
\end{aligned}
\tag{12.47}
$$

Here H_0 is the Hamiltonian for the supersymmetric oscillator (see (12.20)) which we have already studied in the last section and H_I represents all the interactions and has the form

$$
H_I = \frac{1}{2}[(\xi + gq^2)^2 + 2\omega q(\xi + gq^2)] + gq\left[\bar{\psi}, \psi\right].
\tag{12.48}
$$

The interaction Hamiltonian, (12.48), can be rewritten in terms of the creation and the annihilation operators with the identifications in the last section (see (12.21)) and takes the form

$$
\begin{aligned}
H_I &= \frac{1}{2}\Big[(\xi - \frac{g}{2\omega}(a_B - a_B^\dagger)^2)^2 \\
&\quad + i\sqrt{2\omega}(a_B - a_B^\dagger)(\xi - \frac{g}{2\omega}(a_B - a_B^\dagger)^2)\Big] \\
&\quad + \frac{ig}{\sqrt{2\omega}}(a_B - a_B^\dagger)[a_F, a_F^\dagger].
\end{aligned}
\tag{12.49}
$$

The expectation value of this interaction Hamiltonian can now be calculated in the thermal vacuum using the relations in (12.41) and (12.42) and takes the form

$$
\begin{aligned}
\langle 0, \beta | H_I | 0, \beta \rangle &= \frac{1}{2}\left[\xi^2 + \frac{\xi g}{\omega}\frac{1 + e^{-\beta\omega}}{1 - e^{-\beta\omega}} + \frac{3g^2}{4\omega^2}\left(\frac{1 + e^{-\beta\omega}}{1 - e^{-\beta\omega}}\right)^2\right] \\
&= \frac{1}{2}\left[\left(\xi + \frac{g}{2\omega}\frac{1 + e^{-\beta\omega}}{1 - e^{-\beta\omega}}\right)^2 + \frac{g^2}{2\omega^2}\left(\frac{1 + e^{-\beta\omega}}{1 - e^{-\beta\omega}}\right)^2\right].
\end{aligned}
\tag{12.50}
$$

This shows, as we would expect, that the thermal corrections to the ground state energy are explicitly positive, independent of the sign of the parameter ξ. The thermal vacuum, to first order in perturbation, has the energy given by

$$
\begin{aligned}
E(\beta) &= E_0(\beta) + \langle 0, \beta | H_I | 0, \beta \rangle \\
&= \omega \left(\frac{e^{-\beta\omega}}{1 - e^{-\beta\omega}} + \frac{e^{-\beta\omega}}{1 + e^{-\beta\omega}} \right) \\
&\quad + \frac{1}{2} \left[\left(\xi + \frac{g}{2\omega} \frac{1 + e^{-\beta\omega}}{1 - e^{-\beta\omega}} \right)^2 + \frac{g^2}{2\omega^2} \left(\frac{1 + e^{-\beta\omega}}{1 - e^{-\beta\omega}} \right)^2 \right].
\end{aligned} \tag{12.51}
$$

It is now clear that since the thermal effects are additive, interactions do not change our earlier conclusion that supersymmetry is spontaneously broken at finite temperature. We note, in particular from Eq. ((12.51)), that at very hight temperatures, $T \to \infty$ ($\beta \to 0$),

$$
\lim_{T \to \infty} E(\beta) \to \frac{3g^2}{2\beta^2\omega^4} + O(\beta^{-1}). \tag{12.52}
$$

This shows that the dominant term in the ground state energy, at very high temperatures, is independent of the parameter ξ leading to the conclusion that independent of whether supersymmetry is broken at zero temperature or not, it must be broken at high temperatures. Although we have discussed the phenomenon of supersymmetry breaking in the context of a quantum mechanical model, it has also been studied in the context of supersymmetric quantum field theories and the conclusions remain unchanged.

12.4 Nicolai map

The structure of supersymmetric theories is so restrictive that it leads to a very interesting map in such theories known as the Nicolai map. Very briefly, Nicolai map tells us that in a supersymmetric theory, it is possible to integrate out the fermions completely and absorb the resulting determinant into a redefinition of the bosonic fields such that the residual bosonic action can be written as a simple Gaussian. This field redefinition cannot always be written in a closed form except in quantum mechanical models. In higher dimensional field theories, on the other hand, it can be constructed iteratively.

Among other things, Nicolai map shows why the ground state energy in a supersymmetric theory has to vanish for all values of the coupling in the path integral formalism. Such a map exists and is meaningful only if the theory is supersymmetric. As a result, the existence of a Nicolai map becomes a measure of supersymmetry (order parameter). In this section, we will see that even the Nicolai map breaks down at finite temperature consistent with our earlier observation that supersymmetry is broken by thermal effects.

Before discussing the Nicolai map, let us note some identities involving the scalar and fermion propagators in quantum field theories. We know that, for a fermionic and a scalar field of the same mass m,

$$S(p, m) = \frac{1}{\not{p} - m},$$

$$D(p, m) = \frac{1}{p^2 - m^2}. \tag{12.53}$$

From the structure of these Greens functions, it follows now that

$$D(p, m) = -\frac{1}{D} \operatorname{Tr} S(p, m) S(-p, m), \tag{12.54}$$

where D denotes the number of space-time dimensions and "Tr" stands for the trace over the Dirac indices of the fermion fields. It is worth noting here that the identity in (12.53) holds even when $m = 0$ and that for a $0 + 1$ dimensional field theory, namely, quantum mechanics, the fermions do not have any Dirac index and consequently, the identity holds without the trace in the right hand side.

The identity in (12.54) can also be written in several alternate ways in the coordinate space as

$$D(x - x') = -\frac{1}{D} \operatorname{Tr} \int d^D x'' \, S(x - x'') S(x' - x''),$$

$$D^{-1}(x - x') = -\frac{1}{D} \operatorname{Tr} \int d^D x'' \, S^{-1}(x - x'') S^{-1}(x' - x''),$$

$$\delta^D(x - x') = -\frac{1}{D} \operatorname{Tr} \int d^D x_1 d^D x_2 \, S(x_1 - x) D^{-1}(x_1 - x_2)$$

$$\times S(x_2 - x'). \tag{12.55}$$

As we know, a supersymmetric theory necessarily involves bosons and fermions of the same mass and the identities in (12.54) and (12.55) are crucial in establishing the Nicolai map in such theories.

Let us now, for simplicity, look at the supersymmetric quantum mechanical theory of (12.14)

$$
\begin{aligned}
I &= \int \mathrm{d}t \left(\tfrac{1}{2}((\dot{q})^2 - (f(q))^2) + i\bar{\psi}\dot{\psi} - f'(q)\bar{\psi}\psi\right) \\
&= \int \mathrm{d}t \mathrm{d}t' \left(\tfrac{1}{2}(q(t')D^{-1}(t'-t)q(t) - \delta(t-t')(f(q))^2)\right) \\
&\quad + \bar{\psi}(t')S^{-1}(t'-t)\psi(t) - \delta(t-t')f'(q)\bar{\psi}\psi). \qquad (12.56)
\end{aligned}
$$

Here, we have identified

$$
S^{-1}(t-t') = i\frac{\mathrm{d}}{\mathrm{d}t}\,\delta(t-t'),
$$

$$
D^{-1}(t-t') = -\frac{\mathrm{d}^2}{\mathrm{d}t^2}\,\delta(t-t'), \qquad (12.57)
$$

corresponding to massless propagators assuming that if mass terms are present, they are all included in the interactions.

The integration over the fermionic variables can now be done in the path integral (the Lagrangian is bilinear in the fermionic fields) and leads to

$$
\det(S^{-1} - f'(q)) = \det S^{-1} \det(1 - Sf'(q)). \qquad (12.58)
$$

Furthermore, if we now redefine the bosonic variable as

$$
q(t) \to q'(t) = q(t) - \int \mathrm{d}t'\, S(t-t')f(q(t')), \qquad (12.59)
$$

or more compactly as

$$
q \to q' = q - Sf(q), \qquad (12.60)
$$

then, we note that

$$
\begin{aligned}
&\int \mathrm{d}t \mathrm{d}t' q'(t')D^{-1}(t'-t)q'(t) \\
&= q'D^{-1}q' = qD^{-1}q + f(q)SD^{-1}Sf(q) \\
&= \int \mathrm{d}t' \mathrm{d}t\,(q(t')D^{-1}(t'-t)q(t) - \delta(t-t')(f(q))^2). \qquad (12.61)
\end{aligned}
$$

Here we have used the identity involving the bosonic and fermionic propagators in (12.54) and (12.55) as well as the fact that for polynomials, $f(q)$, which do not support instantons, the crossed terms in (12.61) integrate to zero.

We are now ready to show that the Nicolai map exists in this theory. In fact, we note that under the field redefinition in (12.60), the bosonic part of the action in (12.56) does, indeed, become quadratic in the new variables. Moreover, the Jacobian resulting from the change of variables is given by

$$J = \left| \frac{\partial q'}{\partial q} \right| = \det(1 - Sf'(q)), \tag{12.62}$$

which is precisely the Matthews-Salam determinant resulting from the integration of the fermions in (12.58). In other words, the fermion determinant can be absorbed into the bosonic field redefinition which makes the path integral a bosonic Gaussian integral, namely,

$$
\begin{aligned}
Z = e^{iW} &= \det S^{-1} \int \mathcal{D}q \, e^{\frac{i}{2} \int dt \, q' D^{-1} q'} \\
&= \det S^{-1} \left(\det(-D^{-1}) \right)^{-1/2} \\
&= [\det(-SD^{-1}S)]^{-1/2} \\
&= 1.
\end{aligned}
\tag{12.63}
$$

This relation is quite significant in the sense that it implies $W = 0$ which leads to the fact that the ground state energy in supersymmetric theories must vanish for all values of the coupling. As we have seen, in the path integral formalism, this follows from the existence of the Nicolai map which crucially depends on the identities involving the bosonic and the fermionic propagators.

Thus, we see that the Nicolai map, of course, depends on the very special structure of bosonic and fermionic interactions in supersymmetric theories, but it also crucially depends on the identities involving bosonic and fermionic propagators of the same mass. In fact, it is clear that even in the absence of interactions, (12.63) continues to hold because of the identities in (12.54) and (12.55). We will now show that these identities no longer hold at finite temperature leading to a breakdown of the Nicolai map. We can see this, in general, without restricting ourselves to the quantum mechanical case.

We have already seen that in the real time formalism, the propagators separate into a temperature independent part and a temperature dependent part. Thus, for example, we see from (2.50) and (2.62) that the bosonic and the fermionic Greens functions can be written in the matrix form as

$$S(p, m) = (\not{p} + m)$$

$$\times \left[\begin{pmatrix} \frac{1}{p^2 - m^2 + i\epsilon} & -2i\pi\theta(-p^0)\delta(p^2 - m^2) \\ -2i\pi\theta(p^0)\delta(p^2 - m^2) & -\frac{1}{p^2 - m^2 - i\epsilon} \end{pmatrix} \right.$$

$$\left. + 2i\pi n_F(|p^0|)\delta(p^2 - m^2) \begin{pmatrix} 1 & 1 \\ 1 & 1 \end{pmatrix} \right],$$

$$D(p, m) = \begin{pmatrix} \frac{1}{p^2 - m^2 + i\epsilon} & -2i\pi\theta(-p^0)\delta(p^2 - m^2) \\ -2i\pi\theta(p^0)\delta(p^2 - m^2) & -\frac{1}{p^2 - m^2 - i\epsilon} \end{pmatrix}$$

$$- 2i\pi n_B(|p^0|)\delta(p^2 - m^2) \begin{pmatrix} 1 & 1 \\ 1 & 1 \end{pmatrix}. \tag{12.64}$$

Here n_F and n_B represent the fermionic and the bosonic distribution functions respectively, namely,

$$n_B(|p^0|) = \frac{1}{e^{\beta|p^0|} - 1}, \qquad n_F(|p^0|) = \frac{1}{e^{\beta|p^0|} + 1}.$$

It is clear now that the temperature dependent parts of the propagators carry the statistics of the respective fields and since the distribution functions are distinctly different for the two classes of the fields, the zero temperature identities of (12.54) and (12.55) do not carry over to finite temperature.

At finite temperature, the structure of the Lagrangian does not change. Consequently, the special structure of the interactions in a supersymmetric theory continues to hold. However, the only temperature dependence comes in through the propagators which do not obey the identities necessary to show the existence of the Nicolai map. The Nicolai map, therefore, does not exist at finite temperature in a supersymmetric theory, consistent with our earlier conclusion that supersymmetry is spontaneously broken at finite temperature. It is worth noting here that even in the absence of interactions,

(12.63) does not hold at finite temperature because the propagators no longer satisfy the relations in (12.54) and (12.55). This shows that the ground state energy at finite temperature must be nonzero in a supersymmetric theory and, in some sense, clarifies the mechanism behind it.

12.5 Goldstino states

As we have seen in the earlier sections, supersymmetry is spontaneously broken at finite temperature. Furthermore, as we have noted in chapter **6** (see the discussion following (9.20)), the spontaneous breakdown of a continuous global symmetry requires the existence of massless particles (Goldstone particles) in a manifestly Lorentz invariant theory with a positive Hilbert space. In the absence of manifest Lorentz invariance as is the case at finite temperature, on the other hand, we have seen in section **3.8**, that Goldstone theorem only predicts zero energy singularities in the Greens functions. Thus, we extend "Goldstone states" (or "Goldstino states" in the case of the breakdown of a fermionic symmetry) to include such states.

In the case of the supersymmetric oscillator at finite temperature, we have explicitly seen that the "Goldstino states" consist of a "tilde" and a "nontilde" quanta as was already mentioned in chapter **3**. More specifically, we note from (12.45), that the Goldstino states are denoted by $|\chi_1\rangle$ and $|\chi_2\rangle$ and have the form

$$Q|0,\beta\rangle \sim a_B^\dagger(\beta)\widetilde{a}_F^\dagger(\beta)|0,\beta\rangle \sim |\chi_1\rangle,$$

$$\overline{Q}|0,\beta\rangle \sim a_F^\dagger(\beta)\widetilde{a}_B^\dagger(\beta)|0,\beta\rangle \sim |\chi_2\rangle. \tag{12.65}$$

Therefore, we see that the "Goldstino states" consist of a thermal particle and a hole. Such a state is conventionally known as a super thermal pair. When supersymmetry is unbroken at zero temperature, it is a general feature that the "Goldstino states" at finite temperature consist of super thermal pairs.

This, however, raises an interesting question. Namely, as we have noted earlier, when supersymmetry is spontaneously broken at zero temperature, it is not restored at finite temperature, rather it continues to remain spontaneously broken. On the other hand, if supersymmetry is spontaneously broken at zero temperature, then, as we know from the Goldstone theorem, there exists a massless fermionic

state and, consequently, it is not clear whether this zero temperature state continues as the "Goldstino" state or a super thermal pair
replaces it at finite temperature. In fact, at finite temperature, as
we will see now, the zero temperature "Goldstino" state mixes with
super thermal pair states and the true "Goldstino" state at finite
temperature is a linear combination of these states while the orthogonal state becomes a massive state. Let us see this qualitatively. In
general, we note that the supersymmetry generator being fermionic
will have the zero temperature form

$$Q = \phi_B^\dagger \phi_F + v\psi_F^\dagger, \tag{12.66}$$

where $\phi_B^\dagger$ and ϕ_F represent some bosonic creation and fermionic
annihilation operators respectively. The supersymmetry breaking
parameter is denoted by the vacuum expectation value v (which is
in reality the vacuum expectation value of an auxiliary field of the
theory) and $\psi_F^\dagger$ denotes the creation operator for the "Goldstino".

In terms of the thermal operators as is clear from the discussions
of the earlier sections, however, the supersymmetry charge will have
the form

$$Q = \alpha(\beta)\phi_B^\dagger(\beta)\widetilde{\phi}_F^\dagger(\beta) + \gamma(\beta)v\psi_F^\dagger(\beta) + \cdots, \tag{12.67}$$

where $\alpha(\beta)$ and $\gamma(\beta)$ are temperature dependent constants and "$\cdots$"
represent terms which contain at least one annihilation operator. It
is clear now that the "Goldstino" state at finite temperature is given
by

$$|\chi(\beta)\rangle = Q|0, \beta\rangle$$

$$= \alpha(\beta)\phi_B^\dagger(\beta)\widetilde{\phi}_F^\dagger(\beta)|0, \beta\rangle + \gamma(\beta)v\psi_F^\dagger(\beta)|0, \beta\rangle. \tag{12.68}$$

We see, therefore, that if $v = 0$, namely, if supersymmetry is unbroken at zero temperature, the "Goldstino" state at finite temperature
is given by a super thermal pair as we have already seen. On the
other hand, if $v \neq 0$, i.e., if supersymmetry is spontaneously broken
at zero temperature, then the finite temperature "Goldstino" state
is given by a linear combination of the zero temperature "Goldstino"
state and a super thermal pair state. The orthogonal combination
can be shown, with some work, to correspond to a massive state.

12.6 Non-simply connected space-time

We have seen in chapter **1** that, in the imaginary time formalism, a thermal field theory corresponds to defining the theory on a manifold $\mathbf{R}^3 \times S^1$ (see the discussion following (1.90)). In many theories of current interest, it is common to define theories on non-simply connected space-time of this kind. For example, the higher dimensional Kaluza-Klein theories or string theories need to have several dimensions compactified (generally on a torus) and the prototype of such a compactification is on manifolds of the form $\mathbf{R}^n \times S^1$. Such theories are remarkably similar in structure to thermal field theories as we will see shortly. We have also seen earlier, in chapter **1** (see section **1.7**) as well as in chapter **11**, that at finite temperature the two point amplitude (as well as the higher ones) becomes nonanalytic at $p^\mu = 0$. However, we also know that in supersymmetric theories, there are remarkable cancellations among bosonic and fermionic loops and so, it is worth asking if such cancellations in a supersymmetric theory make the amplitudes analytic at finite temperature. There is, however, an obvious difficulty in this investigation because, as we have seen in this chapter, a supersymmetric theory does not remain supersymmetric at finite temperature.

Theories defined on a non-simply connected space-time provide an alternative for such an investigation. Namely, although the structure of theories defined on a space-time $\mathbf{R}^3 \times S^1$ is similar to a thermal field theory, the compactification length L of S^1 is unrelated to temperature and the generating functional does not correspond to a partition function. Consequently, we are not required to impose periodic and anti-periodic boundary conditions for bosons and fermions respectively as we would in order to define a thermal partition function. The breaking of supersymmetry at finite temperature, as we have noted earlier, arises because of the different statistics (equivalently, periodicity conditions) satisfied by the bosons and fermions. In contrast, when we have a theory defined on a non-simply connected space-time, we are free to choose the periodicity conditions for the various fields along the direction being compactified. Consequently, we can choose the same periodicity conditions for bosons and fermions which is likely to preserve supersymmetry. In such a case, it is worth asking whether the amplitudes of the theory become analytic for a supersymmetric theory defined on, say $\mathbf{R}^3 \times S^1$, when the bosons and the fermions satisfy identical periodicity conditions along the compactified direction. In this section, we will first show that supersymmetry is, indeed, preserved when identical periodicity conditions are used for bosons and fermions in the sense that the zero point energy and the tadpoles vanish. Furthermore, we show that the non-analyticity in the amplitudes continues to hold in spite of supersymmetry – supersymmetry merely relates the non-analyticity in the bosonic and the fermionic amplitudes.

Let us consider the Wess-Zumino theory, which describes the supersymmetric interacting theory of Majorana fermions with a scalar and a pseudoscalar, described by the Lagrangian density

$$\mathcal{L} = \frac{1}{2}\partial_\mu A \partial^\mu A + \frac{1}{2}\partial_\mu B \partial^\mu B - \frac{m^2}{2}(A^2 + B^2) + \frac{i}{2}\overline{\psi}\gamma^\mu \partial_\mu \psi$$
$$- \frac{m}{2}\overline{\psi}\psi - g\overline{\psi}(A - i\gamma_5 B)\psi - mgA(A^2 + B^2)$$
$$- \frac{g^2}{2}(A^2 + B^2)^2. \tag{12.69}$$

Here A and B denote, respectively, the scalar and the pseudoscalar fields and ψ represents a Majorana (self charge conjugate) fermion. In $3 + 1$ dimensions, this theory has the special property (following from supersymmetry) that the tadpoles and the vacuum energy vanish. Furthermore, the radiative corrections to the two point functions have the form

$$\frac{k}{A} \!-\!\!\bullet\!\!-\! \frac{k}{A} \quad = a(k^2 + m^2),$$

$$\frac{k}{B} \!-\!\!\bullet\!\!-\! \frac{k}{B} \quad = a(k^2 + m^2),$$

$$\frac{k}{\psi} \!-\!\!\bullet\!\!-\! \frac{k}{\psi} \quad = a\not{k}, \tag{12.70}$$

where a is a constant and the forms of the two point functions signify that the theory only needs a wave function renormalization. This is what we had alluded to earlier in that supersymmetric theories have much better renormalization properties than ordinary theories.

These follow from the nonrenormalization theorem which we will not get into.

Suppose, we would like to study the Wess-Zumino theory on $\mathbf{R}^3 \times S^1$ where we assume that the coordinate x^3 is compactified with a compactification length L. We note here that this is different from the finite temperature case in that here a spatial coordinate is being compactified as opposed to the time coordinate in the Imaginary time formalism. The momentum along this axis (see (1.73)) will be discrete and depending on the boundary condition will have the values

$$
\left.
\begin{array}{ll}
p^3_{(\text{per.})} & = \frac{2\pi n}{L}, \\[2mm]
p^3_{(\text{anti.})} & = \frac{2\pi(n+1/2)}{L},
\end{array}
\right\}, \quad n = 0, \pm 1, \pm 2, \cdots . \tag{12.71}
$$

As discussed in chapter 1 (see (1.99)), the four dimensional momentum integration will now be replaced by

$$
\int \frac{\mathrm{d}^4 p_E}{(2\pi)^4} \to \int \frac{\mathrm{d}^3 p}{(2\pi)^3} \frac{1}{L} \sum_n . \tag{12.72}
$$

However, for simplicity, we will continue to use the notation of four dimensional integration with the understanding that it really stands for a three dimensional integral and a sum as described above.

With these notations, we are now ready to calculate the vacuum energy which, at the lowest order, arises at two loops in this theory

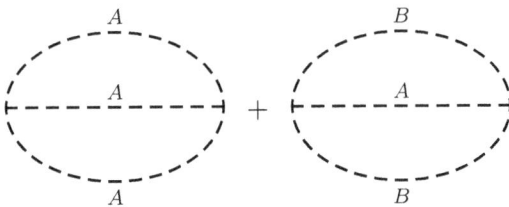

$$
= 4im^2 g^2 \int \frac{\mathrm{d}^4 k}{(2\pi)^4} \frac{\mathrm{d}^4 p}{(2\pi)^4} \frac{1}{(k^2 - m^2)(p^2 - m^2)((k-p)^2 - m^2)},
$$

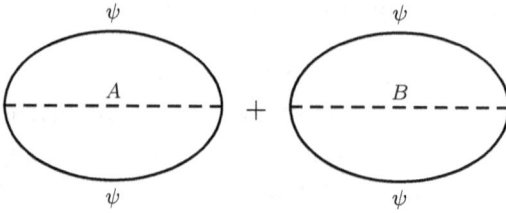

$$= -8ig^2 \int \frac{\mathrm{d}^4k}{(2\pi)^4} \frac{\mathrm{d}^4p}{(2\pi)^4} \frac{k \cdot p}{(k^2 - m^2)(p^2 - m^2)((k-p)^2 - m^2)},$$

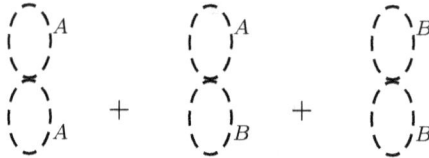

$$= 4ig^2 \int \frac{\mathrm{d}^4k}{(2\pi)^4} \frac{\mathrm{d}^4p}{(2\pi)^4} \frac{1}{(k^2 - m^2)(p^2 - m^2)}. \tag{12.73}$$

It is now an easy matter to check that these three terms algebraically add up to zero (if both bosons and fermions satisfy identical boundary conditions) because of the symmetry properties of the integrals. This shows that the theory does not develop a vacuum energy even on $\mathbf{R}^3 \times S^1$ provided identical boundary conditions are used for the bosons and the fermions along the direction of compactification. Even though we have discussed the vanishing of the zero point energy only at two loops, this analysis can be carried out order by order to show that the theory has a vanishing vacuum energy as is required by supersymmetry.

In a similar manner, we can also show that the theory does not develop any tadpoles when described on $\mathbf{R}^3 \times S^1$ for identical periodicity conditions for bosons and fermions. As a simple example, let us note that

$$= 3mg \int \frac{\mathrm{d}^4k}{(2\pi)^4} \frac{1}{k^2 - m^2},$$

$$= mg \int \frac{d^4k}{(2\pi)^4} \frac{1}{k^2 - m^2},$$

$$= -4mg \int \frac{d^4k}{(2\pi)^4} \frac{1}{k^2 - m^2}. \tag{12.74}$$

Once again, for identical periodicity conditions, it is trivial to check that the three terms add up to zero leading to the conclusion that the theory does not develop any tadpoles, as would be required by supersymmetry.

Next, let us look at the bosonic and the fermionic two point functions of the theory. When the A, B and the ψ fields satisfy identical boundary conditions, the two point function for the A field, at one loop, becomes

$$= -i\pi_A(p) = 4g^2 \int \frac{d^4k}{(2\pi)^4} \frac{2p \cdot (k+p) + m^2}{(k^2 - m^2)((k+p)^2 - m^2)}$$

$$= 4g^2(p^2 + m^2) \int \frac{d^4k}{(2\pi)^4} \frac{1}{(k^2 - m^2)((k+p)^2 - m^2)}. \tag{12.75}$$

Here p^μ represents the external momentum. We write any four momentum, p^μ, as $(\hat{p}, p^3)$ where p^3 takes the values given in (12.71) depending on the periodicity conditions used for the fields.

We can now evaluate the two point function in (12.75) explicitly using (12.71) and (12.72). Rotating to Euclidean space and using periodic boundary conditions, we obtain

$$-i\pi_A^{(\text{per.})}(p) = -4ig^2(p_E^2 - m^2) \int \frac{d^4 k_E}{(2\pi)^4} \frac{1}{(k_E^2 + m^2)}$$

$$\times \frac{1}{((k_E + p_E)^2 + m^2)}$$

$$= -4ig^2(p_E^2 - m^2) \int \frac{d^3 \hat{k}_E}{(2\pi)^3} \frac{1}{L} \sum_n \frac{1}{(\hat{k}_E^2 + (\frac{2\pi n}{L})^2 + m^2)}$$

$$\times \frac{1}{((\hat{k}_E + \hat{p}_E)^2 + (p^3 + \frac{2\pi n}{L})^2 + m^2)}. \qquad (12.76)$$

The sum in (12.76) can be evaluated using (1.102) and we can write (This is, in fact, the same integral as in (1.115) with appropriate identification.)

$$-i\pi_A(p) = -i\pi_A^{(0)}(p) - i\pi_{A,L}^{(\text{per.})}(p), \qquad (12.77)$$

where $\pi_A^{(0)}$ is the part of the two point function independent of L (namely, the $L \to \infty$ value corresponding to the two point function in the four dimensional Wess-Zumino theory) and the L dependent part takes the form

$$-i\pi_{A,L}^{(\text{per.})} = -4ig^2(p_E^2 - m^2)$$

$$\times \int \frac{d^3 \hat{k}_E}{(2\pi)^3} \frac{n_B(L\omega_{\hat{k}})}{\omega_{\hat{k}}} \left[\frac{1}{p_E^2 + 2(\hat{k}_E \cdot \hat{p}_E + ip^3\omega_{\hat{k}})} \right.$$

$$\left. + \frac{1}{p_E^2 + 2(\hat{k}_E \cdot \hat{p}_E - ip^3\omega_{\hat{k}})} \right]$$

$$= -\frac{ig^2}{2\pi^2} \frac{(p_E^2 - m^2)}{\hat{p}} \int_0^\infty d\hat{k} \, \frac{\hat{k} \, n_B(L\omega_{\hat{k}})}{\omega_{\hat{k}}} \ln R. \qquad (12.78)$$

Here, we have defined,

$$\omega_{\hat{k}} = (\hat{k}^2 + m^2)^{1/2},$$

$$R = \frac{[p_E^2 + 2(\hat{k} \cdot \hat{p} + ip^3\omega_{\hat{k}})][p_E^2 + 2(\hat{k} \cdot \hat{p} - ip^3\omega_{\hat{k}})]}{[p_E^2 - 2(\hat{k} \cdot \hat{p} + ip^3\omega_{\hat{k}})][p_E^2 - 2(\hat{k} \cdot \hat{p} - ip^3\omega_{\hat{k}})]}. \qquad (12.79)$$

We could have similarly evaluated the two point function for the A field with anti-periodic boundary condition for all the fields. Here the appropriate identity to use would be

$$\sum_n f(2n+1) = \sum \pi \operatorname{Res} f(z) \tan \pi z, \quad \text{at the poles of } f(z),$$

and without going into details, we simply note here that the result in such a case, has the form

$$-i\pi_A^{(\text{anti.})}(p_E) = -i\pi_A^{(0)} - i\pi_{A,L}^{(\text{anti.})}, \tag{12.80}$$

where

$$-i\pi_{A,L}^{(\text{anti.})}(p_E) = \frac{ig^2}{2\pi^2} \frac{(p_E^2 - m^2)}{\hat{p}} \int_0^\infty d\hat{k} \frac{\hat{k}\, n_F(L\omega_{\hat{k}})}{\omega_{\hat{k}}} \ln R, \tag{12.81}$$

with n_F representing the fermionic distribution function.

The two point functions for the B field can similarly be obtained and we simply note here the results

$$= -i\pi_B(p_E) = -i\pi_B^0(p_E) - i\pi_{B,L}(p_E), \tag{12.82}$$

with

$$-i\pi_{B,L}^{(\text{per.})}(p_E) = -\frac{ig^2}{2\pi^2} \frac{(p_E^2 - m^2)}{\hat{p}} \int_0^\infty d\hat{k} \frac{\hat{k}\, n_B(L\omega_{\hat{k}})}{\omega_{\hat{k}}} \ln R,$$

$$-i\pi_{B,L}^{(\text{anti.})}(p_E) = \frac{ig^2}{2\pi^2} \frac{(p_E^2 - m^2)}{\hat{p}} \int_0^\infty d\hat{k} \frac{\hat{k}\, n_F(L\omega_{\hat{k}})}{\omega_{\hat{k}}} \ln R. \tag{12.83}$$

The two point function for the fermion field can also be calculated in a straightforward manner and the result is

$$= -i\pi_\psi(p_E) = -i\pi_\psi^0(p_E) - i\pi_{\psi,L}(p_E), \tag{12.84}$$

where

$$-i\pi_{\psi,L}^{(\text{per.})}(p_E) = -\frac{ig^2}{2\pi^2}\frac{\not{p}_E}{\hat{p}}\int_0^\infty d\hat{k}\,\frac{\hat{k}n_B\left(L\omega_{\hat{k}}\right)}{\omega_{\hat{k}}}\ln R,$$

$$-i\pi_{\psi,L}^{(\text{anti.})}(p_E) = \frac{ig^2}{2\pi^2}\frac{\not{p}_E}{\hat{p}}\int_0^\infty d\hat{k}\,\frac{\hat{k}n_F\left(L\omega_{\hat{k}}\right)}{\omega_{\hat{k}}}\ln R. \tag{12.85}$$

Thus, we see that, at one loop, the three two point functions can be written as

$$-i\pi_{A,L}(p_E) = (p_E^2 - m^2)\pi_L(p_E),$$

$$-i\pi_{B,L}(p_E) = (p_E^2 - m^2)\pi_L(p_E),$$

$$-i\pi_{\psi,L}(p_E) = \not{p}_E\pi_L(p_E), \tag{12.86}$$

where we have

$$\pi_L^{(\text{per.})}(p_E) = -\frac{ig^2}{2\pi^2\hat{p}}\int_0^\infty d\hat{k}\,\frac{\hat{k}\,n_B(L\omega_{\hat{k}})}{\omega_{\hat{k}}}\ln R,$$

$$\pi_L^{(\text{anti.})}(p_E) = \frac{ig^2}{2\pi^2\hat{p}}\int_0^\infty d\hat{k}\,\frac{\hat{k}\,n_F(L\omega_{\hat{k}})}{\omega_{\hat{k}}}\ln R. \tag{12.87}$$

We note here that, when rotated to Minkowski space, the forms of the two point functions are consistent with the nonrenormalization theorem as discussed in (12.70). However, it is also clear that the two point functions become nonanalytic at the origin in the momentum space. This can be seen from the fact that

$$\lim_{\hat{p}\to 0\,p^3\to 0}\pi_L^{(\text{per.})}(p_E) = -\frac{ig^2}{\pi^2}\int_0^\infty d\hat{k}\,\frac{n_B(L\omega_{\hat{k}})}{\omega_{\hat{k}}},$$

$$\lim_{\hat{p}\to 0\,p^3\to 0}\pi_L^{\text{anti.}}(p_E) = \frac{ig^2}{\pi^2}\int_0^\infty d\hat{k}\,\frac{n_F(L\omega_{\hat{k}})}{\omega_{\hat{k}}}. \tag{12.88}$$

On the other hand, when the limits are reversed, the expressions become

$$\lim_{p^3\to 0\,\hat{p}\to 0}\pi_L^{(\text{per.})}(p_E) = -\frac{ig^2}{\pi^2}\int_0^\infty d\hat{k}\,\frac{\hat{k}^2 n_B(L\omega_{\hat{k}})}{\omega_{\hat{k}}^3},$$

$$\lim_{p^3\to 0\,\hat{p}\to 0}\pi_L^{(\text{anti.})}(p_E) = \frac{ig^2}{\pi^2}\int_0^\infty d\hat{k}\,\frac{\hat{k}^2 n_F(L\omega_{\hat{k}})}{\omega_{\hat{k}}^3}. \tag{12.89}$$

This nonanalytic behavior is precisely the same as seen at finite temperature (see (1.123) and (1.124)). Supersymmetry has not improved this behavior. If anything, it has only related the nonanalyticity in the two point functions for the A, B and the ψ fields. We merely note here without going into details that a periodic boundary condition for the fermionic fields is known to lead to acausal behavior. In such a case, the only consistent boundary condition that can be used for both the bosonic and the fermionic fields appears to be the anti-periodic boundary condition which is what should be used in a consistent compactification of higher dimensional theories. (Different boundary conditions would be incompatible with supersymmetry as we have already seen earlier in this chapter.)

12.7 References

A. Das, Physica **A158**, 1 (1989).

A. Das, *Lectures on Quantum Field Theory* (second edition), World Scientific Publishing (2020).

A. Das and M. Hott, Modern Physics Letters **A10**, 893 (1995).

A. Das and M. Kaku, Physical Review **D18**, 4540 (1978).

A. Das, A. Kharev and V. S. Mathur, Physics Letters **181B**, 299 (1986).

A. Das and V. S. Mathur, Physical Review **D35**, 2053 (1987).

A. Das and V. S. Mathur, Indian Journal of Physics **B61**, 214 (1987).

L. Ford, Physical Review **D20**, 933 (1980).

H. Matsumoto, M. Nakahara, Y. Nakano and H. Umezawa, Physical Review **D29**, 2838 (1984).

H. Nicolai, Nuclear Physics **B170**, 419 (1980).

E. Witten, Nuclear Physics **B202**, 253 (1982).

CHAPTER 13

Effective actions

The one loop effective action for a system of fermions, interacting with background fields – scalars or gauge fields – is an important fundamental quantity in the study of a quantum field theory. It incorporates all the one loop corrections to the action due to the background field interactions and the imaginary part of this effective action determines the stability of the vacuum of the theory. It generates all the one loop n-point amplitudes of the theory (involving the background fields).

At zero temperature, it is known that the (one) loop amplitudes (and, therefore, the effective action) have ultraviolet divergence and need to be regularized. In a gauge theory, the regularization has to respect gauge invariance (it should be gauge invariant). The effective action can, of course, be calculated order by order in perturbation theory by evaluating Feynman diagrams. However, long time ago, Schwinger gave a very elegant way of calculating the complete one loop effective action (not order by order in perturbation theory), at zero temperature, through what is known as the proper time method which we discuss below.

13.1 Proper time method

Let us assume that the tree level Lagrangian density for the fermion theory, interacting with a background field, is given by

$$\mathcal{L} = \overline{\psi}(i\partial \!\!\!/ - m - gA \!\!\!/)\psi, \tag{13.1}$$

where m denotes the fermion mass, g denotes the coupling (constant) strength of the fermions to the background field A which is generic, for example, it can stand for a scalar field or a gauge field A_μ contracted with Dirac matrices and ∂ stands, similarly, for the derivative operator (∂_μ contracted with Dirac matrices (in $0+1$ dimensions, it simply corresponds to ∂_t). Since the Lagrangian (and, therefore, the action) is quadratic in the fermion fields, integrating out the fermions in the path integral formalism, we obtain[1]

$$Z[A] = e^{iW[A]} = N \int \mathcal{D}\overline{\psi}\mathcal{D}\psi \, e^{i\int d^D x \, \overline{\psi}(i\partial - m - gA)\psi}$$

$$= N \det(i\partial - m - gA),$$

$$\text{or,} \quad W[A] = \Gamma[A] = -i\ln\left(N \det(i\partial - m - gA)\right), \qquad (13.2)$$

where N denotes a normalization constant and D stands for the number of space-time dimensions. We also point out here that, when sources (for fields etc.) are not present, the generating functional W coincides with that for the 1PI amplitudes Γ (namely, in such a case, there is no Legendre transformation, see, for example, (5.44)). Furthermore, subtracting out the contribution corresponding to the vanishing external field A (namely, an uninteresting constant), we obtain the genuine field dependent effective action

$$\Gamma_{\text{eff}}[A] = \Gamma[A] - \Gamma[0] = -i\ln \frac{\det(i\partial - m - gA)}{\det(i\partial - m)},$$

$$= -i\text{Tr}\,\ln(i\partial - m - gA) - C = -i\,\text{Tr}\,\ln H - C, \quad (13.3)$$

where C is a divergent constant and we have identified

$$H = (i\partial - m - gA). \qquad (13.4)$$

We have also used the familiar formula that, for any operator (matrix), we can write

$$\det \mathcal{O} = e^{\text{Tr}\,\ln \mathcal{O}}. \qquad (13.5)$$

Therefore, ignoring the divergent constant, we can write (see (13.3))

$$\Gamma_{\text{eff}}[A] = \Gamma[A] = -i\,\text{Tr}\,\ln H. \qquad (13.6)$$

---

[1]See, for example, section **5.5** in A. Das, *Field Theory: A Path Integral Approach* (third edition), World Scientific Publishing (2020).

Furthermore, recalling that the integral representation of the gamma function is given by

$$\int_0^\infty d\tau \, \tau^{\nu-1} e^{-\alpha\tau} = \alpha^{-\nu} \Gamma(\nu), \quad \alpha, \nu \geq 0, \tag{13.7}$$

and using the familiar relation

$$\lim_{\nu \to 0} \alpha^{-\nu} = \lim_{\nu \to 0} e^{-\nu \ln \alpha} = \lim_{\nu \to 0} \left(1 - \nu \ln \alpha + O(\nu^2)\right), \tag{13.8}$$

we obtain

$$\lim_{\nu \to 0} \alpha^{-\nu} \Gamma(\nu) = \lim_{\nu \to 0} \left(1 - \nu \ln \alpha + O(\nu^2)\right) \left(\frac{1}{\nu} + O(\nu^0)\right)$$
$$= C - \ln \alpha,$$

where C is a divergent constant and which leads, from (13.3)-(13.8), to

$$\Gamma_{\text{eff}}[A] = \Gamma[A] = \lim_{\nu \to 0} i \int_0^\infty \frac{d\tau}{\tau^{1-\nu}} \, \text{Tr} \, e^{-\tau H}$$
$$= i \int_0^\infty \frac{d\tau}{\tau} \, \text{Tr} \, e^{-\tau H}. \tag{13.9}$$

As in (13.6) we have neglected an uninteresting divergent constant in (13.9). This (gauge invariant) regularized integral representation for the effective action at zero temperature was derived by Schwinger where τ is known as the "proper time" parameter (sometimes it is denoted by s) for obvious reasons, as we will discuss next.

We note that we can think of the factor $e^{-\tau H}$ in (13.9) as the time evolution operator with Euclidean time τ (with the inverse $e^{\tau H}$ and operators transforming as $\mathcal{O}(\tau) = e^{\tau H} \mathcal{O}(\tau = 0) e^{-\tau H}$). In that case, H would correspond to the Hamiltonian generating translation in the Euclidean time τ and will lead to the (Euclidean) Hamiltonian equations

$$\frac{dx^\mu}{d\tau} = -[x^\mu, H], \qquad \frac{dp_\mu}{d\tau} = -[p_\mu, H]. \tag{13.10}$$

If these equations can be solved exactly, then we can determine the operators $x^\mu(\tau), p_\mu(\tau)$ which, in turn, would lead to determination of the coordinate eigenstates as

$$|x^\mu, \tau\rangle = e^{\tau H} |x, 0\rangle = e^{\tau H} |x\rangle,$$
$$\langle x^\mu, \tau| = \langle x, 0| e^{-\tau H} = \langle x| e^{-\tau H}. \tag{13.11}$$

As a result, we can write the "trace" in (13.9) as

$$\text{Tr}\, e^{-\tau H} = \int d^D x \,\langle x | e^{-\tau H} | x \rangle = \langle x^\mu, \tau | x^\mu, 0 \rangle, \tag{13.12}$$

and the effective action in (13.9) can be determined (or given an integral representation). The D in (13.12) denotes the number of space-time dimensions. When the Hamiltonian equations in (13.10) cannot be solved exactly, but only iteratively, then the effective action in (13.9) can only be determined perturbatively. The proper time method can also be extended to Minkowski time with an appropriate rotation ($\tau \to it$, see Fig. 1.1) with the right $i\epsilon$ prescription for the propagators). Schwinger's proper time method works quite well at zero temperature and there were several attempts, in the past, to generalize this method to finite temperature, leading to conflicting results. We will discuss, in the next section, an alternative method which seems to work well in determining finite temperature effective actions.

13.2 An alternative proposal for finite temperature

The simplifying observation in this context is to recall that there are no ultraviolet divergences at finite temperature and, therefore, a gauge invariant regularization is not needed at finite temperature and there is no necessity of generalizing the complicacies associated with the proper time method to finite temperature. Note also that in calculating effective actions, real time is more appropriate. As we have discussed in earlier chapters (see, particularly, chapter 1), the imaginary time formalism naturally leads to retarded and advanced amplitudes (Green's functions) beyond the two point function (in space-time dimensions $D \geq 2$). Only in the $0 + 1$ dimension, the Feynman and the retarded amplitudes coincide and, therefore, the imaginary time formalism is equally suitable in $0 + 1$ dimension. On the other hand, the effective action that we are interested in should generate Feynman amplitudes and, consequently, the real time formalism is more appropriate. (Retarded and advanced amplitudes can be obtained from the Feynman amplitudes as we will discuss later in this section.)

The proposal is quite simple. We note from (13.4) and (13.6) that, for a massive fermion interacting with an arbitrary background

A, we have

$$\frac{\partial \Gamma_{\text{eff}}[A]}{\partial m} = -i \operatorname{Tr} \frac{(-1)}{H} = \operatorname{Tr} \frac{i}{(i\partial - m - gA)}$$

$$= \int dt d^{D-1}x \, S(t, \mathbf{x}; t, \mathbf{x})[A], \qquad (13.13)$$

where $S(t, \mathbf{x}; t', \mathbf{x}')[A]$ denotes the complete Feynman propagator (including the factor of i) in the presence of the background field A. On the other hand, if the fermion is massless ($m = 0$), then we can write

$$\frac{\delta \Gamma_{\text{eff}}[A]}{\delta A(t, \mathbf{x})} = gS(t, \mathbf{x}; t, \mathbf{x})[A]. \qquad (13.14)$$

Namely, it simply picks out the complete fermion propagator, at coincident coordinates, in the presence of the background field (note that the exact Lorentz structure including factors on the right hand side depends on the nature of the background field, namely, whether it is a scalar or vector etc). So, for example, for a scalar background, the right hand side in (13.14) would represent the complete propagator, at coincident points, while, for a gauge field background, the right hand side would correspond to the current density which is again connected with the complete fermion propagator as $\operatorname{tr} \left(\gamma^\mu S(t, \mathbf{x}; t, \mathbf{x})[A] \right)$ where "tr" stands for the Dirac trace. In either case, it is clear that the complete fermion propagator is the most fundamental quantity in (13.13) or (13.14) if we want to evaluate the effective action. Furthermore, we note that if we were to work in the mixed $(t, \mathbf{p})$ space, (13.14) would take the form

$$\frac{\delta \Gamma_{\text{eff}}[A]}{\delta A(t, -\mathbf{p})} = gS(t, t; \mathbf{p})[A]. \qquad (13.15)$$

To determine the complete fermion propagator at finite temperature directly, let us look for a general propagator function of variables $(t, t'; \mathbf{x}, \mathbf{x}')$ or $(t, t'; \mathbf{p})$ as well as with dependence on the background field A which satisfies

(i) the appropriate equations for the complete propagator of the theory,

(ii) the necessary symmetry properties of the theory such as the Ward identity,

(iii) and the anti-periodicity property associated with the finite
temperature fermion propagator on each of the two time coordinates.

In fact, it is the last requirement that is not there at zero temperature
and is quite crucial for a determination of the complete propagator
at finite temperature. If the theory does not have any ultraviolet divergence at zero temperature, this propagator will turn out to be the
exact complete propagator of the theory leading to the exact and the
correct effective action of the theory including the zero temperature
part. However, if there are ultraviolet divergences at zero temperature which need to be regularized, this propagator will not lead to
the correct zero temperature effective action, but the finite temperature part, which does not have to be regularized, will be exact. Let
us illustrate how the method works with two examples.

13.2.1 0 + 1 dimensional QED.

13.2.1 0 + 1 dimensional QED. The Lagrangian for the $0 + 1$ dimensional QED is given by (see also section **6.5.1**)

$$L = \overline{\psi}(i\partial_t - m - eA(t))\psi, \tag{13.16}$$

where e denotes the electromagnetic coupling, t is the only coordinate (there is no spatial coordinate) and the gauge potential has just
one component (time index) which we have suppressed for simplicity and, therefore, the gauge field background appears like a scalar
background. There are no Dirac matrices so that $\overline{\psi} = \psi^\dagger$. Furthermore, the mass term behaves like a chemical potential since $\overline{\psi}\psi$, for
this theory, corresponds to the number operator which is a conserved
quantity. This simple theory has been studied extensively (we have
already discussed this theory in section **6.5.1** within a different context) and is very useful for us to fix the ideas of the proposal in a
clear manner which will be helpful in going to higher dimensions.

Let us use the closed time path formalism, discussed in chapter
2, where the path in the complex time plane is given in Fig. 2.1 and
on this contour, the dynamical equations satisfied by the complete
fermion propagators take the forms

$$(i\partial_t - m - eA_c(t))S_c(t, t') = i\delta_c(t - t'),$$

$$S_c(t, t')(i\overleftarrow{\partial}_{t'} + m + eA_c(t')) = -i\delta_c(t - t'). \tag{13.17}$$

Here the subscript "c" on the propagator as well as the background field characterizes the branch of the contour and can, in principle, take values $c = +, -, 3$. However, the branch $c = 3$ decouples from amplitudes so that, effectively, we have $c = \pm$ (see the discussion following (2.24)). Following the discussion in section **2.2** we can write the fermion propagator as (see also the appendix **2.6.2**),

$$S_c(t, t') = \left(\theta_c(t - t')C - \theta_c(t' - t)D\right)$$
$$\times\, e^{-im(t-t')-ie\int_{t'}^{t} dt''\, A_c(t'')}, \tag{13.18}$$

where, in order to satisfy the delta function discontinuity across $t = t'$ in (13.17), the constants C and D satisfy

$$C + D = 1, \quad \text{or}, \quad C = 1 - D, \tag{13.19}$$

so that we can write

$$S_c(t, t') = \left(\theta_c(t - t') - D\right)e^{-im(t-t')-ie\int_{t'}^{t} dt''\, A_c(t'')}. \tag{13.20}$$

There are a couple of comments to be made here. First the propagator here has a factor of i incorporated into it (as already pointed out below (13.13), or another way of saying this is that the propagator here $S_c(t, t')$ is really $iS_c(t, t')$ of appendix **2.6.2**). Second, we have changed the constants A, B in **2.6.2** to C, D to avoid any confusion with the background gauge field A. Finally, in **2.6.2** we were discussing the free propagator and, therefore, the exponential involved only the mass term, but here we are determining the complete propagator in the presence of the background gauge field. If we set $A = 0$ in (13.16), the solutions in (13.18)-(13.19) reduce respectively to (2.111)-(2.112). The gauge transformation properties of the propagator follow naturally from (13.20), since under (see section **5.1**)

$$\psi_c(t) \to \psi_c(t)\, e^{-ie\alpha(t)}, \quad \overline{\psi}_c(t) \to \overline{\psi}_c(t)\, e^{ie\alpha(t)},$$
$$A_c(t'') \to A_c(t'') + \partial_{t''}\alpha(t''),$$
$$e^{-ie\int_{t'}^{t} dt''\, A_c(t'')} \to e^{-ie\int_{t'}^{t} dt''\, (A_c(t'')+\partial_{t''}\alpha(t''))}$$
$$= e^{-ie\int_{t'}^{t} dt''\, A_c(t'')} \times e^{-ie(\alpha(t)-\alpha(t'))}. \tag{13.21}$$

Namely, the exponential in (13.18) has the right behavior of transformation for the propagator (recall that the propagator is the thermal average of the time ordered product $\langle T(\psi(t)\overline{\psi}(t'))\rangle_\beta$).

Finally, we come to the last property, namely, the anti-periodicity of the fermion propagator which says that (see (2.113) with $T \rightarrow -\infty$)

$$S_c(-\infty - i\beta, t') = -S_c(-\infty, t'), \tag{13.22}$$

which from (13.18) and (13.19) determines

$$C e^{-im(t-t')-ie\int_{t'}^{t} dt'' \, A_c(t'')}\Big|_{t=-\infty-i\beta}$$
$$= D e^{-im(t-t')-ie\int_{t'}^{t} dt'' \, A_c(t'')}\Big|_{t=-\infty},$$

$$\text{or,} \quad D = C e^{-\beta m - ie\int_{-\infty}^{-\infty-i\beta} dt'' \, A_c(t'')}$$

$$= C e^{-\beta m - ie(a_+ - a_-)}, \tag{13.23}$$

where

$$a_\pm = \int_{-\infty}^{\infty} dt \, A_\pm(t). \tag{13.24}$$

As discussed earlier, only the $\pm$ branches contribute. Furthermore, using the condition (13.19), we immediately determine

$$D = n_F\left(m + \frac{ie}{\beta}(a_+ - a_-)\right),$$

$$C = 1 - n_F\left(m + \frac{ie}{\beta}(a_+ - a_-)\right). \tag{13.25}$$

The propagator now follows from (13.20) to be

$$S_c(t, t')$$
$$= e^{-imt-ie\int d\bar{t}\,\theta_c(t-\bar{t})A_c(\bar{t})}\left(\theta_c(t - t') - n_F\left(m + \frac{ie}{\beta}(a_+ - a_-)\right)\right)$$
$$\times e^{imt'+ie\int d\bar{t}\,\theta_c(t'-\bar{t})A_c(\bar{t})}, \tag{13.26}$$

and it can be checked that the fermion propagator satisfies anti-periodicity on t and t' independently. The exponentials can, of course, be combined into the simpler form given in (13.20), but we have deliberately broken it to this suggestive form which generalizes to higher dimensions. The four components of the propagator $(S_{++}, S_{+-}, S_{-+}, S_{--})$ can now be easily obtained, but this is enough for our purpose.

In addition to satisfying the gauge transformation properties as well as anti-periodicity, this (exact) complete propagator also satisfies the Lippmann-Schwinger equation which is the compact expression (for the propagator) which leads to the perturbative expansion of the propagator around the tree level propagator. To see this, let us consider only the ++ branch for simplicity (we will not write the ++ symbol explicitly). Since the exponential $e^{-im(t-t')}$ will naturally factor out in a chain of propagators, we will factor out this exponential term in the following discussion and with that, the tree level propagator, in this case, can be written as

$$S_0(t, t') = e^{-im(t-t')}\overline{S}_0(t, t'),$$
$$\overline{S}_0(t, t') = \theta(t - t') - n_F(m). \tag{13.27}$$

We note from (13.26) that the complete propagator can, similarly, be written in a compact form (after factoring the mass term in the exponential) as

$$\overline{S}(t, t') = e^{-ie\Phi(t)}\left(\theta(t - t') - n_F\left(m + \frac{ie(a_+ - a_-)}{\beta}\right)\right)e^{ie\Phi(t')}, \tag{13.28}$$

where we have denoted (on the ++ branch with the lower limit of integration is $-\infty$)

$$\Phi(t) = \int dt''\,\theta(t - t'')A(t''), \quad \Phi(-\infty) = 0, \quad \Phi(\infty) = a_+, \tag{13.29}$$

as defined in (13.24). It follows from (13.27)-(13.29) that

$$\frac{de^{-ie\Phi(t)}}{dt} = -ie\frac{d\Phi(t)}{dt}e^{-ie\Phi(t)} = -ieA(t)e^{-ie\Phi(t)},$$

$$\frac{d\overline{S}_0(t, t')}{dt} = \delta(t - t') = -\frac{d\overline{S}_0(t, t')}{dt'},$$

$$\frac{d\overline{S}(t, t')}{dt} = -ieA(t)\overline{S}(t, t') + \delta(t - t'). \tag{13.30}$$

Using these three relations as well as (13.29), we note that we can write

$$ie(\overline{S}_0 A \overline{S})(t,t') = ie \int dt''\, \overline{S}_0(t,t'') A(t'') \overline{S}(t'',t')$$

$$= -\int dt''\, \overline{S}_0(t,t'') \left(\frac{d\overline{S}(t'',t')}{dt''} - \delta(t''-t') \right),$$

which, upon integration by parts the first term and using the second relation in (13.30), leads to

$$= -\overline{S}(t,t') + \overline{S}_0(t,t'). \tag{13.31}$$

This relation can be written either as

$$\overline{S}(t,t') = \overline{S}_0(t,t') - ie(\overline{S}_0 A \overline{S})(t,t'), \tag{13.32}$$

or as

$$\left((1 + ie\overline{S}_0 A)\overline{S} \right)(t,t') = \overline{S}_0(t,t'), \tag{13.33}$$

or equivalently as

$$\overline{S}(t,t') = \left((1 + ie\overline{S}_0 A)^{-1} \overline{S}_0 \right)(t,t'). \tag{13.34}$$

All these three expressions are referred to as the Lippmann-Schwinger equation which expresses the complete propagator $\overline{S}(t,t')$ in terms of the tree level propagator $\overline{S}_0(t,t')$ and, of course, these still hold if we bring back the exponential factors $e^{-im(t-t')}$. This gives us the confidence that our propagator, derived in (13.26), from the three simple requirements given above, is the correct complete propagator for this QED theory in $0+1$ dimension.

All that remains now is to determine the effective action from this complete propagator using, for example, (13.14) or (13.15) which involves the propagator at coincident points. In this $0+1$ dimensional case, the only coordinate is the time coordinate. Furthermore, we note, from (13.26), that the complete propagator on the closed contour can be written, at coincident times $t = t'$ as

$$S_c(t,t) = \frac{1}{2}\left(1 - 2n_F \left(m + \frac{ie}{\beta}(a_+ - a_-) \right) \right)$$

$$= \frac{1}{2} \tanh\left(\frac{ie}{2}(a_+ - a_-) + \frac{\beta m}{2} \right), \tag{13.35}$$

with $a_\pm$ defined in (13.24). We have also taken the symmetric value of the step function at the coincident limit, namely, $\theta_c(0) = \frac{1}{2}$. It is interesting to note that the propagator, at coincident points, is independent of the time coordinate. This, in fact, follows from the Ward identities of the theory (this is a gauge theory, QED in $0 + 1$ dimensions).[2] Of course, there is no momentum in $0 + 1$ dimensions. Therefore, we can easily integrate (13.14) or (13.15) with respect to the gauge field to obtain the normalized effective action to be[3]

$$\Gamma_{\text{eff}}[a_+, a_-] = -i \ln \cosh \left(\frac{ie}{2}(a_+ - a_-) + \frac{\beta m}{2} \right)$$

$$= -i \ln \left[\cos \frac{e(a_+ - a_-)}{2} + i \tanh \frac{\beta m}{2} \sin \frac{e(a_+ - a_-)}{2} \right], \quad (13.36)$$

where in the last step, we have discarded a constant (additive) term for simplicity. This is the complete effective action for this theory at finite temperature which reduces to the well studied action on the C_+ branch of the contour when we set $a_- = 0$. At zero temperature ($\beta \to \infty$), this effective action reduces to

$$\lim_{T \to 0} \Gamma_{\text{eff}}[a_+] \to -i \ln \left(\cos \frac{ea_+}{2} + i \sin \frac{ea_+}{2} \right)$$

$$= -i \ln e^{\frac{iea_+}{2}} = \frac{ea_+}{2}, \quad (13.37)$$

which is the well known Chern-Simons term and is quite well known (see, for example, (6.158)). This explains in some detail how this method works in this simple model. Furthermore, being the complete effective action, (13.36) can lead to retarded/advanced amplitudes as well. For example, the n-point retarded amplitude, which can be expressed as[4]

$$\Gamma_R^{(n)} = \sum_{i_k = \pm} \Gamma_{+i_1 i_2 \cdots i_{n-1}}, \quad (13.38)$$

can be shown, from the form of $\Gamma_{\text{eff}}[a_+, a_-]$ in (13.36), to vanish (since it is a function of the difference $(a_+ - a_-)$) for $n \geq 2$.

---

[2]See, for example, A. Das and G. Dunne, Physical Review **D57**, 5023 (1998); J. Barcelos-Neto and A. Das. Physical Review **D58**, 085022 (1998).

[3]See, for example, I. S. Gradshteyn and I. M. Ryzhik, *Table of Integrals, Series and Products* (1965), formula 2.424.2 (for $n = 0$).

[4]See, for example, F. T. Brandt, A. Das, J. Frenkel and A. J. da Silva, Physical Review **D59**, 065004 (1999); F. T. Brandt, A. Das and J. Frenkel, Physical Review **D60**, 105008 (1999).

13.2.2 Schwinger model. We have already studied the Schwinger model in subsection **7.2.3** as well as in section **10.3** in different contexts. Schwinger model describes massless QED in $1+1$ dimensions. Here we will derive the effective action for the theory at finite temperature using the ideas developed above. Let us recall that the Lagrangian density for this theory is given by

$$\mathcal{L} = \overline{\psi}(t,x)\gamma^{\mu}(i\partial_{\mu} - eA_{\mu}(t,x))\psi(t,x), \quad \mu = 0,1. \tag{13.39}$$

This two dimensional theory is soluble at zero temperature and the interaction of the photons with fermions is known to generate a (gauge invariant) mass for the photon. The effective action for the photon, which results if we integrate out the fermions, has also been studied perturbatively at finite temperature, both for zero chemical potential as well as when the chemical potential is nonzero. Here we will derive the closed form expression for the effective action, at finite temperature, following the ideas developed in this chapter. Let us recall that, at zero temperature, only the two point function for the photon is nonzero in this theory and needs to be regularized (for ultraviolet divergence) and the zero temperature effective action describes a free, massive photon. On the other hand, as we will see, at finite temperature, the effective action will have multi-photon amplitudes which are well behaved (without any need for regularization). Since this method does not involve any regularization, the zero temperature limit of this effective action will not reduce to the correct (regularized) zero temperature value, but, as we have emphasized earlier, our main interest is in the finite temperature part of the effective action.

The simplest and most intuitive way to study the finite temperature part of the effective action is to decompose the coordinates and the fields in the natural basis (for the problem), namely, let us define

$$x^{\pm} = \frac{1}{2}(x^0 \pm x^1), \qquad\qquad p_{\pm} = p_0 \pm p_1,$$

$$\partial_{\pm} = \partial_0 \pm \partial_1, \qquad\qquad A_{\pm} = A_0 \pm A_1,$$

$$\psi_{+} = \frac{1}{2}(\mathbb{1} + \gamma_5)\psi, \qquad\qquad \psi_{-} = \frac{1}{2}(\mathbb{1} - \gamma_5)\psi,$$

$$= \psi_{\mathrm{R}}\begin{pmatrix} 1 \\ 1 \end{pmatrix}, \qquad\qquad = \psi_{\mathrm{L}}\begin{pmatrix} 1 \\ -1 \end{pmatrix} \tag{13.40}$$

where $\gamma_5 = \gamma^0\gamma^1$, in this $1 + 1$ dimensional theory, has the form $\begin{pmatrix} 0 & 1 \\ 1 & 0 \end{pmatrix}$ (see, for example, (7.118)) and

$$\psi_R = \frac{1}{2}(\psi_1 + \psi_2), \qquad \psi_L = \frac{1}{2}(\psi_1 - \psi_2). \tag{13.41}$$

As a result, in this basis, the Lagrangian density in (13.39) decomposes into two decoupled $0 + 1$ dimensional theories, namely, we can write

$$\mathcal{L} = \psi_R^\dagger(i\partial_+ - eA_+)\psi_R + \psi_L^\dagger(i\partial_- - eA_-)\psi_L, \tag{13.42}$$

where ψ_L and ψ_R in (13.42) denote the scalar components given in (13.41) of the two (2 dimensional) spinor fields defined in (13.40). At zero temperature, the regularization mixes the two sectors leading to the chiral anomaly. On the other hand, at finite temperature, we do not need a regularization and, in fact, we have already seen in (7.137) and the discussion following there, as well as in (9.109) and (9.110), that there is no chiral anomaly at finite temperature. Therefore, at finite temperature, the two sectors do not mix and each can be studied as a $0 + 1$ dimensional independent theory as in the earlier subsection **11.2.1**. However, there is one crucial difference which makes the calculations as well as the discussions a bit more involved. Namely, all the field variables in either of the two sectors depend on both the coordinates $x^\pm$ and not just on one coordinate (x^+ or x^-) as is the case in the last subsection. We will discuss how to handle this in some detail in the following.

To begin with, let us concentrate only on the sector of ψ_R where the dynamical coordinate is x^+ (see (13.42)). Therefore, in this sector, we will Fourier transform the x^- coordinate and we will denote the conjugate variable simply as $p_- = p$. Similarly, we will denote any other momentum $k_- = k$ for simplicity. The action for the right handed sector, therefore, can be written as

$$S_R = 2 \int dx^+ dx^-\, \psi_R^\dagger(x^+, x^-)(i\partial_+ - eA_+(x^+, x^-))\psi_R(x^+, x^-)$$

$$= 2 \int dx^+ \frac{dp}{2\pi}\, \psi_R^\dagger(x^+, -p)\Big[i\partial_+\psi_R(x^+, p)$$

$$- e \int \frac{dq}{2\pi} A_+(x^+, p - q)\psi_R(x^+, q)\Big], \tag{13.43}$$

where the overall factor of 2 is from the Jacobian of transformation from (x^0, x^1) to (x^+, x^-). This makes it clear that the equation for the propagator (13.43), in this sector (as well as in the other sector), would involve a convolution which is best studied by the following operator notation. Let us define the operators $\hat{S}(x^+, x'^+)$ and $\hat{A}_+(x^+)$ such that

$$S(x^+, x'^+; p, k) = \langle p | \hat{S}(x^+, x'^+) | k \rangle, \qquad \langle p | k \rangle = 2\pi \delta(p - k),$$

$$A_+(x^+, p - k) = \langle p | \hat{A}_+(x^+) | k \rangle. \tag{13.44}$$

Then, the equation for the fermion propagator, in this sector, (following from (13.43))

$$i\partial_+ S(x^+, x'^+; p, k) - e \int \frac{dq}{2\pi} A_+(x^+, p - q) S(x^+, x'^+; q, k)$$

$$= \frac{i}{2} \delta(x^+ - x'^+) 2\pi \delta(p - k), \tag{13.45}$$

can be written in a simpler form in terms of the operators in (13.44) as

$$(i\partial_+ - e\hat{A}_+(x^+))\hat{S}(x^+, x^-) = \frac{i}{2} \delta(x^+ - x^-). \tag{13.46}$$

Therefore, the equations for the fermion propagator in this sector can be written in the operator form on the closed time path contour as

$$(i\partial_+ - e\hat{A}_{+(c)}(x^+))\hat{S}_c(x^+, x'^+) = \frac{i}{2} \delta_c(x^+ - x'^+),$$

$$\hat{S}_c(x^+, x'^+)(i\overleftarrow{\partial}'_+ + e\hat{A}_{+(c)}(x'^+)) = -\frac{i}{2} \delta_c(x^+ - x'^+). \tag{13.47}$$

The two equations are solved by

$$\hat{S}_c(x^+, x'^+) = \frac{1}{4} e^{-ie \int d\bar{x}^+ \theta_c(x^+ - \bar{x}^+) \hat{A}_{+(c)}(\bar{x}^+)} (\mathrm{sgn}_c(x^+ - x'^+) + \hat{O})$$

$$\times e^{ie \int d\bar{x}'^+ \theta_c(x'^+ - \bar{x}'^+) \hat{A}_{+(c)}(\bar{x}'^+)}, \tag{13.48}$$

where

$$\mathrm{sgn}_c(x^+ - x'^+) = \theta_c(x^+ - x'^+) - \theta_c(x'^+ - x^+), \tag{13.49}$$

and $\hat{\mathcal{O}}$ is an operator independent of the coordinates (x^+, x'^+), but contains all the other essential information resulting from interactions as well as temperature. We can now impose the anti-periodicity condition on the first time coordinate t (the third condition) on this propagator (see also (13.22))

$$\langle p | \hat{S}_c(\frac{-\infty + x}{2}, x'^+) | k \rangle = -e^{-\frac{\beta p}{2}} \langle p | \hat{S}_c(\frac{-\infty - i\beta + x}{2}, x'^+) | k \rangle, \quad (13.50)$$

where the factor $e^{-\frac{\beta p}{2}}$, on the right hand side, arises from the exponential in the Fourier transform of the x^- coordinate to the momentum variables. With some algebra, (13.50) leads to

$$\hat{\mathcal{O}} = 1 - 2(\hat{\mathcal{O}}_+ + 1)^{-1}, \quad (13.51)$$

with

$$\hat{\mathcal{O}}_+ = e^{\frac{ie\left(\hat{a}_{+(+)} - \hat{a}_{+(-)}\right)}{2}} e^{\frac{\beta \hat{P}}{2}} e^{\frac{ie\left(\hat{a}_{+(+)} - \hat{a}_{+(-)}\right)}{2}}, \quad (13.52)$$

where $(\pm)$ denote the two branches $C_\pm$ of the time contour and $\hat{P}$ denotes the momentum operator satisfying

$$\hat{P}|p\rangle = p|p\rangle, \quad \hat{a}_{+(\pm)} = \int_{-\infty}^{\infty} dx^+ \, \hat{A}_{+(\pm)}(x^+). \quad (13.53)$$

This completes the derivation of the complete fermion propagator of the theory satisfying all the three postulates. Let us indicate below how this satisfies, in addition, the Lippmann-Schwinger equation as well.

Let us note, from (13.48) and (13.51)-(13.52), that when there is no background gauge field, $A_\pm = 0$,

$$\hat{\mathcal{O}}_{0,+} = e^{\frac{\beta \hat{P}}{2}}, \quad (13.54)$$

so that we can write

$$\hat{\mathcal{O}}_0 = 1 - 2(\hat{\mathcal{O}}_{0,+} + 1)^{-1} = 1 - 2\left(e^{\frac{\beta \hat{P}}{2}} + 1\right)^{-1}$$

$$= 1 - 2n_F\left(\frac{\hat{P}}{2}\right) = \tanh\frac{\beta \hat{P}}{4}. \quad (13.55)$$

Substituting this into (13.48), we obtain (on the branch of the contour C_+),

$$\hat{S}_0(x^+, x'^+) = \frac{1}{4}\left(\text{sgn}(x^+ - x'^+) + \hat{\mathcal{O}}_0\right)$$

$$= \frac{1}{4}\left(\text{sgn}(x^+ - x'^+) + 1 - 2n_{\text{F}}\left(\frac{\hat{P}}{2}\right)\right)$$

$$= \frac{1}{2}\left(\theta(x^+ - x'^+) - n_{\text{F}}\left(\frac{\hat{P}}{2}\right)\right). \tag{13.56}$$

This structure is very reminiscent of the $0 + 1$ dimensional fermion propagator (13.27) (unlike the $0 + 1$ case, here $m = 0$, but the momentum p is nonzero which was not there in the earlier model). Furthermore, let us define, as in $0 + 1$ dimensional case (and remember that we are on the C_+ branch),

$$\hat{\Phi}(x^+) = \int d\bar{x}^+ \, \theta(x^+ - \bar{x}^+)\hat{A}_+(\bar{x}^+), \tag{13.57}$$

which satisfies

$$\hat{\Phi}(-\infty) = 0, \quad \hat{\Phi}(\infty) = \hat{a}_+ = \int d\bar{x}^+ \, \hat{A}_+(\bar{x}^+), \tag{13.58}$$

so that we can write the complete propagator in (13.48) on C_+ as

$$\hat{S}(x^+, x'^+) = e^{-ie\hat{\Phi}(x^+)}\frac{1}{4}\left(\text{sgn}(x^+ - x'^+) + \hat{\mathcal{O}}\right)e^{ie\hat{\Phi}(x'^+)}. \tag{13.59}$$

Here $\hat{\mathcal{O}}$ is defined in (13.51).

The form of the complete fermion propagator in (13.59) is very similar to the $0 + 1$ dimensional case given in (13.28). Following our earlier derivation, we can now show that the complete fermion propagator $\hat{S}$ can be written in terms of $\hat{S}_0$ as

$$\hat{S}(x^+, x'^+) = \left((1 + 2ie\hat{S}_0\hat{A}_+)^{-1}\hat{S}_0\right)(x^+, x'^+), \tag{13.60}$$

which can be compared with (13.34) and can also be written as

$$\hat{S}(x^+, x'^+) = \hat{S}_0(x^+, x'^+) - 2ie\left(\hat{S}_0\hat{A}_+\hat{S}\right)(x^+, x'^+), \tag{13.61}$$

to be compared with (13.32). These are the Lippmann-Schwinger equations for the right handed fermions and similar relations can be derived in a parallel manner for the left handed fermions as well. We should point out here that the factor of 2 in (13.60) as well as in (13.61) is a consequence of our using light-cone coordinates (see (13.40)).

The fermion propagator in (13.48) involves operators which do not commute in general. However, let us recall from (13.14) that the calculation of the effective action involves the coincidence points, $x^+ = x'^+$ where the complete propagator takes the form

$$\hat{S}_c(x^+, x^+) = \frac{1}{4} e^{-ie\hat{\Phi}(x^+)} \hat{\mathcal{O}} e^{ie\hat{\Phi}(x^+)}. \tag{13.62}$$

In this limit, it is easy to show that the two exponential operators in (13.62) cancel each other. There are various ways in which one can show this, but the simplest is from the equations (13.47) satisfied by the propagator. Namely, we note that

$$i\partial_+ \hat{S}(x^+, x^+) = \lim_{x'^+ \to x^+} \left(i\partial_+ \hat{S}(x^+, x'^+) + i\partial'_+ \hat{S}(x^+, x'^+) \right)$$

$$= e\hat{A}(x^+)\hat{S}(x^+, x^+) + \lim_{x'^+ \to x^+} \frac{i}{2}\delta_c(x^+ - x'^+)$$

$$- e\hat{A}(x^+)\hat{S}(x^+, x^+) - \lim_{x'^+ \to x^+} \frac{i}{2}\delta_c(x^+ - x'^+)$$

$$= 0. \tag{13.63}$$

Namely, at the coincident points, the fermion propagator is independent of the coordinate, as also was the case in the $0 + 1$ dimensional model. This can also be shown from the Ward identities of the theory. Since the operator $\hat{\mathcal{O}}$ is independent of the coordinates, it follows from (13.62) and (13.63) that the two exponential operator commute with $\hat{\mathcal{O}}$ and cancel each other in the limit of coincidence. As a result, in this limit, we can write

$$\hat{S}_c(x^+, x^+) = \frac{1}{4} \hat{\mathcal{O}} = \frac{1}{4} \left(\mathbb{1} - 2(\hat{\mathcal{O}}_+ + \mathbb{1})^{-1} \right), \tag{13.64}$$

where we have used relation (13.51). We see from (13.15) that it is this function that needs to be integrated to determine the effective action in the right handed fermion sector.

To determine the effective action, we note from the definition of $\hat{\mathcal{O}}_+$ in (13.52) that we have

$$\frac{d\hat{\mathcal{O}}_+}{d\hat{a}_+} = ie\hat{\mathcal{O}}_+, \qquad \hat{a}_+ = \hat{a}_{+(+)} - \hat{a}_{+(-)}. \tag{13.65}$$

Therefore, we see that if we write

$$\hat{F} = \frac{1}{2} \ln \hat{O}_{+}, \quad \frac{d\hat{F}}{d\hat{a}_{+}} = \frac{1}{2\hat{O}_{+}} \frac{d\hat{O}_{+}}{d\hat{a}_{+}} = \frac{ie}{2}, \quad \hat{O}_{+} = e^{2\hat{F}}, \qquad (13.66)$$

$$\hat{O} = 1 - 2(\hat{O}_{+} + 1)^{-1} = \tanh \hat{F}. \qquad (13.67)$$

It follows now that

$$\frac{d \ln \cosh \hat{F}}{d\hat{a}_{+}} = \tanh \hat{F} \frac{d\hat{F}}{d\hat{a}_{+}} = \frac{ie}{2} \tanh \hat{F} = \frac{ie}{2} \hat{O}, \qquad (13.68)$$

and the effective action in the right hand sector can be obtained by integrating the propagator (13.64) with respect to $\hat{a}_{+}$ (see, for example, (13.15) or (13.36)), which upon using (13.68), yields

$$\Gamma_{\mathrm{R,eff}} = e \int \frac{dk}{2\pi} \int_{0}^{\hat{a}_{+}} d\hat{a}_{+} \, \langle k | \hat{S}_{c}(x^{+}, x^{+}) | k \rangle$$

$$= e \int \frac{dk}{2\pi} \int_{0}^{\hat{a}_{+}} d\hat{a}_{+} \, \langle k | \frac{1}{4} \hat{O} | k \rangle = -\frac{i}{2} \int \frac{dk}{2\pi} \int_{0}^{\hat{a}_{+}} d\hat{a}_{+} \langle k | \frac{ie}{2} \hat{O} | k \rangle$$

$$= -\frac{i}{2} \int \frac{dk}{2\pi} \, \langle k | \ln \cosh \hat{F} - \ln \cosh \hat{F}_{0} | k \rangle$$

$$= -\frac{i}{2} \int \frac{dk}{2\pi} \, \langle k | \ln \cosh \left(\frac{1}{2} \ln \hat{O}_{+} - \ln \cosh \frac{\beta \hat{P}}{4} \right) | k \rangle$$

$$= -\frac{i}{2} \int \frac{dk}{2\pi} \, \langle k | \ln \cosh \left(\frac{1}{2} \ln \hat{O}_{+} \right) | k \rangle, \qquad (13.69)$$

where we have neglected an unimportant constant term in the last step, as was also done in the $0 + 1$ dimensional case. In a parallel manner, we can also calculate the effective action for the left handed fermions which has the form (neglecting a constant term)

$$\Gamma_{\mathrm{L,eff}} = -\frac{i}{2} \int \frac{dk}{2\pi} \, \langle k | \ln \cosh \left(\frac{1}{2} \ln \hat{O}_{-} \right) | k \rangle, \qquad (13.70)$$

where (see also (13.52))

$$\hat{O}_{-} = e^{\frac{ie\hat{a}_{-}}{2}} e^{\frac{\beta \hat{P}}{2}} e^{\frac{ie\hat{a}_{-}}{2}}, \quad \hat{a}_{-} = \hat{a}_{-(+)} - \hat{a}_{-(-)}, \qquad (13.71)$$

and (see also (13.53))

$$\hat{a}_{-(\pm)} = \int dx^{-} \, A_{-(\pm)}(x^{-}). \qquad (13.72)$$

The perturbative expansion of these effective actions can be compared with the perturbative calculations and we emphasize again that the zero temperature limit of these will not coincide with the regularized calculations at zero temperature. These two examples, the $0 + 1$ and $1 + 1$ dimensional QED, elucidate how the alternative to the proper time method works at finite temperature.

13.3 Derivative expansion

There is yet another method for calculating effective actions known as the derivative expansion. The low energy properties of a fundamental theory can be quite profitably studied through the derivative expansion which we will describe next. The basic idea behind derivative expansion is quite simple. When a theory contains a mass scale which is large compared to the energy scales at which we are interested, we can define an effective theory by integrating out these heavy fields. The resulting effective theory can, then, be expanded in a power series in the ratio of the energy scale and the heavy mass scale. Clearly, if one is interested in processes involving energy scales much smaller than the heavy mass scale, then, only a few terms in the expansion will be able to describe the low energy phenomena quite successfully. This method of expansion in powers of momentum or derivatives is commonly known as the derivative expansion. One can, equivalently, think of this also as an inverse (heavy) mass expansion. The method of derivative expansion has been used, in this way, in the study of decays such as $\pi^0 \rightarrow 2\gamma$ as well as in the study of the properties of Skyrmions, just to name a few. It can also be used to study properties such as superconductivity, superfluidity etc. Clearly, it makes sense to make an expansion in powers of momentum only if there is a heavy mass in the theory. In the absence of such a mass scale, it is known that such an expansion is plagued with infrared divergences which leads to a breakdown of the derivative expansion. However, in some simple two dimensional models, even in the absence of a heavy mass, we will see that the method of derivative expansion works, namely, even though every term in the expansion is infrared divergent, the divergences cancel when the series is summed.

The method of derivative expansion can be best described with the example of a fermion field with a heavy mass interacting with other fields which are not as massive. Thus, for example, we can consider the Lagrangian density for the heavy fermion to be of the form

$$\mathcal{L} = \overline{\psi}\,(i\partial\!\!\!/ - m - K(\phi))\psi = \overline{\psi}\,(S_{\mathrm{f}}^{-1} - K(\phi))\psi, \tag{13.73}$$

where the fermion mass, m, is assumed to be heavy. $K(\phi)$ represents the interaction of some generic bosonic field ϕ with the fermions ($K(\phi)$ is a general functional of the scalar field ϕ) while S_{f} is the Green's function for the fermion defined to be (the Feynman "$i\epsilon$" prescription is to be understood)

$$S_{\mathrm{f}} = \frac{1}{i\partial\!\!\!/ - m}, \tag{13.74}$$

which, in the momentum space, takes the form

$$S_{\mathrm{f}}(p) = \frac{1}{p\!\!\!/ - m}. \tag{13.75}$$

The Lagrangian density for the fermion in (13.73) is bilinear in the fermion fields which can be integrated out in a straightforward manner in the path integral formalism to yield a determinant, namely,

$$Z(\phi) = N \int \mathcal{D}\overline{\psi}\mathcal{D}\psi\, e^{i\int \mathrm{d}^4 x\,\mathcal{L}}$$

$$= N \det\,(S_{\mathrm{f}}^{-1} - K(\phi))$$

$$= N(\det S_{\mathrm{f}}^{-1})\,[\det\,(1 - S_{\mathrm{f}}K(\phi))]$$

$$= N' \det\,(1 - S_{\mathrm{f}}K(\phi)), \quad N' = N\,(\det S_{\mathrm{f}}^{-1}). \tag{13.76}$$

Here N is a normalization constant and in the last line we have absorbed an uninteresting constant into the normalization constant N to write the generating functional in terms of the Matthews-Salam determinant. Therefore, we see that integrating out the fermion fields yields an effective action for the bosonic field ϕ of the form

$$Z(\phi) = N'\, e^{iS_{\mathrm{eff}}(\phi)} = N'\, e^{\mathrm{Tr}\,\ln(1 - S_{\mathrm{f}}K(\phi))}$$

$$\text{or,} \quad S_{\mathrm{eff}}(\phi) = -i\,\mathrm{Tr}\,\ln\,(1 - S_{\mathrm{f}}K(\phi)), \tag{13.77}$$

where we have neglected an uninteresting additive constant $(\ln N')$ in writing the effective action for the bosonic field and have used the fact that for any matrix, A (see also (13.5)),

$$\det A = e^{\text{Tr} \ln A}, \tag{13.78}$$

where "Tr" represents a trace over any matrix index the fermion fields may have (Dirac indices, internal symmetry indices etc.) including an integration over space-time coordinates (indices).

In general, the expression on the right hand side of (13.77) can not be evaluated in a closed form. However, when the fermion field is massive, we can expand the determinant as a series in powers of the derivatives (since the fermion propagator S_f is inversely proportional to the fermion mass) in the following way. We note first that for any matrix, A, we can expand the logarithm as usual as

$$\ln(1 - A) = -A - \frac{A^2}{2} - \frac{A^3}{3} + \cdots . \tag{13.79}$$

Moreover, if we think of the Green's function, $S_f(p)$, as a momentum dependent operator and $K(\phi(x))$ as a coordinate dependent operator, then, "Tr" can be simply written as

$$\text{Tr} A = \text{tr} \int d^4x \, \langle x|A|x \rangle, \tag{13.80}$$

where "tr" stands for the trace over other indices such as the Dirac index and the internal symmetry indices. With these, we can expand the right hand side of (13.77) as

$$\text{Tr} \ln(1 - S_f K(\phi)) = -\text{tr} \int d^4x \, \langle x|[S_f(p)K(\phi(x))$$

$$+ \frac{S_f(p)K(\phi(x))S_f(p)K(\phi(x))}{2} + \cdots]|x \rangle. \tag{13.81}$$

The expression in (13.81) is not yet in a form suitable for direct evaluation because the momentum and the coordinate operators do not commute. However, we can move the momentum dependent factors past the coordinate dependent ones through the use of various identities such as (note from (13.74) and (13.75) the usual identification $p_\mu \to i\partial_\mu$ in the coordinate basis)

$$[p_\mu, g(x)] = i(\partial_\mu g(x)),$$

$$[p^2, g(x)] = (\Box g(x)) + 2ip^\mu(\partial_\mu g(x)), \tag{13.82}$$

$$\left[\frac{1}{p^2}, g(x)\right] = \frac{1}{(p^2)^2}[p^2, g(x)] + \frac{1}{(p^2)^3}[p^2, [p^2, g(x)]] + \cdots.$$

While the first two identities in (13.82) are straightforward, the third can be obtained from the series expansion of the integral representation

$$\frac{1}{p^2} = \int_0^\infty d\beta \, e^{-\beta p^2}. \tag{13.83}$$

With the help of these identities (and others following from these), all the momentum dependent terms in (13.81) can be moved past the coordinate dependent terms so that all the coordinate dependent terms are on the right (or left) and each term in the expression in (13.81) can be written in the form

$$\text{tr} \int d^4x \, \langle x|A(p)B(x)|x\rangle = \text{tr} \int d^4x d^4p \, \langle x|A(p)|p\rangle \langle p|B(x)|x\rangle$$

$$= \text{tr} \int \frac{d^4p}{(2\pi)^4} \, A(p) \int d^4x \, B(x). \tag{13.84}$$

Here we have used the fact that the inner product of the coordinate and the momentum basis states merely gives the phase factor

$$\langle x|p\rangle = \frac{e^{-ip\cdot x}}{(2\pi)^2}. \tag{13.85}$$

Furthermore, the integrands in (13.84) are ordinary functions and the momentum integration would lead to a multiplicative factor while the term with the coordinate integration has exactly the form of an action. Comparing with (13.77) and (13.81), we see now that integrating out the fermion fields leads to an effective action for the bosonic fields which would be a series in powers of the derivatives as is clear from the identities in (13.82). This is the spirit of the derivative expansion and even though we have discussed this within the context of interacting fermions, it works equally well in other theories.

13.3.1 Soluble models. We have discussed the general idea behind the derivative expansion in the last section. In this section, we will describe how it works in the context of simple two dimensional models which are known to be soluble. This would not only clarify how the method works, but would also bring out the interesting feature that in some models without a heavy mass, the derivative expansion still works without any problem of infrared divergence. Let us consider, in $1 + 1$ dimensions, a massless fermion interacting with an Abelian gauge field. This is conventionally known as the Schwinger model where the Lagrangian density for the fermion is given by (see (5.11) and (5.13))

$$\mathcal{L}_{\mathrm{f}} = i\overline{\psi}\gamma^{\mu}D_{\mu}\psi = \overline{\psi}\gamma^{\mu}(i\partial_{\mu} + gA_{\mu})\psi, \tag{13.86}$$

and the action in two dimensions takes the form

$$S_{\mathrm{f}} = \int \mathrm{d}^2x\, \mathcal{L}_{\mathrm{f}}. \tag{13.87}$$

The fermions are massless and it is easy to check that the gauge coupling, g, in (13.86) has the canonical dimension of a mass.

In two dimensions, the Dirac gamma matrices satisfy various identities. Namely, with $\eta^{\mu\nu} = (+, -)$, γ^0 Hermitian and γ^1 anti-Hermitian, if we define a Hermitian $\gamma_5 = \gamma^0\gamma^1$, then it is easy to check that $((\gamma^0)^2 = 1, (\gamma^1)^2 = -1$ and $(\gamma_5)^2 = 1)$

$$\gamma^{\mu}\gamma^{\nu} = \eta^{\mu\nu} + \epsilon^{\mu\nu}\gamma_5,$$

$$\text{or,} \quad \epsilon^{\mu\nu}\gamma_{\nu} = \gamma_5\gamma^{\mu}, \tag{13.88}$$

where the totally anti-symmetric tensor, $\epsilon^{\mu\nu}$, is the two dimensional Levi-Civita tensor, satisfying $\epsilon^{01} = 1$. Furthermore, the dual of a vector in two dimensions is also a vector, defined as

$$A_{\mu}^5 = \epsilon_{\mu\nu}A^{\nu}, \tag{13.89}$$

and, therefore, we can write the Lagrangian density of (13.86) also as

$$\mathcal{L}_{\mathrm{f}} = i\overline{\psi}\gamma^{\mu}D_{\mu}\psi = \overline{\psi}\gamma^{\mu}(i\partial_{\mu} + gA_{\mu})\psi$$

$$= \overline{\psi}\gamma^{\mu}(i\partial_{\mu} + g\xi A_{\mu} + g\eta\gamma_5 A_{\mu}^5)\psi, \qquad \xi + \eta = 1. \tag{13.90}$$

As is well known, different values of the parameters ξ and η correspond to different regularizations used to study the theory in the path integral formalism. In particular, $\xi = 1$ and $\eta = 0$ corresponds to choosing a gauge invariant regularization while $\xi = 0$ and $\eta = 1$ corresponds to the use of a regularization which preserves the axial gauge invariance.

The generating functional, as we have seen in (13.77), is given by

$$Z[A_\mu] = N' e^{iS_{\text{eff}}[A_\mu]} = N \int \mathcal{D}\overline{\psi}\, \mathcal{D}\psi\, e^{iS_f}, \tag{13.91}$$

and is best evaluated by rotating to Euclidean space where the path integrals are well behaved. In rotating to Euclidean space, we treat both ψ and $\overline{\psi}$ as independent fields, $x^0 \to -ix_2$, $\partial_0 \to i\partial_2$ and $\gamma^0 \to i\gamma_2$. Consequently, $\gamma_5 = \gamma^0\gamma^1 \to i\gamma_1\gamma_2$ continues to be Hermitian in Euclidean space although individual γ_μ's are anti-Hermitian. The identity of (13.88), in Euclidean space, takes the form

$$\gamma_\mu\gamma_\nu = -\delta_{\mu\nu} - i\epsilon_{\mu\nu}\gamma_5, \tag{13.92}$$

where the Levi-Civita tensor in the Euclidean space has the value $\epsilon_{12} = 1$. We also note that, for the covariant derivative to transform as the ordinary derivative, namely, for $D_0 \to iD_2$, we must have $A_0 \to iA_2$ as well as $A_0^5 \to iA_2^5$. This shows that it is impossible to maintain the reality of A_μ and A_μ^5 in the Euclidean space simultaneously with the duality between them (see (13.89)). Therefore, we choose to treat them as independent real fields in the Euclidean space. After all the calculations are done, and we have rotated back to Minkowski space, we can use the duality between these two fields, namely, (13.89).

With all these relations, we note that the action for the fermion fields, in Euclidean space, becomes

$$S_f = i \int \mathrm{d}^2 x\, \overline{\psi}\gamma_\mu(i\partial_\mu + g\xi A_\mu + g\eta\gamma_5 A_\mu^5)\psi, \tag{13.93}$$

and following (13.77) we can write down the generating functional to have the form

$$Z[A_\mu] = N' \det(1 + S_f(g\xi \slashed{A} - g\eta\gamma_5 \slashed{A}^5)), \tag{13.94}$$

so that the Euclidean effective action takes the form (see (13.77))

$$S_{\text{eff}}[A_\mu] = \text{Tr}\ln\left(1 + S_{\text{f}}(g\xi\slashed{A} - g\eta\gamma_5\slashed{A}^5)\right)$$

$$= \text{Tr}\sum_{n=0}\frac{(-1)^n}{n+1}\left(S_{\text{f}}(g\xi\slashed{A} - g\eta\gamma_5\slashed{A}^5)\right)^{n+1}. \qquad (13.95)$$

In the present case, the fermion fields are massless and, conse-quently, S_{f} has the simple form

$$S_{\text{f}}(p) = \frac{1}{\slashed{p}} = \frac{\slashed{p}}{p^2}. \qquad (13.96)$$

It is easy to show that in $1+1$ dimensions, the series in (13.95) involves only a finite number of terms. To see this, let us define

$$\slashed{B} = g(\xi\slashed{A} - \eta\gamma_5\slashed{A}^5), \qquad (13.97)$$

and note that in $1+1$ dimension, any vector has the unique decom-position into a longitudinal part and a transverse part as

$$B_\mu = \partial_\mu\phi(x) + \epsilon_{\mu\nu}\partial^\nu\rho(x). \qquad (13.98)$$

As a result, we can write (we will show the vanishing of the higher order terms in Minkowski space, for simplicity)

$$S_{\text{f}}(p)(g\xi\slashed{A} - g\eta\gamma_5\slashed{A}^5) = \frac{1}{\slashed{p}}\slashed{B} = \frac{1}{\slashed{p}}(\slashed{\partial}\phi(x) - \gamma_5\slashed{\partial}\rho(x)), \qquad (13.99)$$

where we have used (13.88). Furthermore, using the commutation relations in (13.82), we can write

$$\frac{1}{\slashed{p}}\slashed{B}(x) = \frac{1}{\slashed{p}}(\slashed{\partial}\phi(x) - \gamma_5\slashed{\partial}\rho(x))$$

$$= \frac{1}{\slashed{p}}(-i)\,[\slashed{p}, \phi(x)] + \gamma_5\frac{1}{\slashed{p}}(-i)\,[\slashed{p}, \rho(x)]$$

$$= -i\phi(x) + i\frac{1}{\slashed{p}}\phi(x)\slashed{p} - i\gamma_5\rho(x) + i\gamma_5\frac{1}{\slashed{p}}\rho(x)\slashed{p}. \qquad (13.100)$$

Therefore, we can write

$$\frac{1}{\not{p}}\mathcal{B}(x)\frac{1}{\not{p}}\mathcal{B}(x) = \left(-i\phi(x)\frac{1}{\not{p}}+i\frac{1}{\not{p}}\phi(x)-i\gamma_5\rho(x)\frac{1}{\not{p}}+i\gamma_5\frac{1}{\not{p}}\rho(x)\right)\mathcal{B}(x)$$

$$= -i\left[\phi(x),\frac{1}{\not{p}}\mathcal{B}(x)\right] - i\gamma_5\left[\rho(x),\frac{1}{\not{p}}\mathcal{B}(x)\right]$$

$$= -i\left[\phi(x)+\gamma_5\rho(x),\frac{1}{\not{p}}\mathcal{B}(x)\right], \qquad (13.101)$$

where we have used the fact that $\rho(x)$ commutes with $\mathcal{B}(x)$ (namely, functions of coordinates commute with each other).

It is clear now, from (13.101), that, for $n \geq 1$,

$$\left(\frac{1}{\not{p}}\mathcal{B}\right)^{n+1} = -\frac{i}{n}\left[\phi(x)+\gamma_5\rho(x),\left(\frac{1}{\not{p}}\mathcal{B}\right)^n\right], \qquad (13.102)$$

which leads to the fact that, for $n \geq 1$,

$$\text{Tr}\left(\frac{1}{\not{p}}\mathcal{B}\right)^{n+1} = -\frac{i}{n}\text{Tr}\left[\phi(x)+\gamma_5\rho(x),\left(\frac{1}{\not{p}}\mathcal{B}\right)^n\right]. \qquad (13.103)$$

From simple power counting, we note that in $1+1$ dimension, the momentum integrals would be divergent only for $n = 0, 1$. For $n > 1$, on the other hand, the integrals are convergent and consequently we can use the cyclicity properties of "Tr" to conclude that

$$\text{Tr}\left(\frac{1}{\not{p}}\mathcal{B}\right)^{n+1} = 0 \qquad \text{for } n > 1. \qquad (13.104)$$

This shows that the series in (13.95) contains only two terms

$$S_{\text{eff}} = \text{Tr}\left(g\frac{\not{p}}{p^2}(\xi\mathcal{A} - \eta\gamma_5\mathcal{A}^5)\right.$$

$$\left. - \frac{g^2}{2}\frac{\not{p}}{p^2}(\xi\mathcal{A} - \eta\gamma_5\mathcal{A}^5)\frac{\not{p}}{p^2}(\xi\mathcal{A} - \eta\gamma_5\mathcal{A}^5)\right). \qquad (13.105)$$

Furthermore, the first term in the series vanishes because it is odd in the momentum variable (and, therefore, the momentum integral would vanish) and, consequently, the series contains only a single term which is quadratic in the field variables. Using the commutation relations in (13.82), we can simplify this to be

$$
S_{\text{eff}} = -\frac{g^2}{2} (\text{tr } \gamma_\mu \gamma_\nu \gamma_\lambda \gamma_\rho) \int d^2x \frac{d^2p}{(2\pi)^2} \frac{p_\mu(p_\lambda - i\partial_\lambda)}{p^2(p - i\partial)^2}
$$
$$
\times (\xi^2 A_\nu(x) A_\rho(x) + \eta^2 A_\nu^5(x) A_\rho^5(x))
$$
$$
- \frac{g^2 \xi \eta}{2} (\text{tr} \gamma_5 \gamma_\mu \gamma_\nu \gamma_\lambda \gamma_\rho) \int d^2x \frac{d^2p}{(2\pi)^2} \frac{p_\mu(p_\lambda - i\partial_\lambda)}{p^2(p - i\partial)^2}
$$
$$
\times (A_\nu(x) A_\rho^5(x) + A_\nu^5(x) A_\rho(x)). \quad (13.106)
$$

Here the derivatives are understood to act only on the first of the two fields. It is worth emphasizing here that normally, in derivative expansion, one expands in powers of the derivatives to obtain an infinite series at every order of expansion. Here, in contrast, we have summed the infinite series to obtain the simple expression above. Furthermore, we have separated the Dirac trace, which will be done after the momentum integral has been evaluated using dimensional regularization (so that the subtleties associated with γ_5 can be avoided), and leads to

$$
S_{\text{eff}} = -\frac{g^2}{2\pi} \int d^2x \left[\xi^2 \left(\delta_{\mu\nu} - \frac{\partial_\mu \partial_\nu}{\Box} \right) A_\mu(x) A_\nu(x) \right.
$$
$$
+ \eta^2 \left(\delta_{\mu\nu} - \frac{\partial_\mu \partial_\nu}{\Box} \right) A_\mu^5(x) A_\nu^5(x)
$$
$$
- \frac{\xi\eta}{2} \left(\epsilon_{\lambda\mu} \frac{\partial_\nu \partial_\lambda}{\Box} + \epsilon_{\lambda\nu} \frac{\partial_\mu \partial_\lambda}{\Box} \right) A_\mu(x) A_\nu^5(x)
$$
$$
\left. - \frac{\xi\eta}{2} \left(\epsilon_{\lambda\mu} \frac{\partial_\nu \partial_\lambda}{\Box} + \epsilon_{\lambda\nu} \frac{\partial_\mu \partial_\lambda}{\Box} \right) A_\mu^5(x) A_\nu(x) \right]. \quad (13.107)
$$

It is understood here that the derivatives act only on the first of the two field variables.

In Euclidean space, A_μ and A_μ^5 are treated as independent. However, now that the integrals are evaluated, we can rotate everything back to Minkowski space where we can use the duality relation,

(13.89), between A_μ and A_μ^5 to write the effective action as

$$S_{\text{eff}} = \frac{g^2}{2\pi} \int d^2x \, A_\mu(x) \left(\xi \eta^{\mu\nu} - \frac{\partial^\mu \partial^\nu}{\Box} \right) A_\nu(x)$$

$$= -\frac{g^2}{2\pi} \int d^2x \, A_\mu(x)(\eta \eta^{\mu\alpha} \eta^{\nu\beta} + \xi \epsilon^{\mu\alpha} \epsilon^{\nu\beta}) \frac{\partial_\alpha \partial_\beta}{\Box} A_\nu(x). \quad (13.108)$$

It is clear now that different values of ξ and η (such that $\xi + \eta = 1$) lead to different effective actions corresponding to use of distinct regularizations. In particular, when $\xi = 1$ and $\eta = 0$, the effective action is obviously gauge invariant.

We can now study various soluble two dimensional models directly from (13.108). Let us consider the massive vector boson model whose Lagrangian density is given by

$$\mathcal{L} = -\frac{1}{4} F_{\mu\nu} F^{\mu\nu} + \frac{\mu^2}{2} A_\mu A^\mu + \mathcal{L}_f$$

$$= \mathcal{L}[A_\mu] + \mathcal{L}_f, \quad (13.109)$$

where μ is the mass of the vector boson and $\mathcal{L}_f$ represents the fermion Lagrangian density of (13.86). In this case, the Euler-Lagrange equation for the vector field gives

$$\partial_\nu F^{\mu\nu} = \mu^2 A^\mu, \quad (13.110)$$

whose consistency requires that

$$\partial_\mu A^\mu = 0. \quad (13.111)$$

Once again, integrating out the fermion fields, subject to the condition of (13.111), would yield an effective Lagrangian density for the gauge fields which can be obtained from (13.108) so that the resulting Lagrangian density for the vector field becomes (remember the two dimensional identity $\epsilon^{\mu\alpha} \epsilon^{\nu\beta} = -\eta^{\mu\nu} \eta^{\alpha\beta} + \eta^{\mu\beta} \eta^{\nu\alpha}$ and $\partial_\mu A^\mu = 0$ from (13.111)).

$$\mathcal{L}_{\text{TOT}} = -\frac{1}{4} F_{\mu\nu} F^{\mu\nu} + \frac{m_R^2}{2} A_\mu A^\mu. \quad (13.112)$$

We see that the fermion interactions merely lead to a renormalization of the vector boson mass as

$$m_R^2 = \mu^2 + \frac{\xi g^2}{2\pi}. \quad (13.113)$$

In particular, we note that in the case of the Schwinger model where the gauge boson is massless, *i.e.*, $\mu = 0$, a gauge invariant regularization ($\xi = 1$) would lead to a gauge boson mass

$$m_{\rm R}^2 = \frac{g^2}{2\pi},\tag{13.114}$$

which is the well known result.

In a similar manner, we can also evaluate the generating functional for the massless Thirring model as follows. The Thirring model represents a self-interacting fermion theory in $1+1$ dimension and is described by the Lagrangian density

$$\mathcal{L} = i\overline{\psi}\slashed{\partial}\psi - \frac{\lambda}{2}\overline{\psi}\gamma^\mu\psi\overline{\psi}\gamma_\mu\psi + \overline{\psi}\gamma^\mu\psi A_\mu$$

$$= \mathcal{L}_{\rm f} - \frac{\lambda}{2}\overline{\psi}\gamma^\mu\psi\overline{\psi}\gamma_\mu\psi.\tag{13.115}$$

Here we have added an external source, A_μ, to the Lagrangian density (to derive correlation functions) and $\mathcal{L}_{\rm f}$ represents the Lagrangian density of (13.86). The generating functional for the Thirring model can now be written as

$$Z_{\rm TH}[A_\mu] = N \int \mathcal{D}\overline{\psi}\mathcal{D}\psi \, e^{i\int {\rm d}^2 x(\mathcal{L}_{\rm f} - \frac{\lambda}{2}\overline{\psi}\gamma^\mu\psi\overline{\psi}\gamma_\mu\psi)}$$

$$= \exp\left(\frac{i\lambda}{2}\int {\rm d}^2 x \, \frac{\delta^2}{\delta A_\mu(x)\delta A^\mu(x)}\right) N \int \mathcal{D}\overline{\psi}\mathcal{D}\psi \, e^{i\int {\rm d}^2 x \, \mathcal{L}_{\rm f}}.\tag{13.116}$$

The generating functional for the fermion fields is already evaluated in (13.108). Substituting this into (13.116) yields the generating functional for the Thirring model to be

$$Z_{\rm TH} = N' \exp\left(-\frac{i}{2\pi}\int {\rm d}^2 x \, A_\mu\left[\frac{\eta\eta^{\mu\alpha}\eta^{\nu\beta}}{1 - \frac{\lambda\eta}{\pi}} + \frac{\xi\epsilon^{\mu\alpha}\epsilon^{\nu\beta}}{1 + \frac{\lambda\xi}{\pi}}\right]\frac{\partial_\alpha\partial_\beta}{\Box}A_\nu\right).\tag{13.117}$$

Other simple two dimensional models can also be solved in this manner and the important thing to note here is that even without a heavy mass for fermions, the derivative expansion works without any problems of infrared divergence.

13.3.2 Derivative expansion at $T \neq 0$. As we have seen earlier, there are novel features that develop at finite temperature. In particular, we have seen, in section **1.7.2**, that at finite temperature, Feynman amplitudes become nonanalytic at the origin in the momentum space. It is, therefore, of interest to investigate whether the derivative expansion continues to hold at finite temperature. To examine this, we will study a toy model (in $3 + 1$ dimensions) of a heavy scalar field interacting with a lighter scalar field through a cubic term described by the Lagrangian density (see, for example, (1.114))

$$\mathcal{L}(B, \phi) = \frac{1}{2}\partial_\mu B \partial^\mu B - \frac{M^2}{2}B^2 - \frac{g}{2}\phi B^2 + \mathcal{L}_0(\phi). \qquad (13.118)$$

We assume that $\mathcal{L}_0(\phi)$ represents the free Lagrangian density for the lighter scalar field, ϕ, and M denotes the heavy mass associated with the scalar field B. Clearly, this is an unrealistic model. However, this has been extensively studied in connection with the nonanalyticity properties at finite temperature and, therefore, it is meaningful to study this model in connection with derivative expansion.

Since the theory is at most quadratic in the heavy B field, we can integrate it out in the path integral to obtain an effective action for the lighter ϕ field as

$$\begin{aligned}
Z &= N' \int \mathcal{D}\phi \mathcal{D}B \, e^{iS(B,\phi)} \\
&= N' \int \mathcal{D}\phi \left(\det(G_B^{-1}(p) - g\phi(x)) \right)^{-\frac{1}{2}} e^{iS_0(\phi)} \\
&= N \int \mathcal{D}\phi \left(\det(1 - gG_B(p)\phi(x)) \right)^{-\frac{1}{2}} e^{iS_0(\phi)} \\
&= N \int \mathcal{D}\phi \, e^{i(S_0(\phi)+S_{\text{eff}}(\phi))} \\
&= N \int \mathcal{D}\phi \, e^{iS_{\text{TOT}}(\phi)}, \qquad (13.119)
\end{aligned}$$

where $G_B(p)$ denotes the Green's function for the B-field and we have absorbed an uninteresting determinant factor into the normalization constant (note also that the power of the scalar determinant is different from that for the fermions). Furthermore, with periodic boundary conditions for the B field, the Green's function for the

B-field at finite temperature has the form (see (2.50))

$$G_B(p) = \frac{1}{p^2 - M^2 + i\epsilon} - 2i\pi n_B(|p^0|)\delta(p^2 - M^2)$$

$$= G_B^{(0)}(p) + G_B^{(\beta)}(p), \tag{13.120}$$

with the bosonic distribution function given by

$$n_B(|p^0|) = \frac{1}{e^{\beta|p^0|} - 1}. \tag{13.121}$$

We note here that for one loop effects, as is the case here, we need only the ++ component of the propagator.

As before, we note that the integration of the heavy B field gives rise to an effective scalar action of the form (the multiplicative factor represents the power of the determinant involved)

$$S_{\text{eff}}(\phi) = \frac{i}{2}\text{Tr}\ln\left(1 - gG_B(p)\phi(x)\right)$$

$$= \frac{i}{2}\text{Tr}\left(-gG_B(p)\phi(x) - \frac{g^2}{2}G_B(p)\phi(x)G_B(p)\phi(x) + \cdots\right), \tag{13.122}$$

where we have expanded, as is done in derivative expansion, the logarithm in powers of the propagator. To compare with the earlier calculations (see section **1.7**) of nonanalyticity in the self-energy, for example, let us consider only the term quadratic in the ϕ field

$$S_{\text{eff}}^Q = -\frac{ig^2}{4}\text{Tr}\left(G_B(p)\phi(x)G_B(p)\phi(x)\right)$$

$$= S_{\text{eff}}^{Q\,(0)} + S_{\text{eff}}^{Q\,(\beta)}, \tag{13.123}$$

where we have separated S_{eff}^Q into its zero temperature part and a part that depends explicitly on temperature.

From the structure of the propagator at finite temperature in (13.120), we see that the temperature dependent part of S_{eff}^Q has the explicit form

$$S_{\text{eff}}^{Q(\beta)} = -\frac{ig^2}{4}\text{Tr}\left(G_B^{(\beta)}(p)\phi(x)G_B^{(0)}(p)\phi(x)\right.$$

$$\left. + G_B^{(0)}(p)\phi(x)G_B^{(\beta)}(p)\phi(x) + G_B^{(\beta)}(p)\phi(x)G_B^{(\beta)}(p)\phi(x)\right). \tag{13.124}$$

Once again, we can move all the momentum dependent factors past coordinate dependent ones through the relation

$$\phi(x)f(p) = \big(f(p - i\partial)\phi(x)\big), \tag{13.125}$$

where the bracket represents the action of the derivatives on the field, namely, the derivatives act only on $\phi(x)$. The derivative expansion (e.g., the last relation in (13.82)) is merely a Taylor series expansion of this formula in powers of derivatives. However, a Taylor series expansion is meaningful only if a function is analytic. Since, we do not know *a priori* whether the temperature dependent part of the action will be analytic, it is much more meaningful to use the relation in (13.125).

Using (13.125), we can move all the momentum dependent factors to the left, which would yield the temperature dependent part of the quadratic action to be

$$
\begin{aligned}
S_{\text{eff}}^{Q\,(\beta)} &= -\frac{ig^2}{4}\text{Tr}\Big(G_B^{(\beta)}(p)\big(G_B^{(0)}(p - i\partial)\phi(x)\big)\phi(x) \\
&\quad + G_B^{(0)}(p)\big(G_B^{(\beta)}(p - i\partial)\phi(x)\big)\phi(x) \\
&\quad + G_B^{(\beta)}(p)\big(G_B^{(\beta)}(p - i\partial)\phi(x)\big)\phi(x)\Big) \\
&= -\frac{g^2}{4}\int\frac{d^4p}{(2\pi)^3}\int d^4x\Big(\delta(p^2 - M^2)\frac{n_B(|p^0|)}{(p - i\partial)^2 - M^2 + i\epsilon}\phi(x) \\
&\quad + \frac{n_B(|p^0 - i\partial^0|)}{p^2 - M^2 + i\epsilon}\delta((p - i\partial)^2 - M^2)\phi(x) \\
&\quad - 2i\pi n_B(|p^0|)n_B(|p^0 - i\partial^0|)\delta(p^2 - M^2) \\
&\quad \times \delta((p - i\partial)^2 - M^2)\phi(x)\Big)\phi(x). \tag{13.126}
\end{aligned}
$$

Here, the derivatives are assumed to act on the first of the two ϕ fields.

The momentum integral in (13.126) can be easily checked to correspond to the integral studied in section **1.7.2** (in the real time formalism) in connection with the nonanalyticity of the self-energy. In fact, this can be evaluated using the modified Feynman parameterization discussed in **11.4** and it is easy to show that

$$\text{Re } S_{\text{eff}}^{Q\,(\beta)} = -\frac{g^2}{32\pi^2} \int d^4 x$$

$$\times \left(\int_0^\infty \frac{pdp}{\omega} \frac{n_B(\omega_p)}{(-\nabla^2)^{\frac{1}{2}}} \text{Re} (\ln R) \phi(x) \right) \phi(x), \quad (13.127)$$

where, as before, we have defined

$$\omega_p = (p^2 + M^2)^{\frac{1}{2}}, \quad p = |\mathbf{p}|, \quad (13.128)$$

and R is as defined in (12.79)

$$R = \frac{[\partial_\mu \partial^\mu + 2i(\omega\partial_0 + p(-\nabla^2)^{\frac{1}{2}})][\partial_\mu \partial^\mu - 2i(\omega\partial_0 - p(-\nabla^2)^{\frac{1}{2}})]}{[\partial_\mu \partial^\mu + 2i(\omega\partial_0 - p(-\nabla^2)^{\frac{1}{2}})][\partial_\mu \partial^\mu - 2i(\omega\partial_0 + p(-\nabla^2)^{\frac{1}{2}})]}.$$
$$(13.129)$$

As we have discussed earlier, in chapter **12**, this expression is manifestly nonanalytic bringing out the nonanalyticity of the self-energy at finite temperature yet in a different fashion. Although we have chosen to study the two point function (self-energy), the cubic, quartic and higher terms in S_{eff} may, in principle, develop similar nonanalyticities at finite temperature. Thus, we see that the derivative expansion breaks down at finite temperature. It is worth noting here, however, that if we had used commutation relations such as in (13.82) (as opposed to the true commutation relation of (13.125)), to move the momentum dependent terms past the coordinate dependent ones, then, we would have a well defined derivative expansion. This, however, would correspond to Taylor expanding the integrand before evaluating the integral and we can check that this corresponds to taking the limit $\partial_0 = 0$ and then $(-\nabla^2)^{\frac{1}{2}} \to 0$. However, there is no *a priori* reason why this should be the preferred limit. In other words, if the effective action is analytic, it should not matter whether we Taylor expand the integrand or the integral. However, when the effective action is nonanalytic, as is the case at finite temperature, expanding the integrand becomes questionable. We also note that effective potential, which is crucial in the study of symmetry breaking, is defined (see, for example, (9.56)) as the part of the effective action without any derivatives (*i.e.*, by setting $\partial_\mu = 0$). However, as we have just seen, at finite temperature, the effective action is nonanalytic precisely at this point and, consequently, the definition of the effective potential also becomes nonunique. This is an issue that is not completely understood, but it is worth pointing out that most physical examples such as the degenerate electron gas or superconductivity can be best understood from the point of view of an effective action in the limit $\partial_0 = 0$ followed by $(-\nabla^2)^{\frac{1}{2}} \to 0$.

13.3.3 Anomaly at $T \neq 0$. As we have seen in the last section, the effective action does become nonanalytic at finite temperature. However, there are still quantities of physical interest in a theory that do not show any nonanalyticity. One such quantity is the chiral anomaly which is known to be at the heart of the solubility of the two dimensional models. In fact, as we have seen earlier in section **6.6** within the context of the Schwinger model, the chiral anomaly is temperature independent. In this section, we will discuss how the anomaly can be derived within the context of derivative expansion and the question of its temperature dependence.

For the sake of comparison with the results of section **6.6**, we will also consider here the example of the Schwinger model with a slightly different notation for clarity. Let us consider the Lagrangian density

$$\mathcal{L}_{\mathrm{f}} = \overline{\psi}(i\partial\!\!\!/ + gV\!\!\!\!/ + \gamma_5 A\!\!\!\!/)\psi, \tag{13.130}$$

where we have represented the vectorial gauge field by V_μ while the external source for the axial vector current is denoted by A_μ. The generating functional, in the presence of this external source, is given by

$$Z[V_\mu, A_\mu] = e^{iS_{\mathrm{eff}}(V_\mu, A_\mu)} = \int \mathcal{D}\psi \mathcal{D}\overline{\psi} \, e^{i \int \mathrm{d}^2x \, \mathcal{L}_{\mathrm{f}}}, \tag{13.131}$$

so that we have

$$\left. \frac{\delta S_{\mathrm{eff}}}{\delta A_\mu(x)} \right|_{A_\mu=0} = \langle \overline{\psi} \gamma_5 \gamma^\mu \psi \rangle = \langle j_5^\mu(x) \rangle$$

$$\text{or,} \quad \langle \partial_\mu j_5^\mu(x) \rangle = \partial_\mu \left. \frac{\delta S_{\mathrm{eff}}}{\delta A_\mu(x)} \right|_{A_\mu=0}. \tag{13.132}$$

This defines the anomaly in the conservation of the axial current.

Following the discussion of the last section (see (13.82)), we see that S_{eff} can be identified with

$$S_{\text{eff}}(V_\mu, A_\mu) = -i \text{Tr} \ln(1 + S_{\text{f}}(p)(g \rlap{/}{V}(x) + \gamma_5 \rlap{/}{A}(x)))$$

$$= -i \sum_{n=0}^{\infty} \text{Tr} \frac{(-1)^n}{n+1} (S_{\text{f}}(p)(g \rlap{/}{V}(x) + \gamma_5 \rlap{/}{A}(x)))^{n+1}. \qquad (13.133)$$

The anomaly, as is clear from (13.132), is obtained only from the term that is linear in A_μ. The first term in the series of (13.133) which contains a term linear in A_μ vanishes following the discussion after (13.105) because the momentum integral is odd. Therefore, the only part of S_{eff} which gives rise to the anomaly comes from the quadratic terms of the series (in particular, from the crossed terms) and has the form

$$S_{\text{eff}}^{\text{anomaly}} = ig \, \text{Tr} \, (S_{\text{f}}(p) \rlap{/}{V}(x) S_{\text{f}}(p) \gamma_5 \rlap{/}{A}(x)), \qquad (13.134)$$

where we have used the cyclicity of the trace.

At zero temperature, the fermion Green's function is that of a massless fermion and has the form ($i\epsilon$ is assumed)

$$S_{\text{f}}(p) = S_{\text{f}}^0(p) = \frac{\rlap{/}{p}}{p^2}. \qquad (13.135)$$

Using this, we can write $S_{\text{eff}}^{\text{anomaly}}$ as

$$S_{\text{eff}}^{\text{anomaly}} = ig \, \text{Tr} \left(\frac{\rlap{/}{p}}{p^2} \rlap{/}{V}(x) \frac{\rlap{/}{p}}{p^2} \gamma_5 \rlap{/}{A}(x) \right)$$

$$= -ig \, \text{Tr}(\gamma_5 \gamma^\mu \gamma^\nu \gamma^\lambda \gamma^\rho) \left(\frac{p_\mu}{p^2} V_\nu(x) \frac{p_\lambda}{p^2} A_\rho(x) \right)$$

$$= -ig \, \text{Tr} \left((\gamma_5 \gamma^\mu \gamma^\nu \gamma^\lambda \gamma^\rho) \left[\frac{p_\mu}{p^2} \frac{(p - i\partial)_\lambda}{(p - i\partial)^2} V_\nu(x) \right] A_\rho(x) \right)$$

$$= -ig \, \text{tr} \, (\gamma_5 \gamma^\mu \gamma^\nu \gamma^\lambda \gamma^\rho)$$

$$\times \int \frac{d^2 p}{(2\pi)^2} d^2 x \left[\frac{p_\mu (p - i\partial)_\lambda}{p^2 (p - i\partial)^2} V_\nu(x) \right] A_\rho(x), \qquad (13.136)$$

where the derivatives are assumed to act on V_ν alone.

The momentum integral, in (13.136), can be done using dimensional regularization in a straightforward manner and leads to (It is worth emphasizing here that the evaluation here corresponds to a gauge invariant regularization.)

$$
S_{\text{eff}}^{\text{anomaly}} = -\frac{g}{2\pi} \int d^2 x \Big(\epsilon^{\mu\nu} V_\mu A_\nu
$$
$$
- (\eta^{\mu\nu} \epsilon^{\lambda\rho} + \eta^{\lambda\rho} \epsilon^{\mu\nu}) \frac{\partial_\mu \partial_\lambda}{\Box} V_\nu A_\rho \Big),
$$

(13.137)

so that from (13.132) and (13.137), we obtain the anomaly in the axial current to be

$$
\langle \partial_\mu j_5^\mu \rangle = \partial_\mu \frac{\delta S_{\text{eff}}^{\text{anomaly}}}{\delta A_\mu}
$$
$$
= -\frac{g}{2\pi} \epsilon^{\mu\nu} \partial_\nu V_\mu + \frac{g}{2\pi} \epsilon^{\mu\nu} \partial_\mu V_\nu
$$
$$
= \frac{g}{\pi} \epsilon^{\mu\nu} \partial_\mu V_\nu.
$$

(13.138)

This is, indeed, the gauge invariant anomaly in the Schwinger model at zero temperature and here we have derived it from the derivative expansion.

Let us next analyze the anomaly at finite temperature within the framework of the derivative expansion. At finite temperature, the fermion Green's function has the form (see (2.62))

$$
S_f(p) = \not{p} \left(\frac{1}{p^2} + 2i\pi n_f(|p^0|)\delta(p^2) \right)
$$
$$
= S_f^{(0)}(p) + S_f^{(\beta)}(p),
$$

(13.139)

where n_f represents the fermion distribution function

$$
n_f(|p^0|) = \frac{1}{e^{\beta|p^0|} + 1}.
$$

(13.140)

Substituting this into (13.134), we see that, in addition to the zero temperature part, the action for the anomaly will now contain a term that is manifestly temperature dependent and is given by (see (13.136))

$$S_{\text{eff}}^{\text{anomaly } (\beta)} = -ig \operatorname{Tr} \left[\gamma_5 (S_f^{(\beta)}(p) \slashed{V}(x) S_f^{(0)}(p) \slashed{A}(x) \right.$$
$$\left. + S_f^{(0)}(p) \slashed{V}(x) S_f^{(\beta)}(p) \slashed{A}(x) + S_f^{(\beta)}(p) \slashed{V}(x) S_f^{(\beta)}(p) \slashed{A}(x)) \right]. \tag{13.141}$$

We note that there are two kinds of terms – linear and quadratic in $S_f^{(\beta)}(p)$. As in (9.107), let us study them separately.

The two terms linear in $S_f^{(\beta)}(p)$ can be written as

$$\text{Linear} := -ig \operatorname{Tr} \left((\gamma_5 \gamma^\mu \gamma^\nu \gamma^\lambda \gamma^\rho) \right.$$
$$\times \left[p_\mu n_f(|p^0|) \delta(p^2) V_\nu \frac{p_\lambda}{p^2} + \frac{p_\mu}{p^2} V_\nu p_\lambda n_f(|p^0|) \delta(p^2) \right] A_\rho \bigg)$$
$$= -ig \operatorname{Tr} \left((\gamma_5 \gamma^\mu \gamma^\nu \gamma^\lambda \gamma^\rho) \left[p_\mu n_f(|p^0|) \delta(p^2) \frac{(p - i\partial)_\lambda}{(p - i\partial)^2} V_\nu \right. \right.$$
$$\left. \left. + \frac{p_\mu}{p^2} (p - i\partial)_\lambda n_f(|(p - i\partial)^0|) \delta((p - i\partial)^2) V_\nu \right] A_\rho \right)$$
$$= -2ig \, (\eta^{\mu\nu} \epsilon^{\lambda\rho} + \eta^{\lambda\rho} \epsilon^{\mu\nu}) \int \frac{d^2 p}{(2\pi)^2} d^2 x \left[\left(\frac{n_f(|p^0|) \delta(p^2)}{(p - i\partial)^2} \right. \right.$$
$$\left. \left. + \frac{n_f(|(p - i\partial)^0|) \delta((p - i\partial)^2)}{p^2} \right) p_\mu (p - i\partial)_\lambda V_\nu \right] A_\rho, \tag{13.142}$$

where we have used (13.125) with the square brackets indicating the action of the derivatives. Furthermore, we note that since the temperature dependent terms in the effective action are finite, we have evaluated the Dirac trace above using two dimensional identities. The contribution to the anomaly from these terms can be easily obtained by taking the functional derivative with respect to A_ρ and applying ∂_ρ to the resulting expression. Thus, we obtain, from the terms linear in $S_f^{(\beta)}$,

$$\text{Anomaly} := -2ig(\eta^{\mu\nu} \epsilon^{\lambda\rho} + \eta^{\lambda\rho} \epsilon^{\mu\nu}) \int \frac{d^2 p}{(2\pi)^2}$$
$$\times \frac{n_f(|p^0|) \delta(p^2)}{(p - i\partial)^2} (p_\mu (p - i\partial)_\lambda + p_\lambda (p - i\partial)_\mu) \partial_\rho V_\nu$$

$$= -2ig \int \frac{d^2p}{(2\pi)^2} \frac{n_f(|p^0|)\delta(p^2)}{(p-i\partial)^2} [-2i(p-i\partial)^2 \epsilon^{\mu\nu} p_\nu] V_\mu$$

$$= -4g\epsilon^{\mu\nu} \int \frac{d^2p}{(2\pi)^2} n_f(|p^0|)\delta(p^2) p_\nu V_\mu = 0. \quad (13.143)$$

In obtaining this result, we have used the substitution

$$p \to -(p-i\partial), \quad (13.144)$$

in the second term in going from (13.142) to (13.143) as well as the two dimensional identity

$$p^\nu \epsilon^{\lambda\rho} p_\lambda \partial_\rho = \epsilon^{\nu\rho}(p^2 \partial_\rho - p \cdot \partial p_\rho). \quad (13.145)$$

This shows that the terms linear in $S_f^{(\beta)}$ do not contribute to the anomaly equation. We can, similarly, analyze the terms quadratic in $S_f^{(\beta)}$ in (13.141).

$$\text{Quadratic} := -ig \text{Tr} \, \gamma_5 S_f^{(\beta)}(p) \not{V} S_f^{(\beta)}(p) \not{A}$$

$$= -2ig(\eta^{\mu\nu}\epsilon^{\lambda\rho} + \eta^{\lambda\rho}\epsilon^{\mu\nu}) \int \frac{d^2p}{(2\pi)^2} d^2x [n_f(|(p-i\partial)^0|)$$

$$\times n_f(|p^0|)\delta(p^2)\delta((p-i\partial)^2) p_\mu (p-i\partial)_\lambda V_\nu] A_\rho. \quad (13.146)$$

The contribution to the anomaly equation, from the above action, can again be obtained by functionally differentiating (13.146) with respect to A_ρ and applying ∂_ρ to the resulting expression. This yields

$$\text{Anomaly} := -2ig(\eta^{\mu\nu}\epsilon^{\lambda\rho} + \eta^{\lambda\rho}\epsilon^{\mu\nu}) \int \frac{d^2p}{(2\pi)^2} n_f(|(p-i\partial)^0|)$$

$$\times n_f(|p^0|)\delta(p^2)\delta((p-i\partial)^2) p_\mu (p-i\partial)_\lambda \partial_\rho V_\nu$$

$$= 2ig \int \frac{d^2p}{(2\pi)^2} n_f(|p^0|)n_f(|(p-i\partial)^0|)\delta(p^2)\delta((p-i\partial)^2)$$

$$\times (-i\epsilon^{\mu\nu}p_\mu(p-i\partial)^2) V_\nu$$

$$= 0. \quad (13.147)$$

This shows that even the terms in S_{eff} which are quadratic in $S_f^{(\beta)}$ do not contribute to the anomaly equation. Consequently, the axial anomaly is temperature independent as we had seen earlier in sections **7.2.3** as well as in **9.6** and, therefore, does not show any nonanalyticity.

13.4 References

I. Aitchison and C. Fraser, Physics Letters **B146**, 63 (1984).

J. Barcelos-Neto and A. Das, Physical Review **D58**, 085022 (1998).

K. S. Babu, A. Das and P. Panigrahi, Physics Letters **B188**, 133 (1987).

K. S. Babu, A. Das and P. Panigrahi, Physical Review **D36**, 3725 (1987).

J. Barcelos-Neto and A. Das, Physical Review **D58**, 085022 (1998).

F. T. Brandt, A. Das, J. Frenkel and A. J. da Silva, Physical Review **D59**, 065004 (1999).

A. Das, *Field Theory: A Path Integral Approach* (third edition), World Scientific Publishing (2020).

A. Das and G. Dunne, Physical Review **D57**, 5023 (1998).

A. Das and J. Frenkel, Physical Review **D80**, 125039 (2009).

A. Das and M. Hott, Physical Review **D50**, 6655 (1994).

A. Das and A. Karev, Physical Review **D36**, 623 (1987); *ibid* **D36**, 2591 (1987).

A. Das and A. J. da Silva, Physical Review **D59**, 105011 (1999).

G. Dunne, K. Lee and C. Lu, Physical Review Letters **78**, 3434 (1997).

A. Karev, *Studies in the Derivative Expansion and Finite Temperature Field Theory*, Ph. D. Thesis, University of Rochester (1987).

B. A. Lippmann and J. Schwinger, Physical Review **79**, 469 (1950).

S. Maciel and S. Perez, Physical Review **D78**, 065005 (2008).

A. M. J. Schakel, *On Broken Symmetries in Fermi Systems*, Ph. D. Thesis, Universiteit van Amsterdam (1989).

J. Schwinger, Physical Review **82**, 664 (1951).

CHAPTER 14

Non-equilibrium phenomena

Our discussion, so far, has been within the context of phenomena in thermal equilibrium. However, there are several physical systems of interest which need not evolve in thermal equilibrium. Most important among these are systems with dynamical phase transitions. Such non-equilibrium phenomena are of great interest, not only from the point of view of the study of physical systems in condensed matter, but also in the cosmological context of the study of the evolution of our physical universe. The question of formation and growth of domains is as important in these contexts as it has also become in systems in high energy physics. In this chapter, we will discuss one such phenomenon, namely, the disoriented chiral condensates and how the study of such systems can be accommodated within the framework of discussions in the earlier chapters.

14.1 Disoriented chiral condensates

To introduce the concept of the disoriented chiral condensates, let us recall that the strong force among elementary particles is described by the Lagrangian density of Quantum chromo dynamics (see (5.20))

$$\mathcal{L}_{QCD} = -\frac{1}{4}F^a_{\mu\nu}F^{\mu\nu\,a} + i\overline{\psi}^i\gamma^\mu(D_\mu\psi)^i - m\overline{\psi}^i\psi^i, \tag{14.1}$$

where $a = 1, 2, \cdots, 8$ and $i = 1, 2, 3$. Thus, we assume that the gauge fields or gluons belong, as usual, to the adjoint representation of the color symmetry group, $SU(3)$, while the quarks or the fundamental building blocks of matter belong to the fundamental representation of the group. Note, that, here we have included a mass term for the fermions as well. In addition to the color index, the fermions also carry a flavor index which we have suppressed for simplicity. (The mass parameter depends on the flavor index, namely, quarks of different flavors have different masses.)

Quarks, as we see from (14.1), carry a color index or color charge. On the other hand, observed hadrons – mesons and baryons – are believed to be made of quarks in color singlet combinations so that they do not carry any color charge. It is much the same way as atoms are made of charged constituents – protons and electrons – and yet are charge neutral. In contrast, however, free quarks are not observed in nature unlike free charged particles. This has led to the postulate that the strong force is so strong that the color charge is confined. Namely, the true physical states of QCD (Quantum chromo dynamics) are the color singlet hadronic states. There is some evidence for this from lattice studies in QCD which also suggests that the hadronic matter undergoes a transition from the confined phase to a deconfined phase of quarks and gluons beyond a temperature of about 150 MeV. In other words, as the temperature of the heat bath is increased, the thermal motion can be thought of as weakening the binding due to the color force leading to a quark-gluon plasma phase beyond a critical temperature of 150 MeV. It is clear that such a phase must have existed at very early stages of the universe when the temperatures were very high. But what is even more interesting is that, the transition temperature is not extremely high so that such a phase of matter can even be produced in the laboratory and its properties studied. This has led to the Heavy Ion Collision experiments where heavy nuclei (typically gold on gold at RHIC, Brookhaven), with relatively high energy per nucleon, are collided in a small region of space. The high energy involved in these experiments is sufficient to heat up the small region of collision to high temperatures which is expected to produce the quark-gluon plasma phase. However, it is clear that the collision region as well as the time scales involved in these processes are so small, that conditions are likely to change very fast so that it would even be inappropriate to take the equilibrium formalism of statistical mechanics as a first approximation. These are very likely transitions far out of equilibrium and should be properly studied, theoretically, as non-equilibrium phenomena.

Experimentally, on the other hand, it is essential to identify a signal which would clearly show that the quark-gluon plasm phase was, indeed, produced during the collision. The main difficulty in this lies in the fact that, at such high energies, the background is nontrivial and that the most common signals of the quark-gluon plasma phase transitions are easily masked by other nuclear effects. A signal which is free from such effects is known as the disoriented chiral condensates and can be understood as follows. As we have noted earlier, different quark flavors have different mass. Experimentally, we know that, of all the flavors, the "up" and the "down" quark flavors have particularly small dynamical masses

$$m_u \sim 5\,\mathrm{MeV}, \quad m_d \sim 10\,\mathrm{MeV}. \tag{14.2}$$

The other quark flavors have relatively higher values for the dynamical mass. (Quarks also have what is known as a constituent mass which is ~ 300 MeV.) It is clear, therefore, that if we are interested in the quark-gluon plasma phase transition, then, the energies (temperatures) we are interested in are much higher than the dynamical masses of the "up" and the "down" quarks and, consequently, for all practical purposes, we can take these two quark flavors to be massless. Restricting ourselves to only these two quark flavors, we note that we can write the dynamical Lagrangian density to be

$$\mathcal{L}_{QCD} = -\frac{1}{4}F^a_{\mu\nu}F^{\mu\nu\,a} + i\overline{\psi}^i\gamma^\mu(D_\mu\psi)^i. \tag{14.3}$$

For the rest of our discussions, we will restrict ourselves to only these two quark flavors and, therefore, (14.3) would represent the dynamical Lagrangian density for our purpose. (Once again, we have suppressed the two flavor indices.)

The important thing to note from (14.3) is the fact that the quarks (fermions) are assumed to be massless. As a result, the fermion Lagrangian density has a larger group of global invariance given by $SU_L(2) \times SU_R(2)$. Namely, the Lagrangian density in (14.3) is invariant under the following two sets of global $SU(2)$ flavor transformations

$$\delta_L\psi^i = \frac{i}{4}(\boldsymbol{\epsilon}_L \cdot \boldsymbol{\tau})(1 - \gamma_5)\psi^i,$$

$$\delta_R\psi^i = \frac{i}{4}(\boldsymbol{\epsilon}_R \cdot \boldsymbol{\tau})(1 + \gamma_5)\psi^i. \tag{14.4}$$

Here $\boldsymbol{\tau}$ stands for three Pauli matrices which are the generators of the two $SU(2)$ flavor transformations and $\boldsymbol{\epsilon}_{L,R}$ are the global parameters associated with these two transformations. (Under these transformations, the color index does not change.)

The transformations in (14.4) are also known as chiral $SU(2)$ transformations since they involve the γ_5 matrix. Quantum mechanically, however, it is known that these chiral symmetries are broken because of anomalies and the true invariance of the theory is the reduced (vector) symmetry

$$SU_L(2) \times SU_R(2) \to SU_V(2). \tag{14.5}$$

The residual vector $SU(2)$ represents the familiar weak isospin symmetry which would be present even for equal, nonzero quark masses. However, the chiral symmetry, as we know, is dynamically broken and the signal for such a breakdown is the appearance of condensates, in our theory, of the form (the repeated index i is being summed)

$$\langle \overline{\psi}\psi \rangle = \langle \overline{\psi}^i \psi^i \rangle \neq 0. \tag{14.6}$$

In other words, we can think of $\langle \overline{\psi}\psi \rangle$ as the order parameter of chiral symmetry breaking. Very roughly speaking, under the $SU(2) \times SU(2)$ transformations of (14.4),

$$\delta(\overline{\psi}\psi) \sim i\overline{\psi} \left(\boldsymbol{\epsilon} \cdot \boldsymbol{\tau} \right) \gamma_5 \psi,$$

$$\delta(\overline{\psi}\boldsymbol{\tau}\gamma_5\psi) \sim i\boldsymbol{\epsilon}\,\overline{\psi}\psi, \tag{14.7}$$

where we can think of $\boldsymbol{\epsilon}$ as the chiral parameter of transformation (it is, in fact, related to $\epsilon_L - \epsilon_R$). From the discussion in chapter **6** (see (9.17)), it is now obvious from (14.7) that the chiral symmetry will be spontaneously broken when the fermion condensate $\langle \overline{\psi}\psi \rangle$ acquires a nonzero value. The associated Goldstone bosons are also known to be the three massless pions (the mass of the pions arise from weak interactions). As the temperature is increased, chiral symmetry is restored and the critical temperature for this phase transition is also estimated to be about 150 MeV. In other words, beyond a temperature of 150 MeV, the fermion condensate has a vanishing vacuum expectation value. It is curious that the critical temperatures for the confining-deconfining transition as well as chiral symmetry restoration are approximately the same. It is not clear, at present, whether there are really two distinct phase transitions, but it is obvious that if we have temperatures high enough to have a quark-gluon plasma phase, then we will necessarily be in a chirally symmetric phase. However, as the small region consisting of the quark-gluon plasma expands, the temperature inside the plasma will decrease and, eventually, chiral symmetry will be broken again. The direction of chiral symmetry breaking, in the isospin space, however, need not be the same as it was originally, before the collision took place. Therefore, if the value of the order parameter is nonzero along the π^0 direction, say, then eventually, when the plasma will hadronize (*i.e.*, come down to the true vacuum with a lower energy), it will produce a large number of neutral pions and not charged ones. This could lead to spectacular events where a large number of pions correlated in isospin space would come out of a single collision signalling clearly that chiral symmetry restoration and, therefore, a quark-gluon plasma phase was, indeed, achieved. It has already been suggested that events of this kind may already have been experimentally observed (the so called Centauro events), but the situation is not very clear.

The difficulty with this picture is as follows. The direction of chiral symmetry breaking need not be the same in the entire plasma. In fact, it is more likely that there will be domains in the plasma with distinct directions of chiral symmetry breaking. Such domain structures are, indeed, quite common in many condensed matter systems as well as in the study of cosmology. If the domain sizes are small (and, therefore, the energy contained in such regions is small) so that only a few pions are produced upon hadronization, then, such an event can hardly be distinguished from that of a random production of pions in a high energy collision. In fact, if the process proceeds through equilibrium, then simple arguments show that this would, indeed, be the case. On the other hand, for a process far out of equilibrium, as is the case here, the domain sizes can, in fact, grow rapidly so that after hadronization, one can have a clear signal of the quark-gluon plasma phase through the appearance of a large number of correlated pions. We will show that this is, indeed, the case in an example where the rapid cooling of the plasma phase is modelled as a sudden quench below the critical temperature.

As we have already pointed out in chapter **2**, the formalism most suitable for study of such non-equilibrium phenomena is within the context of the closed time path formalism. In the next section, we will describe how this method applies to a non-equilibrium situation through a simple example before moving on to the real problem.

14.2 Free Klein-Gordon theory

Let us, as a warm up example, consider a free Klein-Gordon theory, in $3 + 1$ dimensions, described by the Lagrangian density

$$\mathcal{L} = \frac{1}{2}\partial_\mu\phi\partial^\mu\phi - \frac{m_i^2}{2}\phi^2 + \theta(x^0)\frac{\delta m^2}{2}\phi^2, \tag{14.8}$$

where m_i denotes the mass in the initial free theory and m_f to be the mass after interactions (for $x^0 > 0$), namely,

$$m_f^2 = (m_i^2 - \delta m^2). \tag{14.9}$$

This is a time dependent Lagrangian density (and, therefore, the Hamiltonian will also be time dependent) where for $x^0 < 0$, we have a free Klein-Gordon theory with a mass m_i while for positive times ($x^0 \geq 0$), the theory has an imaginary mass (as would be required for a broken symmetry phase described in (9.1)). There is a sudden change of phase at $x^0 = 0$ (prototype of a quenching) and clearly, the evolution of the dynamical system cannot be described through the equilibrium formalism. (We would like to emphasize here that the quench could have happened at an arbitrary time t_0, in which case, the interaction term in (14.8) could, in general, be written with a step function $\theta(x^0) \rightarrow \theta(x^0 - t_0)$. However, for simplicity, we have chosen here $t_0 = 0$.)

Let us assume that, initially, the system is in equilibrium at a temperature described by β. This system described in (14.8) and (14.9) is simple enough that one can solve the dynamical equations explicitly and determine the Green's functions of the theory. In particular, in statistical systems, we are interested in the correlated Green's functions (see (2.55)) which give temperature dependent correlations between various fields. In this section, we will see how the closed time path formalism can be used to derive the same Green's functions. Following our discussion in chapter **2**, we note that the Feynman rules for this theory, for $x^0, y^0 < 0$, at finite temperature are given by the free Green's functions (interaction term vanishes here)

$$\begin{array}{ccc} x & y \\ \underline{\hspace{2cm}} & = & iG^{(0)}_{++}(x-y), \\ + & + \end{array}$$

$$\begin{array}{ccc} x & y \\ \underline{\hspace{2cm}} & = & iG^{(0)}_{+-}(x-y), \\ + & - \end{array}$$

$$\underset{-}{x} \underset{+}{\underline{\qquad\qquad y}} = iG^{(0)}_{-+}(x-y),$$

$$\underset{-}{x} \underset{-}{\underline{\qquad\qquad y}} = iG^{(0)}_{--}(x-y). \tag{14.10}$$

The interaction vertices of the full theory can be taken to be

$$\underset{+\ +}{\underline{\quad\times\quad}} = i\delta m^2 \theta(x^0)\delta(x-y),$$

$$\underset{-\ -}{\underline{\quad\times\quad}} = -i\delta m^2 \theta(x^0)\delta(x-y). \tag{14.11}$$

As we have noted in (2.55)-(2.59), the physical Green's functions, G_R, G_A and G_C can be obtained from the Feynman propagators as

$$\widehat{G} = \begin{pmatrix} 0 & G_A \\ G_R & G_C \end{pmatrix}$$

$$= QGQ^{-1} = Q \begin{pmatrix} G_{++} & G_{+-} \\ G_{-+} & G_{--} \end{pmatrix} Q^{-1}, \tag{14.12}$$

where

$$Q = \frac{1}{\sqrt{2}} \begin{pmatrix} 1 & -1 \\ 1 & 1 \end{pmatrix},$$

$$Q^{-1} = \frac{1}{\sqrt{2}} \begin{pmatrix} 1 & 1 \\ -1 & 1 \end{pmatrix}. \tag{14.13}$$

In the present example, we can treat the $\delta m^2 \theta(x^0)$ term in the Lagrangian density, (14.8), as the interaction term and calculate the complete propagator. Or equivalently, let us define (see (14.11))

$$\Sigma(x,y) = \begin{pmatrix} \delta m^2 & 0 \\ 0 & -\delta m^2 \end{pmatrix} \theta(x^0)\delta(x-y), \tag{14.14}$$

and note that the complete propagator is obtained from the inverse

$$G = \left((G^{(0)})^{-1} + \Sigma \right)^{-1}$$

$$= \left(1 + G^{(0)}\Sigma \right)^{-1} G^{(0)} = G^{(0)} \left(1 + \Sigma G^{(0)} \right)^{-1}. \tag{14.15}$$

Given this, the complete physical Green's functions are obtained from (14.12) and (14.15) to be

$$\widehat{G} = QGQ^{-1}$$

$$= \left(\mathbb{1} + \widehat{G}^{(0)}\widehat{\Sigma}\right)^{-1}\widehat{G}^{(0)} = \widehat{G}^{(0)}\left(\mathbb{1} + \widehat{\Sigma}\widehat{G}^{(0)}\right)^{-1}, \qquad (14.16)$$

where we have defined, consistent with (14.12),

$$\widehat{\Sigma}(x,y) = Q\Sigma(x,y)Q^{-1} = \begin{pmatrix} 0 & \Sigma_{\rm R}(x,y) \\ \Sigma_{\rm A}(x,y) & \Sigma_{\rm C}(x,y) \end{pmatrix}$$

$$= \begin{pmatrix} 0 & \delta m^2 \\ \delta m^2 & 0 \end{pmatrix}\theta(x^0)\delta(x-y), \qquad (14.17)$$

so that we can identify

$$\Sigma_{\rm R}(x,y) = \delta m^2\theta(x^0)\delta(x-y) = \Sigma_{\rm A}(x,y), \quad \Sigma_{\rm C}(x,y) = 0. \quad (14.18)$$

Let us note that each term in (14.16) consists of a 2×2 matrix which can be simplified to give

$$\widehat{G} = \left[\sum_{n=0}^{\infty}(-1)^n\left(\widehat{G}^{(0)}\widehat{\Sigma}\right)^n\right]\widehat{G}^{(0)}. \qquad (14.19)$$

Thus, comparing with (14.12), we obtain the complete physical propagators to be

$$G_{\rm R} = \sum_{n=0}^{\infty}(-1)^n(G_{\rm R}^{(0)}\delta m^2)^n G_{\rm R}^{(0)}$$

$$= \left(1 + G_{\rm R}^{(0)}\delta m^2\right)^{-1}G_{\rm R}^{(0)} = G_{\rm R}^{(0)}\left(1 + \delta m^2 G_{\rm R}^{(0)}\right)^{-1},$$

$$G_{\rm A} = \sum_{n=0}^{\infty}(-1)^n(G_{\rm A}^{(0)}\delta m^2)^n G_{\rm A}^{(0)}$$

$$= \left(1 + G_{\rm A}^{(0)}\delta m^2\right)^{-1}G_{\rm A}^{(0)} = G_{\rm A}^{(0)}\left(1 + \delta m^2 G_{\rm A}^{(0)}\right)^{-1},$$

$$G_{\rm C} = \sum_{n=0}^{\infty}\sum_{m=0}^{n}(-1)^n(G_{\rm R}^{(0)}\delta m^2)^m G_{\rm C}^{(0)}(\delta m^2 G_{\rm A}^{(0)})^{n-m}$$

$$= \left(1 + G_{\rm R}^{(0)}\delta m^2\right)^{-1}G_{\rm C}^{(0)}\left(1 + \delta m^2 G_{\rm A}^{(0)}\right)^{-1}, \qquad (14.20)$$

where the coordinates (as well as the integrations over intermediate coordinates) are suppressed for clarity.

Normally, in a relativistic field theory, inversion of factors like the ones in (14.20) are easily achieved by taking the Fourier transformation. Here, however, the quadratic interaction terms in (14.14) or (14.17), namely, $\delta m^2 \theta(x^0)\delta(x - y)$, are explicitly time dependent and, consequently, the theory is not time translation invariant and going to the Fourier space does not help. The inversion of the factors, in this case, is slightly more involved, but can be done systematically in the following way. Since time translation is not a symmetry of our theory, it is more useful to work in a mixed coordinate basis where the Green's functions are defined as functions of the time coordinate and the spatial momentum. For example, the free Green's functions have the forms, given by (see (2.60) which can also be read out from (2.42) and (2.56))

$$G_R^{(0)}(x^0, y^0, \mathbf{p}) = \int \frac{dp^0}{2\pi} \frac{e^{-ip^0(x^0-y^0)}}{(p^0 + i\epsilon)^2 - \omega_i^2}$$

$$= -\frac{\theta(x^0 - y^0)}{\omega_i} \sin \omega_i(x^0 - y^0),$$

$$G_A^{(0)}(x^0, y^0, \mathbf{p}) = \int \frac{dp^0}{2\pi} \frac{e^{-ip^0(x^0-y^0)}}{(p^0 - i\epsilon)^2 - \omega_i^2}$$

$$= \frac{\theta(y^0 - x^0)}{\omega_i} \sin \omega_i(x^0 - y^0),$$

$$G_C^{(0)}(x^0, y^0, \mathbf{p}) = -2i\pi \int \frac{dp^0}{2\pi} (1 + 2n_B(|p^0|))$$

$$\times \delta(p^2 - m^2) e^{-ip^0(x^0-y^0)}$$

$$= -\frac{i}{\omega_i}(1 + 2n_B(\omega_i)) \cos \omega_i(x^0 - y^0). \quad (14.21)$$

It is to be noted here that even though the complete Green's functions in the interacting theory will not be time translation invariant, the free theory (in the absence of interactions) is time translation invariant and, consequently, the initial physical Green's functions in (14.21) depend only on the difference of the two time coordinates. Furthermore, in writing the Green's functions in (14.21), we have defined

$$\omega_i^2 = \mathbf{p}^2 + m_i^2. \tag{14.22}$$

To derive the complete physical Green's functions, let us note next that

$$(G_{\mathrm{R}}^{(0)}\Sigma_{\mathrm{R}})(x^0, y^0; \mathbf{p}) = \int dz^0\, G_{\mathrm{R}}^{(0)}(x^0, z^0; \mathbf{p})\Sigma_{\mathrm{R}}(z^0, y^0; \mathbf{p})$$

$$= -\delta m^2 \int dz^0\, \theta(x^0 - z^0) \frac{\sin \omega_i(x^0 - z^0)}{\omega_i} \times \theta(z^0)\delta(z^0 - y^0)$$

$$= -\delta m^2\, \theta(x^0 - y^0)\, \theta(y^0) \frac{\sin \omega_i(x^0 - y^0)}{\omega_i}. \tag{14.23}$$

As a result, we can write

$$(1 + G_{\mathrm{R}}^{(0)}\Sigma_{\mathrm{R}})(x^0, y^0; \mathbf{p})$$

$$= \delta(x^0 - y^0) - \delta m^2\, \theta(x^0 - y^0)\theta(y^0) \frac{\sin \omega_i(x^0 - y^0)}{\omega_i}. \tag{14.24}$$

There are two ways we can obtain the inverse of $(1 + G_{\mathrm{R}}^{(0)}\Sigma)$, namely, we can either iterate the result in (14.23) and sum a geometric series to obtain the inverse or a simple inspection suggests that

$$\left(1 + G_{\mathrm{R}}^{(0)}\Sigma_{\mathrm{R}}\right)^{-1}(x^0, y^0; \mathbf{p})$$

$$= \delta(x^0 - y^0) + \delta m^2\, \theta(x^0 - y^0)\, \theta(y^0) \frac{\sin \omega_f(x^0 - y^0)}{\omega_f}, \tag{14.25}$$

where we have identified

$$\omega_f^2 = \omega_i^2 - \delta m^2. \tag{14.26}$$

That (14.25) defines the correct inverse of (14.24) can be easily checked by noting that

$$\int dz^0 \theta(x^0 - z^0) \, \theta(z^0) \, \theta(z^0 - y^0) \, \theta(y^0)$$

$$\times \sin \omega_i(x^0 - z^0) \, \sin \omega_f(z^0 - y^0)$$

$$= \theta(x^0 - y^0)\theta(y^0) \int_{y^0}^{x^0} dz^0 \, \sin \omega_i(x^0 - z^0) \sin \omega_f(z^0 - y^0)$$

$$= \frac{\theta(y^0)\theta(x^0 - y^0)}{\delta m^2} \left(\omega_f \sin \omega_i(x^0 - y^0) - \omega_i \sin \omega_f(x^0 - y^0) \right).$$

$$\tag{14.27}$$

We can now go ahead and obtain the retarded Green's function using (14.20)-(14.22) in a straightforward manner,

$$G_R(x^0, y^0; \mathbf{p}) = \int dz^0 (1 + G_R^{(0)} \Sigma_R)^{-1}(x^0, z^0; \mathbf{p}) G_R^{(0)}(z^0 - y^0; \mathbf{p})$$

$$= \int dz^0 \left(\delta(x^0 - z^0) + \delta m^2 \, \theta(x^0 - z^0) \, \theta(z^0) \, \frac{\sin \omega_f(x^0 - z^0)}{\omega_f} \right)$$

$$\times \left(-\frac{\theta(z^0 - y^0)}{\omega_i} \sin \omega_i(z^0 - y^0) \right)$$

$$= -\frac{\theta(x^0 - y^0)}{\omega_i} \sin \omega_i(x^0 - y^0) - \frac{\delta m^2}{\omega_i \omega_f} \theta(x^0 - y^0)\theta(x^0)$$

$$\times \left[\theta(y^0) \int_{y^0}^{x^0} dz^0 \, \sin \omega_f(x^0 - z^0) \sin \omega_i(z^0 - y^0) \right.$$

$$\left. + \theta(-y^0) \int_0^{x^0} dz^0 \, \sin \omega_f(x^0 - z^0) \sin \omega_i(z^0 - y^0) \right]$$

$$= \theta(-x^0)\theta(-y^0) \, G_R^{(0)}(x^0 - y^0; \omega_i) + \theta(x^0)\theta(y^0) G_R^{(0)}(x^0 - y^0; \omega_f)$$

$$+ \frac{\theta(x^0 - y^0)\theta(x^0)\theta(-y^0)}{2\omega_i \omega_f} \left((\omega_f - \omega_i) \sin(\omega_f x^0 + \omega_i y^0) \right.$$

$$\left. - (\omega_f + \omega_i) \sin(\omega_f x^0 - \omega_i y^0) \right). \tag{14.28}$$

We see that the retarded Green's function has the right (retarded) behavior (namely, $\theta(x^0 - y^0)$) for both $x^0 > 0, x^0 < 0$ and $y^0 > 0, y^0 < 0$. The advanced Green's function can be obtained from

(14.28) simply by letting $x^0 \leftrightarrow y^0$. Furthermore, we note that it is no longer time translation invariant since the mass term δm^2 breaks this symmetry.

We can now obtain the correlated Green's function in a straight-forward manner. We note from (14.20) and (14.21), as well as from the fact that $\Sigma_C = 0$ (see (14.18)) that

$$G_C(x^0, y^0; \mathbf{p}) = (1 + G_R^{(0)}\Sigma_R)^{-1} G_C^0 (1 + \Sigma_A G_A^{(0)})^{-1}(x^0, y^0; \mathbf{p})$$

$$= \int dz^0 dz'^0 \left[\delta(x^0 - z^0) + \theta(x^0 - z^0)\theta(z^0) \frac{\delta m^2}{\omega_f} \sin\omega_f(x^0 - z^0) \right]$$

$$\times \left(-\frac{i}{\omega_i} \right) (1 + 2n_B(\omega_i)) \cos\omega_i(z^0 - z'^0)$$

$$\times \left[\delta(z'^0 - y^0) - \theta(y^0 - z'^0)\theta(z'^0) \frac{\delta m^2}{\omega_f} \sin\omega_f(z'^0 - y^0) \right]$$

$$= -\frac{i}{\omega_i}(1 + 2n_B(\omega_i))\Big\{ \theta(-x^0)\theta(-y^0) \cos\omega_i(x^0 - y^0)$$

$$+ \frac{\theta(-x^0)\theta(y^0)}{2\omega_f}\Big((\omega_f + \omega_i)\cos(\omega_i x^0 - \omega_f y^0)$$

$$+ (\omega_f - \omega_i)\cos(\omega_i x^0 + \omega_f y^0)\Big)$$

$$+ \frac{\theta(x^0)\theta(-y^0)}{2\omega_f}\Big((\omega_f + \omega_i)\cos(\omega_f x^0 - \omega_i y^0)$$

$$+ (\omega_f - \omega_i)\cos(\omega_f x^0 + \omega_i y^0)\Big)$$

$$+ \frac{\theta(x^0)\theta(y^0)}{2\omega_f^2}\Big((\omega_f^2 + \omega_i^2)\cos\omega_f(x^0 - y^0)$$

$$+ (\omega_f^2 - \omega_i^2)\cos\omega_f(x^0 + y^0)\Big)\Big\}. \qquad (14.29)$$

The explicit time dependence of the correlated Green's function is now obvious. Second, we note that, for $x^0 = y^0$, the correlated Green's function becomes

$$G_C(x^0, \mathbf{p}) = -\frac{i}{\omega_i}(1 + 2n_B(\omega_i))$$

$$\times \left[1 + \theta(x^0)\left(-1 + \cos^2 \omega_f x^0 + \left(\frac{\omega_i}{\omega_f}\right)^2 \sin^2 \omega_f x^0\right)\right]. \quad (14.30)$$

We note from (14.30) that, for $x^0 < 0$, the correlated Green's function has the form

$$G_C = -\frac{i}{\omega_i}\left(1 + 2n_B(\omega_i)\right) \quad (14.31)$$

as it should be (see (14.21)). On the other hand, for $x^0 > 0$, the correlated Green's function in (14.30) takes the form

$$G_C(x^0, \mathbf{p}) = -\frac{i}{\omega_i}(1 + 2n_B(\omega_i))\left[\cos^2 \omega_f x^0 + \left(\frac{\omega_i}{\omega_f}\right)^2 \sin^2 \omega_f x^0\right]$$

$$= -\frac{i}{\omega_i}(1 + 2n_B(\omega_i))\left[1 + \frac{1}{2}\left(1 - \left(\frac{\omega_i}{\omega_f}\right)^2\right)(\cos 2\omega_f x^0 - 1)\right]. \quad (14.32)$$

This is, indeed, exactly what is obtained by solving the differential equations for the Green's functions in the present case. Here, we have tried to derive it from the diagrammatic point of view of the closed time path formalism to bring out the flavor of the calculation for the real problem of the disoriented chiral condensates that we are interested in.

14.3 Nambu-Jona-Lasinio model

The study of dynamical breaking of chiral symmetry in QCD is non-trivial and, therefore, we choose a simple model to describe the phenomena of chiral symmetry breaking where calculations are at least manageable. The most familiar of such models is known as the Nambu-Jona-Lasinio model and is described by a Lagrangian density (repeated indices are being summed)

$$\mathcal{L}_{\text{NJL}} = i\overline{\psi}^i \not\partial \psi^i + \frac{g^2}{4N\Lambda^2}\left[(\overline{\psi}^i \psi^i)^2 - (\overline{\psi}^i \tau^a \gamma_5 \psi^i)^2\right]. \quad (14.33)$$

Here $i = 1, 2, \cdots, N$ represent the color indices while the two flavor indices of the fermions are suppressed. The three matrices τ^a,

$a = 1, 2, 3$, (or $\boldsymbol{\tau}$) are the three Pauli matrices which are the generators of the flavor $SU(2)$ symmetry group and which commute with the gamma matrices. In $3 + 1$ dimensions, the Nambu-Jona-Lasinio theory is nonrenormalizable and we should think of this as an effective low energy theory below the cut-off scale Λ. We will study this model in $\frac{1}{N}$ expansion which is known to yield only planar diagrams at the leading order and nonperturbative results in the coupling constant g. This theory, as can be easily checked, is invariant under the global chiral transformations

$$\psi^i \to e^{i\boldsymbol{\alpha}\cdot\boldsymbol{\tau}\gamma_5}\psi^i,$$

$$\overline{\psi}^i \to \overline{\psi}^i e^{i\boldsymbol{\alpha}\cdot\boldsymbol{\tau}\gamma_5}, \tag{14.34}$$

where the three components of $\boldsymbol{\alpha}$ represent the constant parameters of transformation. Infinitesimally, these transformations take the form

$$\delta\psi^i = i\boldsymbol{\epsilon}\cdot\boldsymbol{\tau}\gamma_5\psi^i,$$

$$\delta\overline{\psi}^i = i\overline{\psi}^i\boldsymbol{\epsilon}\cdot\boldsymbol{\tau}\gamma_5. \tag{14.35}$$

We can write the Nambu-Jona-Lasinio model also in an alternate suggestive manner, convenient for our discussion, with the introduction of auxiliary fields as

$$\mathcal{L}_{\text{NJL}} = i\overline{\psi}^i \slashed{\partial}\psi^i - \frac{N\Lambda^2}{2}\left(\sigma^2 + \pi^a\pi^a\right)$$

$$+ \frac{g}{\sqrt{2}}\overline{\psi}^i(\sigma + i\gamma_5\pi^a\tau^a)\psi^i, \tag{14.36}$$

where we can think of σ and π^a as the scalar and the pseudoscalar fields (The three π^a's can be thought of as corresponding to the three pions). The equations of motion for the auxiliary fields are easily seen to be the constraint equations

$$\sigma = \frac{g}{\sqrt{2}}\overline{\psi}^i\psi^i,$$

$$\pi^a = \frac{ig}{\sqrt{2}}\overline{\psi}^i\tau^a\gamma_5\psi^i, \tag{14.37}$$

and if we eliminate the auxiliary fields through their equations of motion, (14.37), then, we get back the Lagrangian density of (14.33). In terms of these variables, the infinitesimal chiral symmetry transformations of the theory take the form (in addition to those in (14.35))

$$\delta \psi^i = i\epsilon \cdot \boldsymbol{\tau} \gamma_5 \psi^i,$$

$$\delta \sigma = 2\epsilon \cdot \boldsymbol{\pi},$$

$$\delta \boldsymbol{\pi} = -2\epsilon \sigma. \tag{14.38}$$

Although chiral symmetry is a symmetry of the tree level Lagrangian, it is, in fact, dynamically broken. This can be seen in the following way. Let us shift the field variables as

$$\sigma \to \sigma + \sigma_c,$$

$$\pi^a \to \pi^a. \tag{14.39}$$

Namely, we are interested in evaluating the effective potential for the scalar field since we expect the vacuum expectation value of the pseudoscalar fields to vanish. Under this shifting, the Lagrangian density of (14.36) becomes

$$\mathcal{L}_{\text{NJL}} = i\overline{\psi}^i \partial\!\!\!/ \psi^i - \frac{N\Lambda^2}{2} (\sigma^2 + \pi^a \pi^a) - N\Lambda^2 \sigma_c \sigma - \frac{N\Lambda^2}{2} \sigma_c^2$$

$$+ \frac{g\sigma_c}{\sqrt{2}} \overline{\psi}^i \psi^i + \frac{g}{\sqrt{2}} \overline{\psi}^i (\sigma + i\pi^a \tau^a \gamma_5) \psi^i. \tag{14.40}$$

Following the discussion in chapter **6** (see *e.g.*, (9.70) and (9.76)), we can now obtain the effective potential including the one loop contribution to be

$$V_{\text{eff}}(\sigma_c) = V^0(\sigma_c) + V^1(\sigma_c)$$

$$= \frac{N\Lambda^2}{2} \sigma_c^2 - 4N \int \frac{\mathrm{d}^4 k_E}{(2\pi)^4} \ln(k_E^2 + m_{\text{eff}}^2), \tag{14.41}$$

where we have identified

$$m_{\text{eff}}^2 = \frac{g^2 \sigma_c^2}{2}. \tag{14.42}$$

We note here that the factors in front of the second term come from tracing over the color indices, the flavor indices as well as the Dirac indices. Furthermore, the difference in sign from that of (9.70) or (9.76) is because of the fact that for fermions, the power of the determinant is positive (as opposed to bosons where it is negative). (It is also worth emphasizing that there is a factor of $\frac{1}{2}$ coming from the fact that the fermion two point function is given by ($\not{k} + m_{\text{eff}}$) and not ($k^2 - m^2$) as would be the case for a scalar field.) Finally, we also note that since the Nambu-Jona-Lasinio theory is a low energy effective theory, the momentum integrals should be cut off at Λ. With all this information, we can, of course, evaluate the effective potential as in chapter **6**. However, let us note that we are not really interested in the exact form of the effective potential – rather, we would like to know if there exists a nontrivial minimum of the effective potential. Namely, we would like to know if a nontrivial solution of the equation

$$\frac{\mathrm{d}V_{\text{eff}}}{\mathrm{d}\sigma_c} = 0. \tag{14.43}$$

From (14.41) and (14.42), we see that this is equivalent to

$$N\sigma_c \left[\Lambda^2 - 4g^2 \int \frac{\mathrm{d}^4 k_E}{(2\pi)^4} \frac{1}{k_E^2 + m_{\text{eff}}^2} \right] = 0. \tag{14.44}$$

The minimum equation, (14.44), is commonly known as the gap equation and we note here that the integral, in (14.44), can be trivially evaluated with a four-dimensional cut-off. However, we would evaluate it slightly differently so that it can be generalized to finite temperature. Let us first do the k^4 integral in the above followed by the spatial momentum integral which will be cut-off at Λ. In such a case, the gap equation becomes, (k stands for the radial momentum)

$$N\sigma_c \left[\Lambda^2 - 4g^2 \int \frac{\mathrm{d}^3 k}{(2\pi)^3} \frac{1}{\sqrt{k^2 + m_{\text{eff}}^2}} \right] = 0,$$

$$\text{or,}\ \ \sigma_c \left[\Lambda^2 - \frac{g^2}{\pi^2} \int_0^\Lambda \mathrm{d}k \frac{k^2}{\sqrt{k^2 + m_{\text{eff}}^2}} \right] = 0. \tag{14.45}$$

The radial integral in (14.45) can be easily done to give

$$\sigma_c \left[\Lambda^2 - \frac{g^2}{2\pi^2} \left\{ \Lambda \sqrt{\Lambda^2 + m_{\text{eff}}^2} \right. \right.$$

$$- m_{\text{eff}}^2 \ln \frac{\Lambda + \sqrt{\Lambda^2 + m_{\text{eff}}^2}}{m_{\text{eff}}} \Bigg\} \Bigg] = 0. \tag{14.46}$$

We note here that, for large values of Λ, this equation takes the form

$$\sigma_c \left[\left(1 - \frac{g^2}{2\pi^2} \right) \Lambda^2 + \frac{g^4 \sigma_c^2}{4\pi^2} \ln \frac{2\sqrt{2}\Lambda}{g\sigma_c} \right] = 0, \tag{14.47}$$

where we have used (14.42).

We see, from (14.47), that the logarithm is a positive quantity and, consequently, a nontrivial solution of the minimum equation (gap equation) would exist, if

$$\frac{g^4 \sigma_c^2}{4\pi^2} \ln \frac{2\sqrt{2}\Lambda}{g\sigma_c} = - \left(1 - \frac{g^2}{2\pi^2} \right) \Lambda^2, \tag{14.48}$$

and such a solution is possible only if

$$- \left(1 - \frac{g^2}{2\pi^2} \right) > 0,$$

$$\text{or,} \quad g^2 > 2\pi^2. \tag{14.49}$$

On the other hand, if (14.49) does not hold, then, the minimum of the potential will occur at

$$\sigma_c = 0. \tag{14.50}$$

In other words, for sufficiently strong coupling, given by (14.49), the σ field picks up a vacuum expectation value which according to (14.37) also implies that the fermion condensate $\overline{\psi}^i \psi^i$ has a nonvanishing vacuum expectation value. Furthermore, from (14.38) as well as our discussions in chapter **6**, we see that in this case, chiral symmetry is dynamically broken. On the other hand, for weak coupling (when (14.49) is not true), chiral symmetry is a true symmetry of the theory.

There is an alternative but equivalent method to see this result and goes under the name of vanishing tadpoles. We note that if there is symmetry breaking and we have shifted the fields by the true vacuum expectation value of the field, then, there cannot be a nonvanishing one point function for the scalar field (see *e.g.*, (9.20)).

The reason is quite clear. A nonvanishing one point function would imply that the scalar field has a nonvanishing vacuum expectation value which would be inconsistent with the fact that, by shifting, we have defined our quantum theory where the vacuum expectation values of all fields are zero. Thus, we expect that in a quantum theory, all one point functions (or tadpoles) must vanish. If we apply this to the theory in (14.40), then, the one point functions up to one loop must add to zero.

In other words, we must have (remember the trace over the two flavors and that a factor $-i$ arises in rotating to Euclidean space)

$$-iN\Lambda^2\sigma_c - \frac{ig}{\sqrt{2}}\int\frac{d^4k}{(2\pi)^4}\,\mathrm{Tr}\,\frac{i}{\slashed{k}+\frac{g\sigma_c}{\sqrt{2}}} = 0,$$

or, $\quad -iN\Lambda^2\sigma_c - 4g^2N\sigma_c\int\frac{d^4k}{(2\pi)^4}\frac{1}{k^2-\frac{g^2\sigma_c^2}{2}} = 0,$

or, $\quad -iN\sigma_c\left[\Lambda^2 - 4g^2\int\frac{d^4k_E}{(2\pi)^4}\frac{1}{k_E^2+m_{\mathrm{eff}}^2}\right] = 0.$ $\qquad$ (14.51)

Thus, we see that the condition for the vanishing of the one point function is identical to the equation for the minimum of the potential, namely, (14.44).

The tadpole method is quite useful if we want to estimate the critical temperature for the restoration of chiral symmetry breaking (in the case that the coupling is strong enough to lead to a breakdown of the chiral symmetry at zero temperature). For example, we note that at finite temperature, there would be an additional temperature dependent term in the second diagram coming from the temperature dependent part of the fermion propagator. From (2.62), we note that this additional term would lead to a contribution, at high temperatures, of the form

$$-\frac{ig}{\sqrt{2}}\int\frac{d^4k}{(2\pi)^4}\,\mathrm{Tr}\,iS_{++}^{\beta}(k)$$

$$= -\frac{ig^2 N\sigma_c}{2\pi^3} \int d^4k \, n_F(|k^0|)\delta(k^2 - g^2\sigma_c^2/2)$$

$$= -\frac{ig^2 N\sigma_c}{2\pi^3} \int d^3k \, \frac{n_F(\omega)}{\omega}$$

$$= -\frac{ig^2 NC^{-1}\sigma_c}{2\pi^3 \beta^2} + O(1/\beta), \tag{14.52}$$

where C is a positive constant whose explicit value is not essential for our discussions. Adding this to the temperature independent part in (14.47), we see that the condition for the vanishing of the tadpoles at finite temperature becomes

$$\sigma_c \left[\left\{ \left(1 - \frac{g^2}{2\pi^2}\right) \Lambda^2 + \frac{g^2 C^{-1}}{2\pi^3 \beta^2} \right\} + \frac{g^4 \sigma_c^2}{4\pi^2} \ln \frac{2\sqrt{2}\Lambda}{g\sigma_c} \right] = 0. \tag{14.53}$$

It is now clear that even if the coupling is strong enough so that (14.49) holds at zero temperature, as the temperature is raised (equivalently, β decreased), beyond a certain temperature, the terms in the curly brackets will become positive and, consequently, at temperatures larger than this critical temperature, the minimum of the potential will occur at $\sigma_c = 0$ (see (14.50)) indicating that chiral symmetry, if it is broken at zero temperature, will be restored beyond this critical temperature. We can determine the critical temperature from (14.53) to be

$$T_c^2 = -2\pi^3 C \left(\frac{1}{g^2} - \frac{1}{2\pi^2}\right) \Lambda^2. \tag{14.54}$$

This shows that the critical temperature for the restoration of chiral symmetry is, in fact, dependent on the magnitude of the coupling constant, namely,

$$T_c = T_c(g). \tag{14.55}$$

Thus, we see that the Nambu-Jona-Lasinio model displays all the features of chiral symmetry breaking and restoration that we expect from QCD. Furthermore, this model is simple calculationally – at least in $\frac{1}{N}$ expansion – so that we can take it as a toy model for the study of the disoriented chiral condensates. To model the quark-gluon plasma phase and the sudden quenching of the expanding plasma, let us assume the coupling constant of the Nambu-Jona-Lasinio model to have a simple time dependent form

$$g(t) = \theta(t)g, \tag{14.56}$$

where we assume that the time independent coupling, g, is strong enough so that the Nambu-Jona-Lasinio model displays chiral symmetry breaking for positive times. We can also assume that the initial temperature of the system is

$$T < T_c(g). \tag{14.57}$$

Thus, our model with the time dependent coupling would, initially $(t < 0)$ describe a gas of free massless fermions in a chirally symmetric phase much like the quark-gluon plasma. At $t = 0$, interactions are introduced such that the system suddenly finds itself at temperatures much below the critical temperature. This makes the initial chirally symmetric state unstable and chiral symmetry will be broken completely in parallel with an expanding quark-gluon plasma with a sudden quench. It is worth noting here that since the initial state of the system is chirally symmetric and since the dynamics is also chirally symmetric, $\langle \sigma \rangle$ will have a vanishing average value in the ensemble. However, within any given element of the ensemble, different regions of space will have different directions for symmetry breaking and we are interested in calculating the growth of the correlations of σ with t. We will carry out this calculation in the next section.

14.4 Domain growth

As we have discussed earlier, we are interested in finding how the size of correlated regions grows with time in the expanding plasma phase. In the present example, this would correspond to determining the time evolution of the σ correlated (or π correlated) regions for positive times. In other words, we are interested in calculating the time dependence of correlated Green's functions, for example, of the kind

$$\langle \sigma(t, \mathbf{x})\sigma(t, 0) \rangle.$$

This calculation is, therefore, similar to the one we had already performed earlier in connection with the free Klein-Gordon theory. However, here, we are dealing with an interacting theory and, therefore, the calculations will be slightly more involved. First, we note that because of interactions, the σ Green's functions will have radiative corrections of the form

$$G_{Ab}^{-1}(x, y) = G_{Ab}^{(0)\,-1}(x, y) + \Sigma_{ab}(x, y), \qquad a, b = \pm, \qquad (14.58)$$

where the self-energy, Σ_{ab}, can be calculated diagrammatically. Here, we will define Σ_{+-} and Σ_{-+} with a relative minus sign since one of the external legs would correspond to the $-$ branch. With this, then, the physical sigma propagators can be obtained as in (14.16) and we have

$$\Sigma = \begin{pmatrix} \Sigma_{++} & -\Sigma_{+-} \\ -\Sigma_{-+} & \Sigma_{--} \end{pmatrix},$$

$$\widehat{\Sigma} = Q\Sigma Q^{-1} = \begin{pmatrix} \Sigma_C & \Sigma_A \\ \Sigma_R & 0 \end{pmatrix},$$

$$G_R = G_R^{(0)}(1 + \Sigma_R G_R^{(0)})^{-1},$$

$$G_A = G_A^{(0)}(1 + \Sigma_A G_A^{(0)})^{-1},$$

$$G_C = G_R \Sigma_C G_A + (1 + G_R^{(0)} \Sigma_R)^{-1} G_C^{(0)}(1 + \Sigma_A G_A^{(0)})^{-1}. \quad (14.59)$$

The relevant quantities for our calculation are the Green's functions for the fermions given in (2.62) (with $m = 0$ and the color and flavor diagonal indices suppressed) which would allow us to calculate the self-energy for the σ field. The tree level physical Green's functions for the σ field are also easily obtained to be

$$G_R^{(0)} = -\frac{1}{N\Lambda^2},$$

$$G_A^{(0)} = -\frac{1}{N\Lambda^2},$$

$$G_C^{(0)} = 0. \qquad (14.60)$$

This unusual form of the tree level Green's functions can be understood as follows. The σ field is an auxiliary field and, consequently, does not correspond to real propagating particles which can thermalize. Another way of saying this is to note that we can think of the σ field as a massive field with a mass of the order of the cut-off, Λ, and, therefore, they are Boltzmann suppressed at any temperature much lower than the cut-off scale. We note here that because of the vanishing of $G_C^{(0)}$, the correlated Green's function in (14.59) simplifies and has only one term.

Let us next discuss the strategy for calculating G_C before doing the actual calculation. We note that because $g(t) = \theta(t)g$, the self-energies (Feynman or physical) have the generic form

$$\Sigma(x, y) = \theta(x^0)\overline{\Sigma}(x, y)\theta(y^0). \tag{14.61}$$

It follows, therefore, that all self-energies including Σ_R, Σ_A and Σ_C vanish if $x^0 < 0$ or $y^0 < 0$. (Another way of saying this is that for $t < 0$, we have a free theory and, therefore, there cannot be any self-energy graph.) Furthermore, $\overline{\Sigma}$ is a quantity that is calculated with a constant g, much the same way the equilibrium calculation is done (The equilibrium quantities will depend on the difference in the coordinates.). As a result, evaluation of factors such as $(1 + \Sigma_R G_R^{(0)})^{-1}$, which can have nontrivial time dependence, becomes simpler for positive times. Thus, for example,

$$(1 + \Sigma_R G_R^{(0)})^{-1}(1 + \Sigma_R G_R^{(0)})(x, y) = 1,$$

$$\text{or,} \quad \int d^4z \, (1 + \Sigma_R G_R^{(0)})^{-1}(x, z)$$

$$\times \left[\delta^4(z - y) - \frac{\theta(z^0)}{N\Lambda^2}\theta(z^0)\overline{\Sigma}_R(z - y)\theta(y^0)\right] = \delta^4(x - y). \tag{14.62}$$

Since, by definition, $\Sigma_R(z-y)$ vanishes for $z^0 < y^0$, if we restrict ourselves to positive times, namely, $y^0 > 0$, the two step functions in the second term can be ignored and the inversion can be carried out much like in the equilibrium case. The knowledge of $(1 + \Sigma_R G_R^{(0)})^{-1}(x, y)$, for positive times $x^0, y^0 > 0$ is enough to determine the retarded Green's function as

$$G_R(x, y) = \int d^4z \, G_R^{(0)}(x, z)(1 + \Sigma_R G_R^{(0)})^{-1}(z, y)$$

$$= \overline{G}_R(x - y). \tag{14.63}$$

We can easily see that the same argument also goes through for the advanced Green's function. But it is, in general, not true for all the Green's function. In particular, we note here that the Feynman Green's function describes propagation both forwards and backwards in time and, therefore, the determination of the Feynman Green's function would be much more involved. As we will see shortly, we only need G_R and G_A for positive times for the purpose of our calculation, we can use

$$\overline{G}_R(x,y) = \int \frac{d^4k}{(2\pi)^4} \frac{e^{-ik\cdot(x-y)}}{-N\Lambda^2 + \overline{\Sigma}_R(k)},$$

$$\overline{G}_A(x,y) = \int \frac{d^4k}{(2\pi)^4} \frac{e^{-ik\cdot(x-y)}}{-N\Lambda^2 + \overline{\Sigma}_A(k)}, \tag{14.64}$$

$$G_C(x,y) = \int d^4z d^4z' \, \overline{G}_R(x-z)\Sigma_C(z,z')\overline{G}_A(z'-y)$$

$$= \int d^4z d^4z' \overline{G}_R(x-z)\theta(z^0)\overline{\Sigma}_C(z-z')\theta(z'^0)\overline{G}_A(z'-y).$$

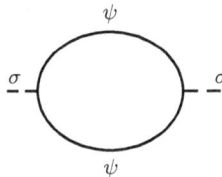

The self-energy, $\overline{\Sigma}_{ab}$, $a,b = \pm$, can be evaluated diagrammatically, in the $\frac{1}{N}$ expansion where the leading term is the fermion bubble, with the Feynman rules of the theory and are determined to be

$$i\overline{\Sigma}_{ab}(p) = (ab)(-)\left(\frac{ig}{\sqrt{2}}\right)^2 2N \int \frac{d^4k}{(2\pi)^4} iS_{ab}(k)iS_{ba}(k+p)$$

$$= (-ab)g^2 N \int \frac{d^4k}{(2\pi)^4} S_{ab}(k)S_{ba}(k+p), \tag{14.65}$$

where the indices a,b are not summed. The color and the flavor indices of the propagator are explicitly taken out as well as the negative sign coming from the fermion loop. With these, the self-energy takes the form

$$\overline{\Sigma}_{++}(p) = 4g^2 N \int \frac{\mathrm{d}^3 k}{(2\pi)^3}$$

$$\times \left[\left(\frac{1 - 2n(\omega_k)}{2\omega_k} \frac{p^0 \omega_k + \mathbf{p} \cdot \mathbf{k}}{(p^0 - \omega_k)^2 - \omega_{k+p}^2} - \frac{i\pi}{\omega_k}(p^0 \omega_k - \mathbf{p} \cdot \mathbf{k}) \right. \right.$$

$$\times \delta((p^0 + \omega_k)^2 - \omega_{k+p}^2) \left(\frac{\epsilon(p^0)}{2} + n(\omega_k) - n(\omega_k)n(\omega_{k+p}) \right)$$

$$\left. + p^0 \to -p^0 \right]$$

$$= \overline{\Sigma}_{--}^*(p^*),$$

$$\overline{\Sigma}_{+-}(p) = -4i\pi g^2 N \int \frac{\mathrm{d}^3 k}{(2\pi)^3}$$

$$\times \frac{1}{\omega_k} \left[(p^0 \omega_k - \mathbf{p} \cdot \mathbf{k})\delta((p^0 + \omega_k)^2 - \omega_{k+p}^2) \right.$$

$$\times (n(\omega_k) - 1)(n(p^0 + \omega_k) - \theta(-p^0 - \omega_k))$$

$$+ (-(p^0)^2 - \mathbf{p} \cdot \mathbf{k})\delta((p^0 - \omega_k)^2 - \omega_{k+p}^2)$$

$$\times n(\omega_k)(n(p^0 - \omega_k) - \theta(-p^0 + \omega_k)) \right],$$

$$\overline{\Sigma}_{-+}(p) = -4i\pi g^2 N \int \frac{\mathrm{d}^3 k}{(2\pi)^3}$$

$$\times \frac{1}{\omega_k} \left[(p^0 \omega_k - \mathbf{p} \cdot \mathbf{k})\delta((p^0 + \omega_k)^2 - \omega_{k+p}^2) \right.$$

$$\times n(\omega_k)(n(p^0 - \omega_k) - \theta(-p^0 + \omega_k))$$

$$+ (-(p^0)^2 - \mathbf{p} \cdot \mathbf{k})\delta((p^0 - \omega_k)^2 - \omega_{k+p}^2)$$

$$\times (n(\omega_k) - 1)(n(p^0 + \omega_k) - \theta(-p^0 - \omega_k)) \right]. \qquad (14.66)$$

From these, we can construct $\overline{\Sigma}_R$, $\overline{\Sigma}_A$ and $\overline{\Sigma}_C$ (see, for example, (14.59)), which after some simplification take the forms

$$\overline{\Sigma}_R(p) = 4g^2 N \int \frac{\mathrm{d}^3 k}{(2\pi)^3} \frac{1 - 2\omega_k}{2\omega_k}$$

$$\times \left[\frac{p^0 \omega_k + \mathbf{p} \cdot \mathbf{k}}{(p^0 - \omega_k + i\epsilon)^2 - \omega_{k+p}^2} + (p^0, \epsilon) \to (-p^0, -\epsilon) \right]$$

$$= \overline{\Sigma}_A^*(p^*),$$

$$\overline{\Sigma}_C(p) = 4i\pi g^2 N((p^0)^2 - \mathbf{p}^2) \int \frac{d^3k}{(2\pi)^3} \frac{1}{\omega_k} \Big[\delta((p^0 + \omega_k)^2 - \omega_{k+p}^2)$$

$$\times \left(\frac{\epsilon(p^0)}{2} + n(\omega_k) - n(\omega_k)n(p^0 + \omega_k) \right) + p^0 \to -p^0 \Big].$$

$$(14.67)$$

In these expressions, $\epsilon(p^0)$ represents the alternating step function which can also be identified with $\text{sgn}(p^0)$.

We can now obtain $\overline{G}_{R(A)}$ in a straightforward manner from (14.64). Furthermore, if we know the locations of the poles of the retarded and advanced Green's functions, $\overline{G}_{R(A)}(p^0, \mathbf{p})$, namely, α_p (α_p^*), then, we can also evaluate the correlated function, $\overline{G}_C$, in a simple manner. The location of the poles can be determined in the following manner. For example, we see from (14.64) that at the location of the poles of $\overline{G}_R$, we have

$$-N\Lambda^2 + \overline{\Sigma}_R(p^0, \mathbf{p}) = 0, \tag{14.68}$$

where $\overline{\Sigma}_R$ is given in (14.67). Let us Taylor expand this relation about $\mathbf{p} = 0$ and look at the lowest order term. In other words, let us set $\mathbf{p} = 0$ in the above equation. In this case, (14.68) becomes (after doing the k integration)

$$\Lambda^2 \left(1 - \frac{g^2}{2\pi^2} \right) + \frac{g^2}{8\pi^2} (p^0)^2 \ln \left(\frac{-4\Lambda^2}{(p^0)^2} \right)$$

$$- \frac{16g^2}{\pi^2} \int_0^\Lambda dk \frac{1}{e^{\beta k} + 1} \frac{k^3}{(p^0)^2 - 4k^2} = 0. \tag{14.69}$$

A simple graphical analysis shows that this equation, for $g^2 > 2\pi^2$ and $T < T_c$ always has a solution for $(p^0)^2 < 0$. Let us call this root as $-M^2$. To find the root of (14.68) when the next term in the Taylor expansion is retained, we substitute

$$(p^0)^2 \to -M^2 + b\mathbf{p}^2, \tag{14.70}$$

and retain terms up second order in $\mathbf{p}$. Here b is an arbitrary constant to be determined. After some tedious algebra, we obtain that, at the location of the pole to this order, we have

$$(\frac{7}{3} - b)I_3 + (\frac{5}{3} - b)I_5 = 0, \tag{14.71}$$

where $I_{3(5)}$ stand for the integrals

$$I_3 = 4M^4 \int_0^\Lambda dk \, \frac{k^3 \tanh(\beta k/2)}{(M^2 + 4k^2)^3},$$

$$I_5 = 16M^2 \int_0^\Lambda dk \, \frac{k^5 \tanh(\beta k/2)}{(M^2 + 4k^2)^3}. \tag{14.72}$$

Once again, with a little bit of analysis, it is easy to see that the solution of (14.71) lies within the range

$$\frac{5}{3} < b < \frac{7}{3}. \tag{14.73}$$

The position of the pole to this order can, therefore, be written as

$$(p^0)^2 = \alpha_p^2 = -M^2 + b\mathbf{p}^2$$

$$\text{or,} \quad \alpha_p \sim \pm i \left(M - \frac{b\mathbf{p}^2}{2M} \right), \tag{14.74}$$

where b takes the value as given in (14.73). The exact values of M and b are not important for our discussion. Rather, what is important is the wrong sign of the mass term, $-(M^2)$ which signals the instability of the initial state. Furthermore, we can carry out our discussion to higher orders, but as we will see shortly, this is enough for our discussion. As we have noted earlier, the location of the pole for the advanced Green's function is simply given by the complex conjugate.

We can now calculate, using (14.64), the equal time correlated Green's function, namely, for $x = (t, \mathbf{x})$ and $y = (t, 0)$ (and $t > 0$)

$$G_C(x, y) = \int d^4 z \, d^4 z' \, \theta(z^0) \theta(z'^0) \int \frac{d^4 k}{(2\pi)^4} \frac{d^4 p}{(2\pi)^4} \frac{d^4 q}{(2\pi)^4}$$

$$\times \, e^{-ik\cdot(x-z) - ip\cdot(z-z') - iq\cdot(z'-y)} \, \overline{G}_R(k) \overline{\Sigma}_C(p) \overline{G}_A(q)$$

$$= \int \frac{d^4 k}{(2\pi)^4} \frac{dp^0}{2\pi} \frac{dq^0}{2\pi} \frac{i}{k^0 - p^0 + i\epsilon} \frac{i}{p^0 - q^0 + i\epsilon}$$

$$\times \, e^{i\mathbf{k}\cdot\mathbf{x} - i(k^0 - q^0)t} \, \overline{G}_R(k^0, \mathbf{k}) \overline{\Sigma}_C(p^0, \mathbf{k}) \overline{G}_A(q^0, \mathbf{k}). \tag{14.75}$$

There are several things to note here. First of all, it is the limited range of integrations of z^0 and z'^0 which produce the first two factors (as opposed to the usual δ functions which ensure conservation of energy). Second, from (14.74), we note that the system is unstable and, consequently, $\overline{G}_R$ grows with time. As a result, one should be very careful in defining the Fourier transform of such functions. It is well known that, for functions growing with time as $\overline{G}_R \sim e^{\alpha t}$ for large t, the Fourier transform, $\overline{G}_R(\omega)$, will be analytic only above the line Im $\omega = \alpha$ and, consequently, the inverse Fourier transform is defined along a line slightly above Im $\omega = \alpha$. The opposite holds for the advanced Green's function where the region of analyticity is Im $\omega < \alpha$. Keeping these in mind, we can perform the integrations over k^0 and q^0 in (14.75) which yields

$$\int \frac{dk^0}{2\pi} \frac{e^{-ik^0 t} \overline{G}_R(k^0, \mathbf{k})}{k^0 - p^0 + i\epsilon}$$

$$= -i\left(\overline{g}_R(\alpha_k, \mathbf{k}) \frac{e^{-i\alpha_k t}}{\alpha_k - p^0} + e^{-ip^0 t} \overline{G}_R(p^0, \mathbf{k})\right),$$

$$\int \frac{dq^0}{2\pi} \frac{e^{iq^0 t} \overline{G}_A(q^0, \mathbf{k})}{q^0 - p^0 - i\epsilon}$$

$$= i\left(\overline{g}_A(\alpha_k^*, \mathbf{k}) \frac{e^{i\alpha_k^* t}}{\alpha_k^* - p^0} + e^{ip^0 t} \overline{G}_A(p^0, \mathbf{k})\right), \qquad (14.76)$$

where $\overline{g}_{R(A)}$ represent the residues at the poles α_k (α_k^* respectively).

As we have noted earlier, α_k is complex, leading to both exponentially growing as well as damped behavior. For large times, it is the exponentially growing term which would dominate. Furthermore, we note that the terms independent of α_k do not grow with time and are the ones that would have been present in an equilibrium situation where the coupling constant would be constant. Thus, keeping only the dominant terms, we obtain, for $t > 0$,

$$G_C(t, \mathbf{x}) = \int \frac{dp^0}{2\pi} \frac{d^3k}{(2\pi)^3} e^{i\mathbf{k}\cdot\mathbf{x}} |\overline{g}_R(\alpha_k, \mathbf{k})|^2 \frac{e^{-i(\alpha_k - \alpha_k^*)t}}{|p^0 - \alpha_k|^2} \overline{\Sigma}_C(p^0, \mathbf{k})$$

$$= \int \frac{d^3k}{(2\pi)^3} e^{i\mathbf{k}\cdot\mathbf{x}} e^{-i(\alpha_k - \alpha_k^*)t} f(\mathbf{k}^2, T), \qquad (14.77)$$

where

$$f(\mathbf{k}^2, T) = g^2 N \int \frac{d^3k}{(2\pi)^3} \frac{|\overline{g}_R(\alpha_k, \mathbf{k})|^2}{\omega_k \omega_{k+p}}$$

$$\times \left[\frac{(\omega_k + \omega_{k+p})^2 - \mathbf{k}^2}{|\omega_p + \omega_{k+p} + \alpha_k|^2} \left(\frac{1}{2} - n(\omega_p) + n(\omega_p)n(\omega_{k+p}) \right) \right.$$

$$+ \frac{(\omega_k - \omega_{k+p})^2 - \mathbf{k}^2}{|\omega_p - \omega_{k+p} - \alpha_k|^2}$$

$$\times \left(\frac{\epsilon(\omega_p - \omega_{k+p})}{2} - n(\omega_p) + n(\omega_p)n(\omega_{k+p}) \right) \tag{14.78}$$

$$+ \frac{(\omega_k - \omega_{k+p})^2 - \mathbf{k}^2}{|\omega_{k+p} - \omega_p - \alpha_k|^2}$$

$$\times \left(\frac{\epsilon(\omega_p - \omega_{k+p})}{2} - n(\omega_p) + n(\omega_p)n(\omega_{k+p}) \right)$$

$$\left. + \frac{(\omega_k + \omega_{k+p})^2 - \mathbf{k}^2}{|\omega_p + \omega_{k+p} - \alpha_k|^2} \left(-\frac{1}{2} - n(\omega_p) + n(\omega_p)n(\omega_{k+p}) \right) \right].$$

With a little bit of analysis, it is easy to see that f vanishes for $\mathbf{k}^2 \to 0$ and, for small $k = |\mathbf{k}|$, we can write

$$f(\mathbf{k}^2, T) \to T^3 d(k/T), \tag{14.79}$$

where $d(x)$ vanishes for $x \to 0$ as a positive power of x, namely, $d(x) \sim x^\alpha$ with $\alpha > 0$. Now, we can evaluate G_C in (14.77) for large positive times using the saddle point approximation. The saddle point occurs at

$$k_s = \frac{2iM|\mathbf{x}|}{bt}. \tag{14.80}$$

Therefore, we see that $k \to 0$ as $t \to \infty$ and this justifies our earlier assumption of retaining only the low order terms in the Taylor expansion in (14.69). Using this, we obtain that for large positive times

$$G_C(t, |\mathbf{x}|) = C(T) \left(\frac{|\mathbf{x}|}{t} \right)^\alpha e^{Mt} e^{-\frac{M|\mathbf{x}|^2}{2bt}}$$

$$= C(T) \left(\frac{|\mathbf{x}|}{t} \right)^\alpha e^{Mt} e^{-\frac{|\mathbf{x}|^2}{L^2(t)}}, \tag{14.81}$$

where $C(T)$ is a function independent of $(t, \mathbf{x})$ and $L(t)$ is the typical size of the domain given by

$$L(t) = \sqrt{\frac{b}{2M}} t. \tag{14.82}$$

The size of the domain, of course, depends on the details of the theory and is encoded in the values of the parameters b and M. But the power law dependence on t is universal in the approximation that we have used. Equation (14.82) shows that the size of the domains, indeed, grow with time in the quenched approximation and, therefore, if a quark-gluon plasma phase is created and there is sudden quenching, the plasma is likely to produce a large number of coherent pions upon hadronization. We see from our analysis that the domain size appears to increase indefinitely. This is because in the $\frac{1}{N}$ expansion there is no collision. The plasma would thermalize once collisions are taken into account and we have to go beyond the leading order to see that. This analysis shows how the closed time path formalism can be used to study non-equilibrium phenomena.

14.5 References

A. Anselm, Physics Letters **B217**, 169 (1989).

A. Anselm and M. G. Ryskin, Physics Letters **B266**, 482 (1991).

P. F. Bedaque and A. Das, Modern Physics Letters **A8**, 3151 (1993) and unpublished notes.

J. D. Bjorken, International Journal of Modern Physics **A7**, 4189 (1992).

D. Boyanovsky, D. S. Lee and A. Singh, Physical Review **D48**, 800 (1993).

Y. Nambu and G. Jona-Lasinio, Physical Review **122**, 345 (1961).

K. Rajagopal and F. Wilczek, Nuclear Physics **B204**, 577 (1993).

J. Schwinger, *Lecture Notes of Brandeis University Summer Institute in Theoretical Physics*, (1960).

CHAPTER 15

Fluctuation-dissipation theorem

While studying the phenomenon of Brownian motion colloidal particles, Einstein found an unexpected relation between two seemingly distinct quantities. Namely, the coefficient of diffusion of the particles was inversely proportional to the coefficient of friction with temperature as the constant of proportionality, or, explicitly,

$$D = \frac{kT}{\gamma m},$$

(15.1)

where D denotes the coefficient of diffusion, γ the coefficient of friction, m the mass of the particle, k the Boltzmann constant and T the temperature. Independently, a little later, Johnson had observed experimentally that, in the absence of an applied current, the mean square voltage in a conductor was proportional to its resistance and the temperature, which was theoretically explained by Nyquist. These early observations led to the formulation of the fluctuation-dissipation theorem which has become a powerful tool in the study of equilibrium statistical mechanics. Very roughly, the fluctuation-dissipation theorem relates the statistical fluctuation in a variable in the medium to the imaginary part of its (medium's) response function to a weak external perturbation, when linear response theory is valid, through a temperature dependent factor. In fact, Einstein had already pointed out that such an unexpected relation can possibly be understood as follows, namely, part of the same random force which leads to the statistical fluctuation in the variable (motion of the Brownian particles) also causes a drag (friction) and, therefore, the two must be related. We will elaborate on both these points in the next section and then probe deeper into its origin.

15.1 Linear response theory

We note here that the fluctuation-dissipation theorem holds both classically as well as quantum mechanically. Here we will discuss a simple classical model of electric susceptibility to elucidate the linear response theory as well as the FD (fluctuation-dissipation) theorem. Then, we will indicate how it is formulated quantum mechanically so that we can use it profitably in the following sections.

Let us consider the motion of an electron with charge $(-e)$ and mass $m = 1$ (for simplicity) subject to an oscillating (in time) external electric field $E(t)$ along the z-axis. If we assume that the electron has a natural frequency (of oscillation) Ω_0, the equation of motion, along the z-axis, will take the form

$$\left(\frac{d^2}{dt^2} + \gamma\frac{d}{dt} + \Omega_0^2\right)z(t) = -eE(t), \quad \gamma, \Omega_0 > 0, \qquad (15.2)$$

where γ denotes the coefficient of friction. Since the external electric field is along the z-direction, the other components of motion will not be affected by the presence of the external electric field. The simplest way to solve the second order equation (15.2) is through the method of Green's functions which, in this case, satisfies the equation (in high energy physics, conventionally, one uses an overall negative sign on the right hand side)

$$\left(\frac{d^2}{dt^2} + \gamma\frac{d}{dt} + \Omega_0^2\right)G(t - t') = \delta(t - t'), \qquad (15.3)$$

where we have used the fact that, since the coefficients γ, Ω_0 are time independent, the Green's function can depend only on the difference in the time coordinates, $G(t, t') = G(t - t')$, because of time translation invariance. If we can determine the Green's function, the solution of the dynamical equation (15.2) can be written as

$$z(t) = -e\int_{-\infty}^{\infty} dt'\, G(t - t')E(t'). \qquad (15.4)$$

The determination of the Green's function is easily carried out in the Fourier transformed space. For example, defining

$$\delta(t - t') = \int_{-\infty}^{\infty}\frac{d\omega}{2\pi}\, e^{-i\omega(t-t')},$$

$$G(t - t') = \int_{-\infty}^{\infty}\frac{d\omega}{2\pi}\, e^{-i\omega(t-t')}\, \widetilde{G}(\omega), \qquad (15.5)$$

and substituting these into (15.3), we obtain

$$\left(-\omega^2 - i\gamma\omega + \Omega_0^2\right)\widetilde{G}(\omega) = 1,$$

$$\text{or,} \quad \widetilde{G}(\omega) = -\frac{1}{(\omega + \frac{i\gamma}{2})^2 - (\Omega_0^2 - \frac{\gamma^2}{4})}$$

$$= -\frac{1}{(\omega + \Omega + \frac{i\gamma}{2})(\omega - \Omega + \frac{i\gamma}{2})}, \qquad (15.6)$$

where we have identified

$$\Omega^2 = \Omega_0^2 - \frac{\gamma^2}{4}, \qquad \gamma \ll \Omega_0 \quad \text{so that} \quad \Omega > 0. \qquad (15.7)$$

Fourier transforming this back to the t space, we obtain

$$G(t - t') = \int_{-\infty}^{\infty} \frac{d\omega}{2\pi} e^{-i\omega(t-t')} \widetilde{G}(\omega)$$

$$= -\int_{-\infty}^{\infty} \frac{d\omega}{2\pi} \frac{e^{-i\omega(t-t')}}{(\omega + \Omega + \frac{i\gamma}{2})(\omega - \Omega + \frac{i\gamma}{2})}. \qquad (15.8)$$

We note, from (15.8), that the integrand for the Green's function has two poles at

$$\omega_\pm = \pm\Omega - \frac{i\gamma}{2}, \qquad (15.9)$$

namely, both the poles are below the real axis. Therefore, a Feynman ($i\epsilon$) prescription is not necessary in this case and a nonzero contribution is obtained only if the contour is enclosed in the lower half of the complex ω plane in the clockwise direction (in order to enclose the poles) as well as only if $t - t' > 0$ (so that the exponential is damped). In other words, the Green's function is a retarded Green's function as we will expect in a classical physical problem where cause has to precede the effect. In fact, the integral in (15.8) can be easily evaluated and has the explicit form

$$G(t - t') = G_R(t - t')$$

$$= \theta(t - t') (-2\pi i) \times \left(-\frac{1}{2\pi}\right) \left(\frac{e^{-i\omega_+(t-t')}}{2\Omega} + \frac{e^{-i\omega_-(t-t')}}{(-2\Omega)}\right)$$

$$= \theta(t-t') \frac{ie^{-\frac{\gamma}{2}(t-t')}}{2\Omega} \left(e^{-i\Omega(t-t')} - e^{i\Omega(t-t')} \right)$$

$$= \theta(t-t') e^{-\frac{\gamma}{2}(t-t')} \frac{\sin \Omega(t-t')}{\Omega}. \tag{15.10}$$

As a result, from (15.4) we note that, as a consequence of the applied electric field, the electric charge gets displaced by an amount along the z axis given by

$$z(t) = -e \int_{-\infty}^{\infty} dt' \, G(t-t') E(t'), \tag{15.11}$$

and the electron acquires a dipole moment along the z axis of amount

$$p(t) = (-e)z(t) = e^2 \int_{-\infty}^{\infty} dt' \, G(t-t') E(t'). \tag{15.12}$$

Let us assume that we have a polarizable medium with N electrons per unit volume subject to this oscillating electric field Then, the medium, as a whole, will acquire a time dependent polarization given by

$$P(t) = Np(t) = Ne^2 \int_{-\infty}^{\infty} dt' \, G(t-t') E(t')$$

$$= Ne^2 \int_{-\infty}^{\infty} dt' \int_{-\infty}^{\infty} \frac{d\omega}{2\pi} e^{-i\omega(t-t')} \widetilde{G}(\omega) \int_{-\infty}^{\infty} \frac{d\omega'}{2\pi} e^{-i\omega't'} \widetilde{E}(\omega')$$

$$= Ne^2 \iint_{-\infty}^{\infty} \frac{d\omega d\omega'}{(2\pi)^2} \widetilde{G}(\omega) \widetilde{E}(\omega') e^{-i\omega t} \int_{-\infty}^{\infty} dt' e^{i(\omega-\omega')t'}$$

$$= Ne^2 \iint_{-\infty}^{\infty} \frac{d\omega d\omega'}{(2\pi)^2} \widetilde{G}(\omega) \widetilde{E}(\omega') e^{-i\omega t} (2\pi)\delta(\omega-\omega')$$

$$= Ne^2 \int_{-\infty}^{\infty} \frac{d\omega}{2\pi} \widetilde{G}(\omega) \widetilde{E}(\omega) e^{-i\omega t}, \tag{15.13}$$

which, upon taking the Fourier transform of the left hand side, leads to

$$\widetilde{P}(\omega) = Ne^2 \, \widetilde{G}(\omega)\widetilde{E}(\omega) = -\frac{Ne^2}{(\omega+\Omega+\frac{i\gamma}{2})(\omega-\Omega+\frac{i\gamma}{2})} \widetilde{E}(\omega)$$

$$= \frac{Ne^2}{(\Omega+\omega+\frac{i\gamma}{2})(\Omega-\omega-\frac{i\gamma}{2})} \widetilde{E}(\omega) = \chi(\omega)\widetilde{E}(\omega), \tag{15.14}$$

where

$$\chi(\omega) = -\frac{Ne^2}{(\omega + \Omega + \frac{i\gamma}{2})(\omega - \Omega + \frac{i\gamma}{2})}$$

$$= \frac{Ne^2}{(\Omega + \omega + \frac{i\gamma}{2})(\Omega - \omega - \frac{i\gamma}{2})}, \quad (15.15)$$

is known as the electric susceptibility of the system.

This is an overly simplified example of linear response theory. The applied external electric field $\widetilde{E}(\omega)$ (perturbation) induces a polarization $\widetilde{P}(\omega)$ in the medium which changes linearly with the perturbation and the electric susceptibility $\chi(\omega)$ measures the response of the system to the external perturbation. It is known as the response function and we note that we can write (15.15) also has

$$\chi(\omega) = Ne^2 \, \widetilde{G}(\omega) = \frac{Ne^2}{\Omega^2 - (\omega + \frac{i\gamma}{2})^2} = \frac{Ne^2}{\Omega_0^2 - \omega^2 - i\gamma\omega}, \quad (15.16)$$

where we have used the relation (15.7) in the final step. Several interesting properties of the response function follow from (15.16), namely,

- $\chi(0) = \frac{Ne^2}{\Omega_0^2} > 0$, static response is positive.

- $\chi(\omega) \to$ large, for $\omega \to \Omega_0$, namely, the response is larger near the resonance.

- $\chi(\omega) \to 0$, for $\omega \to \infty$.

- $\mathrm{Im}\, \chi(\omega) = \frac{Ne^2(\gamma\omega)}{(\Omega_0^2 - \omega^2)^2 + (\gamma\omega)^2} \geq 0$.

Furthermore, since the response function is a retarded quantity (it is related to the retarded Green's function, see (15.16)), it also satisfies the Kramers-Kronig relation which says that, for any retarded function $f(\omega)$,

$$\mathrm{Re}\, f(\omega) = \frac{1}{\pi} \int d\omega' \, \mathrm{P}\!\left(\frac{1}{\omega - \omega'}\right) \mathrm{Im}\, f(\omega'),$$

$$\mathrm{Im}\, f(\omega) = -\frac{1}{\pi} \int d\omega' \, \mathrm{P}\!\left(\frac{1}{\omega - \omega'}\right) \mathrm{Re}\, f(\omega'), \quad (15.17)$$

where

$$P\left(\frac{1}{x}\right) = \lim_{\epsilon \to 0} \frac{1}{2}\left(\frac{1}{x+i\epsilon} + \frac{1}{x-i\epsilon}\right) = \lim_{\epsilon \to 0} \frac{x}{x^2 + \epsilon^2}, \qquad (15.18)$$

denotes the principal value. In fact, from (see (15.15)), we note that we can write

$$\operatorname{Re}\chi(\omega) = -\frac{Ne^2}{2}\left(\frac{1}{(\omega + \Omega + \frac{i\gamma}{2})(\omega - \Omega + \frac{i\gamma}{2})}\right.$$
$$+ \left.\frac{1}{(\omega + \Omega - \frac{i\gamma}{2})(\omega - \Omega - \frac{i\gamma}{2})}\right),$$

$$\operatorname{Im}\chi(\omega) = \frac{iNe^2}{2}\left(\frac{1}{(\omega + \Omega + \frac{i\gamma}{2})(\omega - \Omega + \frac{i\gamma}{2})}\right.$$
$$- \left.\frac{1}{(\omega + \Omega - \frac{i\gamma}{2})(\omega - \Omega - \frac{i\gamma}{2})}\right). \qquad (15.19)$$

Using these as well as (15.18), we can check explicitly that the Kramers-Kronig relations, (15.17), hold in the present case.

15.1.1 Fluctuation-dissipation theorem.

The simple model above clarifies that the response function to a weak external perturbation is proportional to the retarded Green's function of the theory. Let us recall, from (2.55), that there are three independent physical Green's functions in a theory, namely, the retarded, advanced and the correlated Green's functions. In the case of bosonic fields, two of them (retarded and advanced) are described in terms of commutators of field operators (the difference between the two lies in the multiplying theta function specifying the time ordering of the two fields) while the correlated Green's function is defined in terms of an anti-commutator. As we have shown above, in a given medium, the response function to a (weak) external perturbation is given by the retarded Green's function. On the other hand, any statistical fluctuation in the system is described by the correlated Green's function. We will now derive the classical result derived above in a quantum mechanical setting and formulate the fluctuation-dissipation theorem.

The point to note here is that, while there are two independent operator products we can construct from two noncommuting field operators, say, for example, either $\phi(x)\phi(0)$ and $\phi(0)\phi(x)$ or $[\phi(x), \phi(0)]$

and $[\phi(x), \phi(0)]_+$, there is only one independent thermal average because of the KMS (Kubo-Martin-Schwinger) condition (see, for example, section **1.3** or **3.6** or **4.5.3**) which imposes the periodicity (in real time formalisms)

$$\text{Tr}\left(e^{-\beta H}\phi(x^0,\mathbf{x})\phi(0)\right) = \text{Tr}\left(e^{-\beta H}\phi(0)\phi(x^0+i\beta,\mathbf{x})\right), \quad (15.20)$$

and follows from the cyclicity of the trace as well as from the fact that H denotes the Hamiltonian operator of the system. Let us define the thermal average (we suppress the space coordinates which are not relevant for our discussion here)

$$S(t) = \langle \phi(t)\phi(0)\rangle_\beta = \frac{1}{Z(\beta)}\,\text{Tr}\left(e^{-\beta H}\phi(t)\phi(0)\right)$$

$$= \frac{1}{Z(\beta)}\,\text{Tr}\left(e^{-\beta H}\phi(0)\phi(t+i\beta)\right) = \langle \phi(0)\phi(t+i\beta)\rangle_\beta$$

$$= S(-t-i\beta), \qquad (15.21)$$

which, in the Fourier transformed space, leads to

$$\tilde{S}(\omega) = \int dt\, e^{i\omega t} S(t) = \int dt\, e^{i\omega t} S(-t-i\beta)$$

$$= \int dt\, e^{i\omega(-t-i\beta)} S(t) = e^{\beta\omega}\int dt\, e^{-i\omega t} S(t)$$

$$= e^{\beta\omega}\tilde{S}(-\omega), \qquad (15.22)$$

which can also be written, equivalently, as

$$\tilde{S}(-\omega) = e^{-\beta\omega}\tilde{S}(\omega). \qquad (15.23)$$

Using these relations, we now see that the retarded Green's function can be written as

$$G_{\mathrm{R}}(t) = -i\theta(t)\langle[\phi(t),\phi(0)]\rangle_\beta = (G_{\mathrm{R}}(t))^* = \theta(t)G_\rho(t), \quad (15.24)$$

where we have identified the spectral Green's function as

$$G_\rho(t) = -i\langle[\phi(t),\phi(0)]\rangle_\beta = -i\big(S(t)-S(-t)\big)$$
$$= -G_\rho(-t) = (G_\rho(t))^*, \qquad (15.25)$$

which, in the Fourier transformed space, leads to

$$\widetilde{G}_\rho(\omega) = -i\big(\widetilde{S}(\omega) - \widetilde{S}(-\omega)\big) = -i(1 - e^{-\beta\omega})\widetilde{S}(\omega). \qquad (15.26)$$

Furthermore, we note that

$$\chi(\omega) = \widetilde{G}_R(\omega) = \int dt\, e^{i\omega t}\, G_R(t) = \int dt\, e^{i\omega t}\, \theta(t)\, G_\rho(t), \quad (15.27)$$

so that

$$\chi^*(\omega) = \widetilde{G}_R^*(\omega) = \int dt\, e^{-i\omega t}\, \theta(t)\, G_\rho(t) = \int dt\, e^{i\omega t}\, \theta(-t)\, G_\rho(-t)$$

$$= -\int dt\, e^{i\omega t}\, (1 - \theta(t))\, G_\rho(t) = -\widetilde{G}_\rho(\omega) + \chi(\omega), \quad (15.28)$$

where we have used (15.25) and (15.27) and this leads to

$$\widetilde{G}_\rho(\omega) = \chi(\omega) - \chi^*(\omega) = 2i\,\mathrm{Im}\,\chi(\omega) = 2i\,\mathrm{Im}\,\widetilde{G}_R(\omega). \qquad (15.29)$$

On the other hand, the correlated Green's function has the form

$$C(t) = \frac{1}{2}\langle[\phi(t),\phi(0)]_+\rangle_\beta = \frac{1}{2}\big(S(t) + S(-t)\big), \qquad (15.30)$$

and, in the Fourier transformed space, this leads to

$$\widetilde{C}(\omega) = \frac{1}{2}\big(\widetilde{S}(\omega) + \widetilde{S}(-\omega)\big) = \frac{1}{2}(1 + e^{-\beta\omega})\widetilde{S}(\omega),$$

which, upon using (15.22), (15.26) and (15.29) yields

$$\widetilde{C}(\omega) = \frac{i}{2}\frac{1 + e^{-\beta\omega}}{1 - e^{-\beta\omega}}\,\widetilde{G}_\rho(\omega) = \frac{i}{2}\coth\frac{\beta\omega}{2}\,\widetilde{G}_\rho(\omega) \qquad (15.31)$$

$$= -\coth\frac{\beta\omega}{2}\,\mathrm{Im}\,\chi(\omega) = -\coth\frac{\beta\omega}{2}\,\mathrm{Im}\,\widetilde{G}_R(\omega). \qquad (15.32)$$

This describes the fluctuation-dissipation (FD) theorem and says that the statistical fluctuations in a theory in equilibrium (given by $C(\omega)$) are related to the imaginary (dissipative) part of the response function ($\chi(\omega)$) through a temperature factor. Let us also recall that the Feynman propagator (at finite temperature) is given by

$$iG_F(x) = \langle(\theta(x^0)\phi(x)\phi(0) + \theta(-x^0)\phi(0)\phi(x))\rangle_\beta$$

$$= \frac{1}{2}\langle([\phi(x),\phi(0)]_+ + \mathrm{sgn}(x^0)\,[\phi(x),\phi(0)])\rangle_\beta$$

$$= C(x) + \frac{i}{2}\big(G_R(x) + G_R(-x)\big), \qquad (15.33)$$

where we have used the definitions in (15.30) as well as (15.24). This ties in with Einstein's observation, namely, since $C(x)$ and $iG_R(x)$ correspond to two parts of the Feynman propagator (within the field theoretic context), the FD theorem only gives a relation between these two components. In fact, since the anti-commutator (correlated function) is Hermitian while the commutator (retarded function) is anti-Hermitian, it follows from (15.33) that

$$\text{Im}\, G_F = -C, \tag{15.34}$$

which holds in both coordinate as well as momentum space ($\text{Im}\, G_F(x)$ and $C(x)$ are real and even functions of x) and, as a result, we can write (see (15.32))

$$\text{Im}\, G_F(p) = -C(p) = \coth \frac{\beta p_0}{2} \,\text{Im}\, G_R(p). \tag{15.35}$$

This relation (describing fluctuation-dissipation) can be compared with (2.78) which follows simply from the analyticity properties of these functions. However, the relation (15.33) (or (15.35)) does not explain the origin of such a relation which we will do next.

15.2 Unitarity and the FD theorem

Conservation of probability in a quantum mechanical process is expressed by the unitarity of the scattering matrix in a quantum field theory, namely,

$$SS^\dagger = \mathbb{1} = S^\dagger S, \tag{15.36}$$

where S represents the scattering matrix (operator). As we have already discussed in chapter **8** (in particular, in (8.1) and (8.2)), if we represent the scattering matrix as

$$S = \mathbb{1} + iT, \quad S^\dagger = \mathbb{1} - iT, \tag{15.37}$$

unitarity, (15.36), can be expressed in the operatorial form

$$-i(T - T^\dagger) = TT^\dagger = T^\dagger T, \quad 2\text{Im}\, T = TT^\dagger = T^\dagger T. \tag{15.38}$$

We note here that the $\mathbb{1}$, in the S matrix in (15.37), simply corresponds to particles scattering without scattering (interacting) and

T represents the "true" scattering matrix (because of interactions). The equation for the imaginary part of the T matrix can be given in terms of a cutting description of Feynman graphs known as the Cutkosky rules, as we have explained in detail in chapter **8**.

Let us summarize here the main results from chapter **8** which will be useful in our present discussion. We saw that the imaginary part of any Feynman graph, at zero temperature, can be given a cutting description and can be expressed as a sum of all possible cuttings of the graph with the cut towards the left so that the external vertices on the left (where momentum enters the diagram) are circled. This is summarized in (8.38) with the graphical representation

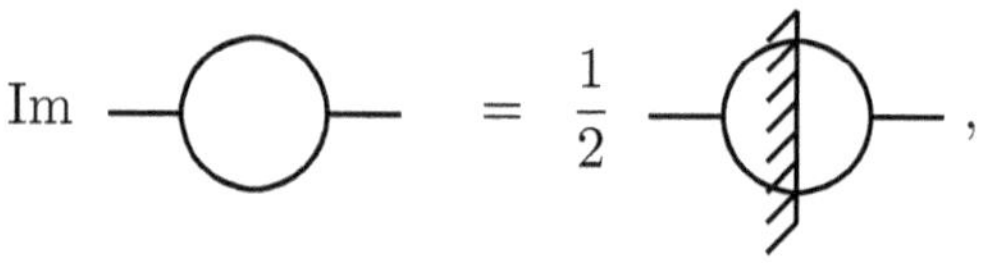

$$\text{Im} \quad \underset{}{-\bigcirc-} \quad = \quad \frac{1}{2} \quad -\bigoplus- \quad ,$$

whereas graphs with the cut to the right, namely, with circled vertices to the right of the cut, including external vertices (where momentum leaves the diagram) vanish. We want to emphasize that a graph by graph cutting description is not required for unitarity to hold, only the amplitude, as a whole, should have a cutting description, but this is a special feature at zero temperature. At finite temperature, on

$$\Pi^L_{ab}(p) = \underset{p}{\overset{a}{\longrightarrow}}\bigoplus\underset{p}{\overset{b}{\longrightarrow}} \quad , \quad \Pi^R_{ab}(p) = \underset{p}{\overset{a}{\longrightarrow}}\bigoplus\underset{p}{\overset{b}{\longrightarrow}}$$

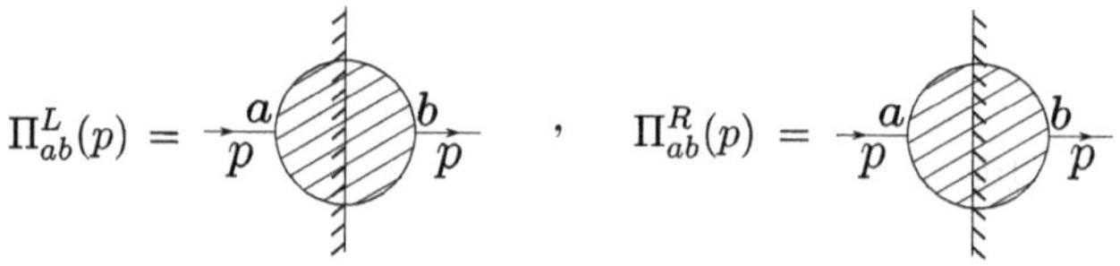

Figure 15.1: Self-energy diagrams with cuts to the left and to the right.

the other hand, the cutting description holds only at the amplitude level (and not graph by graph). This has to do with the fact that a cut propagator, at zero temperature, depends only on positive or negative energies (see (8.39)). At finite temperature, on the other hand, a cut propagator depends on both positive and negative energies (see, for example, (8.54)). This also leads to the consequence that graphs with a cut to the right do not vanish, in general. The vanishing happens only for retarded amplitudes (see (8.85)). As a

result, at finite temperature, the cutting rule for the self-energy has
the form

$$2 \operatorname{Im} \Pi_{ab}(p) = \Pi_{ab}^{L}(p) + \Pi_{ab}^{R}(p), \quad a, b = \pm, \tag{15.39}$$

where $\Pi_{ab}^{L}(p)$ and $\Pi_{ab}^{R}(p)$ are shown in Fig. 15.1.

The retarded self-energy, on the other hand, is given by (see, for
example, (8.90))

$$\Pi_{\mathrm{R}}(p) = \Pi_{++}(p) + \Pi_{+-}(p),$$

so that

$$\begin{aligned}
2 \operatorname{Im} \Pi_{\mathrm{R}}(p) &= 2 \operatorname{Im} \Pi_{++}(p) + 2 \operatorname{Im} \Pi_{+-}(p) \\
&= \Pi_{++}^{L}(p) + \Pi_{++}^{R}(p) + \Pi_{+-}^{L}(p) + \Pi_{+-}^{R}(p) \\
&= \left(\Pi_{++}^{L}(p) + \Pi_{+-}^{L}(p) \right) + \left(\Pi_{++}^{R}(p) + \Pi_{+-}^{R}(p) \right) \\
&= \Pi_{++}^{L}(p) + \Pi_{+-}^{L}(p) + 2 \operatorname{Im} \Pi_{\mathrm{R}}^{R}(p) \\
&= \Pi_{++}^{L}(p) + \Pi_{+-}^{L}(p), \tag{15.40}
\end{aligned}$$

where we have used (8.85), namely, the fact that, for the retarded self-
energy, the diagrams with the cut to the right vanish. Furthermore,
from Fig. 15.1, we note that

$$\Pi_{ab}^{L}(-p) = \Pi_{ba}^{R}(p), \quad \Pi_{ab}^{R}(-p) = \Pi_{ba}^{L}(p), \tag{15.41}$$

which, in fact, holds graph by graph, but it is sufficient, for our
purpose, that it holds at the amplitude level. It can also be concluded
from the detailed analysis in section **8.2** that the sum of all cut graphs
(in an amplitude) are real, namely, $(a, b = \pm)$

$$\left(\Pi_{ab}^{L}(p) \right)^{*} = \Pi_{ab}^{L}(p), \qquad \left(\Pi_{ab}^{R}(p) \right)^{*} = \Pi_{ab}^{R}(p),$$

as well as

$$\left(\Pi_{ab}^{L}(p) \right)^{*} = \Pi_{(-b)(-a)}^{R}(p), \quad \left(\Pi_{ab}^{R}(p) \right)^{*} = \Pi_{(-b)(-a)}^{L}(p),$$

which together lead to

$$\Pi_{ab}^{L}(p) = \Pi_{(-b)(-a)}^{R}(p), \qquad \Pi_{ab}^{R}(p) = \Pi_{(-b)(-a)}^{L}(p). \tag{15.42}$$

Furthermore, it follows, from the relations in (8.58) (as well as the fact that the vertex changes sign in the sum as $+ \leftrightarrow -$), that

$$\sum_{a=\pm} \Pi^L_{ab}(p) = 0 = \sum_{b=\pm} \Pi^R_{ab}(p), \qquad (15.43)$$

which allows us to conclude from (15.39) and (15.40) that

$$2\operatorname{Im}\Pi_{++}(p) = \Pi^L_{++}(p) + \Pi^R_{++}(p),$$

$$2\operatorname{Im}\Pi_R(p) = \Pi^L_{++}(p) + \Pi^L_{+-}(p) = \Pi^L_{++}(p) - \Pi^R_{++}(p). \qquad (15.44)$$

With all this information, let us next consider the Feynman diagram shown in Fig. 15.2 which can be written as

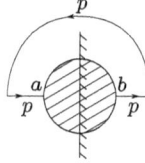

Figure 15.2: Self-energy diagrams with cuts to the left and to the right.

$$I_{ab} = \int d^4p \, \Pi^R_{ab}(p) \, G^{(0)}_{a(-a)}(-p), \qquad a,b, \quad \text{fixed}, \qquad (15.45)$$

where $G^{(0)}_{a(-a)}(p)$ denotes the tree level propagator and we have used the fact that the index of a propagator (Green's function), in the cut region, is determined by the index in the uncut region and is opposite to it (see (8.58)). Furthermore, if we let $p \to -p$ in (15.45) and use (15.41), we can write

$$I_{ab} = \int d^4p \, \Pi^R_{ab}(-p) \, G^{(0)}_{a(-a)}(-p) = \int d^4p \, \Pi^L_{ba}(p) G^{(0)}_{a(-a)}(p)$$

$$= \int d^4p \, \Pi^L_{ba}(p) \, e^{-a\beta p_0} \, G^{(0)}_{a(-a)}(-p), \qquad (15.46)$$

where, in the last step, we have used (8.60) which follows from the KMS condition (see (8.59)). As a result, comparing (15.45) and (15.46), we conclude that

$$\Pi^R_{ab}(p) = e^{-a\beta p_0} \, \Pi^L_{ba}(p), \qquad a,b, \quad \text{fixed},$$

which, in particular, gives

$$\Pi^R_{++}(p) = e^{-\beta p_0}\, \Pi^L_{++}(p), \quad \Pi^R_{+-}(p) = e^{-\beta p_0}\, \Pi^L_{-+}(p), \qquad (15.47)$$

so that, at zero temperature $(\beta \to \infty)$

$$\Pi^R_R(p) = \Pi^R_{++}(p) - \Pi^R_{+-}(p) = 0, \quad \text{for} \quad p_0 > 0, \qquad (15.48)$$

as is well known from the cutting rules at zero temperature (see (8.85)). Furthermore, it now follows from (15.44) and (15.47) that

$$\operatorname{Im}\Pi_{++}(p) = \frac{1}{2}\left(\Pi^L_{++}(p) + \Pi^R_{++}(p)\right) = \frac{1}{2}\left(1 + e^{-\beta p_0}\right)\Pi^L_{++}(p),$$

$$\operatorname{Im}\Pi_R(p) = \frac{1}{2}\left(\Pi^L_{++}(p) - \Pi^R_{++}(p)\right) = \frac{1}{2}\left(1 - e^{-\beta p_0}\right)\Pi^L_{++}(p),$$

so that we have

$$\operatorname{Im}\Pi_{++}(p) = \coth\frac{\beta p_0}{2}\operatorname{Im}\Pi_R(p), \quad p_0 > 0. \qquad (15.49)$$

The connection of this result, which follows from the unitarity of the theory, to the FD (fluctuation-dissipation) theorem is now straightforward. Let us recall that the complete Feynman Green's functions (propagators) as well as the physical Green's functions (propagators) of the theory can be defined as matrices in the closed time path formalism as (see (2.49) and (2.57))

$$G(p) = \begin{pmatrix} G_{++}(p) & G_{+-}(p) \\ G_{-+}(p) & G_{--}(p) \end{pmatrix}, \quad \widehat{G}(p) = \begin{pmatrix} 0 & G_A(p) \\ G_R(p) & G_C(p) \end{pmatrix}, \qquad (15.50)$$

and correspondingly, the complete self-energies are defined as

$$\Pi(p) = \begin{pmatrix} \Pi_{++}(p) & \Pi_{+-}(p) \\ \Pi_{-+}(p) & \Pi_{--}(p) \end{pmatrix}, \quad \widehat{\Pi}(p) = \begin{pmatrix} \Pi_C(p) & \Pi_A(p) \\ \Pi_R(p) & 0 \end{pmatrix}. \qquad (15.51)$$

Furthermore, the complete self-energy is expressed in terms of the Green's functions (propagators) as

$$G^{-1}(p) = (G^{(0)}(p))^{-1} - \Pi(p),$$
$$\widehat{G}^{-1}(p) = (\widehat{G}^{(0)}(p))^{-1} - \widehat{\Pi}(p), \qquad (15.52)$$

where $G^{(0)}(p), \widehat{G}^{(0)}(p)$ are the tree level propagators. (We point out here that the difference in the relative sign between the two terms in (14.15) and (15.52) is because of the chosen sign for the mass term in the Lagrangian density (14.8) simply because we are studying a phase transition there.) The inverse of the Green's functions in (15.50) can be obtained easily to correspond to

$$
G^{-1}(p) = \frac{1}{\det G(p)} \begin{pmatrix} G_{--}(p) & -G_{+-}(p) \\ -G_{-+}(p) & G_{++}(p) \end{pmatrix},
$$

$$
\widehat{G}^{-1}(p) = \frac{1}{\det \widehat{G}(p)} \begin{pmatrix} G_{\mathrm{C}}(p) & -G_{\mathrm{A}}(p) \\ -G_{\mathrm{R}}(p) & 0 \end{pmatrix}. \tag{15.53}
$$

Let us also recall from (2.58) and (2.59) that $G(p)$ $(G^{(0)}(p))$ and $\widehat{G}(p)$ $(\widehat{G}^{(0)}(p))$ are related through a unitary matrix so that

$$
\det G^{(0)}(p) = \det \widehat{G}^{(0)}(p), \quad \det G(p) = \det \widehat{G}(p). \tag{15.54}
$$

We now see, from (15.50)-(15.54) that

$$
\operatorname{Im} \Pi(p) = -\operatorname{Im}\left(G^{-1}(p)\right) + \operatorname{Im}\left((G^{(0)}(p))^{-1}\right).
$$

This leads to

$$
\begin{aligned}
\operatorname{Im} \Pi_{++}(p) &= -\frac{1}{\det G(p)} \operatorname{Im} G_{--}(p) + \frac{1}{\det G^{(0)}(p)} \operatorname{Im} G^{(0)}_{--}(p) \\
&= \frac{1}{\det G(p)} \operatorname{Im} G_{++}(p) - \frac{1}{\det G^{(0)}(p)} \operatorname{Im} G^{(0)}_{++}(p),
\end{aligned} \tag{15.55}
$$

where we have used $G_{--}(p) = (G_{++}(p))^*, G^{(0)}_{--}(p) = (G^{(0)}_{++}(p))^*$. On the other hand,

$$
\operatorname{Im} \widehat{\Pi}(p) = -\operatorname{Im}\left(\widehat{G}^{-1}(p)\right) + \operatorname{Im}\left((\widehat{G}^{(0)}(p))^{-1}\right),
$$

leads to

$$
\operatorname{Im} \Pi_{\mathrm{R}}(p) = \frac{1}{\det G(p)} G_{\mathrm{R}}(p) - \frac{1}{\det G^{(0)}(p)} \operatorname{Im} G^{(0)}_{\mathrm{R}}(p). \tag{15.56}
$$

Therefore, we conclude from (15.49), (15.55) and (15.56) that, for

$p_0 > 0$,

$$\left(\operatorname{Im}\Pi_{++}(p) - \coth\frac{\beta p_0}{2}\operatorname{Im}\Pi_{\mathrm{R}}(p)\right)$$

$$= \frac{1}{\det G(p)}\left(\operatorname{Im}G_{++}(p) - \coth\frac{\beta p_0}{2}\operatorname{Im}G_{\mathrm{R}}(p)\right)$$

$$- \frac{1}{\det G^{(0)}(p)}\left(\operatorname{Im}G_{++}^{(0)}(p) - \coth\frac{\beta p_0}{2}\operatorname{Im}G_{\mathrm{R}}^{(0)}(p)\right) = 0. \quad (15.57)$$

On the other hand, we see explicitly, from (2.50) and (2.60), that the second bracket in (15.57) identically vanishes. As a result, it follows from (15.57) that

$$\operatorname{Im}G_{++}(p) = \coth\frac{\beta p_0}{2}\operatorname{Im}G_{\mathrm{R}}(p), \quad p_0 > 0, \qquad (15.58)$$

which can be compared with the FD (fluctuation-dissipation) theorem derived in (15.35) (recall that $G_{\mathrm{F}}(p) = G_{++}(p)$). However, here the result is derived from the unitarity of the S-matrix in a quantum field theory which is a statement about the conservation of probability.

15.3 Generalized FD theorem in an out of equilibrium model

As we have seen, FD (fluctuation-dissipation) theorem is a powerful relation which is very useful and works in systems in thermal equilibrium. However, such a relation becomes meaningless in a system not in equilibrium simply because a temperature can only be defined for a system in equilibrium. One can extend the FD theorem to a system, very near equilibrium, through what is known as a gradient expansion if the fluctuations in the temperature are small (slowly varying systems). However, for systems truly out of equilibrium, the FD relation needs to be generalized. The reason, in some sense, is quite clear mathematically. For a system, in equilibrium, the Green's functions depend only on the difference of coordinates, in particular, they depend on the difference of time coordinates $(x^0 - y^0)$. In contrast, when things are out of equilibrium, as we have seen in the last chapter (chapter **14**), the Green's functions depend on the coordinates x^0 and y^0 independently. Alternatively, we can say that, in addition to depending on $t = x^0 - y^0$, the Green's functions depend also on $T = \frac{x^0 + y^0}{2}$.

For an out of equilibrium system, there are two rather well known proposed extensions of the FD theorem which we will discuss below. Let us recall from (15.35) that the FD theorem, in equilibrium where the coordinate dependence is on $x - y$, expresses the relation (in the Fourier transformed space incorporating the KMS condition)

$$C(p) = -\coth \frac{\beta|p_0|}{2} \operatorname{Im} G_{\mathrm{R}}(p), \tag{15.59}$$

where, p_μ is the momentum conjugate to $x^\mu - y^\mu$ and

$$\coth \frac{\beta|p_0|}{2} = 1 + 2n_{\mathrm{B}}(|p_0|), \quad n_{\mathrm{B}}(|p_0|) = \frac{1}{e^{\beta|p_0|} - 1}. \tag{15.60}$$

Furthermore, as defined in (15.24) and (15.30)

$$G_{\mathrm{R}}(x, y) = -i\theta(x^0 - y^0)\langle[\phi(x), \phi(y)]\rangle,$$
$$G_{\mathrm{C}}(x, y) = -2iC(x, y) = -i\langle[\phi(x), \phi(y)]_+\rangle, \tag{15.61}$$

and we note from (15.24) and (15.25) that, if we define a Green's function,

$$G_\rho(x, y) = -i\langle[\phi(x), \phi(y)]\rangle_\beta = -G_\rho(y, x), \tag{15.62}$$

known as the spectral Green's function, we can write

$$G_{\mathrm{R}}(x, y) = \theta(x^0 - y^0) G_\rho(x, y). \tag{15.63}$$

For a system, near equilibrium, Kadanoff and Baym propose a generalization of the FD relation given in (15.59) as ($T = \frac{x^0 + y^0}{2}$)

$$C(p, T) = -(1 + 2f(p_0, T)) \operatorname{Im} G_{\mathrm{R}}(p, T), \tag{15.64}$$

where the conventional bosonic distribution function $n_B(p_0)$ has been generalized to a new distribution function, namely, $f(p_0, T)$, which can be systematically determined from the self-consistency of of the relation in (15.64) perturbatively. This proposed generalization is intuitively quite clear. Namely, when a system is near equilibrium, it obviously is being acted on by a very weak perturbation and, in this case, as we have discussed in section **15.1**, the effect of the perturbation can be described by the linear response theory which allows us to calculate the change in the distribution function perturbatively.

In some special cases, this proposal can also be extended to out of equilibrium systems.

On the other hand, if a system is truly out of equilibrium, such a simple proposal is unlikely to work. Therefore, Kadanoff-Baym had a second proposal for a class of systems known as *glassy materials* where the relaxation time is long compared to any other observable time scale. As a result, the system takes a long time to come to true equilibrium and, at any time before the true equilibrium, it is in a quasi-equilibrium state. There is a wide class of such materials which have been studied experimentally and, from various considerations, the second generalization of the FD theorem was proposed to have the form

$$\frac{1}{\beta_{\text{eff}}} G_{\text{R}}(x^0, y^0; \mathbf{p}) = -\theta(x^0 - y^0) \frac{\partial C(x^0, y^0; \mathbf{p})}{\partial y^0}, \qquad (15.65)$$

where β_{eff} denotes an effective inverse temperature to be determined consistently. There are various things to note from the relation in (15.65). First, this relation is in a mixed space of time coordinates (x^0, y^0) and the spatial components of the momenta $\mathbf{p}$ which are conjugate to the spatial coordinate differences $(\mathbf{x} - \mathbf{y})$. Second, since the left hand side contains a retarded Green's function, the theta function on the right hand side simply guarantees the consistency of the equation since $C(x, y)$ is not a retarded function (see (15.61)). Furthermore, since x^0 and y^0 are independent variables (namely, $C(x^0, y^0) \neq C(x^0 - y^0)$ so that $\frac{\partial}{\partial x^0} \neq -\frac{\partial}{\partial y^0}$), it is important to be careful with the derivative on the right hand side. Finally, since G_{R} and C are Green's functions with the same canonical dimensions, the overall factor on the left, $\frac{1}{\beta_{\text{eff}}} \sim T_{\text{eff}}$ which is related to an effective temperature during an observation time interval and the derivative on the right, $\frac{\partial}{\partial y^0}$ make the equation dimensionally consistent since both correspond to a unit of energy (with $\hbar = k = 1$). In some simpler models, relation (15.65) can also be written as

$$\frac{\Omega_{\text{eff}}}{2} \coth \frac{\beta_{\text{eff}} \Omega_{\text{eff}}}{2} G_{\text{R}}(x^0, y^0; \mathbf{p}) = -\theta(x^0 - y^0) \frac{\partial C(x^0, y^0; \mathbf{p})}{\partial y^0},$$

$$(15.66)$$

which reduces to (15.65) in the limit $\beta_{\text{eff}} \Omega_{\text{eff}} \to 0$. Both β_{eff} and Ω_{eff} can be determined from the consistency of the relation order by order.

Let us check these ideas in a simple and soluble model. Let us consider a quenched scalar field model described by the Lagrangian density

$$\mathcal{L} = \frac{1}{2}\partial_\mu\phi\partial^\mu\phi - \frac{m_i^2}{2}\phi^2 - \frac{\delta m^2}{2}\theta(x^0)\phi^2 = \mathcal{L}_0 + \mathcal{L}_I, \qquad (15.67)$$

where the interaction term

$$\mathcal{L}_I = -\frac{\delta m^2}{2}\theta(x^0)\phi^2, \qquad (15.68)$$

describes a sudden quenching which also breaks time translation invariance. This indeed makes this model a genuine out of equilibrium model for any nontrivial value of the parameter δm^2. This simple model reflects the behavior of a scalar plasma with a sudden expansion for $\delta m^2 < 0$ or a sudden contraction for $\delta m^2 > 0$. Correspondingly, the system cools or heats up (respectively) on time scales much smaller than the relaxation time scale of the system. This is a simple and soluble model which we have already studied in section **14.2** of the last chapter (see (14.8) for the Lagrangian density) with

$$\delta m^2 \to -\delta m^2,$$

so that we can define

$$m_f^2 = m_i^2 + \delta m^2, \qquad (15.69)$$

which can be compared with (14.9). As a result, the Lagrangian density in (15.67) describes a free scalar theory with mass m_i for x^0 where the mass suddenly changes to m_f given by (15.69) for $x^0 \geq 0$. The time, where the sudden change occurs, can be arbitrary, but we have chosen it to be at $x^0 = 0$ for simplicity as was also the case in section **14.2**.

We have worked out the complete physical Green's functions for this model in (14.28) and (14.29) (the advanced Green's function is obtained from the retarded one by letting $x^0 \leftrightarrow y^0$). Putting in the value of $G_{\mathrm{R}}^{(0)}(x^0, y^0; \mathbf{p})$ from (14.21), we obtain

$$G_{\mathrm{R}}(x^0, y^0; \mathbf{p}) = \theta(x^0 - y^0) \times$$

$$\left[- \frac{\theta(-x^0)\theta(-y^0)}{\omega_i} \sin \omega_i (x^0 - y^0) - \frac{\theta(x^0)\theta(y^0)}{\omega_f} \sin \omega_f (x^0 - y^0) \right.$$

$$+ \frac{\theta(x^0)\theta(-y^0)}{2\omega_i} \left(\left(1 - \frac{\omega_i}{\omega_f}\right) \sin(\omega_f x^0 + \omega_i y^0) \right.$$

$$\left. - \left(1 + \frac{\omega_i}{\omega_f}\right) \sin(\omega_f x^0 - \omega_i y^0) \right) \right], \tag{15.70}$$

while (14.29) and (15.61) lead to

$$C(x^0, y^0; \mathbf{p}) = \frac{i}{2} G_{\mathrm{C}}(x^0, y^0; \mathbf{p}) = \frac{\coth \frac{\beta \omega_i}{2}}{2\omega_i} \times$$

$$\left[\theta(-x^0)\theta(-y^0) \cos \omega_i (x^0 - y^0) \right.$$

$$+ \frac{1}{2}\theta(-x^0)\theta(y^0) \left(\left(1 + \frac{\omega_i}{\omega_f}\right) \cos(\omega_i x^0 - \omega_f y^0) \right.$$

$$\left. + \left(1 - \frac{\omega_i}{\omega_f}\right) \cos(\omega_i x^0 + \omega_f y^0) \right)$$

$$+ \frac{1}{2}\theta(x^0)\theta(-y^0) \left(\left(1 + \frac{\omega_i}{\omega_f}\right) \cos(\omega_f x^0 - \omega_i y^0) \right.$$

$$\left. + \left(1 - \frac{\omega_i}{\omega_f}\right) \cos(\omega_f x^0 + \omega_i y^0) \right)$$

$$+ \frac{1}{2}\theta(x^0)\theta(y^0) \left(\left(\frac{\omega_f^2 + \omega_i^2}{\omega_f^2}\right) \cos \omega_f (x^0 - y^0) \right.$$

$$\left. + \left(\frac{\omega_f^2 - \omega_i^2}{\omega_f^2}\right) \cos \omega_f (x^0 + y^0) \right) \right]. \tag{15.71}$$

Here we have identified (see, for example, (14.22))

$$\omega_i^2 = \mathbf{p}^2 + m_i^2, \quad \omega_f^2 = \mathbf{p}^2 + m_f^2. \tag{15.72}$$

As discussed earlier, let us define new time coordinates

$$t = x^0 - y^0, \quad T = \frac{x^0 + y^0}{2}, \tag{15.73}$$

and the conjugate momentum to t as p_0. Then, we can Fourier transform the Green's functions in (15.70) and (15.71) with respect to t to obtain

$$\widetilde{G}_R(p_0, T; \mathbf{p}) = \int_{-\infty}^{\infty} dt\, e^{ip_0 t}\, G_R(t, T; \mathbf{p}),$$

$$C(p_0, T; \mathbf{p}) = \int_{-\infty}^{\infty} dt\, e^{ip_0 t}\, C(t, T; \mathbf{p}), \tag{15.74}$$

which are the relevant functions in the FD theorem (see, for example, (15.64)). The integrations in (15.74) can be done in a straightforward manner and yield

$$C(p_0, T; \mathbf{p}) = \frac{\coth \frac{\beta \omega_i}{2}}{2\omega_i} \times$$

$$\left[-\theta(-T)\left(\frac{\sin 2(p_0 + \omega_i)T}{p_0 + \omega_i} + \frac{\sin 2(p_0 - \omega_i)T}{p_0 - \omega_i} \right) \right.$$

$$+ \theta(T)\left(\frac{\omega_f^2 + \omega_i^2}{2\omega_f^2} \left(\frac{\sin 2(p_0 + \omega_f)T}{p_0 + \omega_f} + \frac{\sin 2(p_0 - \omega_f)T}{p_0 - \omega_f} \right) \right.$$

$$\left. + \frac{\omega_f^2 - \omega_i^2}{\omega_f^2} \cos 2\omega_f T \frac{\sin 2p_0 T}{p_0} \right)$$

$$+ \frac{\omega_+}{\omega_f}\left(-\theta(T)\left(\frac{\sin 2(p_0 + \omega_f)T}{p_0 + \omega_+} + \frac{\sin 2(p_0 - \omega_f)T}{p_0 - \omega_+} \right) \right.$$

$$\left. + \theta(-T)\left(\frac{\sin 2(p_0 + \omega_i)T}{p_0 + \omega_+} + \frac{\sin 2(p_0 - \omega_i)T}{p_0 - \omega_+} \right) + (\omega_i \leftrightarrow -\omega_i) \right)$$

$$\left. + \pi\left(\frac{\omega_+}{\omega_f} \cos 2\omega_- T\left(\delta(p_0 + \omega_+) + \delta(p_0 - \omega_+) \right) + (\omega_i \leftrightarrow -\omega_i) \right) \right], \tag{15.75}$$

where we have defined

$$\omega_\pm = \frac{\omega_f \pm \omega_i}{2}, \quad \text{such that} \quad \omega_i \to -\omega_i \Rightarrow \omega_+ \leftrightarrow \omega_-. \tag{15.76}$$

Similarly, we have

$$\text{Im}\, G_R(p_0, T; \mathbf{p}) = \frac{\theta(T)}{2\omega_f}\left(\frac{\sin 2(p_0 + \omega_f)T}{p_0 + \omega_f} - \frac{\sin 2(p_0 - \omega_f)T}{p_0 - \omega_f} \right)$$

$$-\frac{\theta(-T)}{2\omega_i}\left(\frac{\sin 2(p_0+\omega_i)T}{p_0+\omega_i}-\frac{\sin 2(p_0-\omega_i)T}{p_0-\omega_i}\right)$$

$$+\left[\frac{\omega_+}{2\omega_i\omega_f}\left(-\theta(T)\left(\frac{\sin 2(p_0+\omega_f)T}{p_0+\omega_+}+\frac{\sin 2(p_0-\omega_f)T}{p_0-\omega_+}\right)\right.\right.$$

$$\left.\left.+\theta(-T)\left(\frac{\sin 2(p_0+\omega_i)T}{p_0+\omega_+}+\frac{\sin 2(p_0-\omega_i)T}{p_0-\omega_+}\right)\right)+(\omega_i\leftrightarrow-\omega_i)\right]$$

$$+\pi\left[\frac{\omega_+}{2\omega_i\omega_f}\cos 2\omega_-T\left(\delta(p_0+\omega_+)-\delta(p_0-\omega_+)\right)+(\omega_i\leftrightarrow-\omega_i)\right].$$

$$(15.77)$$

A few comments are in order here. First, we note that there are both finite terms as well as pole terms (at $p_0=\pm\omega_+$) in both (15.75) and (15.77). The finite contributions arise from a bounded region $|t|\leq 2|T|$ from terms where x^0,y^0 have the same sign (namely, from the $\theta(x^0)\theta(y^0)$ and $\theta(-x^0)\theta(-y^0)$ terms, see (15.73)). On the other hand, the pole terms come from the unbounded region $2|T|<|t|\leq\infty$ when x^0 and y^0 have opposite signs (namely, terms with $\theta(x^0)\theta(-y^0)$ and $\theta(-x^0)\theta(y^0)$ in (15.75) and (15.77)). The $i\epsilon$ regularization terms, needed for convergence at temporal infinity, give rise to the delta function terms. We see that the finite terms in (15.75) and (15.77) seem to have some simple relationship. However, let us concentrate only on the pole terms at $p_0=\pm\omega_+$ where see from (15.75) and (15.77) that these terms are simply related as

$$C(p_0,T;\mathbf{p})=\mp\coth\left(\frac{\beta\omega_i p_0}{2\omega_\pm}\right)\mathrm{Im}\,G_R(p_0,T;\mathbf{p})\qquad(15.78)$$

$$=\mp\left(1+2n_B\left(\frac{\beta\omega_i p_0}{\omega_\pm}\right)\right)\mathrm{Im}\,G_R(p_0,T;\mathbf{p}).\qquad(15.79)$$

Here the overall negative sign is for the poles at $p_0=\omega_\pm$ while the positive overall sign is for the poles at $p_0=-\omega_\pm$. This can be compared with (15.64).

We note here that this simple soluble model allows us to see what happens at short times after the quench, when the system is far from equilibrium, in which case there is no generalized FD theorem of the Kadanoff-Baym kind. Such a relation exists only a long time after the quench. Let us look at the full Green's functions in the mixed space given in (15.70)-(15.71) and consider the behavior at large times after the quench which is relevant in the case of a glassy material. Namely, let us consider $x^0, y^0 \gg 0$ so that $x^0 + y^0 \gg 0$, but we choose the observational interval $x^0 - y^0$ to be finite. In this case, we can set $\theta(-x^0) = 0 = \theta(-y^0)$ and the two Green's functions in (15.70) and (15.71) take the forms

$$G_{\mathrm{R}}(x^0, y^0; \mathbf{p}) = -\frac{\theta(x^0 - y^0)}{\omega_f} \sin \omega_f (x^0 - y^0), \tag{15.80}$$

$$C(x^0, y^0; \mathbf{p}) = \frac{\coth \frac{\beta \omega_i}{2}}{4\omega_i} \left(\left(\frac{\omega_f^2 + \omega_i^2}{\omega_f^2} \right) \cos \omega_f (x^0 - y^0) \right.$$
$$\left. + \left(\frac{\omega_f^2 - \omega_i^2}{\omega_f^2} \right) \cos \omega_f (x^0 + y^0) \right). \tag{15.81}$$

A comment is in order here. In writing the Green's functions in (15.70) and (15.71) (and, therefore, in (15.80) and (15.81)), we have suppressed the regularization factor coming from the Feynman $i\epsilon$ factor. This, however, becomes relevant in studying large time behaviors. If we were to restore these regularization factors, the first terms in the two equations above would be multiplied by a factor of $e^{-\epsilon|x^0 - y^0|}$ while the second term (in the second equation) will have a regularization factor of $e^{-\epsilon(x^0 + y^0)}$. Since $x^0 + y^0 \gg 0$ and $|x^0 - y^0|$ is a finite interval, the second term in the second functions would be strongly suppressed. With all these information, let us calculate

$$\theta(x^0 - y^0) \frac{\partial C(x^0, y^0; \mathbf{p})}{\partial y^0}$$
$$\simeq \theta(x^0 - y^0) \coth \frac{\beta \omega_i}{2} \frac{\omega_f^2 + \omega_i^2}{4\omega_i \omega_f} \sin \omega_f (x^0 - y^0), \tag{15.82}$$

while

$$G_{\mathrm{R}}(x^0, y^0; \mathbf{p}) \simeq -\frac{\theta(x^0 - y^0)}{\omega_f} \sin \omega_f (x^0 - y^0), \tag{15.83}$$

so that we can write

$$-\theta(x^0 - y^0) \frac{\partial C(x^0, y^0; \mathbf{p})}{\partial y^0} = \frac{(\omega_f^2 + \omega_i^2) \coth \frac{\beta \omega_i}{2}}{4\omega_i} G_{\mathrm{R}}(x^0, y^0; \mathbf{p}). \tag{15.84}$$

Comparing with (15.66) and identifying $\Omega_{\text{eff}} = \omega_f$, we can write

$$\frac{\omega_f}{2} \coth \frac{\beta_{\text{eff}}\omega_f}{2} = \frac{\omega_f^2 + \omega_i^2}{4\omega_i} \coth \frac{\beta\omega_i}{2}, \tag{15.85}$$

which can be solved to obtain the effective temperature as

$$\beta_{\text{eff}} = \frac{2}{\omega_f} \text{Arcth}\left(\frac{\omega_f^2 + \omega_i^2}{2\omega_i\omega_f} \coth \frac{\beta\omega_i}{2}\right) = \frac{4\omega_i}{\omega_f^2 + \omega_i^2} \tanh \frac{\beta\omega_i}{2} + \cdots$$

which, for $\beta\omega_i \ll 1$ leads to

$$\frac{1}{\beta_{\text{eff}}} = \frac{\omega_f^2 + \omega_i^2}{2\omega_i^2} \frac{1}{\beta} = \left(1 + \frac{\delta m^2}{2\omega_i^2}\right) \frac{1}{\beta}, \tag{15.86}$$

where $\frac{1}{\beta}$ denotes the temperature. This result demonstrates, as we have noted earlier also (in the paragraph following (15.68)), that, for $\delta m^2 > 0$, there is a sudden contraction leading to a higher temperature while, for $\delta m^2 < 0$, there is a sudden expansion leading to a lower temperature. There is another simple soluble model that one can study given by the Lagrangian density

$$\mathcal{L} = \frac{1}{2}\partial_\mu\phi\partial^\mu\phi - \frac{m^2}{2}\phi^2 - \Delta m\, \delta(x^0). \tag{15.87}$$

This is also an out of equilibrium model where one can also work out and verify the generalized FD (fluctuation-dissipation) theorem in detail.[1] We do not go into the details here.

15.4 References

A. Altland and B. Simons, *Condensed Matter Field Theory* (Cambridge University Press, Cambridge, 2010).

P. F. Bedaque, Physics Letters **B344**, 23 (1995).

P. F. Bedaque and A. Das, Modern Physics Letters **A8**, 3151 (1993).

P. F. Bedaque, A. Das and S. Naik, Modern Physics Letters **A12**, 2481 (1997).

---

[1]See, for example, the appendix in A. L. M. Britto, A. Das and J. Frenkel, Physical Review **D93**, 105034 (2016).

A. Berera, M. Gleiser and R. O. Ramos, Physical Review **D58**, 123508 (1998).

A. L. M. Britto, A. Das and J. Frenkel, Physical Review **92**, 025020 (2015).

A. L. M. Britto, A. Das and J. Frenkel, Physical Review **93**, 105034 (2016).

H. B. Callen and T. A. Welton, Physical Review **83**, 34 (1951).

A. Cristiani and F. Ritort, Journal of Physics **A36**, R181 (2003).

R. Cutkosky, Journal of Mathematical Physics **1**, 429 (1960).

A. Das and J. Frenkel, Modern Physics Letters **A30**, 1550163 (2015).

A. Das and J. Frenkel, Physical Review **D89**, 087701 (2014).

A. Einstein, Annalen der Physik **17**, 549 (1905).

M. Gleiser and R. O. Ramos, Physical Review **D50**, 244 (1994).

J. B. Johnson, Physical Review **32**, 97 (1928).

L. P. Kadanoff and G. Baym, *Quantum Statistical Mechanics: Green's Function Methods in Equilibrium and Non-equilibrium Problems* (Benjamin, New York, 1962).

R. Kubo, Journal of Physical Society of Japan **12**, 570 (1957).

R. Kubo, Reports on Progress in Physics **29**, 255 (1966).

L. Leuzzi and T. M. Nieuwenhuizen, *Thermodynamics of the Glassy State* (Taylor and Francis, London, 2007).

M. Lindner and M. M. Muller, Physical Review **D73**, 125002 (2006).

P. Lipavsky, V. Spicka and B. Velicky, Physical Review **B34**, 6933 (1986).

E. Marinari, G. Parisi, F. Ricci-Tersenghi and J. Ruiz-Lorenzo, Journal of Physics **A31**, 2611 (1998).

P. C. Martin and J. Schwinger, Physical Review **115**, 1342 (1959).

P. Millington, A. Pilaftsis, Physical Review **D88**, 8 (2013).

H. Nyquist, Physical Review **32**, 110 (1928).

K. Rajagopal and F. Wilczek, Nuclear Physics **B204**, 577 (1993).

R. Zwanzig, *Nonequilibrium Statistical Mechanics* (Oxford University Press, New York, 2001).

Index

This book discusses all three formalisms used in the study of finite temperature field theory, namely the imaginary time formalism, the closed time formalism and thermofield dynamics. In addition, the finite temperature description on an arbitrary path in the complex t-plane is also described in detail. Gauge field theories and symmetry restoration at finite temperature are among the practical examples discussed in depth. The thermal operator representation relating the zero temperature Feynman graphs to the finite temperature ones are also explained in depth. Applications of the formalisms are worked out in detail. The consistent generalization of light-front field theories to finite temperature is systematically explained as well as the phenomenon of Unruh radiation. Cutting (Cutcosky) rules for the imaginary parts of amplitudes at finite temperature are discussed in careful detail and examples are worked out. Spontaneous and dynamical symmetry breaking and possible symmetry restoration at finite temperature are described. The question of gauge dependence of the effective potential as well as physical parameters (like mass) and the Nielsen identities are explained with examples. The methods for calculating effective actions at finite temperature are described with examples. The subtleties which arise at finite temperature are pointed out in detail also with examples. The nonrestoration of some of the symmetries at high temperature (such as supersymmetry) and theories on nonsimply connected space-times are described thoroughly. Examples of nonequilibrium phenomena are discussed with the disoriented chiral condensates as an illustration. Fluctuation-dissipation theorem is explained in detail and is worked out systematically for glassy materials. Several appendices are added at the end of some of the chapters to help the readers appreciate the discussions of the individual chapters.

This book is a very useful tool for graduate students, teachers and researchers in theoretical physics.

World Scientific
www.worldscientific.com
13308 sc

ISBN 978-981-127-293-6 (pbk)